Van De Graaff's Photographic Atlas

for the

Biology Laboratory

SEVENTH EDITION

Byron J. Adams
Brigham Young University

John L. Crawley

925 W. Kenyon Avenue, Unit 12
Englewood, CO 80110

www.morton-pub.com

To our teachers, colleagues, friends, and students who share with us a mutual love for biology.

The Galapagos marine iguana, *Amblyrhynchus cristatus,* is unique among lizards due to its ability to live and forage in the sea. Adult iguanas can dive up to 30 feet. It lives throughout the Galapagos Archipelago.

ISBN: 978-1-61731-058-4

10 9 8 7 6 5 4

Printed in the United States of America

Cover: Sally lightfoot crab, *Grapsus grapsus.*

Preface

Biology is an exciting, dynamic, and challenging science. It is the study of life. Students are fortunate to be living at a time when insights and discoveries in almost all aspects of biology are occurring at a very rapid pace. Much of the knowledge learned in a biology course has application in improving humanity and the quality of life. An understanding of biology is essential in establishing a secure foundation for more advanced courses in the biological sciences or health sciences.

Biology is a visually oriented science. *Van De Graaff's Photographic Atlas for the Biology Laboratory* is intended to provide you with quality photographs of animals similar to those you may have the opportunity to observe in a biology laboratory. It is designed to accompany any biology text or laboratory manual you may be using in the classroom. In certain courses *Van De Graaff's Photographic Atlas for the Biology Laboratory* could serve as the laboratory manual.

An objective of this atlas is to provide you with a balanced visual representation of the major kingdoms of biological organisms. Great care has been taken to construct completely labeled, informative figures that are depicted clearly and accurately. The micrographs are representative of what students will actually be looking at in their labs, not amazing one-of-a-kind photo contest winners. The terms used in this atlas are in agreement with those appearing in the more commonly used college biology texts.

Numerous dissections of plants and invertebrate and vertebrate animals were completed and photographed in the preparation of this atlas. These images are included for those students who have the opportunity to do similar dissections as part of their laboratory requirement.

Chapter 9 of this atlas is devoted to the biology of the human organism, which is emphasized in many biology textbooks and courses. In this chapter, you are provided with a complete set of photographs for each of the human body systems. Human cadavers have been carefully dissected and photographed to clearly depict each of the principal organs from each of the body systems. Selected radiographs (X-rays), CT scans, and MR images depict structures from living persons and thus provide an applied dimension to this portion of the atlas.

Preface to Seventh Edition

The success of the previous editions of *Van De Graaff's Photographic Atlas for the Biology Laboratory* provided opportunities to make changes to enhance the value of this new edition in aiding students in learning about living organisms. The revision of this atlas presented in its seventh edition required planning, organization, and significant work. As authors we have the opportunity and obligation to listen to the critiques and suggestions from students and faculty who have used this atlas. This constructive input is appreciated and has resulted in a greatly improved atlas.

One objective in preparing this edition of the atlas was to create an inviting pedagogy. The page layout was improved by careful selection of updated, new, and replacement photographs. All new illustrations were added, including key cladograms making the connections between taxonomy, morphology, and evolutionary history more intuitive. Each image in this atlas was carefully evaluated for its quality, effectiveness, and accuracy. Quality photographs of detailed dissections were updated enhancing the value of this edition. Reformatting of the pedagogy enabled more photographs, photomicrographs, enlarged images in certain chapters, and additional photographs of representative organisms. Micrographs were chosen that would closely approximate what students would see in the lab. Perhaps most important to this seventh edition was Dr. Byron Adams, Brigham Young University. Byron has brought important professional input and rounded out the team.

About the Authors

Byron J. Adams

Byron grew up on a small farm in rural northeastern California, where his parents and schoolteachers nurtured his love of the natural world. He completed his undergraduate degree in Zoology in 1993 from Brigham Young University with an emphasis in marine biology and his Ph.D. in Biological Sciences from the University of Nebraska in 1998. Following a short stint as a postdoctoral fellow at the University of California-Davis, Byron took his first faculty position at the University of Florida prior to returning to Brigham Young University.

Byron's approach to understanding biology involves inferring evolutionary processes from patterns in nature. His research programs in biodiversity, evolution, and ecology have had the continuous support of the National Science Foundation as well as other agencies, including the United States Department of Agriculture and the National Human Genome Research Institute. His most recent projects involve fieldwork in Antarctica, where he and his colleagues are studying the relationship between biodiversity, ecosystem functioning, and climate change. When he's not freezing his butt off in the McMurdo Dry Valleys or southern Transantarctic Mountains, he makes his home in Woodland Hills, Utah.

John L. Crawley

John spent his early years growing up in Southern California, where he took every opportunity to explore nature and the outdoors. He currently resides in Provo, Utah, where he enjoys the proximity to the mountains, desert, and local rivers and lakes.

He received his degree in Zoology from Brigham Young University in 1988. While working as a researcher for the National Forest Service and Utah Division of Wildlife Resources in the early 1990s, John was invited to work on his first project for Morton Publishing, *A Photographic Atlas for the Anatomy and Physiology Laboratory*. After completion of that title John started work on *A Photographic Atlas for the Zoology Laboratory*. To date John has completed five titles with Morton Publishing.

John has spent much of his life observing nature and taking pictures. His photography has provided the opportunity for him to travel widely, allowing him to observe and learn about other cultures and lands. His photos have appeared in national ads, magazines, and numerous publications. He has worked for groups such as Delta Airlines, *National Geographic*, Bureau of Land Management, U.S. Forest Service, and many others. His projects with Morton Publishing have been a great fit for his passion for photography and the biological sciences.

Byron on the plane making his way back from the Transantarctic Mountains heading for McMurdo Station.

John snorkeling with green sea turtles in the Galapagos.

Prelude

Scientists work to determine accuracy in understanding the relationship of organisms even when it requires changing established concepts. DNA sequences, developmental pathways, and morphological structures, along with the fossil record and geological dating, are used to recover the evolutionary history of life (phylogeny) and represent this in a hierarchical classification (taxonomy). New methods for generating and analyzing evolutionary hypotheses continue to improve our understanding of phylogenetic relationships. Because classification schemes that reflect phylogenetic relationships have so much more explanatory power than simple lists of organisms, scientists are constantly updating their classification schemes to reflect these advances in knowledge.

In 1758 Carolus Linnaeus, a Swedish naturalist, assigned all known kinds of organisms into two kingdoms—plants and animals. For over two centuries, this dichotomy of plants and animals served biologists well but has been replaced by the hypothesis of shared common ancestry by three major evolutionary lineages (see exhibit 1). This hypothesis is based primarily on DNA sequence data but corroborates numerous other lines of evidence as well.

Exhibit 1 Domains, Kingdoms, and Representative Examples

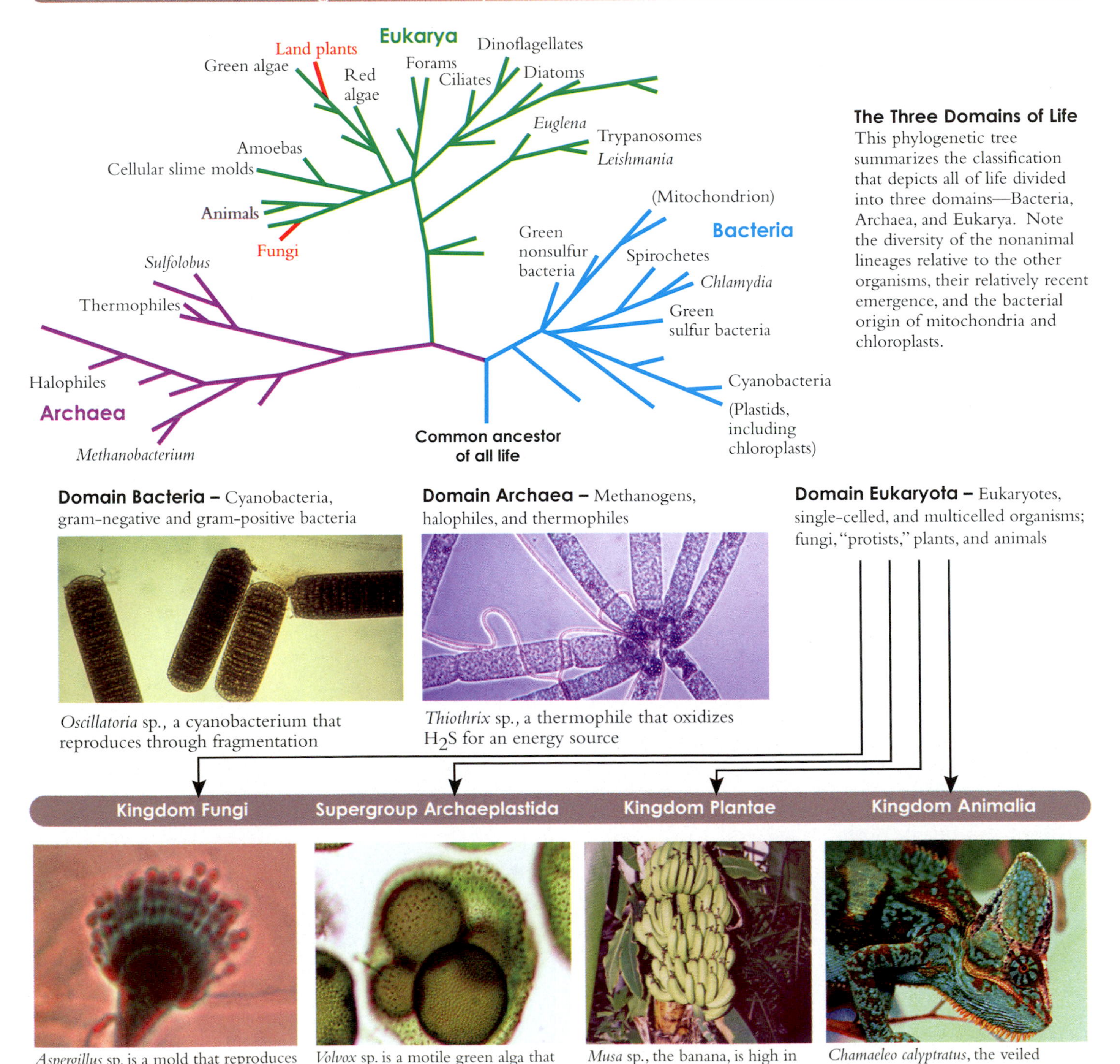

The Three Domains of Life
This phylogenetic tree summarizes the classification that depicts all of life divided into three domains—Bacteria, Archaea, and Eukarya. Note the diversity of the nonanimal lineages relative to the other organisms, their relatively recent emergence, and the bacterial origin of mitochondria and chloroplasts.

Domain Bacteria – Cyanobacteria, gram-negative and gram-positive bacteria

Oscillatoria sp., a cyanobacterium that reproduces through fragmentation

Domain Archaea – Methanogens, halophiles, and thermophiles

Thiothrix sp., a thermophile that oxidizes H_2S for an energy source

Domain Eukaryota – Eukaryotes, single-celled, and multicelled organisms; fungi, "protists," plants, and animals

Aspergillus sp. is a mold that reproduces asexually and sometimes sexually

Volvox sp. is a motile green alga that reproduces asexually or sexually

Musa sp., the banana, is high in nutritional value.

Chamaeleo calyptratus, the veiled chameleon, is known for its ability to change colors according to its mood

Basic Characteristics of Domains

Domain	Characteristics
Domain Bacteria — Bacteria	Prokaryotic cell; single circular chromosome; cell wall containing peptidoglycan; chemosynthetic autotrophs, chlorophyll-based photosynthesis, photosynthetic autotrophs, and heterotrophs; gram-negative and gram-positive forms; lacking nuclear envelope; lacking organelles and cytoskeleton
Domain Archaea — Archaea	Prokaryotic cell; single circular chromosome; cell wall; unique membrane lipids, ribosomes, and RNA sequences; lacking nuclear envelope; some with chlorophyll-based photosynthesis; with organelle and cytoskeleton
Domain Eukaryota — Eukarya	Single-celled and multicelled organisms; nuclear envelope enclosing more than one linear chromosome; membrane-bound organelles in most; some with chlorphyll-based photosynthesis

Common Classification System of Some Groups of Living Eukaryotes

Eukaryote Supergroups

Excavata – Diplomonads, Parabasalids, and Euglenozoans
Chromalveolata
- Alveolates – Dinoflagellates, Apicomplexans, and Ciliates
- Stramenopiles – Diatoms, Golden algae, Brown algae, and Oomycetes

Rhizaria – Cercozoans, Forams, and Radiolarians
Archaeplastida – Red algae, Green algae, Chlorophytes, Charophytes, and Land plants
Unikonta
- Amoebozoans – Slime molds, Gymnamoebas, and Entamoebas
- Opisthokonts – Nuclearids, Fungi, Choanoflagellates, Animals

* Single-Celled Eukaryote Supergroup Phyla – heterotrophic and phototrophic "protists"
- Phylum Amoebozoa – amoebas and slime molds
- Phylum Heterokontophyta – water molds, diatoms, golden algae
- Phylum Euglenozoa – euglenoids
- Phylum Cryptophyta – cryptomonads
- Phylum Rhodophyta – red algae
- Phylum Dinoflagellata – dinoflagellates
- Phylum Haptophyta – haptophytes

Kingdom Fungi
- Phylum Chytridiomycota – chytrids
- Phylum Zygomycota – zygomycetes
- Phylum Glomeromycota – glomeromycetes
- Phylum Ascomycota – ascomycetes
- Phylum Basidiomycota – basidiomycetes

Kingdom Plantae – bryophytes and vascular plants
- Phylum Hepatophyta – liverworts
- Phylum Anthocerophyta – hornworts
- Phylum Bryophyta – mosses
- Phylum Lycophyta (= Lycopodiophyta) – club moss, ground pines, and spike mosses
- Phylum Pteridophyta – whisk ferns, horsetails, ferns
- Phylum Cycadophyta – cycads
- Phylum Ginkgophyta – Ginkgo
- Phylum Pinophyta (= Coniferophyta) – conifers
- Phylum Gnetophyta – gnetophytes
- Phylum Magnoliophyta (= Anthophyta) – angiosperms (flowering plants)

** Kingdom Animalia – invertebrate and vertebrate animals
- Phylum Ctenophora – comb jellies
- Phylum Porifera – sponges
- Phylum Cnidaria – coral, hydra, and jellyfish
- Phylum Chordata – lancelets, tunicates, and vertebrates
- Phylum Echinodermata – sea stars and sea urchins
- Phylum Hemichordata – acorn worms
- Phylum Nematoda – roundworms
- Phylum Nematomorpha – horsehair worms
- Phylum Tardigrada – water bears
- Phylum Arthropoda – crustaceans, insects, and spiders
- Phylum Kinorhyncha – spiny-crown worms
- Phylum Bryozoa – moss animals
- Phylum Entoprocta – goblet worm
- Phylum Annelida – segmented worms
- Phylum Mollusca – clams, snails, and squids
- Phylum Nemertea – proboscis worms
- Phylum Brachiopoda – lamp shells
- Phylum Phoronida – horseshoe worms
- Phylum Gastrotricha – hairy backs
- Phylum Platyhelminthes – flatworms
- Phylum Rotifera – rotifers

* Historically considered a Kingdom, protists are no longer recognized as such in modern taxonomy. For convenient reference to earlier classification schemes, protist phyla are presented here, but note that each of these is depicted more accurately within the Eukaryote Supergroups.

** Some minor and/or poorly known phyla are not covered in this atlas. Where Phyla are grouped by chapter, they are done so to reflect phylogenetic relationships (with the exception of chapter 3, the unicellular microeukaryotes ["protists"], and chapter 9, the pseudocoelomates).

Acknowledgments

Many professionals have assisted in the preparation of *Van De Graaff's Photographic Atlas for the Biology Laboratory,* seventh edition, and have shared our enthusiasm about its value for students of biology. We are especially appreciative of Chrissy Simmons from Southern Illinois University Edwardsville, Heidi Richter from University of the Fraser Valley, Heather Brient-Johnson from Inver Hills Community College, Pam Dobbins from Shelton State Community College, and Matthew McClure from Lamar State College for their detailed review of this atlas. Drs. Ronald A. Meyers, John F. Mull, and Samuel I. Zeveloff of the Department of Zoology at Weber State University and Dr. Samuel R. Rushforth and Dr. Robert R. Robbins at Utah Valley University were especially helpful and supportive of this project. The radiographs, CT scans, and MR images have been made possible through the generosity of Gary M. Watts, M.D., and the Department of Radiology at Utah Valley Regional Medical Center.

We thank Jake Christiansen, James Barrett, and Austen Slade for their specimen dissections. Others who aided in specimen dissections were Nathan A. Jacobson, D.O., R. Richard Rasmussen, M.D., and Sandra E. Sephton, Ph.D. We are indebted to Douglas Morton and the personnel at Morton Publishing Company for the opportunity, encouragement, and support to prepare this atlas.

Photo Credits

Many of the photographs of living plants and animals were made possible because of the cooperation and generosity of the San Diego Zoo, San Diego Wild Animal Park, Sea World (San Diego, CA), Hogle Zoo (Salt Lake City, UT), and Aquatica (Orem, UT). We are especially appreciative to the professional biologists at these fine institutions.

We are appreciative of Dr. Wilford M. Hess and Dr. William B. Winborn for their help in obtaining photographs and photomicrographs. The electron micrographs are courtesy of Scott C. Miller and James V. Allen.

Figure 1.2 Leica Inc.

Figures 1.13, 4.22, 4.24, 4.25, 4.26, 4.27, 4.28, and **4.34** from *A Photographic Atlas for the Microbiology Laboratory, 3rd Edition,* by Michael J. Leboffe and Burton E. Pierce. © 2001 Morton Publishing.

Figures 6.139, 6.140, 6.162, and **6.288** Champion Paper Co.

Figures 6.255 (c) Craig K. Lorenz / PhotoResearchers.com

Figures 7.12 and **7.101** NOAA (National Oceanic and Atmospheric Administration)

Figure 7.198 Ari Pani

Figure 7.200 NOAA Okeanos Explorer Program, INDEX-SATAL 2010

Figure 7.220 (f) Linda Snook, NOAA

Figure 7.247 (a) Louis Porras

Figure 7.250 (k) U.S. Fish and Wildlife Service

Figures 7.212, 7.215, 7.217, 7.218, 7.219, 8.4, 8.11, and **8.12** from *Comparative Anatomy: Manual of Vertebrate Dissection, 2nd Edition,* by Dale W. Fishbeck and Aurora Sebastiani. © 2008 Morton Publishing.

Figures 8.110, 8.111, 8.112, 8.113, 8.114, 8.115, 8.116, 8.117, 8.118, 8.119, and **8.120** from *Mammalian Anatomy: The Cat, 2nd Edition,* by Aurora Sebastiani and Dale W. Fishbeck. © 2005 Morton Publishing.

Book Team

Publisher: Douglas N. Morton
President: David M. Ferguson
Acquisitions Editor: Marta R. Martins
Typography and Text Design: John L. Crawley
Project Manager: Melanie Stafford
Editorial Assistant: Rayna Bailey
Illustrations: Imagineering Media Services, Inc.
Cover Design: Joanne Saliger & Will Kelley

Table of Contents

Cells and Tissues

Chapter 1

All organisms are composed of one or more cells. *Cells* are the basic structural and functional units of organisms. A cell is a minute, membrane-enclosed, protoplasmic mass consisting of chromosomes surrounded by cytoplasm. Specific organelles are contained in the cytoplasm that function independently but in coordination with one another. Prokaryotic cells (fig. 1.1) and eukaryotic cells (figs. 1.3 and 1.18) are the two basic types.

Prokaryotic cells lack a membrane-bound nucleus, instead containing a single strand of *nucleic acid*. These cells contain few organelles. A rigid or semirigid cell wall provides shape to the cell outside the *cell (plasma) membrane*. Bacteria are examples of prokaryotic, single-celled organisms.

Eukaryotic cells contain a true *nucleus* with multiple chromosomes, have several types of specialized organelles, and have a differentially permeable cell membrane. Organisms consisting of eukaryotic cells include protozoa, fungi, algae, plants, and invertebrate and vertebrate animals.

Plant cells differ in some ways from other eukaryotic cells in that their cell walls contain *cellulose* for stiffness (fig. 1.3). Plant cells also contain vacuoles for water storage and membrane-bound *chloroplasts* with photosynthetic pigments for photosynthesis.

The *nucleus* is the large, spheroid body within the eukaryotic cell that contains the genetic material of the cell. The nucleus is enclosed by a double membrane called the *nuclear membrane*, or *nuclear envelope*. The *nucleolus* is a dense, nonmembranous body composed of protein and RNA molecules. The chromatin are fibers of protein and DNA molecules that make up a eukaryotic chromosome. Prior to cellular division, the chromatin shortens and coils into rod-shaped *chromosomes*. Chromosomes consist of DNA and structural proteins called *histones*.

The *cytoplasm* of the eukaryotic cell is the medium between the nuclear membrane and the cell membrane. *Organelles* are small membrane-bound structures within the cytoplasm. The cellular functions carried out by organelles are referred to as *metabolism*. The structure and function of the nucleus and principal organelles are listed in table 1.1. In order for cells to remain alive, metabolize, and maintain homeostasis, they must have access to nutrients and respiratory gases, be able to eliminate wastes, and be in a constant, protective environment.

The *cell membrane* is composed of phospholipid, protein, and carbohydrate molecules. The cell membrane gives form to a cell and controls the passage of material into and out of a cell. More specifically, the proteins in the cell membrane provide:

1. structural support;
2. a mechanism of molecule transport across the membrane;
3. enzymatic control of chemical reactions;
4. receptors for hormones and other regulatory molecules; and
5. cellular markers (antigens), which identify the blood and tissue type.

The carbohydrate molecules:

1. repel negative objects due to their negative charge;
2. act as receptors for hormones and other regulatory molecules;
3. form specific cell markers that enable like cells to attach and aggregate into tissues; and
4. enter into immune reactions.

Tissues are groups of similar cells that perform specific functions (see fig. 1.9). A flowering plant, for example, is composed of three tissue systems:

1. the *ground tissue system*, providing support, regeneration, respiration, photosynthesis, and storage;
2. the *vascular tissue system*, providing conduction passageways through the plant; and
3. the *dermal tissue system*, providing protection to the plant.

The tissues of the body of a multicellular animal are classified into four principal types (see fig. 1.36):

1. *epithelial tissue* covers body and organ surfaces, lines body cavities and lumina (hollow portions of body tubes), and forms various glands;
2. *connective tissue* binds, supports, and protects body parts;
3. *muscle tissue* contracts to produce movements; and
4. *nervous tissue* initiates and transmits nerve impulses.

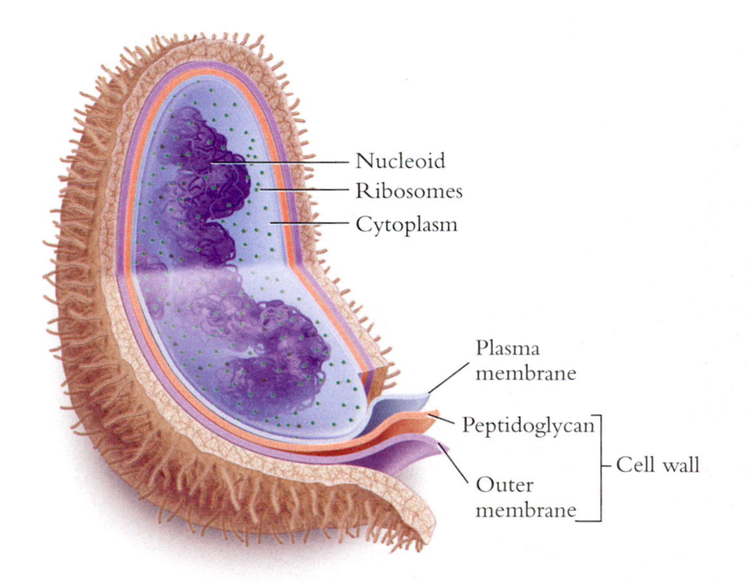

Figure 1.1 A generalized prokaryotic cell.

Table 1.1 Structure and Function of Eukaryotic Cellular Components

Component	Structure	Function
Cell (plasma) membrane	Composed of protein and phospholipid molecules	Provides form to cell; controls passage of materials into and out of cell
Cell wall	Cellulose fibrils	Provides structure and rigidity to plant cell
Cytoplasm	Fluid to jellylike substance	Serves as suspending medium for organelles and dissolved molecules
Endoplasmic reticulum	Interconnecting membrane-lined channels	Enables cell transport and processing of metabolic chemicals
Ribosome	Granules of nucleic acid (RNA) and protein	Synthesizes protein
Mitochondrion	Double-membraned sac with cristae (chambers)	Assembles ATP (cellular respiration)
Golgi complex	Flattened membrane-lined chambers	Synthesizes carbohydrates and packages molecules for secretion
Lysosome	Membrane-surrounded sac of enzymes	Digests foreign molecules and worn cells
Centrosome	Mass of protein that may contain rodlike centrioles	Organizes spindle fibers and assists mitosis and meiosis
Vacuole	Membranous sac	Stores and excretes substances within the cytoplasm, regulates cellular turgor pressure
Microfibril and microtubule	Protein strands and tubes	Forms cytoskeleton, supports cytoplasm, and transports materials
Cilium and flagellum	Cytoplasmic extensions from cell; containing microtubules	Movements of particles along cell surface, or cell movement
Nucleus	Nuclear envelope (membrane), nucleolus, and chromatin (DNA)	Contains genetic code that directs cell activity; forms ribosomes
Chloroplast	Inner (grana) membrane within outer membrane	Involved in photosynthesis

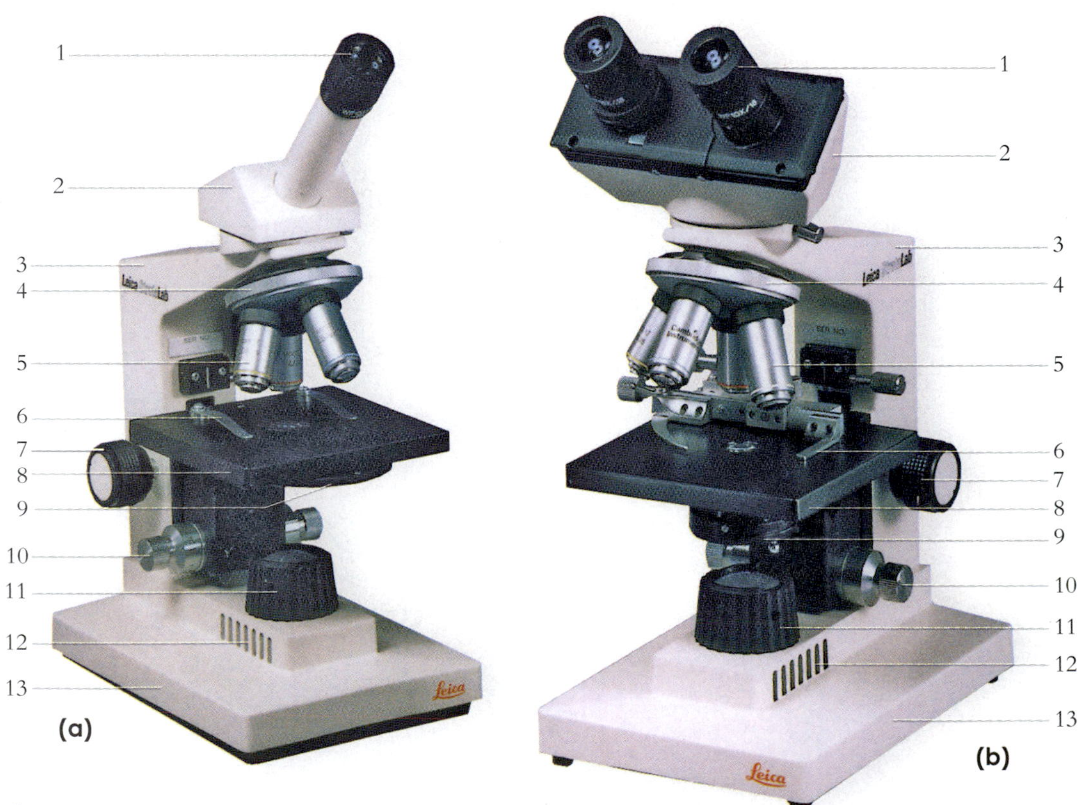

Figure 1.2 (a) A compound monocular microscope, and (b) a compound binocular microscope.

1. Eyepiece (ocular)
2. Head
3. Arm
4. Nosepiece
5. Objective
6. Stage clip
7. Coarse focus adjustment knob
8. Stage
9. Condenser
10. Fine focus adjustment knob
11. Collector lens with iris
12. Illuminator (inside)
13. Base

Plant Cells and Tissues

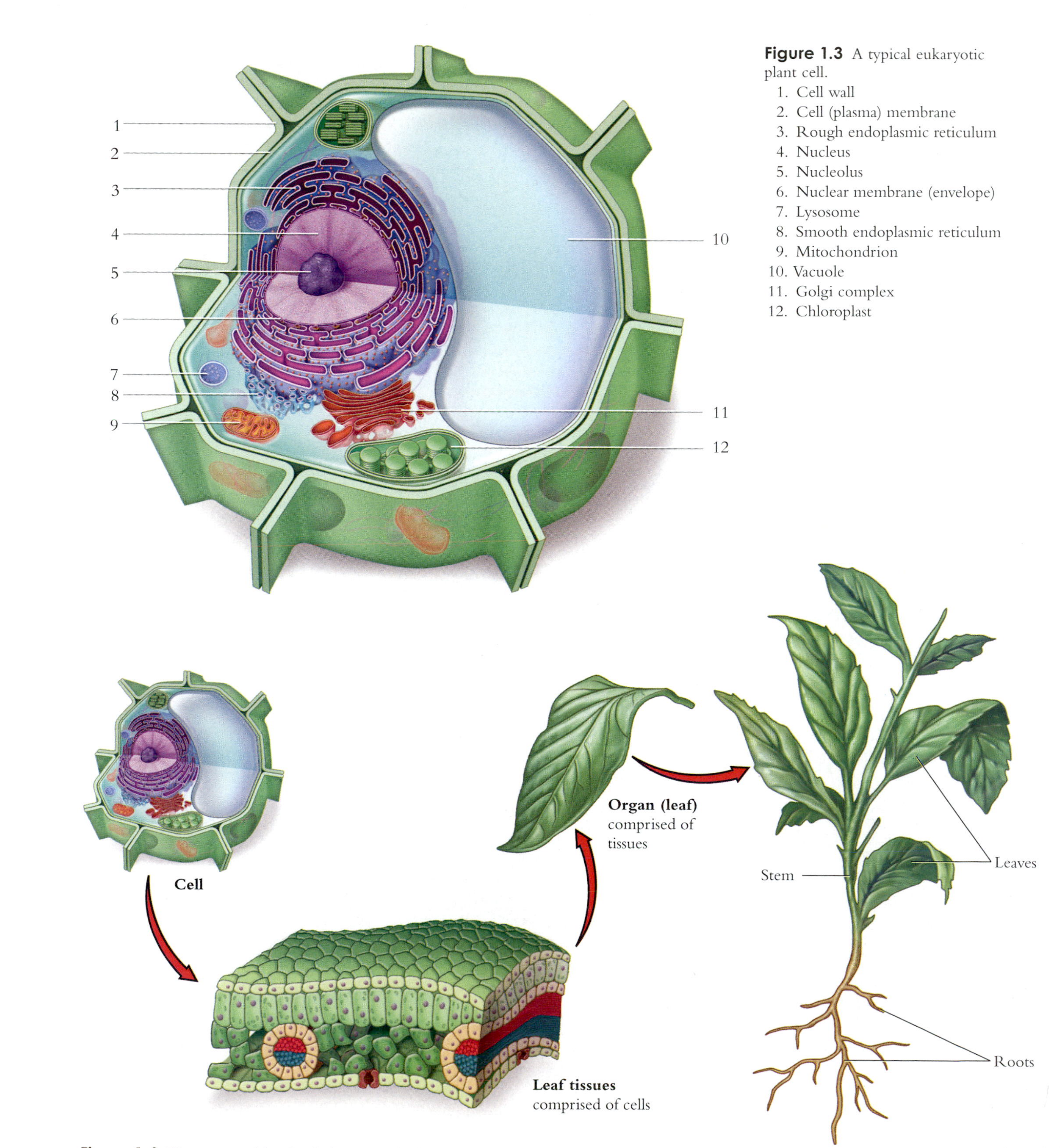

Figure 1.3 A typical eukaryotic plant cell.

1. Cell wall
2. Cell (plasma) membrane
3. Rough endoplasmic reticulum
4. Nucleus
5. Nucleolus
6. Nuclear membrane (envelope)
7. Lysosome
8. Smooth endoplasmic reticulum
9. Mitochondrion
10. Vacuole
11. Golgi complex
12. Chloroplast

Figure 1.4 The structural levels of plant organization.

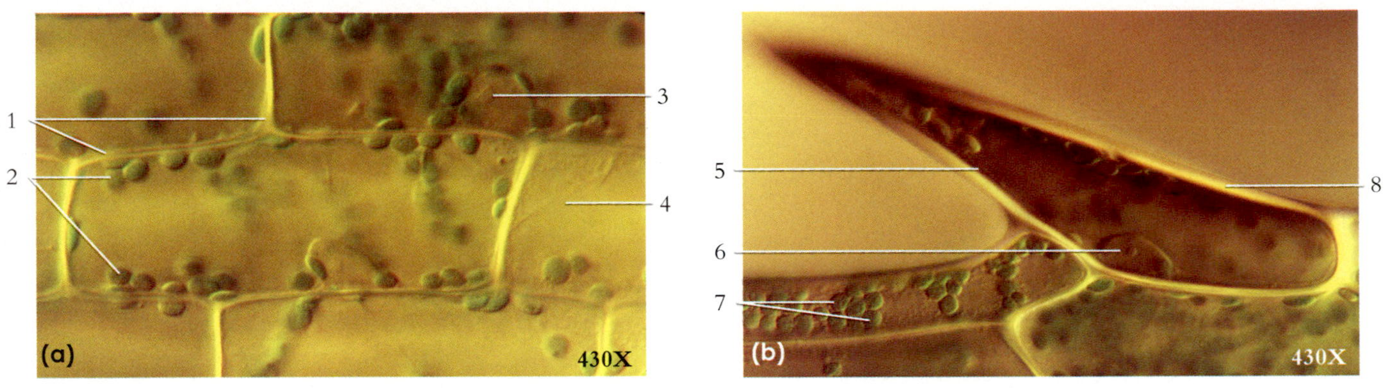

Figure 1.5 A live *Elodea* sp. leaf cells (a) photographed at the center of the leaf and (b) at the edge of the leaf.

1. Cell wall
2. Chloroplasts
3. Nucleus
4. Vacuole
5. Spine-shaped cell on exposed edge of leaf
6. Nucleus
7. Chloroplasts
8. Cell wall

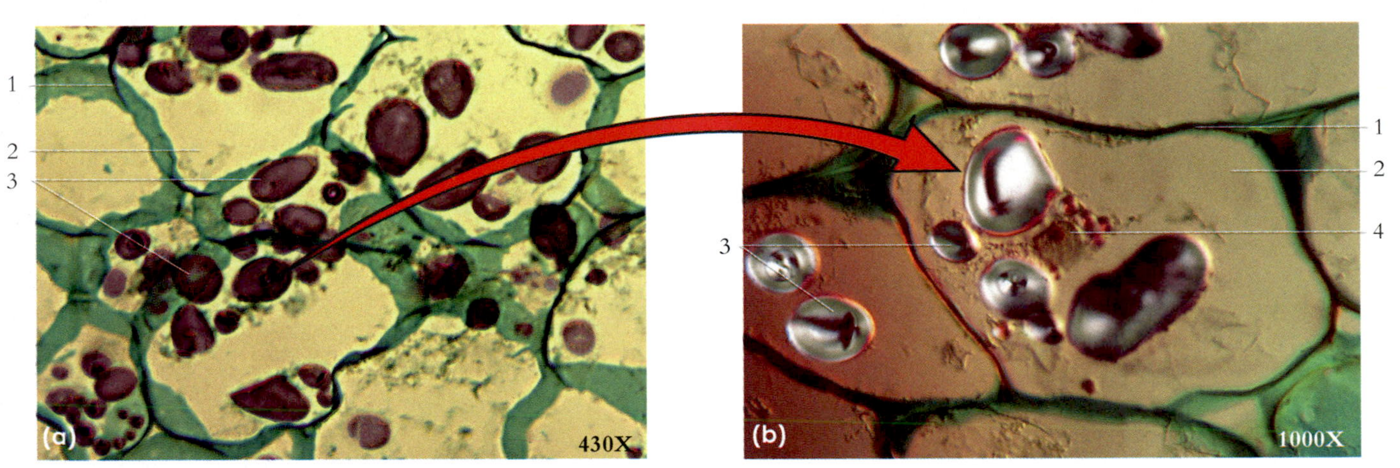

Figure 1.6 (a) Cells of a potato, *Solanum tuberosum*, showing starch grains at a low magnification, and (b) at a high magnification. Food is stored as starch in potato cells, which is deposited in organelles called amyloplasts.

1. Cell wall
2. Cytoplasm
3. Starch grains
4. Nucleus

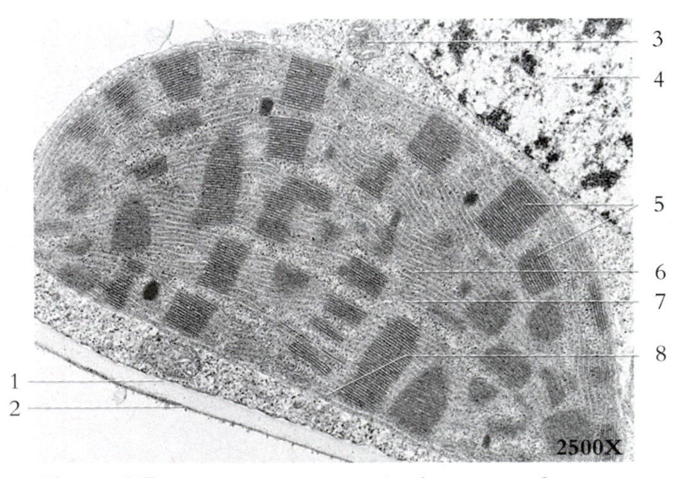

Figure 1.7 An electron micrograph of a portion of a sugarcane leaf cell.

1. Cell membrane
2. Cell wall
3. Mitochondrion
4. Nucleus
5. Grana
6. Stroma
7. Thylakoid membrane
8. Chloroplast envelope (outer membrane)

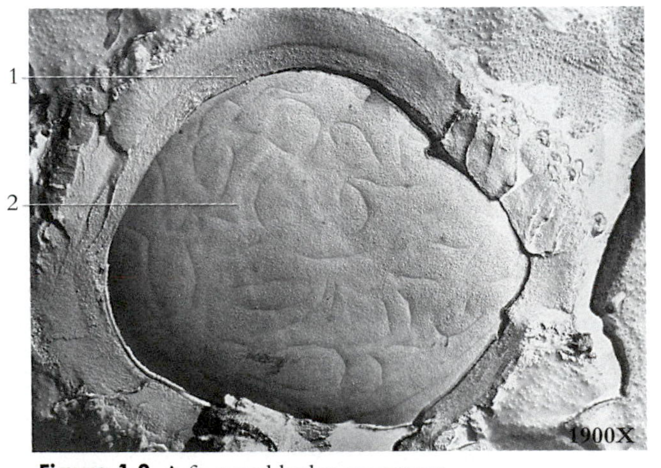

Figure 1.8 A fractured barley smut spore.

1. Cell wall
2. Cell membrane

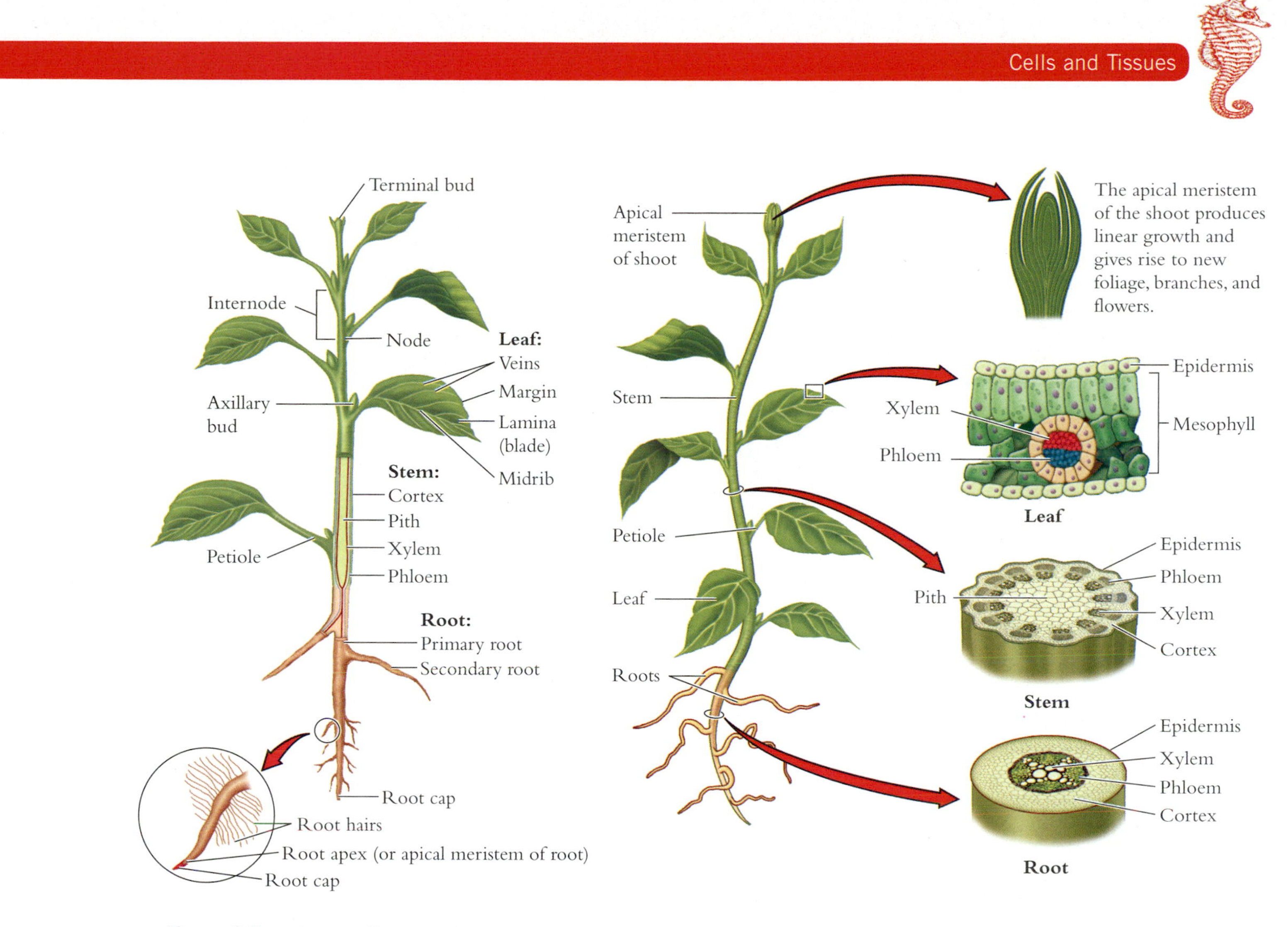

Figure 1.9 A diagram illustrating the anatomy and the principal organs and tissues of a typical dicot.

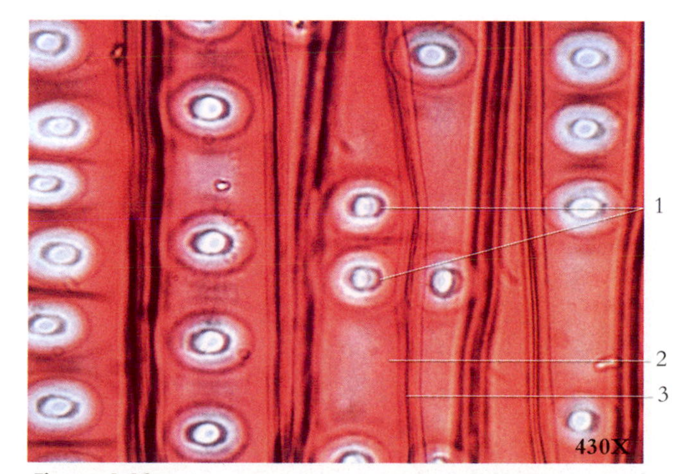

Figure 1.10 A longitudinal section through the xylem of a pine, *Pinus*, showing tracheid cells with prominent bordered pits.

1. Bordered pits
2. Tracheid cell
3. Cell wall

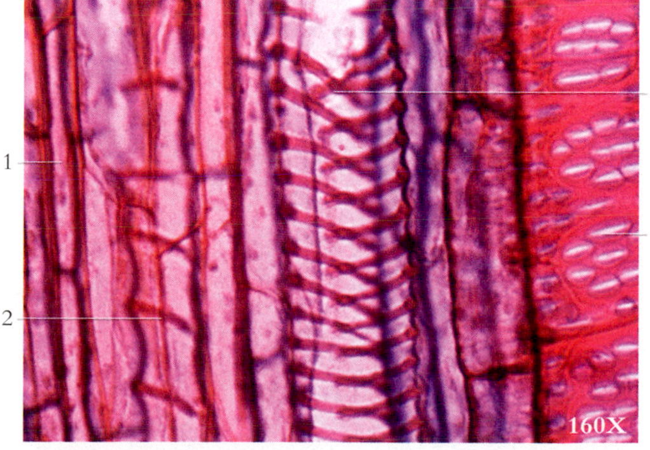

Figure 1.11 Longitudinal section through the xylem of a squash stem, *Cucurbita maxima*. The vessel elements shown here have several different patterns of wall thickenings.

1. Parenchyma
2. Annular vessel elements
3. Helical vessel elements
4. Pitted vessel elements

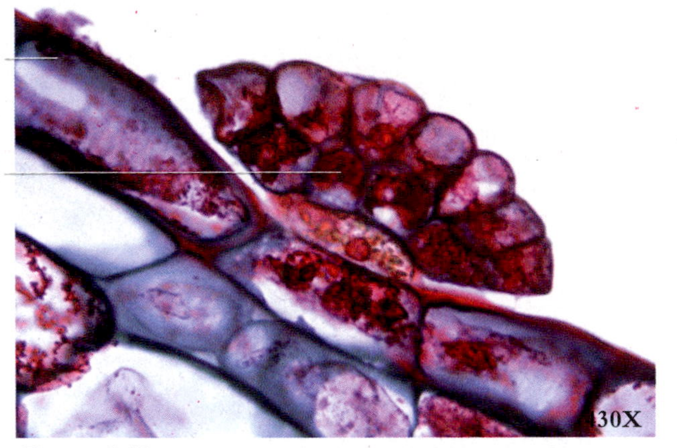

Figure 1.12 A section through a leaf of the venus flytrap, *Dionaea muscipula*, showing epidermal cells with a digestive gland. The gland is composed of secretory parenchyma cells.

1. Epidermis
2. Gland

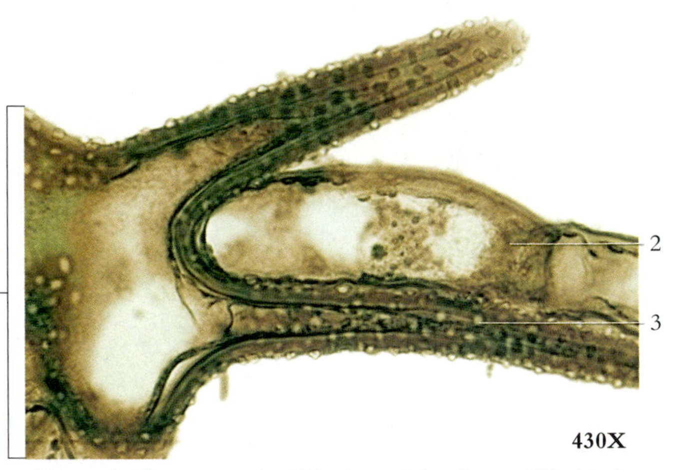

Figure 1.13 An astrosclereid in the petiole of a pond lily, *Nuphar*.

1. Astrosclereid
2. Parenchyma cell
3. Crystals in cell wall

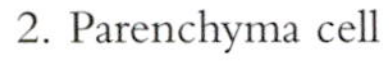

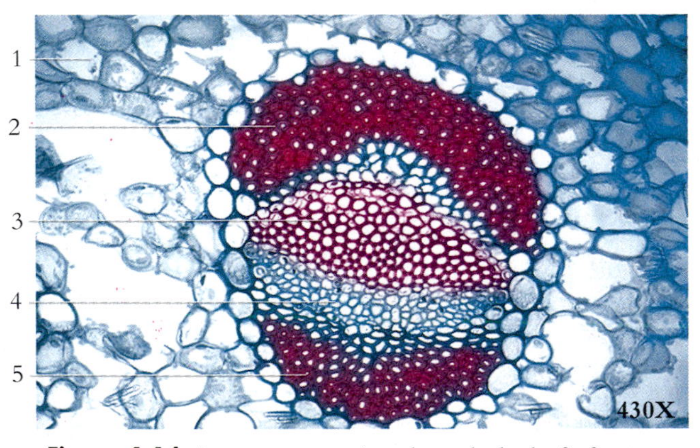

Figure 1.14 A transverse section through the leaf of a yucca, *Yucca brevifolia*, showing a vascular bundle (vein). Note the prominent sclerenchyma tissue forming caps on both sides of the bundle.

1. Leaf parenchyma
2. Leaf sclerenchyma (bundle cap)
3. Xylem
4. Phloem
5. Bundle cap

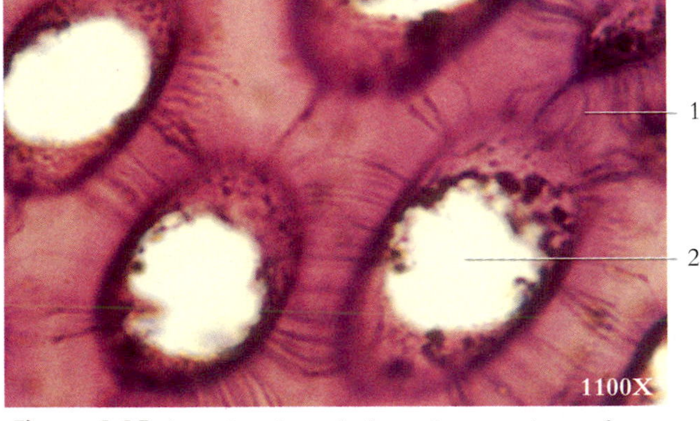

Figure 1.15 A section through the endosperm tissue of a persimmon, *Diospyros virginiana*. These thick-walled cells are actually parenchyma cells. Cytoplasmic connections, or plasmodesmata, are evident between cells.

1. Plasmodesmata
2. Cell lumen (interior space)

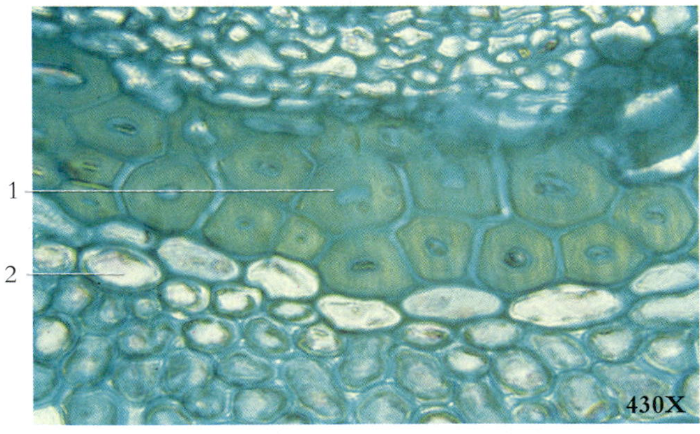

Figure 1.16 A transverse section through the stem of flax, *Linum*. Note the thick-walled fibers as compared to the thin-walled parenchyma cells.

1. Fibers
2. Parenchyma cell

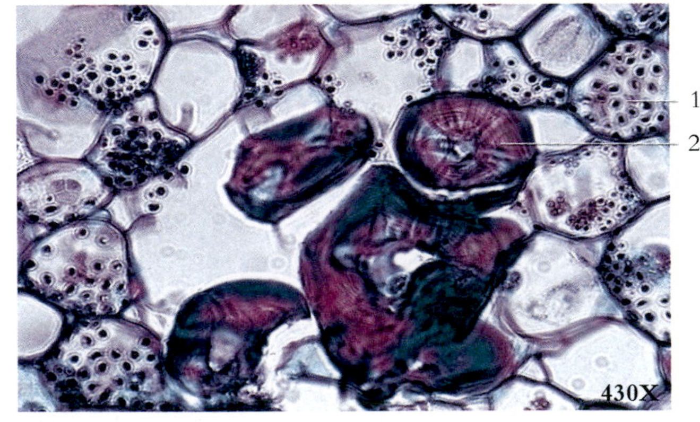

Figure 1.17 A section through the stem of a wax plant, *Hoya carnosa*. Thick-walled sclereids (stone cells) are evident.

1. Parenchyma cell containing starch grains
2. Sclereid (stone cell)

Animal Cells and Tissues

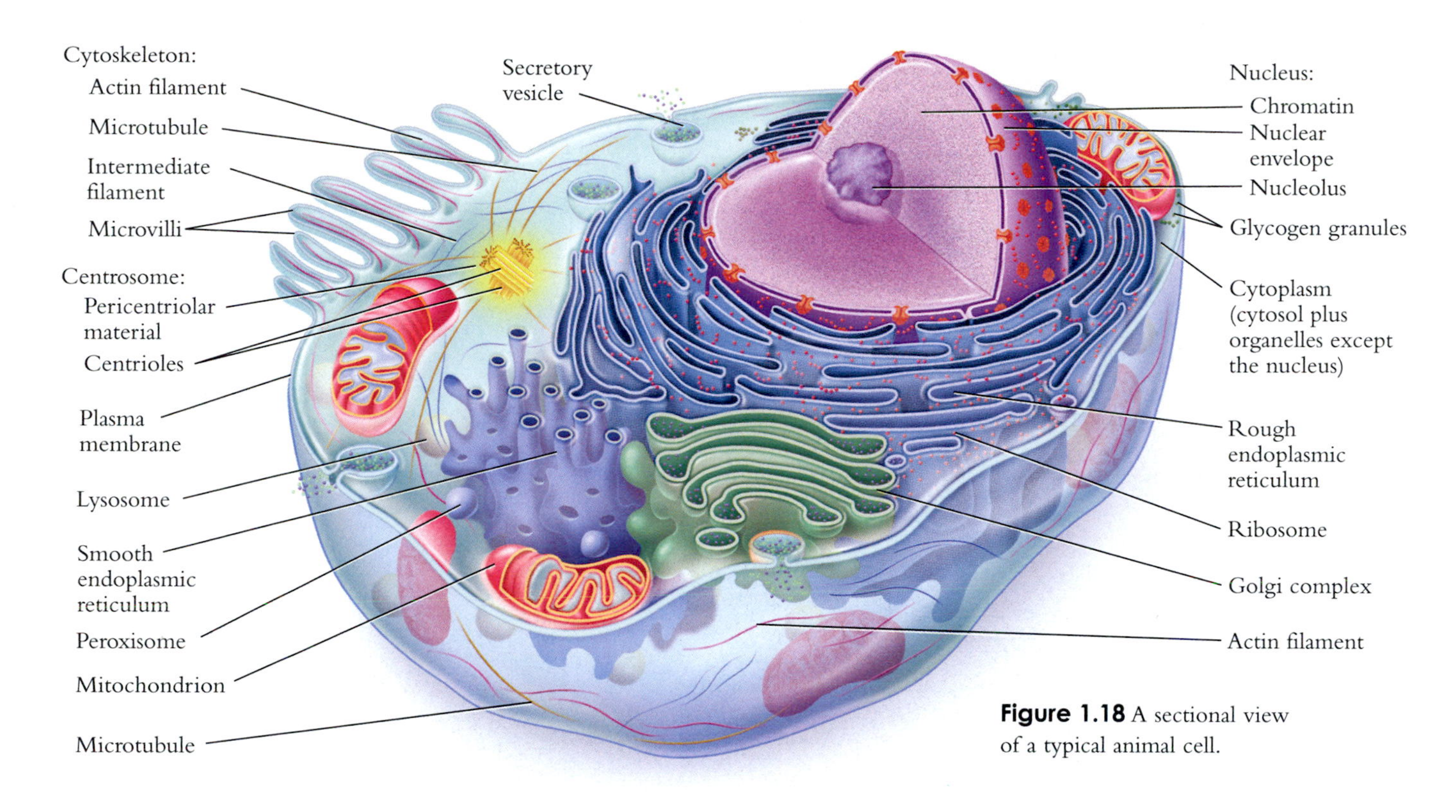

Figure 1.18 A sectional view of a typical animal cell.

Figure 1.19 An electron micrograph of a freeze-fractured nuclear envelope showing the nuclear pores.

1. Nuclear pores

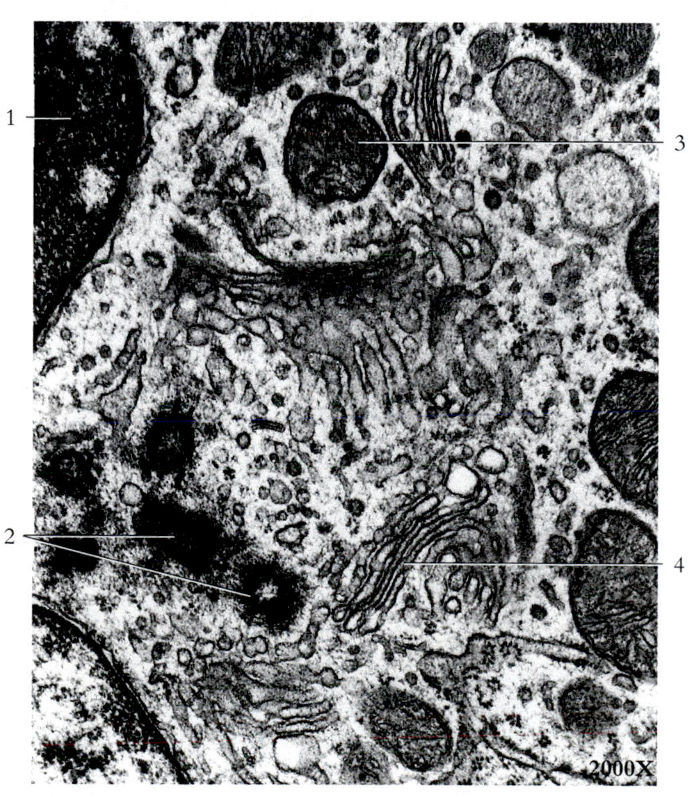

Figure 1.20 An electron micrograph of various organelles.

1. Nucleus
2. Centrioles
3. Mitochondrion
4. Golgi complex

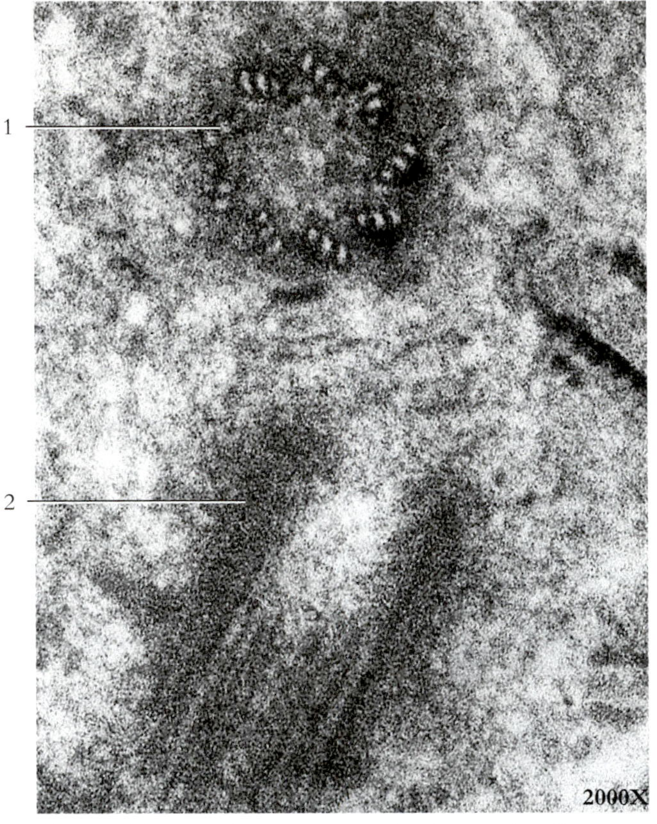

Figure 1.21 An electron micrograph of centrioles. The centrioles are positioned at right angles to one another.
1. Centriole (shown in transverse section)
2. Centriole (shown in longitudinal section)

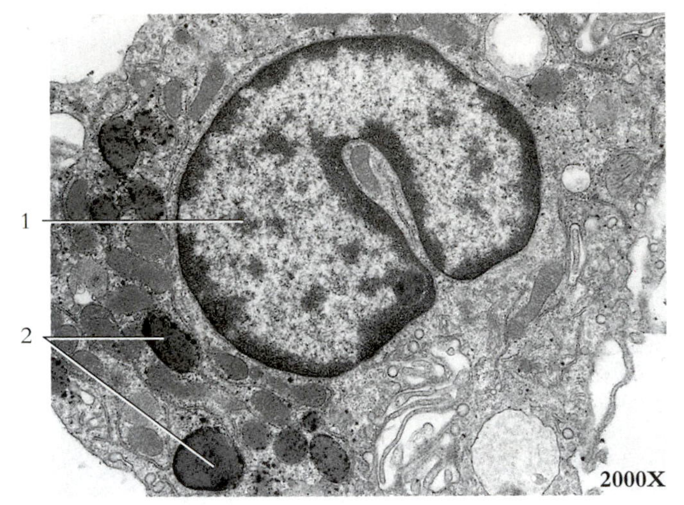

Figure 1.22 An electron micrograph of lysosomes.
1. Nucleus
2. Lysosomes

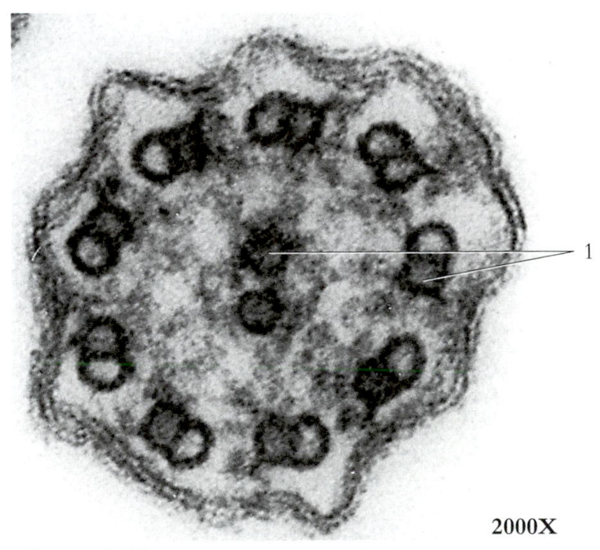

Figure 1.24 An electron micrograph of cilia (transverse section) showing the characteristic "9 + 2" arrangement of microtubules in the transverse sections.
1. Microtubules

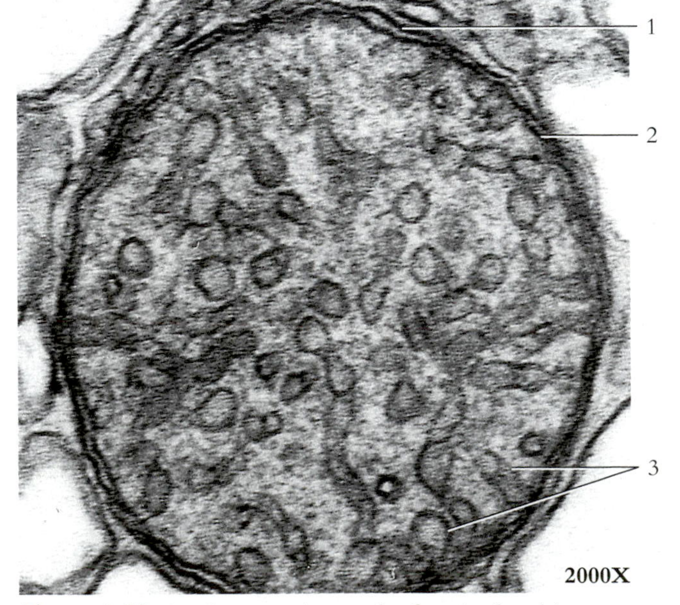

Figure 1.23 An electron micrograph of a mitochondrion.
1. Outer membrane
2. Inner membrane
3. Crista

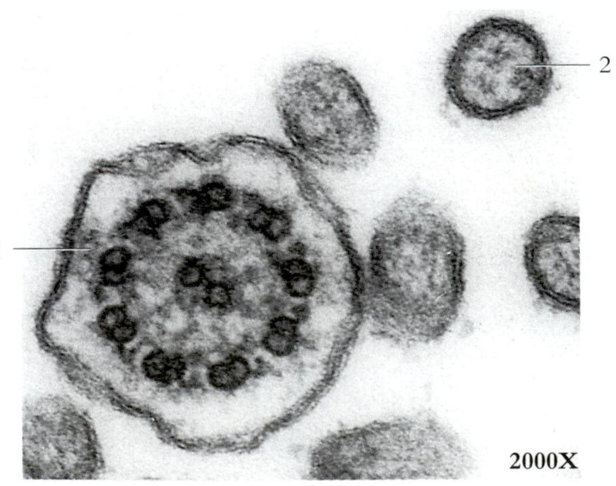

Figure 1.25 An electron micrograph showing the difference between a microvillus and a cilium.
1. Cilium
2. Microvillus

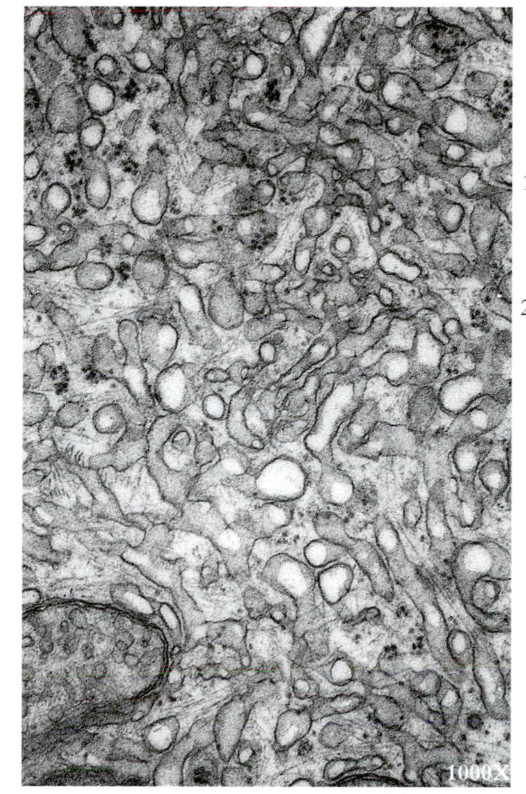

Figure 1.26 An electron micrograph of smooth endoplasmic reticulum from the testis.

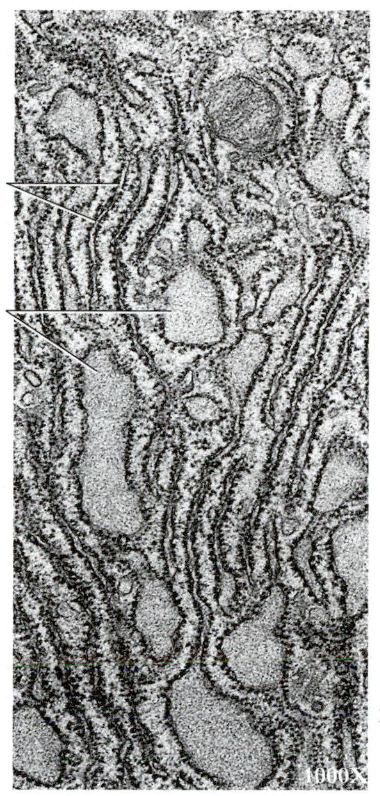

Figure 1.27 An electron micrograph of rough endoplasmic reticulum.

1. Ribosomes
2. Cisternae

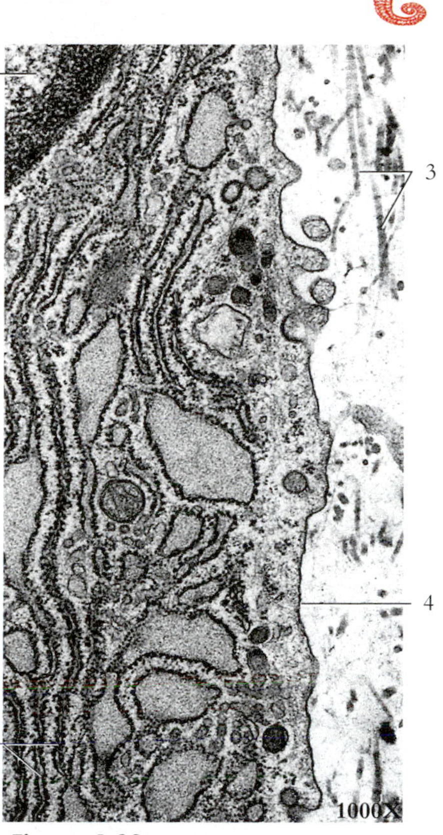

Figure 1.28 An electron micrograph of rough endoplasmic reticulum secreting collagenous filaments to the outside of the cell.

1. Nucleus
2. Rough endoplasmic reticulum
3. Collagenous filaments
4. Cell membrane

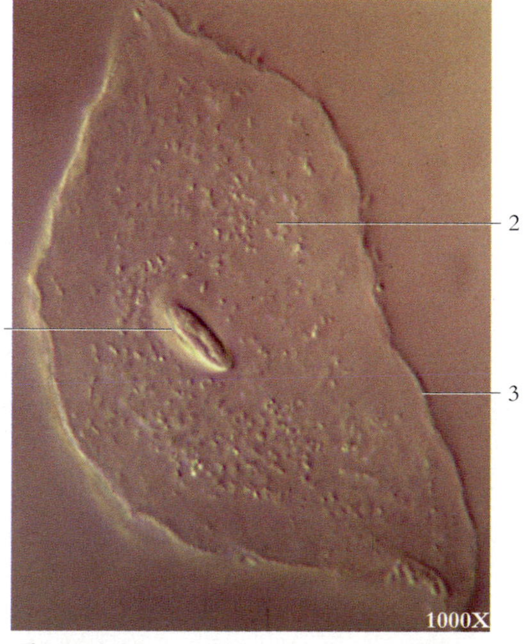

Figure 1.29 An epithelial cell from a cheek scraping.

1. Nucleus
2. Cytoplasm
3. Cell membrane

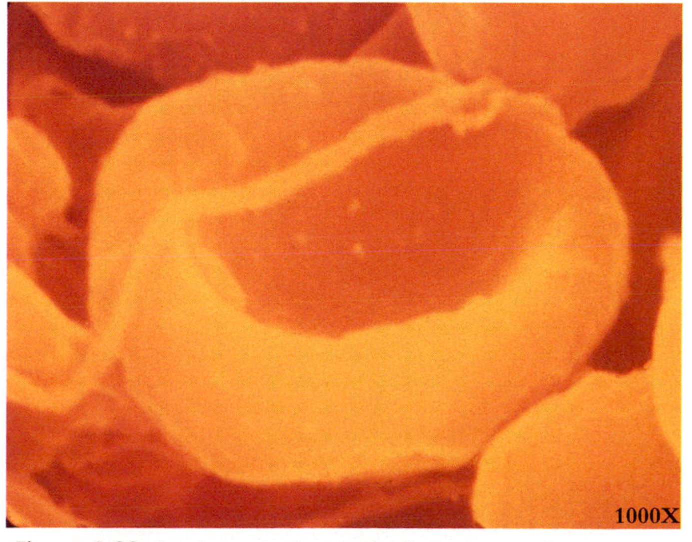

Figure 1.30 An electron micrograph of a human erythrocyte (red blood cell).

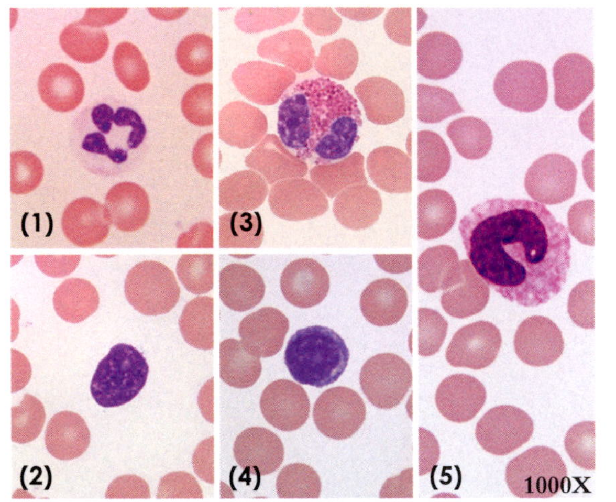

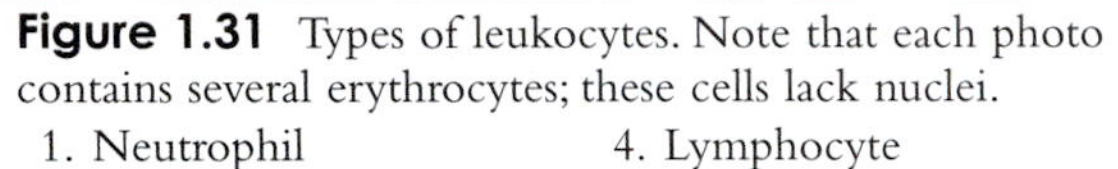

Figure 1.31 Types of leukocytes. Note that each photo contains several erythrocytes; these cells lack nuclei.

1. Neutrophil
2. Basophil
3. Eosinophil
4. Lymphocyte
5. Monocyte

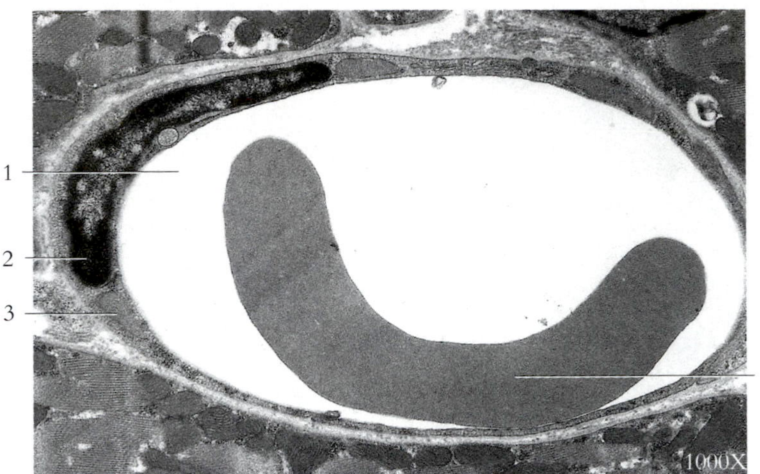

Figure 1.32 An electron micrograph of a capillary containing an erythrocyte.

1. Lumen of capillary
2. Nucleus of endothelial cell
3. Endothelial cell
4. Erythrocyte

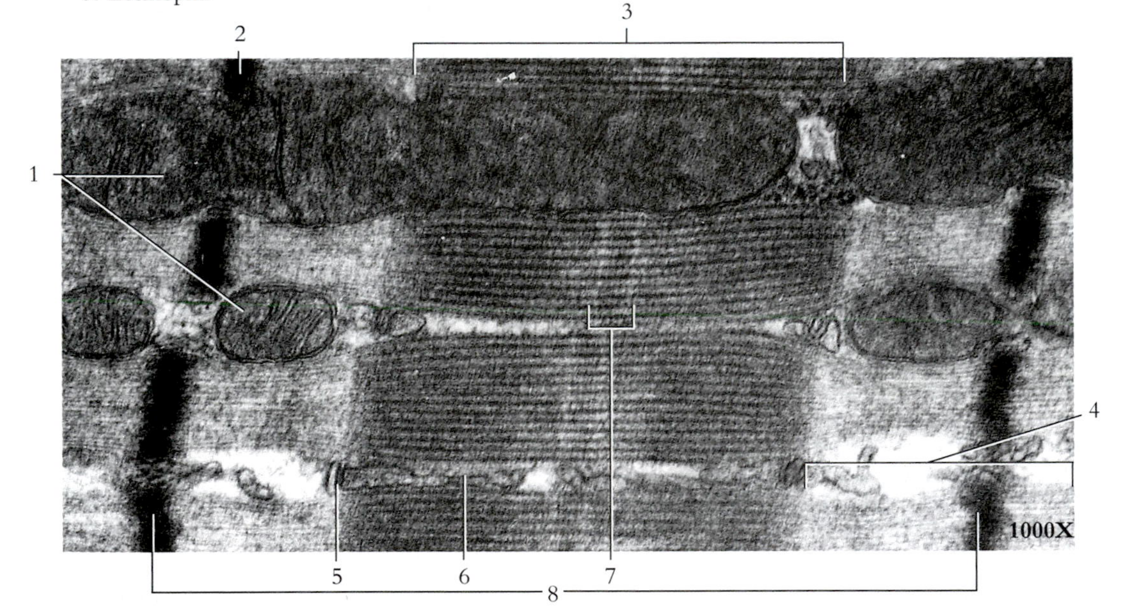

Figure 1.33 An electron micrograph of a skeletal muscle myofibril, showing the striations.

1. Mitochondria
2. Z line
3. A band
4. I band
5. T-tubule
6. Sarcoplasmic reticulum
7. M line
8. Sarcomere

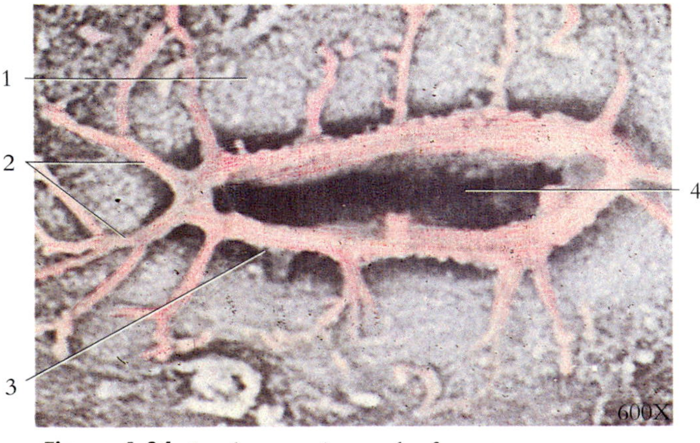

Figure 1.34 An electron micrograph of an osteocyte (bone cell) in cortical bone matrix.

1. Bone matrix
2. Canaliculi
3. Lacuna
4. Osteocyte

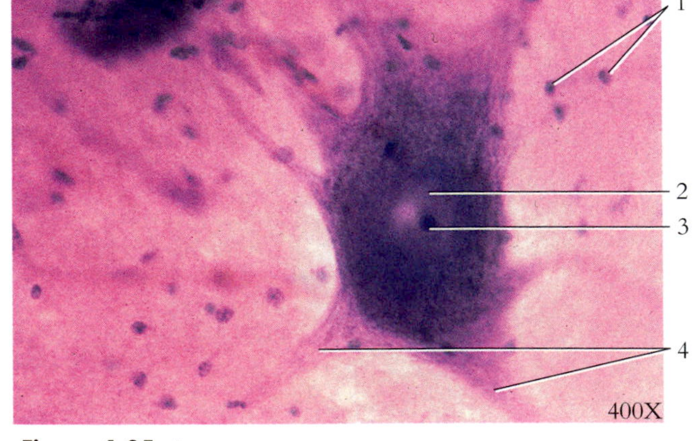

Figure 1.35 A neuron smear.

1. Nuclei of surrounding neuroglial cells
2. Nucleus of neuron
3. Nucleolus of neuron
4. Dendrites of neuron

Epithelial Tissue

Epithelial tissue covers the outside of the body and lines all organs. Its primary function is to provide protection.

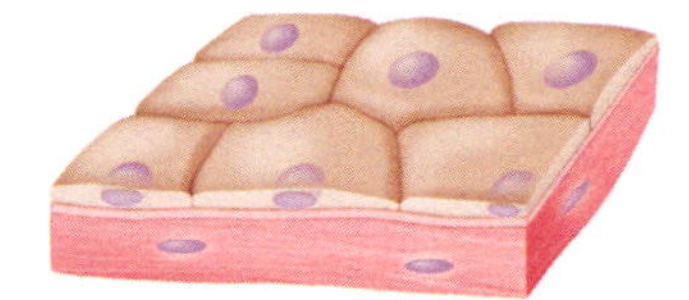

Simple squamous epithelium

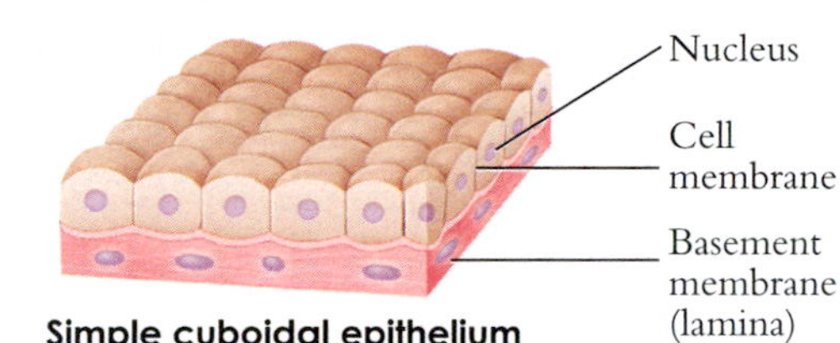

Simple cuboidal epithelium

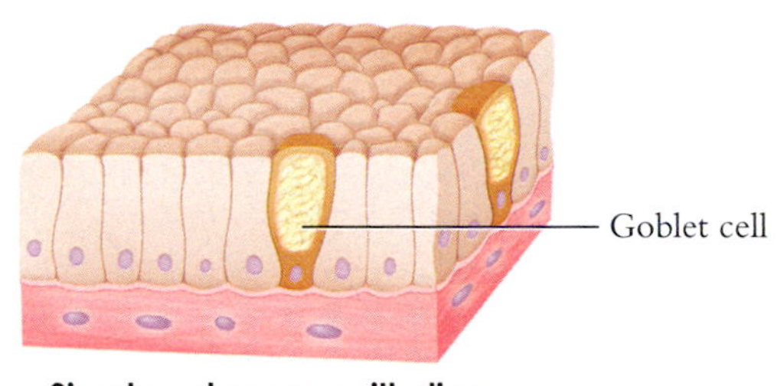

Simple columnar epithelium

Connective Tissue

Connective tissue functions as a binding and supportive tissue for all other tissues in the organism.

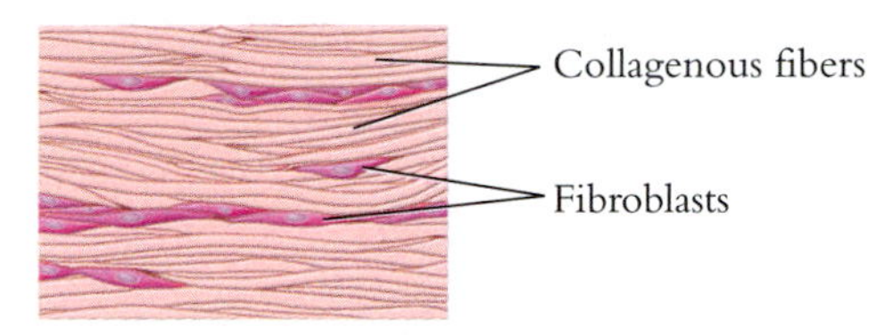

Dense regular connective tissue

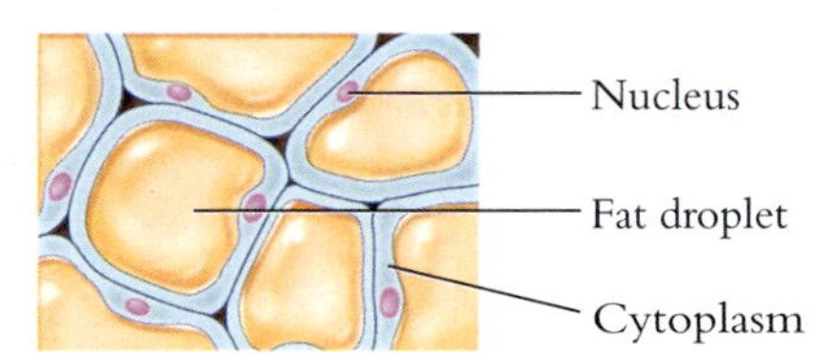

Adipose tissue

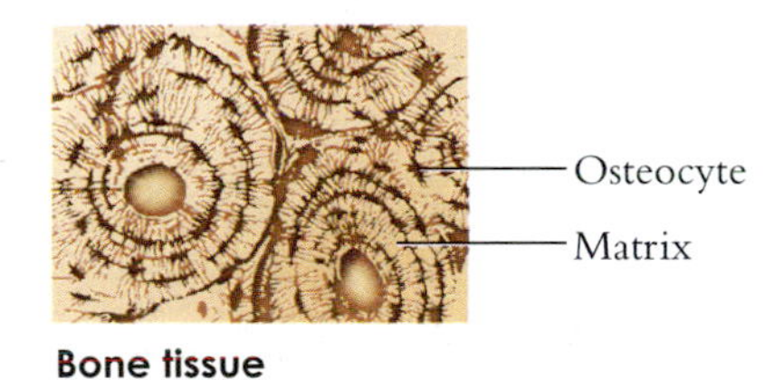

Bone tissue

Muscle Tissue

Muscle tissue is a tissue adapted to contract. Muscles provide movement and functionality to the organism.

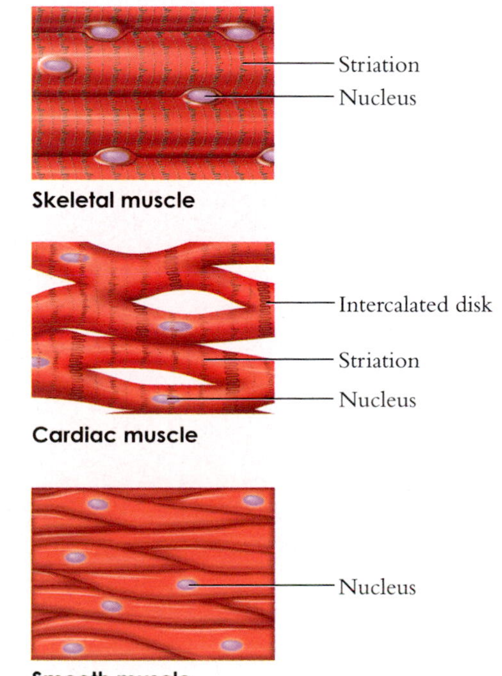

Nervous Tissue

Nervous tissue functions to receive stimuli and transmits signals from one part of the organism to another.

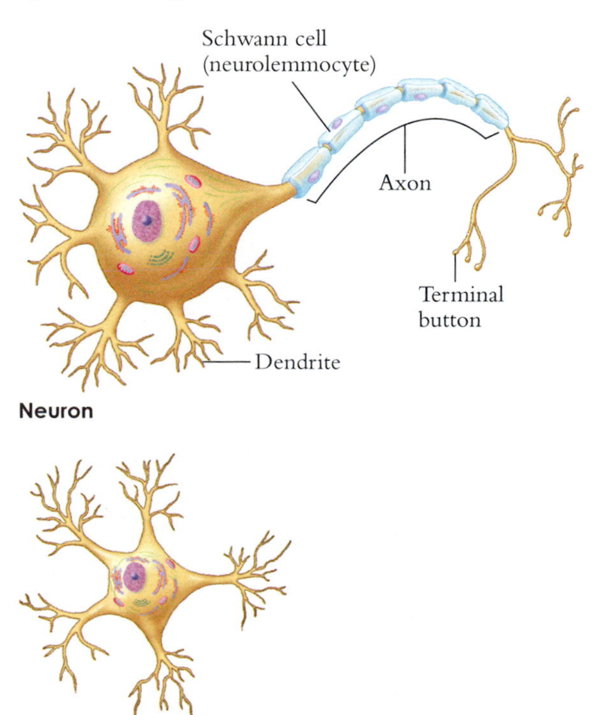

Figure 1.36 Some examples of animal tissues.

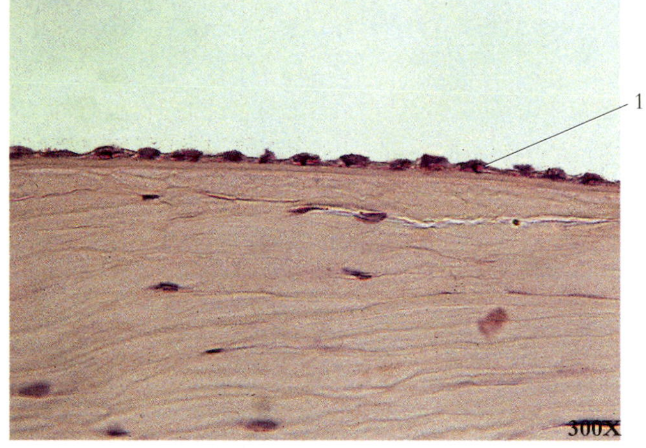

Figure 1.37 Simple squamous epithelium.
1. Single layer of flattened cells with elliptical nuclei

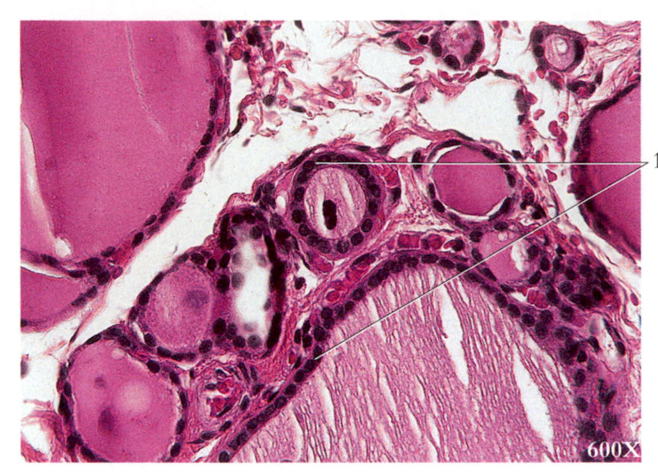

Figure 1.38 Simple cuboidal epithelium.
1. Single layer of cells with round nuclei

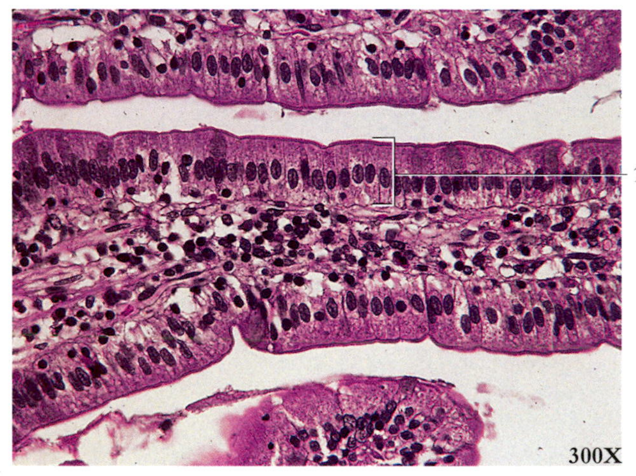

Figure 1.39 Simple columnar epithelium.
1. Single layer of cells with oval nuclei

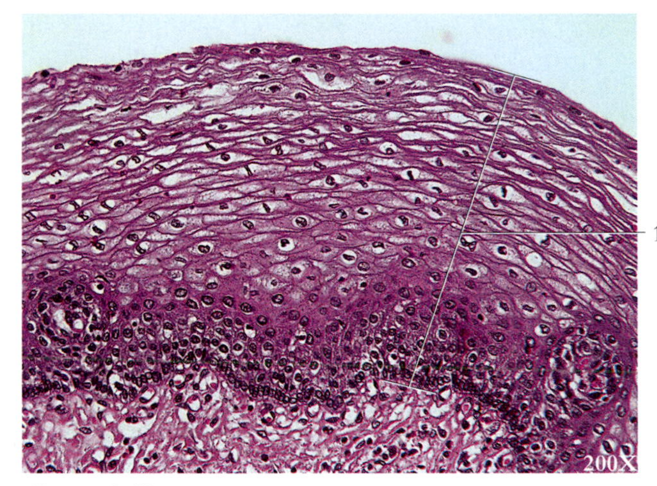

Figure 1.40 Stratified squamous epithelium.
1. Multiple layers of cells that are flattened at the upper layer

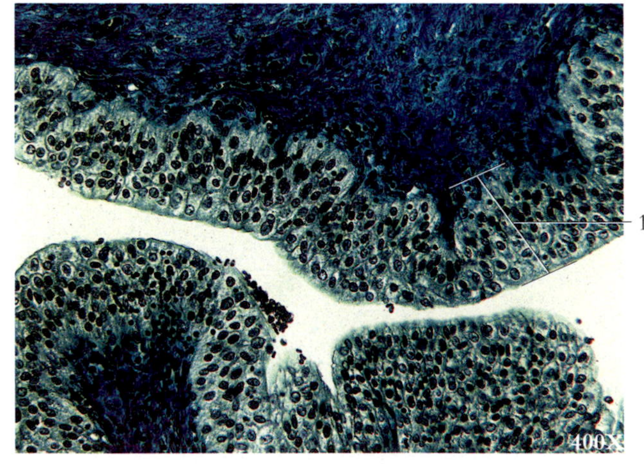

Figure 1.41 Stratified columnar epithelium.
1. Cells are balloon-like at surface

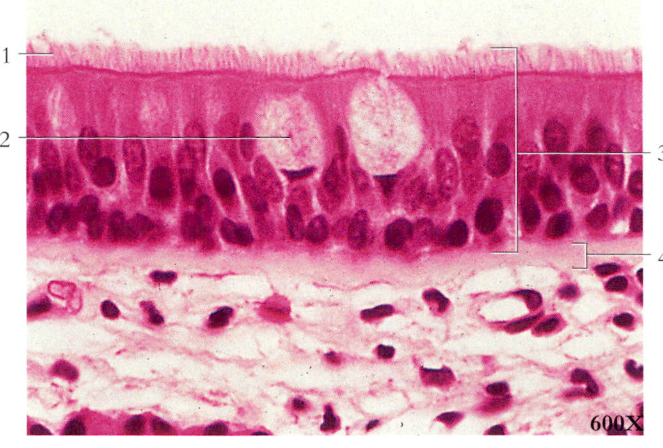

Figure 1.42 Pseudostratified columnar epithelium.
1. Cilia
2. Goblet cell
3. Pseudostratified columnar epithelium
4. Basement membrane

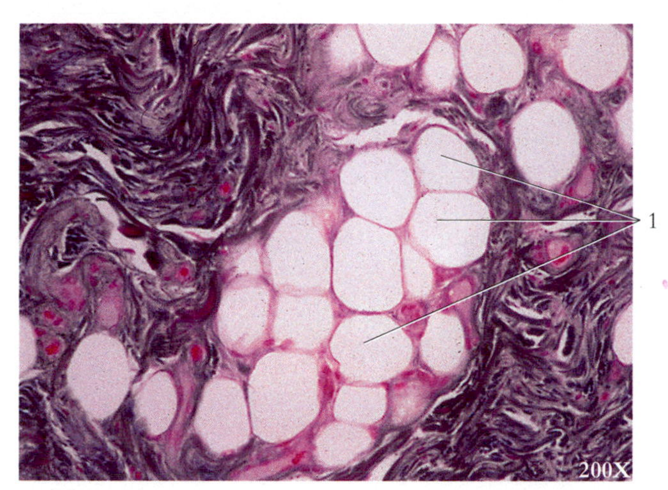

Figure 1.43 Adipose connective tissue.
1. Adipocytes (adipose cells)

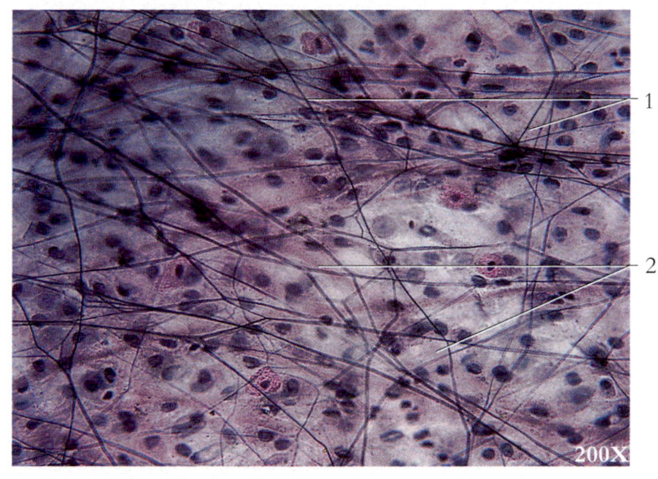

Figure 1.44 Loose connective tissue stained for fibers.
1. Elastic fibers (black)
2. Collagen fibers (pink)

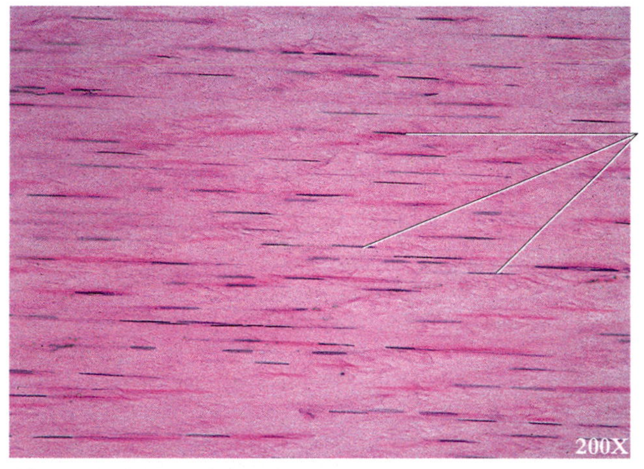

Figure 1.45 Dense regular connective tissue.
1. Nuclei of fibroblasts arranged in parallel rows between pink-stained collagen fibers

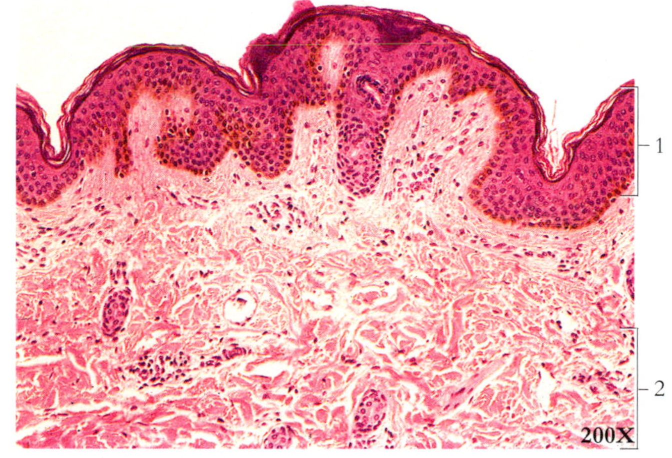

Figure 1.46 Dense irregular connective tissue.
1. Epidermis
2. Dense irregular connective tissue (reticular layer of dermis)

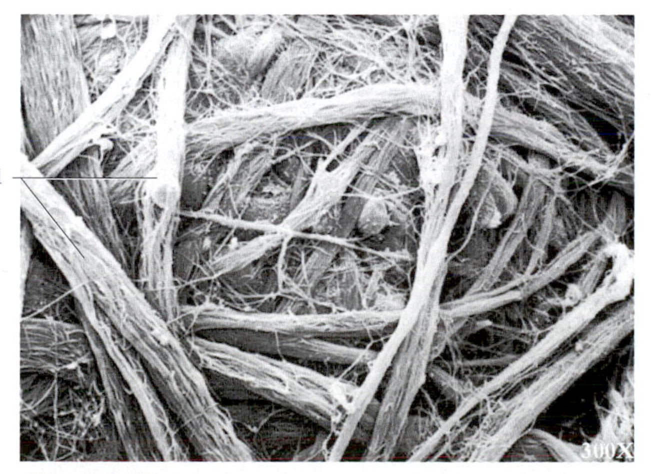

Figure 1.47 An electron micrograph of dense irregular connective tissue.
1. Collagenous fibers

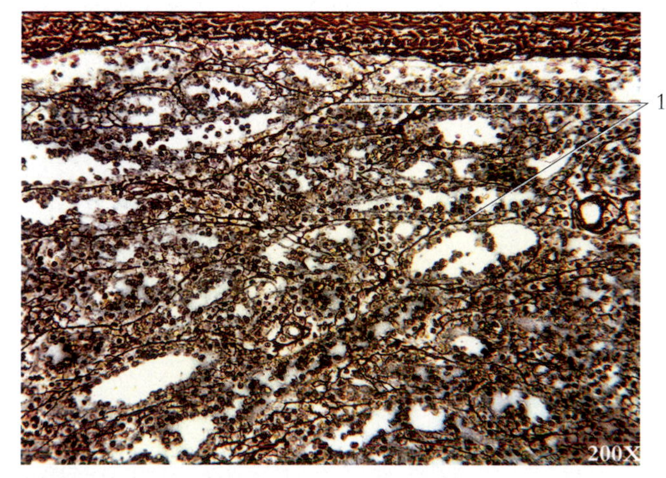

Figure 1.48 Reticular connective tissue.
1. Reticular fibers

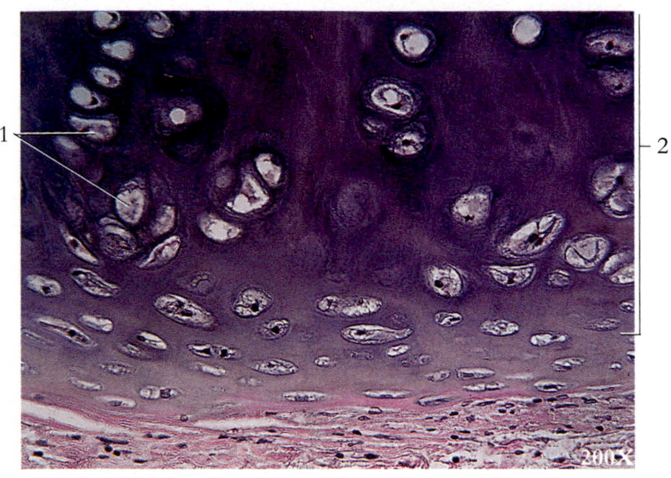

Figure 1.49 Hyaline cartilage.
1. Chondrocytes
2. Hyaline cartilage

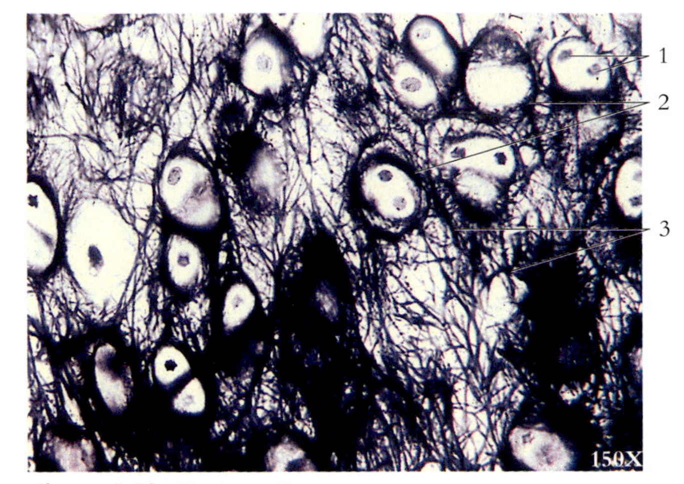

Figure 1.50 Elastic cartilage.
1. Chondrocytes
2. Lacunae
3. Elastic fibers

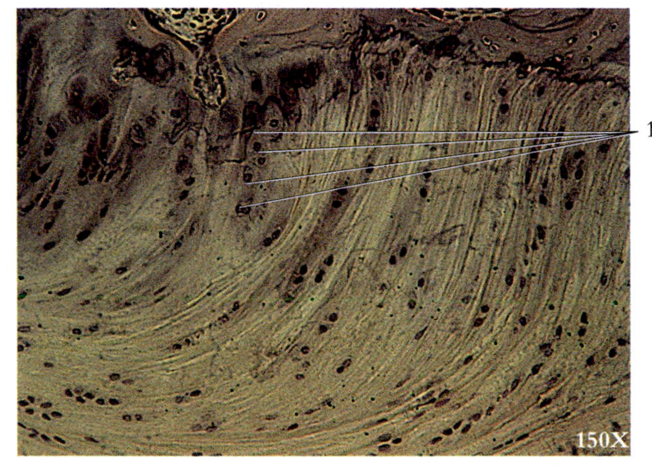

Figure 1.51 Fibrocartilage.
1. Chondrocytes arranged in a row

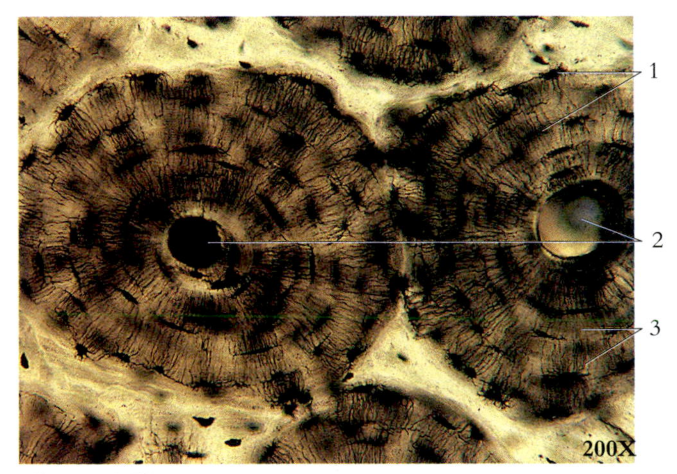

Figure 1.52 A transverse section of two osteons in compact bone tissue.
1. Lacunae containing osteocytes
2. Central (Haversian) canals
3. Lamellae

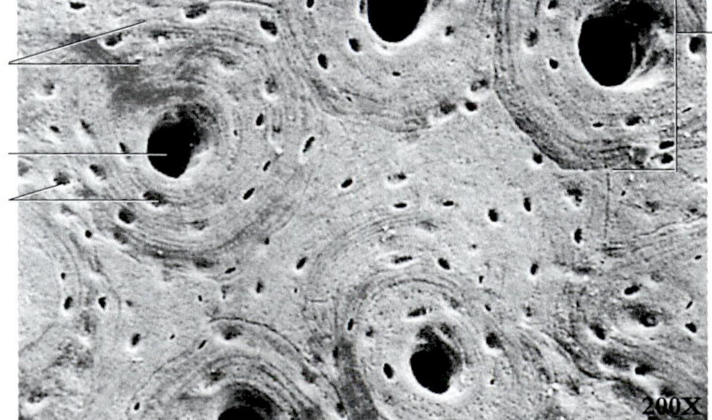

Figure 1.53 An electron micrograph of bone tissue.
1. Interstitial lamellae
2. Lamellae
3. Central canal (Haversian canal)
4. Lacunae
5. Osteon (Haversian system)

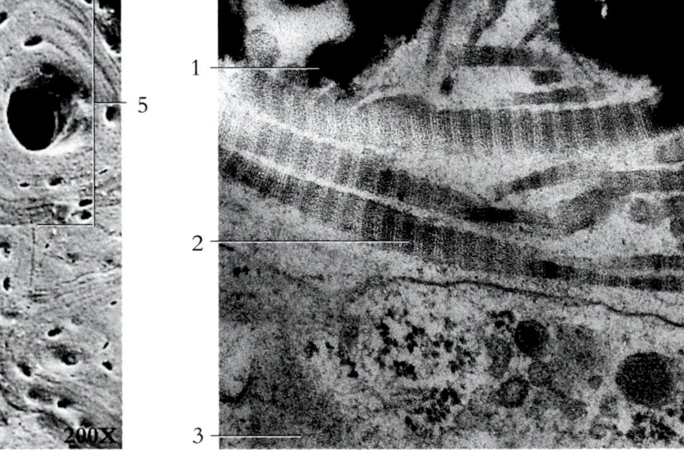

Figure 1.54 An electron micrograph of bone tissue formation.
1. Bone mineral (calcium salts stain black)
2. Collagenous filament (distinct banding pattern)
3. Collagen-secreting osteoblasts

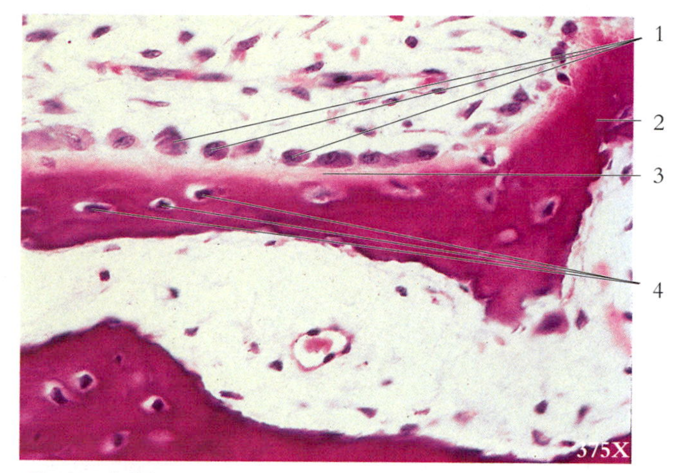

Figure 1.55 Osteoblasts.
1. Osteoblasts
2. Bone
3. Osteoid
4. Osteocytes

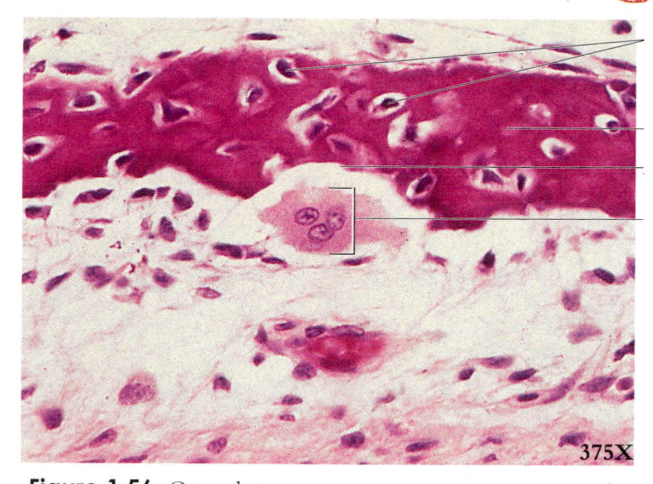

Figure 1.56 Osteoclast.
1. Osteocytes
2. Bone
3. Howship's lacuna
4. Osteoclast in Howship's lacuna

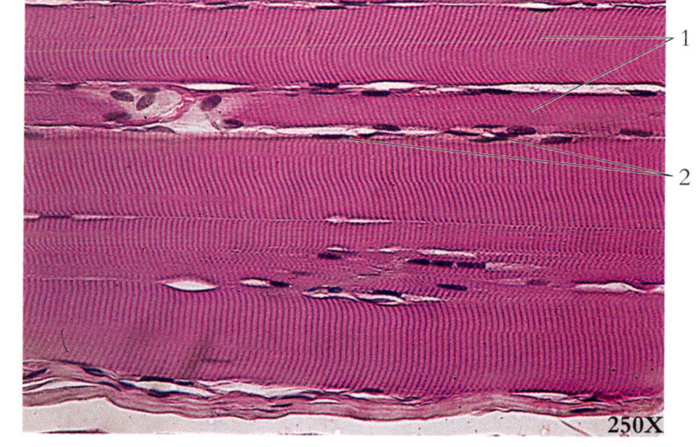

Figure 1.57 A longitudinal section of skeletal muscle tissue.
1. Skeletal muscle cells (note striations)
2. Multiple nuclei in periphery of cell

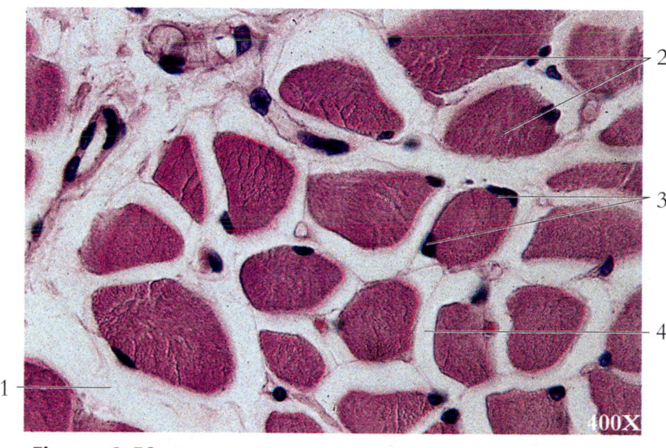

Figure 1.58 A transverse section of skeletal muscle tissue.
1. Perimysium (surrounds bundles of cells)
2. Skeletal muscle cells
3. Nuclei in periphery of cell
4. Endomysium (surrounds cells)

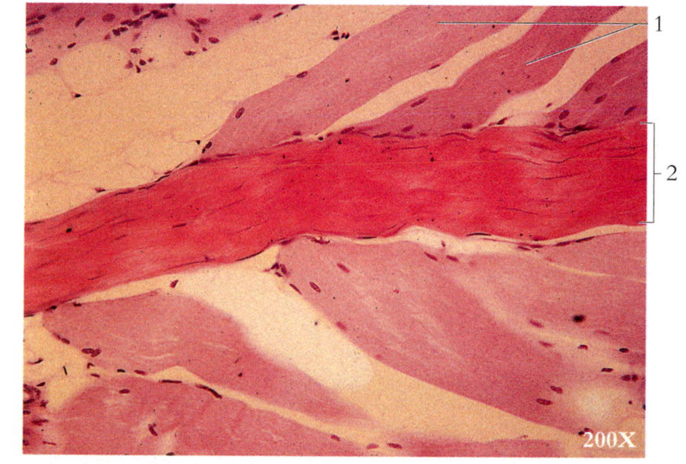

Figure 1.59 The attachment of skeletal muscle to tendon.
1. Skeletal muscle
2. Dense regular connective tissue (tendon)

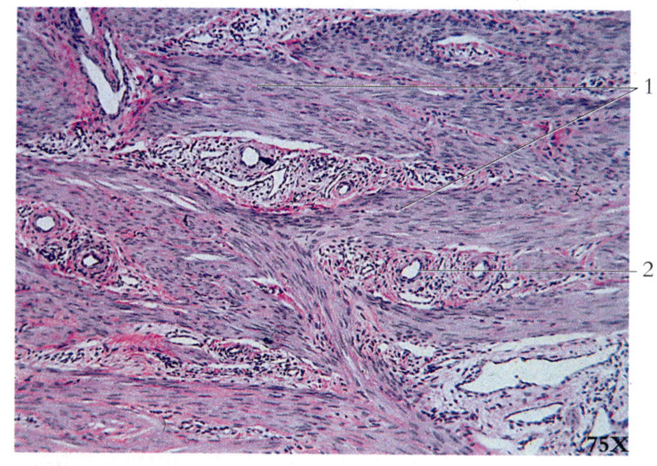

Figure 1.60 Smooth muscle tissue.
1. Smooth muscle
2. Blood vessel

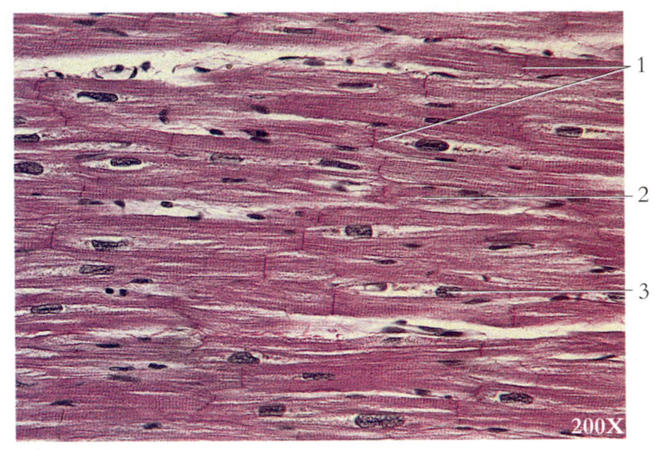

Figure 1.61 Cardiac muscle tissue.
1. Intercalated disks
2. Light-staining perinuclear sarcoplasm
3. Nucleus in center of cell

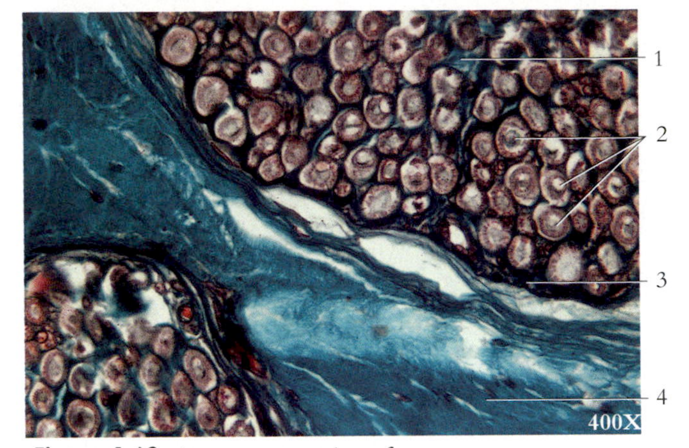

Figure 1.62 A transverse section of a nerve.
1. Endoneurium
2. Axons
3. Perineurium
4. Epineurium

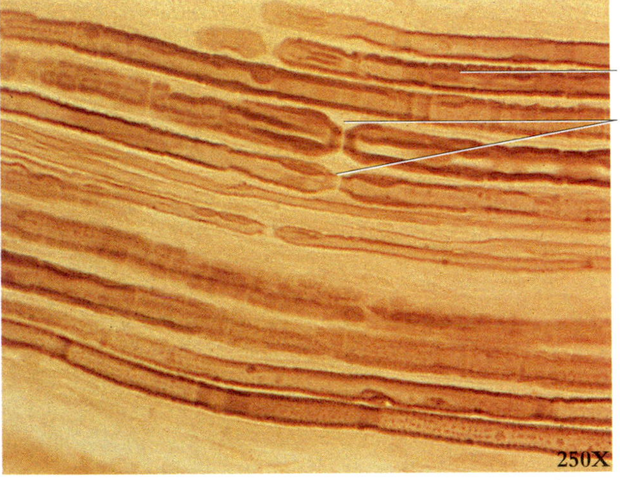

Figure 1.63 A longitudinal section of axons.
1. Myelin sheath
2. Neurofibril nodes (nodes of Ranvier)

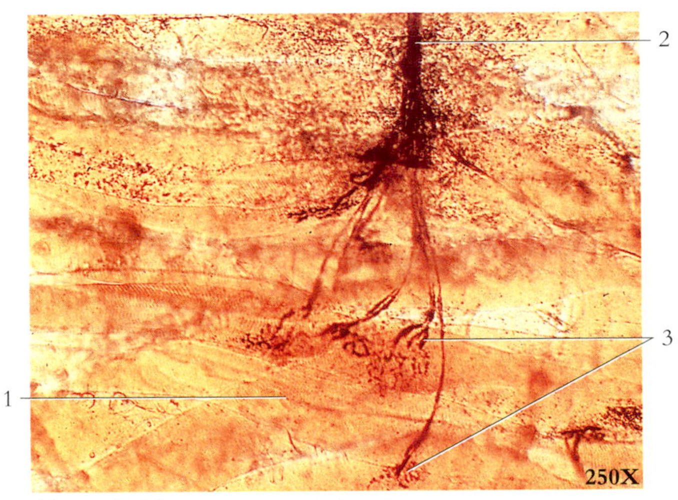

Figure 1.64 A neuromuscular junction.
1. Skeletal muscle fiber
2. Motor nerve
3. Motor end plates

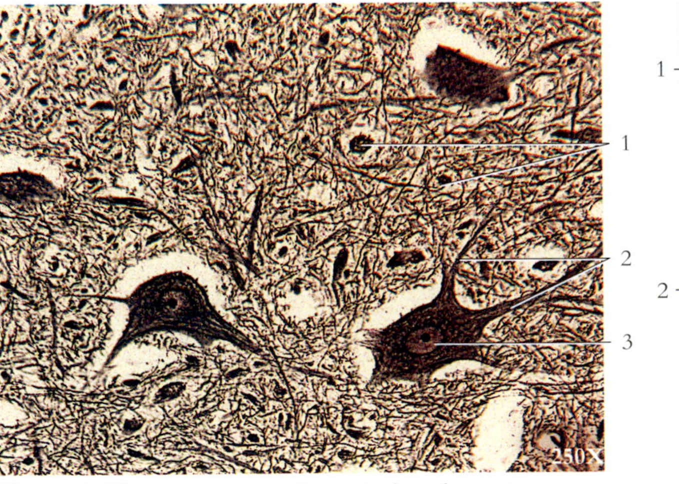

Figure 1.65 Motor neurons from spinal cord.
1. Neuroglia cells
2. Dendrites
3. Nucleus

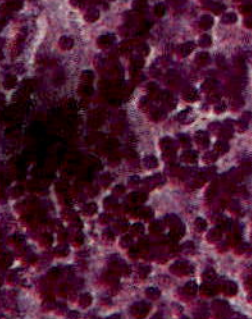

Figure 1.66 Purkinje neurons from the cerebellum.
1. Molecular layer of cerebellar cortex
2. Granular layer of cerebellar cortex
3. Dendrites of Purkinje cell
4. Purkinje cell body

Perpetuation of Life

Chapter 2

The term *cell cycle* refers to how a multicellular organism develops, grows, and maintains and repairs body tissues. In the cell cycle, each new cell receives a complete copy of all genetic information in the parent cell and the cytoplasmic substances and organelles to carry out hereditary instructions.

The animal cell cycle (see fig. 2.3) is divided into: 1) interphase, which includes Gap 1 (G1), Synthesis (S), and Gap 2 (G2) phases; and 2) mitosis, which includes prophase, metaphase, anaphase, and telophase. *Interphase* is the interval between successive cell divisions during which the cell is metabolizing and the chromosomes are directing RNA synthesis. The *G1 phase* is the first growth phase, the *S phase* is when DNA is replicated, and the *G2 phase* is the second growth phase. *Mitosis* (also known as karyokinesis) is the division of the nuclear parts of a cell to form two daughter nuclei with the same number of chromosomes as the original nucleus.

Like the animal cell cycle, the plant cell cycle consists of growth, synthesis, mitosis, and cytokinesis. *Growth* is the increase in cellular mass as the result of metabolism; *synthesis* is the production of DNA and RNA to regulate cellular activity; mitosis is the splitting of the nucleus and the equal separation of the chromatids; and cytokinesis is the division of the cytoplasm that accompanies mitosis.

Unlike animal cells, plant cells have a rigid cell wall that does not cleave during cytokinesis. Instead, a new cell wall is constructed between the daughter cells. Furthermore, many land plants do not have centrioles for the attachment of spindles. The microtubules in these plants form a barrel-shaped anastral spindle at each pole. Mitosis and cytokinesis in plants occur in basically the same sequence as these processes in animal cells.

Asexual reproduction is propagation without sex; that is, the production of new individuals by processes that do not involve *gametes* (sex cells). Asexual reproduction occurs in a variety of microorganisms, fungi, plants, and animals, wherein a single parent produces offspring with characteristics identical to itself. Asexual reproduction is not dependent on the presence of other individuals. No egg or sperm is required. In asexual reproduction, all the offspring are genetically identical (except for mutants). Types of asexual reproduction include:

1. *fission*—subdivision of a cell (or organism, population, species, etc.) into to separate parts. Binary fission produces two separate parts; multiple fission produces more than two separate parts (cells, populations, species, etc.);
2. *sporulation*—multiple fission: many cells are formed and join together in a cyst-like structure (protozoans and fungi);
3. *budding*—buds develop organisms like the parent and then detach themselves (hydras, yeast, certain plants); and
4. *fragmentation*—organisms break into two or more parts, and each part is capable of becoming a complete organism (algae, flatworms, echinoderms).

Sexual reproduction is propagation of new organisms through the union of genetic material from two parents. Sexual reproduction usually involves the fusion of haploid gametes (such as sperm and egg cells) during fertilization to form a zygote.

The major biological difference between sexual and asexual reproduction is that sexual reproduction produces genetic variation in the offspring. The combining of genetic material from the gametes produces offspring that are different from either parent and contain new combinations of characteristics. This may increase the ability of the species to survive environmental changes or to reproduce in new habitats. The only genetic variation that can arise in asexual reproduction comes from mutations.

Figure 2.1 Sexual reproduction. A pair of cinnamon teal, *Anas cyanoptera*, in early spring.

Vegetative propagation

A plant produces external stems, or runners. Simple vegetative propagation occurs in a number of flowering plants, such as strawberries.

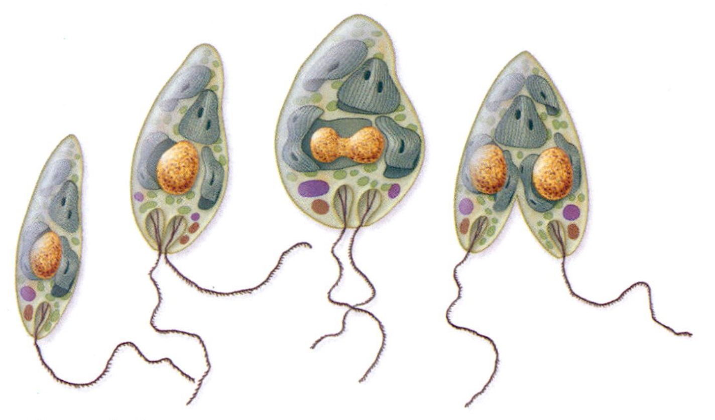

Binary fission

A single cell divides, forming two separate cells. Fission occurs in bacteria, protozoans, and other single-celled organisms.

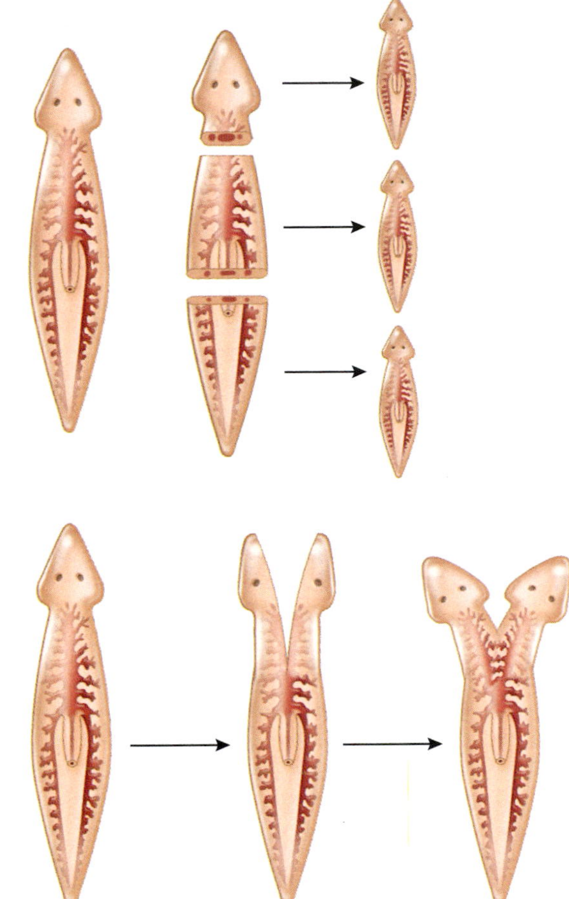

Fragmentation

An organism breaks into two or more parts, each capable of becoming a complete organism. Fragmentation occurs in flatworms and echinoderms.

Figure 2.2 Types of asexual reproduction: vegetative propagation, binary fission, and fragmentation.

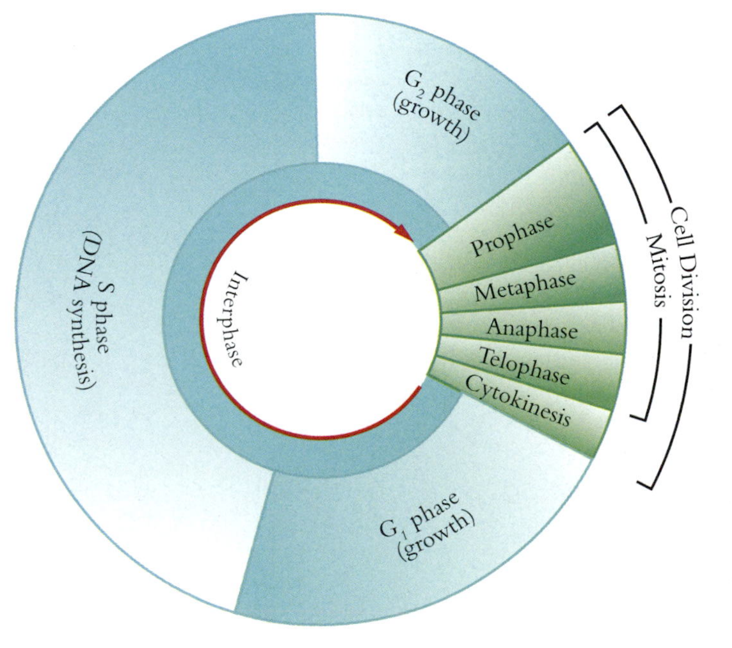

Figure 2.3 The animal cell cycle.

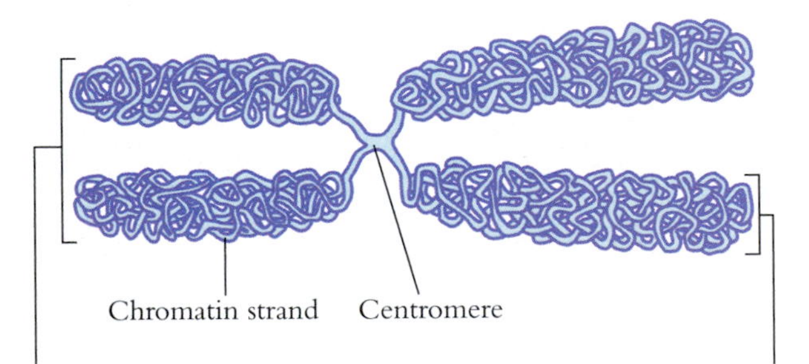

Figure 2.4 Each duplicated chromosome consists of two identical chromatids attached at the centrally located and constricted centromere.

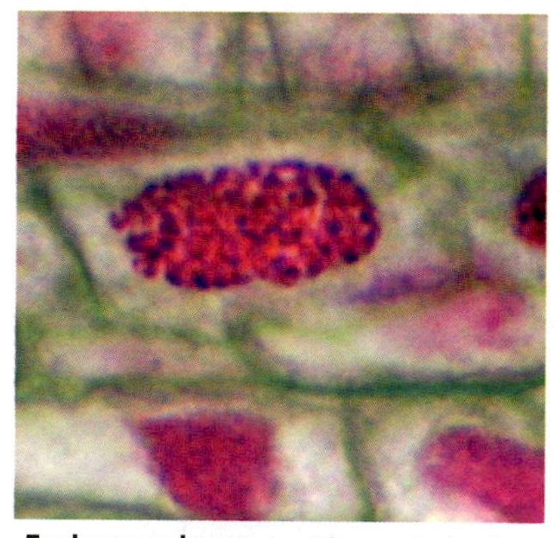

Early prophase — Chromatin begins to condense to form chromosomes.

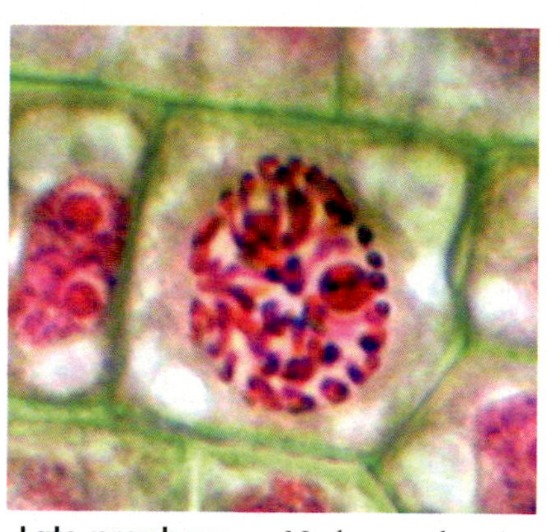

Late prophase — Nuclear envelope is intact, and chromatin condenses into chromosomes.

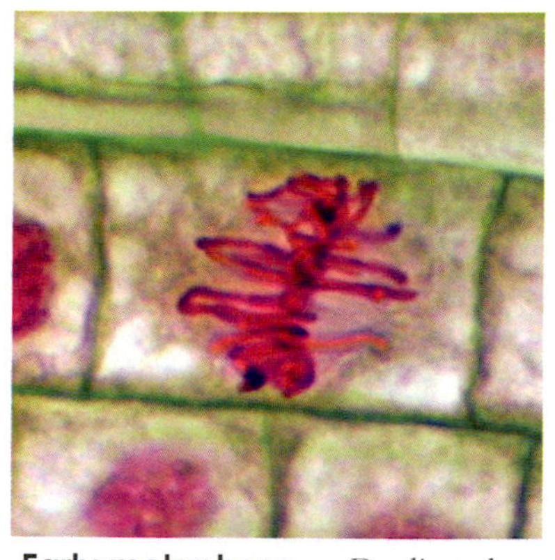

Early metaphase — Duplicated chromosomes are each made up of two chromatids, at equatorial plane.

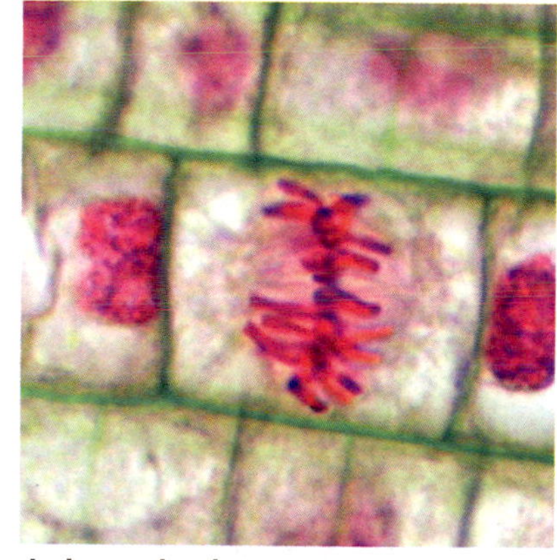

Late metaphase — Duplicated chromosomes are each made up of two chromatids, at equatorial plane.

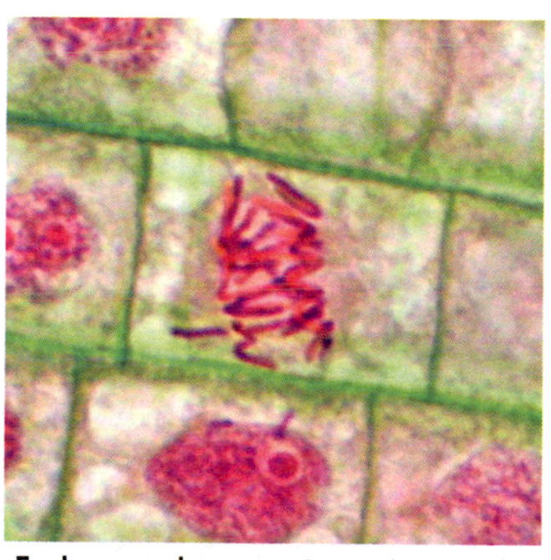

Early anaphase — Sister chromatids are beginning to separate into daughter chromosomes.

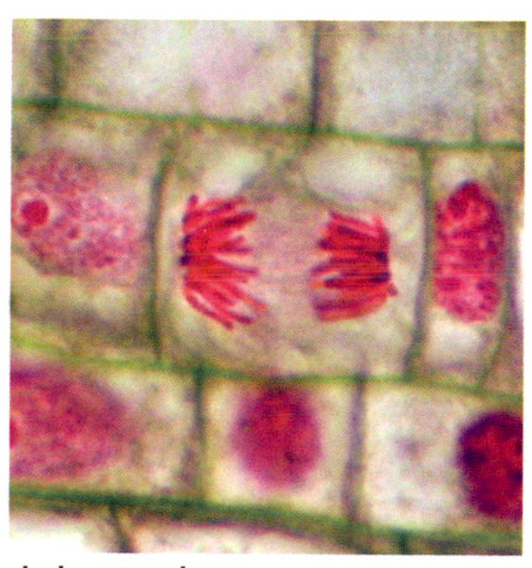

Late anaphase — Daughter chromosomes are nearing poles.

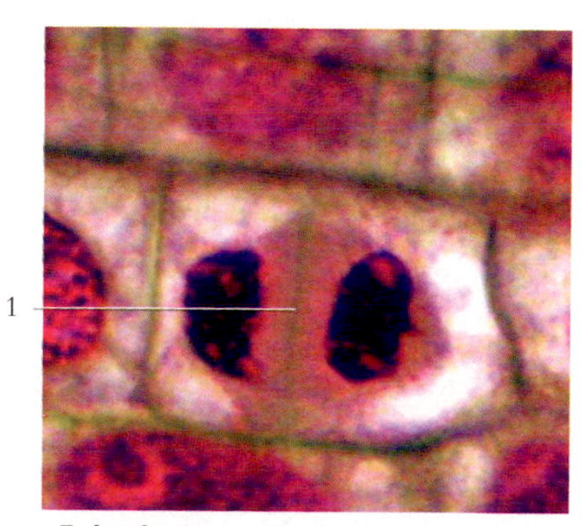

Telophase — Daughter chromosomes are at poles, and cell plate is forming.
1. Cell plate

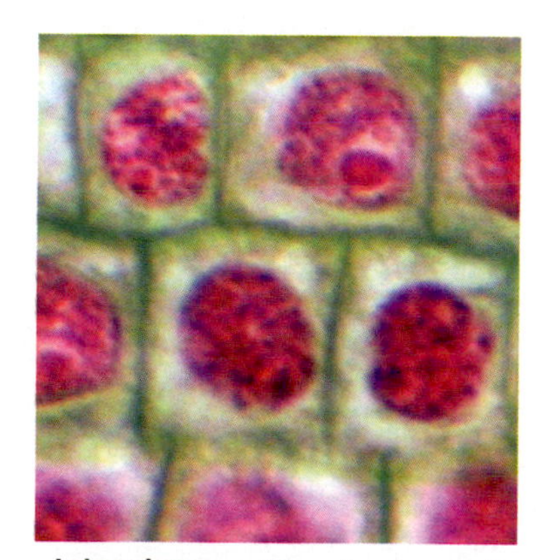

Interphase — Two daughter cells result from cytokinesis.

Figure 2.5 The stages of mitosis in Hyacinth, *Hyacinthus*, root tip. (all 430X)

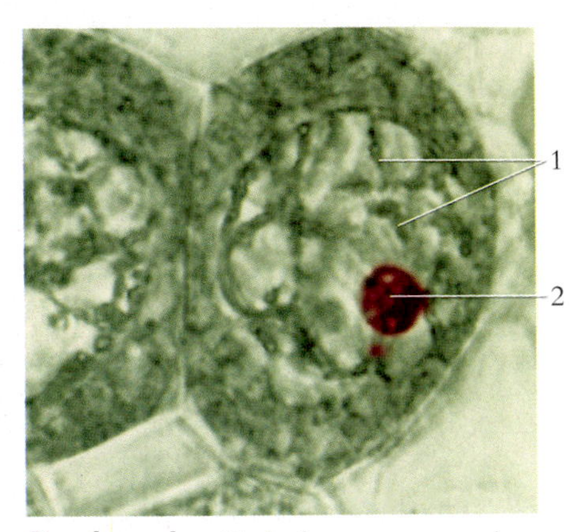

Prophase I — Each chromosome consists of two chromatids joined by a centromere.
1. Chromatids
2. Nucleolus

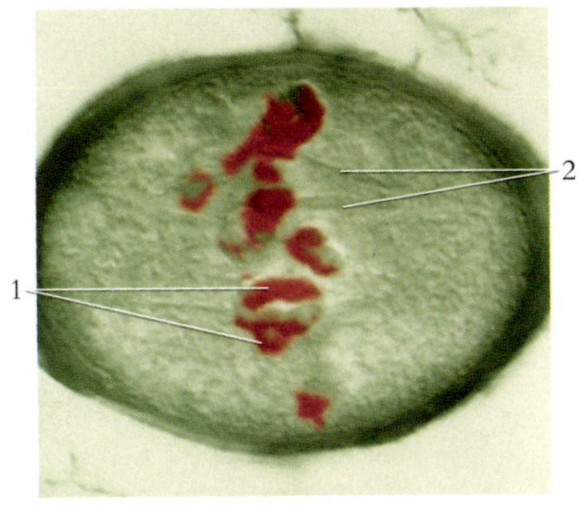

Metaphase I — Chromosome pairs align at the equator.
1. Chromosome pairs at equator
2. Spindle fibers

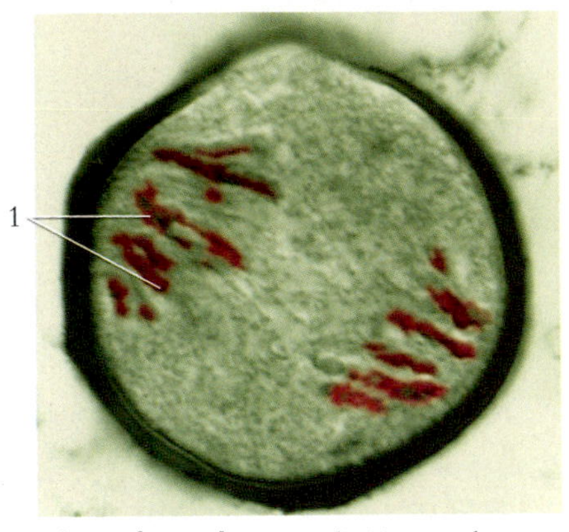

Anaphase I — No division at the centromeres occurs as the chromosomes separate, so one entire chromosome goes to each pole.
1. Chromosomes (two chromatids each)

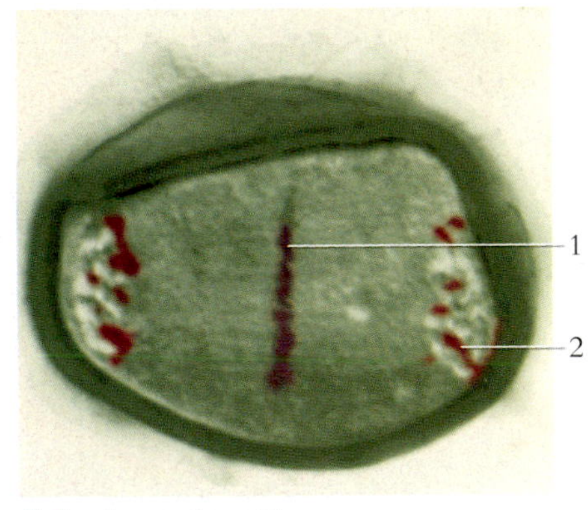

Telophase I — Chromosomes lengthen and become less distinct. The cell plate (in some plants) forms between forming cells.
1. Cell plate (new cell wall)
2. Chromosome

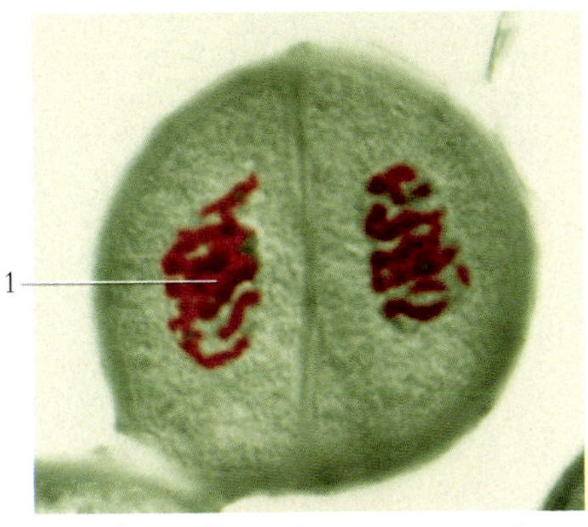

Prophase II — Chromosomes condense as in prophase I.
1. Chromosomes

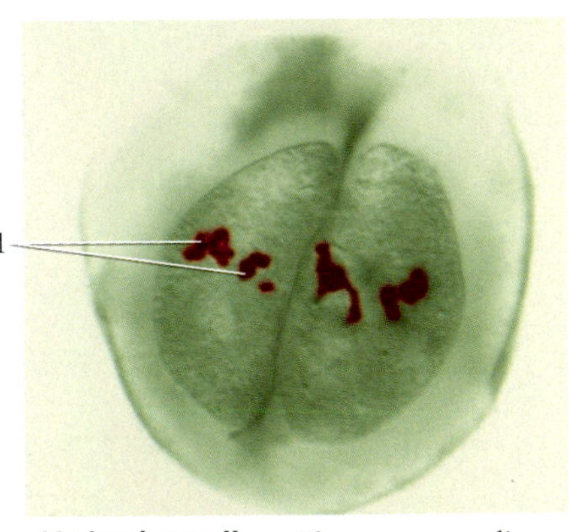

Metaphase II — Chromosomes align on the equator, and spindle fibers attach to the centromeres. This is similar to metaphase in mitosis.
1. Chromosomes

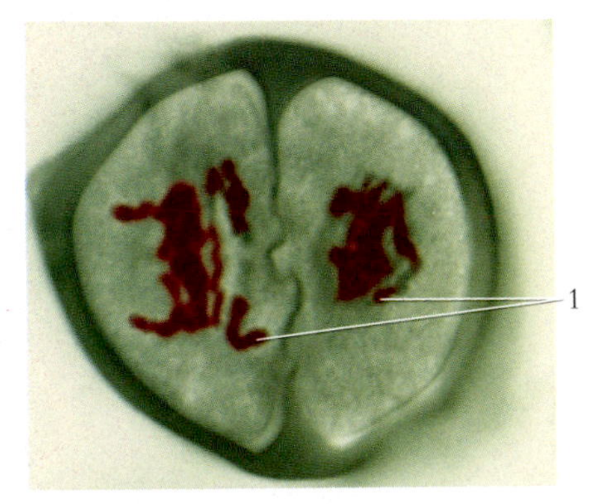

Anaphase II — Chromatids separate, and each is pulled to an opposite pole.
1. Chromatids

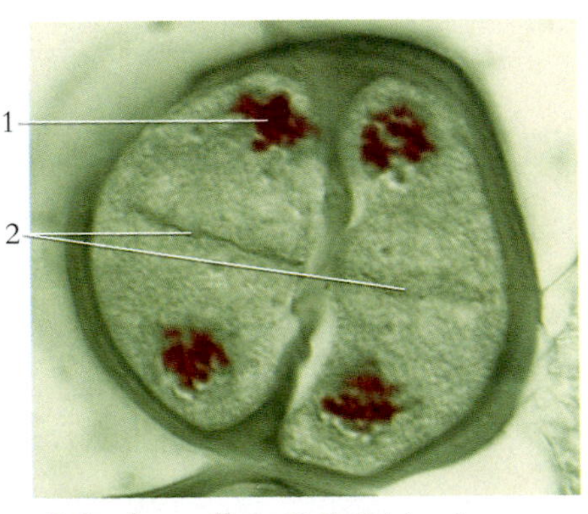

Telophase II — Cell division is complete, and cell walls of four haploid cells are formed.
1. Chromatids
2. New cell walls (cell plates)

Figure 2.6 The stages of meiosis in lily microsporocytes to form microspores. 1000X

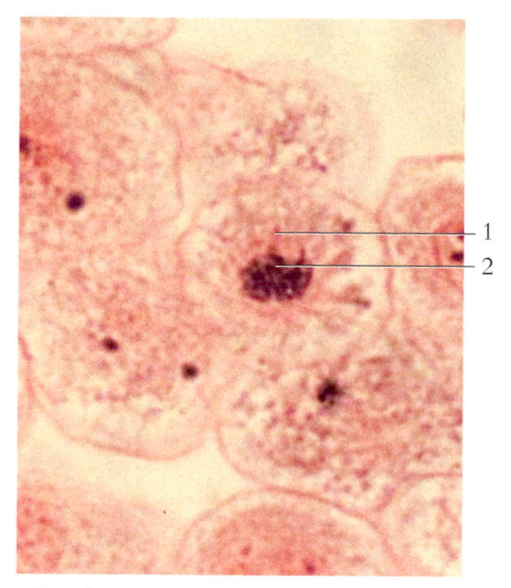

Prophase
Each chromosome consists of two chromatids joined by a centromere. Spindle fibers extend from each centriole.

1. Aster around centriole
2. Chromosomes

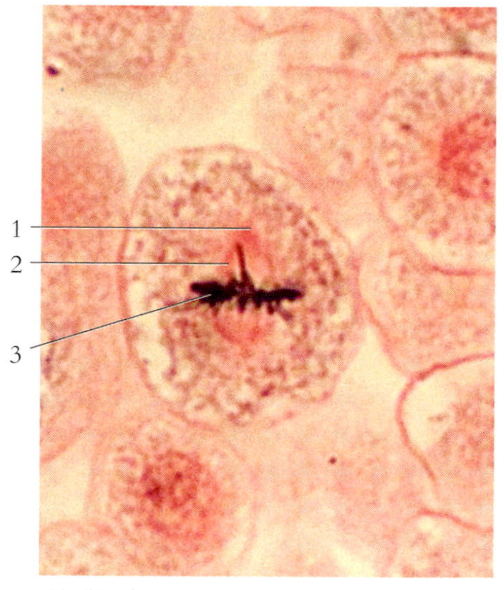

Metaphase
The chromosomes are positioned at the equator. The spindle fibers from each centriole attach to the centromeres.

1. Aster around centriole
2. Spindle fibers
3. Chromosomes at equator

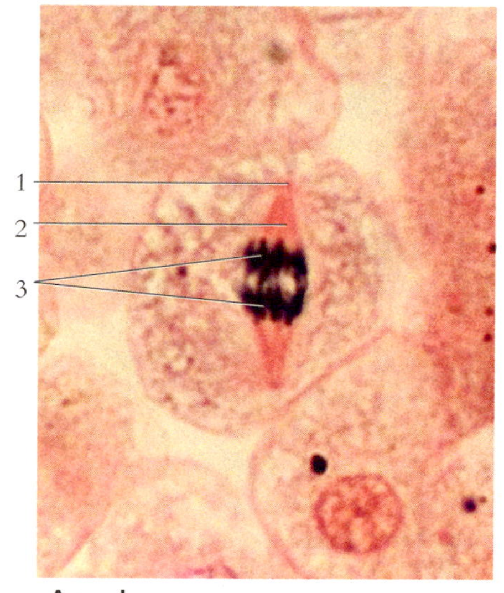

Anaphase
The centromeres split, and the sister chromatids separate as each is pulled to an opposite pole.

1. Aster around centriole
2. Spindle fibers
3. Separating chromosomes

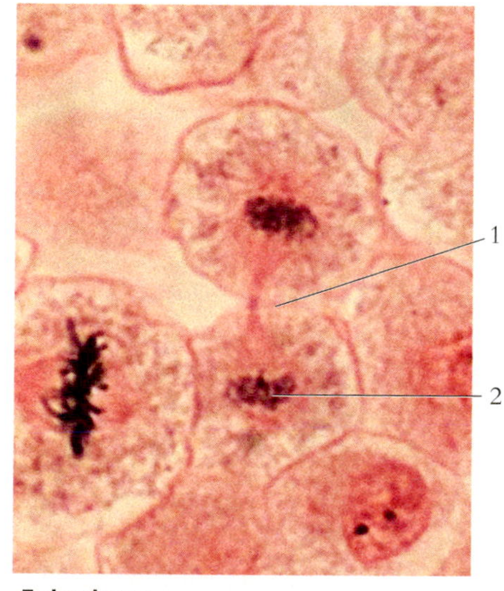

Telophase
The chromosomes lengthen (decondense) and become less distinct. The cell membrane forms between the forming daughter cells.

1. New cell membrane
2. Newly forming nucleus

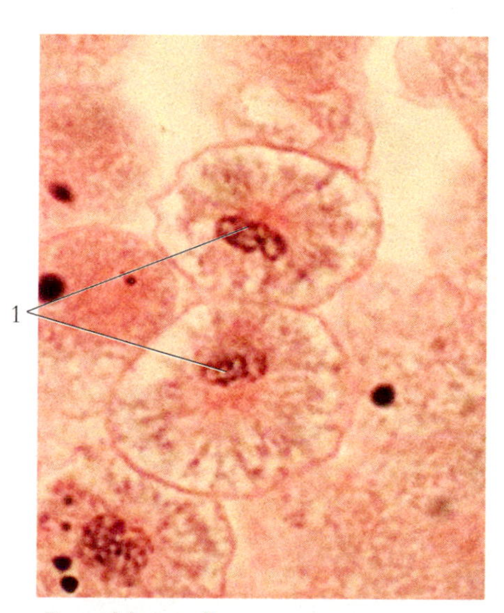

Daughter cells
The single chromosomes (former chromatids—see anaphase) continue to lengthen (decondense) as the nuclear membrane reforms. Cell division is complete, and the newly formed cells grow and mature.

1. Daughter nuclei

Figure 2.7 The stages of animal cell mitosis followed by cytokinesis. Whitefish blastula. 500X

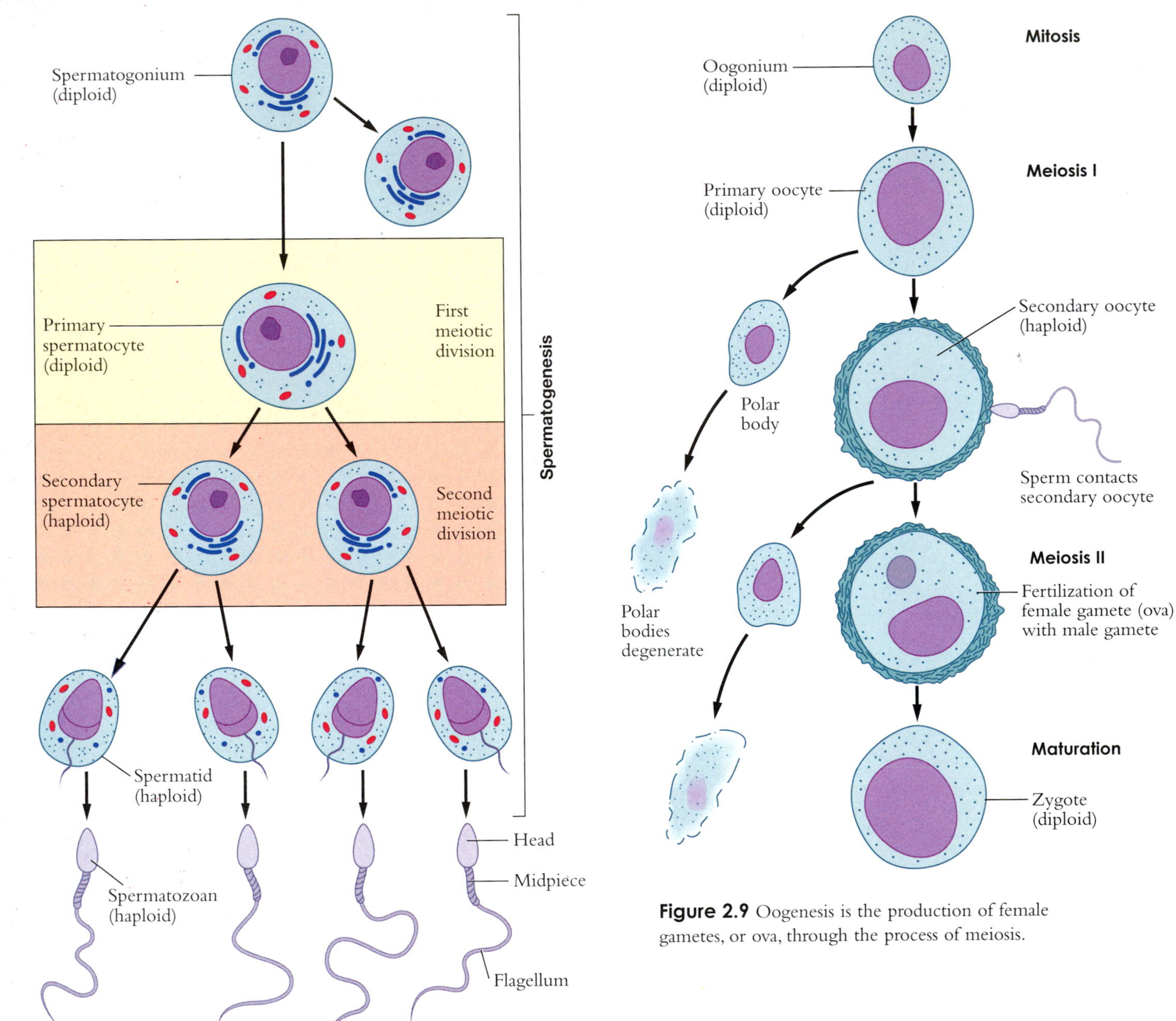

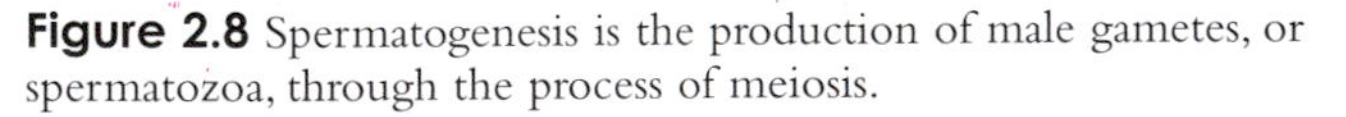

Figure 2.8 Spermatogenesis is the production of male gametes, or spermatozoa, through the process of meiosis.

Figure 2.9 Oogenesis is the production of female gametes, or ova, through the process of meiosis.

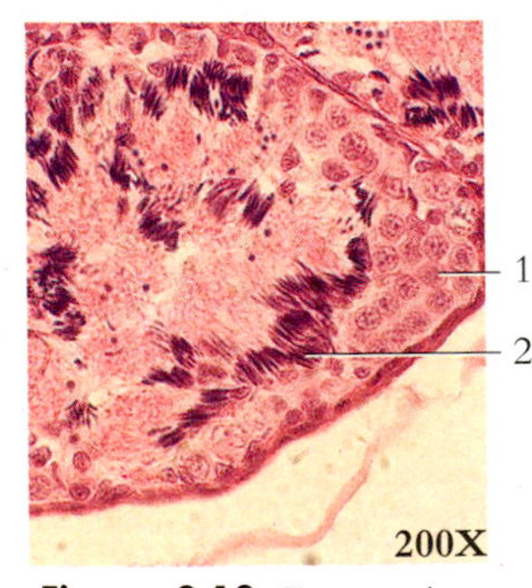

Figure 2.10 Frog testis.
1. Spermatocytes
2. Developing sperm

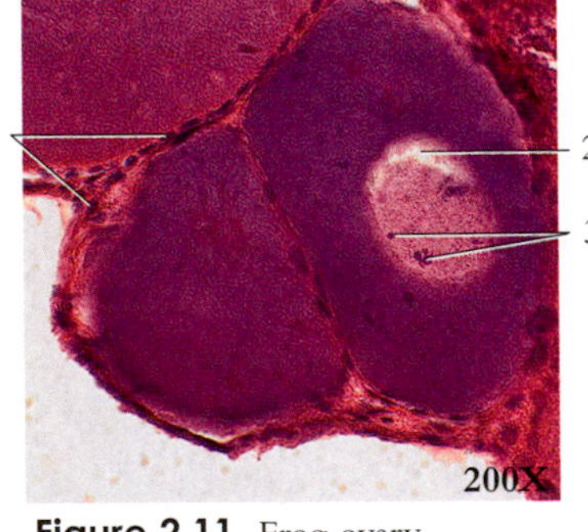

Figure 2.11 Frog ovary.
1. Follicle cells
2. Germinal vesicle
3. Nucleoli

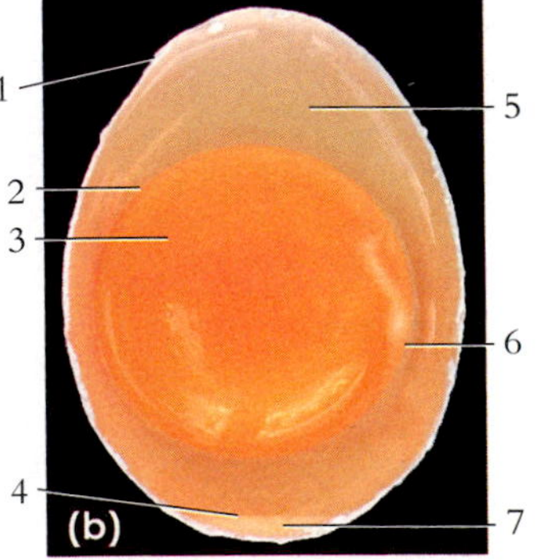

Figure 2.12 (a) An intact chicken egg and (b) a portion of the shell is removed exposing the internal structures.
1. Shell
2. Vitelline membrane
3. Yolk
4. Shell membrane
5. Albumen (egg white)
6. Chalaza (dense albumen)
7. Air space

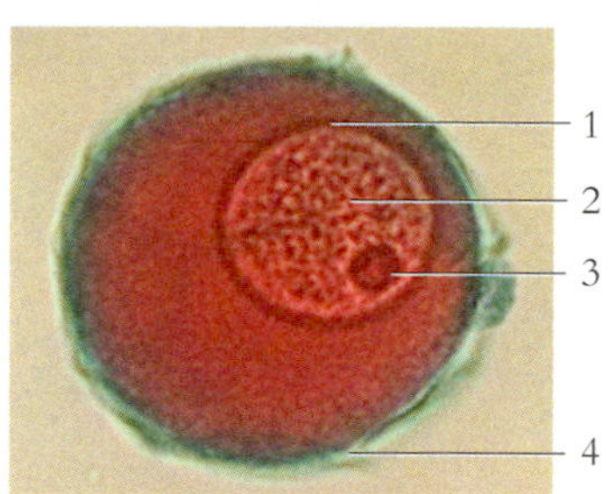

Unfertilized egg
1. Nuclear membrane
2. Nucleus
3. Nucleolus
4. Cell membrane

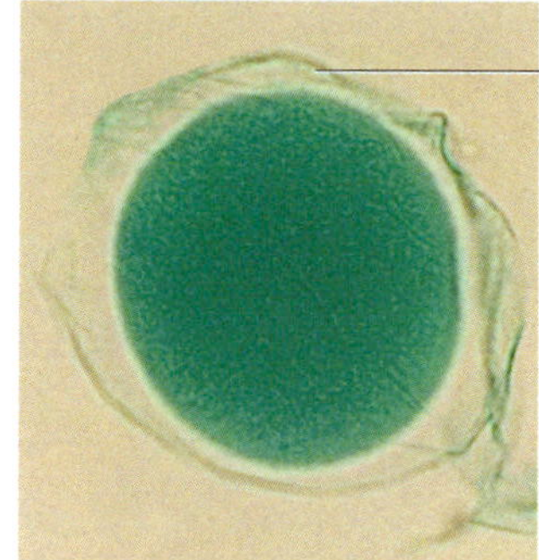

Fertilized egg
1. Fertilization membrane

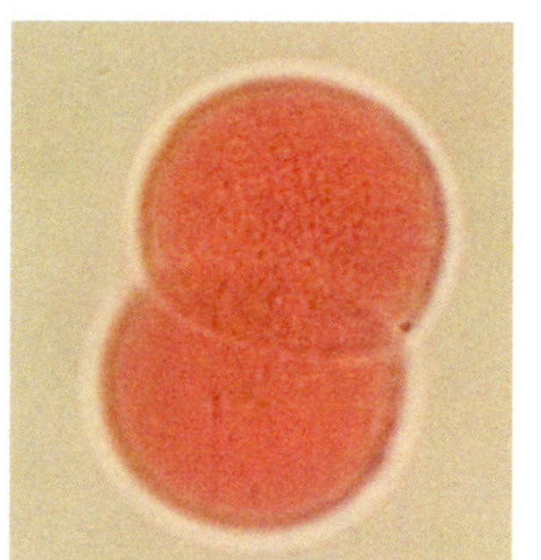

2-cell stage

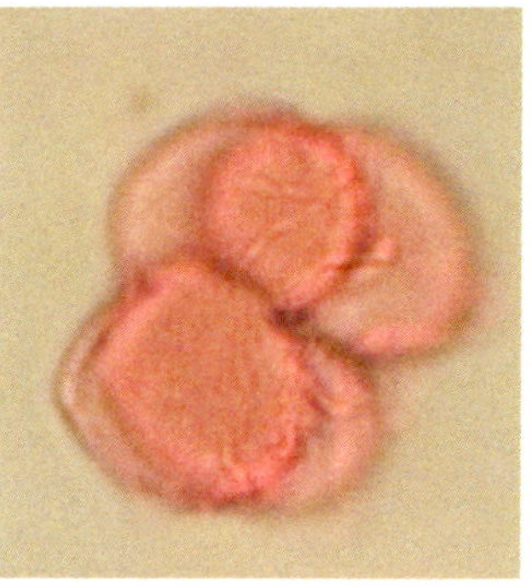

4-cell stage

8-cell stage

16-cell stage

32-cell stage

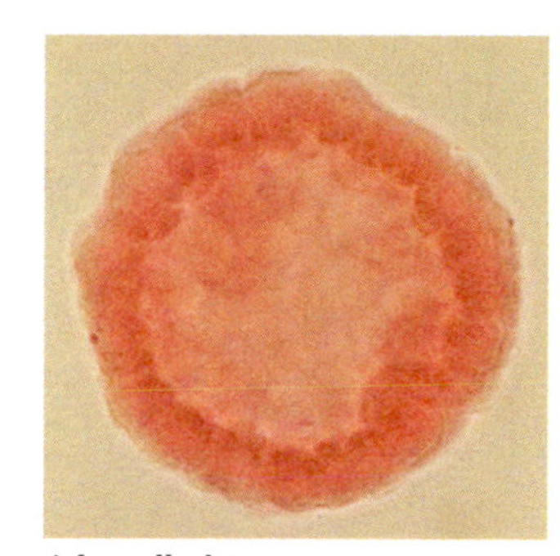

64-cell stage

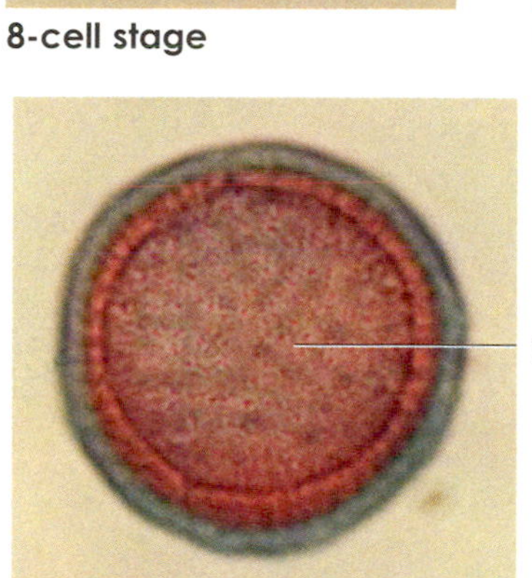

Blastula
1. Blastocoel

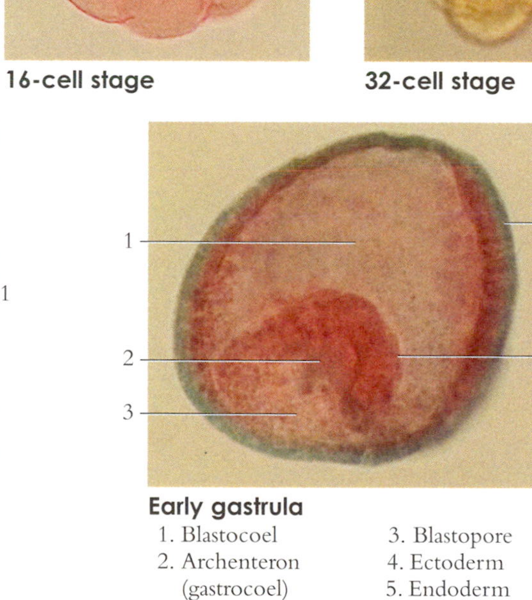

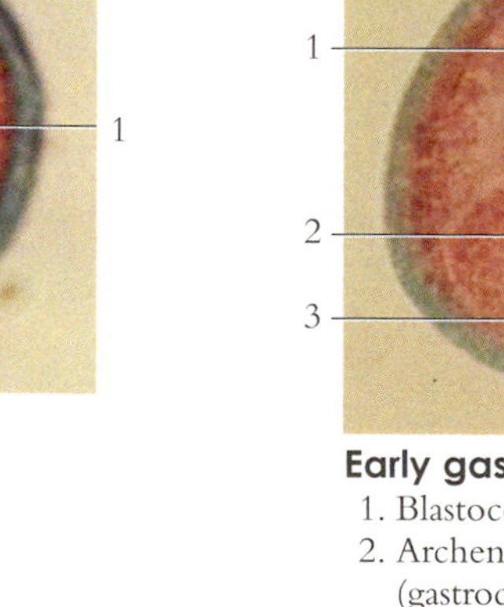

Early gastrula
1. Blastocoel
2. Archenteron (gastrocoel)
3. Blastopore
4. Ectoderm
5. Endoderm

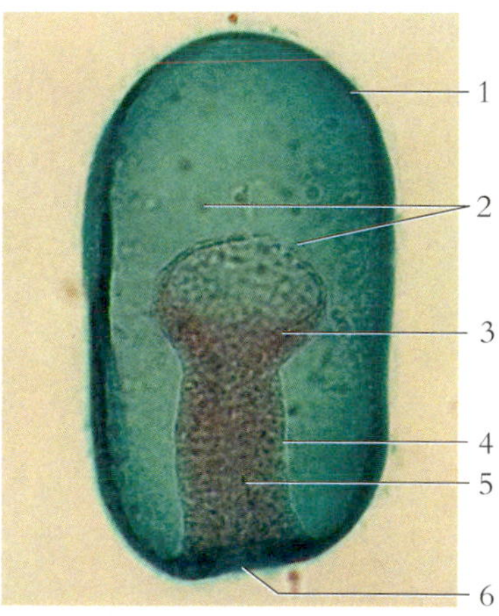

Late gastrula
1. Ectoderm
2. Mesenchyme cells
3. Coelomic sac
4. Endoderm
5. Archenteron (gastrocoel)
6. Blastopore

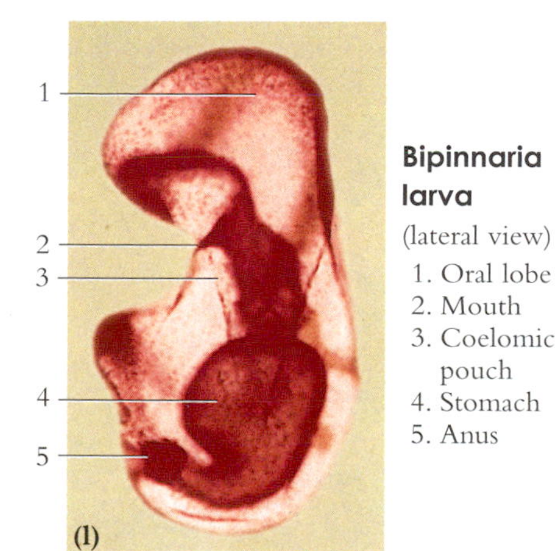

(l)

Bipinnaria larva
(lateral view)
1. Oral lobe
2. Mouth
3. Coelomic pouch
4. Stomach
5. Anus

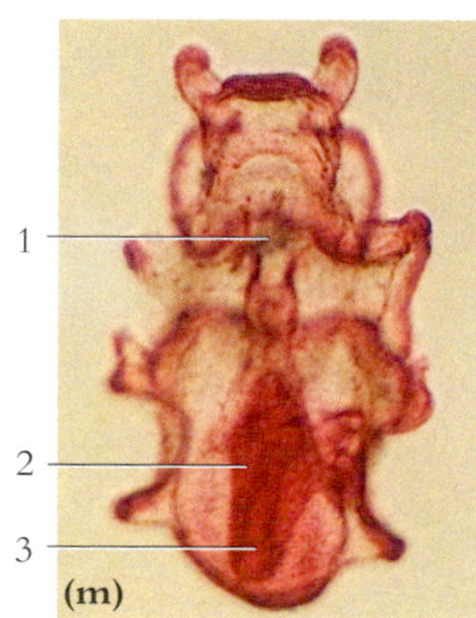

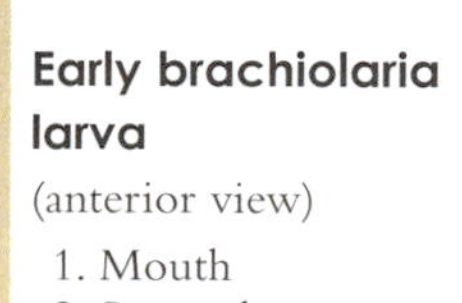

(m)

Early brachiolaria larva
(anterior view)
1. Mouth
2. Stomach
3. Anus

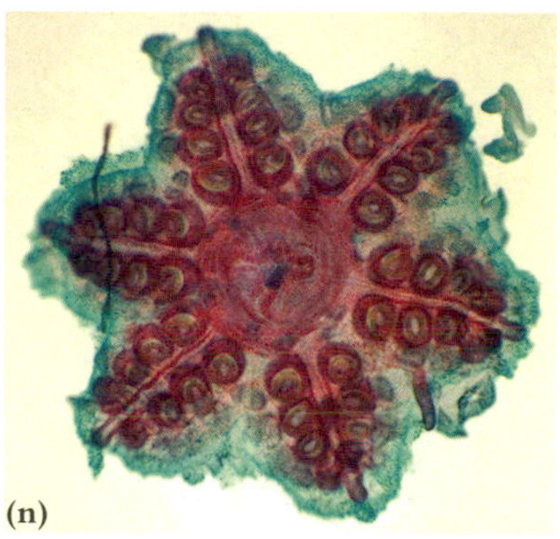

(n)

Young sea star

Figure 2.13 The development of the sea star, *Asterias* sp. 100X.

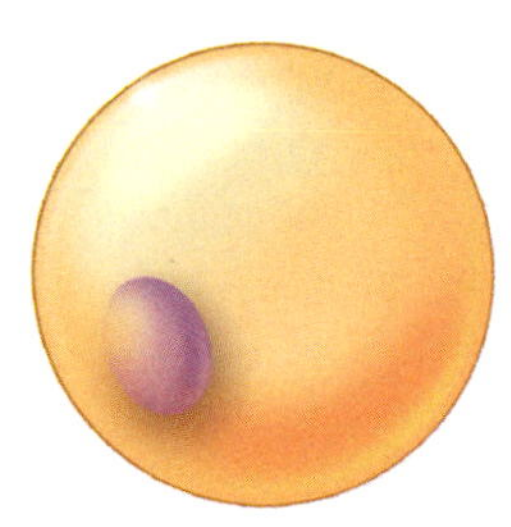

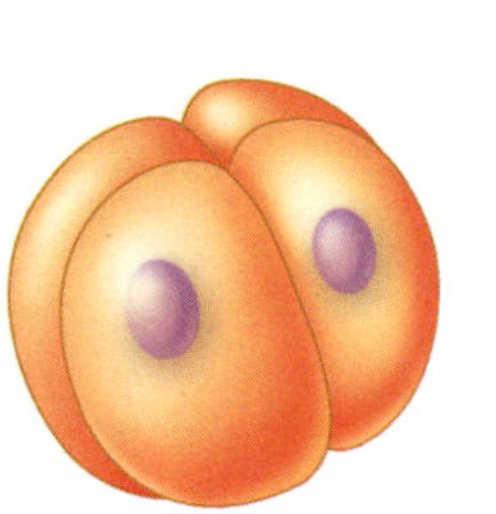

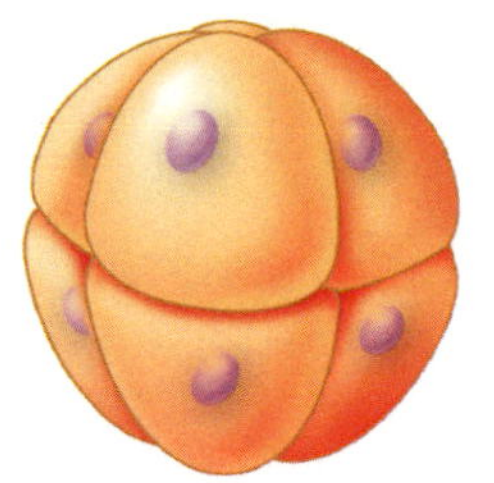

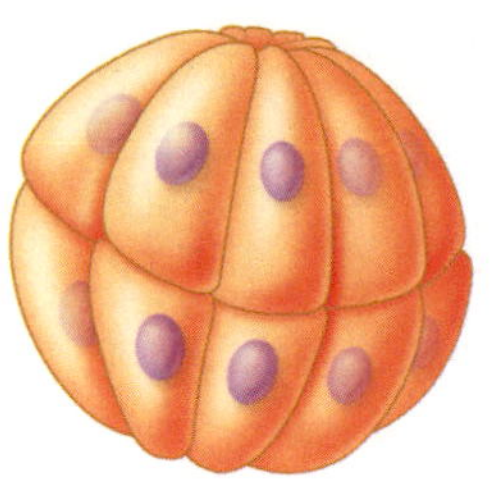

Fertilized egg

4-cell stage

8-cell stage

16-cell stage

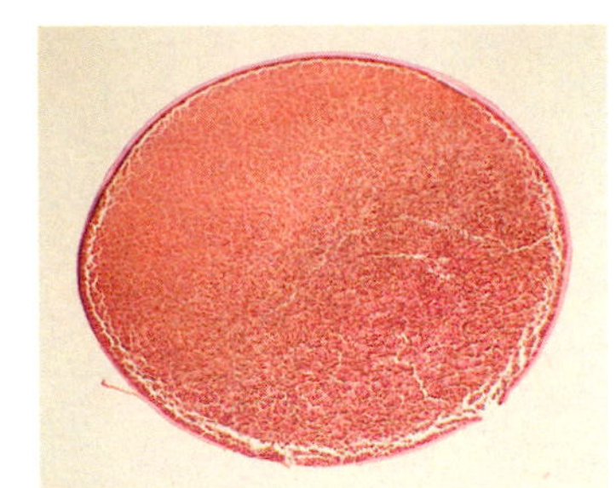

Fertilized egg
(transverse section)

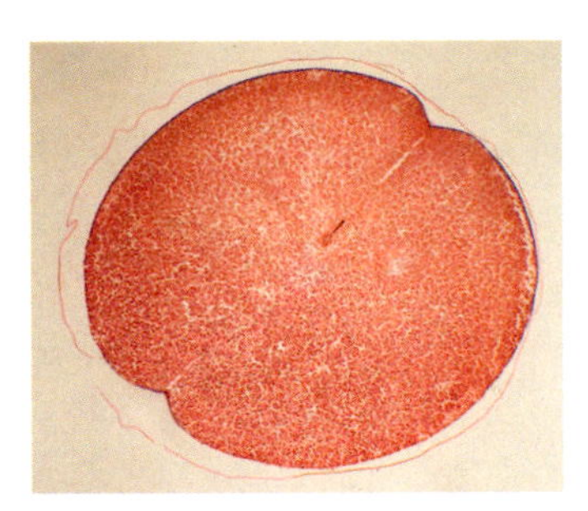

2-cell stage
(transverse section)

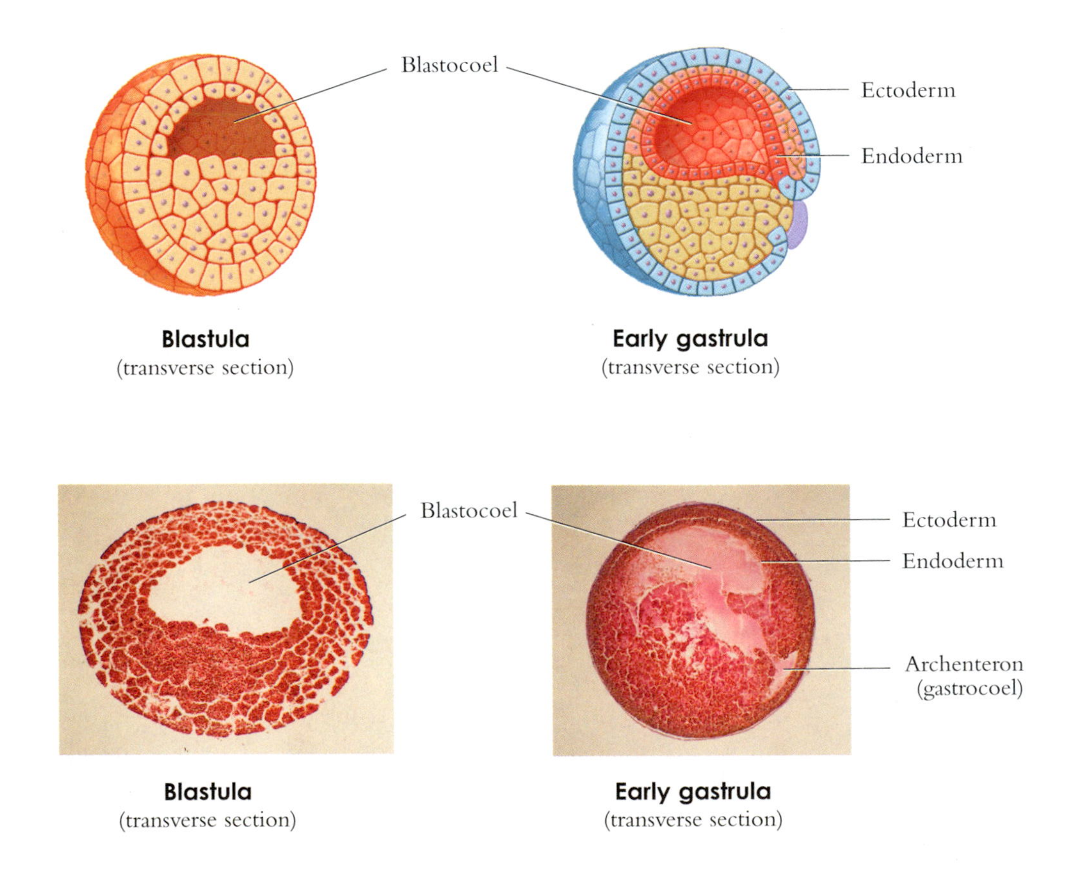

Blastula
(transverse section)

Early gastrula
(transverse section)

Blastula
(transverse section)

Early gastrula
(transverse section)

Figure 2.14 Frog development from fertilized egg to early gastrula, shown in diagram and photomicrographs. 100X.

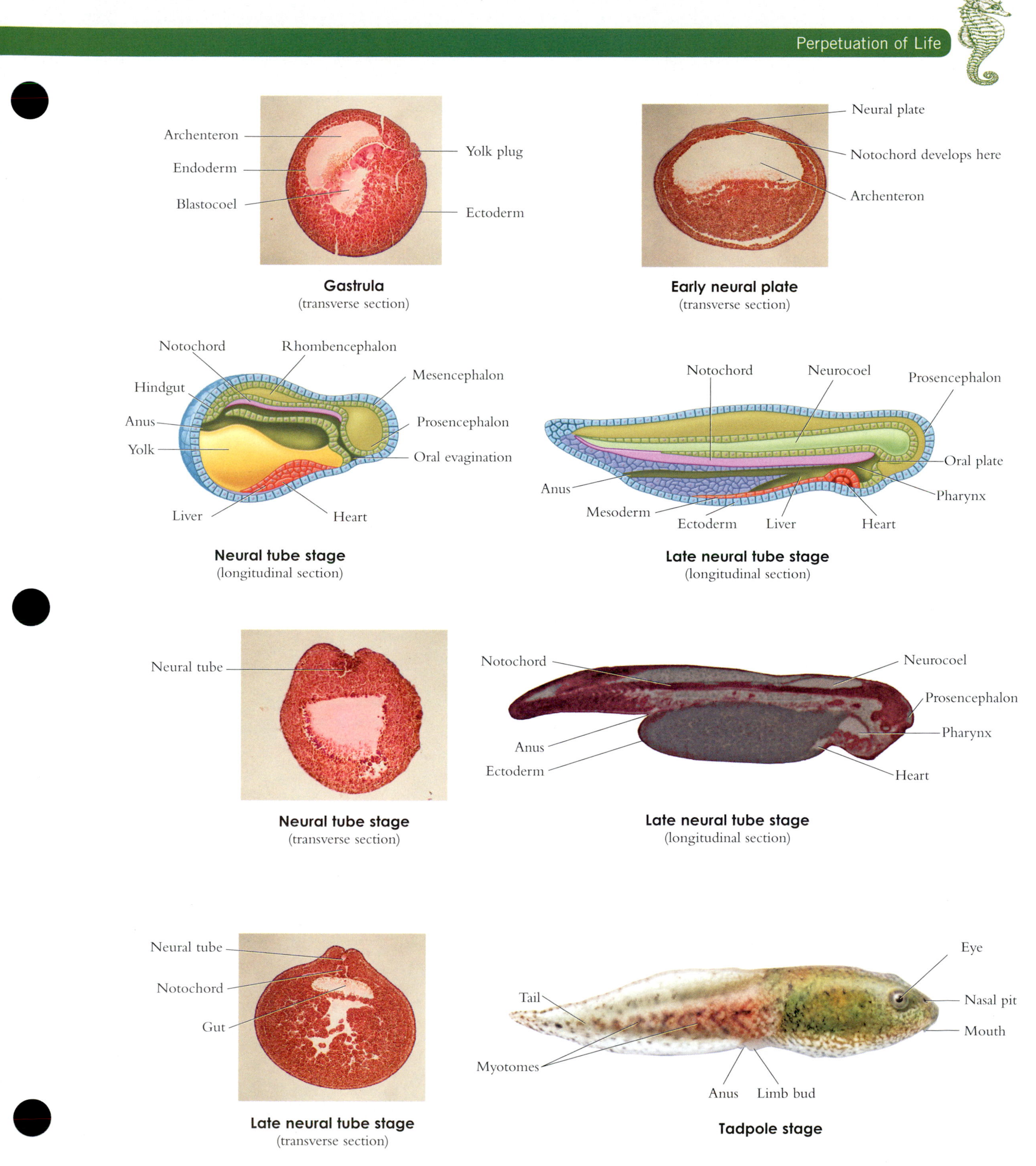

Figure 2.15 Frog development from gastrula to tadpole, shown in diagram and photomicrographs. 100X.

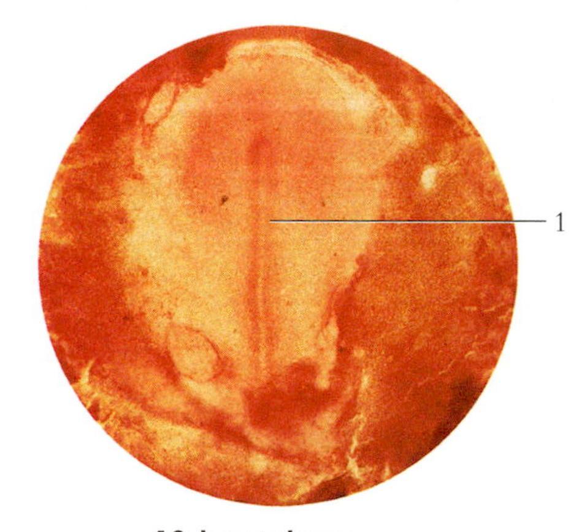

13-hour stage
1. Embryo main body formation

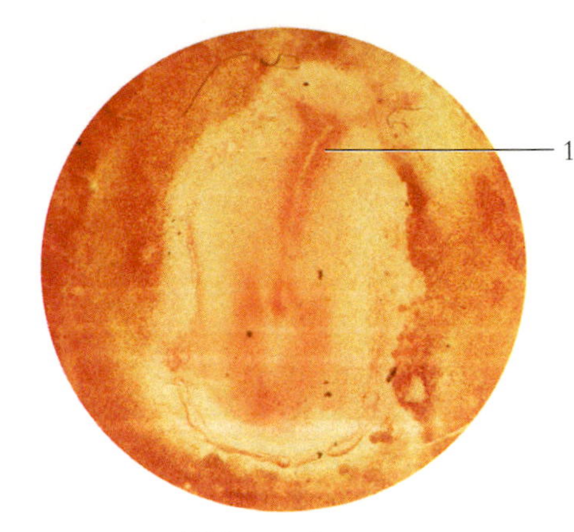

18-hour stage
1. Neurulation beginning

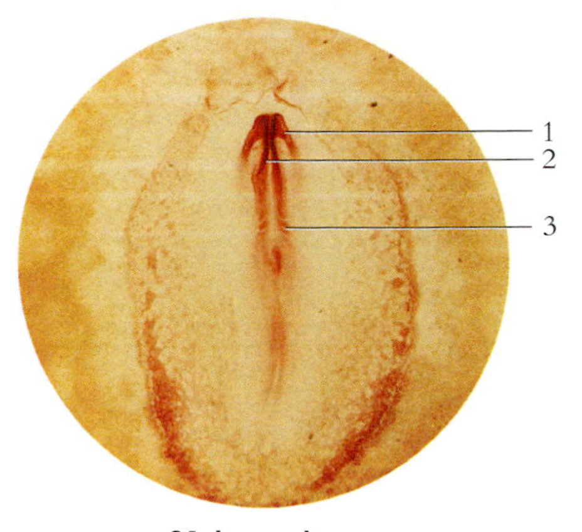

21-hour stage
1. Head fold
2. Neural fold
3. Muscle plate (somites)

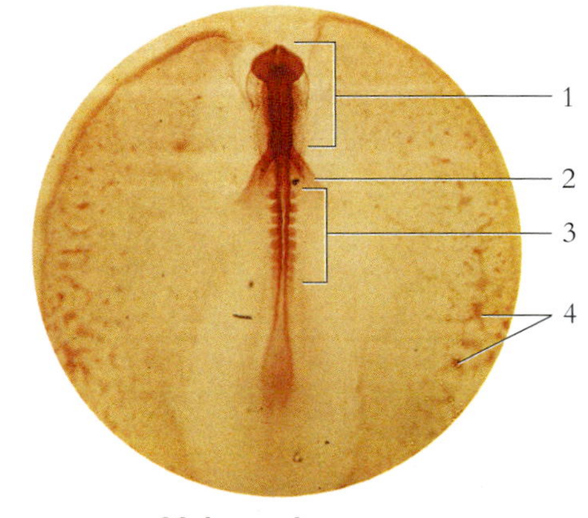

28-hour stage
1. Head fold and brain
2. Artery formation
3. Muscle plate (somites)
4. Blood vessel formation

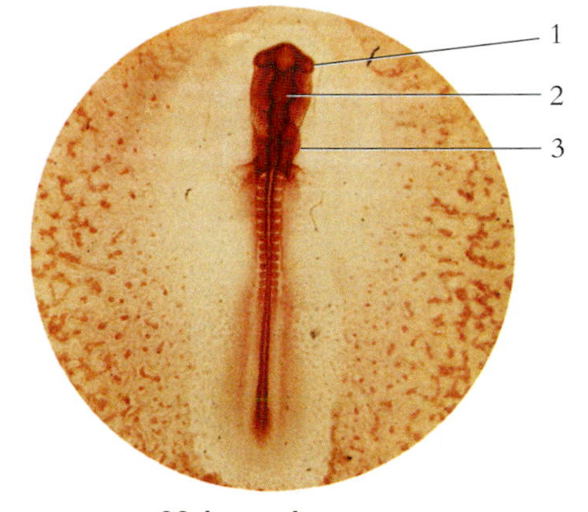

38-hour stage
1. Optic vesicle
2. Brain with five regions
3. Heart

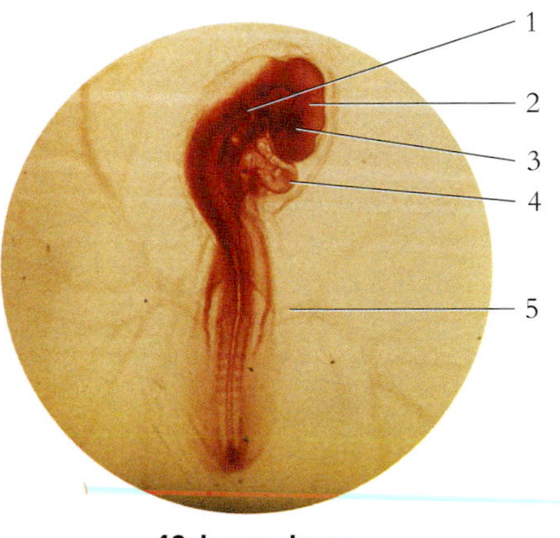

48-hour stage
1. Ear
2. Brain
3. Eye
4. Heart
5. Artery

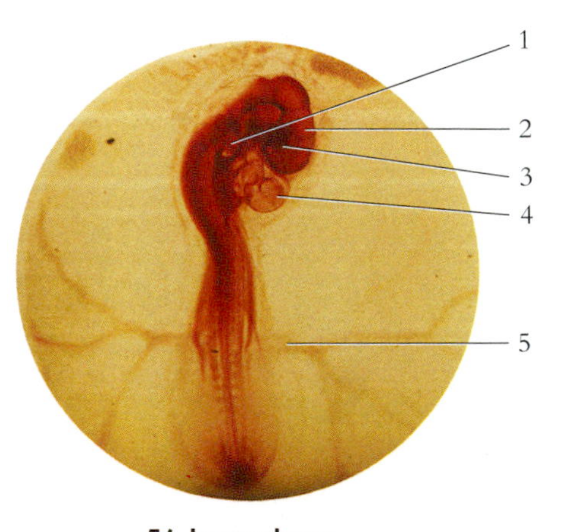

56-hour stage
1. Ear
2. Brain
3. Eye
4. Heart
5. Artery

96-hour stage
1. Eye
2. Mesencephalon
3. Heart
4. Wing formation
5. Fecal sac (allantois)
6. Leg formation

Figure 2.16 The stages of chick development. 20X

Bacteria and Archaea

Chapter 3

Bacteria range between 1 and 50 μm in width or diameter. The morphological appearance of bacteria may be spiral (spirillum), spherical (coccus), or rod-shaped (bacillus). Cocci and bacilli frequently form clusters or linear filaments and may have bacterial flagella. Relatively few species of bacteria cause infection. Hundreds of species of nonpathogenic bacteria live on the human body and within the gastrointestinal (GI) tract. Those in the GI tract constitute a person's gut fauna and are biologically critical to humans.

Photosynthetic bacteria contain chlorophyll and release oxygen during photosynthesis. Some bacteria are *obligate aerobes* (require O_2 for metabolism) and others are *facultative anaerobes* (indifferent to O_2 for metabolism). Some are *obligate anaerobes* (oxygen may poison them). Most bacteria are heterotrophic *saprophytes*, which secrete enzymes to break down surrounding organic molecules into absorbable compounds.

Archaea are adapted to a limited range of extreme conditions. The cell walls of Archaea lack peptidoglycan (characteristic of bacteria). Archaea have distinctive RNAs and RNA polymerase enzymes. They include methanogens, typically found in swamps and marshes, and thermoacidophiles, found in acid hot springs, acidic soil, and deep oceanic volcanic vents.

Methanogens exist in oxygen-free environments and subsist on simple compounds such as CO_2, acetate, or methanol. As their name implies, Methanogens produce methane gas as a by-product of metabolism. These organisms are typically found in organic-rich mud and sludge that often contain fecal wastes.

Thermoacidophiles are resistant to hot temperatures and high acid concentrations. The cell membrane of these organisms contains high amounts of saturated fats, and their enzymes and other proteins are able to withstand extreme conditions without denaturation. These microscopic organisms thrive in most hot springs and hot, acid soils.

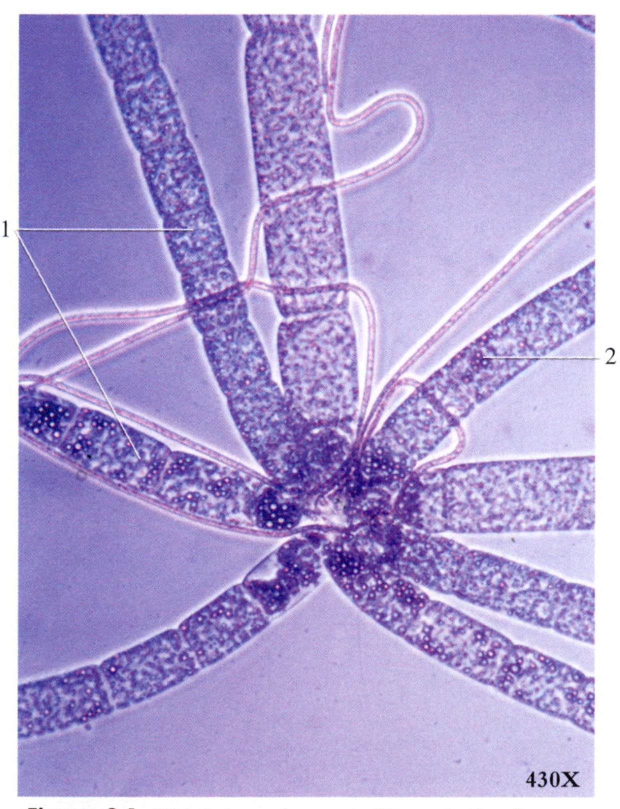

Figure 3.1 *Thiothrix* sp., a genus of bacteria that forms sulfur granules in its cytoplasm. These organisms obtain energy from oxidation of H_2S.

1. Filaments 2. Sulfur granules

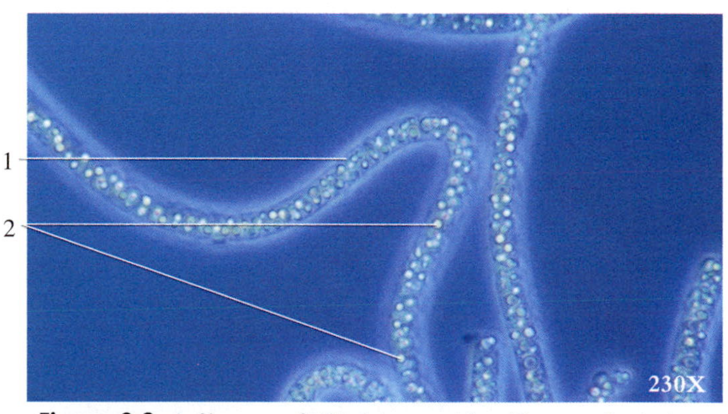

Figure 3.2 A filament of *Thiothrix* sp. with sulfur granules in its cytoplasm.

1. Filament 2. Sulfur granules

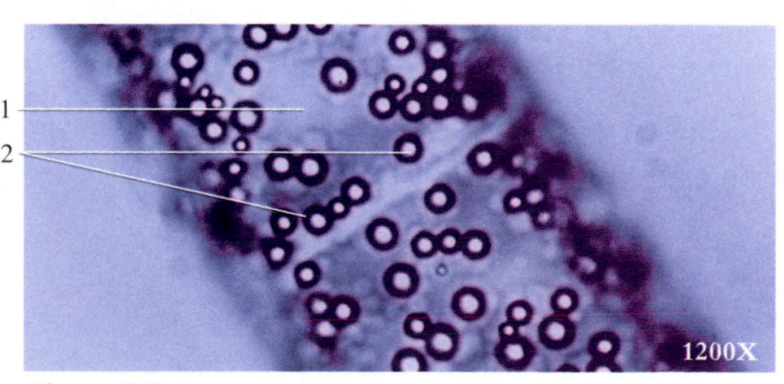

Figure 3.3 A magnified *Thiothrix* sp. filament with sulfur granules in its cytoplasm.

1. Cytoplasm 2. Sulfur granules

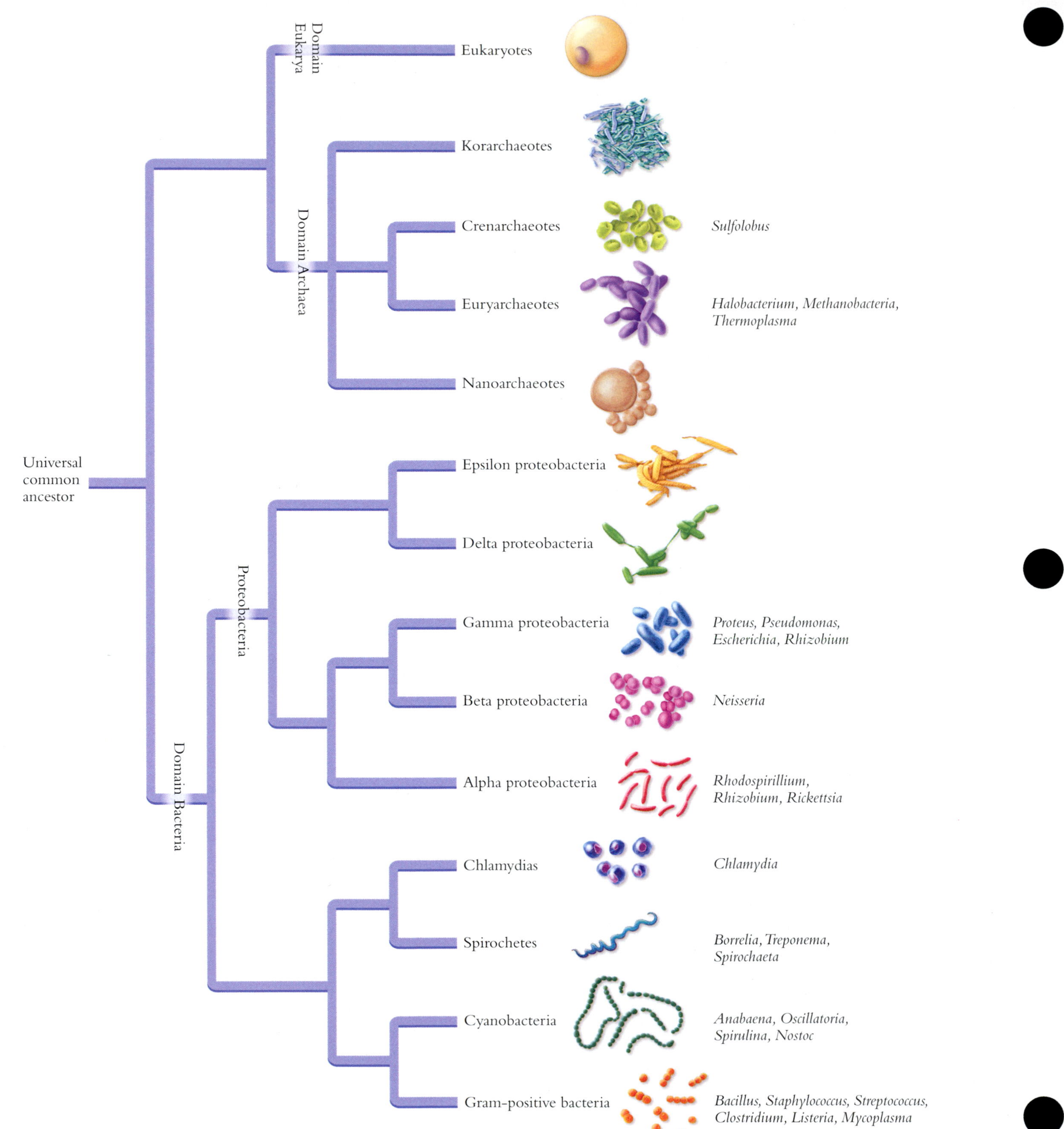

Figure 3.4 Phylogenetic relationships and classification of major bacteria and archaea lineages.

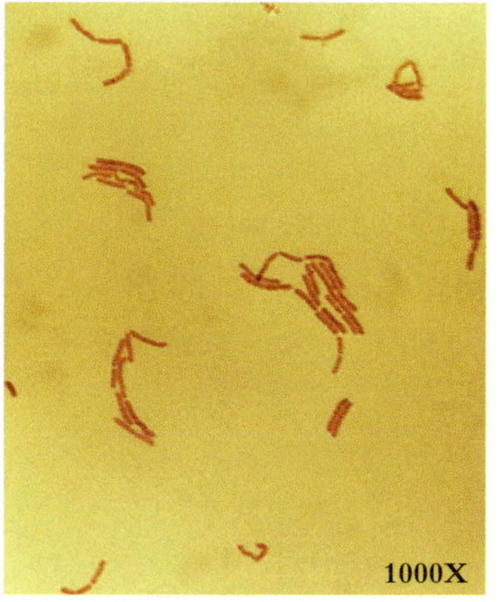

Figure 3.5 The bacterium *Bacillus megaterium*. *Bacillus* is capable of producing endospores. This species of *Bacillus* generally remains in chains after it divides.

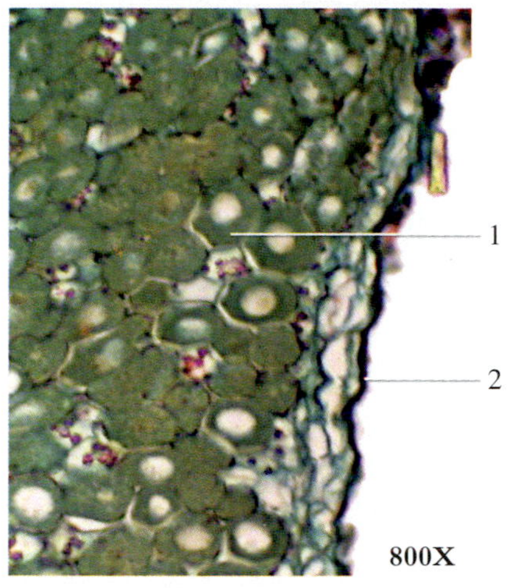

Figure 3.6 Transverse section through the root nodule of clover showing intracellular nitrogen-fixing bacteria.

1. Cell with bacteria
2. Epidermis

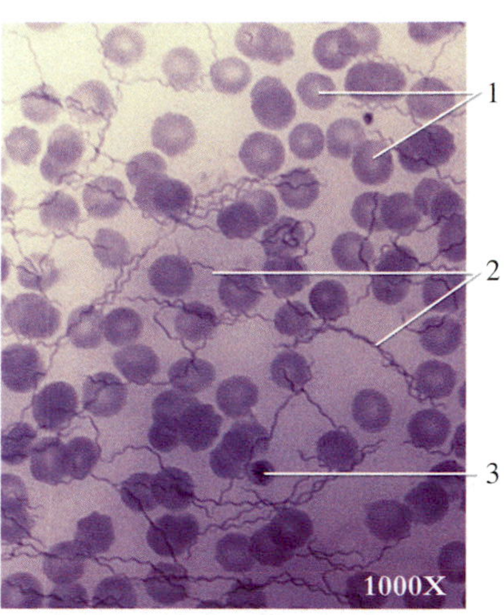

Figure 3.7 The spirochete, *Borella recurrentis*. Spirochetes are flexible rods twisted into helical shapes. This species causes relapsing fever.

1. Red blood cells
2. Spirochete
3. White blood cell

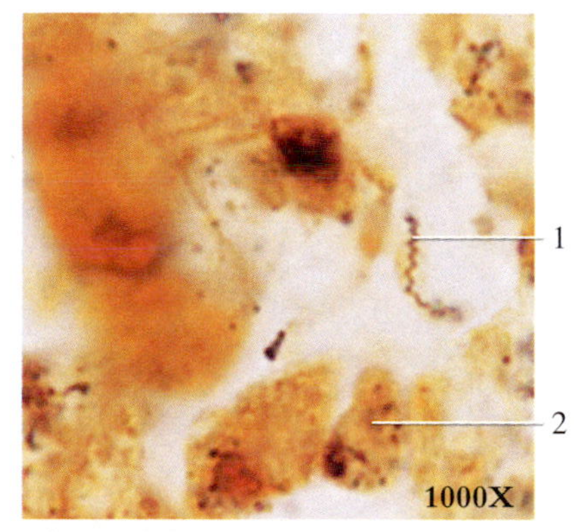

Figure 3.8 The spirochete *Treponema pallidum*. This species causes syphilis.

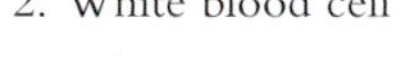

1. *Treponema pallidum*
2. White blood cell

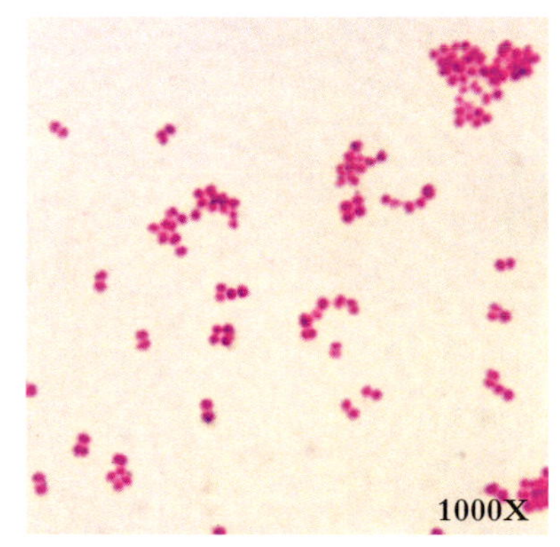

Figure 3.9 *Neisseria gonorrhoeae*. This is a diplococcus that causes gonorrhea.

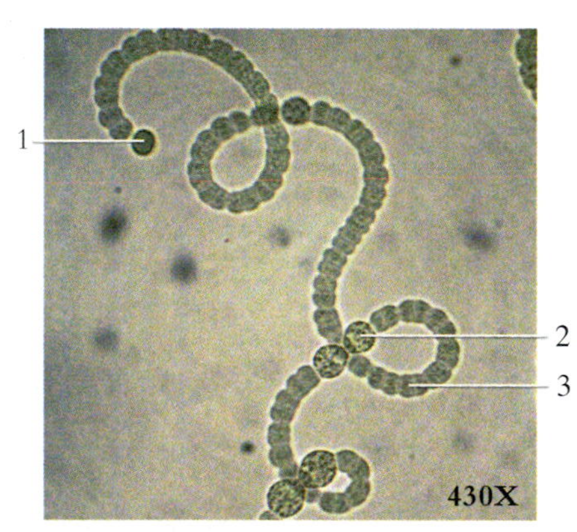

Figure 3.10 An *Anabaena* sp. filament. This organism is a nitrogen-fixing cyanobacterium. Nitrogen fixation takes place within the heterocyst cells.

1. Heterocyst
2. Spore (akinete)
3. Vegetative cell

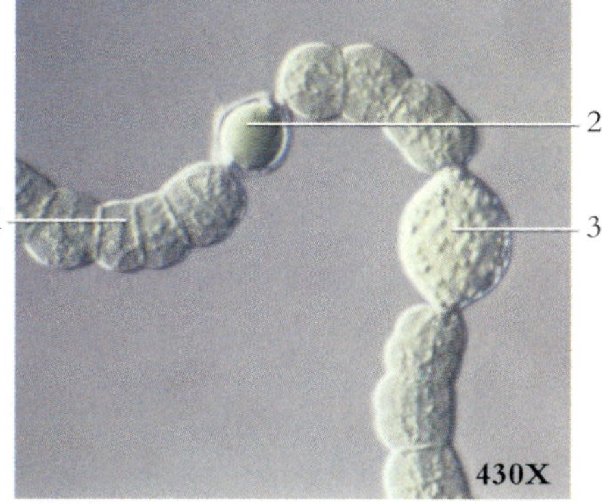

Figure 3.11 An *Anabaena* sp. filament. This is a nitrogen-fixing cyanobacterium. Nitrogen fixation takes place within the heterocyst cells.

1. Vegetative cell
2. Heterocyst
3. Spore

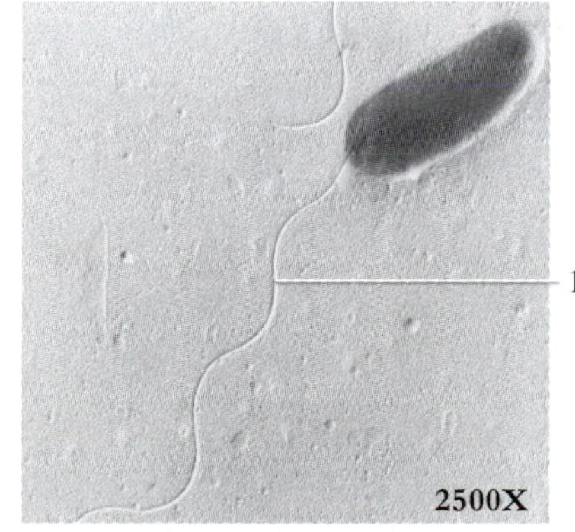

Figure 3.12 The flagellated bacterium, *Pseudomonas* sp.

1. Flagellum

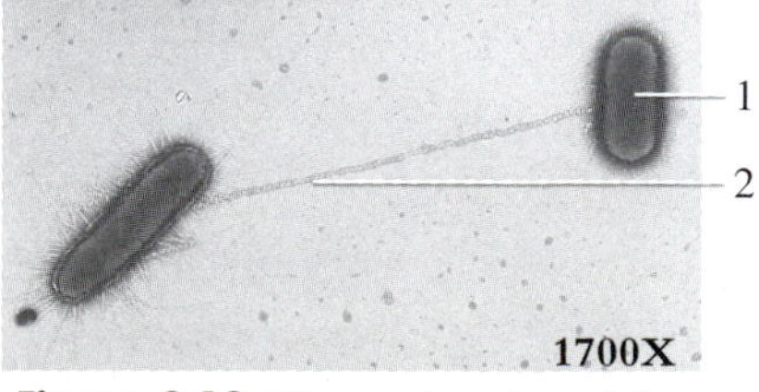

Figure 3.13 The conjugation of the bacterium *Escherichia coli*. By this process of conjugation, genetic material is transferred through the conjugation tube from one cell to the other allowing genetic recombination.

1. Bacterium
2. Conjugation tube

Table 3.1 Some Representatives of Bacteria and Archaea

Categories	Representative Genera
Bacteria	
Photosynthetic bacteria	
Cyanobacteria	*Anabaena, Oscillatoria, Spirulina, Nostoc*
Green bacteria	*Chlorobium*
Purple bacteria	*Rhodospirillum*
Gram-negative bacteria	*Proteus, Pseudomonas, Escherichia, Rhizobium, Neisseria*
Gram-positive bacteria	*Bacillus, Staphylococcus, Streptococcus, Clostridium, Listeria*
Spirochetes	*Spirochaeta, Treponema*
Actinomycetes	*Streptomyces*
Rickettsias and Chlamydias	*Rickettsia, Chlamydia*
Mycoplasmas	*Mycoplasma*
Archaea	
Methanogens	*Halobacterium, Methanobacteria*
Thermoacidophiles	*Thermoplasma, Sulfolobus*

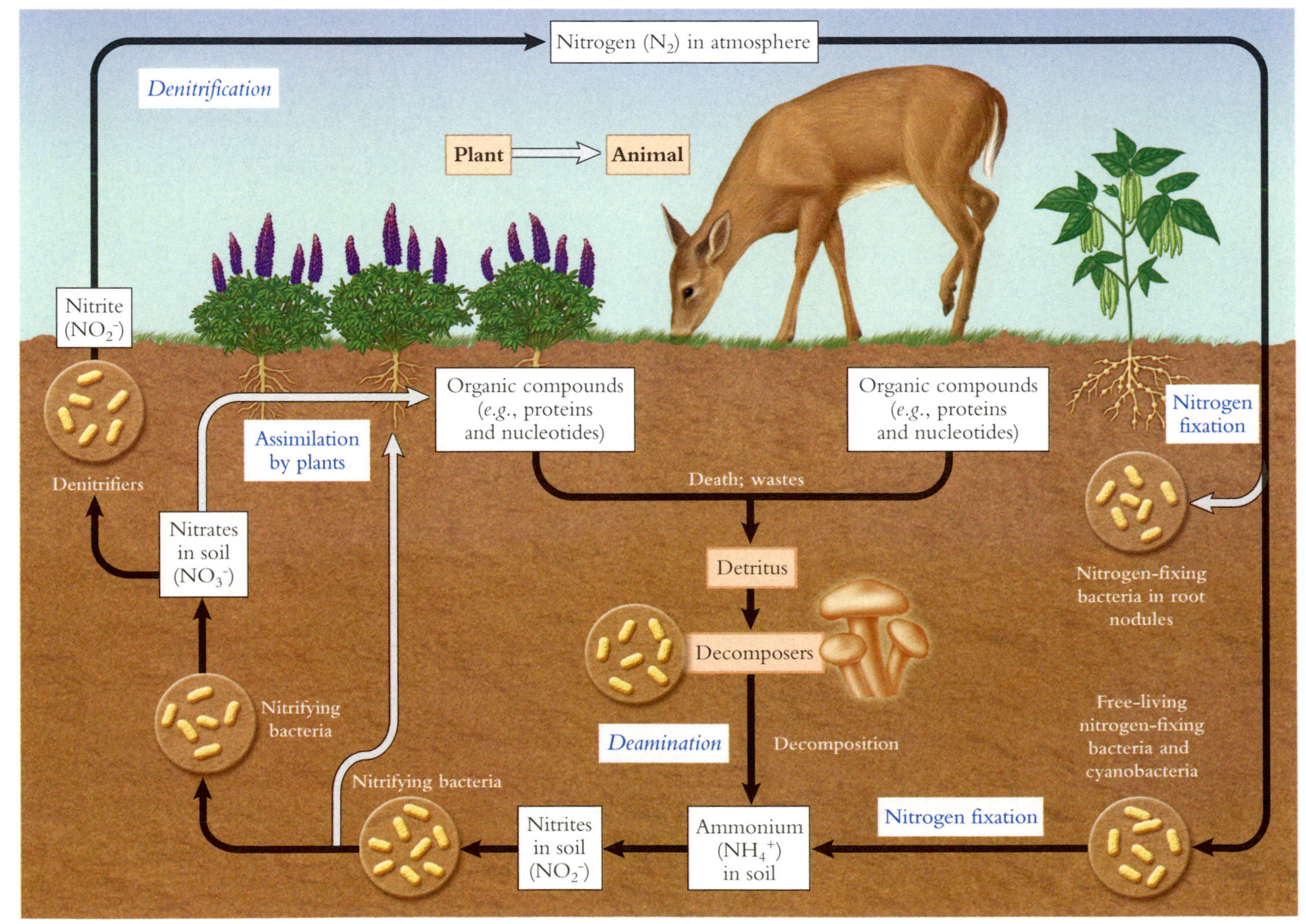

Figure 3.14 Few organisms have the ability to utilize atmospheric nitrogen. Nitrogen-fixing bacteria within the root nodules of legumes (and some free-living bacteria) provide a usable source of nitrogen to plants.

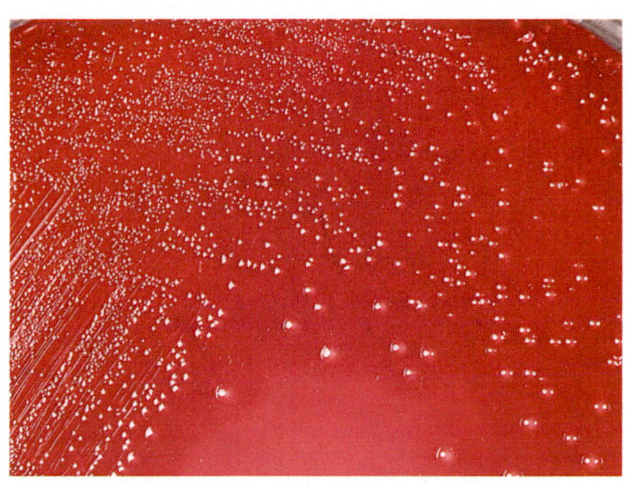

Figure 3.15 Colonies of *Streptococcus pyogenes* cultured on a sheep blood agar plate. *S. pyogenes* causes strep throat and rheumatic fever in humans. This agar plate is approximately 10 cm in diameter.

Figure 3.16 Cyanobacteria living in hot springs and hot streams, such as this 40 meter effluent from a geyser in Yellowstone National Park.

1. Mats of *Cyanophyta*

Figure 3.17 Cyanobacteria of several species growing in the effluent from a geyser. The different species are temperature-dependent and form the bands of color.

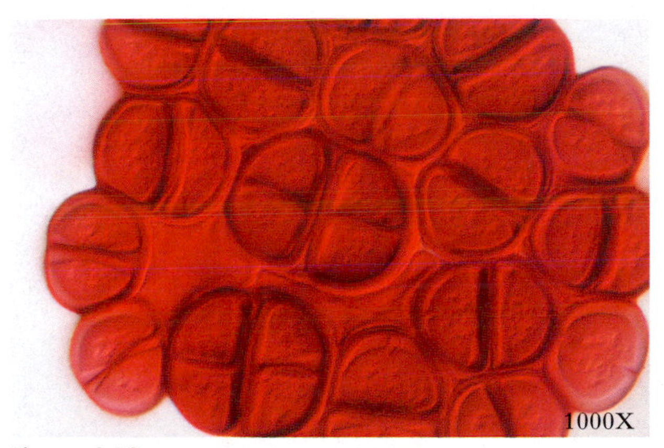

Figure 3.18 A magnified view of the cyanobacterium *Chroococcus* sp. shown with a red biological stain.

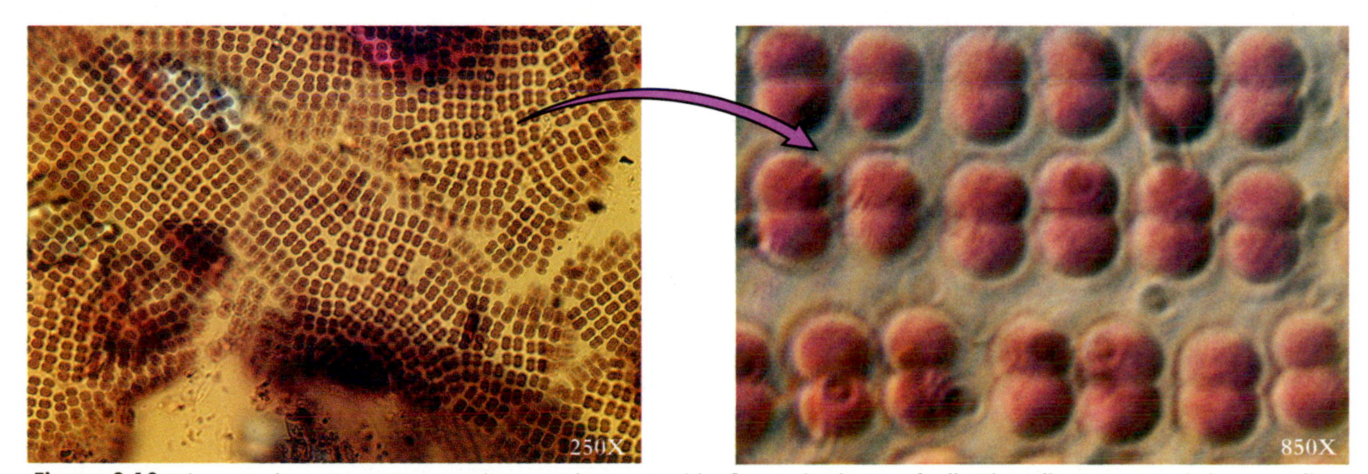

Figure 3.19 The cyanobacterium, *Merismopedia* sp., is characterized by flattened colonies of cells. The cells are in a single layer, usually aligned into groups of two or four.

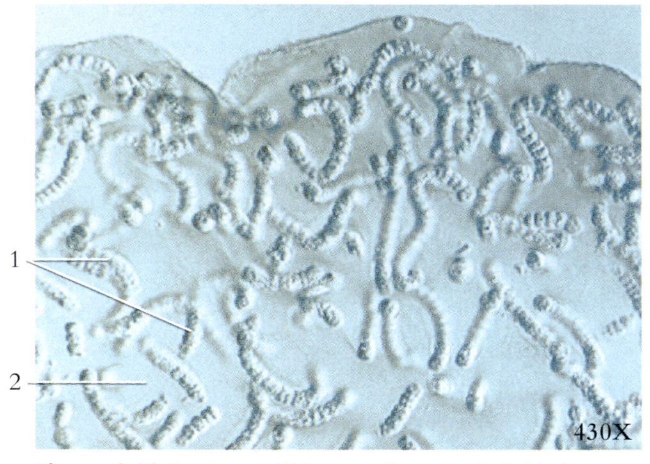

Figure 3.20 A colony of *Nostoc* sp. filaments. Individual filaments secrete mucilage, which forms a gelatinous matrix around all filaments.
1. Filaments 2. Gelatinous matrix

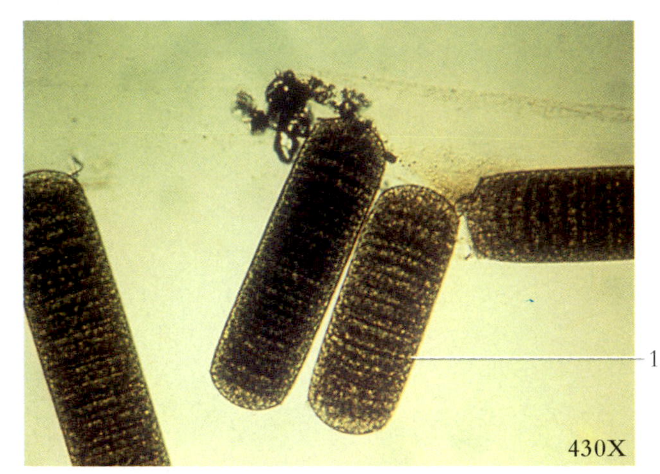

Figure 3.21 The filaments of *Oscillatoria* sp. The only way this cyanobacterium can reproduce is through fragmentation of a filament. Fragments are known as hormogonia.
1. Hormogonium

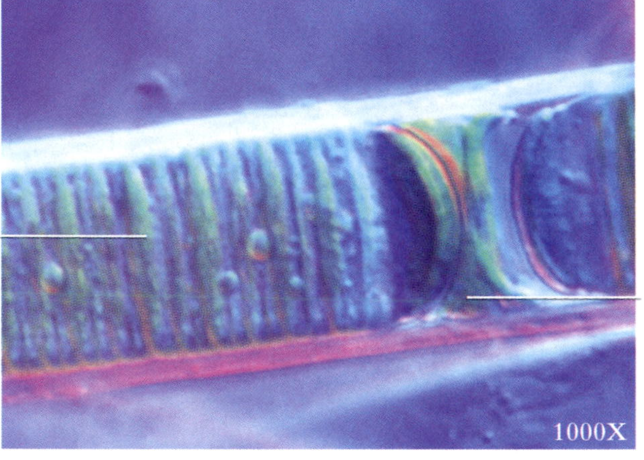

Figure 3.22 A portion of a cylindrical filament of *Oscillatoria* sp. This cyanobacterium is common in most aquatic habitats.
1. Filament segment (hormogonium) 2. Separation disk (necridium)

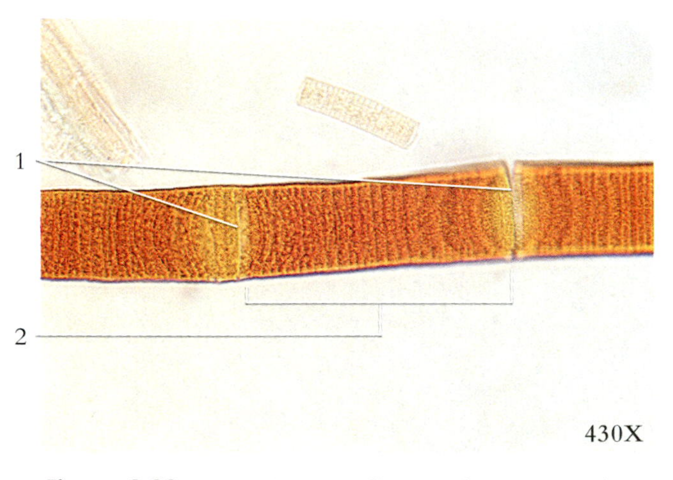

Figure 3.23 An *Oscillatoria* sp. filament showing necridia.
1. Necridia 2. Hormogonium

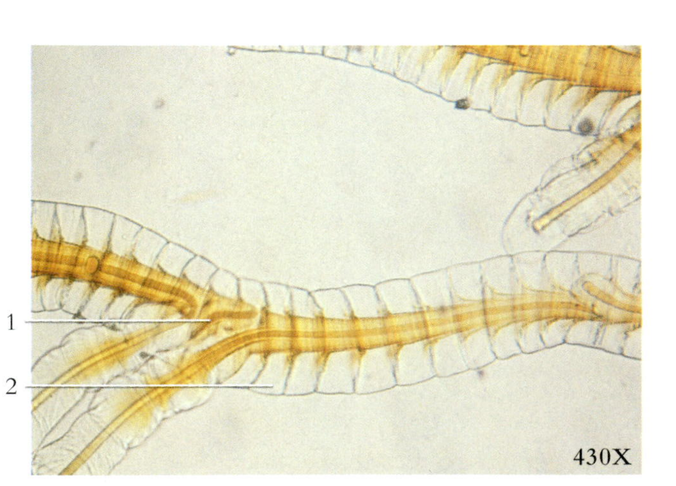

Figure 3.24 *Scytonema* sp., a cyanobacterium, common on moistened soil. Notice the falsely branched filament typical of this genus. This species also demonstrates "winged" sheaths.
1. False branching 2. "Winged" sheath

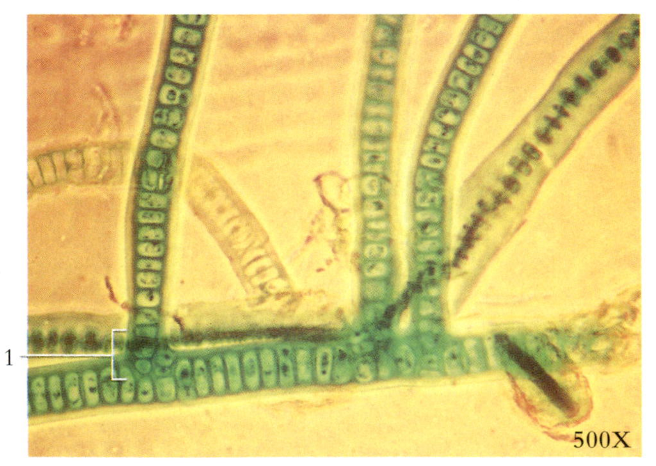

Figure 3.25 The cyanobacterium, *Stigonema* sp. This species has true-branched filaments caused from cell division in two separate planes.
1. True branching

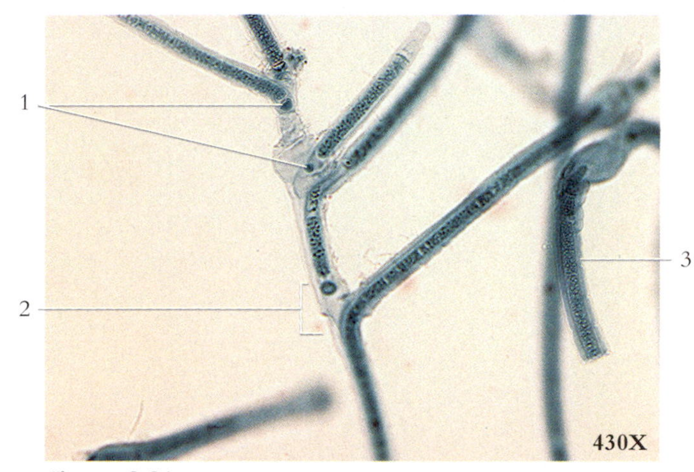

Figure 3.26 *Tolypothrix* sp., a cyanobacterium with a single false-branched filament.
1. Heterocysts 2. False branching 3. Sheath

Figure 3.27 Longitudinal section of a fossilized stromatolite two billion years old. Layering indicates the communities of bacteria and cyanobacteria mixed with sediments. This specimen originates from Australia (scale in mm).

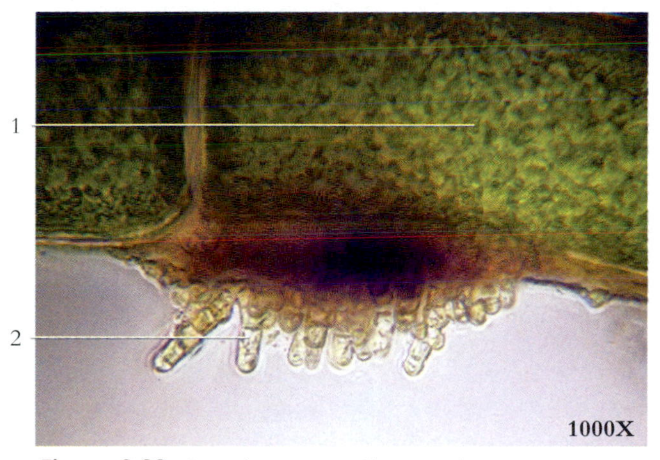

Figure 3.28 Cyanobacterium, *Chamaesisphon* sp., growing as an epiphyte on green algae, *Cladophora* sp.
1. *Cladophora* sp. 2. *Chamaesisphon* sp.

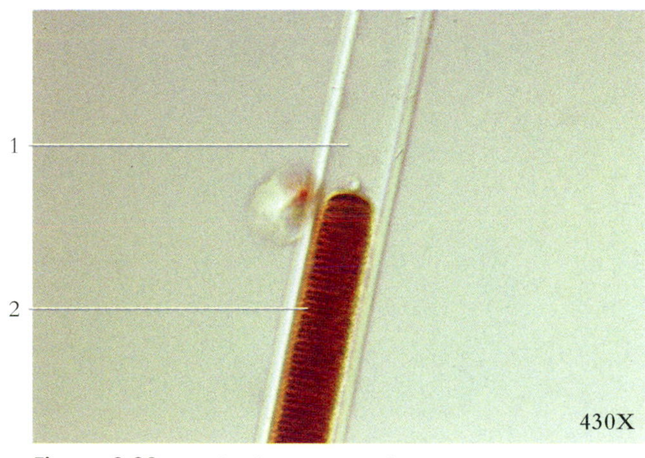

Figure 3.29 *Lyngbya birgeii*, a cyanobacterium, is common in eutrophic water throughout North America.
1. Extended sheath 2. Filament of living cells

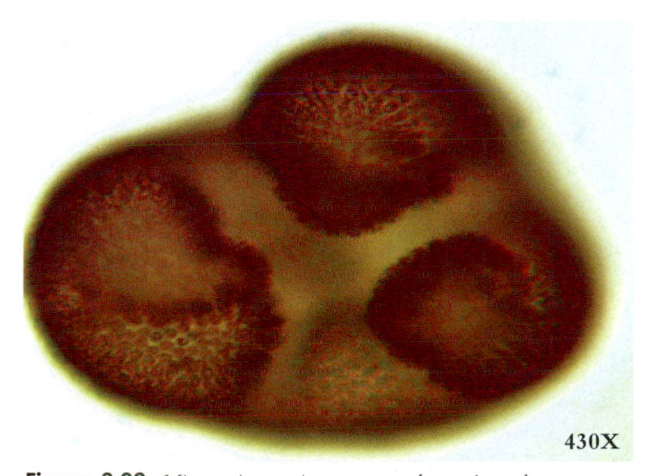

Figure 3.30 *Microcystis aeruginosa*, a cyanobacterium that can cause toxic water "blooms."

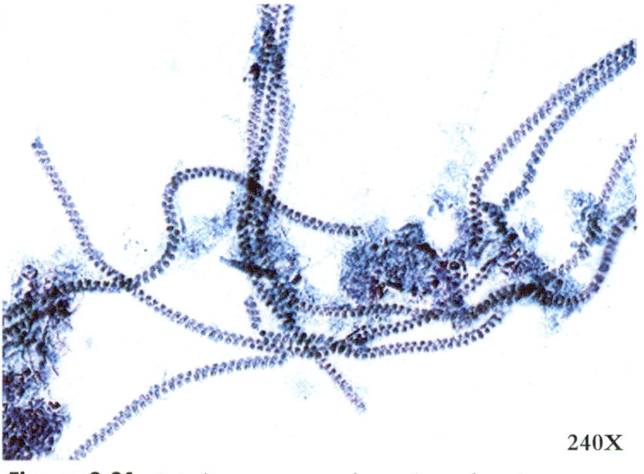

Figure 3.31 *Spirulina* sp., a cyanobacterium, showing characteristic spiral trichomes.

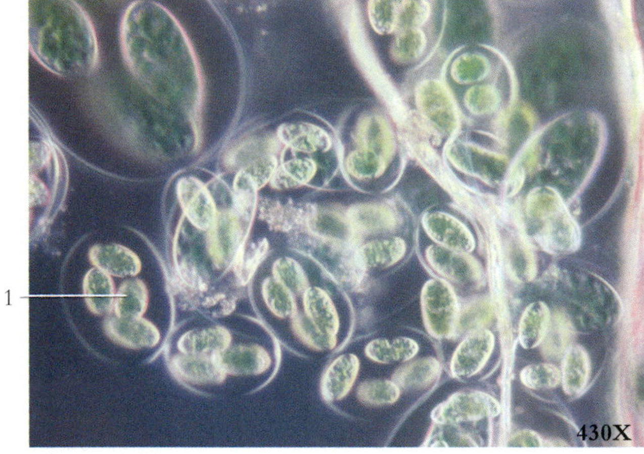

Figure 3.32 *Glaucocystis* sp., a green alga with cyanobacteria as endosymbionts.
1. Cyanobacteria endosymbiont

Figure 3.33 *Microcoleus* sp., one of the most common cyanobacteria in and on soils throughout the world. It is characterized by several filaments in a common sheath.

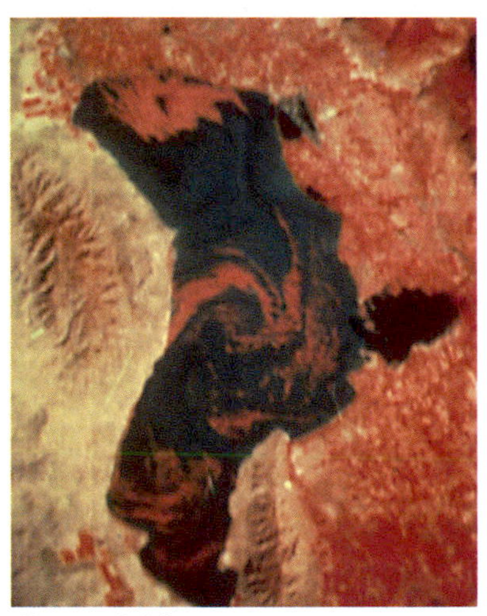

Figure 3.34 A satellite image of a large lake. The circular pattern in the water is composed of dense growths of cyanobacteria.

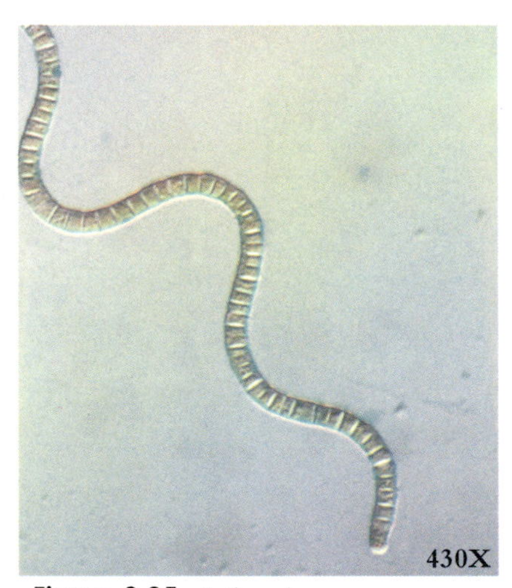

Figure 3.35 *Arthrospira* sp., a common cyanobacterium.

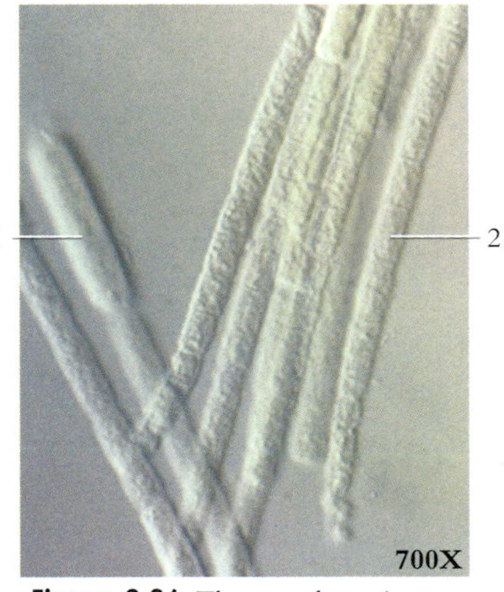

Figure 3.36 The cyanobacterium *Aphanizomenon* sp., common in nutrient-rich (often polluted) waters around the world.
1. Spore (akinete) 2. Filament

Figure 3.37 A spring seep in Zion National Park, Utah.
1. Mat of cyanobacteria.

Figure 3.38 A researcher examining cyanobacterial growths on soil in Canyonlands National Park, Utah.

Figure 3.39 A close-up photo of cryptobiotic soil crust. These crusts are composed of cyanobacteria, fungi, lichens, and other organisms.

Select Single-Celled Eukaryote Supergroup Phyla ("Protists")

Chapter 4

All animals are eukaryotes—their cells contain a membrane-bound nucleus that contains their genetic material. Most eukaryotic cells also contain membrane-bound organelles, such as mitochondria, chloroplasts, and digestive vacuoles and are even capable of meiosis and sexual reproduction. Eukaryotes are most closely related to Archaea but acquired their organelles from Bacteria by way of endosymbiosis (see exhibit 1 on page vi).

We easily recognize the majority of multicellular animals—the Metazoa—and distinguish these from plants and fungi. But there is a tremendous diversity of eukaryotes that aren't metazoans, fungi, or plants. Some contain chloroplasts, some don't. Most are single-celled, but some aren't. Most are microscopic, but some, like giant kelp, are very large. These organisms, which do not constitute a natural, or monophyletic group, are defined more by what they aren't than by what they are. But because they play an important role in understanding the evolutionary transitions that took place between prokaryotes and metazoans over a billion years ago, they are crucial components of any serious study of zoology.

Historically, the Linnean classification system ranked taxa according to morphological similarity. As phylogenetic analyses have become increasingly sophisticated and accurate, some of the well-known Linnean taxa have turned out to be evolutionary grades (as opposed to clades), united by primitive (plesiomorphic), as opposed to derived (apomorphic) characters. Such is the case for many independent evolutionary lineages of eukaryotes that are either unicellular or multicellular but without specialized tissues. Heretofore known as "protists," in this chapter we present them in a phylogenetic context that more accurately reflects their evolutionary history and current taxonomic status.

Most of the unicellular taxa in fig. 3.1 are abundant in aquatic habitats, and many are important constituents of plankton. Plankton are communities of organisms that drift passively or swim slowly in ponds, lakes, and oceans. Plankton are a major source of food for other aquatic organisms. Phototrophic (plantlike) microeukaryotes are major food producers in aquatic ecosystems. Key members of this group are from the Phylum Heterokontophyta, which includes the diatoms and golden algae. The cell wall of a diatom is composed largely of silica rather than cellulose. Some diatoms move in a slow, gliding way as cytoplasm glides through slits in the cell wall.

The Phylum Dinoflagellata also constitutes a large component of the phototrophic plankton. In most species of dinoflagellates, the cell wall is formed of armor-like plates of cellulose. Dinoflagellates are motile, having two flagella. Generally, one encircles the organism in a transverse groove, and the other projects to the posterior.

Among the unicellular microeukaryotes, or 'protozoan' (animallike) phyla are the Amoebozoa, Apicomplexa, Euglenozoa, Metamonada and Ciliophora. Locomotion of these heterotrophs is by way of flagella, cilia, or pseudopodia of various sorts. In feeding upon other organisms or organic particles, they use simple diffusion, pinocytosis, active transport, or phagocytosis. Although most of these organisms reproduce asexually, some species may also reproduce sexually during a portion of their life cycle. Most protozoa are harmless, although some are parasitic and may cause human disease, including African sleeping sickness and malaria.

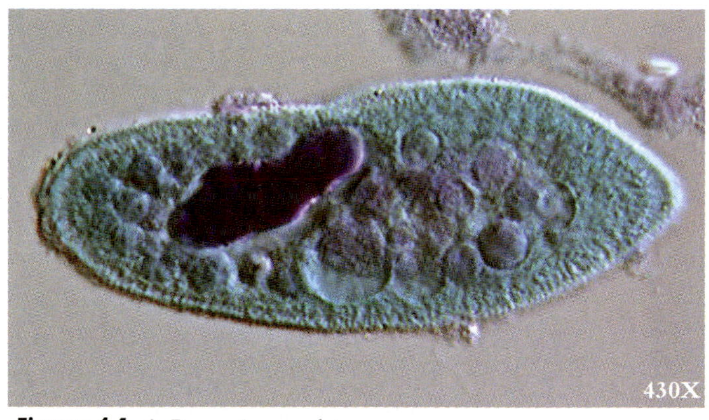

Figure 4.1 A *Paramecium caudatum*.

Table 4.1 Representative Single-Celled Eukaryote Supergroup Phyla

	Phyla and Representative Kinds	Characteristics
Plantlike	**Heterokontophyta** — diatoms and golden algae	Diatom cell walls composed of or impregnated with silica, often with two halves; plastids often golden in Chrysophyceae due to chlorophyll composition
Animallike	**Dinoflagellata** — dinoflagellates	Two flagella in grooves of wall; brownish-gold plastids
	Amoebozoa — amoebozoa	Cytoskeleton of microtubules and microfilaments; amoeboid locomotion
	Apicomplexa — sporozoa and *Plasmodium*	Lack locomotor capabilities and contractile vacuoles; mostly parasitic
	Euglenozoa — flagellated protozoa	Use flagella or pseudopodia for locomotion; mostly parasitic
	Metamonada — trichomonadas	Flagellate protozoan, *Trichomonas* sp.
	Ciliophora — ciliates and *Paramecium*	Use cilia to move and feed

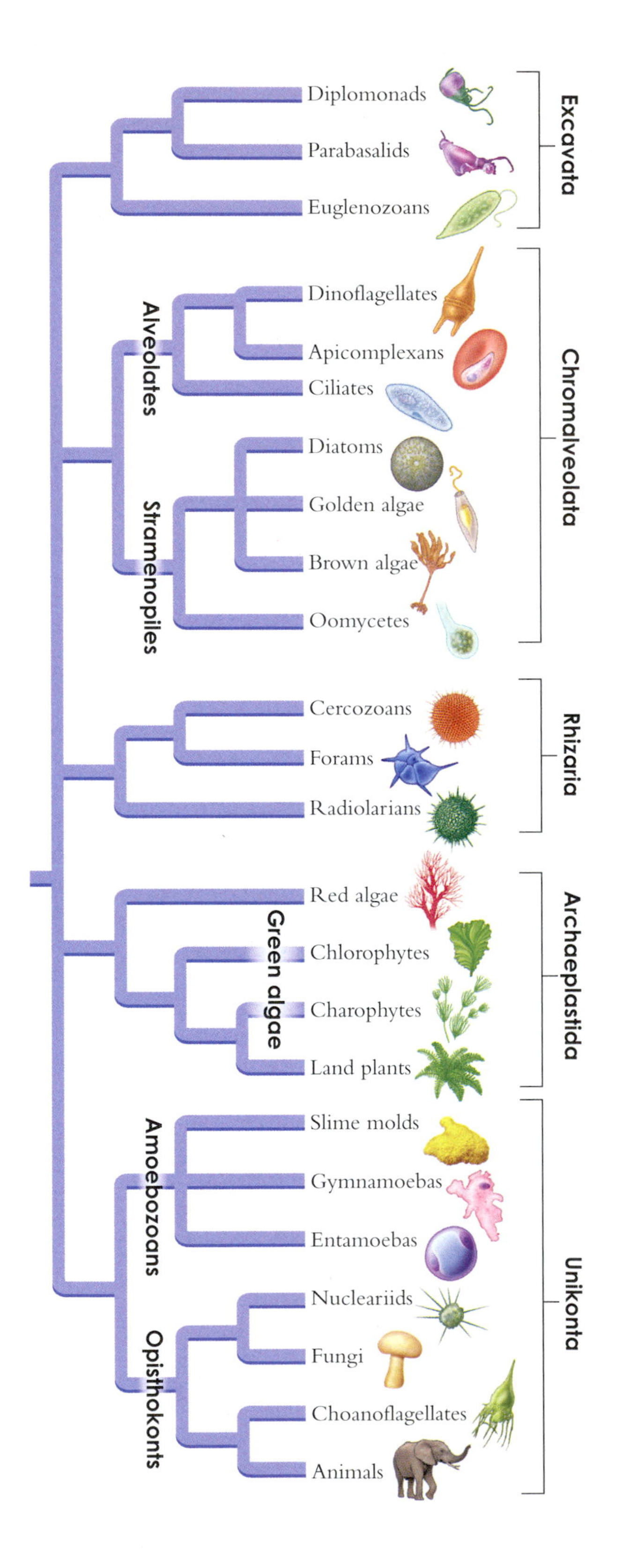

Figure 4.2 The phylogenetic relationships and classification of major eukaryote lineages.

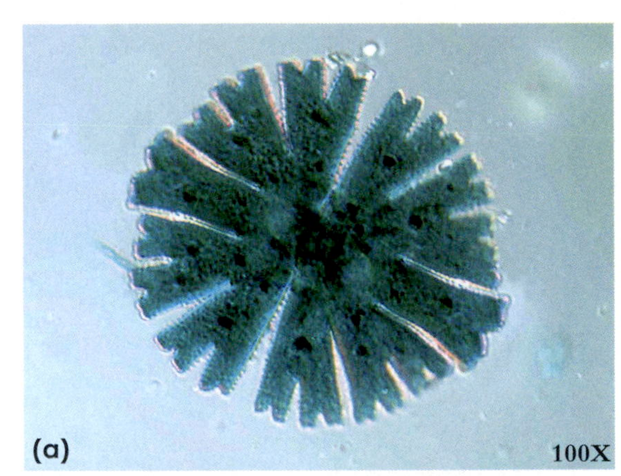

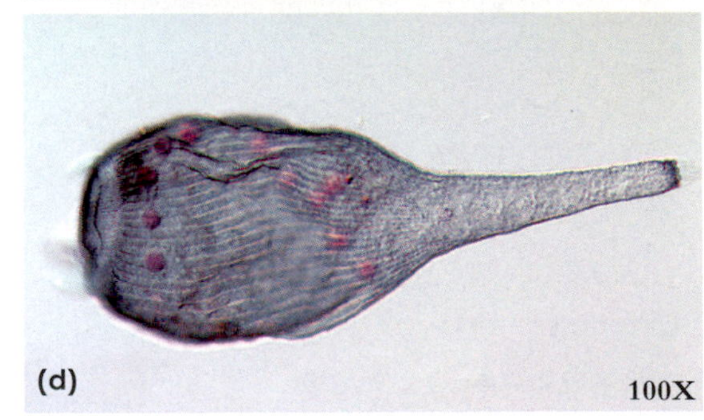

Figure 4.3 Example Protista include: (a) Desmid, *Micrasterias* sp., (b) kelp, *Macrocystis* sp., (c) a slime mold, *Physarum cinerea*, and (d) a protozoa, *Stentor* sp.

Phylum Heterokontophyta - diatoms and golden algae

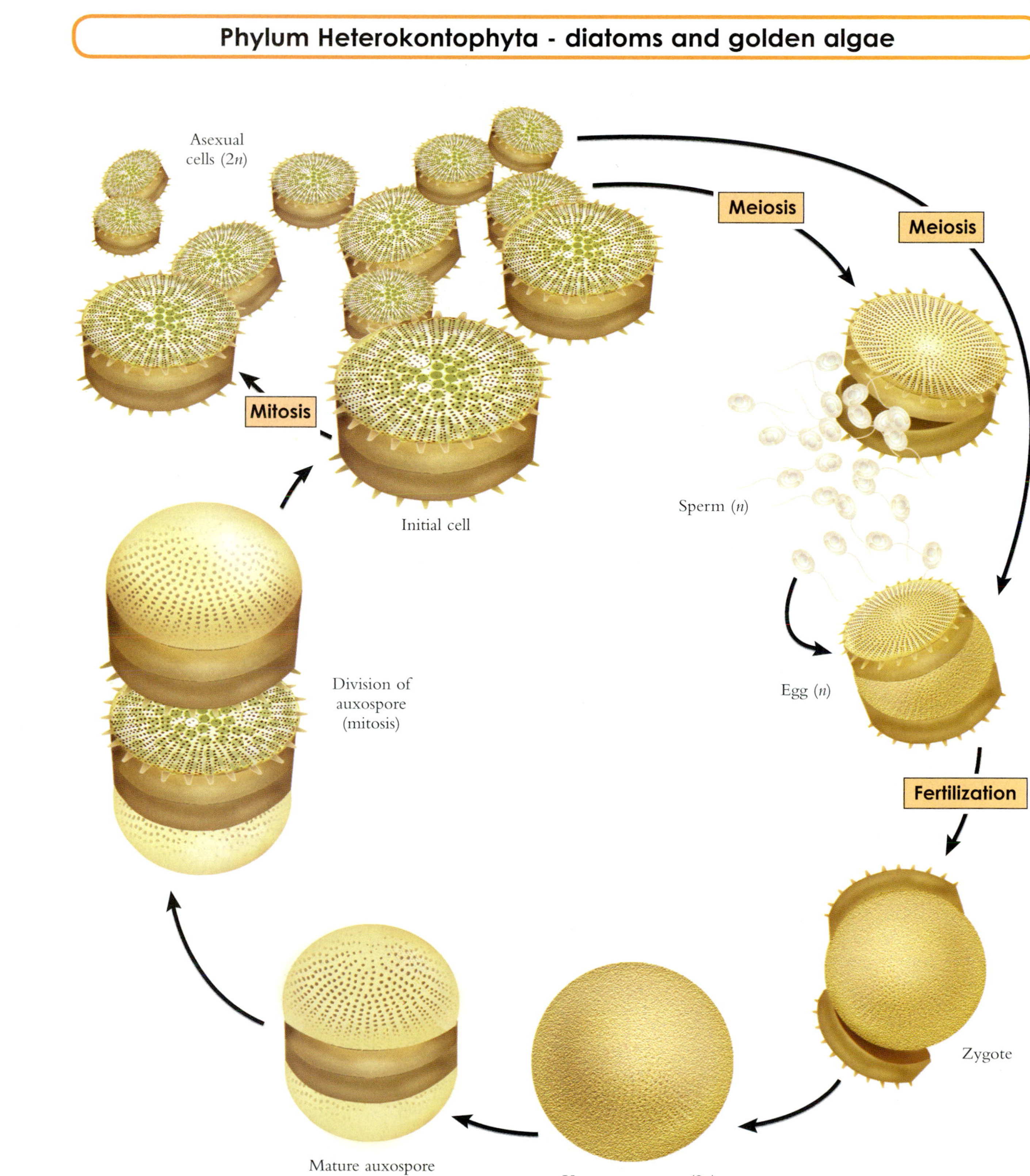

Figure 4.4 Life cycle of a centric diatom.

Figure 4.5 *Biddulphia* sp., a diatom forming colonies. These cells are beginning cell division.

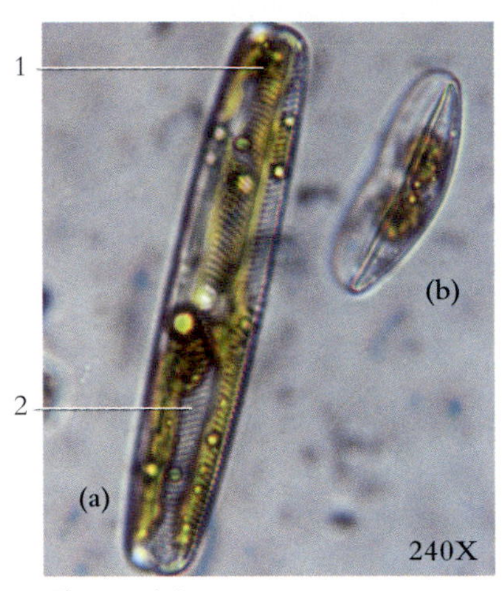

Figure 4.6 Live specimens of pennate (bilaterally symmetrical) diatoms. (a) *Navicula* sp., and (b) *Cymbella* sp.

1. Chloroplast
2. Striae

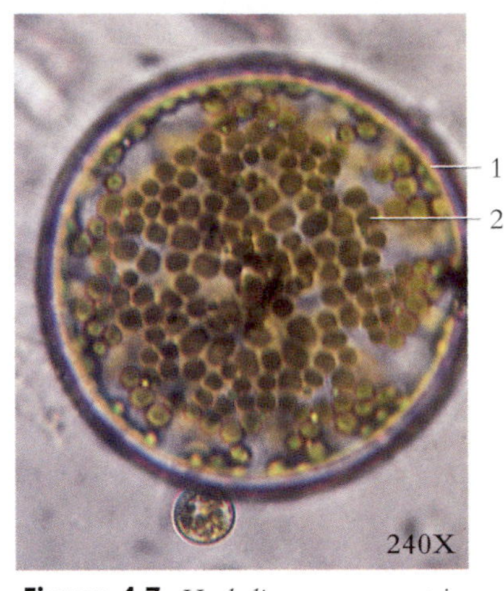

Figure 4.7 *Hyalodiscus* sp., a centric (radially symmetrical) diatom, from a freshwater spring in Nevada.

1. Silica cell wall
2. Chloroplasts

Figure 4.8 *Epithemia* sp., a distinctive pennate freshwater diatom.

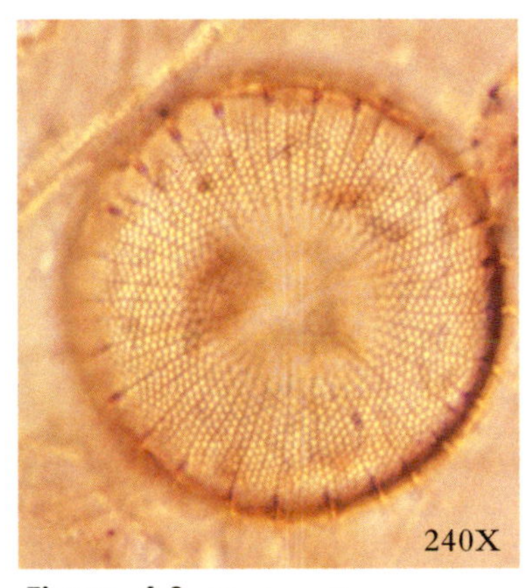

Figure 4.9 *Stephanodiscus* sp., a centric diatom.

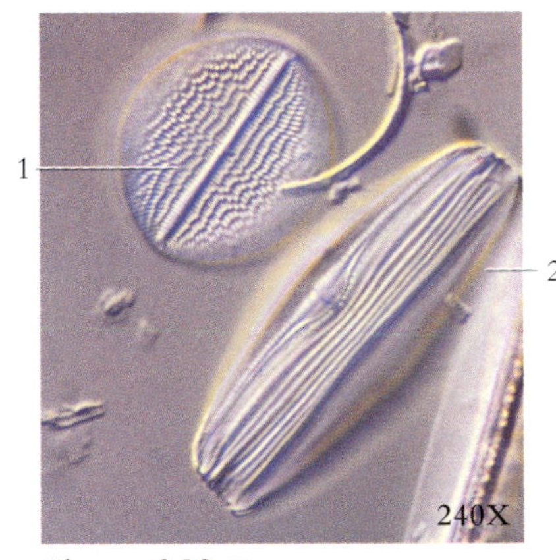

Figure 4.10 Two common freshwater diatoms.

1. *Cocconeis*
2. *Amphora*

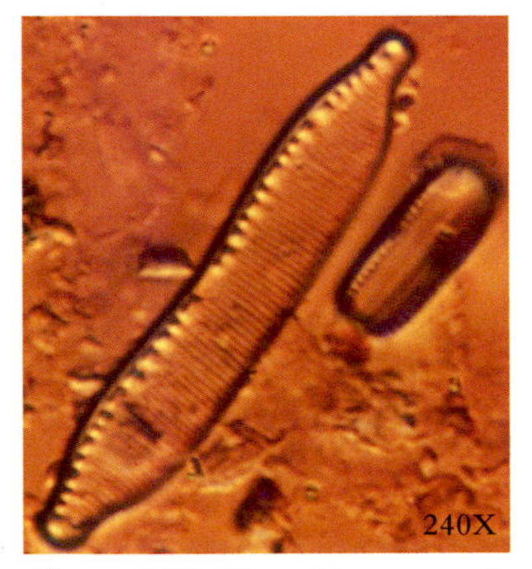

Figure 4.11 *Hantzschia* sp., one of the most common soil diatoms.

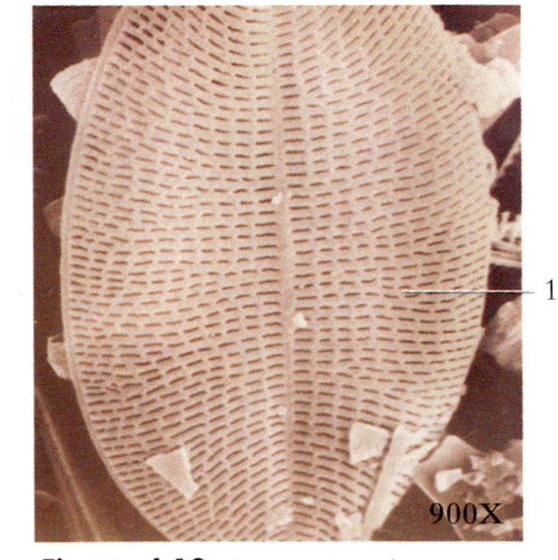

Figure 4.12 A scanning electron micrograph of *Cocconeis* sp., a common freshwater diatom.

1. Striae containing pores, or punctae, in the frustule (silicon cell wall).

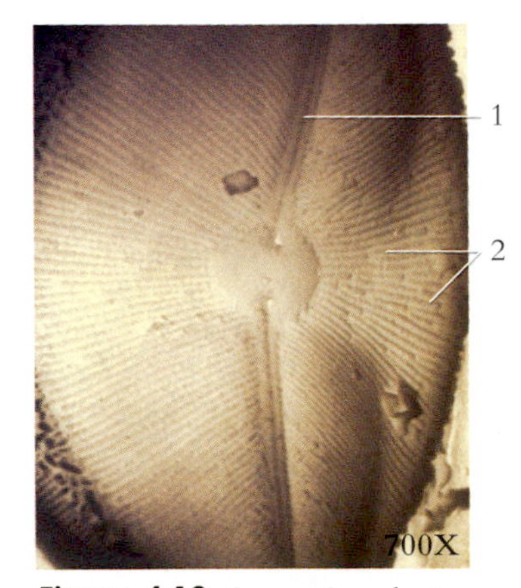

Figure 4.13 A scanning electron micrograph of the diatom *Achnanthes flexella*.

1. Raphe
2. Striae

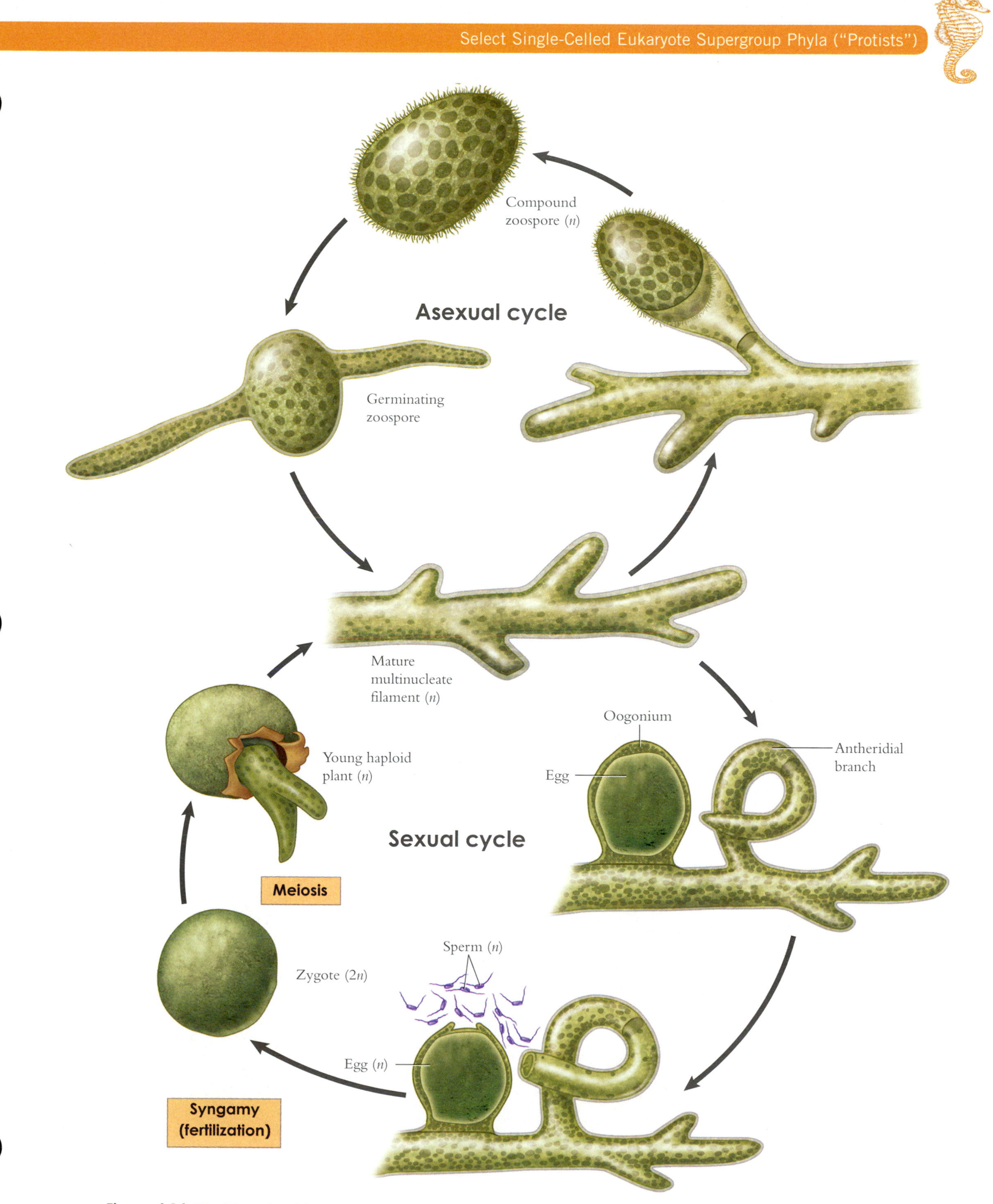

Figure 4.14 The life cycle of the "water felt alga," *Vaucheria* sp.

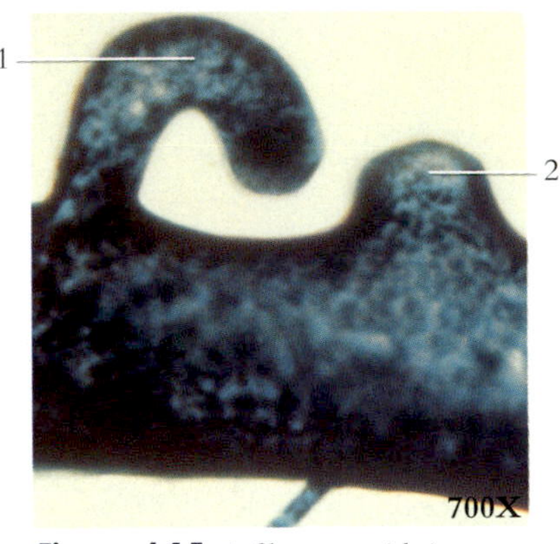

Figure 4.15 A filament with immature gametangia of the "water felt" alga, *Vaucheria* sp. *Vaucheria* is a chrysophyte that is widespread in freshwater and marine habitats. It is also found in the mud of brackish areas that periodically become submerged and then exposed to air.

1. Antheridium
2. Developing oogonium

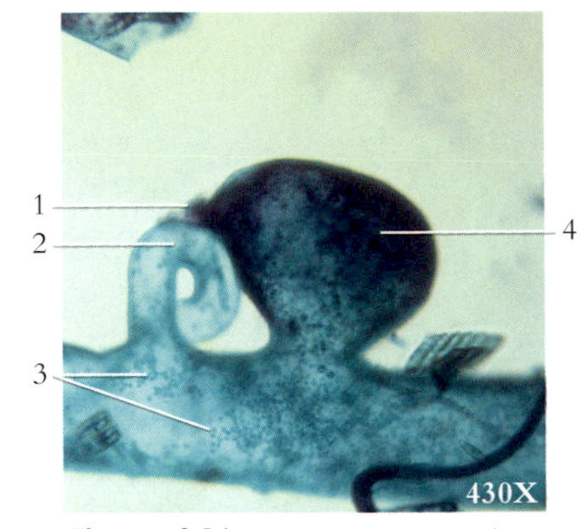

Figure 4.16 A *Vaucheria* sp., with mature gametangia.

1. Fertilization pore
2. Antheridium
3. Chloroplasts
4. Developing oogonium

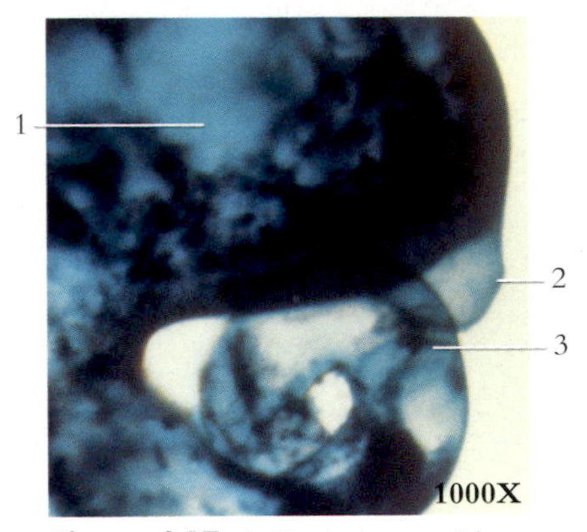

Figure 4.17 A *Vaucheria* sp., with mature gametangia.

1. Oogonium
2. Fertilization pore
3. Antheridium

Phylum Dinoflagellata - dinoflagellates

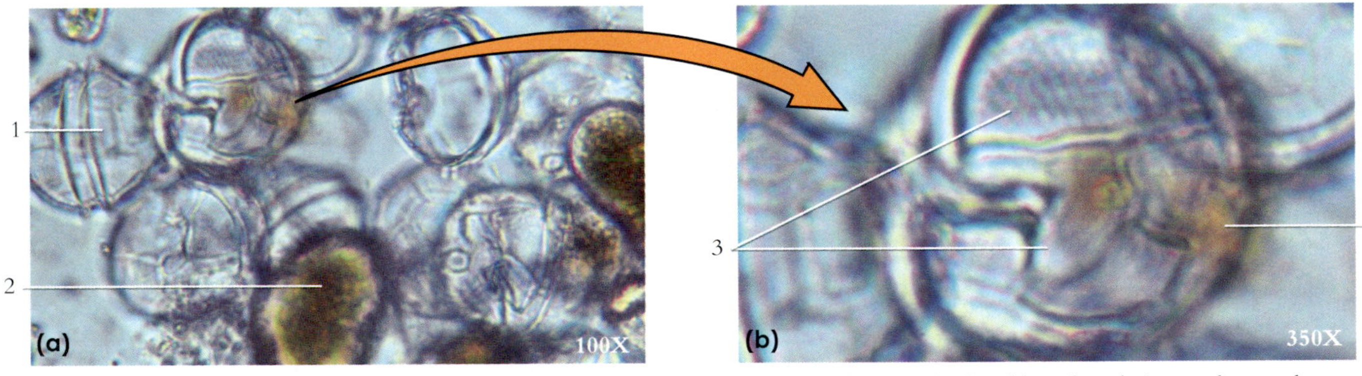

Figure 4.18 The dinoflagellates, *Peridinium* sp. (a) Some organisms are living; (b) others are dead and have lost their cytoplasm and consist of resistant cell walls.

1. Dead dinoflagellate
2. Living dinoflagellate
3. Cellulose plates
4. Remnant of cytoplasm

Figure 4.19 A giant clam with bluish coloration due to endosymbiont dinoflagellates.

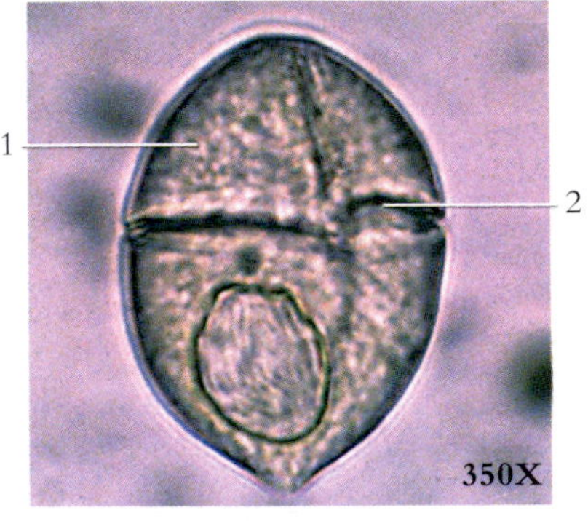

Figure 4.20 A photomicrograph of *Peridinium* sp. The cell wall of many dinoflagellates is composed of overlapping plates of cellulose.

1. Wall of cellulose plates
2. Transverse groove

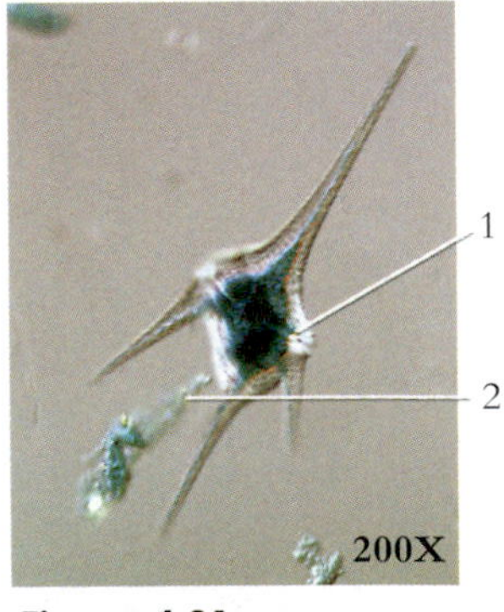

Figure 4.21 *Ceratium* sp. is a common freshwater dinoflagellate.

1. Transverse groove
2. Trailing flagellum

Phylum Amoebozoa - amoebas

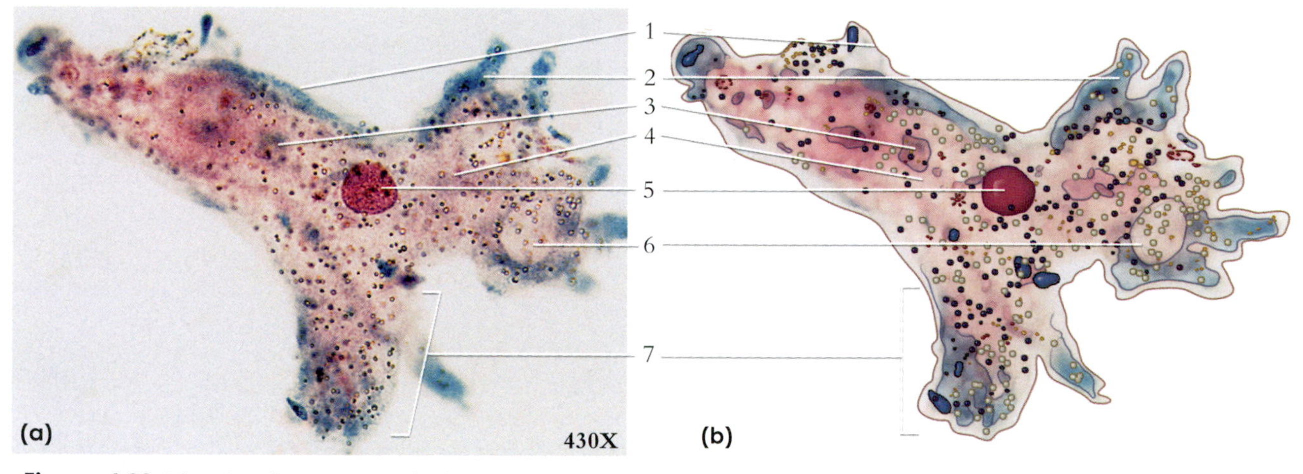

Figure 4.22 The *Amoeba proteus* is a freshwater protozoan that moves by forming cytoplasmic extensions called pseudopodia. (a) Stained cell, and (b) diagram.

1. Cell membrane
2. Ectoplasm
3. Food vacuole
4. Endoplasm
5. Nucleus
6. Contractile vacuole
7. Pseudopodia

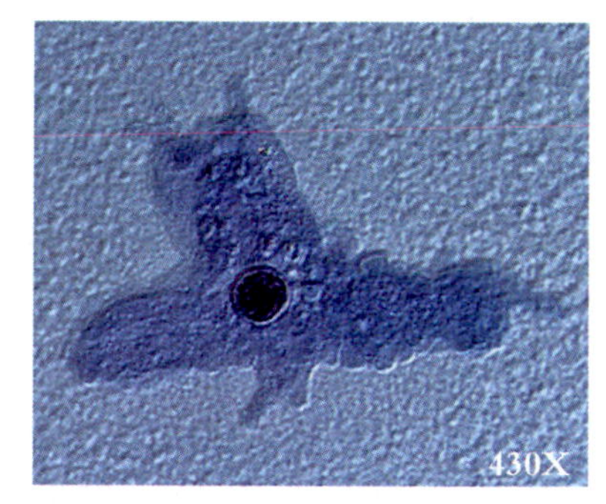

Figure 4.23 *Amoeba proteus* (stained blue).

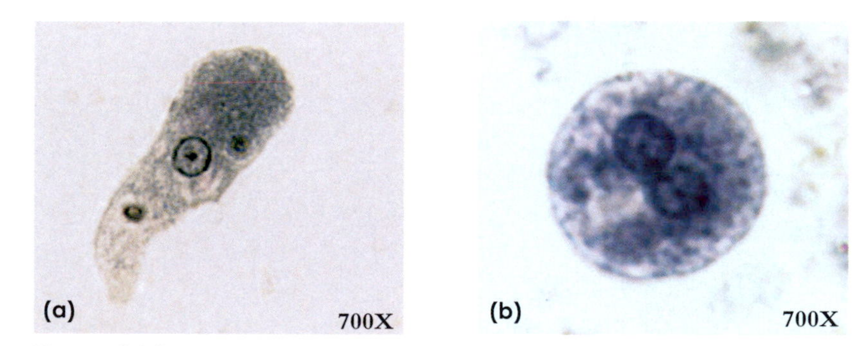

Figure 4.24 Protozoan *Entamoeba histolytica* is the causative agent of amoebic dysentery, a disease most common in areas with poor sanitation. (a) A trophozoite, and (b) a cyst.

Phylum Apicomplexa - *Plasmodium*

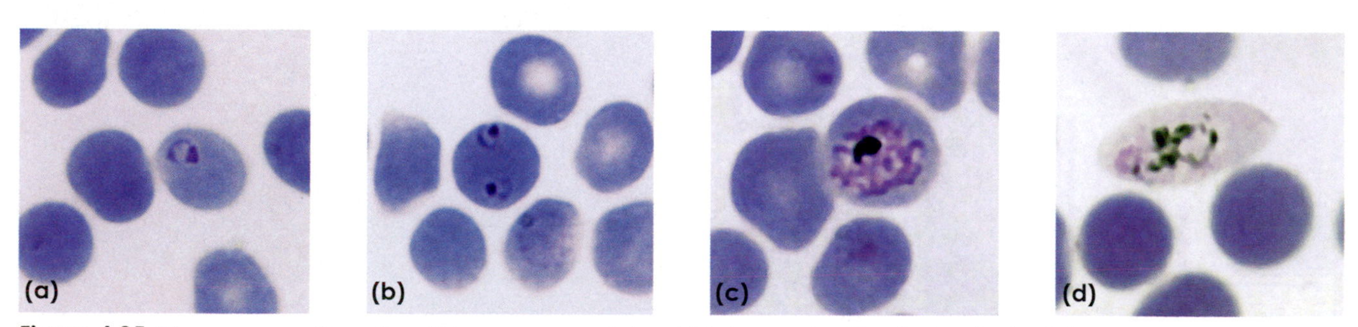

Figure 4.25 The protozoan *Plasmodium falciparum* causes malaria, which is transmitted by the female *Anopheles* mosquito. (a) The ring stage in a red blood cell, (b) a double infection, (c) a developing schizont, and (d) a gametocyte.

Phylum Metamonada - (Trichomonas) and Phylum Euglenozoa - (Leishmania and Trypanosoma): flagellated protozoans

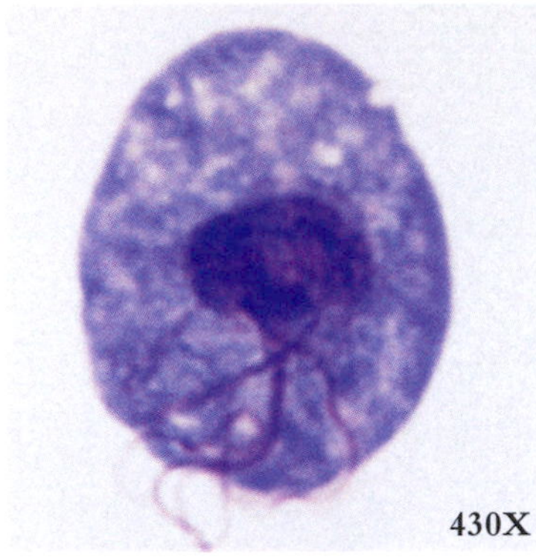

Figure 4.26 The protozoan *Trichomonas vaginalis* is the causative agent of trichomoniasis. Trichomoniasis is an inflammation of the genitourinary mucosal surfaces—the urethra, vulva, vagina, and cervix in females and the urethra, prostate, and seminal vesicles in males.

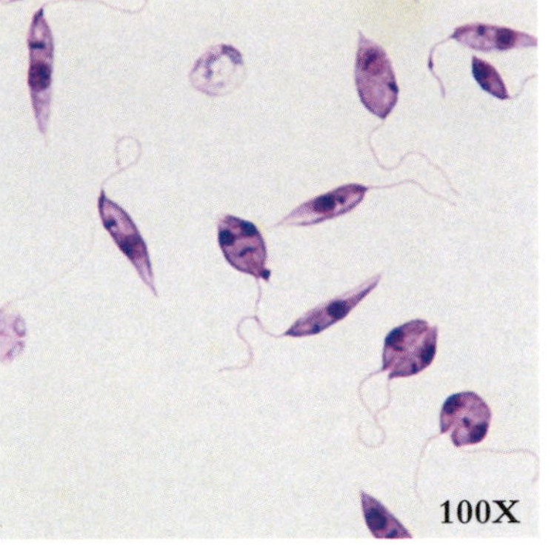

Figure 4.27 The protozoan *Leishmania donovani* is the causative agent of leishmaniasis, or kala-azar disease, in humans. The sandfly is the infectious host of this disease.

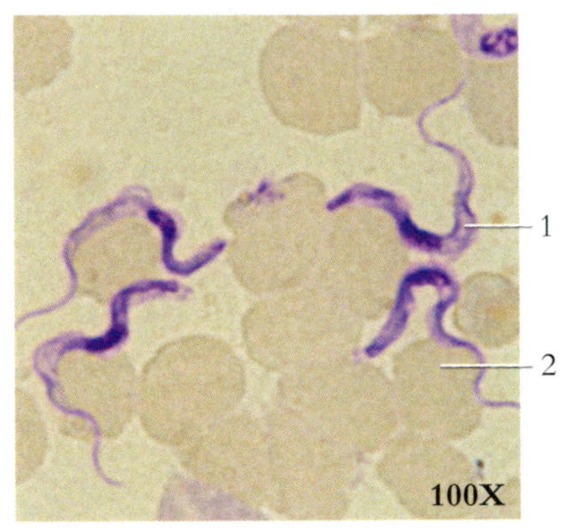

Figure 4.28 A flagellated protozoan, *Trypanosoma brucei*, is the causative agent of trypanosomiasis, or African sleeping sickness. The tsetse fly is the infectious host of this disease in humans.

1. *Trypanosoma brucei*
2. Red blood cell

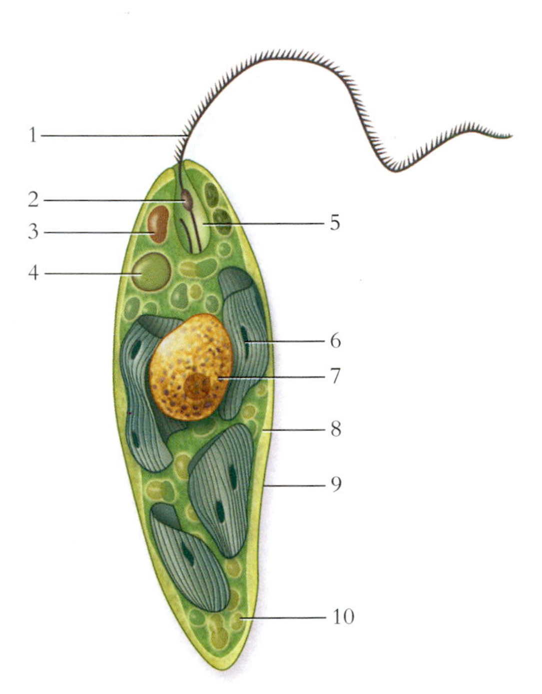

Figure 4.29 A diagram of *Euglena*, a genus of flagellates that contain chloroplasts. They are freshwater organisms that have a flexible pellicle rather than a rigid cell wall.

1. Long flagellum
2. Photoreceptor
3. Eyespot
4. Contractile vacuole
5. Reservoir
6. Chloroplast
7. Nucleus
8. Pellicle
9. Cell membrane
10. Paramylon granule

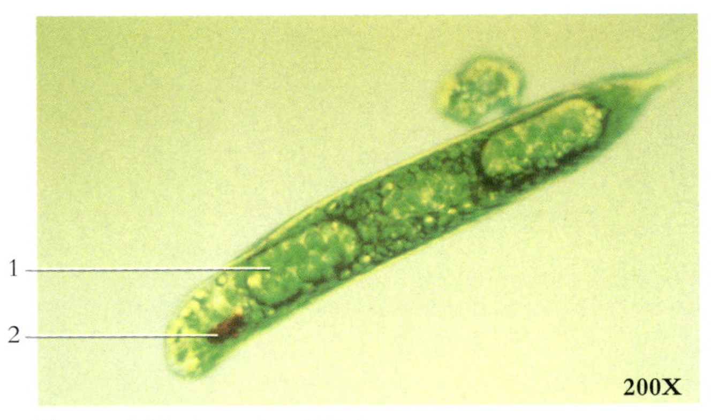

Figure 4.30 A species of *Euglena*.

1. Paramylum body
2. Photoreceptor

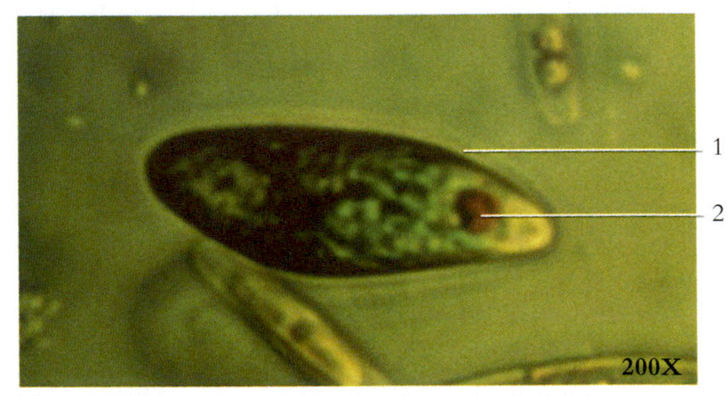

Figure 4.31 A species of *Euglena* from a brackish lake.

1. Pellicle
2. Photoreceptor

Phylum Ciliophora - ciliates and paramecia

Figure 4.32 *Paramecium caudatum* is a ciliated protozoan. The poisonous trichocysts of these unicellular organisms are used for defense and capturing prey.

1. Pellicle
2. Contractile vacuole
3. Macronucleus
4. Cilia
5. Trichocyst
6. Food vacuole
7. Forming food vacuole
8. Gullet
9. Oral groove
10. Micronucleus

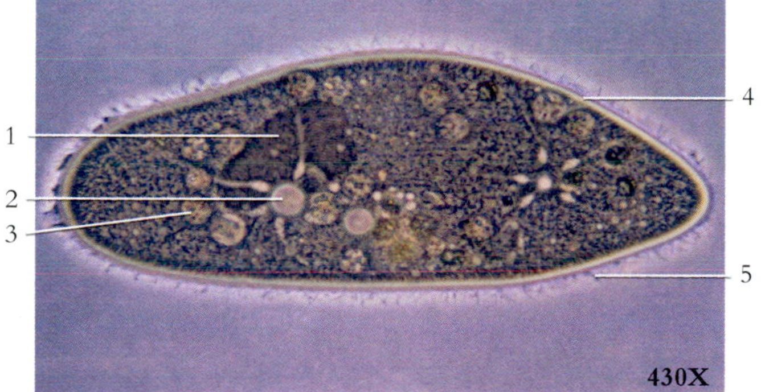

Figure 4.33 *Paramecium caudatum* is a ciliated protozoan. Paramecia are usually common in ponds containing decaying organic matter.

1. Macronucleus
2. Contractile vacuole
3. Micronucleus
4. Pellicle
5. Cilia

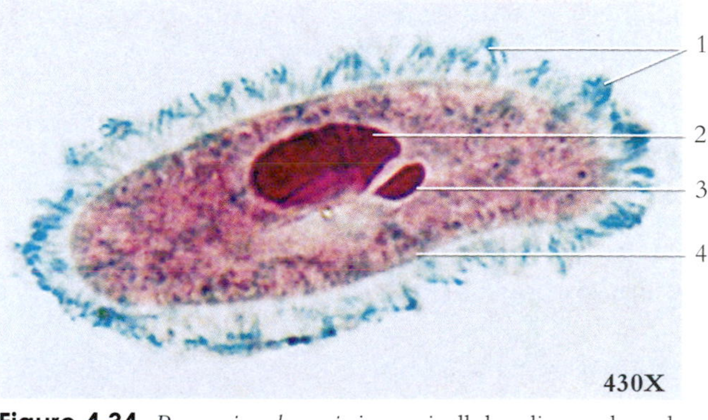

Figure 4.34 *Paramecium bursaria* is a unicellular, slipper-shaped organism. When disturbed or threatened, they release spear-like trichocysts as a defense.

1. Trichocysts
2. Macronucleus
3. Micronucleus
4. Pellicle

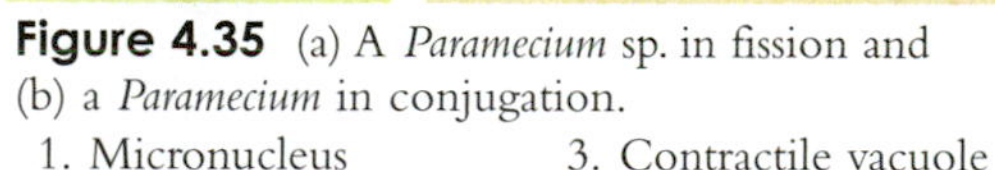

Figure 4.35 (a) A *Paramecium* sp. in fission and (b) a *Paramecium* in conjugation.

1. Micronucleus
2. Macronucleus
3. Contractile vacuole

Figure 4.36 A prepaired slide showing a group of *Paramecium* sp.

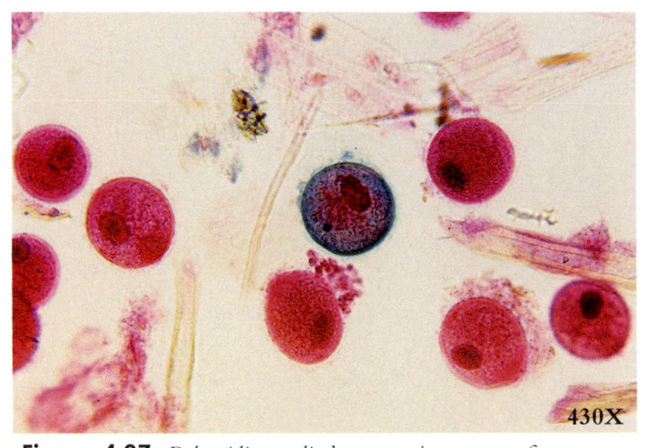

Figure 4.37 *Balantidium coli*, the causative agent of balantidiasis. Cysts in sewage-contaminated water are the infective form.

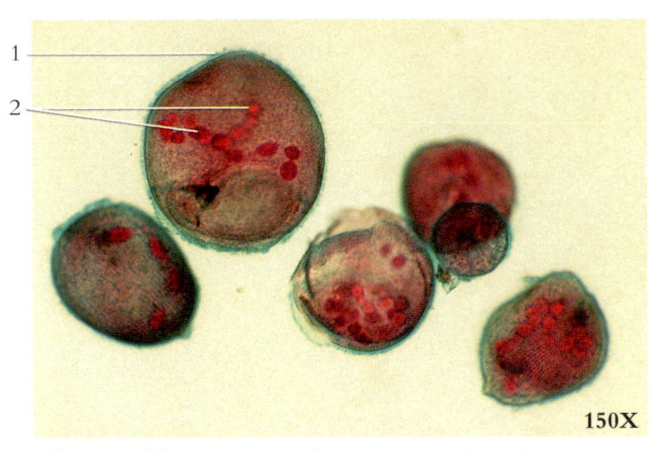

Figure 4.38 *Stentor* sp., a free-swimming form that has adopted an oval shape.
1. Cilia
2. Macronucleus (monoliform)

Table 4.2 Some Representatives of Protista: Primarily Multicellular Organisms

Phylum and Representative Kinds	Characteristics
Algae	
Phylum Chlorophyta—green algae	Unicellular, colonial, filamentous, and multicellular platelike forms; mosly freshwater; reproduce asexually and sexually
Phylum Phaeophyta—brown algae, giant kelp	Multicellular, mostly marine often in the intertidal zone; most with alternation of generations
Phylum Rhodophyta—red algae	Multicellular, mostly marine; sexual reproduction but with no flagellated cells; alternation of generations common
Protists Resembling Fungi	
Phylum Myxomycota—plasmodial slime molds	Multinucleated continuum of cytoplasm without cell membranes; amoeboid plasmodium during feeding stage; produce asexual fruiting bodies; gametes produced by meiosis
Phylum Dictyosteliomycota—cellular slime molds	Solitary cells during feeding stage; cells aggregate when food is scarce; produce asexual fruiting bodies
Phylum Oomycota—water molds, white rusts, and downy mildews	Decomposers or parasitic forms; walls of cellulose, dispersal by nonmotile spores or flagellated zoospores, gametes produced by meiosis

Phylum Chlorophyta - green algae

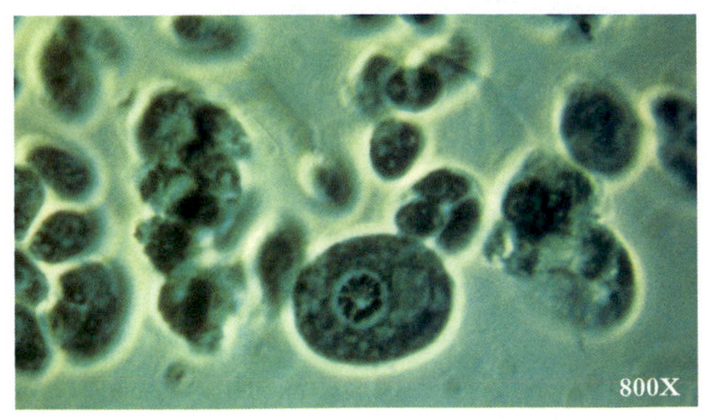

Figure 4.39 *Chlamydomonas* sp., a common unicellular green alga.

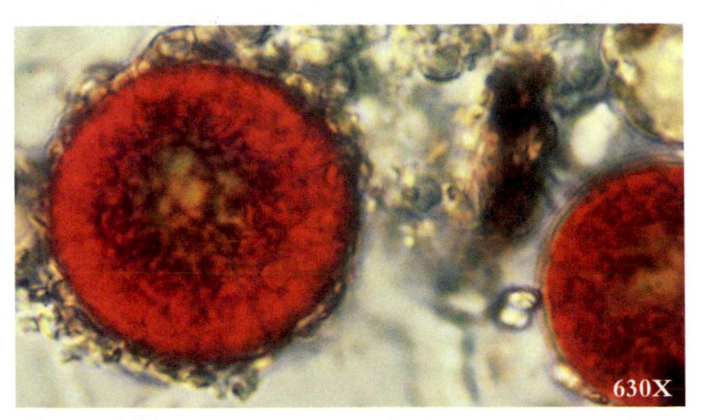

Figure 4.40 *Chlamydomonas nivalis*, the common snow alga.

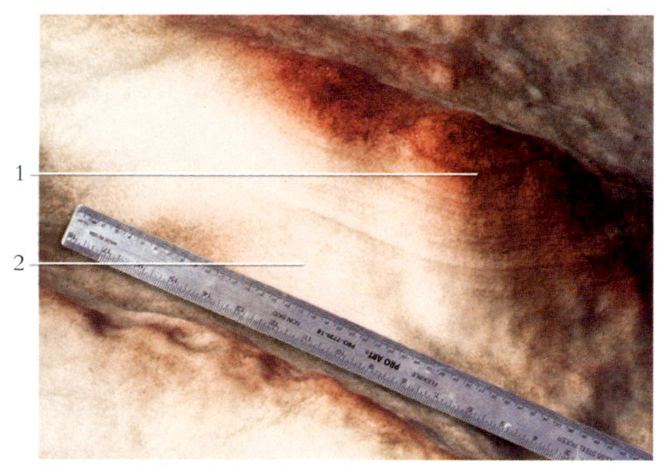

Figure 4.41 A habitat shot of *Chlamydomonas nivalis* creating "red snow."

1. *Chlamydomonas nivalis*
2. Snow

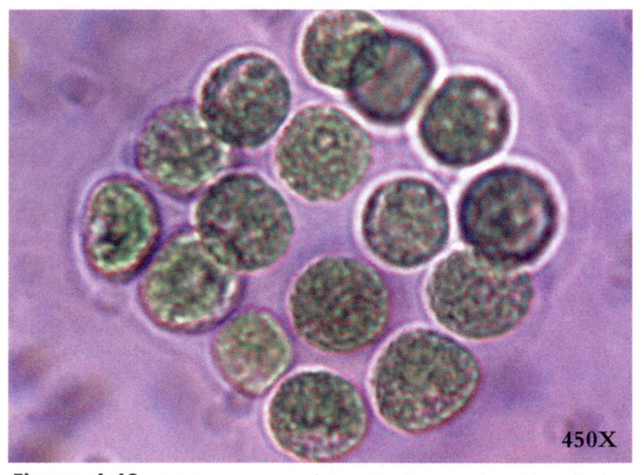

Figure 4.42 A *Gonium* sp. colony. *Gonium* sp. is a 16-celled flat colony of *Chlamydomonas*-like cells.

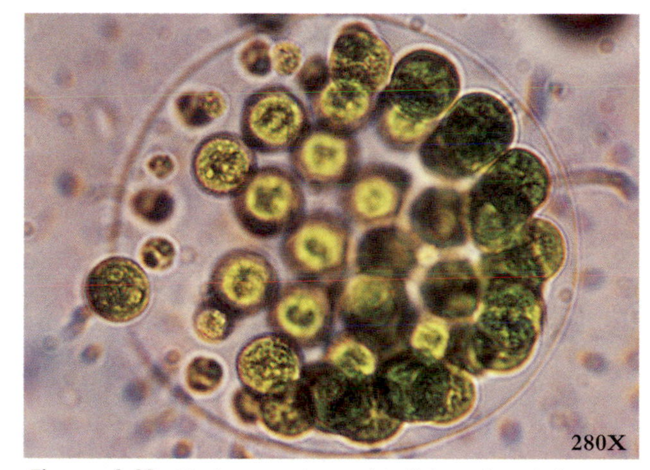

Figure 4.43 *Pleodorina* sp., is a multicellular colony (often 64-celled) relative of *Chlamydomonas* and *Volvox*.

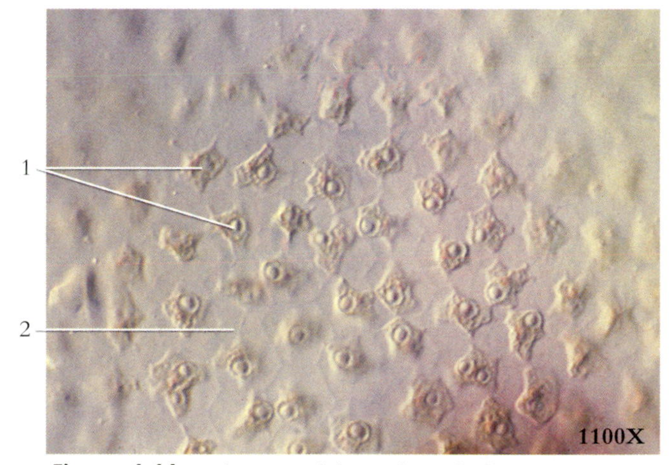

Figure 4.44 A close-up of the surface of *Volvox* sp. showing the interconnections between cells.

1. Vegetative cells
2. Cytoplasmic connection between cells

Figure 4.45 A *Volvox* sp. Three separate organisms are shown in this photomicrograph, each containing daughter colonies of various ages.

1. Immature daughter colony
2. Daughter colonies
3. Vegetative cells

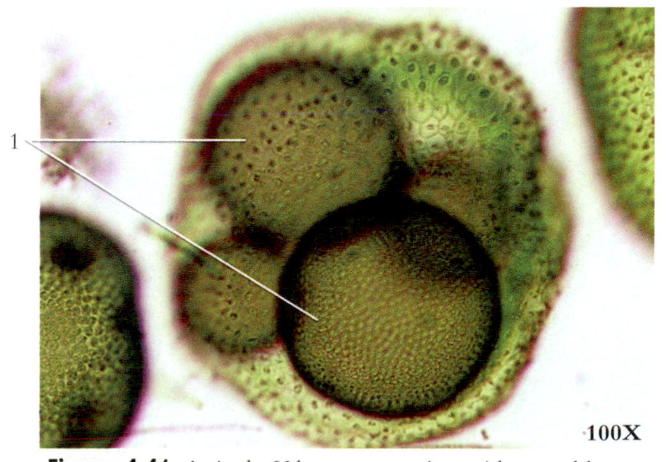

Figure 4.46 A single *Volvox* sp., organism with several large daughter colonies.

1. Daughter colonies

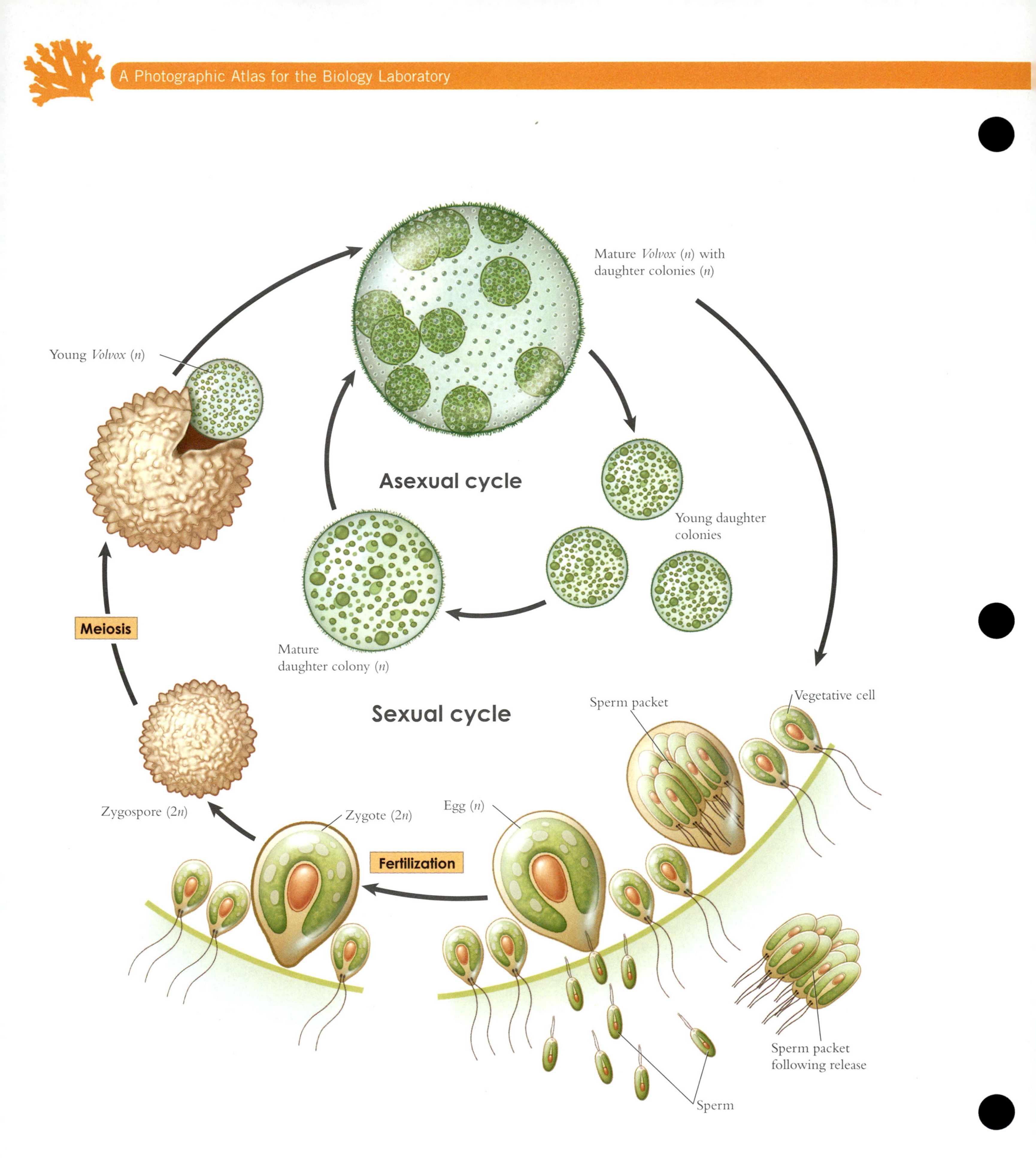

Figure 4.47 The life cycle of *Volvox*, a common freshwater chlorophyte. *Volvox* is considered by some to be a colony and by others to be a single, integrated plant.

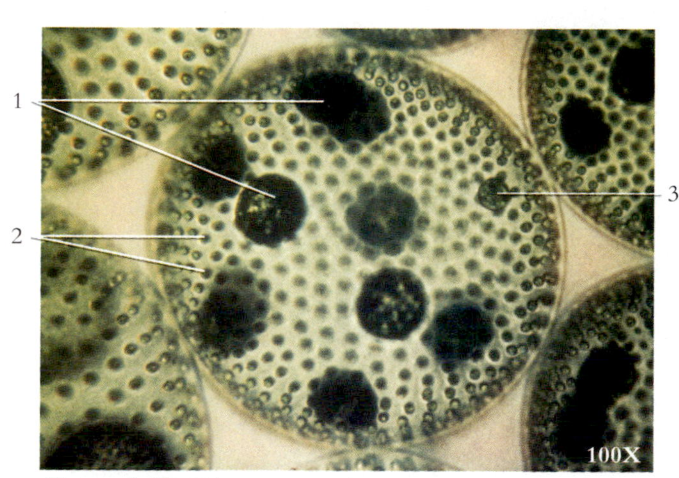

Figure 4.48 *Volvox* sp., a single mature specimen with several eggs and zygotes.
1. Zygotes 2. Vegetative cells 3. Egg

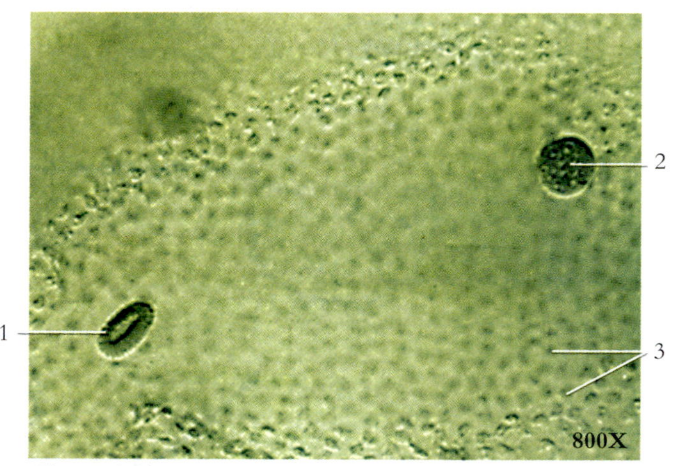

Figure 4.49 A single mature specimen of *Volvox* sp. This photomicrograph is a highly magnified view of a single organism showing gametes.
1. Sperm packet
2. Egg
3. Vegetative cells

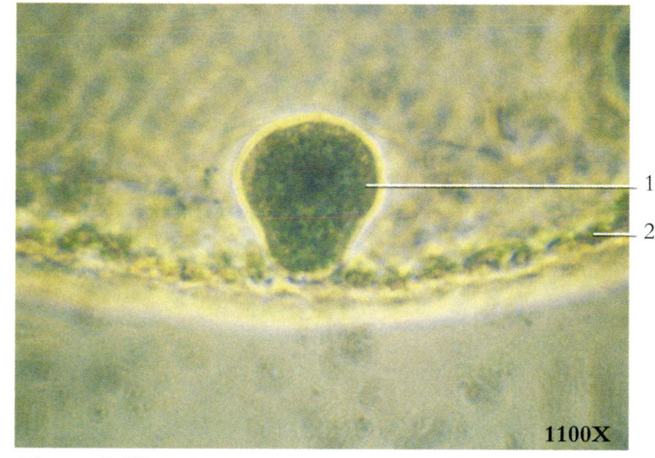

Figure 4.50 *Volvox* sp., showing a prominent egg at the edge of the organism. This egg will be fertilized to develop a zygote and then a zygospore.
1. Egg 2. Vegetative cells

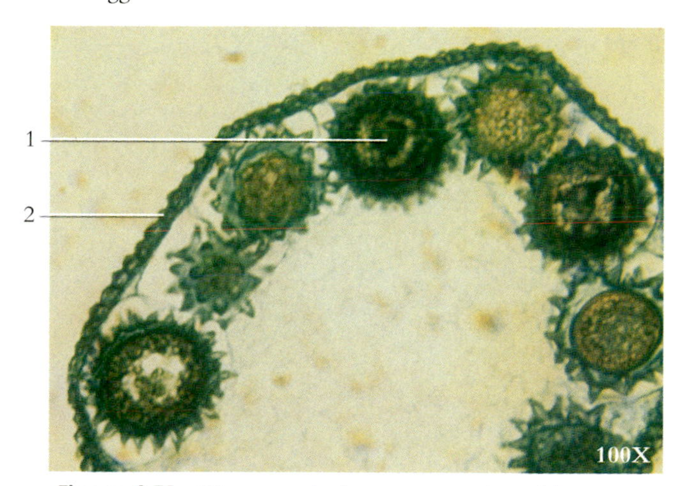

Figure 4.51 *Volvox* sp., a single mature organism with zygospores.
1. Zygospore
2. Vegetative cells

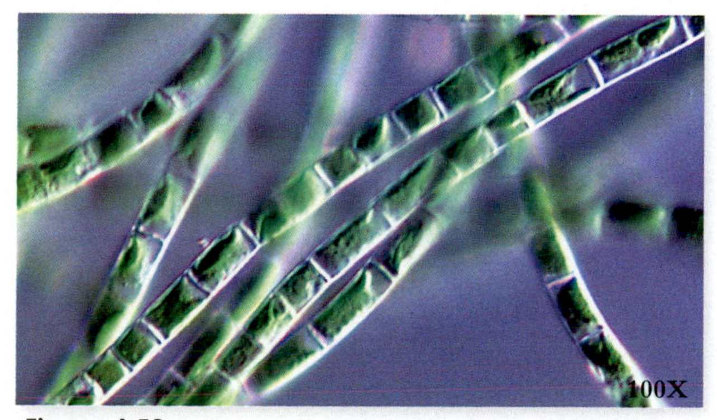

Figure 4.52 A live specimens of *Ulothrix* sp., an unbranched, filamentous green alga.

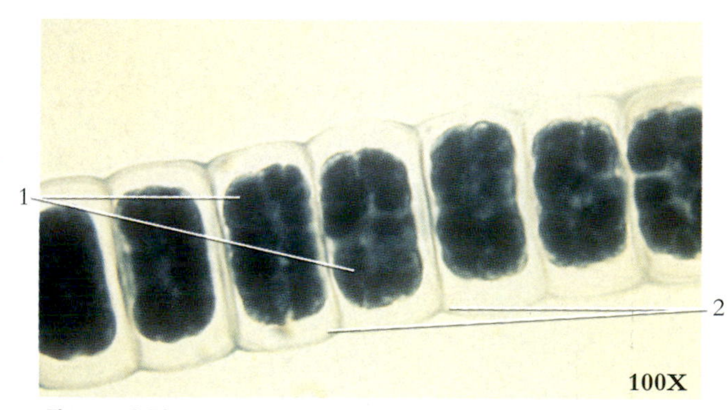

Figure 4.53 *Ulothrix* sp., an unbranched, filamentous green alga.
1. Zoospores 2. Individual cells (known as sporangia when they produce spores)

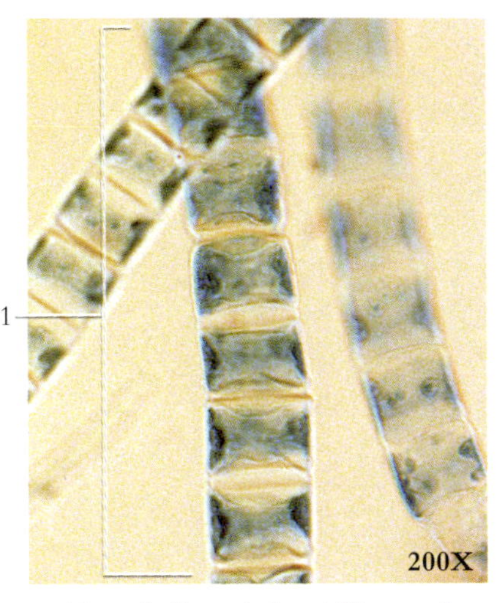

Vegetative stained filament

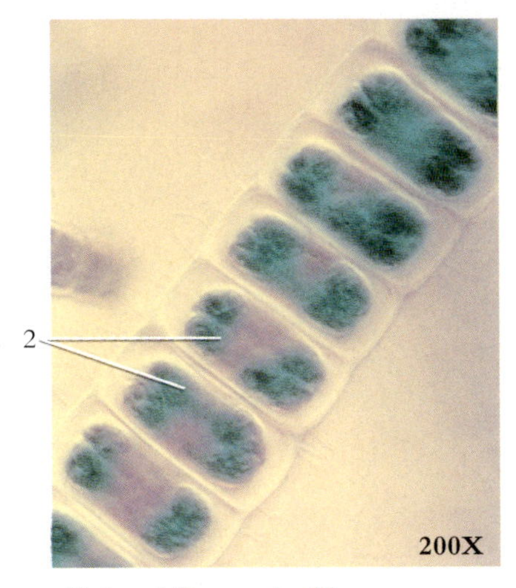

Stained filament with zoospores

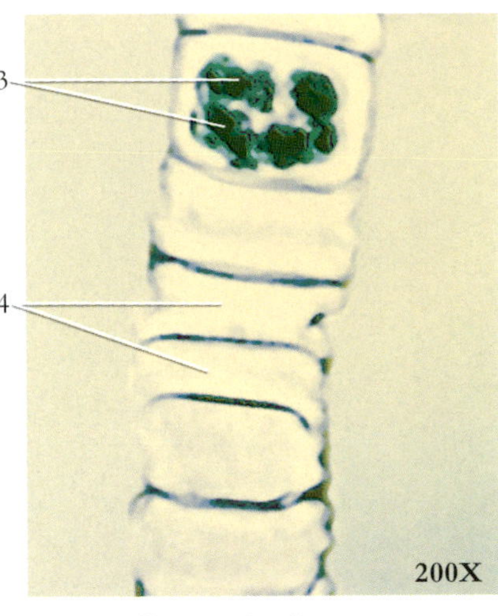

Empty filament, after zoospores have been released

Figure 4.54 The production and release of zoospores in the green alga *Ulothrix* sp.
1. Filament 2. Young zoospores 3. Mature zoospores 4. Empty cells following zoospore release

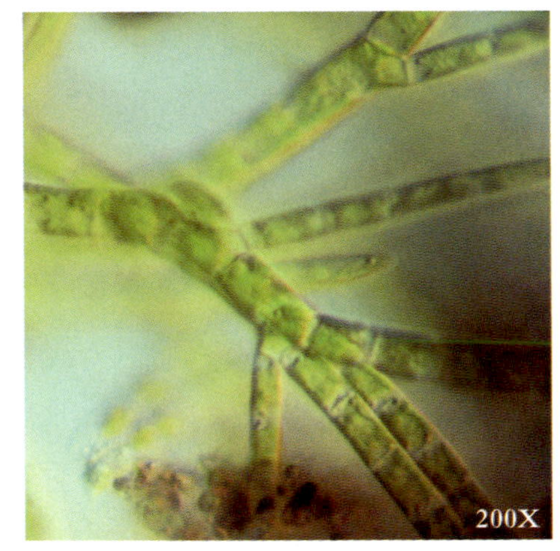

Figure 4.55 *Stigeocolonium* sp., a close relative of *Ulothrix*, showing a branched thallus.

Figure 4.56 *Draparnaldia* sp., a relative of *Ulothrix*, showing different cell sizes in the thallus and a characteristic branching pattern.

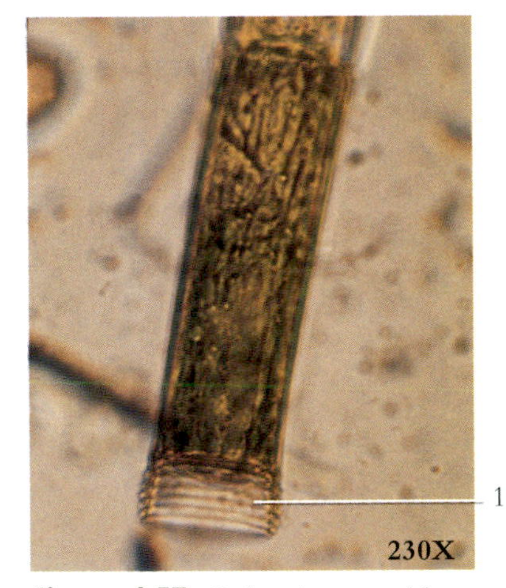

Figure 4.57 *Oedogonium* sp. with distinct "apical caps" that accrue from cell division in this genus.
1. Apical caps

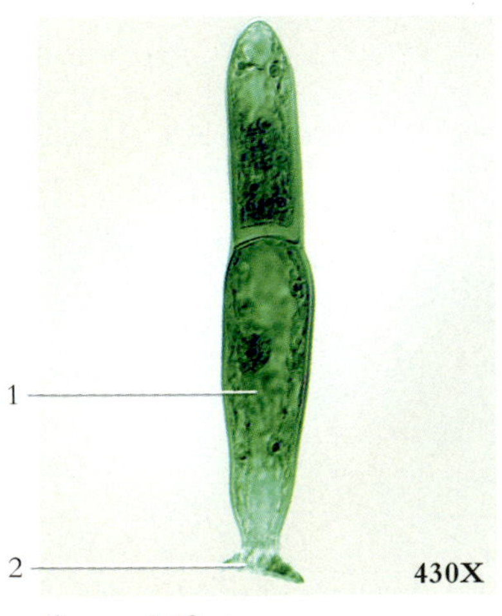

Figure 4.58 A young filament of *Oedogonium* sp.
1. Basal cell
2. Holdfast

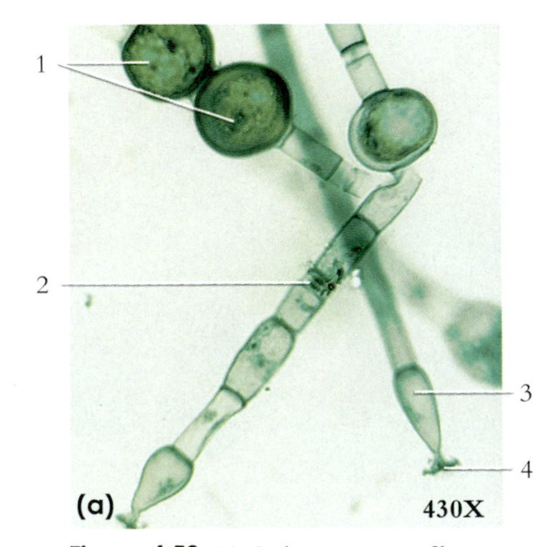

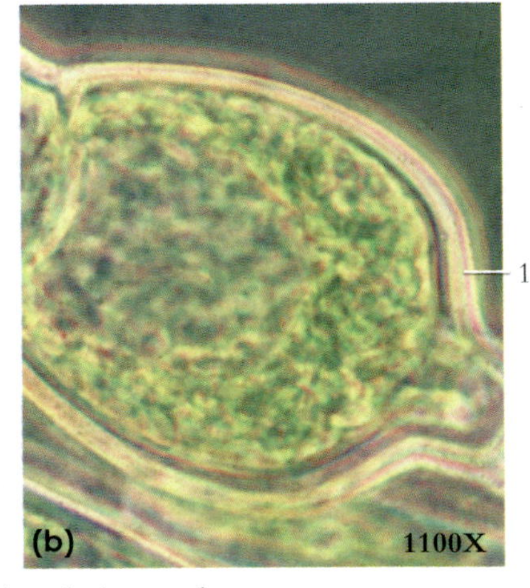

Figure 4.59 (a) *Oedogonium* sp., a filamentous, unbranched, green alga. (b) Close-up of an oogonium.
1. Oogonia
2. Antheridium
3. Basal holdfast cell
4. Holdfast

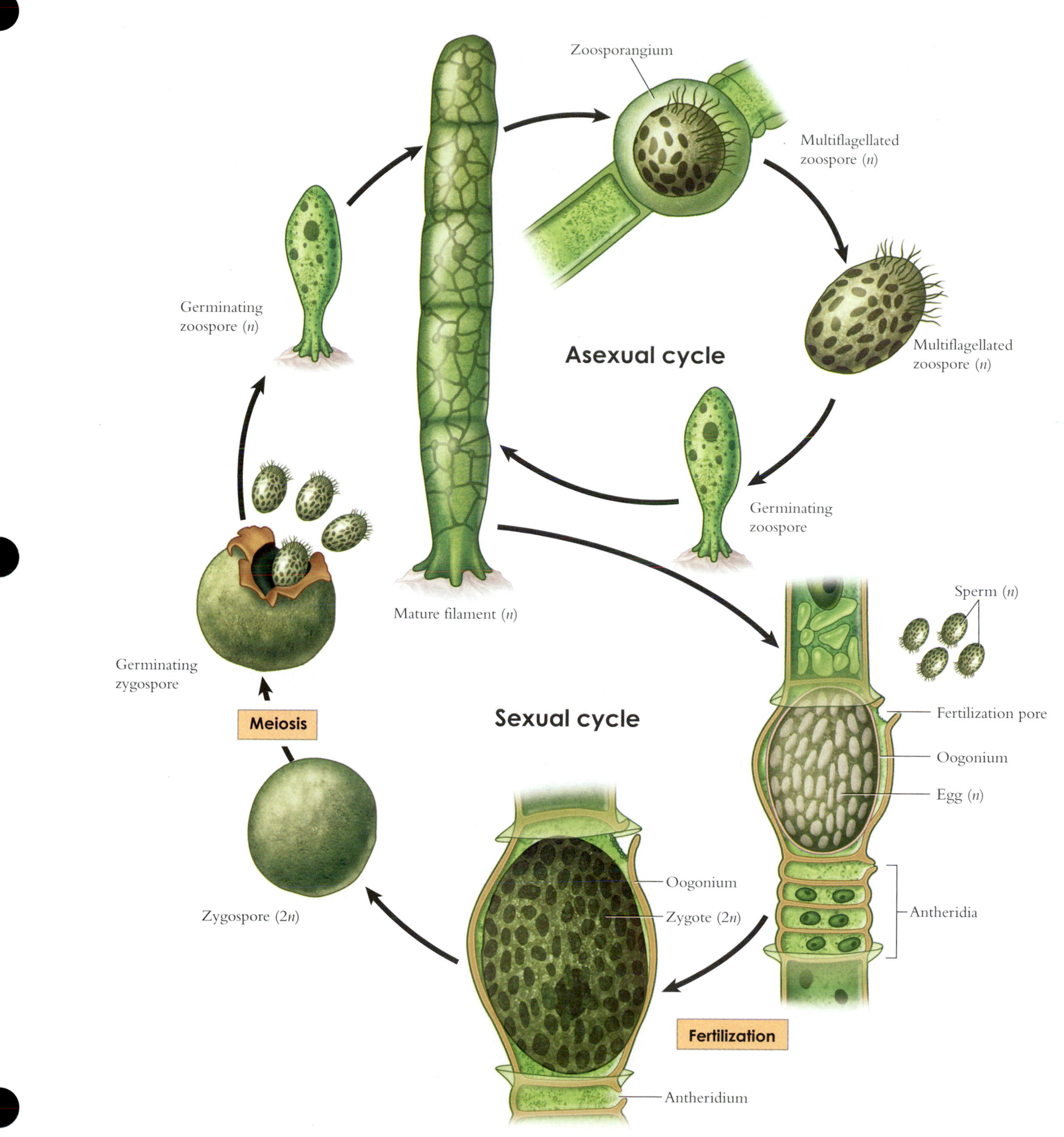

Figure 4.60 The life cycle of *Oedogonium* sp., an unbranched, filamentous green alga.

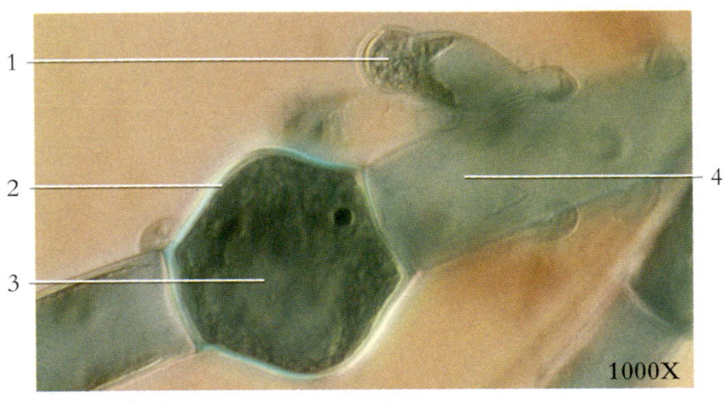

Figure 4.61 The oogonium of the unbranched green alga, *Oedogonium* sp.

1. Dwarf male filament
2. Oogonium
3. Developing egg
4. Vegetative cell

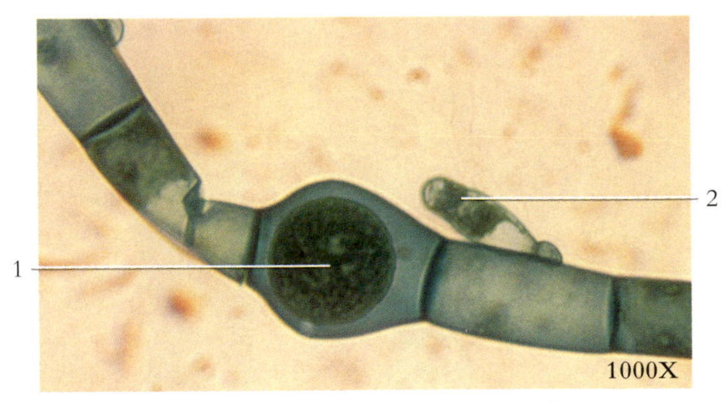

Figure 4.62 An oogonium with mature egg and dwarf male filament.

1. Egg
2. Dwarf male filament

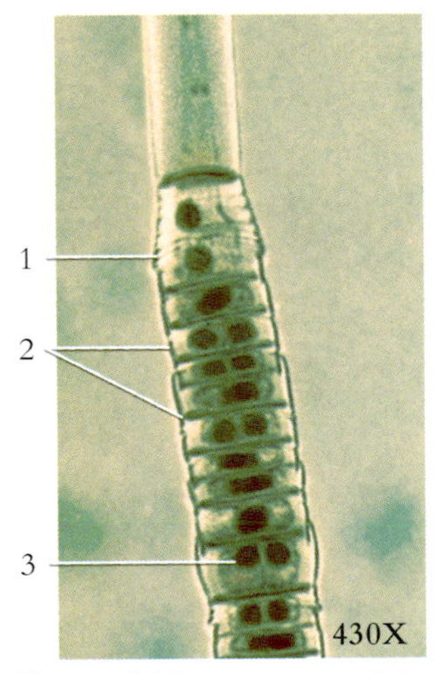

Figure 4.63 A filament of the green alga, *Oedogonium* sp.

1. Annular scars from cell division
2. Antheridia
3. Sperm

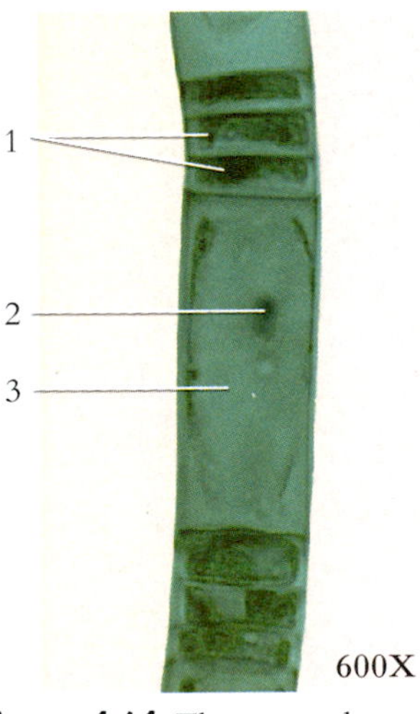

Figure 4.64 The green alga, *Oedogonium* sp., showing antheridia between vegetative cells.

1. Sperm within antheridia
2. Nucleus of vegetative cell
3. Vegetative cell

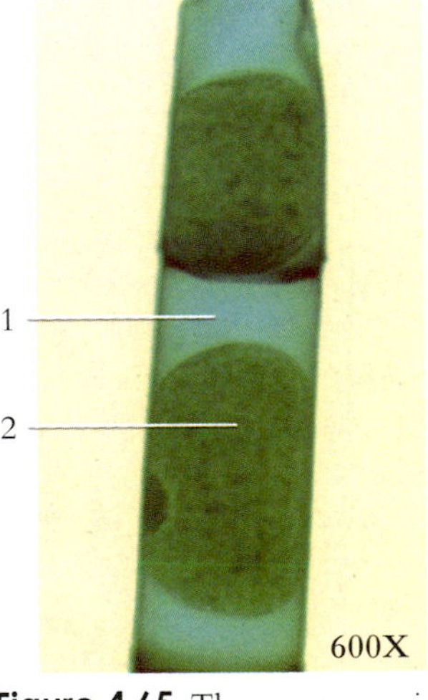

Figure 4.65 The zoosporangium of the unbranched green alga, *Oedogonium* sp.

1. Zoosporangium
2. Zoospore

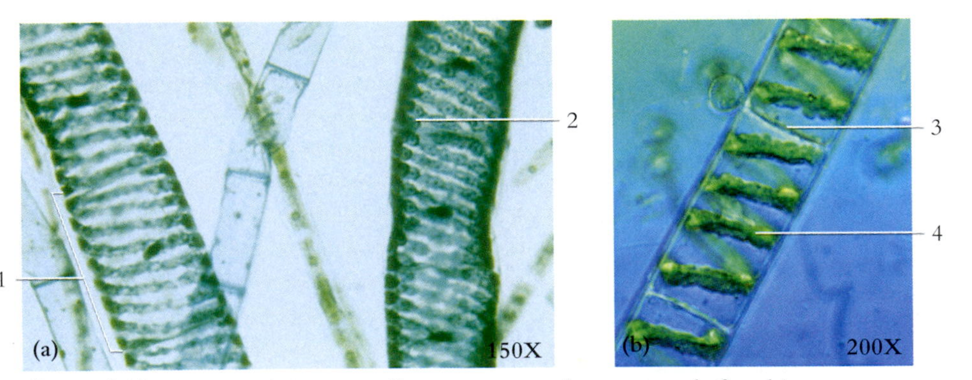

Figure 4.66 The genus *Spirogyra* are filamentous green algae commonly found in green masses on the surfaces of ponds and streams. Their chloroplasts are arranged as a spiral within the cell. (a) Several cells compose a filament. (b) A magnified view of a single filament comprised of several cells.

1. Single cell
2. Filaments
3. Cell wall
4. Chloroplast

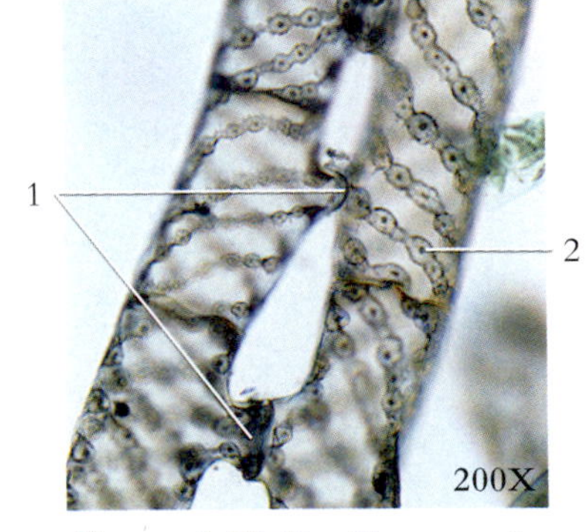

Figure 4.67 The filaments of *Spirogyra* sp. showing initial contact of conjugation tubes.

1. Conjugation tube
2. Pyrenoid in chloroplast

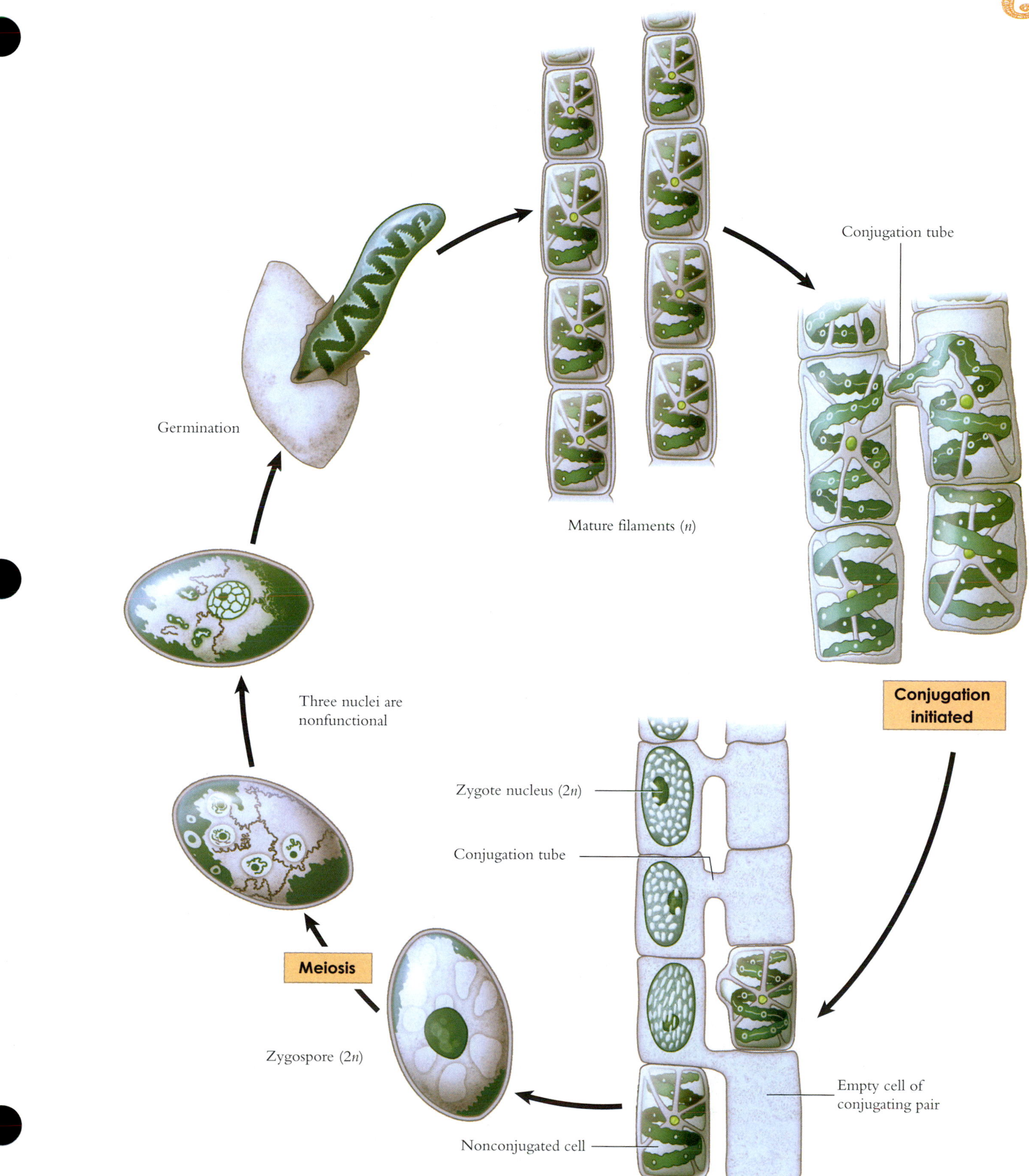

Figure 4.68 The life cycle of *Spirogyra*, a common freshwater green alga.

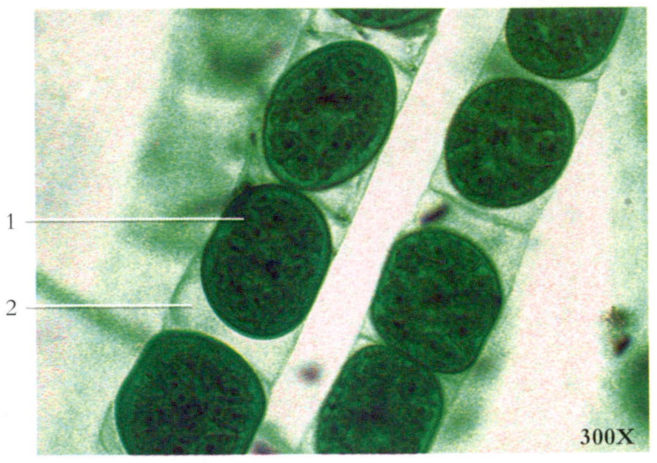

Figure 4.69 Two filaments of *Spirogyra* sp. with aplanospores.
1. Aplanospore 2. Cell wall

Figure 4.70 *Spirogyra* sp. in a small freshwater pond.

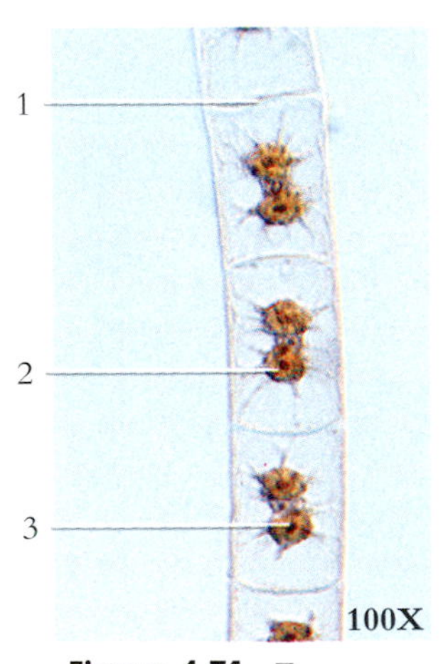

Figure 4.71 *Zygnema* sp. filament showing the star-shaped chloroplasts.
1. Cell wall
2. Chloroplast
3. Pyrenoid

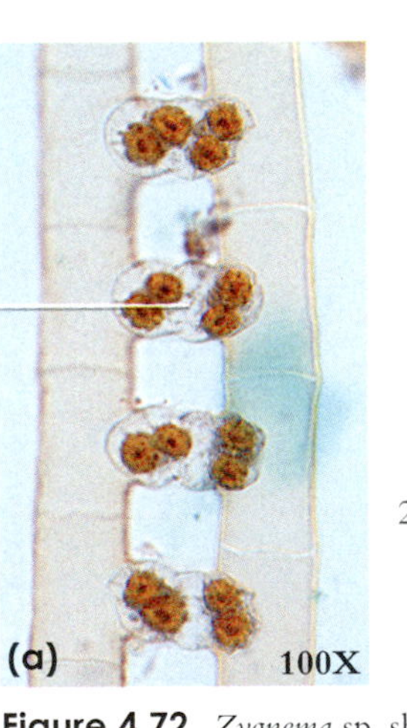

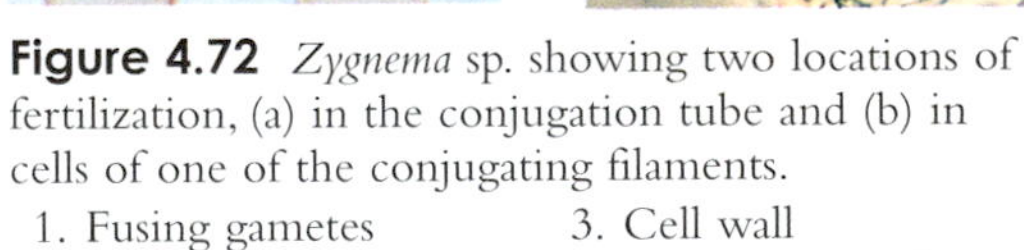

Figure 4.72 *Zygnema* sp. showing two locations of fertilization, (a) in the conjugation tube and (b) in cells of one of the conjugating filaments.
1. Fusing gametes
2. Zygote
3. Cell wall
4. Conjugation tube

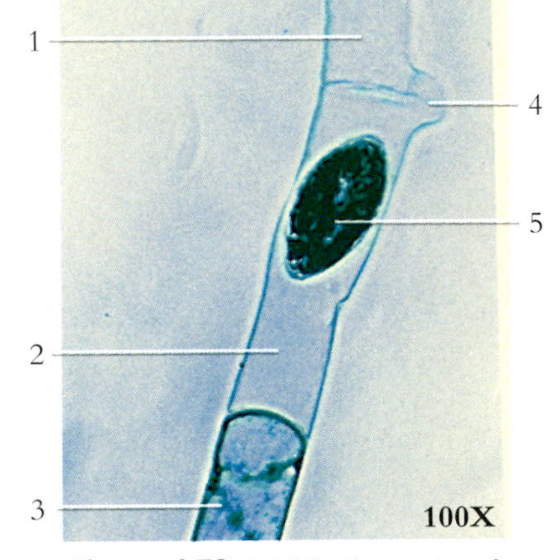

Figure 4.73 Self-fertile species of *Spirogyra* sp. A gamete has migrated from the upper cell to form a zygote in the lower cell.
1. Upper cell
2. Lower cell
3. Chloroplast
4. Conjugation tube
5. Zygote

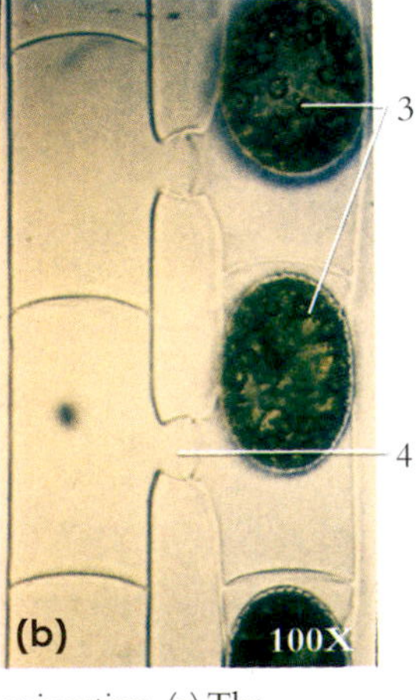

Figure 4.74 *Zygonema* sp. undergoing conjugation. (a) The filament is just forming conjugation tubes; and (b) two conjugated filaments.
1. Developing gametes
2. Developing conjugation tubes
3. Zygotes
4. Conjugation tube

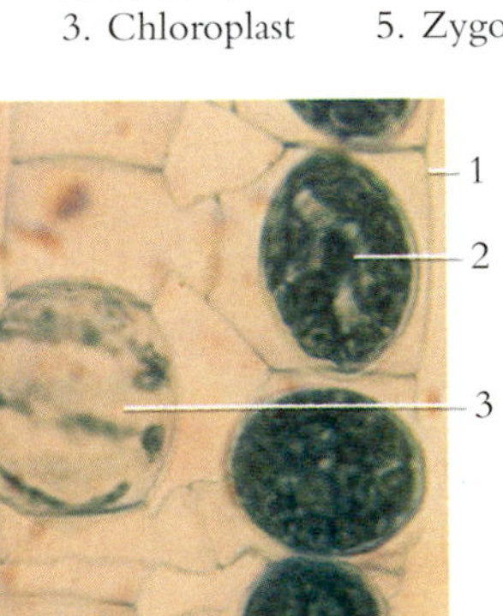

Figure 4.75 Conjugation in *Spirogyra* sp.
1. Cell-bearing zygote
2. Zygote
3. Cell that did not conjugate

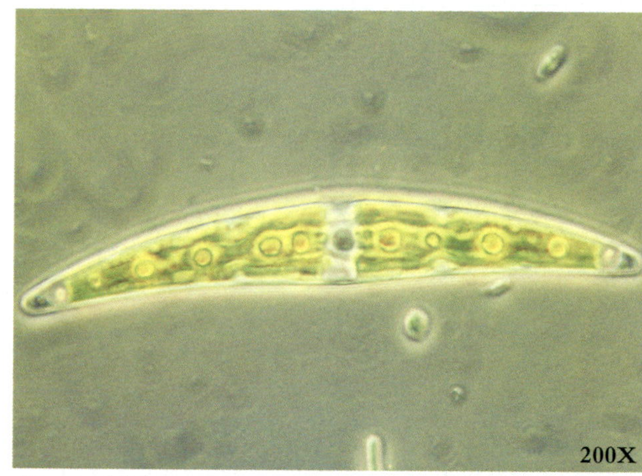

Figure 4.76 Desmid *Closterium* sp. Desmids are unicellular, freshwater chlorophyta that reproduce sexually by conjugation.

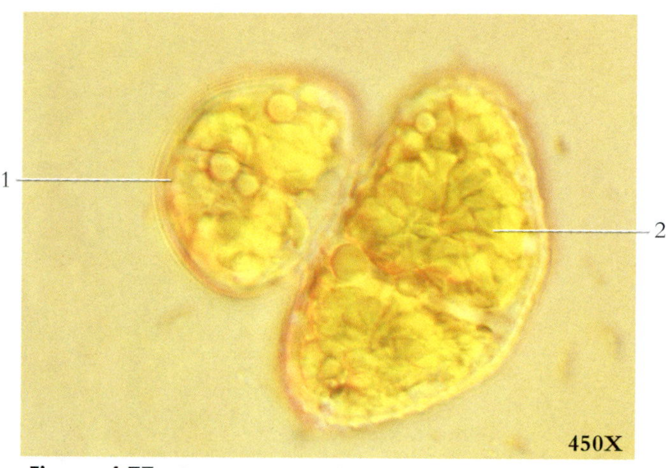

Figure 4.77 *Cosmarium* sp., a desmid, soon after cell division forming a new semicell.

1. New semicell
2. Dividing cell

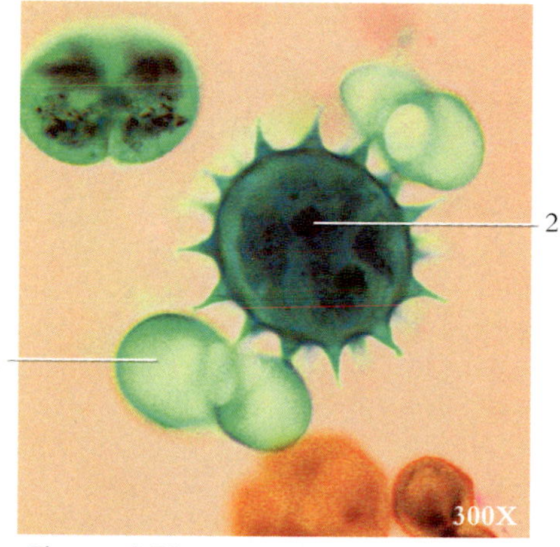

Figure 4.78 Zygospore of the desmid *Cosmarium* sp.

1. Empty cell that has been involved in conjugation
2. Zygospore

Figure 4.79 Desmid *Micrasterias* sp.

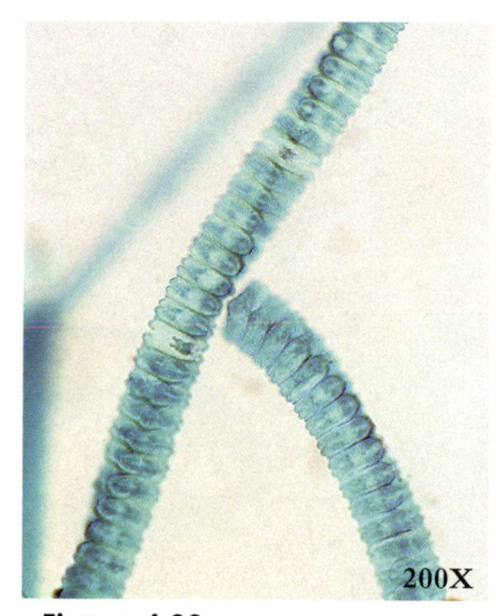

Figure 4.80 *Desmidium* sp., a filamentous (colonial) desmid.

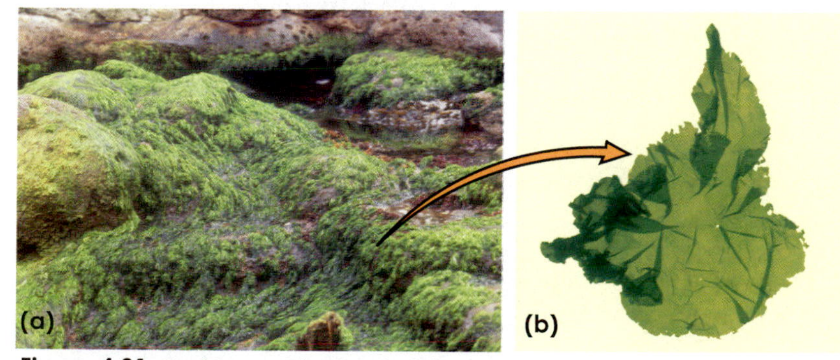

Figure 4.81 Sea lettuce, *Ulva* sp., which lives as a flat membranous chlorophyte in marine environments.

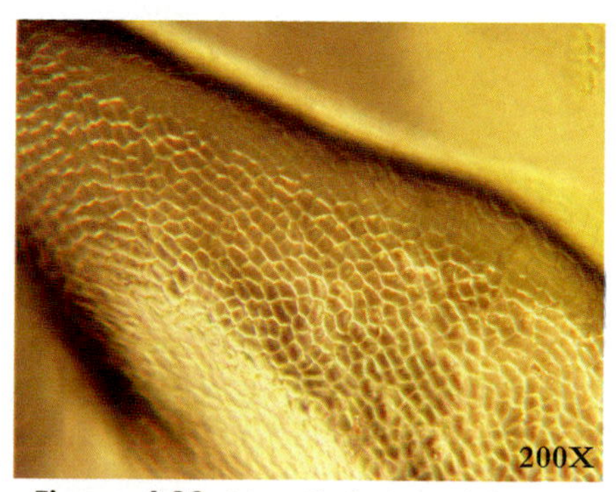

Figure 4.82 Magnified view of the surface of *Enteromorpha intestinalis*. *Enteromorpha is* closely related to *Ulva*.

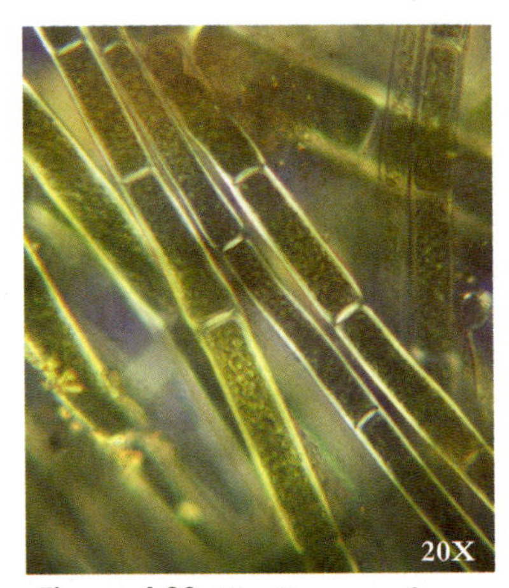

Figure 4.83 The filaments of *Cladophora* sp. This member of class Ulvophyceae is found in both freshwater and marine habitats.

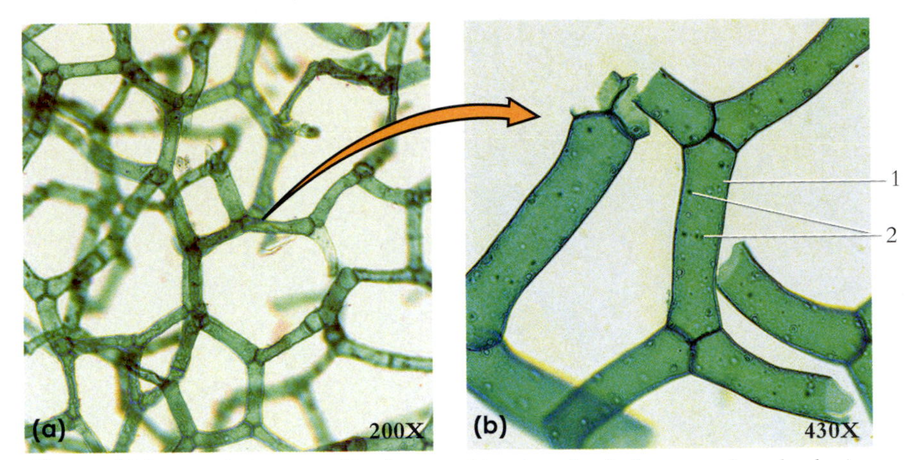

Figure 4.84 A *Hydrodictyon* sp. The large, multinucleated cells form net-shaped colonies.
1. Individual cell
2. Nuclei of cell

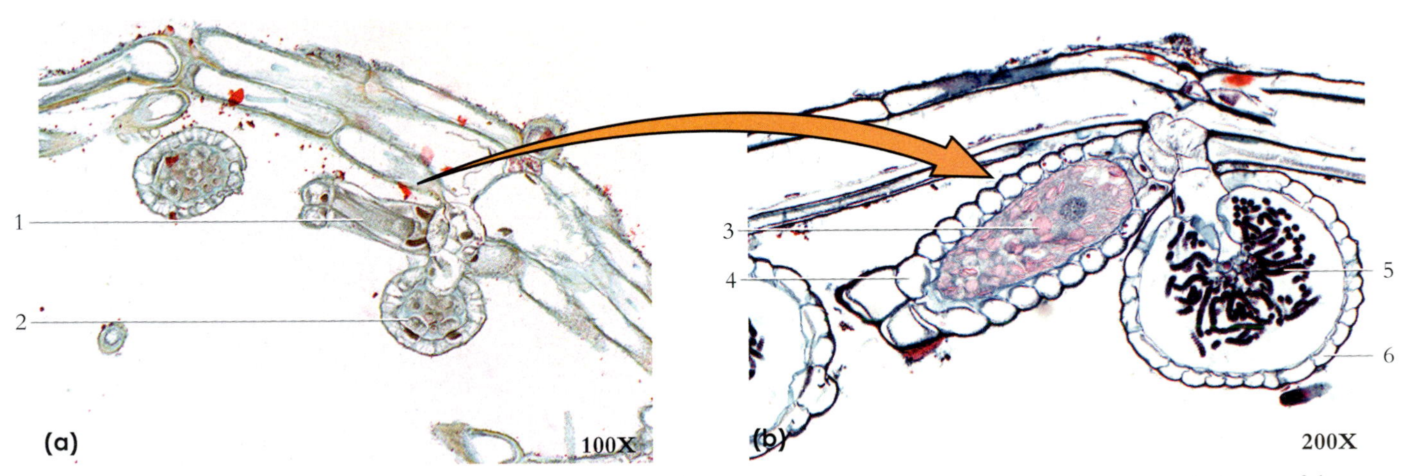

Figure 4.85 (a) *Chara* sp. inhabits marshes or shallow, temperate lakes, showing characteristic gametangia. (b) A magnified view of the gametangia.
1. Oogonium
2. Antheridium
3. Egg
4. Oogonium
5. Sperm-producing cells (filaments)
6. Antheridium

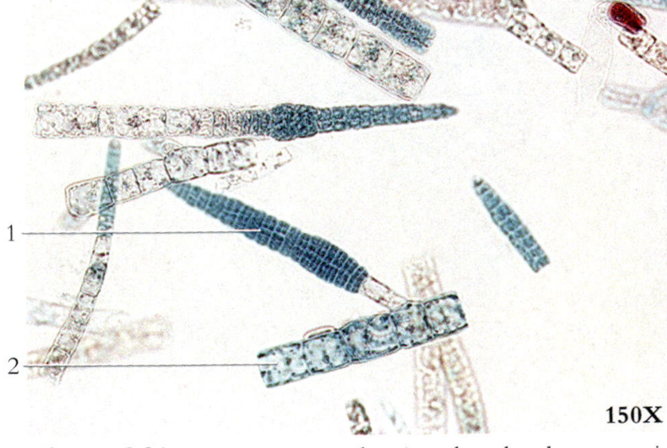

Figure 4.86 An *Ectocarpus* sp. showing pleurolocular sporangia.
1. Pleurolocular sporangium
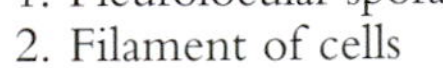
2. Filament of cells

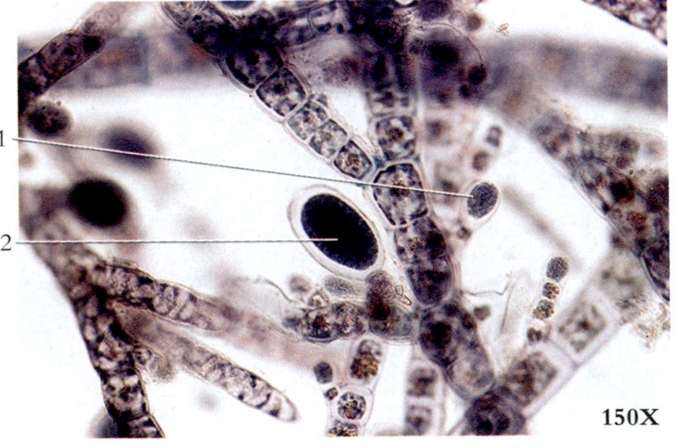

Figure 4.87 An *Ectocarpus* sp. showing unilocular sporangia.
1. Immature unilocular sporangium
2. Mature unilocular sporangium

Phylum Phaeophyta - brown algae and giant kelp

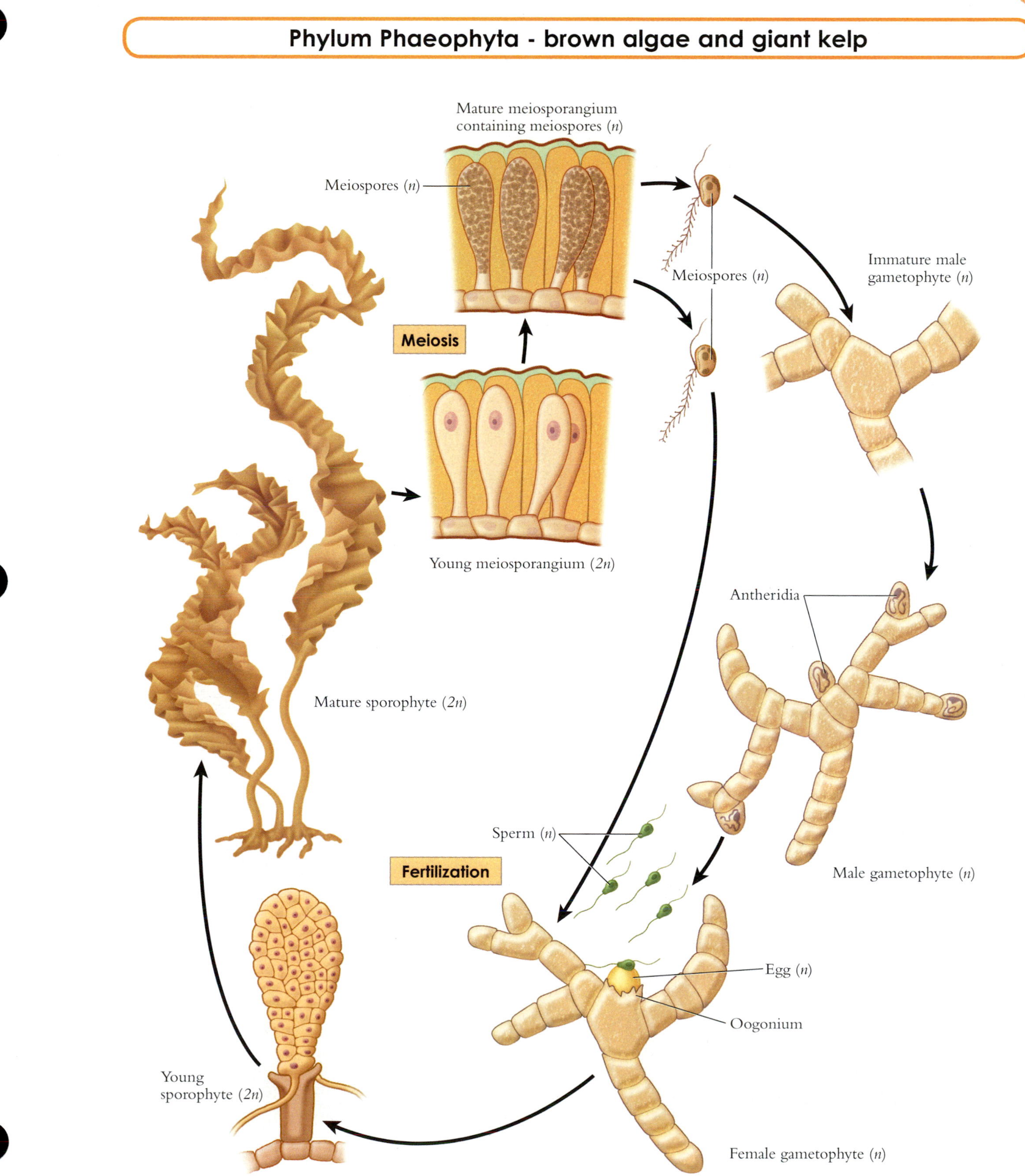

Figure 4.88 The life cycle of *Laminaria*, a common kelp.

Figure 4.89 Rocky coast of southern Alaska showing dense growths of the brown alga, *Fucus* sp.

Figure 4.90 "Sea palm," *Postelsia palmaeformis*, a common brown alga found on the western coast of North America.

Macrocystis sp.

Macrocystis sp.

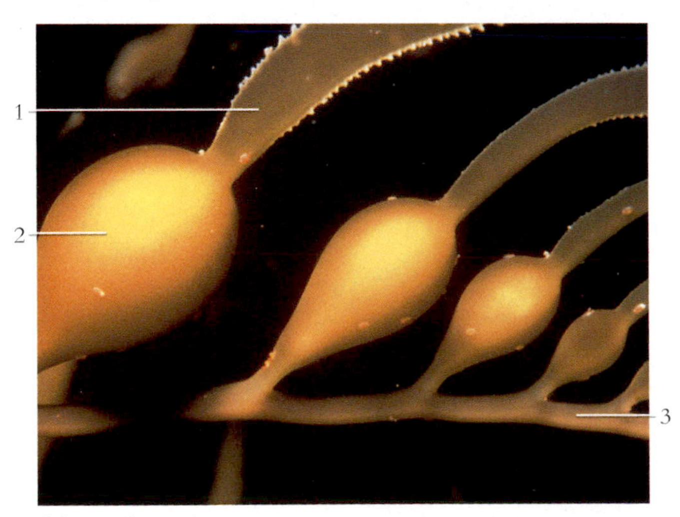

Macrocystis sp.

Egregia sp.

Figure 4.91 Some examples of brown algae, Phaeophyta. These large species are commonly known as kelps.
1. Blade 2. Float (air-filled bladder) 3. Stipe

Figure 4.92 The kelp, *Laminaria* sp., one of the common "seaweeds" found along many rocky coasts.

Figure 4.93 A tidal pool with green, brown, and red algae.

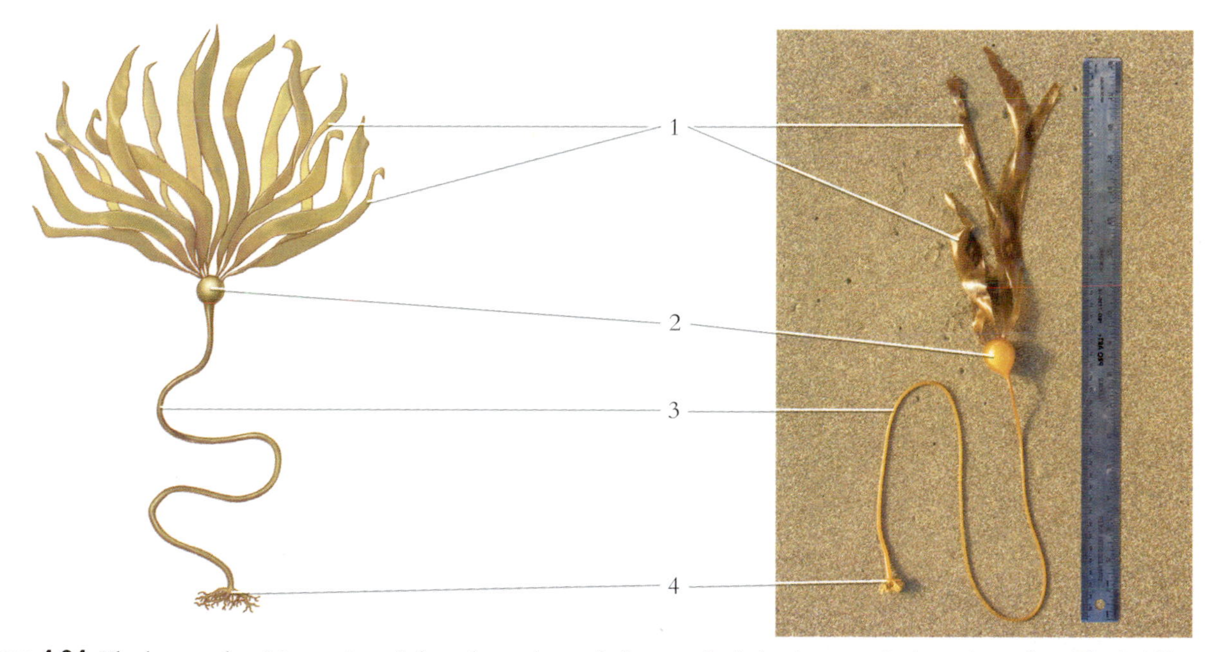

Figure 4.94 The brown alga, *Nereocystis* sp. It has a long stipe and photosynthetic laminae attached to a large float. The holdfast anchors the alga to the ocean floor. This alga and others can grow to lengths of several meters.

1. Lamina
2. Floats (air-filled bladders)
3. Stipe
4. Holdfasts

Figure 4.95 *Sargassum* sp., a brown alga common in the Sargasso sea.

1. Floats
2. Blade
3. Stipe

Figure 4.96 A mixture of kelps washed onto shore to form "windrows" of *Phaeophyta* sp.

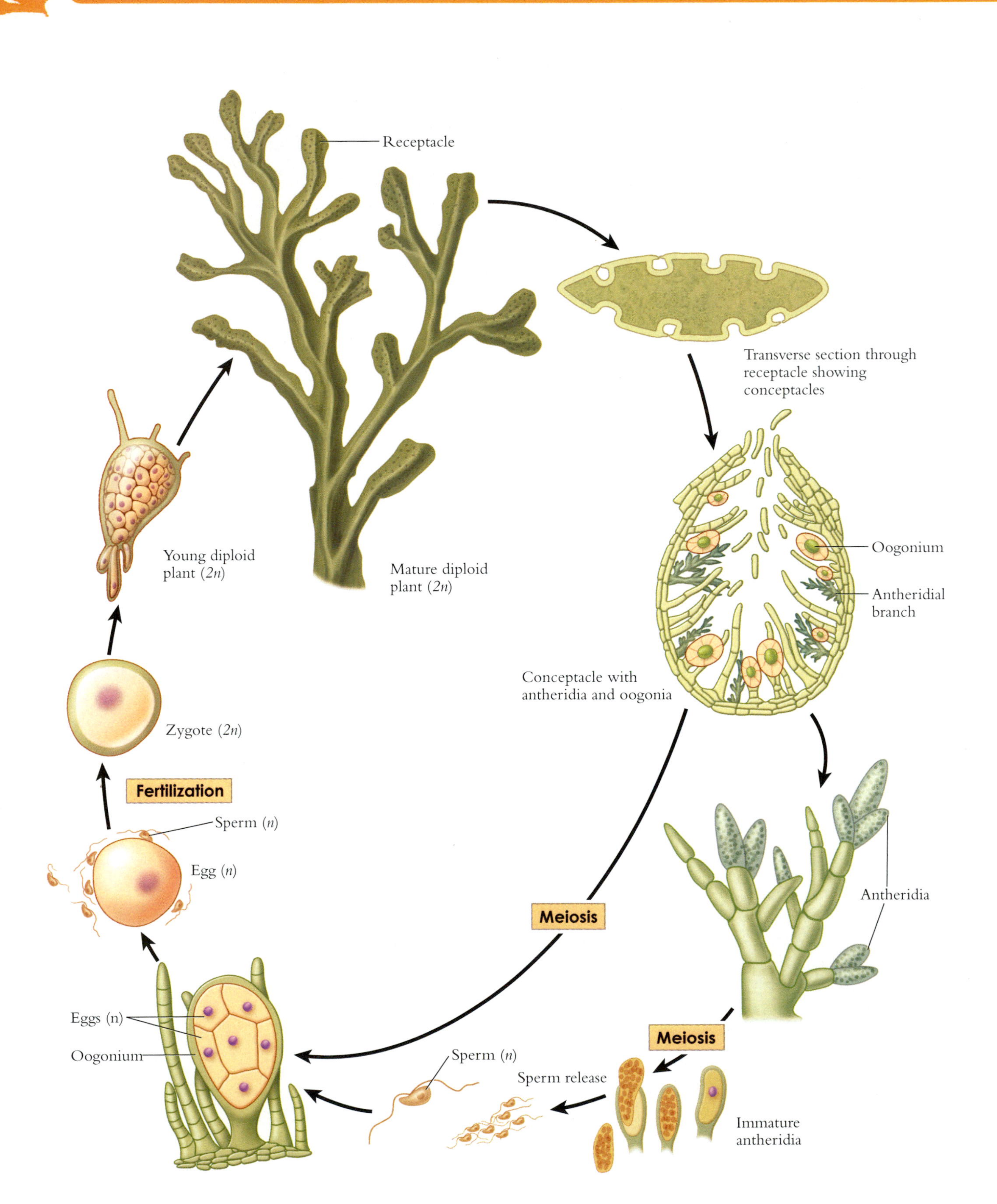

Figure 4.97 The life cycle of *Fucus*, a common brown alga.

Figure 4.98 (a) *Fucus* sp., a brown alga, commonly called rockweed. (b) An enlargement of a blade supporting the receptacles.

1. Blade
2. Receptacle
3. Conceptacles (spots) are chambers embedded in the receptacles
4. Blade

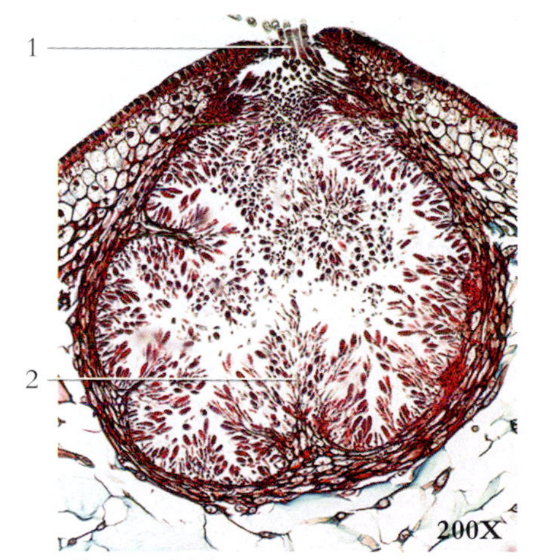

Figure 4.99 A conceptacle of *Fucus* sp.

1. Sterile paraphases
2. Antheridial branches

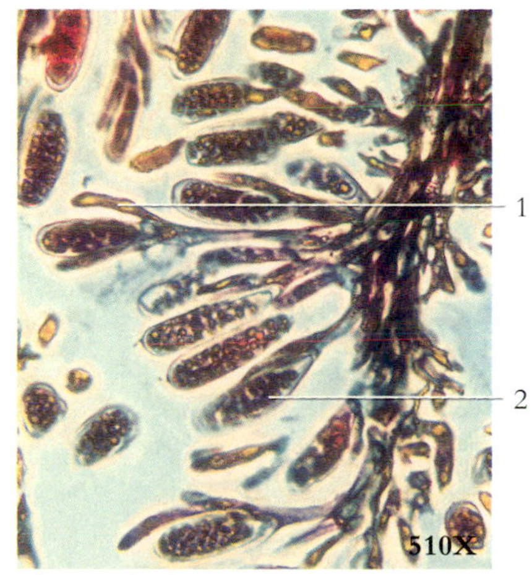

Figure 4.100 A close-up of antheridial branch of *Fucus* sp.

1. Antheridial branch
2. Antheridium

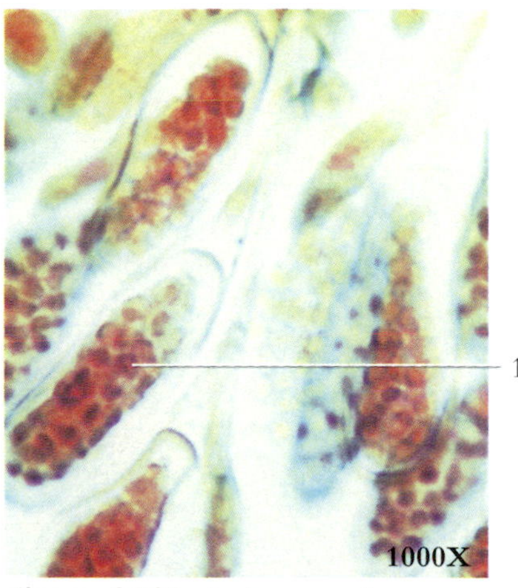

Figure 4.101 A close-up of antheridium of *Fucus* sp.

1. Sperm within antheridium

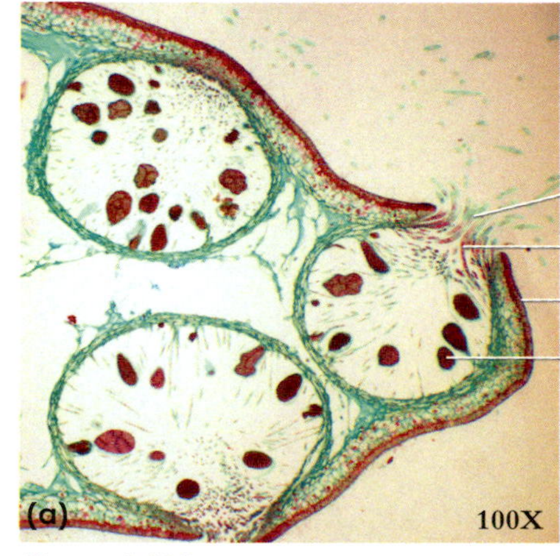

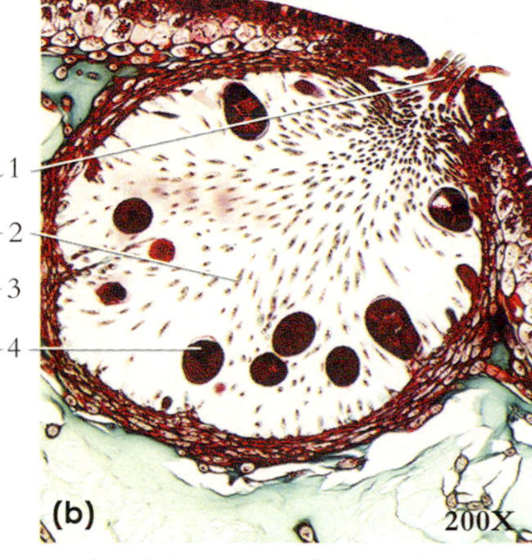

Figure 4.102 A section through a *Fucus* sp. receptacle. (a) Low magnification showing three conceptacles and (b) higher magnification of a single conceptacle with oogonia.

1. Ostiole
2. Paraphyses (sterile hairs)
3. Surface of receptacle
4. Oogonium

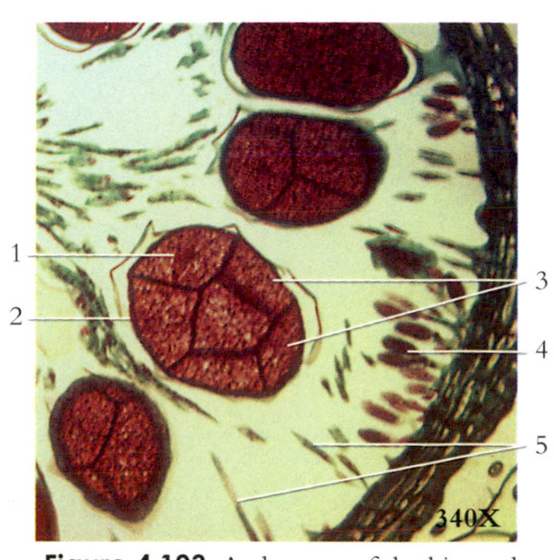

Figure 4.103 A close-up of the bisexual conceptacle of *Fucus* sp.

1. Nucleus of egg
2. Oogonium
3. Eggs
4. Antheridium
5. Paraphyses

Phylum Rhodophyta - red algae

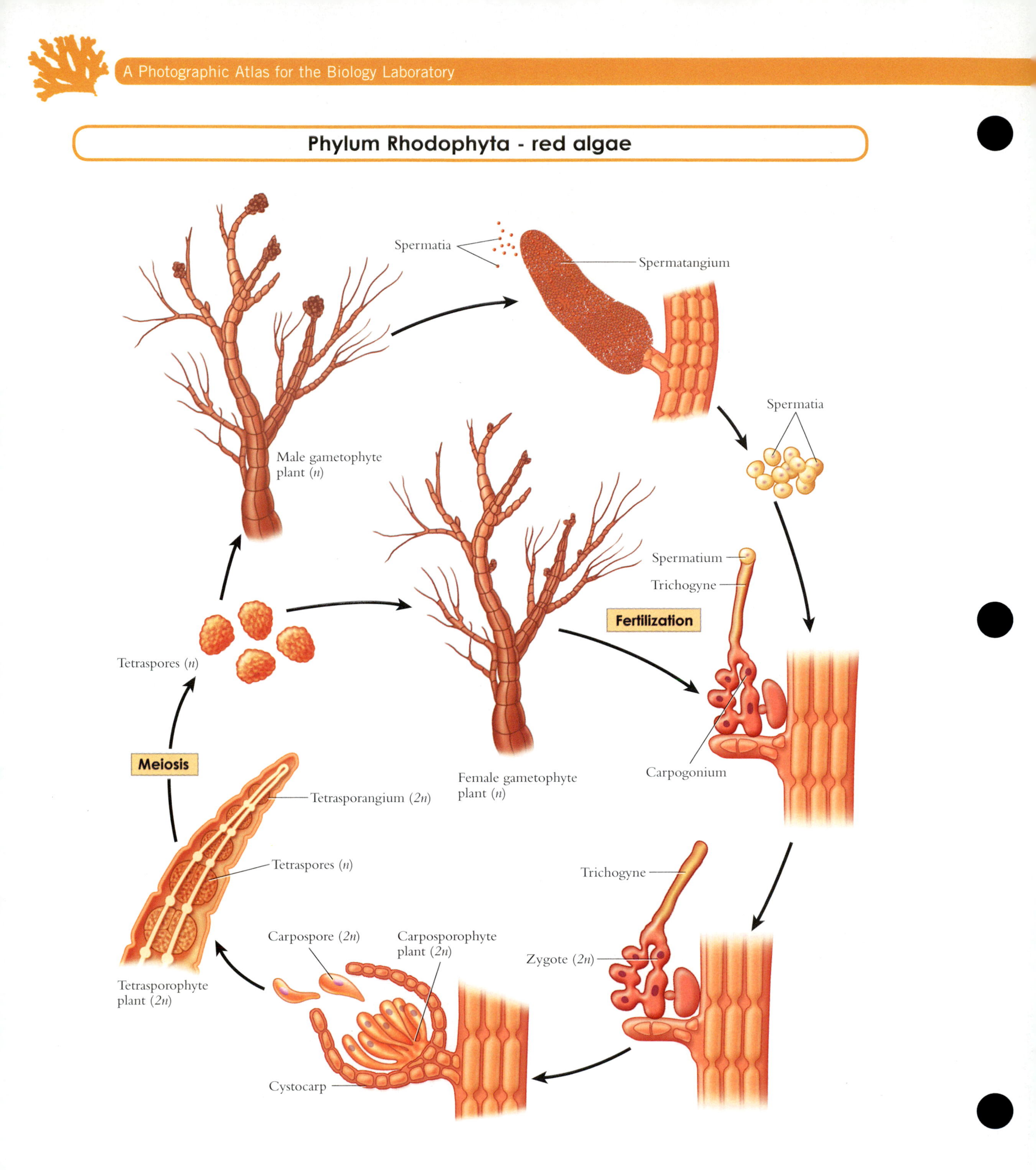

Figure 4.104 The life cycle of the red alga, *Polysiphonia*.

Figure 4.105 Intertidal zone showing a colony of red alga, *Bangia* sp.

Figure 4.106 Mature plant of the red alga, *Rhodymenia* sp.

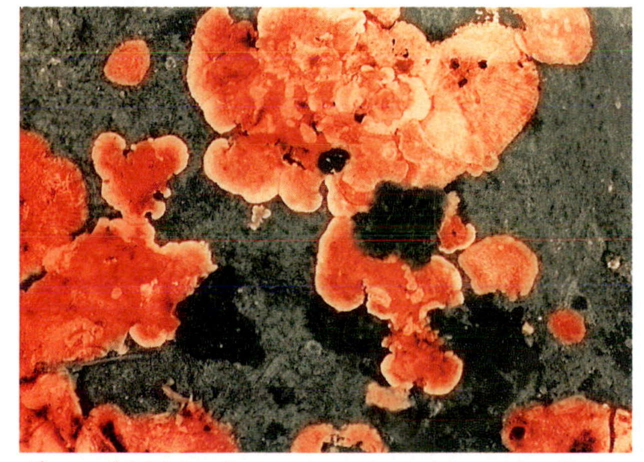

Figure 4.107 Small encrusting colonies of a species of red alga on a stone. The colonies shown are bright red and are only a few millimeters in size.

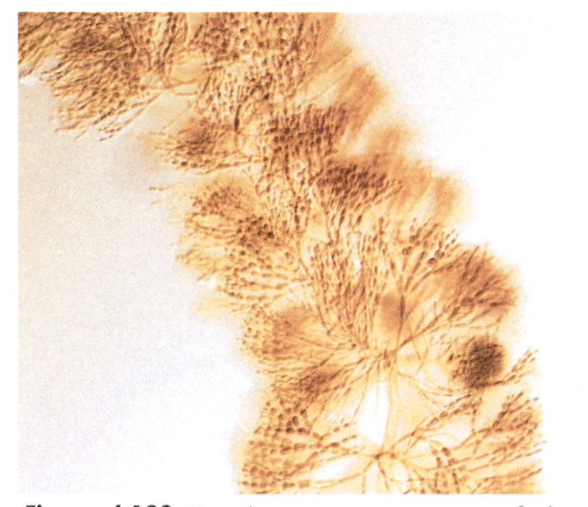

Figure 4.108 *Batrachospermum* sp., a common freshwater red alga.

Figure 4.109 *Audouinella* sp. is a freshwater member of Rhodophyta. This organism was collected from a coldwater spring.

Figure 4.110 Mature plant of the common red alga, *Polysiphonia* sp.

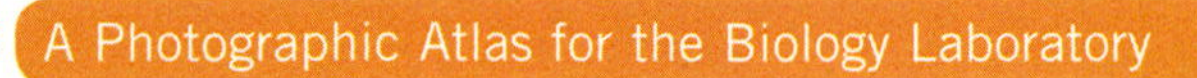

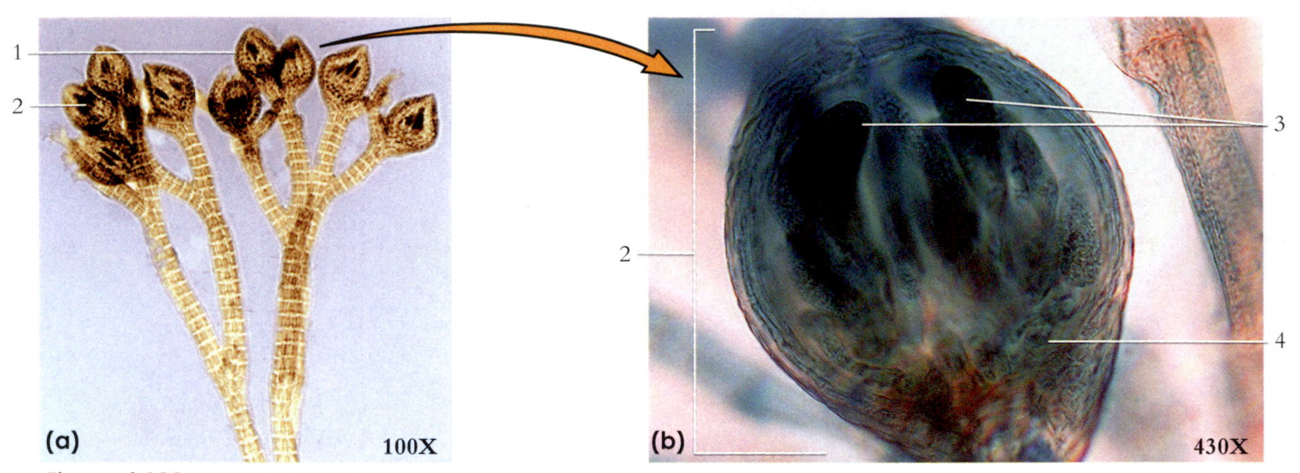

Figure 4.111 The red alga, *Polysiphonia* sp. It has alternation of three generations. (a) Female gametophyte with attached carposporophyte generation. (b) A close-up of the cystocarp plant.

1. Pericarp 2. Cystocarp 3. Carpospores 4. Carposporophyte

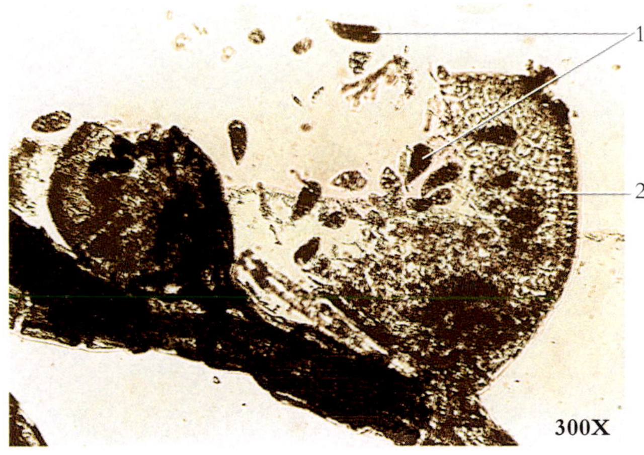

Figure 4.112 *Polysiphonia* sp., showing the release of carpospores.

1. Carpospores (2*n*) 2. Ruptured cystocarp

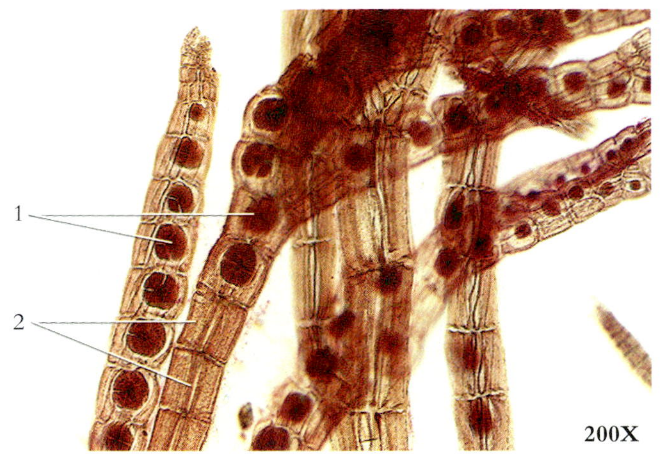

Figure 4.113 A tetrasporophyte generation of *Polysiphonia* sp. showing tetraspores (meiospores).

1. Tetraspores 2. Cells of tetrasporophyte plant

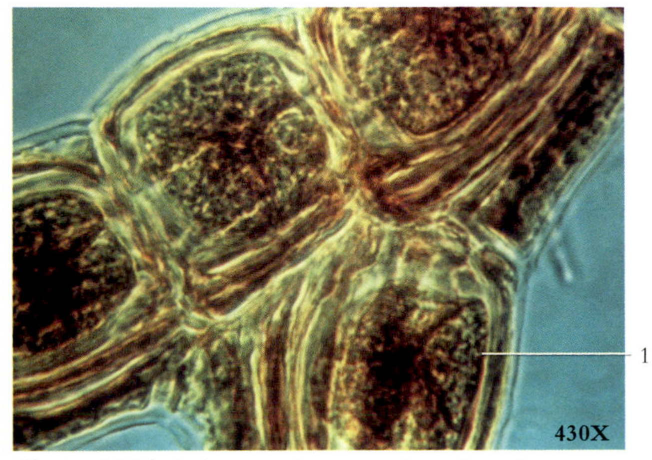

Figure 4.114 A close-up of tetrasporophyte plant of *Polysiphonia* sp.

1. Tetraspore (meiospore)

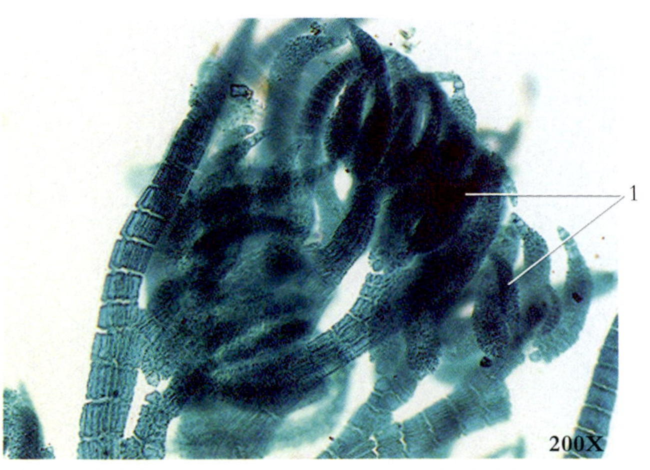

Figure 4.115 A male gametophyte plant of *Polysiphonia* sp., showing spermatangia (green stain).

1. Spermatangia with spermatia

Phylum Myxomycota - plasmodial slime molds

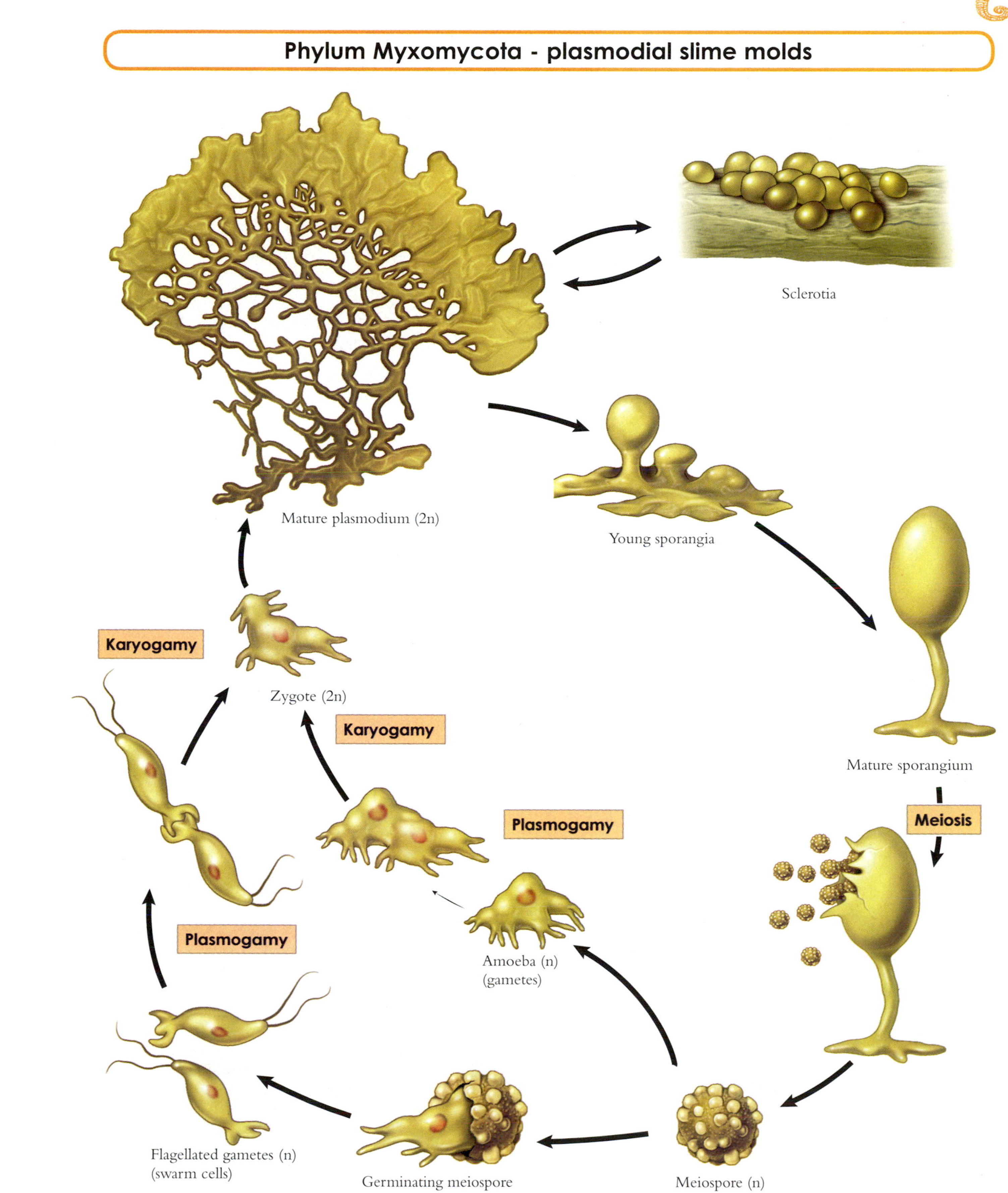

Figure 4.116 The life cycle of a plasmodial slime mold.

Figure 4.117 The sporangia of the slime mold *Comatricha typhoides*.

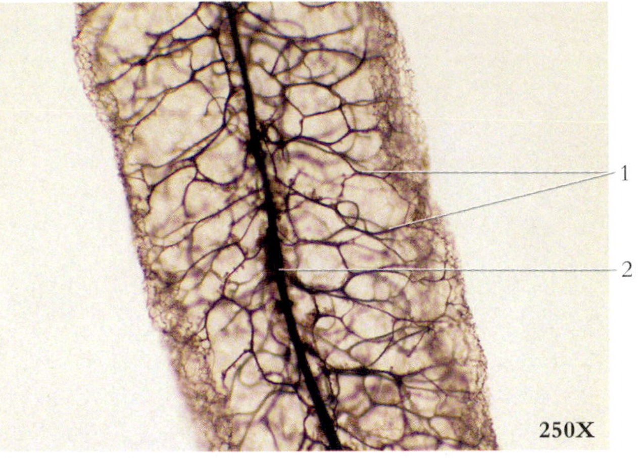

Figure 4.118 A longitudinal section through the sporangium of *Stemonitis* sp.
1. Cellular filaments (capillitum)
2. Columella

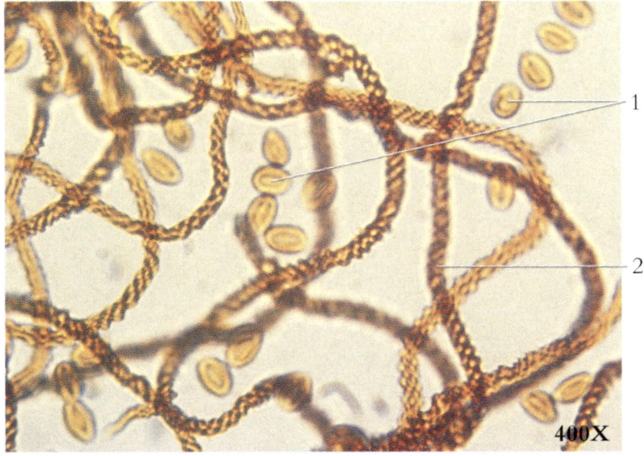

Figure 4.119 A close-up through the sporangium of *Stemonitis* sp.
1. Spores
2. Capillitum

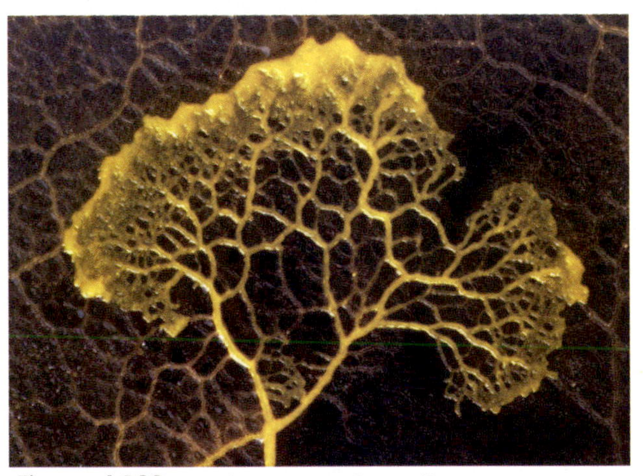

Figure 4.120 A *Physarum* sp. plasmodium.

Figure 4.121 The sporangia of slime mold. These vary considerably in size and shape. One species of *Lycogala* sp. is shown here.

Figure 4.122 A slime mold specimen from a high-mountain locality.

Phylum Oomycota - water molds, white rusts, and downy mildews

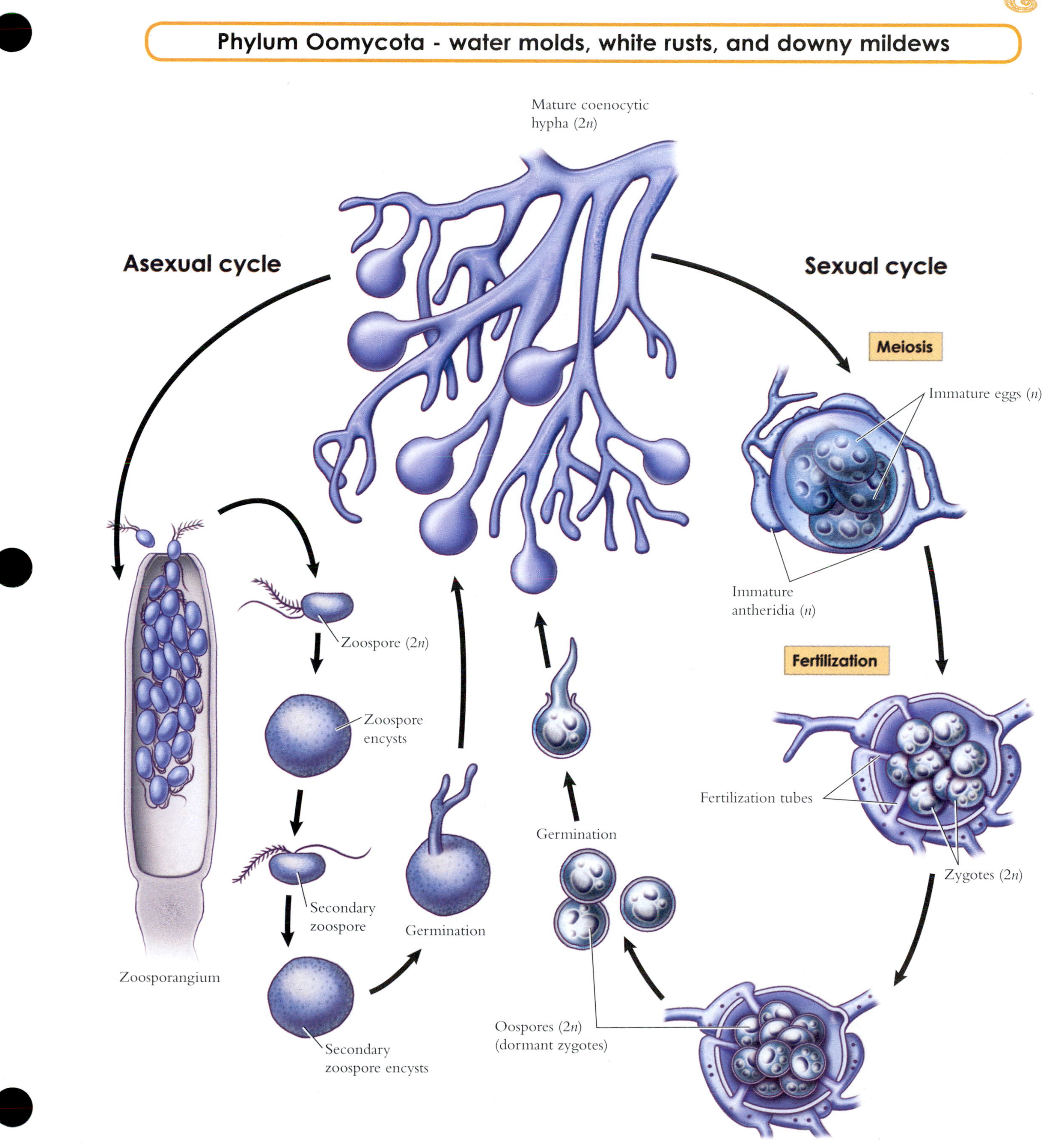

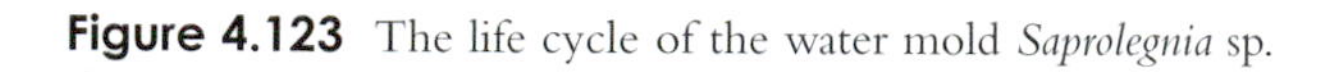

Figure 4.123 The life cycle of the water mold *Saprolegnia* sp.

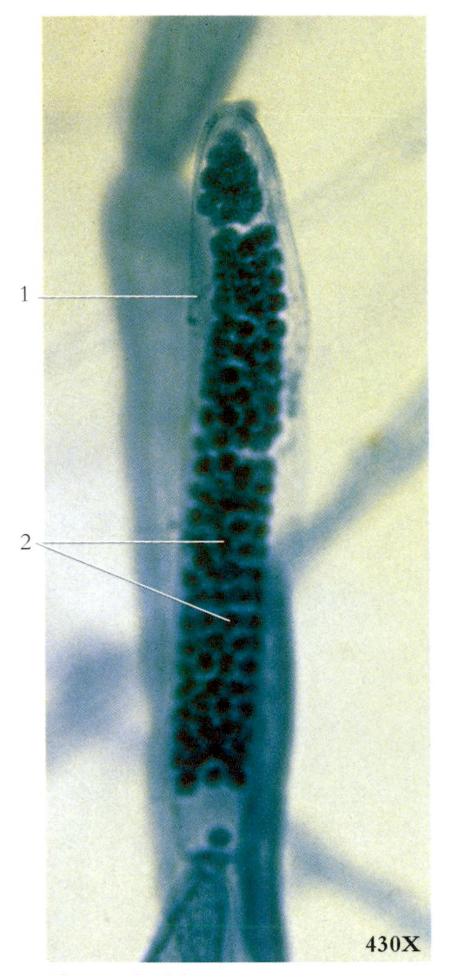

Figure 4.124 The zoosporangium of the water mold *Saprolegnia* sp.
1. Zoosporangium
2. Zoospores

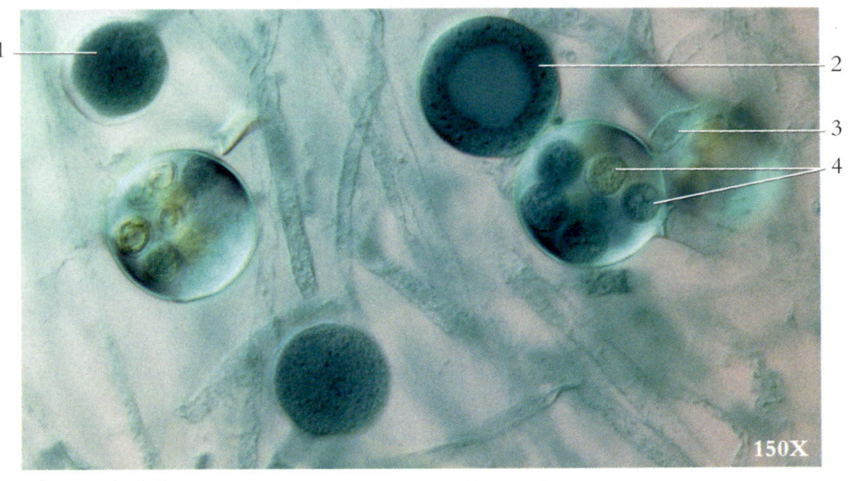

Figure 4.125 The oogonia of the water mold *Saprolegnia* sp.
1. Young oogonium
2. Developing oogonium
3. Young antheridium
4. Eggs

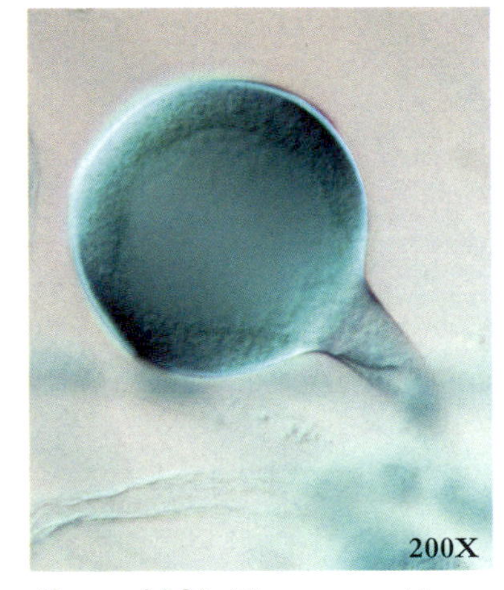

Figure 4.126 The water mold, *Saprolegnia* sp., showing a young oogonium before eggs have been formed.

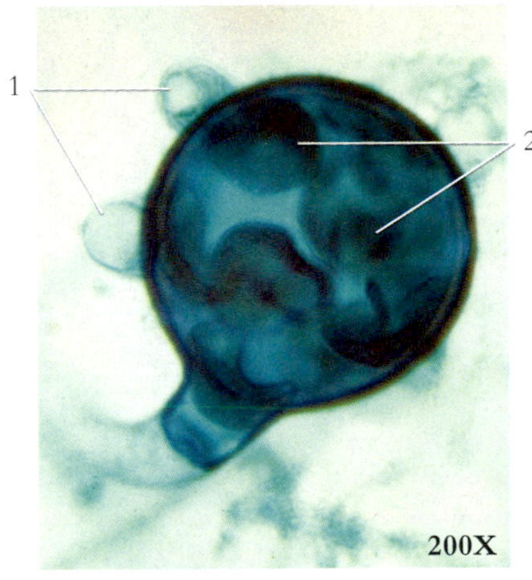

Figure 4.127 The oogonia of the water mold *Saprolegnia* sp.
1. Empty antheridia
2. Zygotes

Figure 4.128 The skin of this brown trout has been infected by the common water mold, *Saprolegnia* sp.

Fungi

Chapter 5

About 250,000 species of fungi are currently extant on Earth. All fungi are heterotrophs; they absorb nutrients through their cell walls and cell membranes. The kingdom Fungi includes the conjugation fungi, yeasts, mushrooms, toadstools, rusts, and lichens. Most are saprobes, absorbing nutrients from dead organic material. Some are parasitic, absorbing nutrients from living hosts. Fungi decompose organic material, helping to recycle nutrients essential for plant growth.

Except for the unicellular yeasts, fungi consist of elongated filaments called *hyphae*. Hyphae begin as cellular extensions of spores that branch as they grow to form a network of hyphae called a *mycelium*. Even the body of a mushroom consists of a mass of tightly packed hyphae attached to an underground mycelium. Fungi are nonmotile and reproduce by means of spores produced sexually or asexually.

Many species of fungi are commercially important. Some are used as food, such as mushrooms; or in the production of foods, such as bread, cheese, beer, and wine. Other species are important in medicine, for example, in the production of the antibiotic penicillin. Many other species of fungi are of medical and economic concern because they cause plant and animal diseases and destroy crops and stored goods.

Table 5.1 Some Representatives of Fungi

Phyla and Representative Kinds	Characteristics
Zygomycota — bread molds, fly fungi	Hyphae lack cross walls along filaments; sexual reproduction by conjugation
Ascomycota — yeasts, molds, morels, and truffles	Septate hyphae; reproductive structure contains ascospores within asci on a fruiting body known as ascoma (ascocarp); asexual reproduction by budding or conidia
Deuteromycota — conidial molds	Fungi that repoduce only by asexual spores (conidia); sexual stages are lacking or unknown
Basidiomycota — mushrooms, toadstools, rusts, and smuts	Septate hyphae; 4 meiospores produced externally on cells called basidia formed on basidioma (basidiocarp)
Lichens — not a phylum, but a symbiotic association of an alga and a fungus	Algal component (usually a green alga) provides food from photosynthesis; fungal component (usually an ascomycete) may provide anchorage, water retention, and/or nutrient absorbance

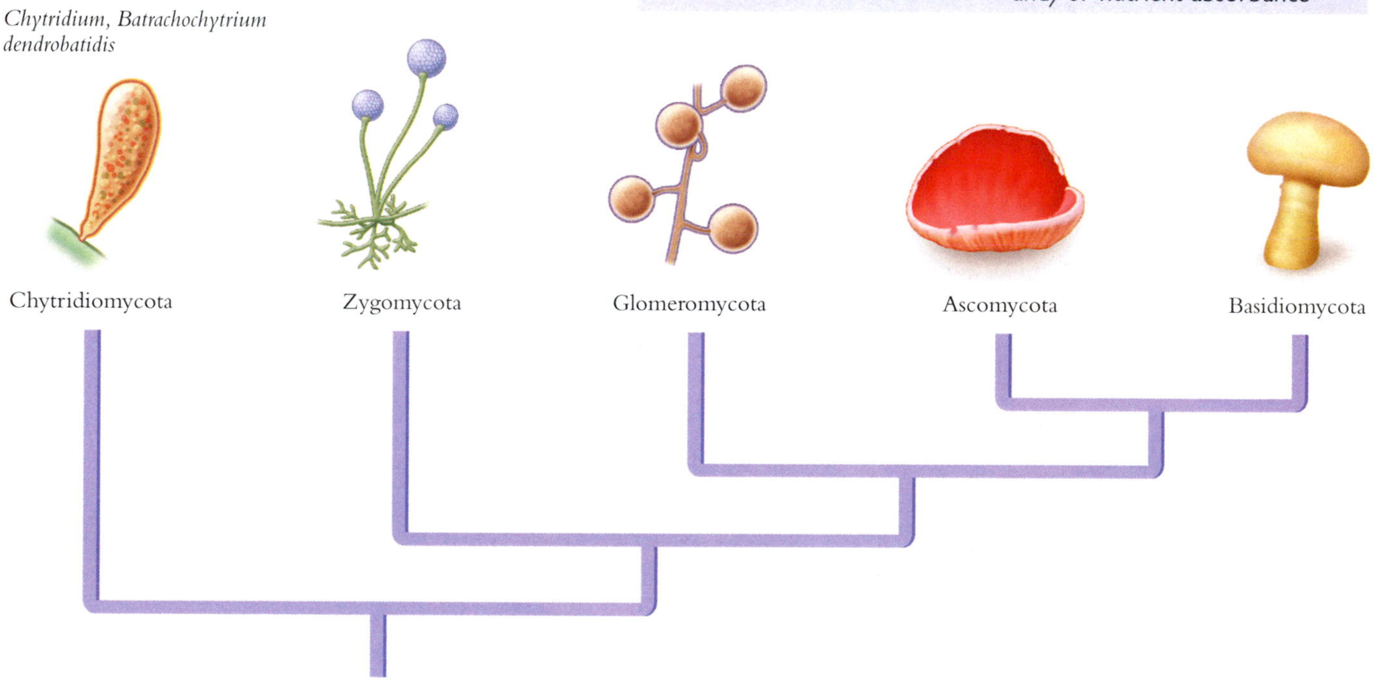

Chytridium, Batrachochytrium dendrobatidis

Chytridiomycota

Zygomycota

Glomeromycota

Ascomycota

Basidiomycota

Figure 5.1 The phylogenetic relationships and classification of major fungi lineages.

Phylum Zygomycota - conjugation fungi

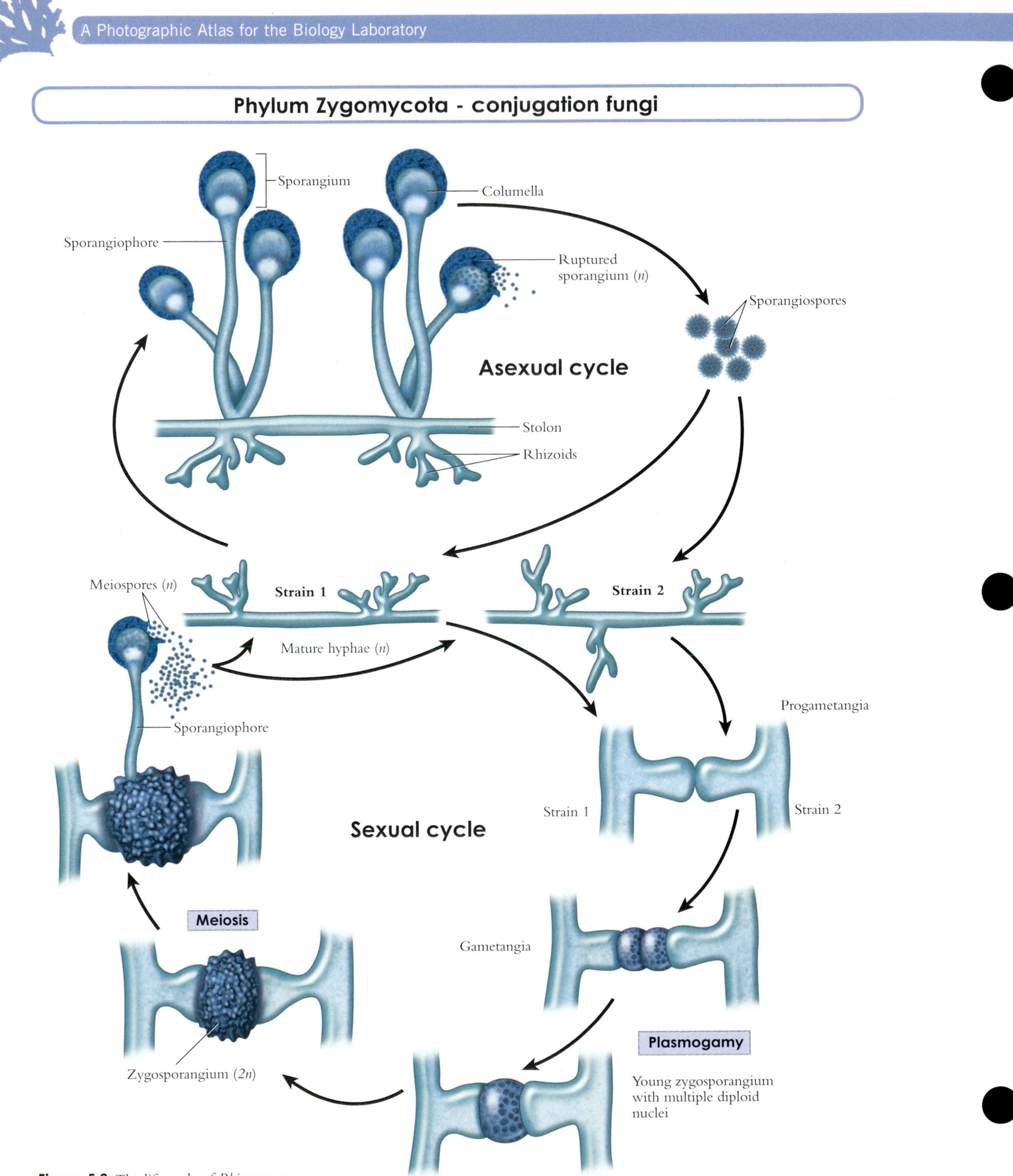

Figure 5.2 The life cycle of *Rhizopus* sp., the common bread mold.

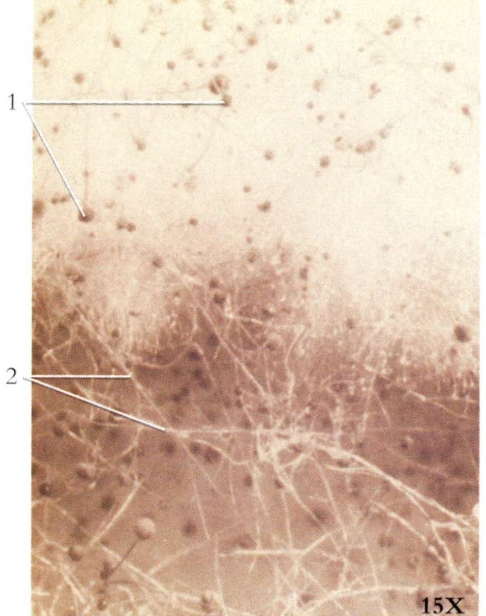

Figure 5.3 A *Rhizopus* species growing on a slice of bread.
1. Sporangia
2. Hyphae (stolon)

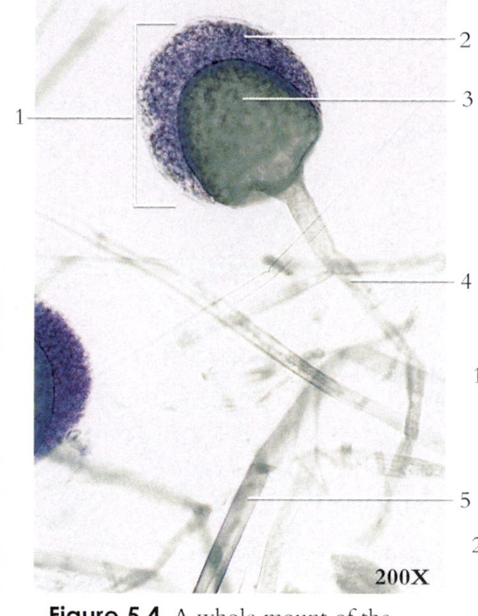

Figure 5.4 A whole mount of the bread mold, *Rhizopus* sp.
1. Sporangium
2. Spores
3. Columella
4. Sporangiophore
5. Hyphae

Figure 5.5 A mature sporangium in the asexual reproductive cycle of the bread mold, *Rhizopus* sp.
1. Sporangium
2. Sporangiophore
3. Spores
4. Columella

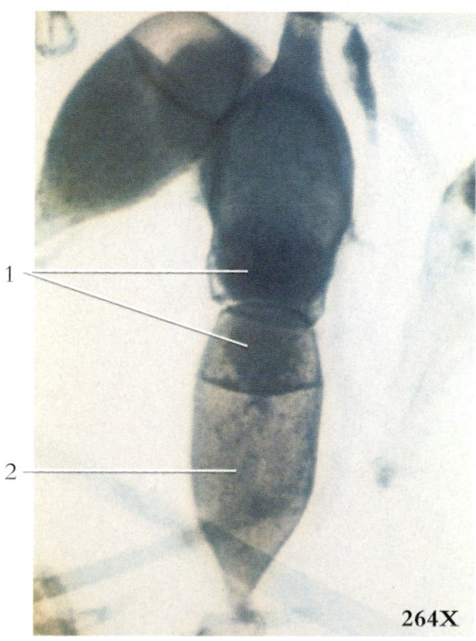

Figure 5.6 A young gametangia of *Rhizopus* sp. contacting prior to plasmogamy.
1. Immature gametangia
2. Suspensor cell

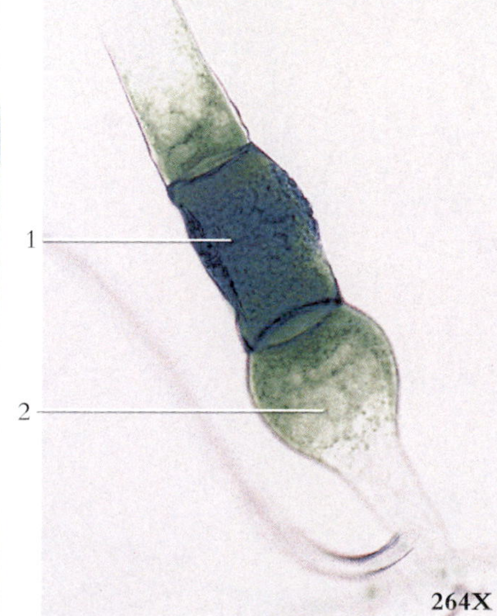

Figure 5.7 An immature *Rhizopus* sp. zygospore following plasmogamy.
1. Immature zygosporangium
2. Suspensor cell

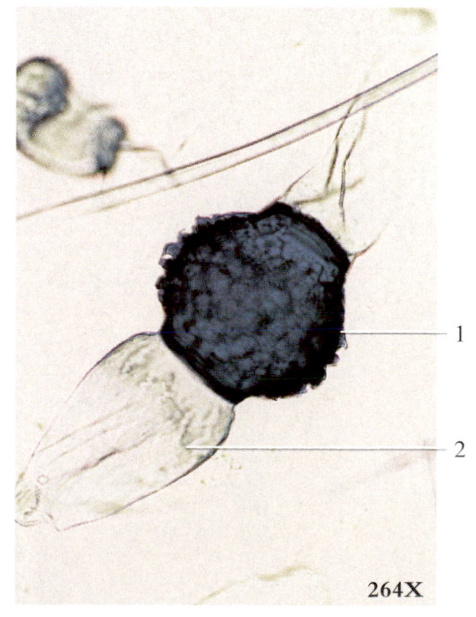

Figure 5.8 A mature *Rhizopus* sp. zygospore.
1. Zygosporangium
2. Suspensor cell

Phylum Ascomycota - yeasts, molds, morels, and truffles

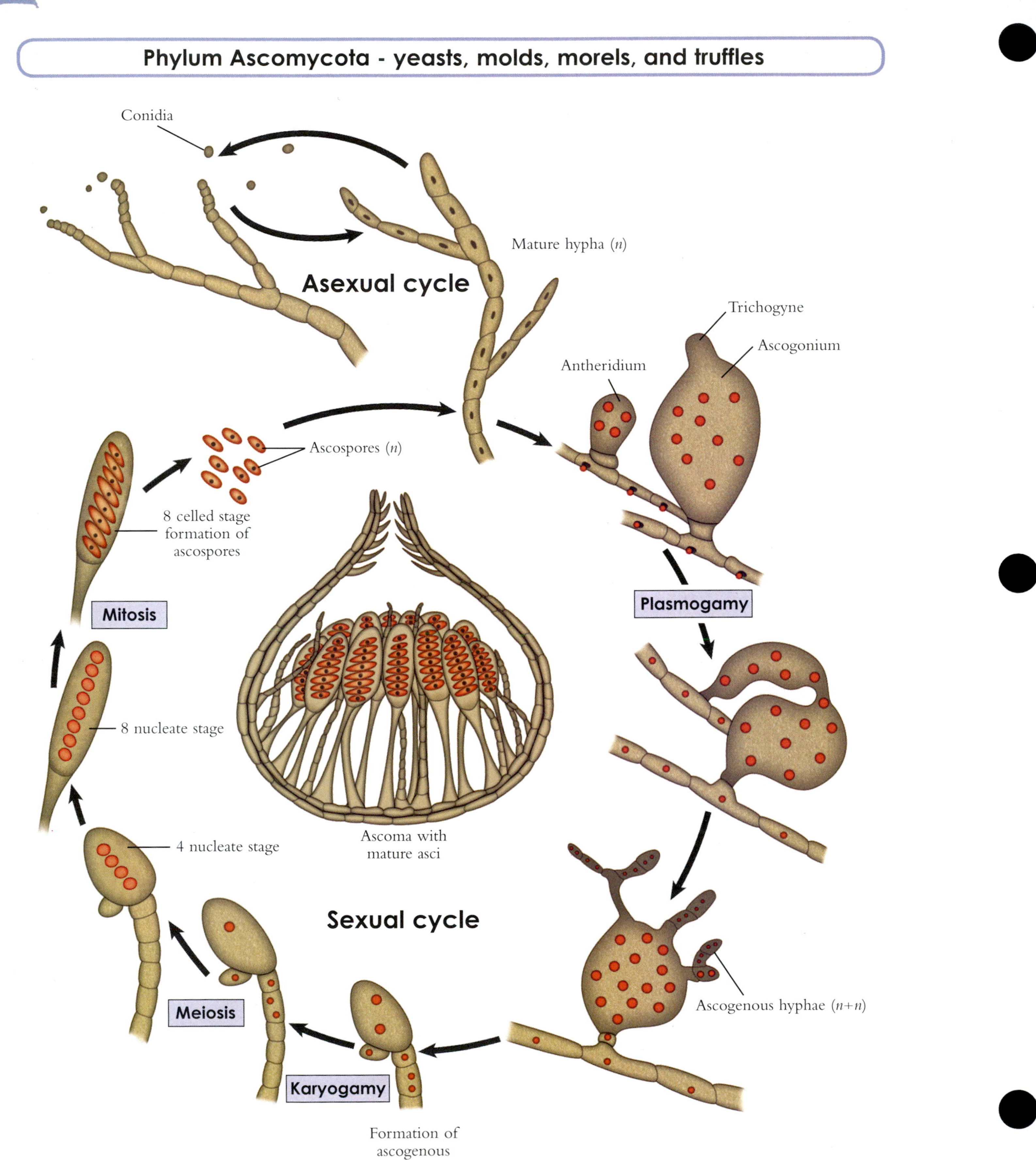

Figure 5.9 The life cycle of an ascomycete.

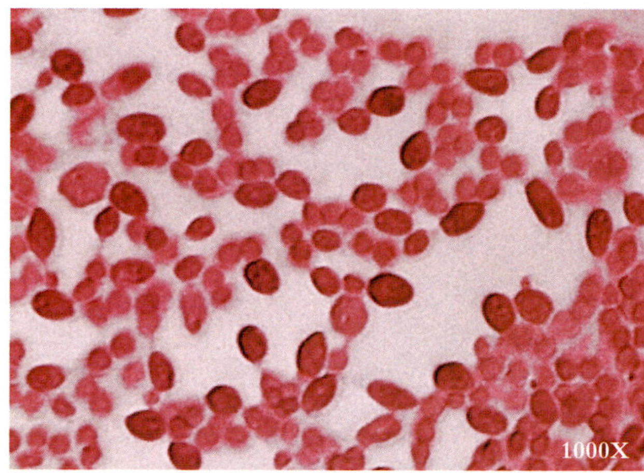

Figure 5.10 Baker's yeast, *Saccharomyces cerevisiae*. The ascospores of this unicellular ascomycete are characteristically spheroidal or ellipsoidal in shape.

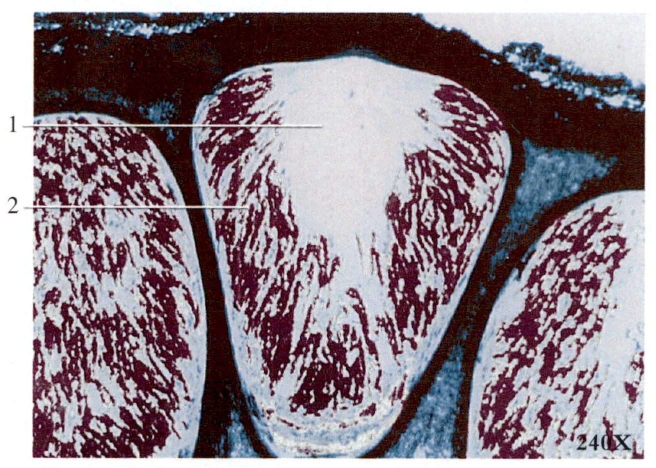

Figure 5.11 A close-up of the parasitic ascomycete, *Hypoxylon* sp., showing embedded perithecia.
1. Perithecium
2. Hymenium

Figure 5.12 The parasitic ascomycete, *Dibotryon morbosum*, on a branch of a chokecherry, *Prunus virginiana*.
1. Fungus
2. Chokecherry stem

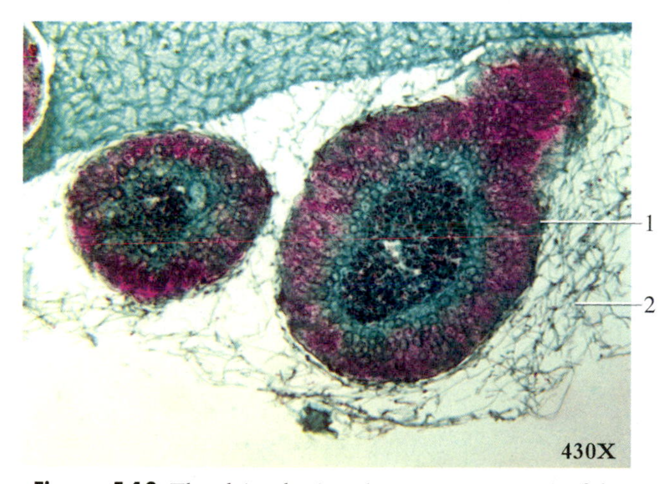

Figure 5.13 The cleistothecium (ascocarp or ascoma) of the ascomycete *Penicillium* sp.
1. Cleistothecium
2. Hyphae

Peziza repanda *Scutellinia scutellata* *Morchella* sp. *Helvella* sp.

Figure 5.14 Fruiting bodies (ascocarps or ascoma) of common ascomycetes. *Peziza repanda* is a common woodland cup fungus. *Scutellinia scutellata* is commonly called the eyelash cup fungus. *Morchella esculenta* is a common edible morel. *Helvella* is sometimes known as a saddle fungus since the fruiting body is thought by some to resemble a saddle.

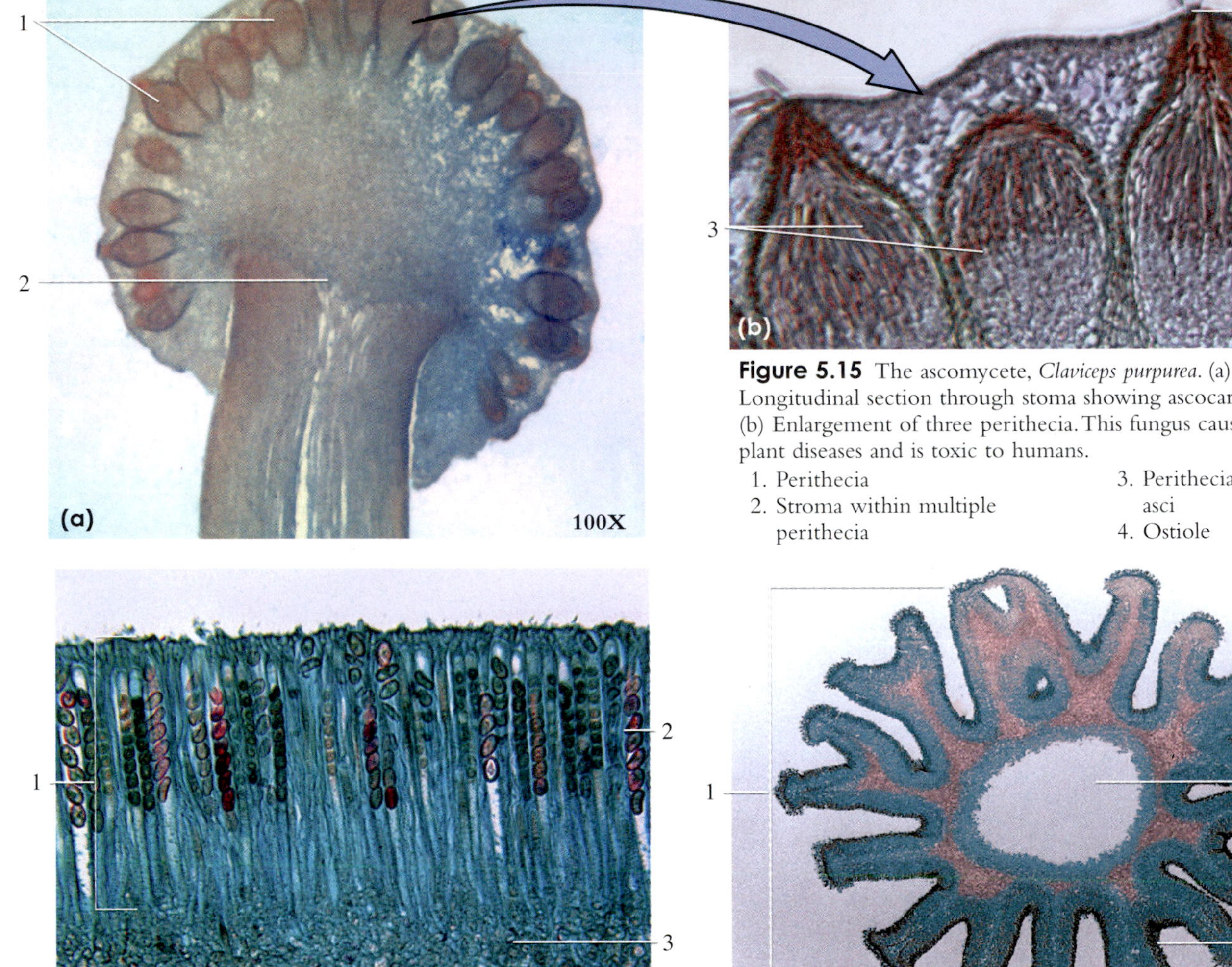

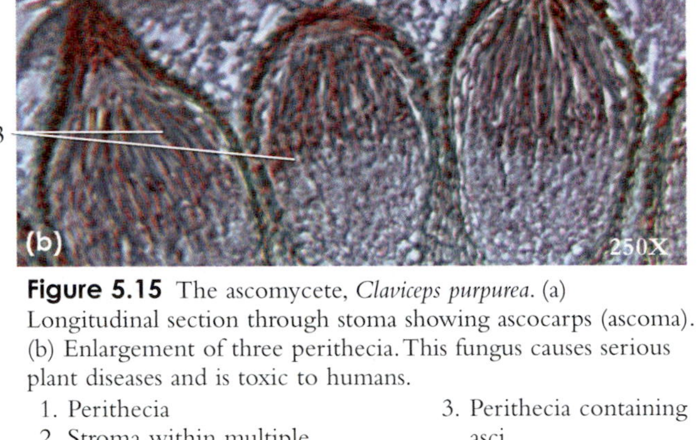

Figure 5.15 The ascomycete, *Claviceps purpurea*. (a) Longitudinal section through stoma showing ascocarps (ascoma). (b) Enlargement of three perithecia. This fungus causes serious plant diseases and is toxic to humans.

1. Perithecia
2. Stroma within multiple perithecia
3. Perithecia containing asci
4. Ostiole

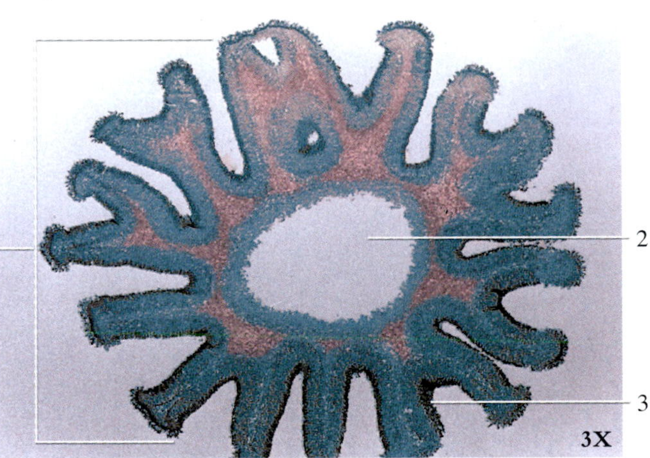

Figure 5.16 A section through the hymenial layer of the apothecium of *Peziza* sp., showing asci with ascospores.

1. Hymenial layer
2. Ascus with ascospores
3. Ascocarp (ascoma) mycelium

Figure 5.17 A section through an ascocarp (ascoma) of the morel, *Morchella* sp. True morels are prized for their excellent flavor.

1. Convoluted fruiting body
2. Hollow "stalk"
3. Hymenium

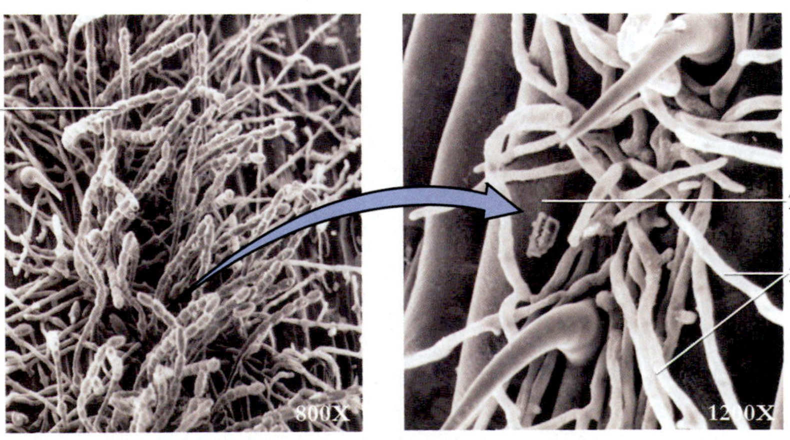

Figure 5.18 Scanning electron micrographs of the powdery mildew, *Erysiphe graminis*, on the surface of wheat. As the mycelium develops, it produces spores (conidia) that give a powdery appearance to the wheat.

1. Conidia
2. Wheat host
3. Hyphae of the fungus

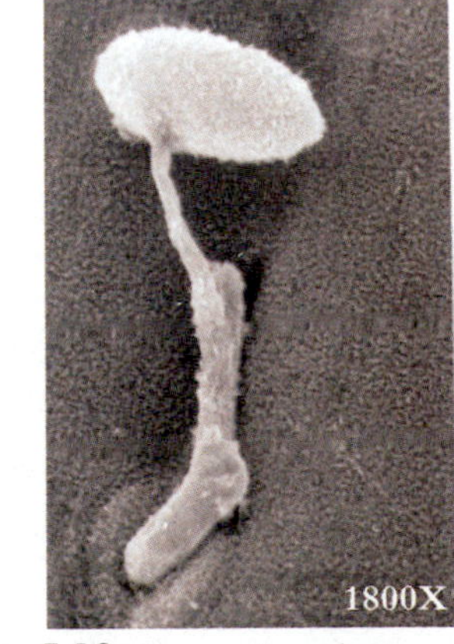

Figure 5.19 A scanning electron micrograph of a germinating spore (conidium) of the powdery mildew, *Erysiphe graminis*. The spore develops into a mycelium that penetrates the epidermis and then spreads over the host plant, producing a powdery appearance.

Phylum Deuteromycota - conidial molds

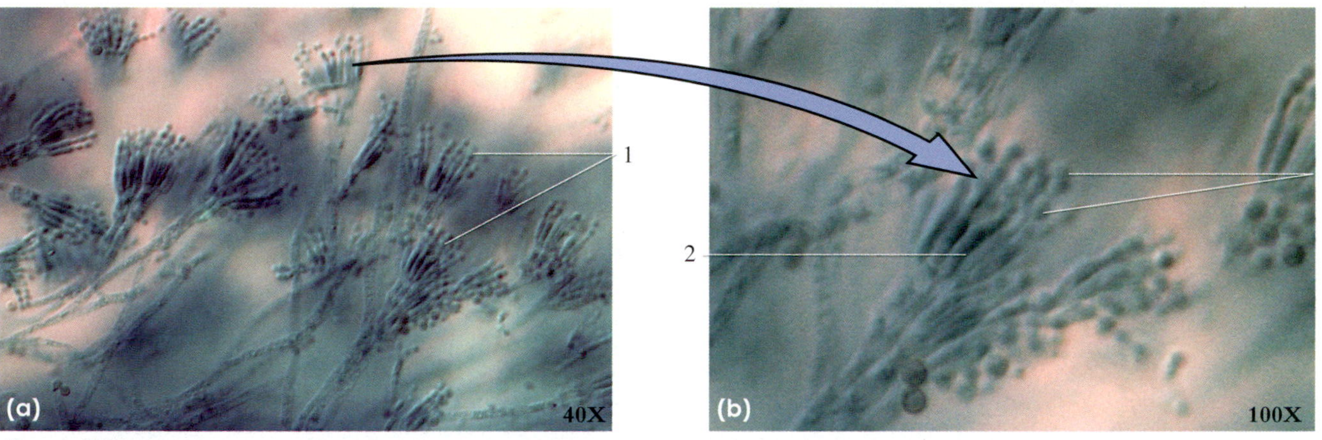

Figure 5.20 The fungus *Penicillium* sp. causes economic damage as a mold but is also the source of important antibiotics. (a) A colony of *Penicillium* sp., and (b) a close-up of a conidiophore with chains of asexual spores (conidia) at the end.
1. Conidia 2. Conidiophore 3. Conidia

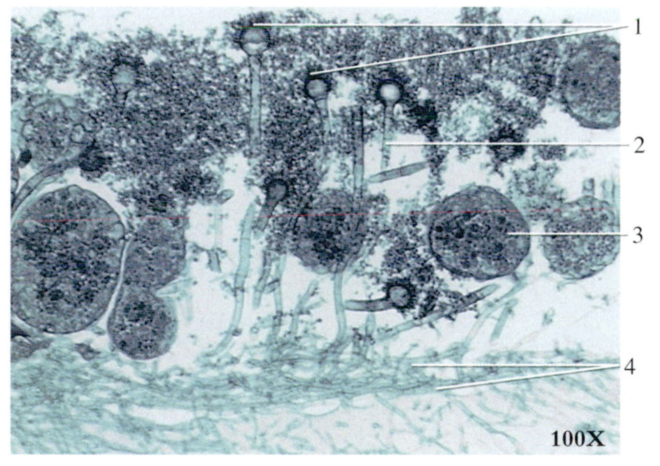

Figure 5.21 A common mold, *Aspergillus* sp.
1. Conidia (spores)
2. Conidiophore
3. Cleistothecium
4. Hyphae

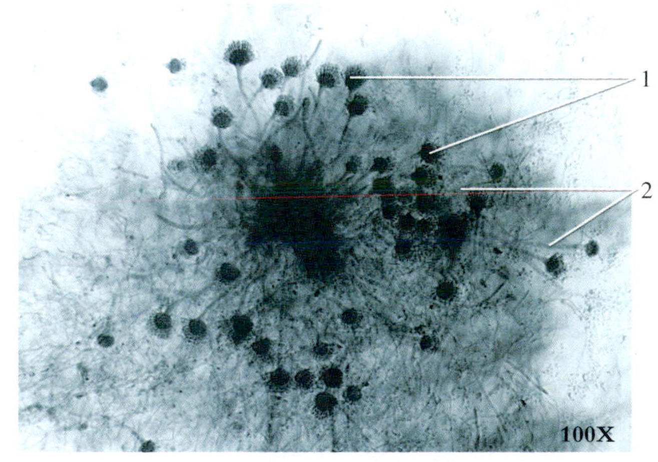

Figure 5.22 A common mold, *Aspergillus* sp.
1. Conidia 2. Conidiophores

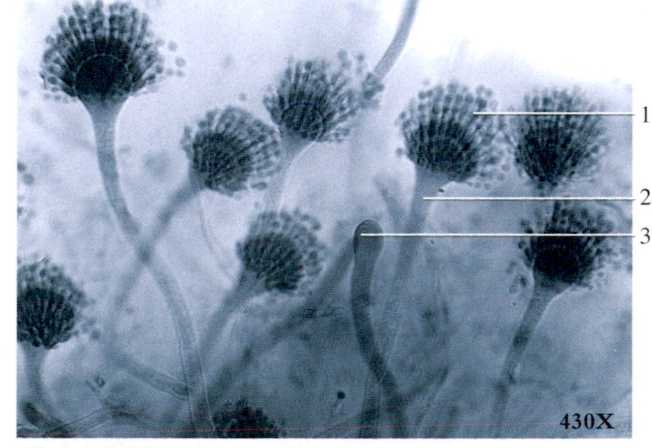

Figure 5.23 A close-up of sporangia of the mold, *Aspergillus* sp. The conidia, or spores, of this genus are produced in a characteristic radiate pattern.
1. Conidia (spores)
2. Conidiophore
3. Developing conidiophore

Figure 5.24 An electron micrograph of an *Aspergillus* sp. spore. Note the rodlet pattern on the spore wall.

Phylum Basidiomycota - mushrooms, toadstools, rusts, and smuts

Pleurotus sp.

Hericium sp.

Coriolus sp.

Astreus sp.

Coprinus sp.

Amanita sp.

Chantarella sp.

Amanita sp.

Nidularia sp.

Boletus sp.

Figure 5.25 Some representative basidocarps (basidiomas or fruiting bodies) of basidiomycetes.

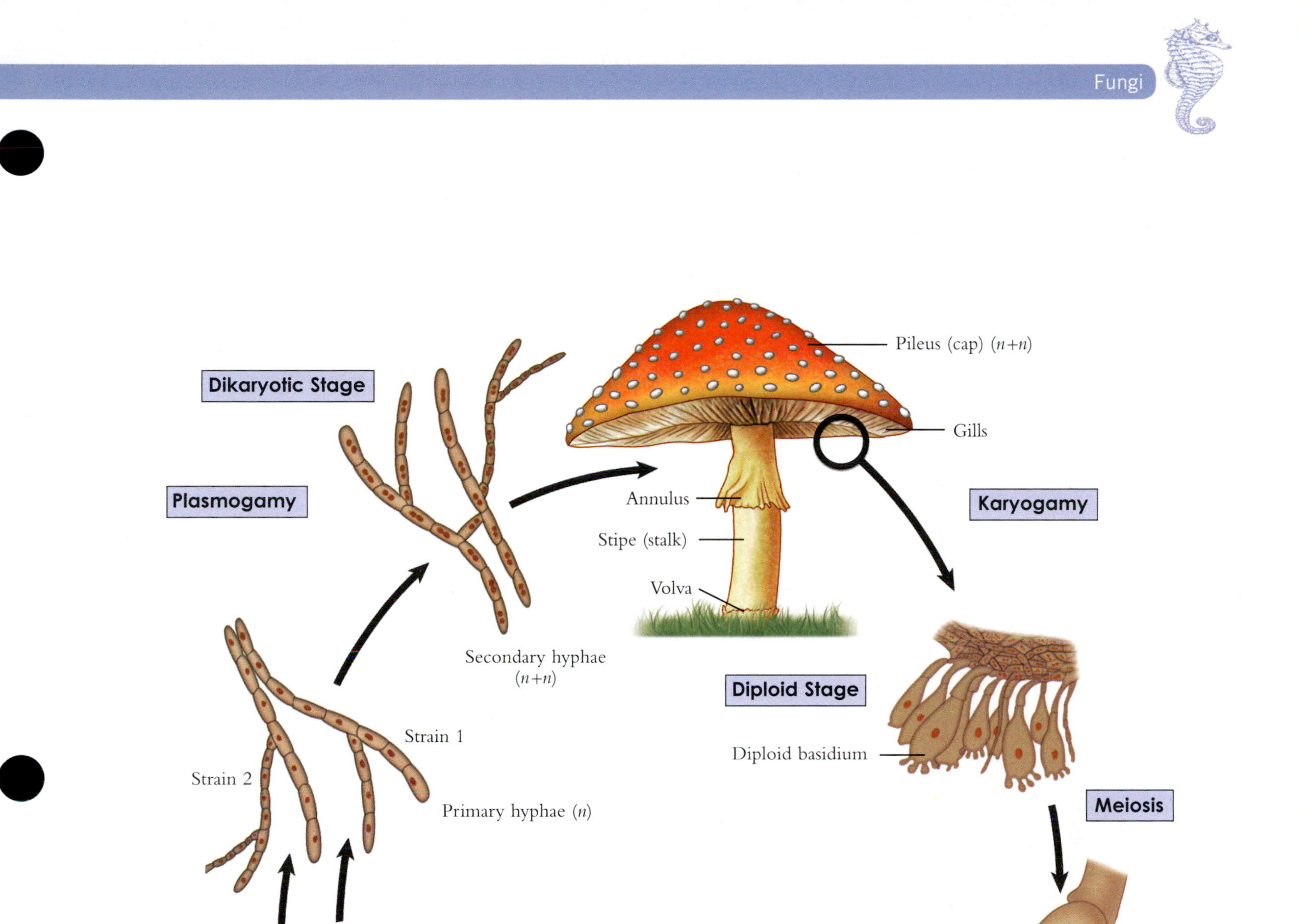

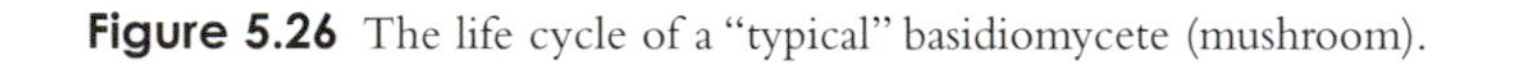

Figure 5.26 The life cycle of a "typical" basidiomycete (mushroom).

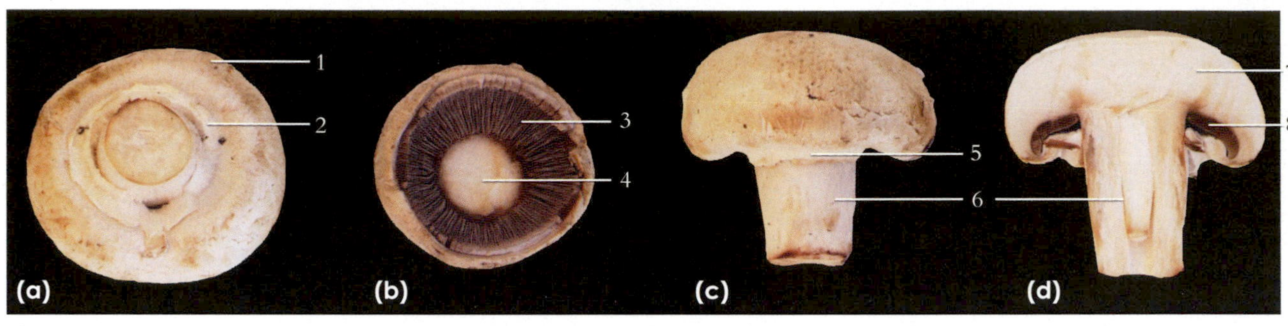

Figure 5.27 Structure of a mushroom. (a) An inferior view with the annulus intact, (b) an inferior view with the annulus removed to show the gills, (c) a lateral view, and (d) a longitudinal section.

1. Pileus (cap)
2. Veil
3. Gills
4. Stipe (stalk)
5. Annulus
6. Stipe (stalk)
7. Pileus (cap)
8. Gills

Figure 5.28 Basidiomycete puffballs growing on a decaying log.

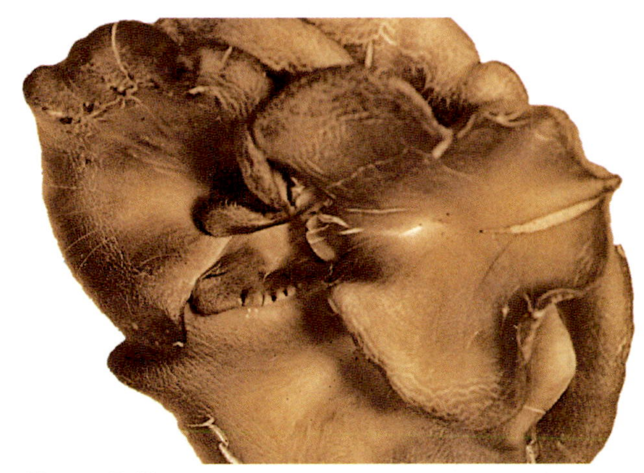

Figure 5.29 Herbarium specimen of the wood fungus, *Stropharia semiglobata*. Growing on decaying wood and other organic matter, basidiomycetes are important decomposers in forest communities.

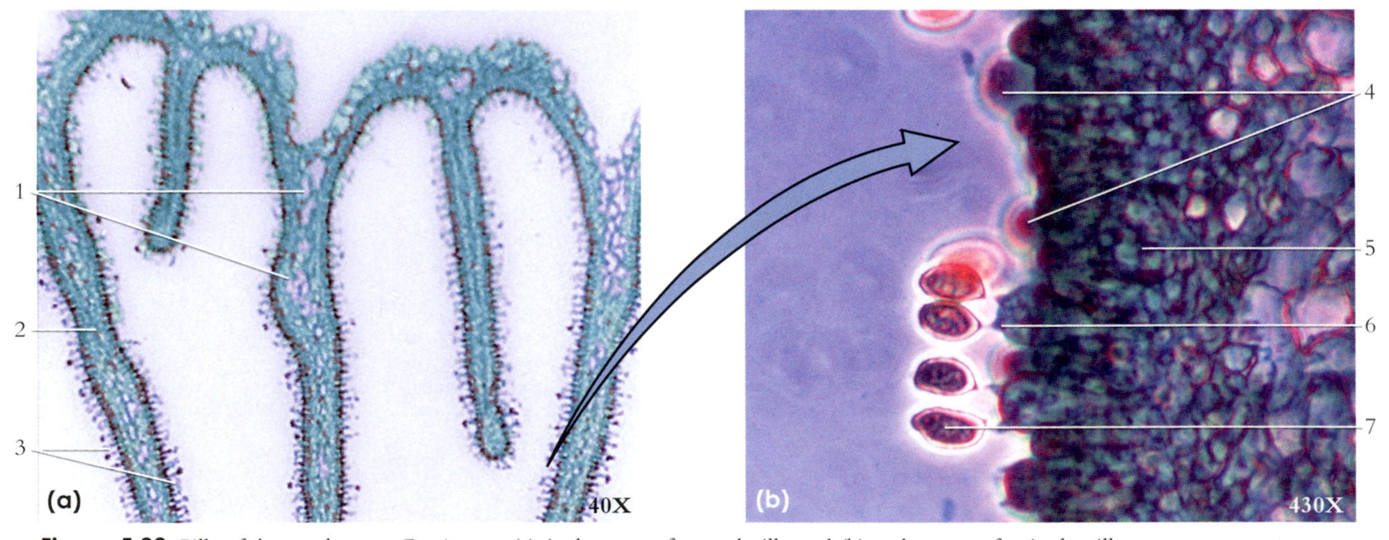

Figure 5.30 Gills of the mushroom *Coprinus* sp. (a) A close-up of several gills, and (b) a close-up of a single gill.

1. Hyphae composing the gills
2. Gill
3. Basidiospores
4. Immature basidia
5. Gill (composed of hyphae)
6. Sterigma
7. Basidiospore

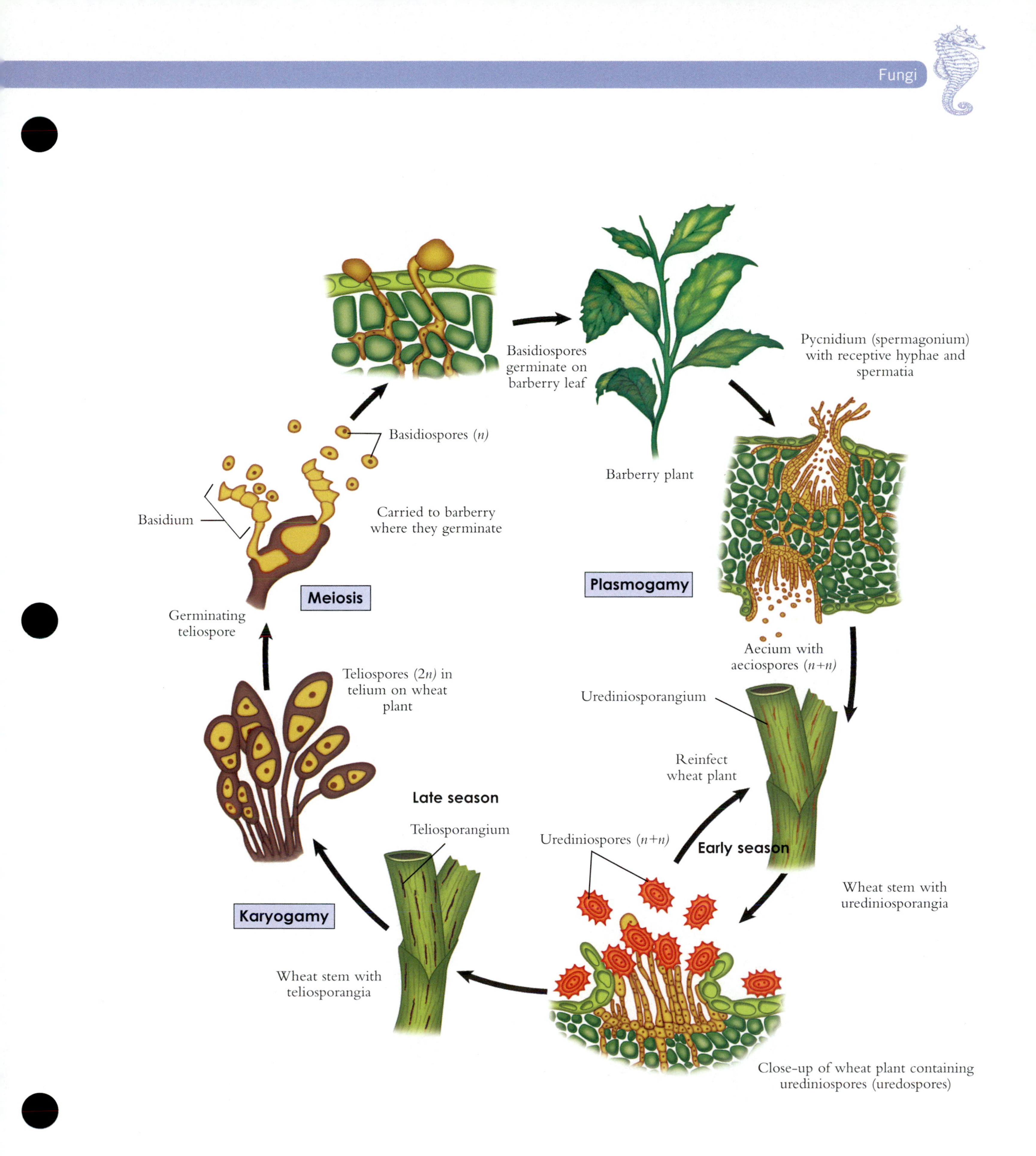

Figure 5.31 The life cycle of wheat rust, *Puccinia graminis.*

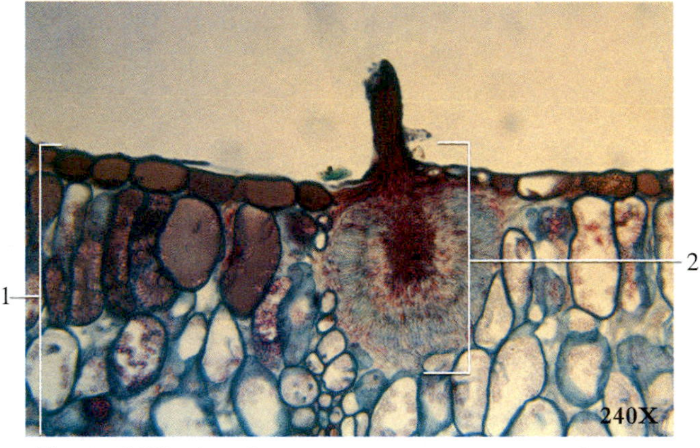

Figure 5.32 The wheat rust, *Puccinia graminis,* pycnidium on barberry leaf.
1. Barberry leaf 2. Pycnidium

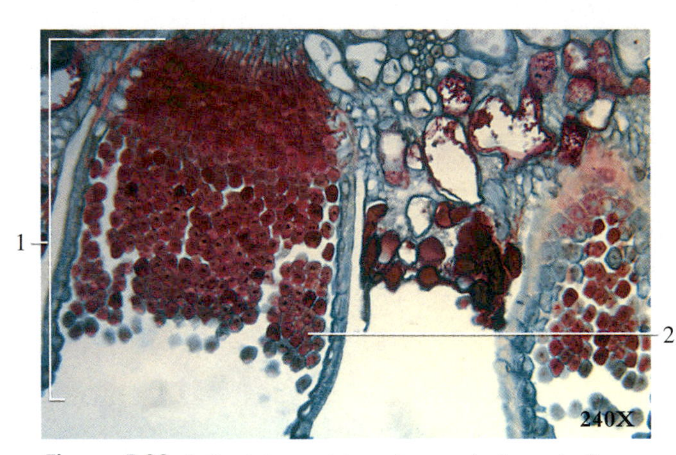

Figure 5.33 A *Puccinia graminis* aecium on barberry leaf.
1. Aecium 2. Aeciospores

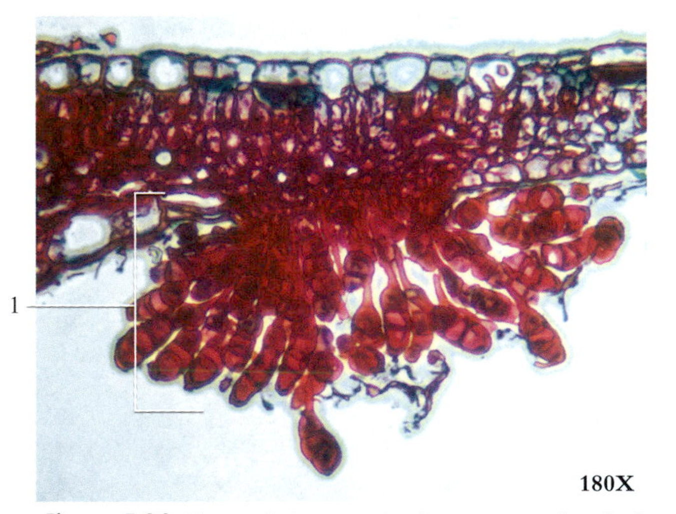

Figure 5.34 The urediniosporangia of *Puccinia* on wheat leaf.
1. Urediniosporangia

Figure 5.35 Black stem wheat rust, *Puccinia graminis,* on the lower surface of barberry leaves.
1. Clusters of aecia

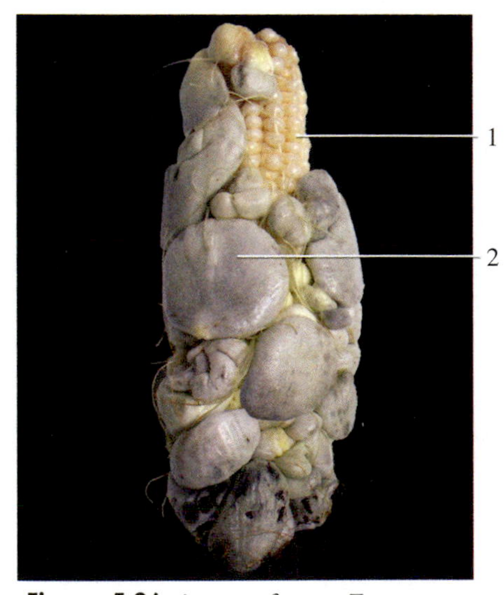

Figure 5.36 An ear of corn, *Zea mays*, infected by the smut *Ustilago maydis,* which is destroying the fruit (ear).
1. Corn ear 2. Fungus

Figure 5.37 Smut-infected brome grass. The grains have been destroyed by the smut fungus.

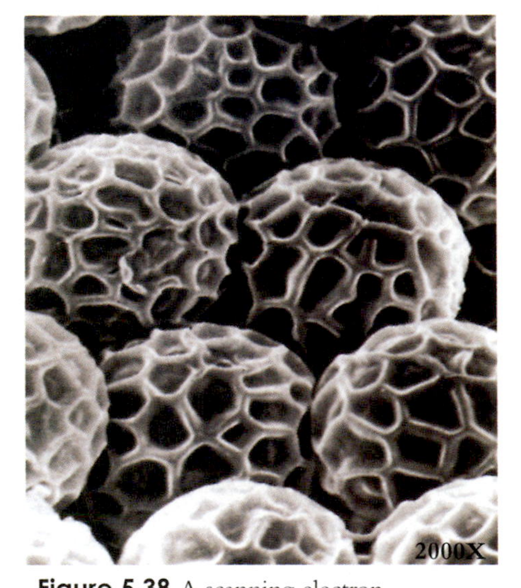

Figure 5.38 A scanning electron micrograph of teliospores of a wheat smut fungus.

Lichens (symbiotic associations of fungi and algae)

Figure 5.39 Lichens are often separated informally on the basis of their form. (a) Crustose lichen, (b) foliose lichen, and (c) fruticose lichen.

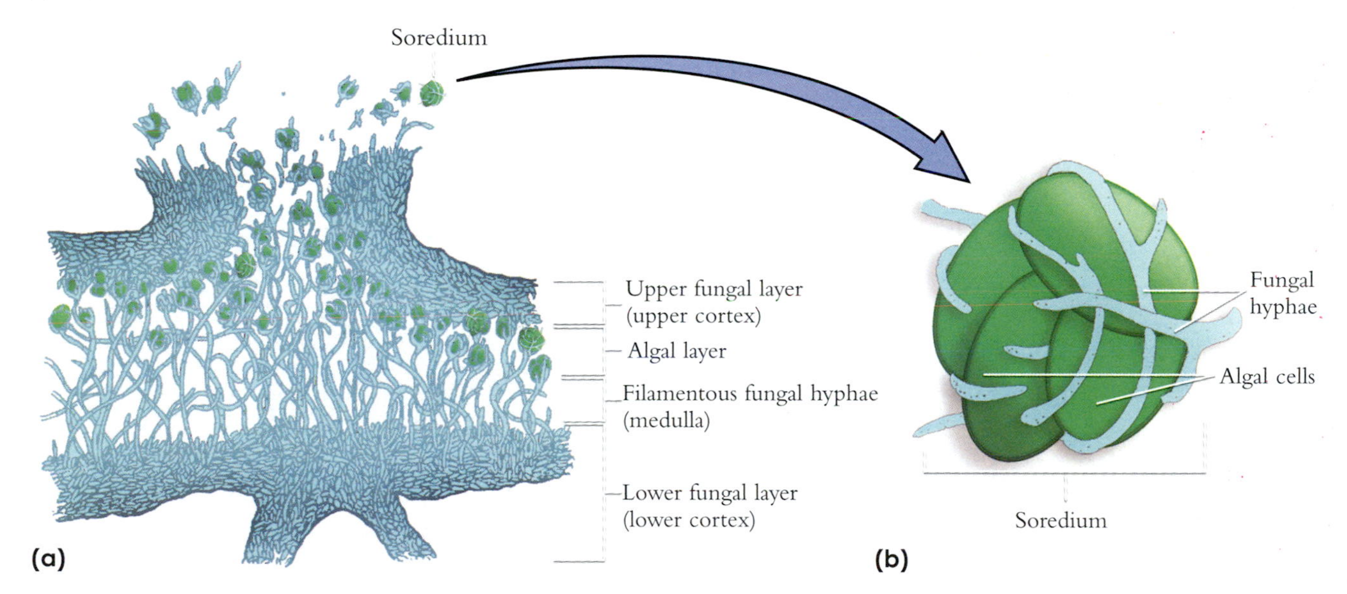

Figure 5.40 Many lichens reproduce by producing soredia, which are small bodies containing both algal and fungal cells. (a) Lichen thallus, and (b) soredium.

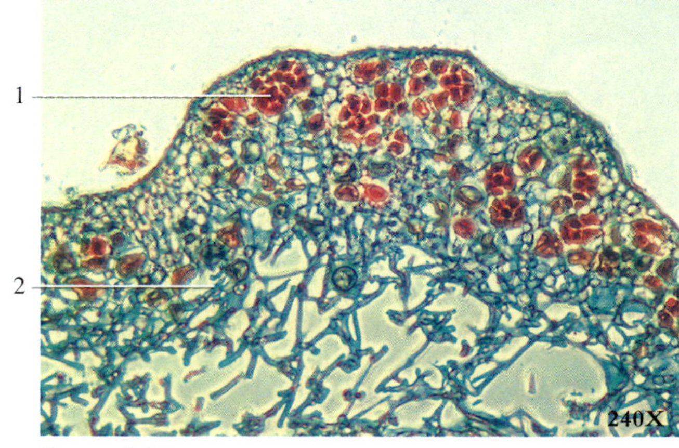

Figure 5.41 Transverse section through a lichen thallus.
1. Algal cells 2. Fungal hyphae

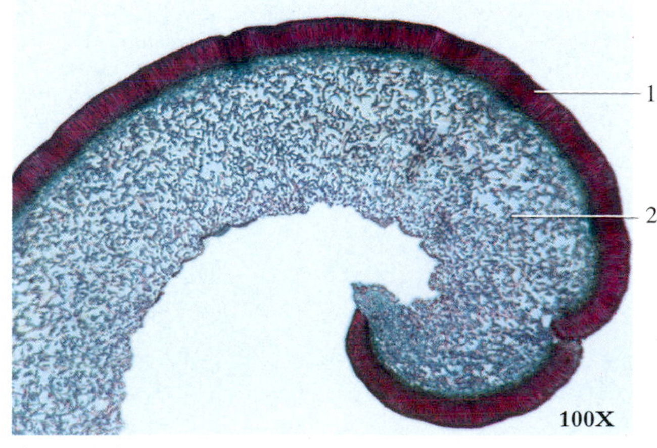

Figure 5.42 Ascomycete lichen thallus demonstrating a surface layer of asci.
1. Asci 2. Loose fungal filaments

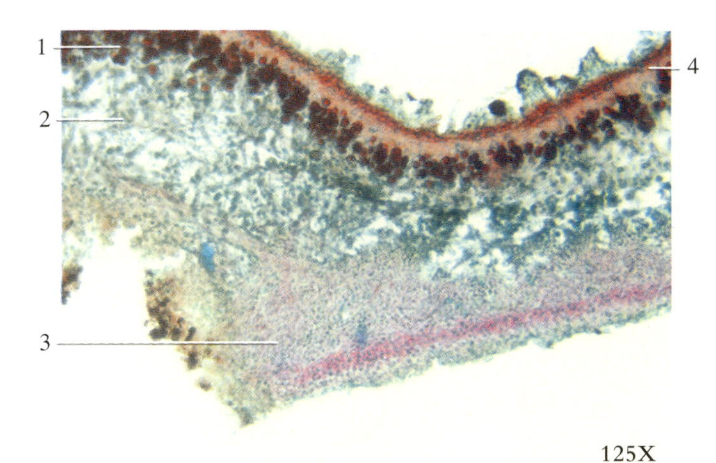

Figure 5.43 A transverse section through a lichen thallus.
1. Algal cells
2. Medulla
3. Lower cortex
4. Fungal layer (upper cortex)

Figure 5.44 The foliose lichen *Xanthoria* sp. growing on the bark of a tree.
1. Lichen
2. Bark

Figure 5.45 The crustose lichen *Lecanora* sp. growing on sandstone in an arid southern Utah environment.

Figure 5.46 The foliose lichen *Hypogymnia* sp. growing on a pine branch in the Northwest.

Figure 5.47 Fruticose lichen, British soldier, *Cladonia cristatella*, growing in Alaska.

Figure 5.48 The foliose and fruticose lichens in the Pacific Northwest.
1. Foliose lichen *Hypogymnia* sp.
2. Foliose lichen *Evernia* sp.
3. Fruticose lichen *Usnea* sp.

Plantae

Chapter 6

Plants are photosynthetic, multicellular eukaryotes. *Cellulose* in their cell walls provides protection and rigidity, and the *pores* or *stomata,* and *cuticle* of stems and leaves regulate gas exchange. Mitosis and meiosis are characteristic of all plants. Jacketed sex organs, called *gametangia*, protect the gametes and embryos from desiccation. All land plants have heteromorphic alternation of generations with distinctive haploid *gametophyte* and diploid *sporophyte* forms. Photosynthetic cells within plants contain *chloroplasts* with the pigments chlorophyll *a*, chlorophyll *b*, and a variety of carotenoids. Carbohydrates are produced by plants and stored in the form of starch.

Reproduction in seed plants is well adapted to a land existence. The conifers produce their seeds in protective *cones*, and the angiosperms produce their seeds in protective *fruits*. In the life cycle of a conifer, such as a pine, the mature *sporophyte* (tree) has female cones that produce *megaspores* that develop into the female gametophyte generation, and male cones that produce *microspores* that develop into the male gametophyte generation (mature pollen grains).

Following fertilization, immature sporophyte generations are present in seeds located on the female cones. The female cone opens and the *seeds* (pine nuts) disperse to the ground and germinate if the conditions are right. Reproduction in angiosperms is similar to reproduction in gymnosperms except that the angiosperm pollen and ovules are produced in flowers rather than in cones, and a fruit is formed.

Table 6.1 Some Representatives of Plantae

Phyla (Division) and Representative Kinds	Characteristics
Bryophytes — liverworts, hornworts, and mosses	Lack vascular tissue; rhizoids; homosporous (bisexual gametophyte)
VASCULAR PLANTS	
Lycopodiophyta — clubmosses, spike mosses, and quillworts	Sporangia borne on sporophylls; homosporous or heterosporous (unisexual gametophyte); many are epiphytes
Pteridophyta Psilotopsida — whisk ferns	True roots and leaves are absent, but vascular tissue present; rhizome and rhizoids present
Pteridophyta Equisetopsida — horsetails	Epidermis embedded with silica; tips of stems bear cone-like structures containing sporangia; most homosporous
Pteridophyta Polypodiopsida — ferns	Fronds as leaves; underground roots coming off rhizomes; most homosporous
SEED PLANTS	
Cycadophyta — cycads	Heterosporous; pollen and seed cones borne of different plants; palmlike leaves
Ginkgophyta — ginkgo	Heterosporous; seed-producing; deciduous, fan-shaped leaves
Pinophyta (= Coniferophyta) — conifers	Heterosporous; pollen and seed cones same plant; needlelike or scalelike leaves
ANGIOSPERMS	
Magnoliophyta (= Anthophyta) — flowering plants	Heterosporous; flowering plants that produce their seeds enclosed in fruit; most are free-living, some are saprophytic or parasitic

Table 6.2 Some Representative Bryophytes

Phyla and Representative Kinds	Characteristics
Marchantiophyta (= Hepatophyta) — liverworts	Flat or leafy gametophytes; single-celled rhizoids; simple sporophytes and elaters present; stomata and columella absent
Anthocerophyta — hornworts	Flat, lobed gametophytes; more complex sporophytes with stomata; pseudoelaters and columella present
Bryophyta — mosses	Leafy gametophytes, multicellular rhizoids; sporophytes with stomata, columella, peristome teeth and/or operculum present

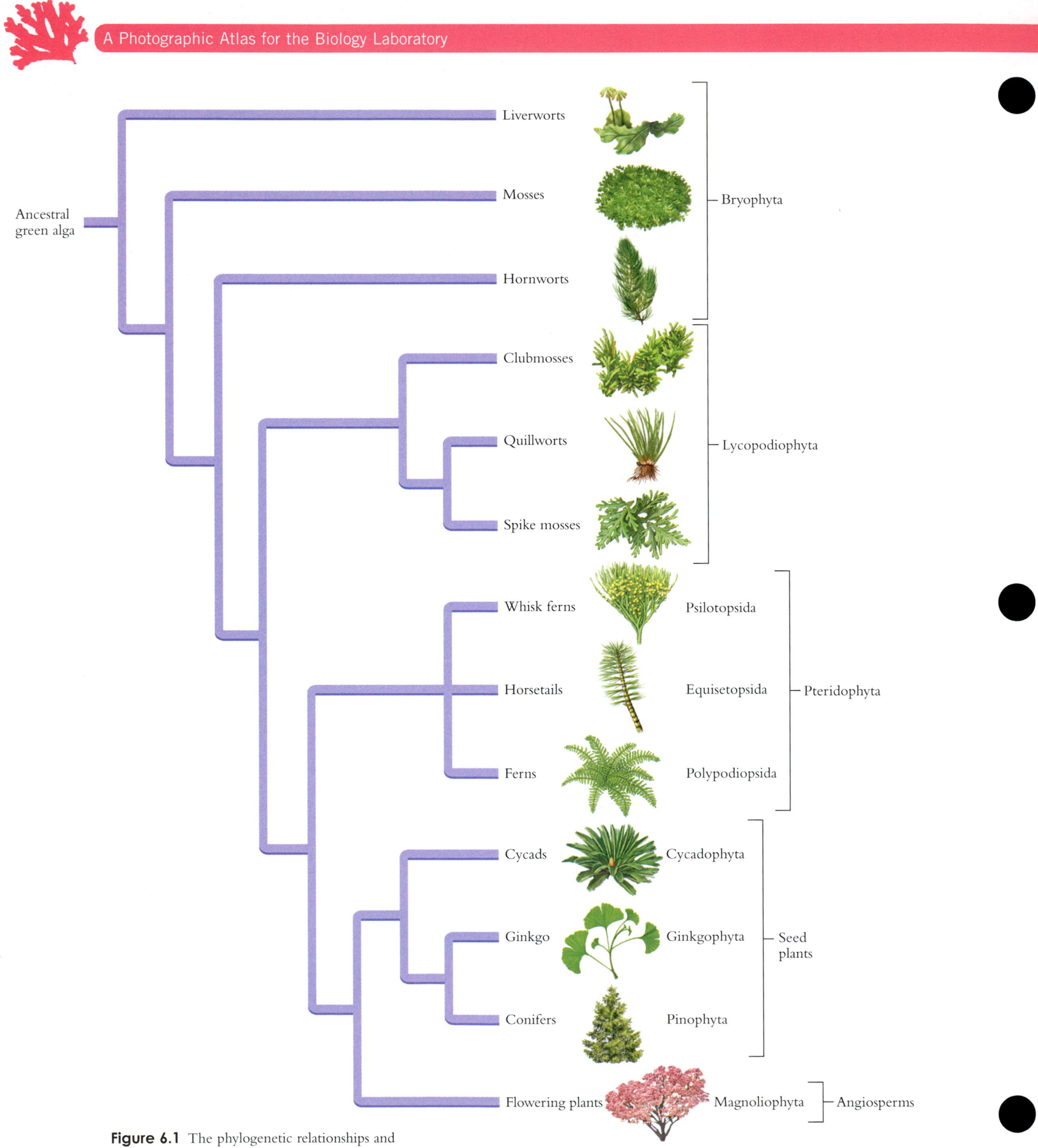

Figure 6.1 The phylogenetic relationships and classification of Plantae.

Phylum Hepatophyta - liverworts

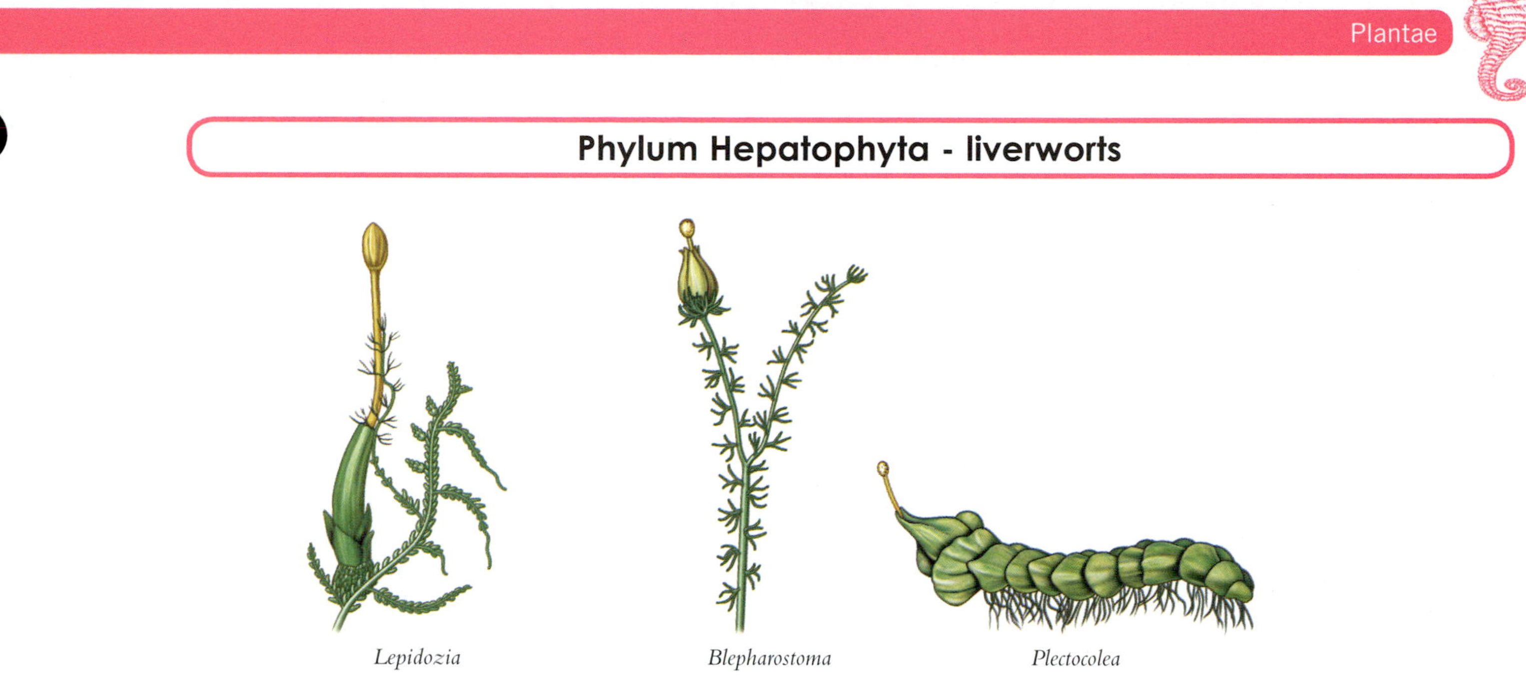

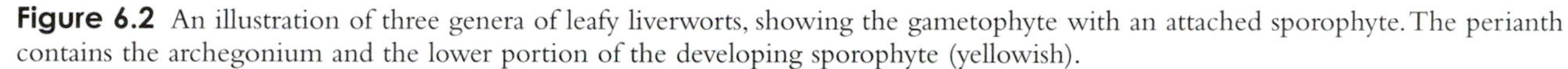

Figure 6.2 An illustration of three genera of leafy liverworts, showing the gametophyte with an attached sporophyte. The perianth contains the archegonium and the lower portion of the developing sporophyte (yellowish).

Calopegia sp.

Conocephalum sp.

Bazzania sp.

Porella sp.

Riccia sp.

Scapania sp.

Figure 6.3 Some examples of liverworts (scale in mm).

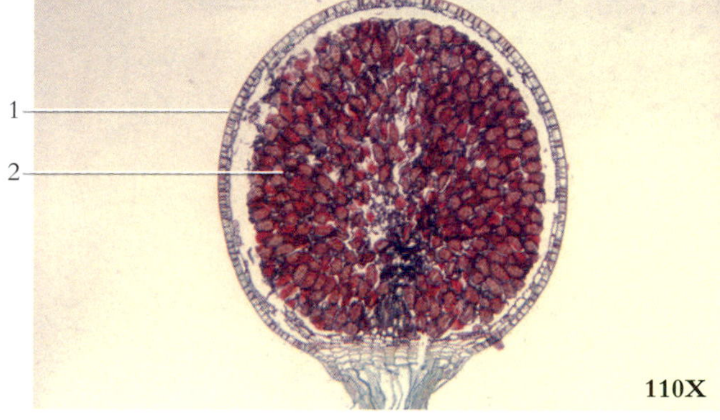

Figure 6.4 A sporophyte (capsule) of the leafy liverwort, *Pelia* sp.
1. Capsule 2. Sporogenous tissue

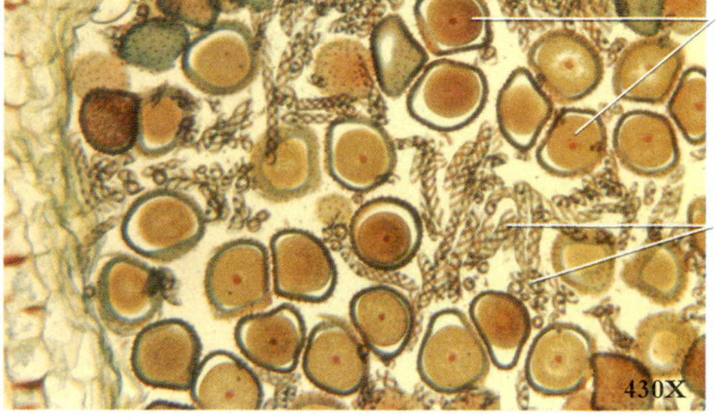

Figure 6.5 A capsule from the leafy liverwort, *Pelia* sp., in longitudinal view.
1. Spores 2. Elaters

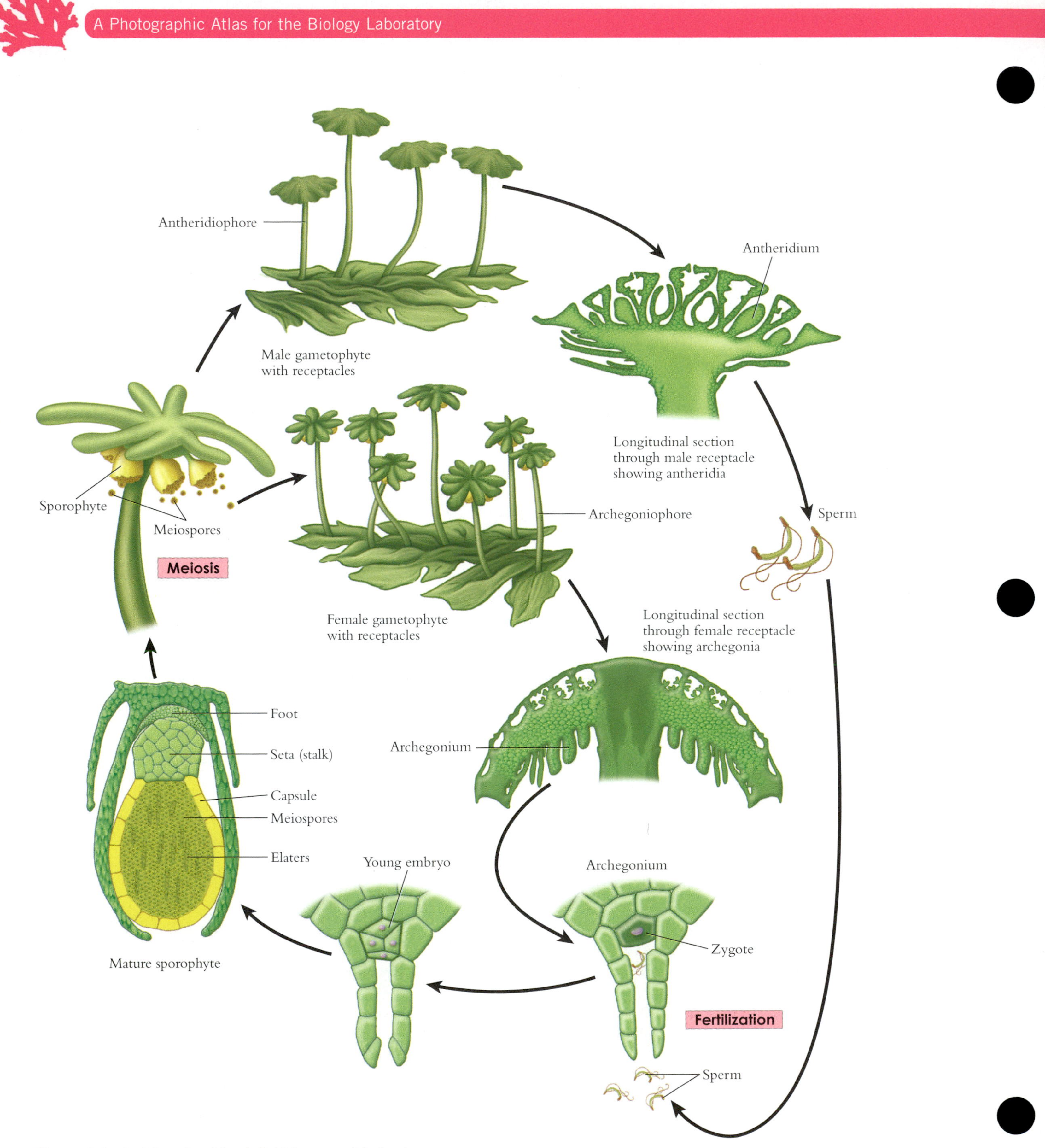

Figure 6.6 The life cycle of the thalloid liverwort, *Marchantia* sp.

Figure 6.7 A detail of *Marchantia* sp. with prominent male antheridial receptacles.

1. Antheridial receptacles
2. Gametophyte thallus

Figure 6.8 The liverwort *Marchantia* sp., showing archegonial receptacles.

1. Archegonial receptacles

Figure 6.9 A detail of *Marchantia* sp. gametophyte plants with prominent gemmae cupules.

1. Gemmae cupules with gemmae

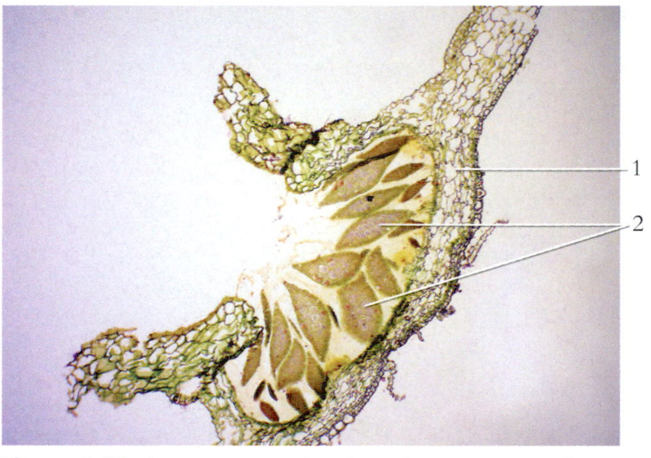

Figure 6.10 A transverse section through a gemma cupule of *Marchantia* sp.

1. Gemmae cupule
2. Gemmae

Figure 6.11 The liverwort *Marchantia* sp., showing rhizoids.

1. Rhizoids

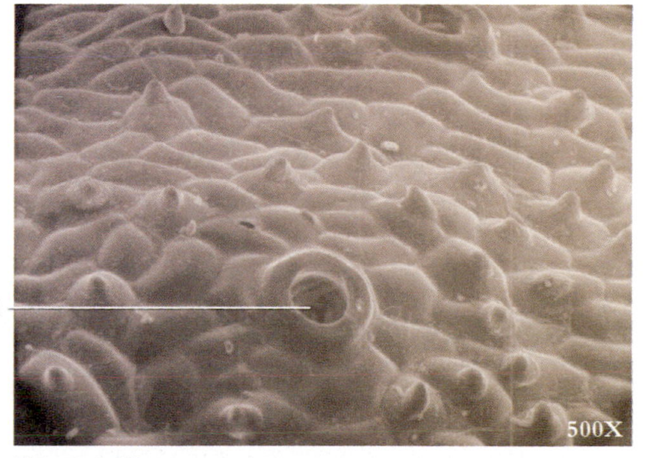

Figure 6.12 A scanning electron micrograph of the thallus of *Marchantia* sp.

1. Air pore

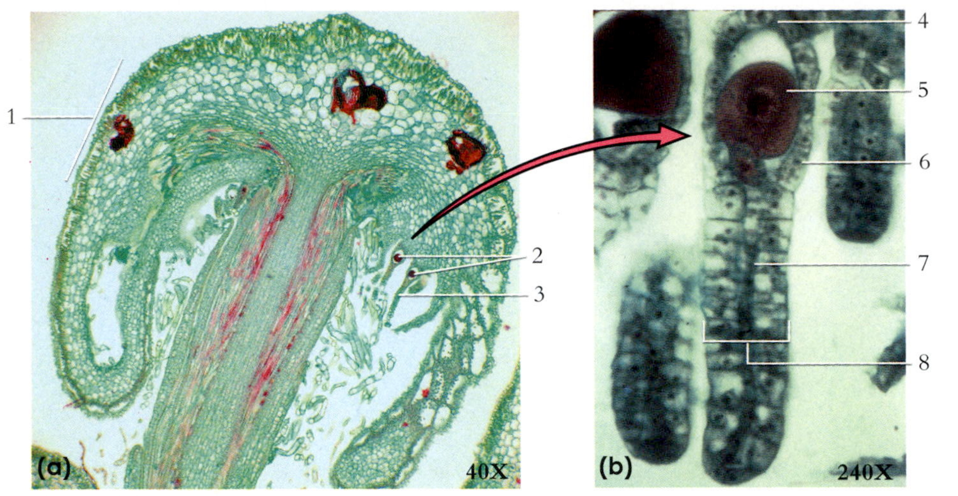

Figure 6.13 (a) The archegonial receptacle of a liverwort, *Marchantia* sp., in a longitudinal section. (b) Archegonium with egg.

1. Archegonial receptacle
2. Eggs
3. Neck of archegonium
4. Base of archegonium
5. Egg
6. Venter of archegonium
7. Neck canal
8. Neck of archegonium

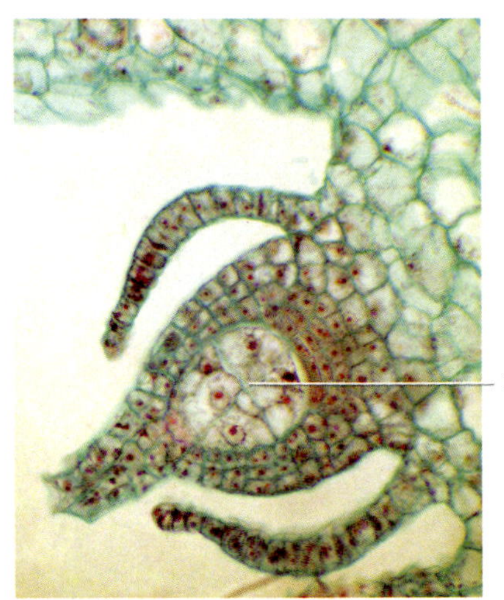

Figure 6.14 A young sporophyte of *Marchantia* sp.

1. Young embryo

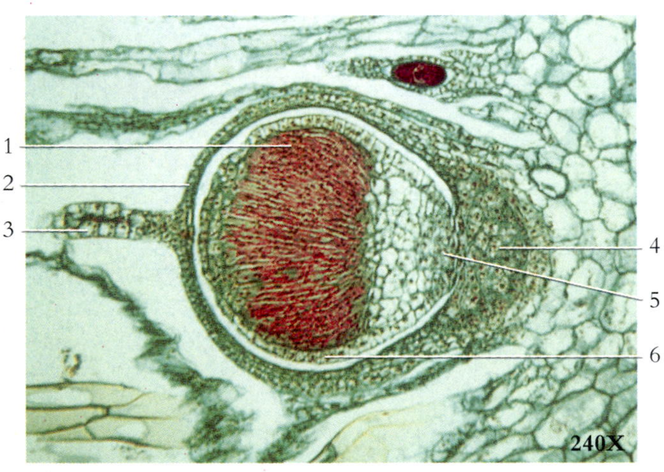

Figure 6.15 A young sporophyte of *Marchantia* sp., in longitudinal section.

1. Sporogenous tissue ($2n$)
2. Enlarged archegonium (calyptra)
3. Neck of archegonium
4. Foot
5. Seta (stalk)
6. Capsule

Figure 6.16 Immature and mature sporophytes.

1. Foot
2. Seta (stalk)
3. Sporangium (capsule)
4. Spores (n) and elaters ($2n$)

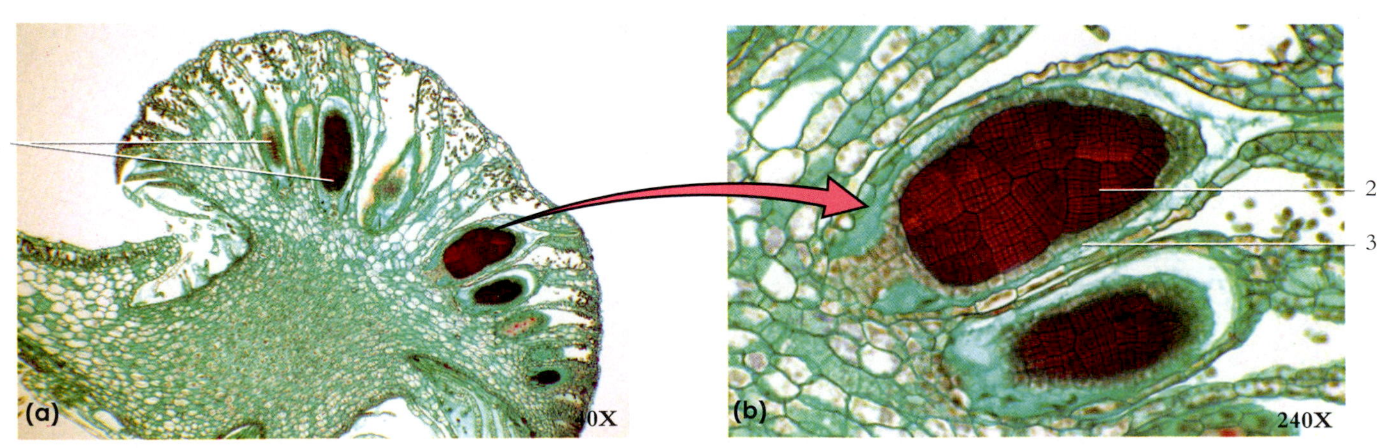

Figure 6.17 (a) A male receptacle with antheridia of a liverwort, *Marchantia* sp., in a longitudinal section. (b) Antheridial head showing a developing antheridium.

1. Antheridia
2. Spermatogenous tissue
3. Antheridium

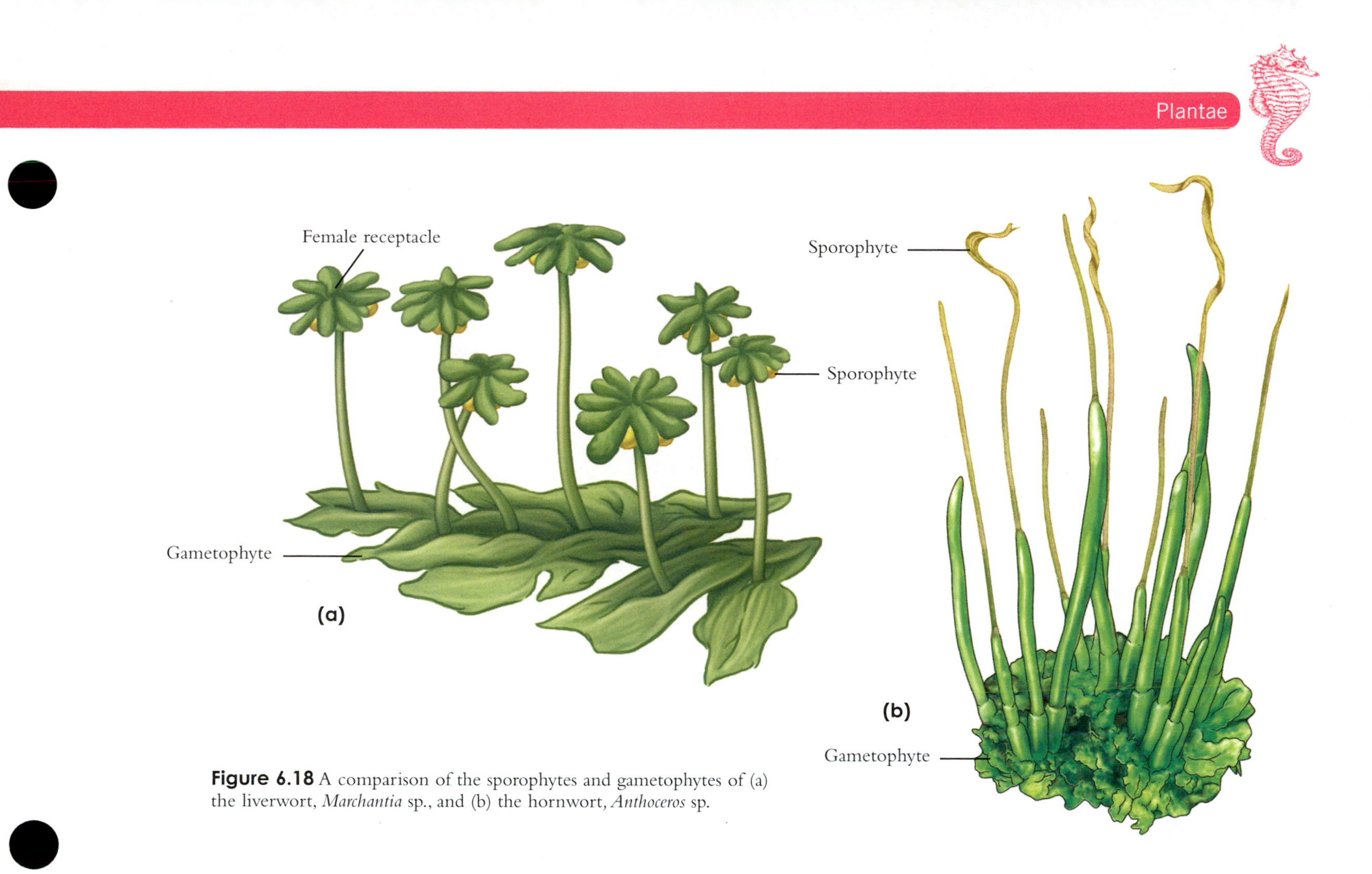

Figure 6.18 A comparison of the sporophytes and gametophytes of (a) the liverwort, *Marchantia* sp., and (b) the hornwort, *Anthoceros* sp.

Phylum Anthocerophyta - hornworts

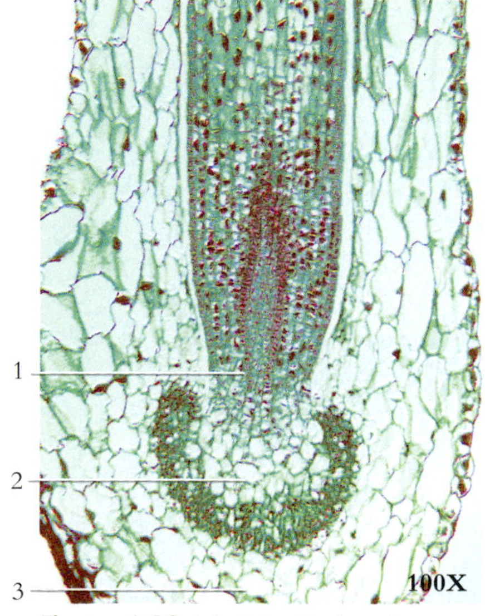

Figure 6.19 A longitudinal section of a portion of the sporophyte of the hornwort, *Anthoceros* sp.

1. Meristematic region of sporophyte
2. Foot
3. Gametophyte

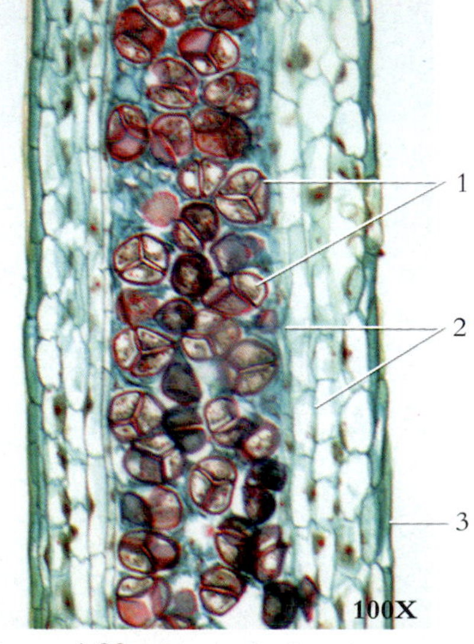

Figure 6.20 A longitudinal section of the sporangium of a sporophyte from the hornwort, *Anthoceros* sp.

1. Spores
2. Elater-like structures (pseudoelaters)
3. Capsule

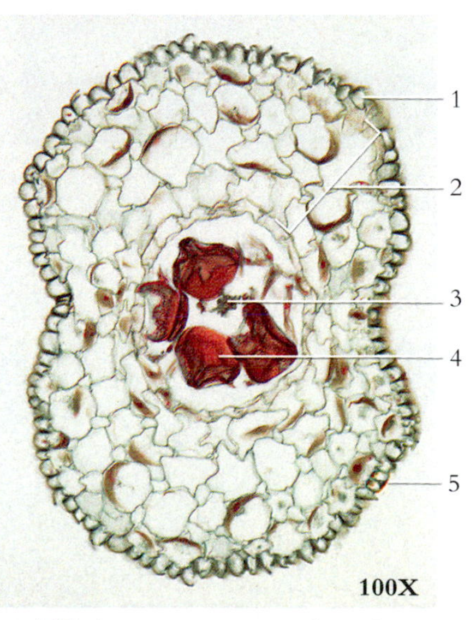

Figure 6.21 A transverse section through the capsule of a sporophyte of the hornwort, *Anthoceros* sp.

1. Epidermis
2. Photosynthetic tissue
3. Columella
4. Tetrad of spores
5. Pore (stomate)

Phylum Bryophyta - mosses

Figure 6.22 A *Sphagnum* sp. bog in the high Rocky Mountains. This lake has nearly been filled in with dense growths of *Sphagnum* sp.

Figure 6.23 A detail of *Sphagnum* sp. bog showing gametophyte plants.

Figure 6.24 A detail of gametophyte plants of peat moss, *Sphagnum* sp.

Figure 6.25 A gametophyte plant of peat moss, *Sphagnum* sp., showing attached sporophytes (scale in mm).

1. Sporophyte
2. Pseudopodium
3. Gametophyte

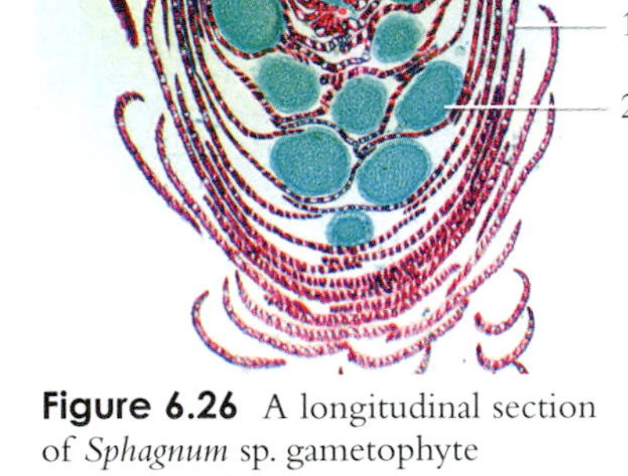

Figure 6.26 A longitudinal section of *Sphagnum* sp. gametophyte showing antheridia.

1. "leaf"
2. Antheridium

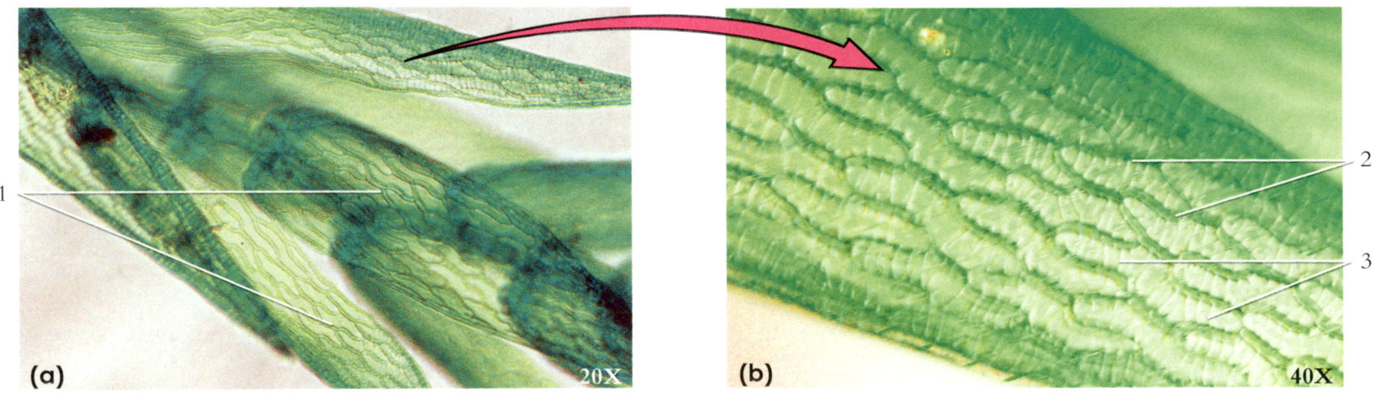

Figure 6.27 (a) A gametophyte of peat moss, *Sphagnum* sp. (b) A magnified view of a "leaf" showing the dead cell chambers that aid in water storage.

1. "Leaves"
2. Photosynthetic cells
3. Dead cells

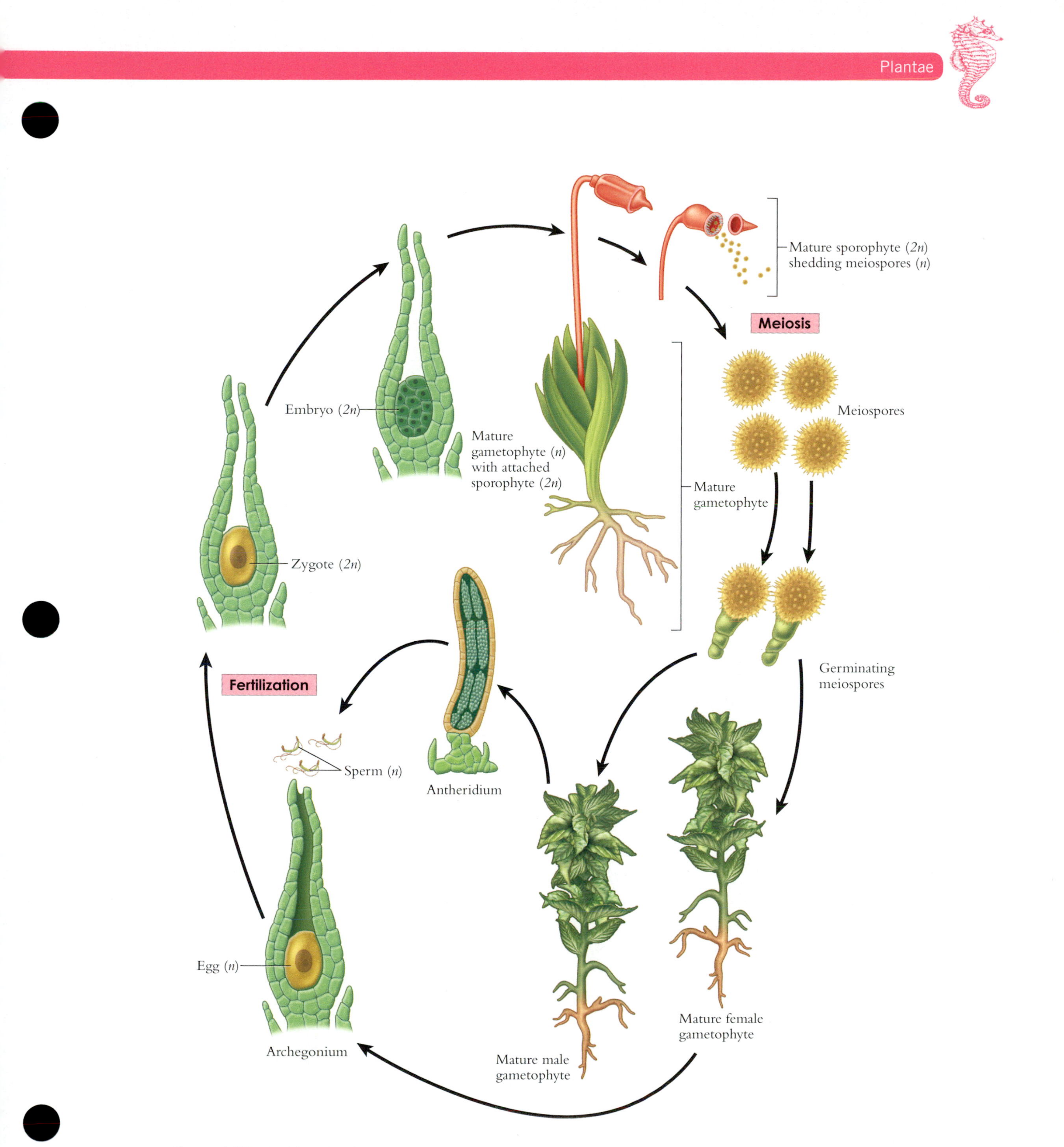

Figure 6.28 The life cycle of a moss (Bryophyta).

Figure 6.29 A habitat shot of a moss growing in a wooded environment.
1. Moss
2. Vascular plants

Figure 6.30 A moss-covered sandstone. Under dry conditions, mosses may become dormant and lose their intense green.
1. Stone
2. Moss

Figure 6.31 Four common mosses often used in course work, (a) *Polytrichum* sp., (b) *Mnium* sp., (c) *Hypnum* sp., and (d) *Dicranum* sp.

Figure 6.32 Gametophyte plants with sporophyte plant attached.
1. Calyptera
2. Capsule of sporophyte (covered by calyptera)
3. Stalk (seta)
4. Gametophyte

Figure 6.33 A sporophyte plant and capsule.
1. Operculum
2. Capsule of sporophyte (with calyptera absent)
3. Stalk (seta)

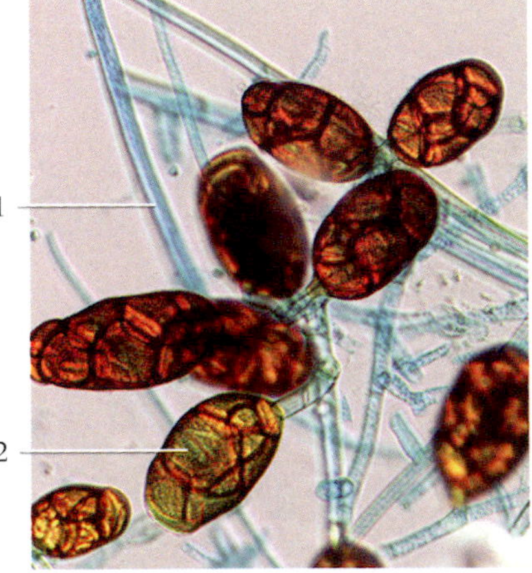

Figure 6.34 The protonemata and bulbils of a moss. The bulbils will grow to become a new gametophyte plant.
1. Protonema
2. Bulbil

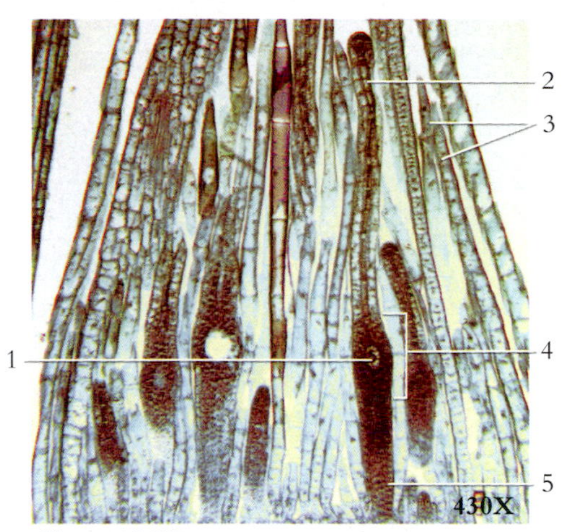

Figure 6.35 A longitudinal section of the archegonial head of the moss *Mnium* sp. The paraphyses are nonreproductive filaments that support the archegonia.

1. Egg
2. Neck
3. Paraphyses
4. Venter
5. Stalk

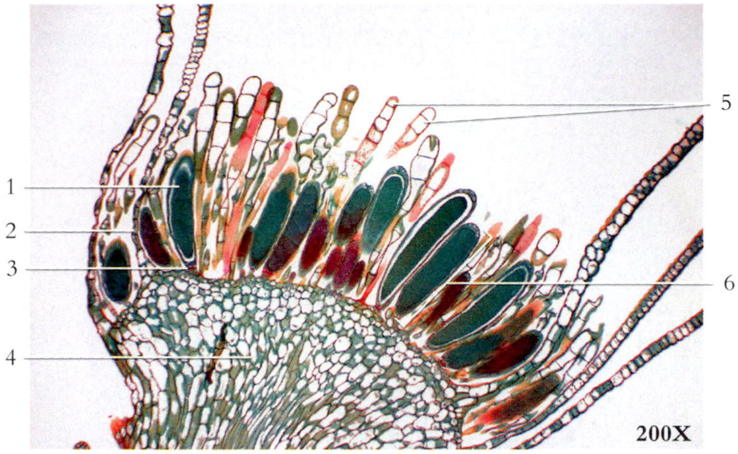

Figure 6.36 A longitudinal section of the antheridial head of the moss *Mnium* sp.

1. Spermatogenous tissue
2. Sterile jacket layer
3. Stalk
4. Male gametophyte (*n*)
5. Paraphyses (sterile filaments)
6. Antheridium (*n*)

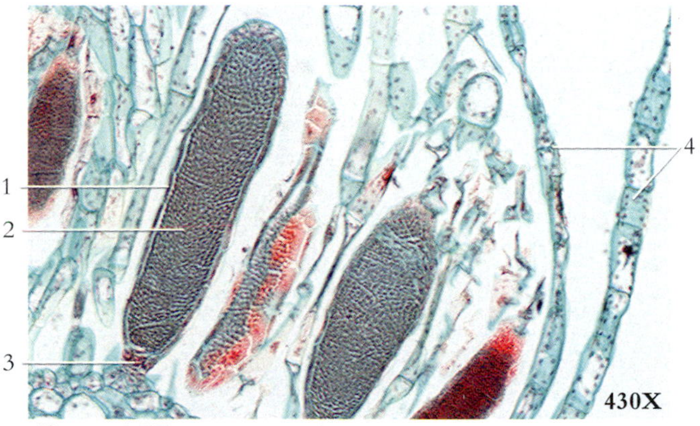

Figure 6.37 A close-up of *Mnium* sp.

1. Antheridium (*n*)
2. Spermatogenous tissue
3. Stalk
4. Paraphyses

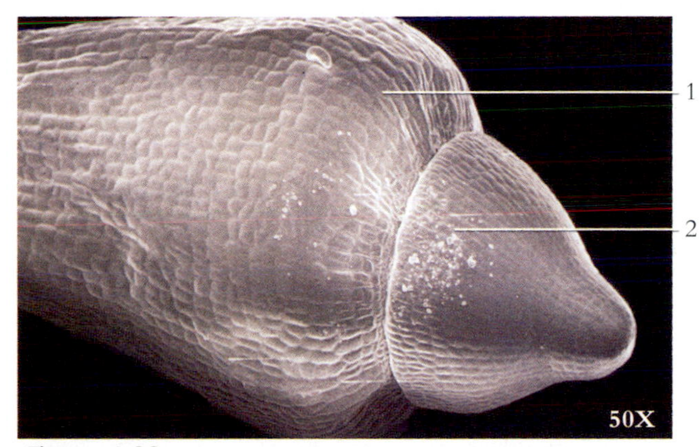

Figure 6.38 A scanning electron micrograph of the sporophyte capsule of the moss *Mnium* sp.

1. Capsule
2. Operculum

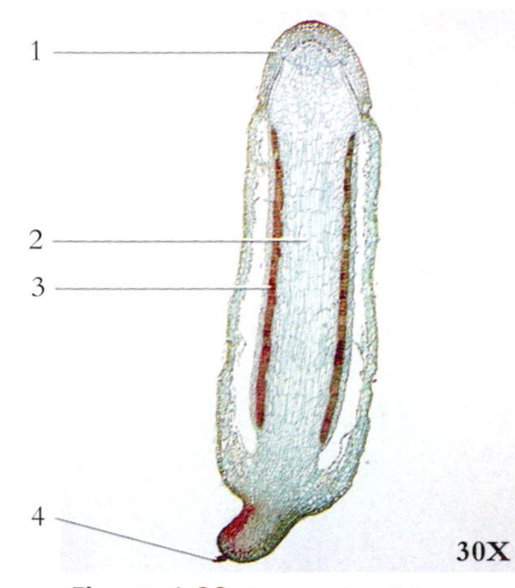

Figure 6.39 A capsule of the moss *Mnium* sp.

1. Operculum
2. Columella
3. Spores
4. Seta

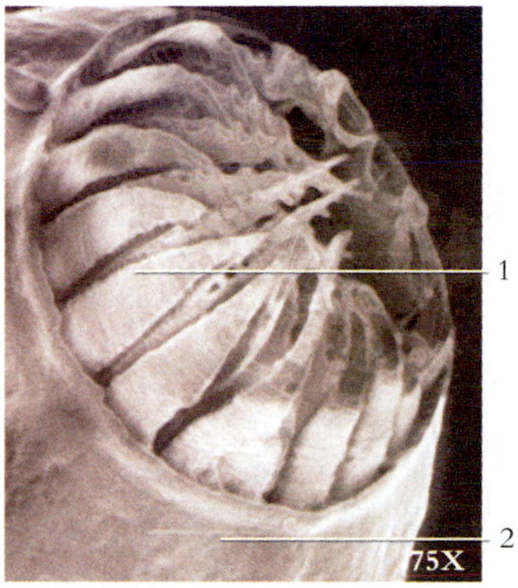

Figure 6.40 A scanning electron micrograph of the peristome of the moss *Mnium* sp. The operculum is absent in the specimen.

1. Peristome
2. Capsule

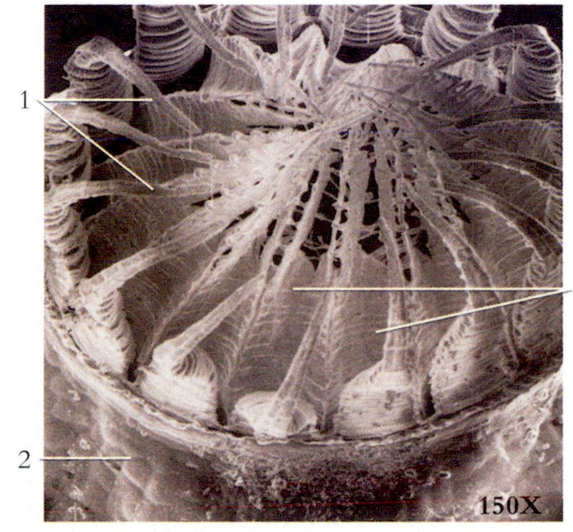

Figure 6.41 A scanning electron micrograph of the peristome of the moss *Mnium* sp.

1. Outer teeth of peristome
2. Capsule
3. Inner teeth of peristome

Lycophyta (= Lycopodiophyta) - club mosses, quillworts, and spike mosses

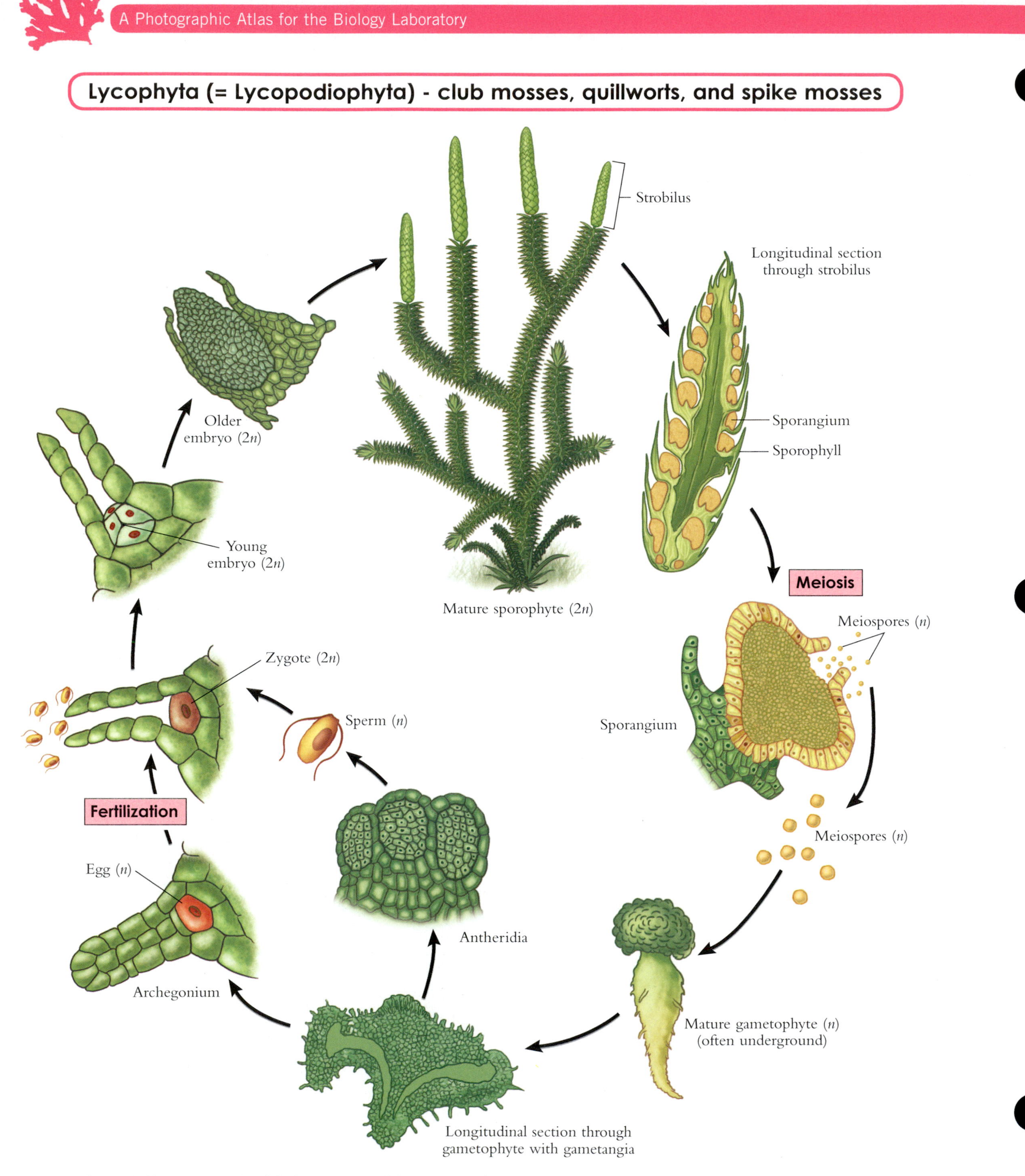

Figure 6.42 The life cycle of the homosporous clubmoss, *Lycopodium* sp.

Figure 6.43 A specimen of a lycopod, *Lycopodium clavatum*, (a) plant and (b) strobilus. *Lycopodium* occurs from the arctic to the tropics (scale in mm).
1. Strobilus
2. Stem

Figure 6.44 An enlargement of a specimen of *Lycopodium* sp., showing branch tip with sporangia on the upper surface of sporophylls (scale in mm).
1. Sporangia
2. Sporophylls (leaves with attached sporangia)

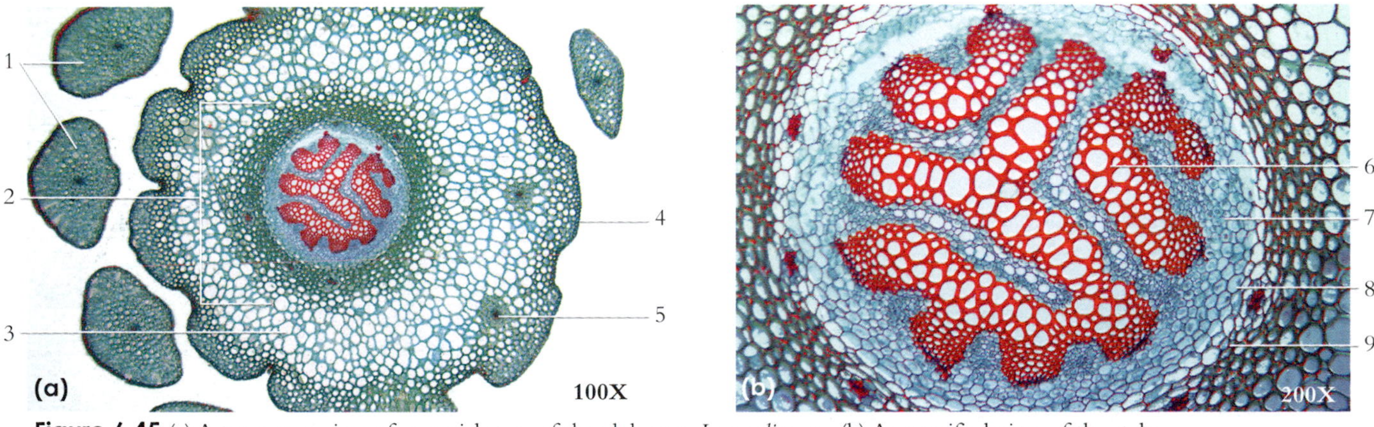

Figure 6.45 (a) A transverse view of an aerial stem of the clubmoss, *Lycopodium* sp. (b) A magnified view of the stele.
1. Leaves (microphylls)
2. Stele
3. Cortex
4. Epidermis
5. Leaf trace
6. Xylem
7. Phloem
8. Pericycle
9. Endodermis

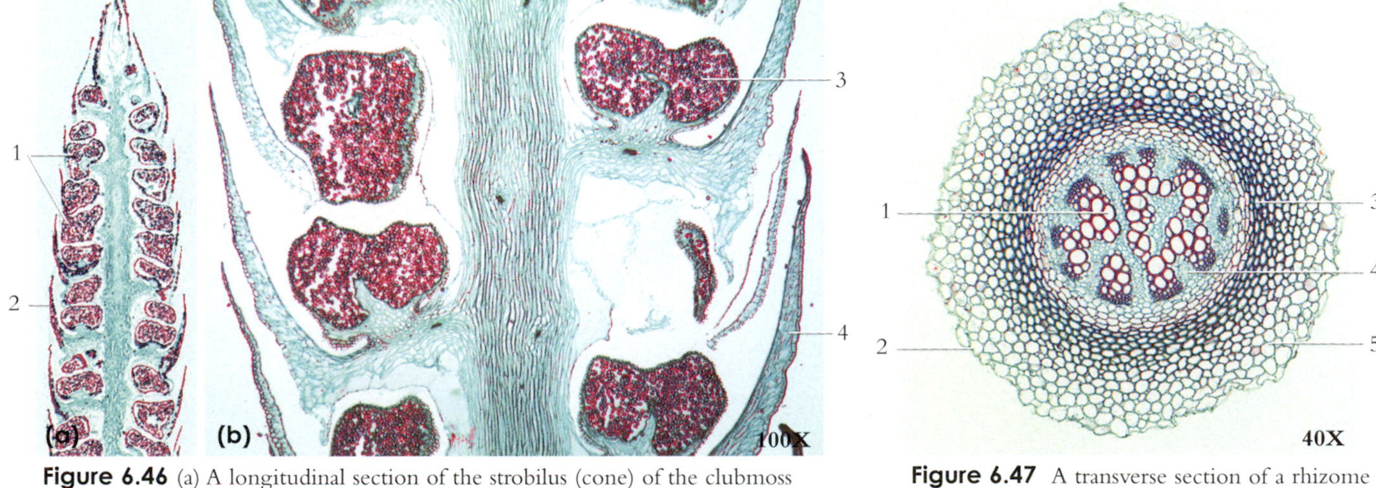

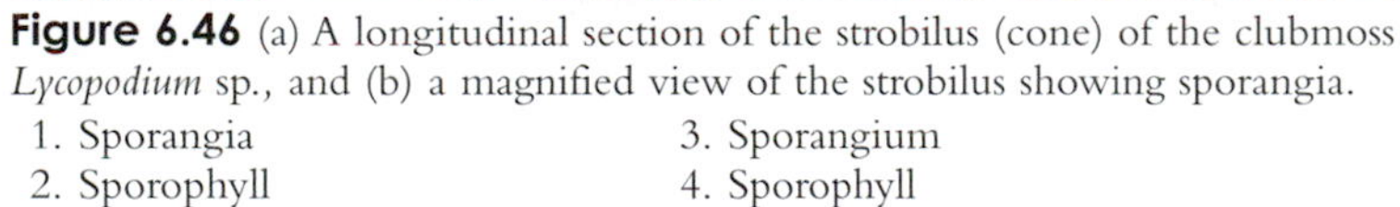

Figure 6.46 (a) A longitudinal section of the strobilus (cone) of the clubmoss *Lycopodium* sp., and (b) a magnified view of the strobilus showing sporangia.
1. Sporangia
2. Sporophyll
3. Sporangium
4. Sporophyll

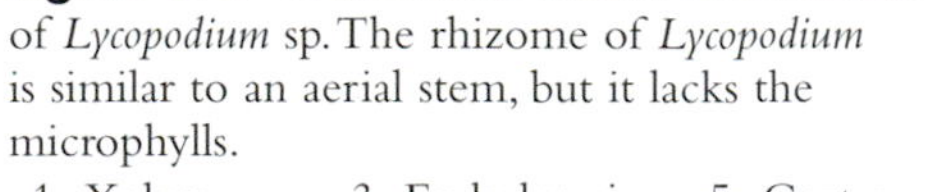

Figure 6.47 A transverse section of a rhizome of *Lycopodium* sp. The rhizome of *Lycopodium* is similar to an aerial stem, but it lacks the microphylls.
1. Xylem
2. Epidermis
3. Endodermis
4. Phloem
5. Cortex

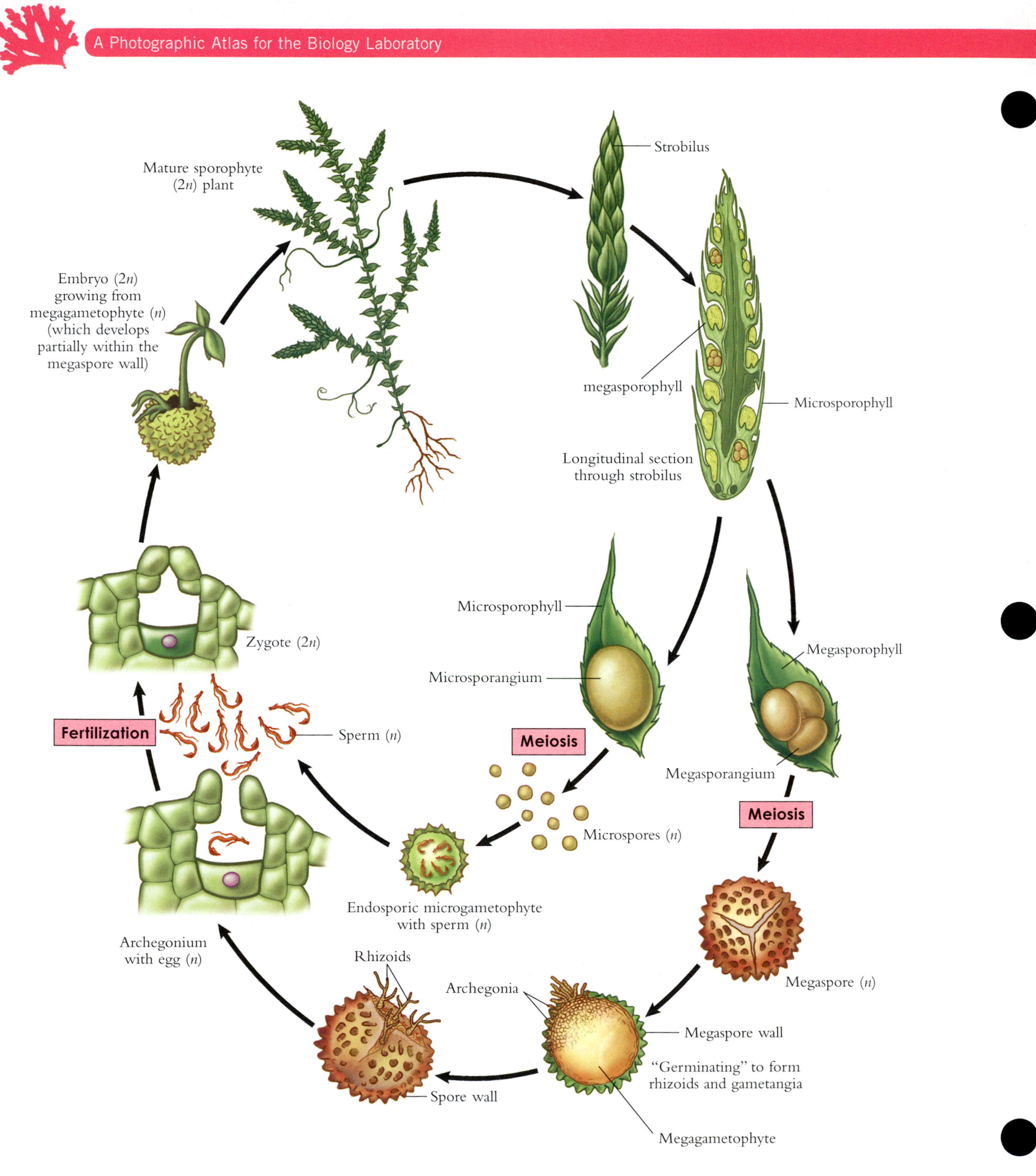

Figure 6.48 The life cycle of *Selaginella* sp., which is heterosporous.

Figure 6.49 The spike moss, *Selaginella kraussiana* (a) growth habit and (b) strobili (cones).
1. Strobili (cones)
2. Sporophyll with sporangium

Figure 6.50 The spike moss, *Selaginella pulcherrima*.

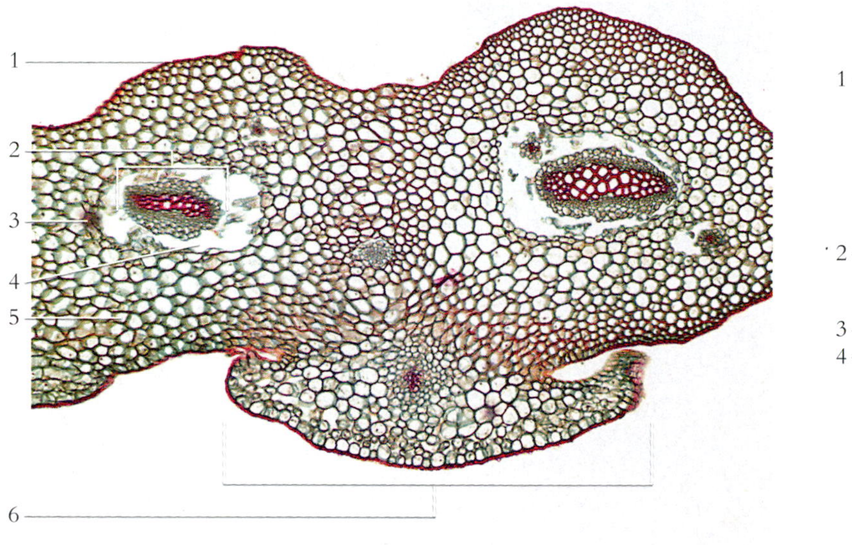

Figure 6.51 A transverse section through stem of *Selaginella* sp. immediately above dichotomous branching.
1. Epidermis
2. Protostele (surrounded by endodermis)
3. Root trace
4. Air cavity
5. Cortex
6. Leaf base

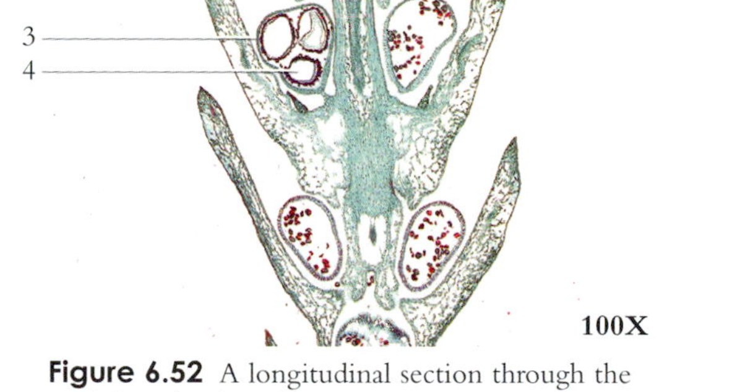

Figure 6.52 A longitudinal section through the strobilus of *Selaginella* sp.
1. Ligule
2. Megasporophyll
3. Megasporangium
4. Megaspore
5. Microsporophyll
6. Microsporangium
7. Microspore
8. Cone axis

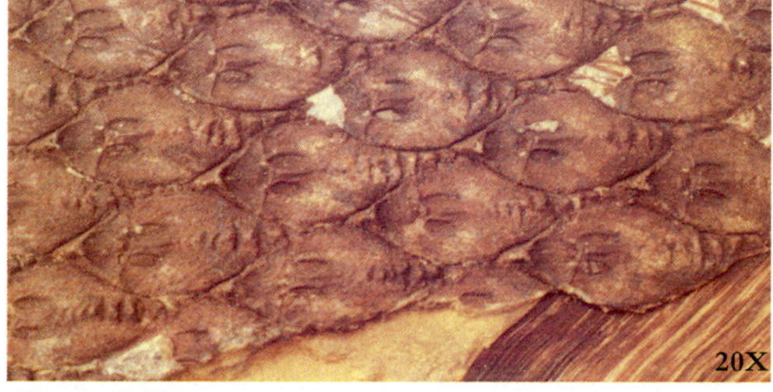

Figure 6.53 A longitudinal view of the surface of the fossil lycophyte *Lepidodendron* sp., a common lycopod from perhaps 300 million years ago.

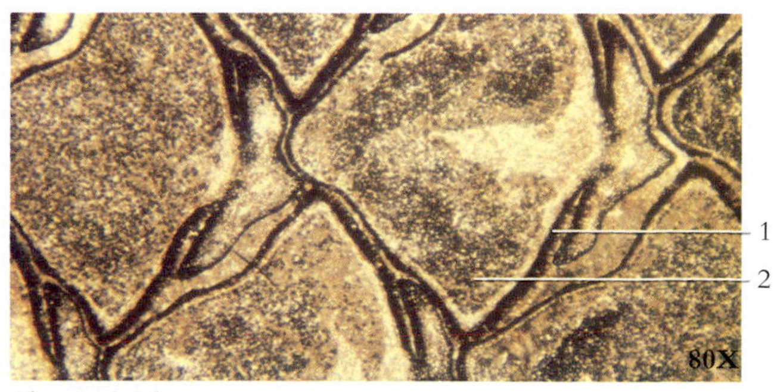

Figure 6.54 A longitudinal section through a fossil strobilus of the lycophyte *Lepidostrobus* sp., from approximately 300 million years ago.
1. Sporangium
2. Sporogenous tissue

Phylum Psilotophyta (= Psilophyta) - whisk ferns

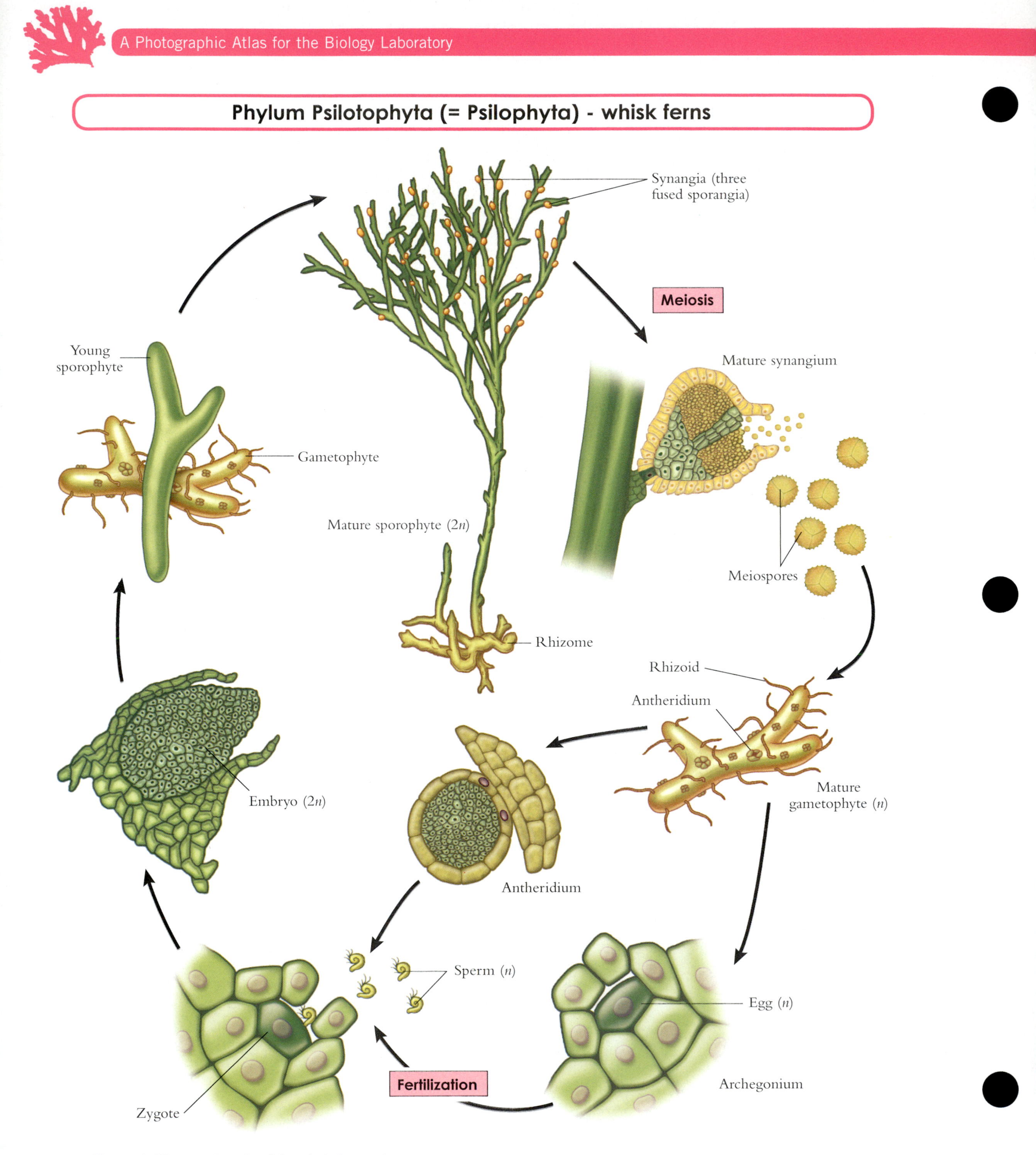

Figure 6.55 The life cycle of the whisk fern, *Psilotum* sp.

Figure 6.56 A *Tmesipteris* sp., growing as an epiphyte on a tree fern in Australia.

Figure 6.57 A whisk fern, *Psilotum nudum,* is a simple vascular plant lacking true leaves and roots.

Figure 6.58 The branches (axes) of *Psilotum nudum* (scale in mm).
1. Aerial axis
2. Rhizome

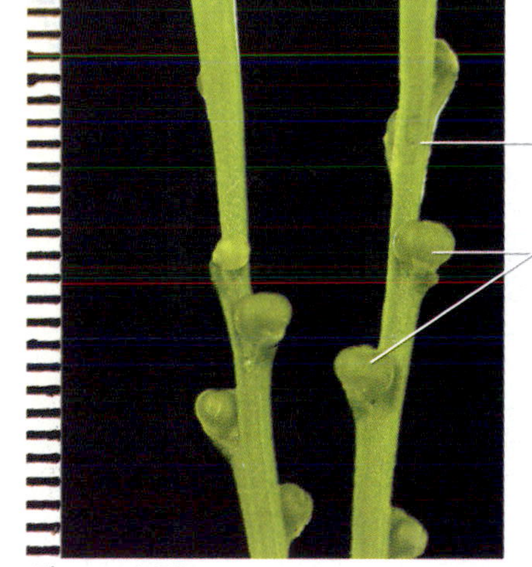

Figure 6.59 A sporophyte of the whisk fern, *Psilotum nudum.* The axes of the sporophyte support sporangia (synangia), which produce spores (scale in mm).
1. Branch (axis)
2. Sporangia (synangia)

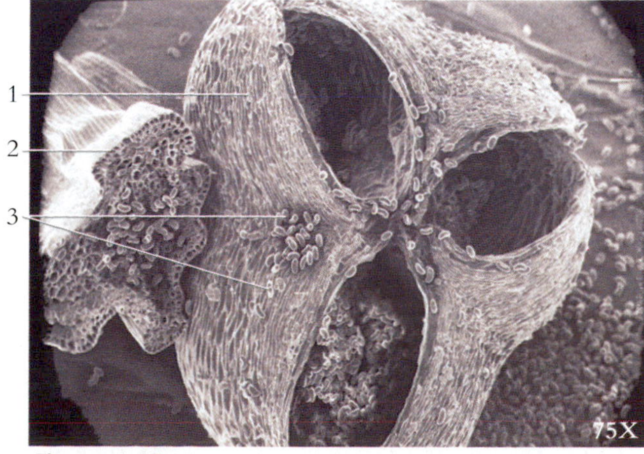

Figure 6.60 A scanning electron micrograph of a ruptured synangium (three fused sporangia) of *Psilotum* sp. that is spilling spores.
1. Sporangium (often called synangia)
2. Axis
3. Spores

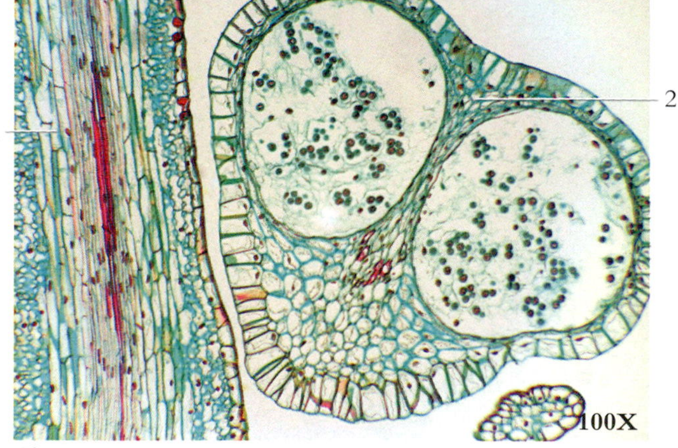

Figure 6.61 A longitudinal section through a stem and sporangium (synangium) of *Psilotum* sp.
1. Axis
2. Sporangia (synangium)

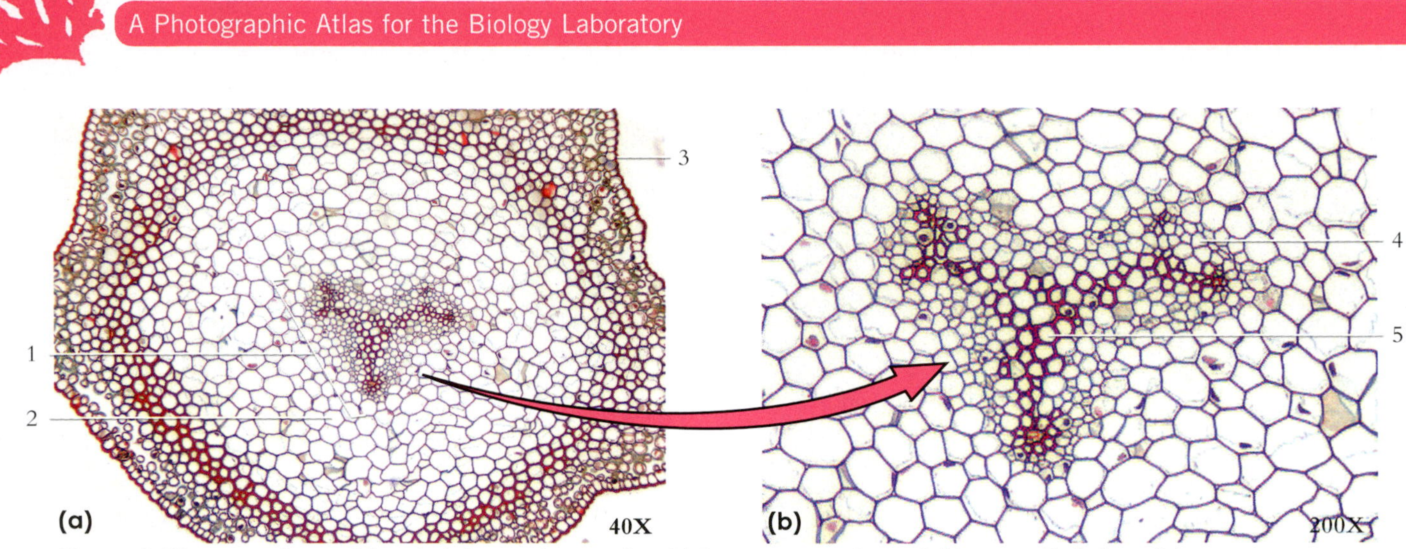

Figure 6.62 An aerial axis of the whisk fern, *Psilotum nudum*. (a) A transverse section and (b) a magnified view of the vascular cylinder (stele).

1. Stele 2. Cortex 3. Epidermis 4. Phloem 5. Xylem

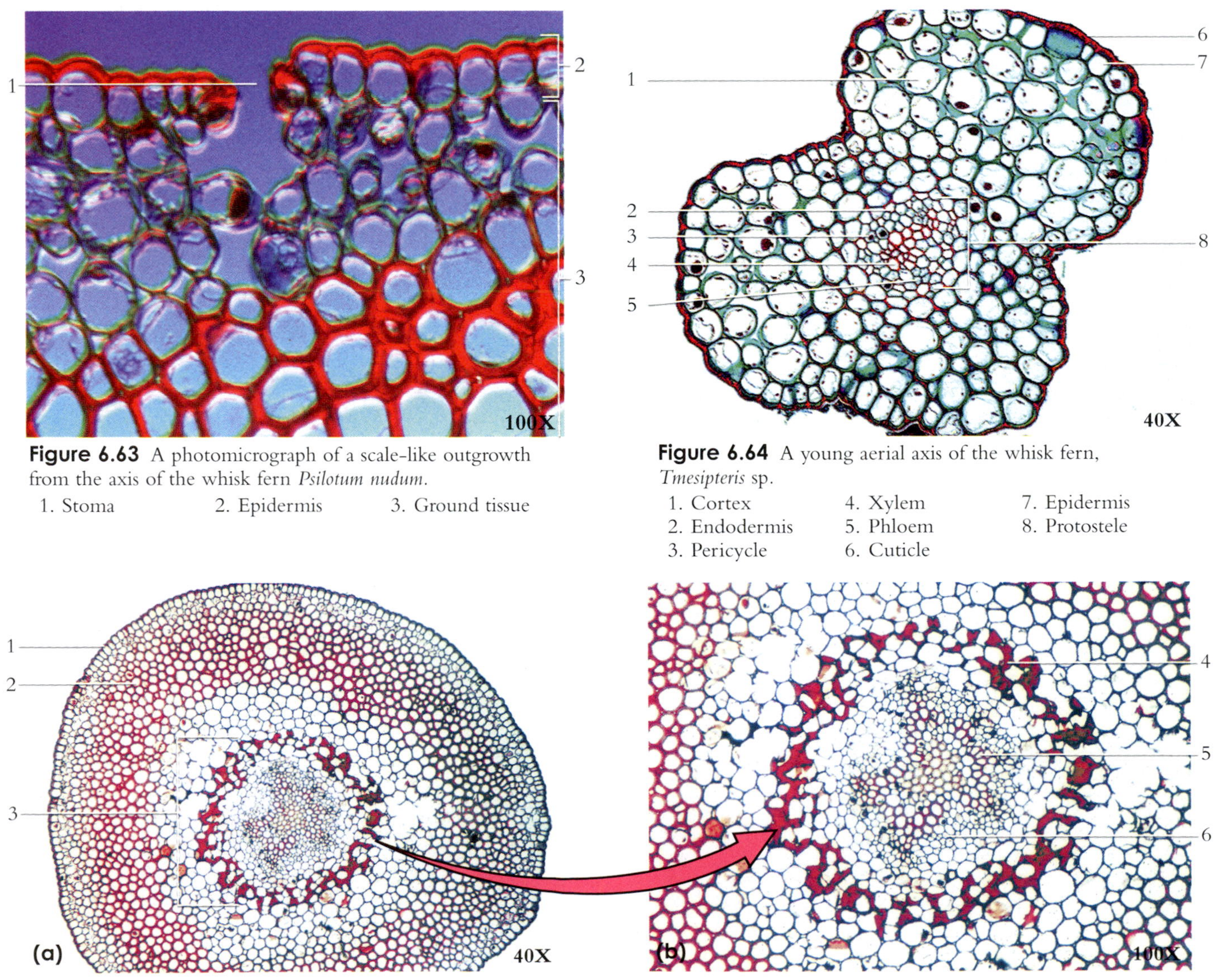

Figure 6.63 A photomicrograph of a scale-like outgrowth from the axis of the whisk fern *Psilotum nudum*.

1. Stoma 2. Epidermis 3. Ground tissue

Figure 6.64 A young aerial axis of the whisk fern, *Tmesipteris* sp.

1. Cortex
2. Endodermis
3. Pericycle
4. Xylem
5. Phloem
6. Cuticle
7. Epidermis
8. Protostele

Figure 6.65 An older aerial axis of the whisk fern, *Tmesipteris* sp. The genus *Tmesipteris* is restricted to distribution in Australia, New Zealand, New Caledonia, and other South Pacific islands. (a) Axis arising from the aerial axis and (b) a magnified view of the stele.

1. Epidermis 2. Cortex 3. Stele 4. Endodermis 5. Xylem 6. Phloem

Phylum Sphenophyta (= Equisetophyta) - horsetails

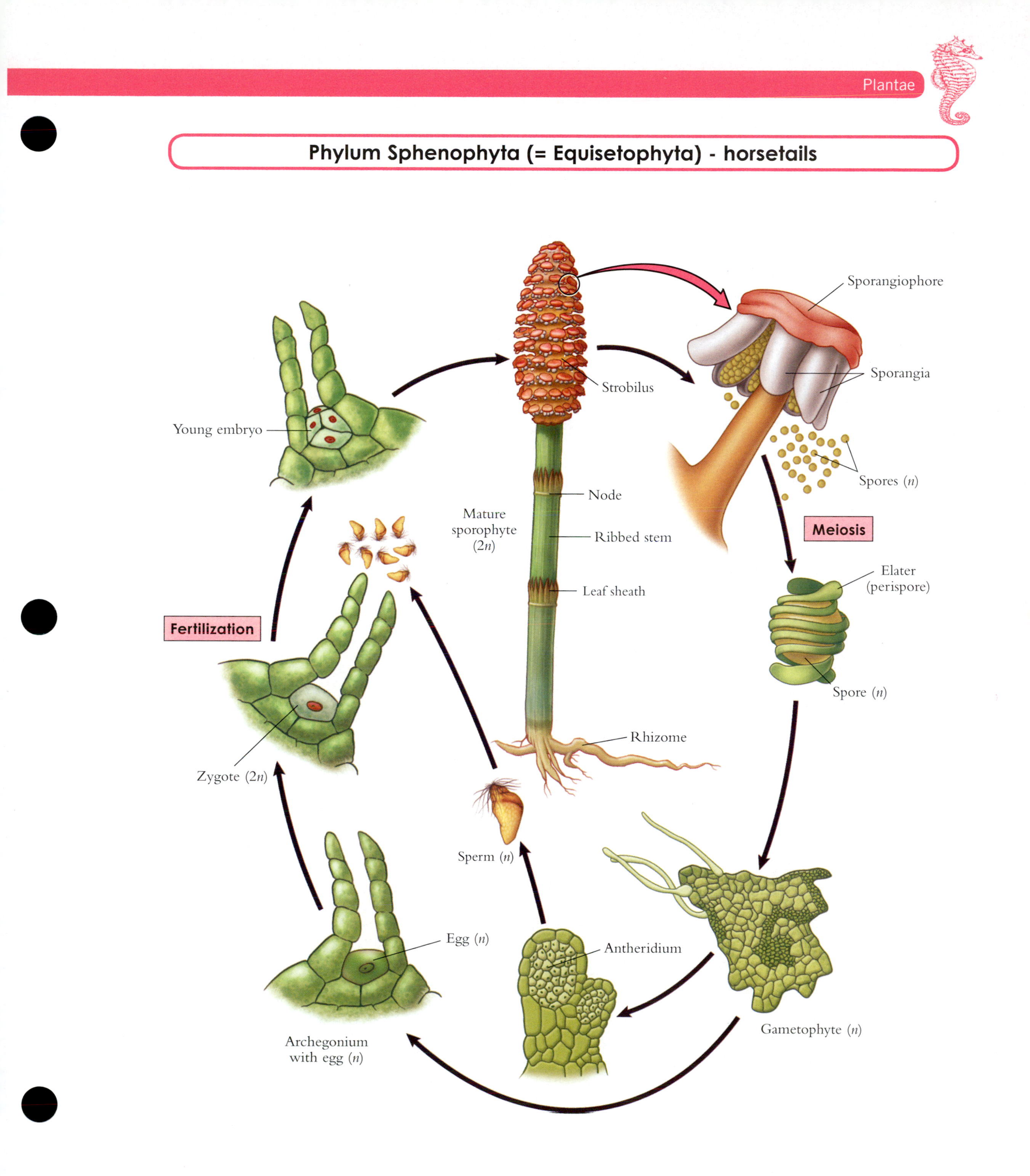

Figure 6.66 The life cycle of the horsetail, *Equisetum* sp.

Figure 6.67 An *Equisetum telmateia* showing lateral branching.

Figure 6.68 A close-up of *Equisetum telmateia* showing lateral branches growing through leaf sheath.

Figure 6.69 The stems of *Equisetum* sp. without lateral branching and showing a prominent leaf sheath at the node.
1. Stem
2. Leaf sheath

Figure 6.70 The horsetail, *Equisetum* sp. Numerous species of Equisetophyta were abundant throughout tropical regions during the Paleozoic Era, some 300 million years ago. Currently, Equisetophyta are represented by this single genus. The meadow horsetail, *Equisetum* sp., showing (a) an immature strobilus, (b) mature strobilus, shedding spores, (c) an open strobilus, and (d) a sporangiophore with its spores released.
1. Sporangiophores
2. Separated sporangiophores revealing sporangia
3. Sporangiophores after spores are shed
4. Open sporangia with spores shed

1
2
20X

Figure 6.71 A young gametophyte of *Equisetum* sp.
1. Rhizoids
2. Antheridium

1
2
400X

Figure 6.72 A longitudinal section of *Equisetum* sp. shoot apex.
1. Apical cell
2. Leaf primordium

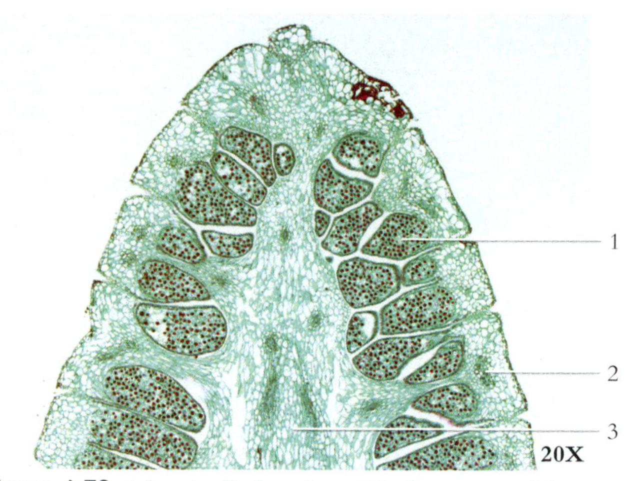

Figure 6.73 A longitudinal section of *Equisetum* sp. strobilus.
1. Sporangium 2. Sporangiophore 3. Strobilus axis

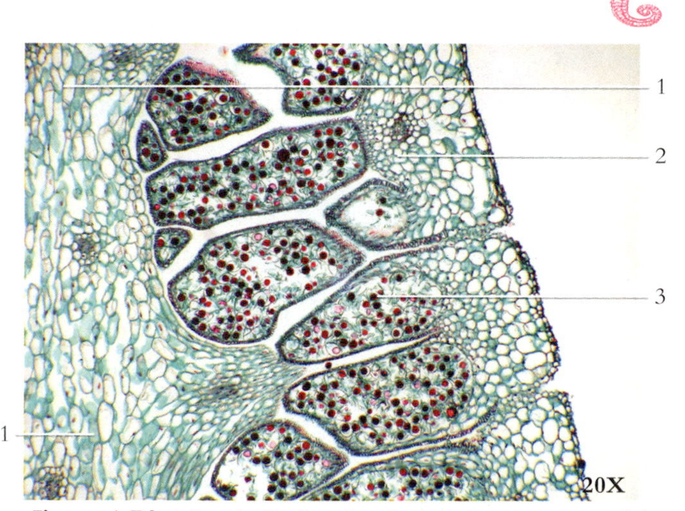

Figure 6.74 A longitudinal section through *Equisetum* sp. strobilus.
1. Axis of the strobilus 2. Sporangiophore 3. Sporangium

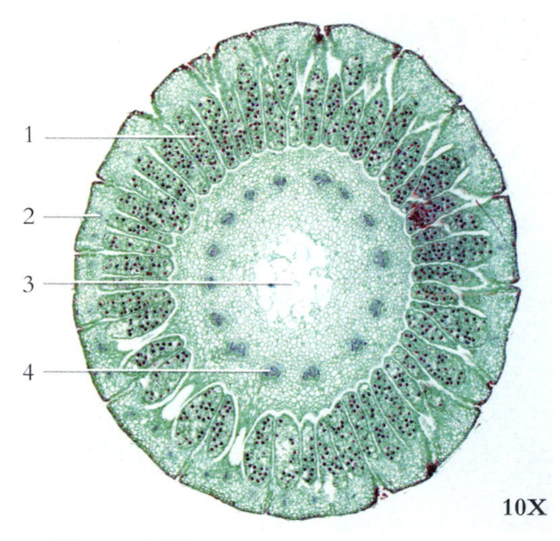

Figure 6.75 A transverse section of the strobilus of *Equisetum* sp.
1. Sporangium
2. Sporangiophore
3. Strobilus axis
4. Vascular bundle

Figure 6.76 A transverse section of the stem of *Equisetum* sp. just above a node.
1. Leaf sheath 2. Main stem 3. Branch

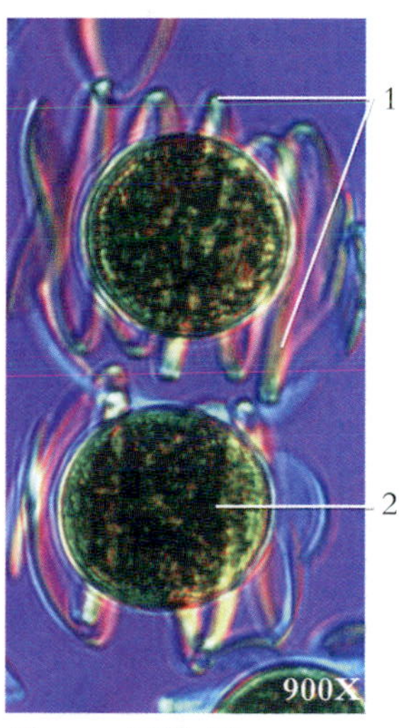

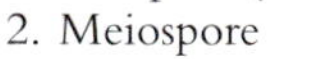

Figure 6.77 The meiospores of *Equisetum* sp.
1. Perispore (elater)
2. Meiospore

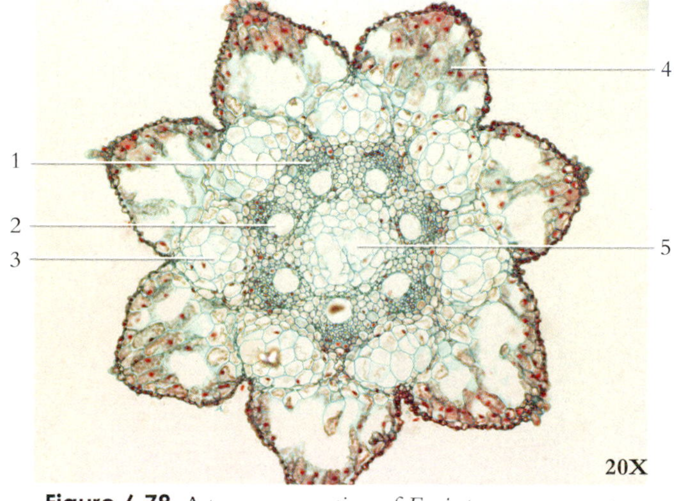

Figure 6.78 A transverse section of *Equisetum* sp. young stem.
1. Vascular tissue 2. Air canal 3. Future air canal 4. Cortex 5. Pith

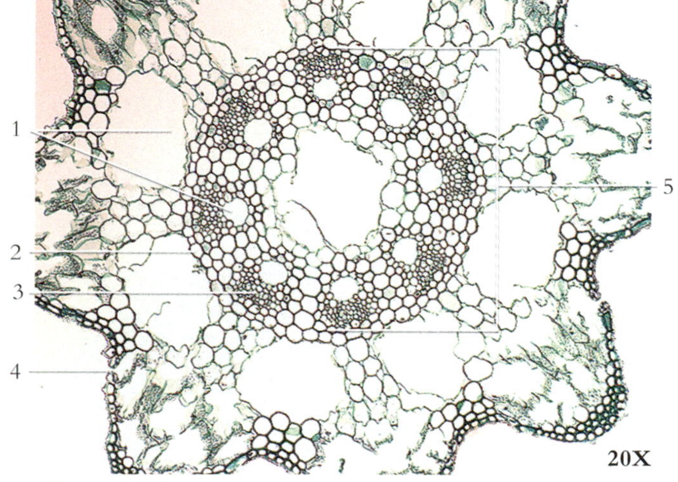

Figure 6.79 A transverse section of *Equisetum* sp. older stem.
1. Air canals 2. Endodermis 3. Vascular tissue 4. Stomate 5. Eustele

Phylum Pteridophyta (= Polypodiophyta) - ferns

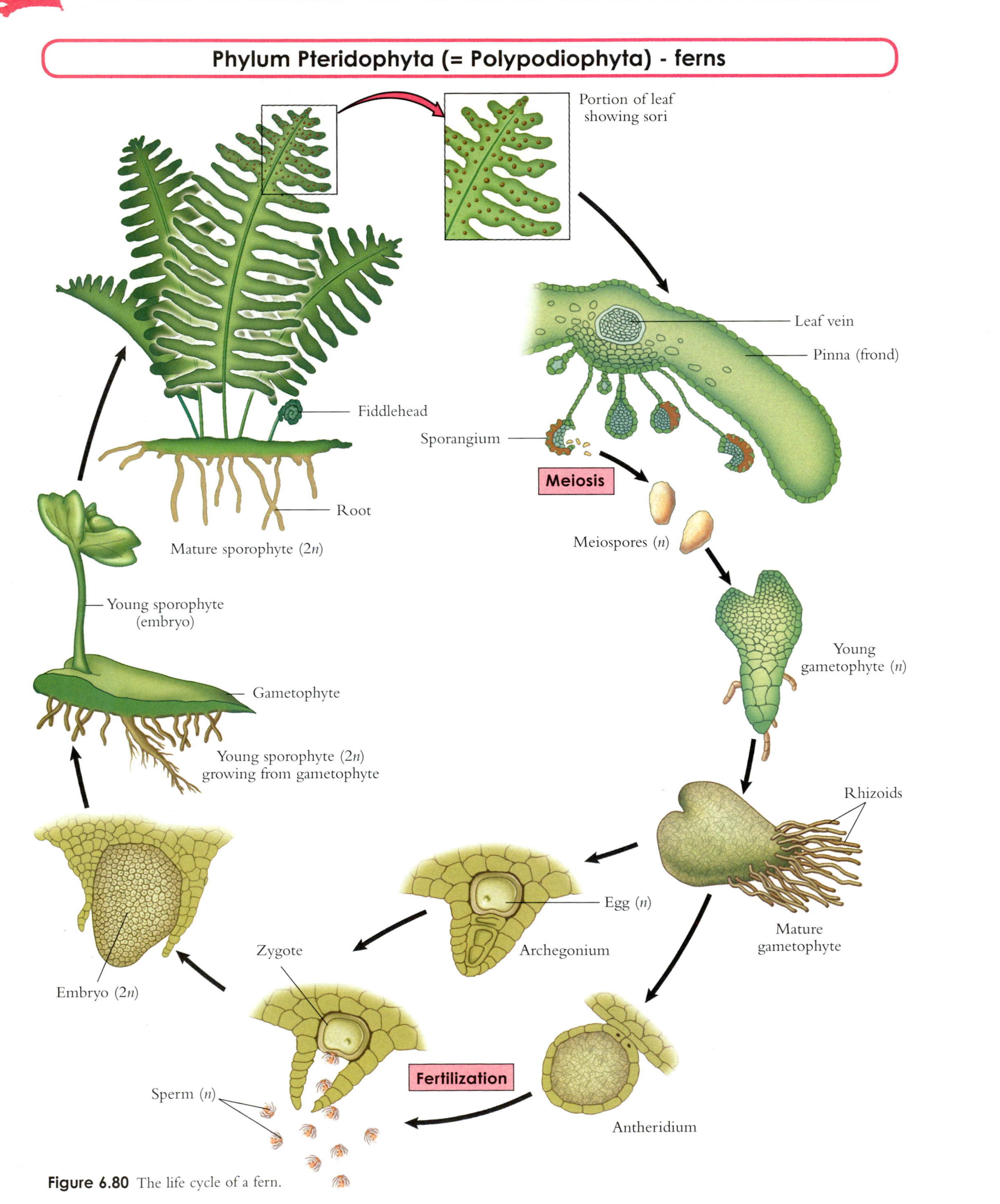

Figure 6.80 The life cycle of a fern.

Figure 6.81 The water fern, *Azolla* sp., is a floating freshwater plant found throughout Europe and the United States.

Figure 6.82 A view of a new (a) compound and (b) simple fern leaf showing circinate vernation forming a fiddlehead.

Figure 6.83 The fronds of the staghorn fern, *Platycerium alcicorne*.

Figure 6.84 A pinnate leaf showing pinnate venation in the leaflets of a fern.

1. Leaf
2. Pinnae
3. Venation

Figure 6.85 A leaf of the fern *Phanerophlebia* sp., or holly fern.

Figure 6.86 A leaf of the fern *Phanerophlebia* sp., showing sori (groups of sporangia).

1. Pinna
2. Sori

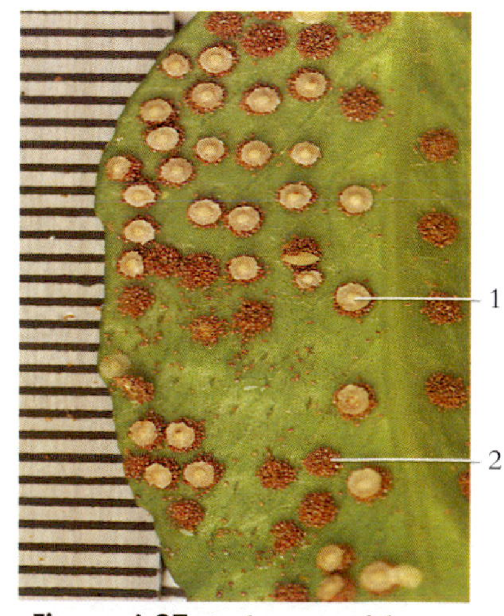

Figure 6.87 A close-up of the fern leaf of *Phanerophlebia* sp. (scale in mm).

1. Sorus with indusium
2. Sorus with indusium shed

Figure 6.88 The leaf of the fern *Polypodium virginianum*.

Figure 6.89 The leaf of the fern *Polypodium virginianum*, showing sori (groups of sporangia).
1. Pinna 2. Sori

Figure 6.90 A close-up of the fern pinna of *Polypodium virginianum* (scale in mm).
1. Sorus

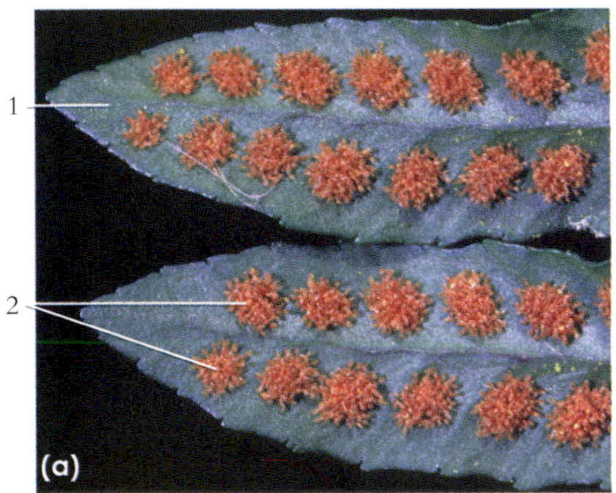

Figure 6.91 The fern *Polypodium* sp. (a) Sori on the undersurface of the pinnae, and (b) a scanning electron micrograph of a sorus.
1. Pinna 2. Sori 3. Annulus 4. Sporangium

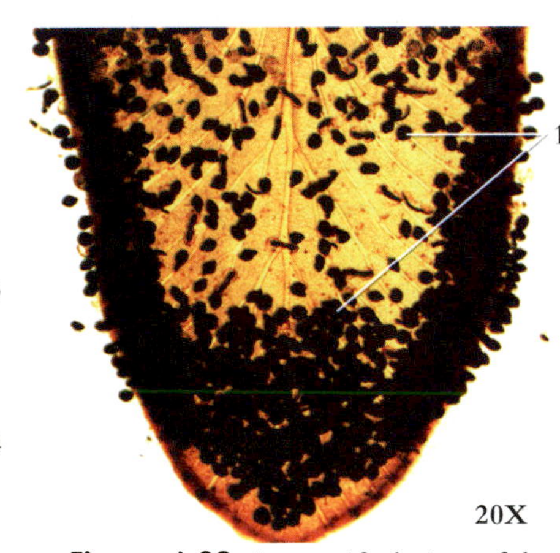

Figure 6.92 A magnified view of the fern pinna of *Pteridium* sp. showing numerous scattered sporangia.
1. Sporangia

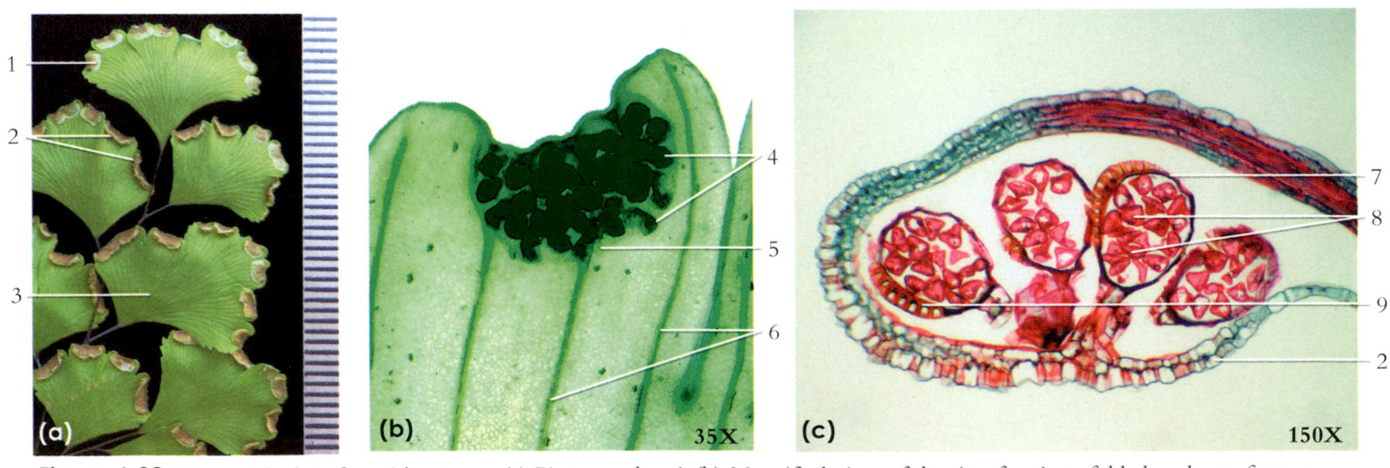

Figure 6.93 The maidenhair fern *Adiantum* sp. (a) Pinnae and sori. (b) Magnified view of the tip of a pinna folded under to form a false indusium that encloses the sorus. (c) Sorus with sporangia containing spores (scale in mm).
1. False indusium 2. Sori 3. Pinna 4. Sporangia with spores 5. False indusium enclosing a sorus 6. Vascular tissue (veins) of the pinna 7. Sporangium 8. Spores 9. Annulus

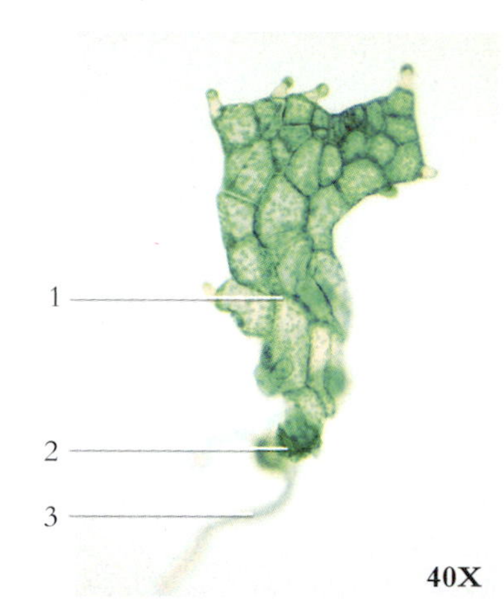

Figure 6.94 A young fern gametophyte.
1. Gametophyte
2. Spore cell wall
3. Rhizoid

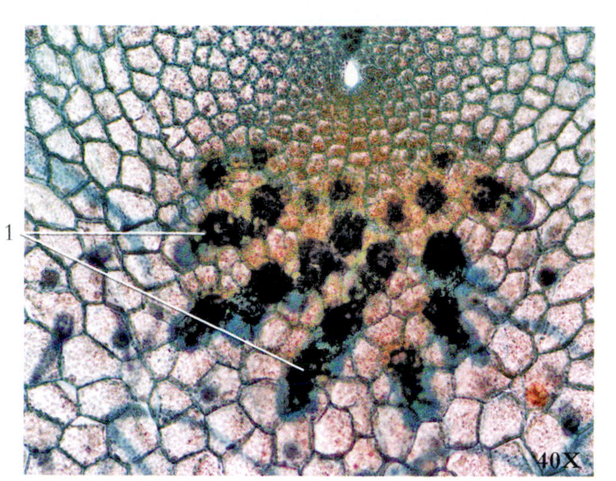

Figure 6.95 A fern gametophyte with archegonia.
1. Archegonia

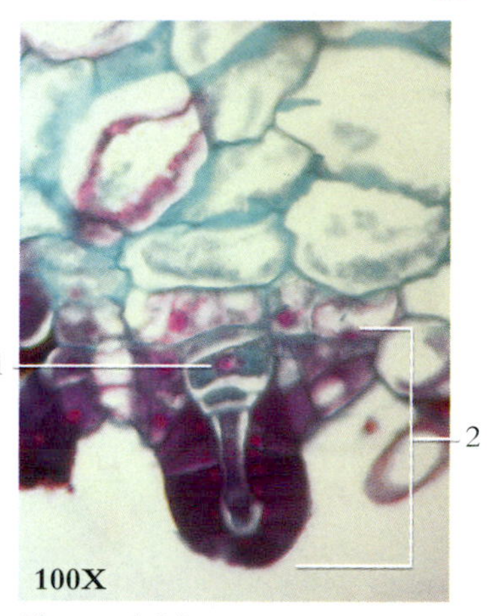

Figure 6.96 A fern gametophyte showing archegonium.
1. Egg
2. Archegonium

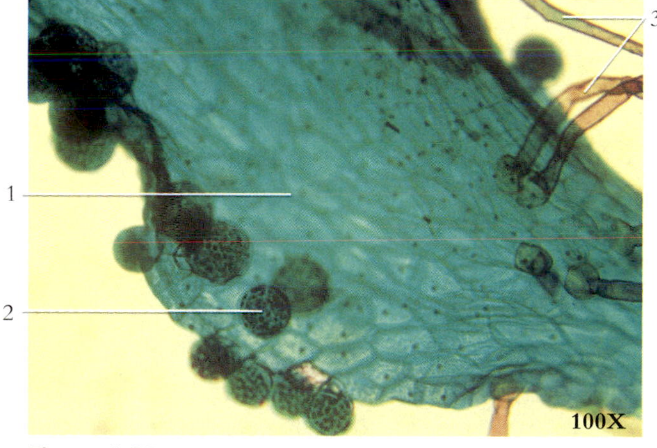

Figure 6.97 A fern gametophyte showing antheridia.
1. Gametophyte (prothallus)
2. Antheridium with sperm
3. Rhizoids

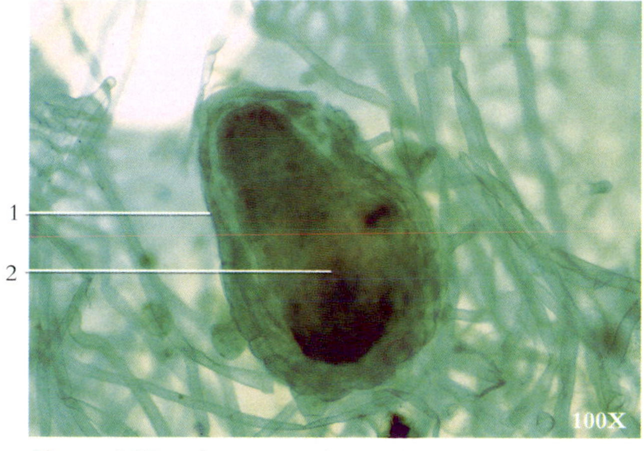

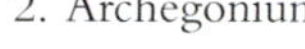

Figure 6.98 A fern gametophyte with a young sporophyte attached.
1. Expanded archegonium
2. Young sporophyte

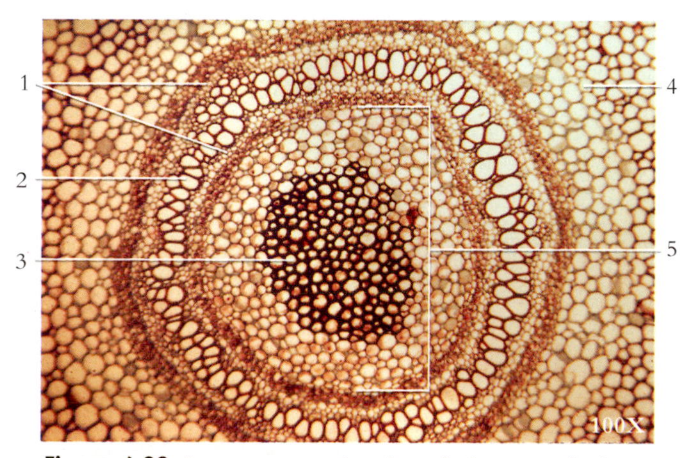

Figure 6.99 A transverse section through the stem of a fern, *Dicksonia* sp. showing a siphonostele.
1. Phloem
2. Xylem
3. Sclerified pith
4. Cortex
5. Pith

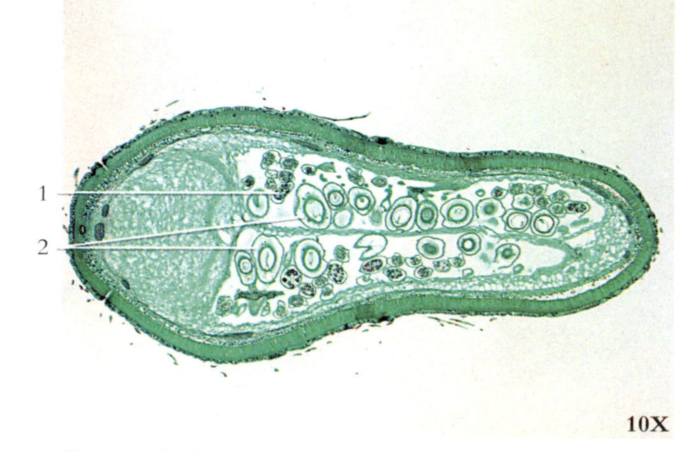

Figure 6.100 A transverse section of a sporocarp of the water fern, *Marsilea* sp., which is one of the two living orders of heterosporous ferns.
1. Microsporangium with microspores
2. Megasporangia with megaspores

Phylum Cycadophyta - cycads

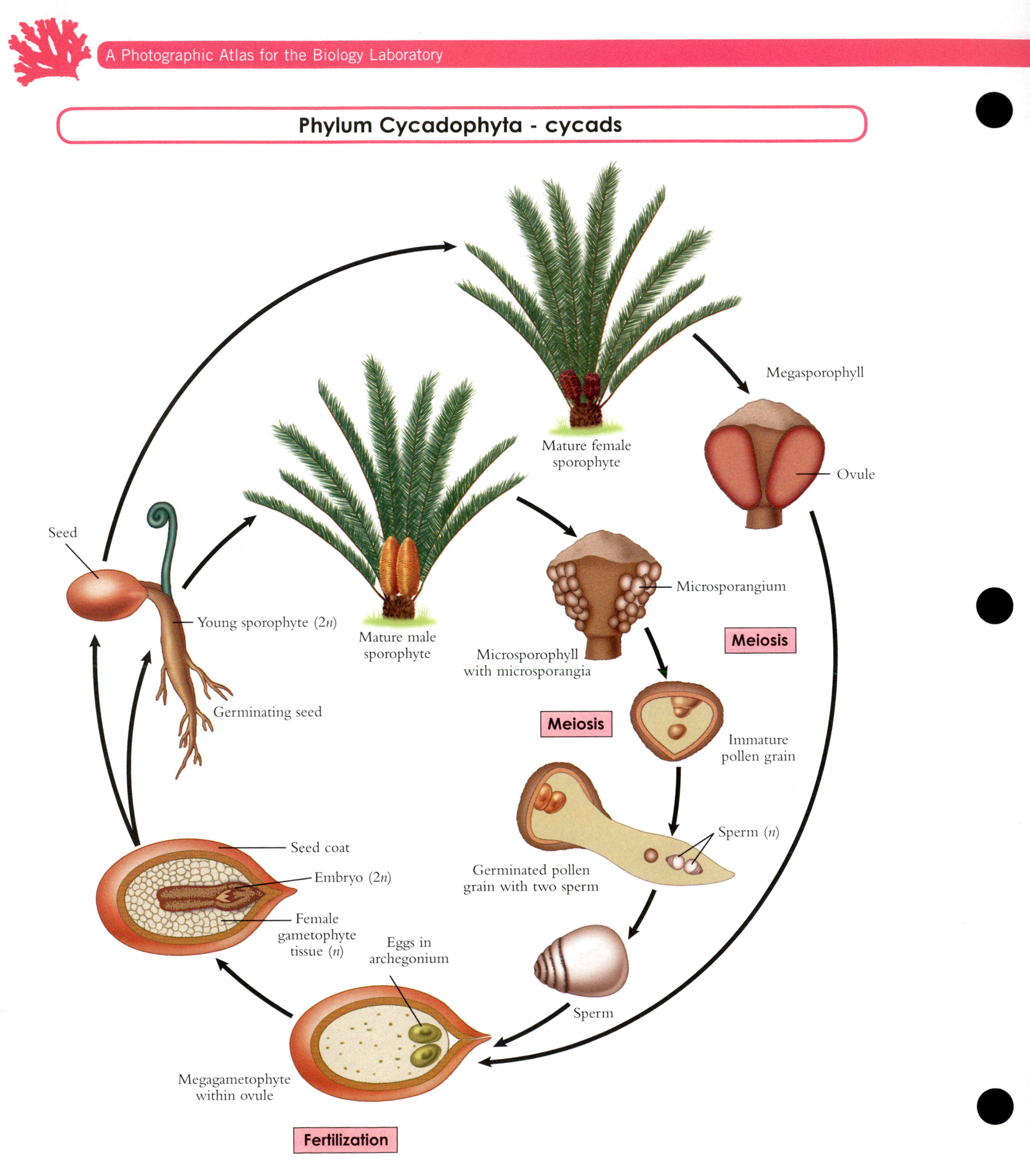

Figure 6.101 The life cycle of a cycad.

Figure 6.102 A *Cycas revoluta*. Cycads were abundant during the Mesozoic Era. Currently, there are 10 living genera, with about 100 species, that are found mainly in tropical and subtropical areas. The trunk of many cycads is densely covered with petioles of shed leaves.

Figure 6.103 A *Cycas revoluta* showing a female cone.
1. Cone

Figure 6.104 A *Cycas revoluta* showing a close-up view of a female cone with developing seeds.
1. Seeds 2. Megasporophyll

Figure 6.105 A *Cycas revoluta* showing a close-up view of a female cone during seed dispersal.
1. Seeds

Figure 6.106 A male cone of *Cycas revoluta*.
1. Cone

Figure 6.107 A male cone of *Cycas revoluta* after release of pollen.

Figure 6.108 A young plant of the cycad *Zamia pumila*. Found in Florida, this cycad is the only species native to the United States. The rootstocks and stems of this plant were an important source of food for some Native Americans.

Figure 6.109 Microsporangiate cones of the cycad *Zamia* sp.

Figure 6.110 The *Encephalartos villosus* is a nonthreatened species of cycad native to southeastern Africa.

Figure 6.111 A maturing female cone of *Encephalartos villosus*.

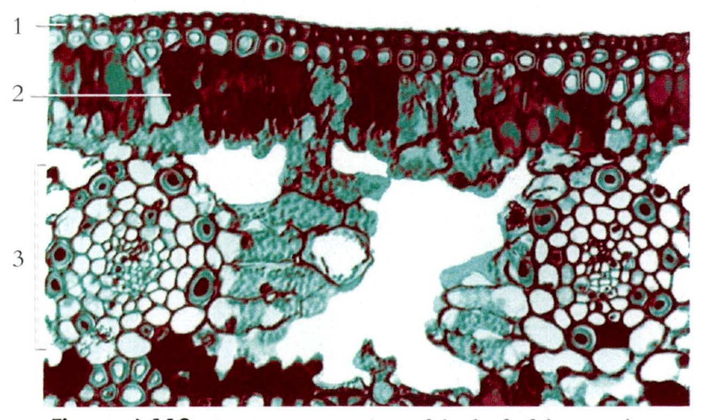

Figure 6.112 A transverse section of the leaf of the cycad *Zamia* sp.

1. Upper epidermis
2. Palisade mesophyll
3. Vascular bundle (vein)

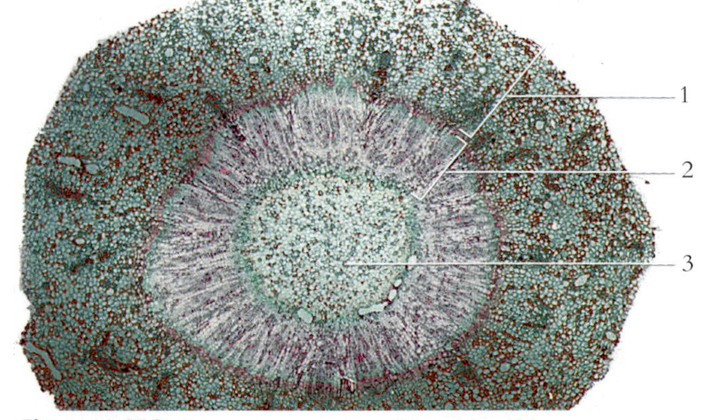

Figure 6.113 A transverse section of the stem of the cycad *Zamia* sp.

1. Cortex
2. Vascular tissue
3. Pith

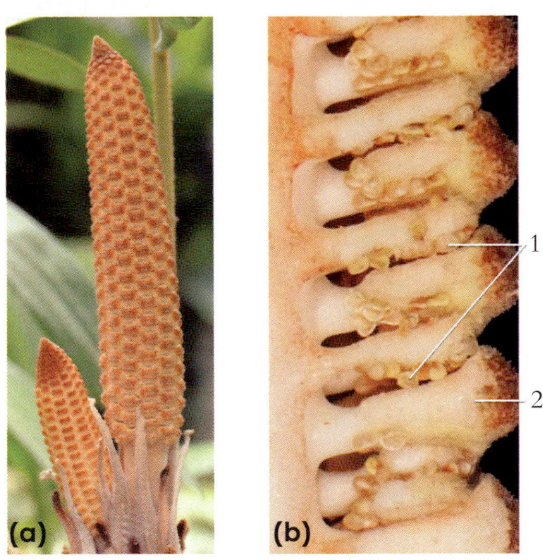

Figure 6.114 A microsporangiate cone of the cycad, *Zamia* sp. The cone on the right (b) is longitudinally sectioned.
1. Microsporangia 2. Microsporophyll

Figure 6.115 A microsporangiate cone of a cycad showing microsporangia on microsporophylls.
1. Microsporangium 2. Microsporophyll

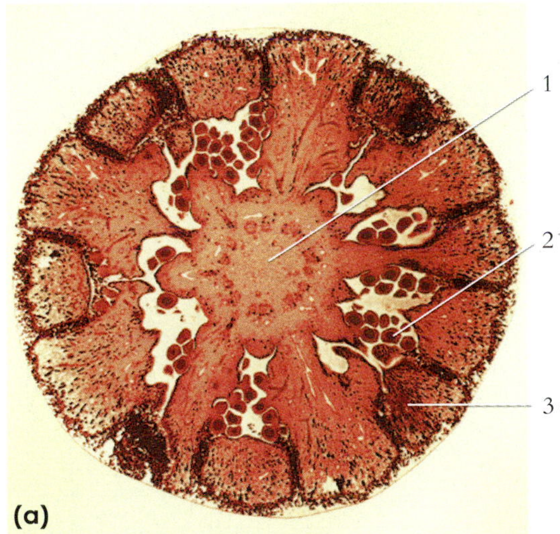

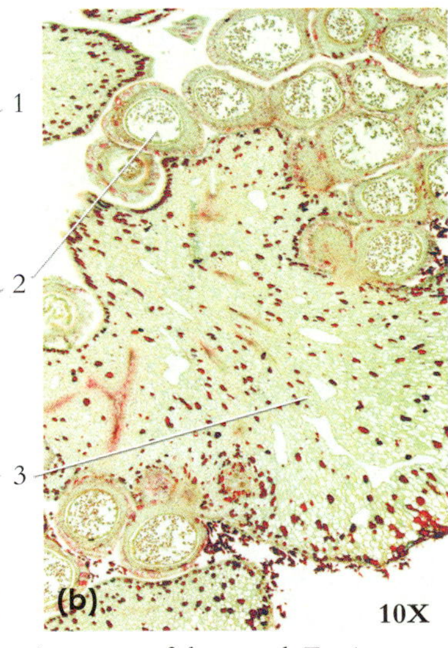

Figure 6.116 A transverse section of a microsporangiate cone of the cycad *Zamia* sp. (a) A low magnification, and (b) a magnified view.
1. Cone axis 2. Microsporangia 3. Microsporophyll

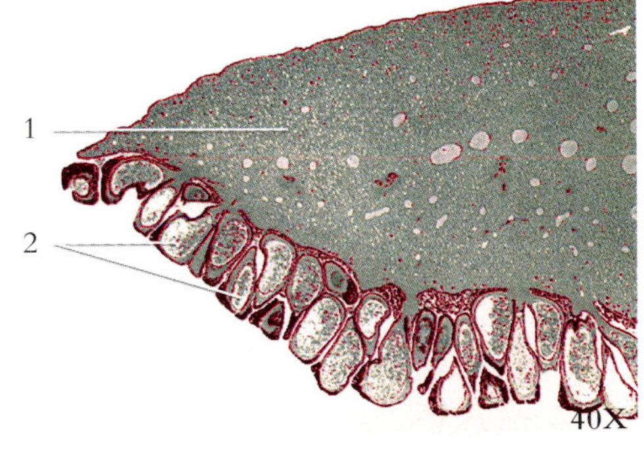

Figure 6.117 A longitudinal section of a microsporophyll of the cycad *Cycas* sp. Note that the microsporangia develop on the undersurface of the microsporophyll.
1. Microsporophyll
2. Microsporangia

Figure 6.118 A megasporangiate cone of *Cycas revoluta* showing ovules on leaflike megasporophylls near the time of pollination.

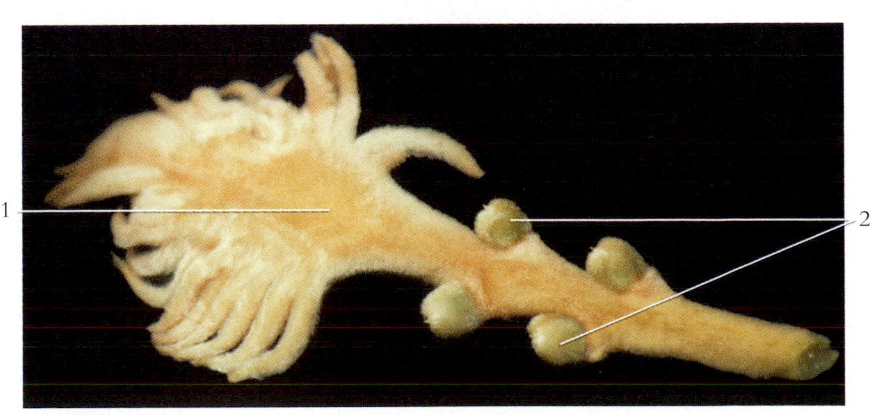

Figure 6.119 The megasporophyll and ovules of *Cycas revoluta*.
1. Megasporophyll 2. Ovules

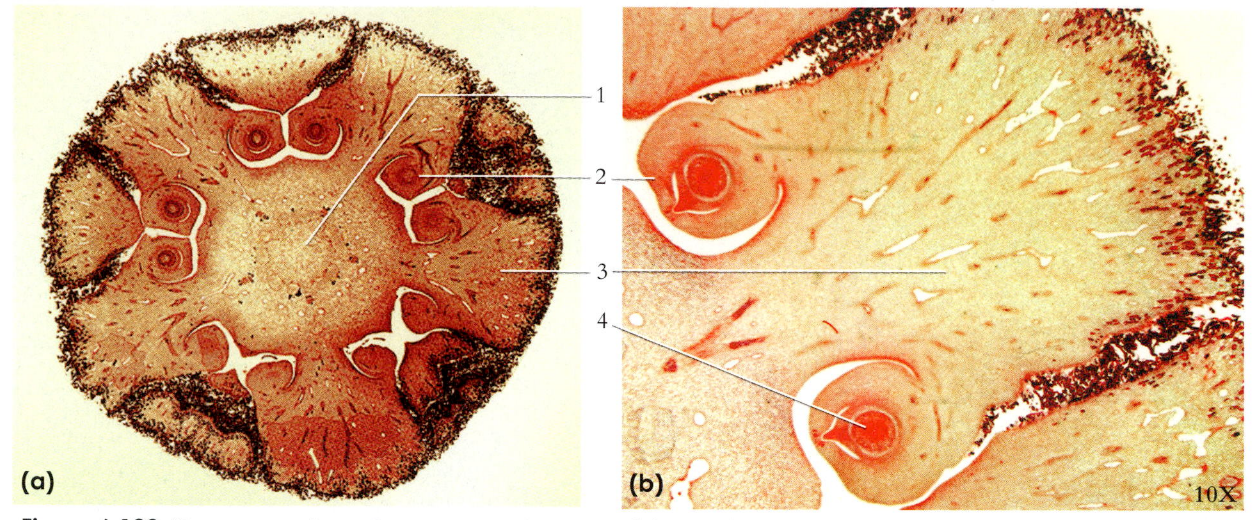

Figure 6.120 Transverse sections of a megasporangiate cone of the cycad *Zamia* sp. (a) A low magnification, and (b) a magnified view.
1. Cone axis
2. Ovule
3. Megasporophyll
4. Megasporocyte

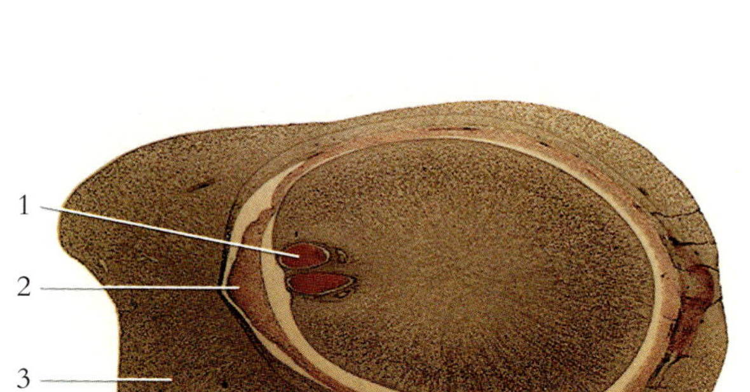

Figure 6.121 An ovule of the cycad *Zamia* sp. The ovule has two archegonia and is ready to be fertilized.
1. Archegonium
2. Megasporangium (nucellus)
3. Integument (will become seed coat)

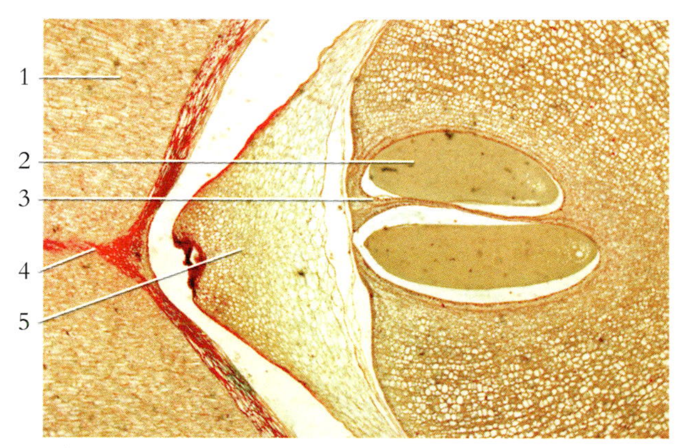

Figure 6.122 A magnified view of the ovule of the cycad *Zamia* sp. showing eggs in archegonia.
1. Integument
2. Egg
3. Archegonium
4. Micropyle area
5. Megasporangium

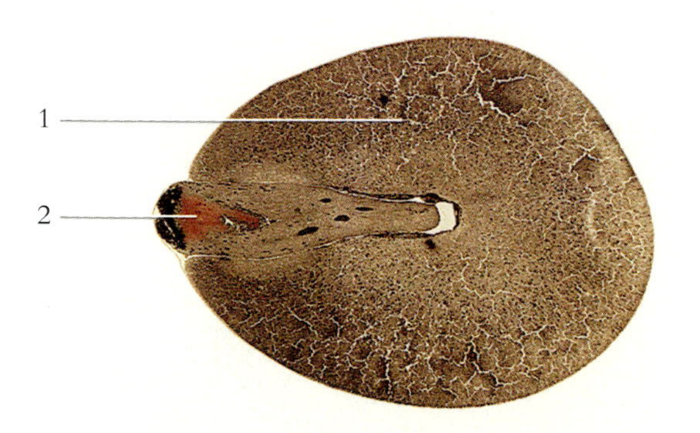

Figure 6.123 An ovule of the cycad *Zamia* sp. The ovule has been fertilized and contains an embryo. The seed coat has been removed from this specimen.
1. Female gametophyte
2. Embryo

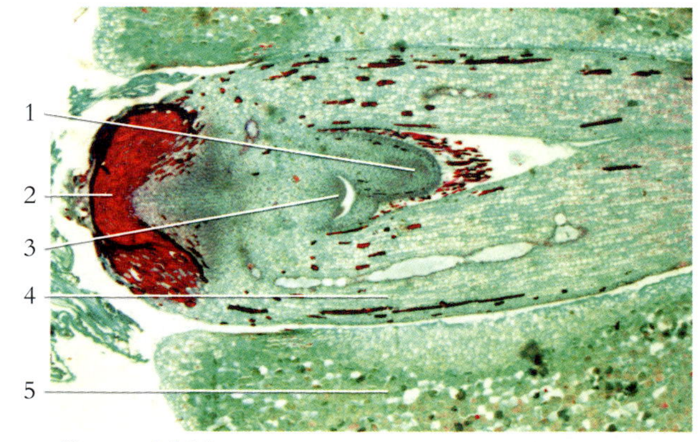

Figure 6.124 A magnified view of the ovule of the cycad *Zamia* sp. showing the embryo.
1. Leaf primordium
2. Root apex
3. Shoot apex
4. Cotyledon
5. Female gametophyte

Phylum Ginkgophyta - *Ginkgo*

Figure 6.125 The *Ginkgo biloba*, or maidenhair tree. Consisting of a central trunk with lateral branches, a mature *Ginkgo* grows to 100 feet tall. Native to China, *Ginkgo biloba* has been introduced into countries with temperate climates throughout the world as an interesting and hardy ornamental tree.

Figure 6.127 A fossil *Ginkgo biloba* leaf impression from Paleocene sediment. This specimen was found in Morton County, North Dakota.

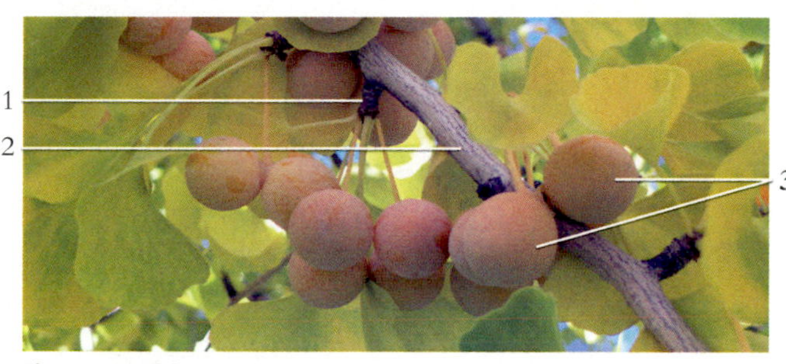

Figure 6.129 A branch of a *Ginkgo biloba* tree supporting a mature seed.

1. Short shoot (spur)
2. Long shoot
3. Mature seeds

Figure 6.126 A leaf from the *Ginkgo biloba* tree. The fan-shaped leaf is characteristic of this species.

Figure 6.128 As the sole member of the phylum Ginkgophyta, *Ginkgo biloba* is able to withstand air pollution. Ginkgos are often used as ornamental trees within city parks. *Ginkgo biloba* may have the longest genetic lineage among seed plants.

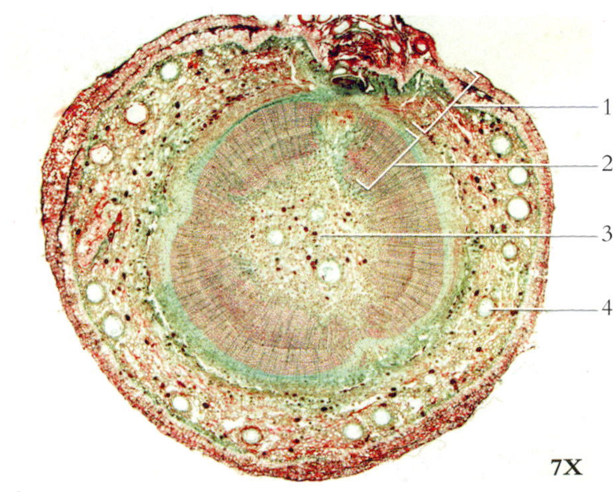

Figure 6.130 A transverse section of a short branch from *Ginkgo biloba*.

1. Cortex
2. Vascular tissue
3. Pith
4. Mucilage duct

Figure 6.131 The leaves and immature ovules on a short shoot of the ginkgo tree, *Ginkgo biloba*.
1. Leaf
2. Immature ovules
3. Short shoot
4. Long shoot

Figure 6.132 The pollen strobili of the ginkgo tree, *Ginkgo biloba*.
1. Leaf
2. Pollen strobilus
3. Long shoot

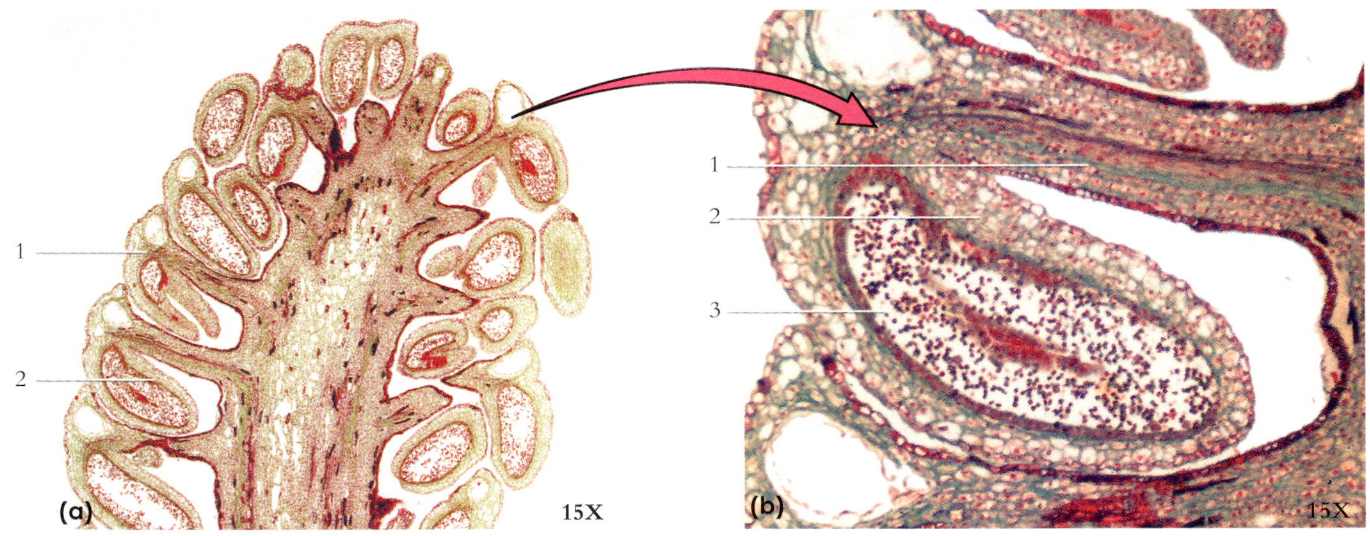

Figure 6.133 A microsporangiate strobilus of *Ginkgo biloba*. (a) A longitudinal section and (b) a magnified view showing a microsporangium.
1. Sporophyll
2. Microsporangium
3. Pollen

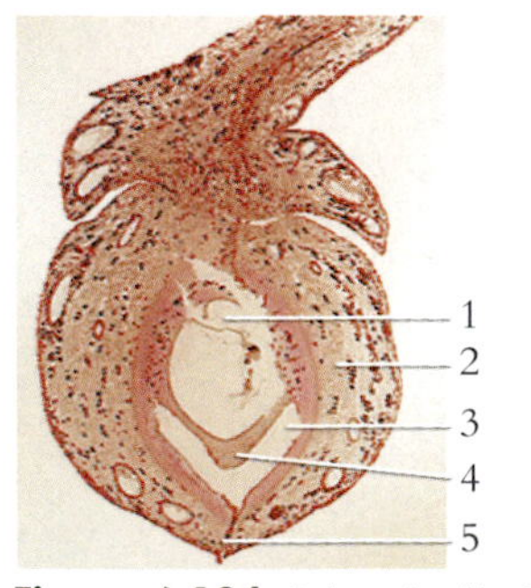

Figure 6.134 A longitudinal section of an ovule of *Ginkgo biloba* prior to fertilization.
1. Megagametophyte
2. Integument
3. Pollen chamber
4. Nucellus
5. Micropyle

Figure 6.135 Transverse and longitudinal sections through a living immature seed of *Ginkgo biloba* showing the green megagametophyte.
1. Fleshy layer of integument
2. Megagametophyte
3. Stony layer of integument

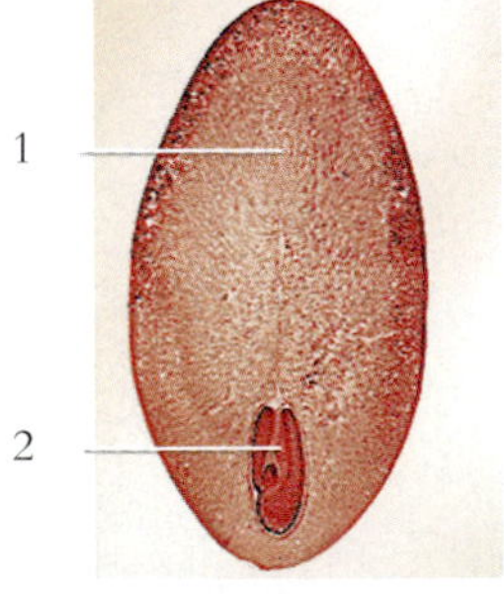

Figure 6.136 A longitudinal section of a seed of *Ginkgo biloba* with the seed coat removed.
1. Megagametophyte
2. Developing embryo

Figure 6.137 A magnified view of the ovule of *Ginkgo biloba* showing the embryo.
1. Leaf primordium
2. Shoot apex
3. Root apex
4. Megagametophyte

Phylum Pinophyta (= Coniferophyta) - conifers

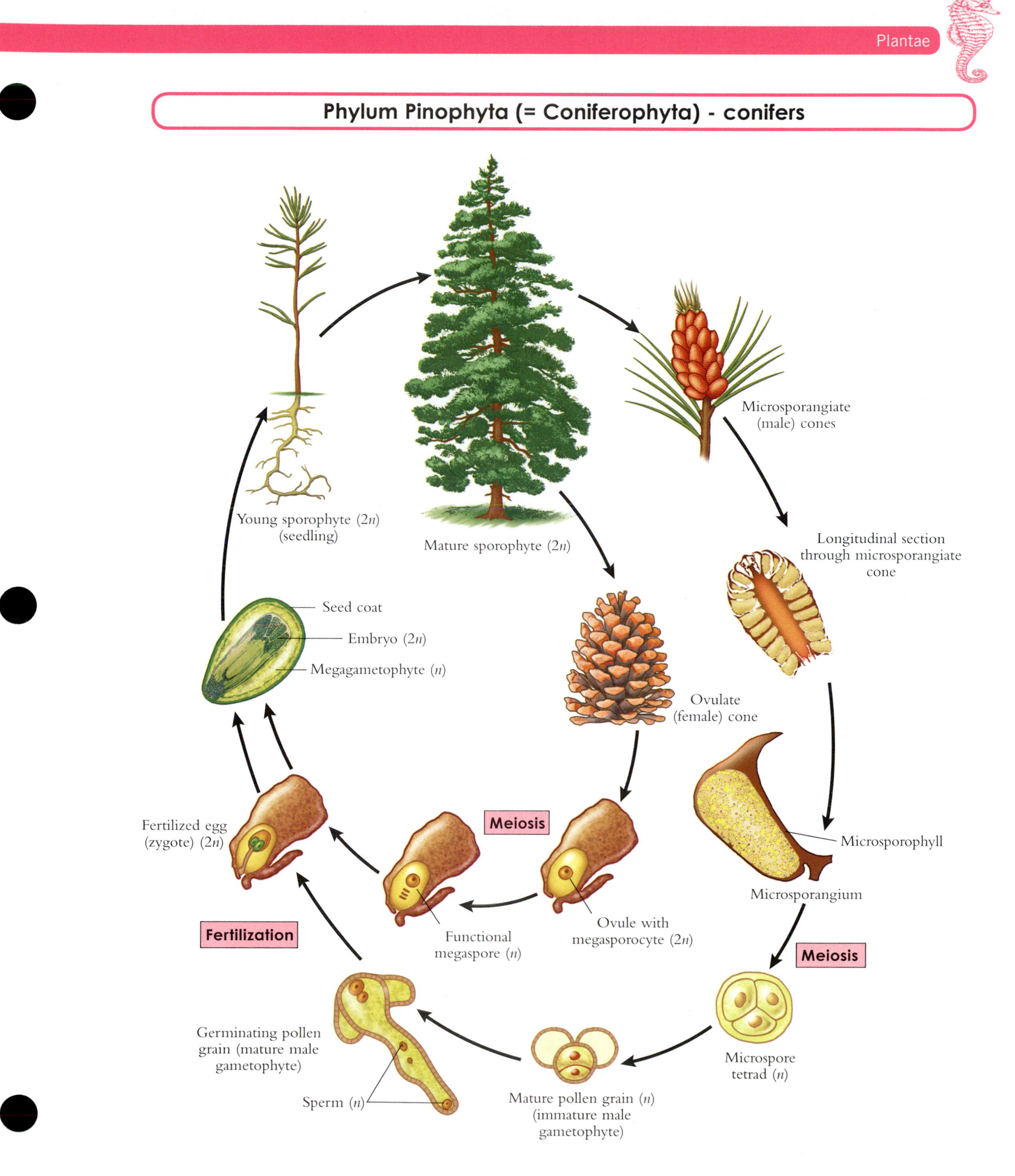

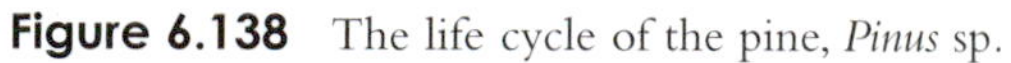

Figure 6.138 The life cycle of the pine, *Pinus* sp.

Figure 6.139 A diagram of the tissues in the stem (trunk) of a conifer. The periderm and dead secondary phloem (outer bark) protects the tree against water lost and the infestation of insects and fungi. The cells of the phloem (inner bark) compress and become nonfunctional after a relatively short period. The vascular cambium annually produces new phloem and xylem and accounts for the growth rings in the wood. The secondary xylem is a water-transporting layer of the stem and provides structural support to the tree.

1. Outer bark
2. Phloem
3. Vascular cambium
4. Secondary xylem

Figure 6.140 The stem (trunk) of a pine tree that was harvested in the year 2000 when the tree was 62 years old. The growth rings of a tree indicate environmental conditions that occurred during the tree's life.

1. 1939—A pine seedling.
2. 1944—Healthy, undisturbed growth indicated by broad and evenly spaced rings.
3. 1949—Growth disparity probably due to the falling of a dead tree onto the young healthy six-year-old tree. The wider "reaction rings" on the lower side help support the tree.
4. 1959—The tree is growing straight again, but the narrow rings indicate competition for sunlight and moisture from neighboring trees.
5. 1962—The surrounding trees are harvested, thus permitting rapid growth once again.
6. 1965—A burn scar from a fire that quickly scorched the forest.
7. 1977—Narrow growth rings resulting from a prolonged drought.
8. 1992—Narrow growth rings resulting from a sawfly insect infestation, whose larvae eat the needles and buds of many kinds of conifers.

Figure 6.141 The leaves of most species of conifers are needle-shaped such as those of the blue spruce, *Picea pungens* (a). *Araucaria heterophyla,* Norfolk Island pine, however, (b) has awl-shaped leaves, and *Podocarpus* sp. (c) has strap-shaped leaves.

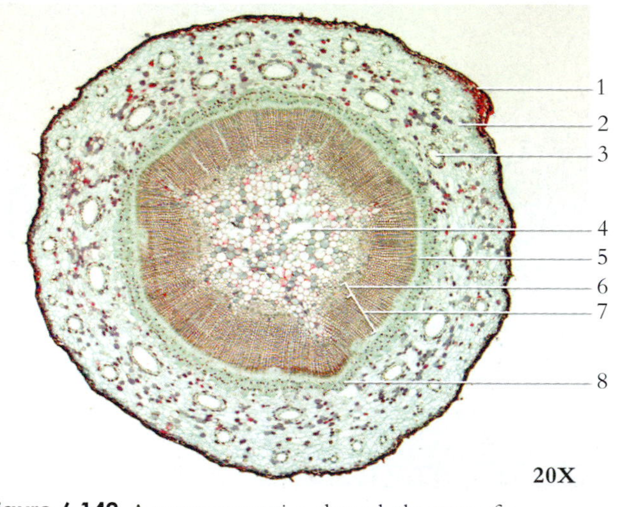

Figure 6.142 A transverse section through the stem of a young conifer showing the arrangement of the tissue layers.

1. Epidermis
2. Cortex
3. Resin duct
4. Pith
5. Cambium
6. Primary xylem
7. Spring wood of secondary xylem
8. Primary phloem

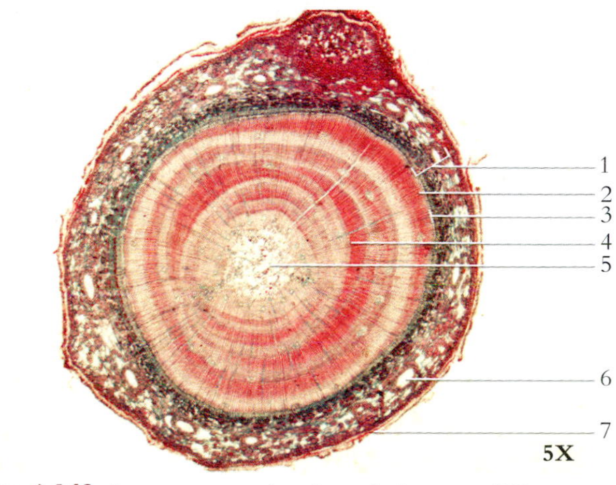

Figure 6.143 A transverse section through the stem of *Pinus* sp., showing secondary stem growth.

1. Bark (cortex and periderm)
2. Secondary phloem
3. Vascular cambium
4. Secondary xylem
5. Pith
6. Resin duct
7. Epidermis

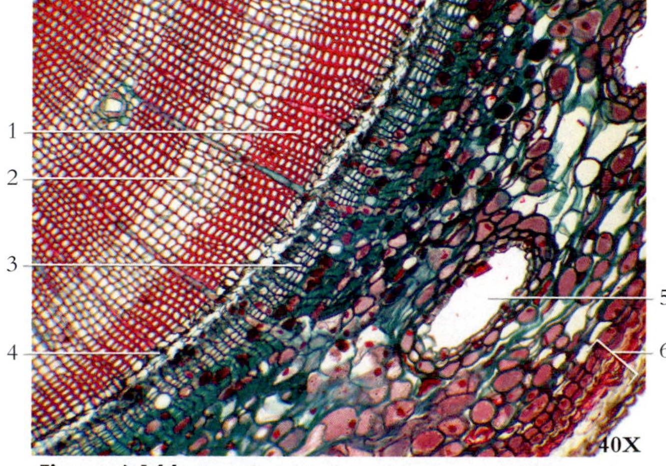

Figure 6.144 An enlarged view of the stem of *Pinus* sp. showing tissues following secondary growth.

1. Late secondary xylem (wood)
2. Early secondary xylem (wood)
3. Secondary phloem
4. Vascular cambium
5. Resin duct
6. Periderm

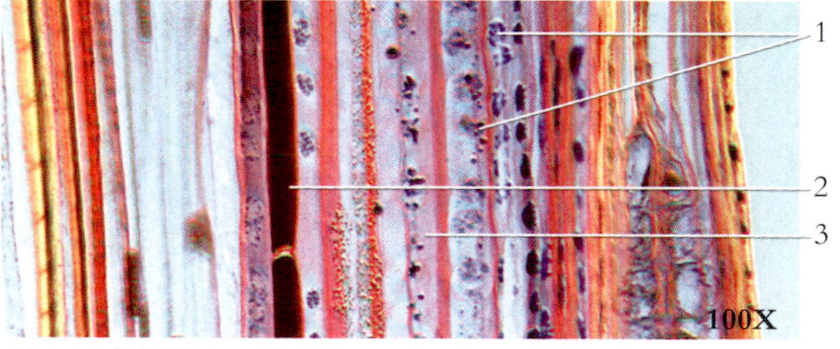

Figure 6.145 A radial longitudinal section through the phloem of *Pinus* sp.

1. Sieve areas on a sieve cell
2. Storage parenchyma
3. Sieve cell

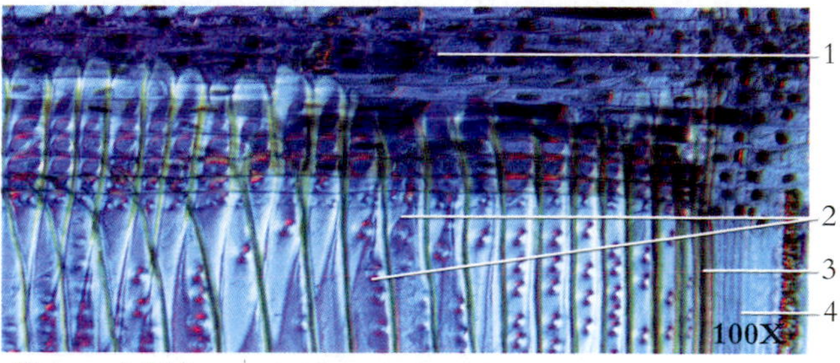

Figure 6.146 A radial longitudinal section through a stem of *Pinus* sp., cut through the xylem tissue.

1. Ray parenchyma
2. Tracheids
3. Vascular cambium
4. Sieve cells

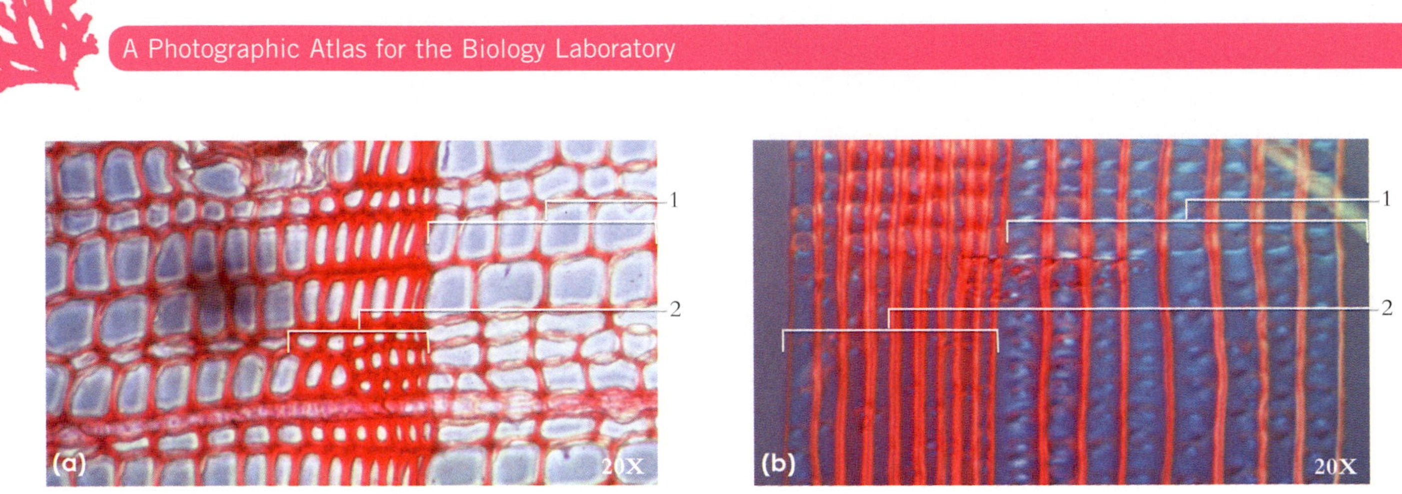

Figure 6.147 The growth rings in *Pinus* sp. (a) Transverse section through a stem; and (b) radial longitudinal section through a stem.

1. Early wood
2. Late wood

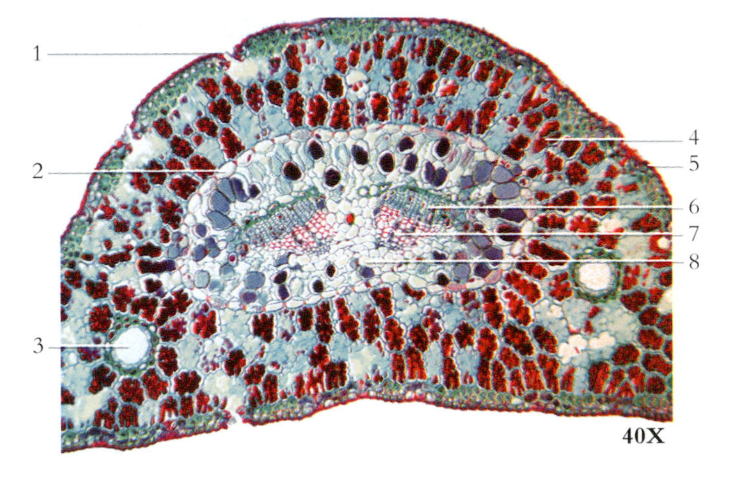

Figure 6.148 The transverse section of a leaf (needle) of *Pinus* sp.

1. Stoma
2. Endodermis
3. Resin duct
4. Photosynthetic mesophyll
5. Epidermis
6. Phloem
7. Xylem
8. Transfusion tissue

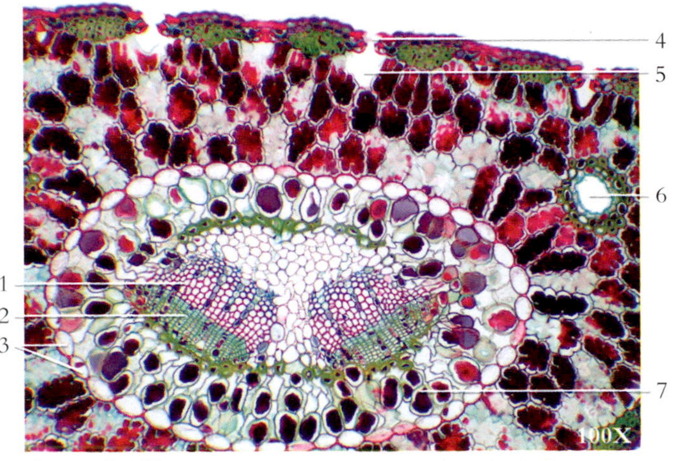

Figure 6.149 The transverse section through the leaf (needle) of *Pinus* sp.

1. Xylem
2. Phloem
3. Endodermis
4. Sunken stoma
5. Substomatal chamber
6. Resin duct
7. Transfusion tissue (surrounding vascular tissue)

Aibes sp. *Taxodium* sp. *Araucaria* sp. *Taxus* sp. *Pinus* sp.

Figure 6.150 The megasporangiate cones from various species of conifers.

Figure 6.151 A first-year ovulate cone in *Pinus* sp.
1. Pollen cones 2. First-year ovulate cone

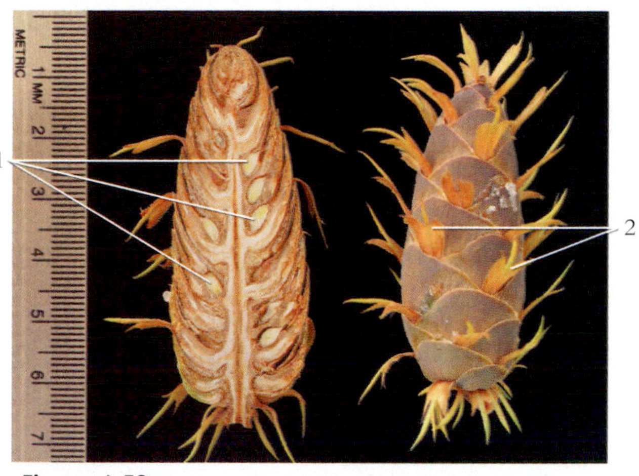

Figure 6.52 A transverse section through a first-year ovulate cone in *Pseudotsuga* sp. (scale in mm).
1. Immature ovules 2. Cone scale bracts

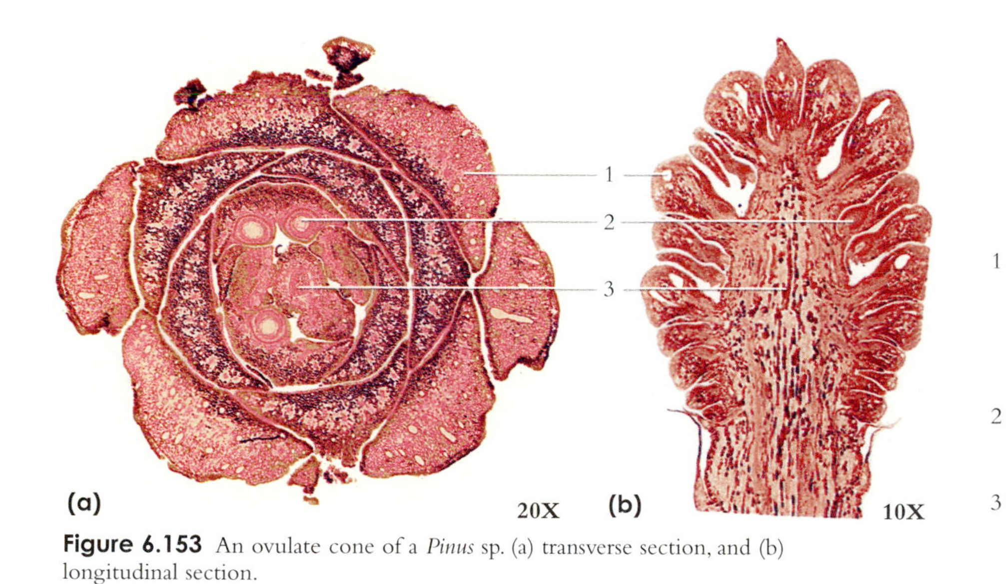

Figure 6.153 An ovulate cone of a *Pinus* sp. (a) transverse section, and (b) longitudinal section.
1. Ovuliferous scale 2. Ovule 3. Cone axis

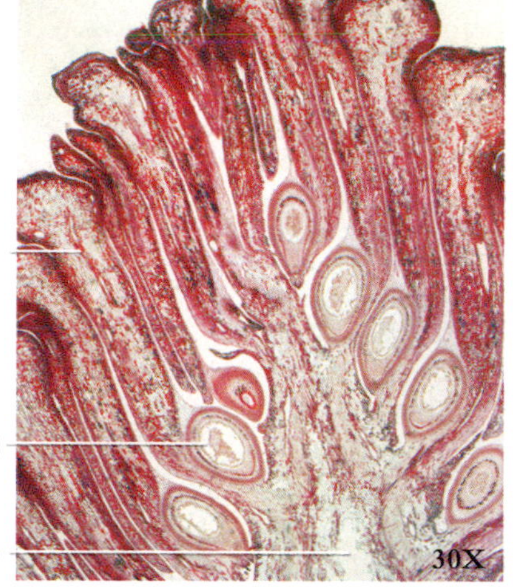

Figure 6.154 A magnified view of a *Pinus* sp. ovulate cone (longitudinal view).
1. Ovuliferous scale 2. Ovule 3. Cone axis

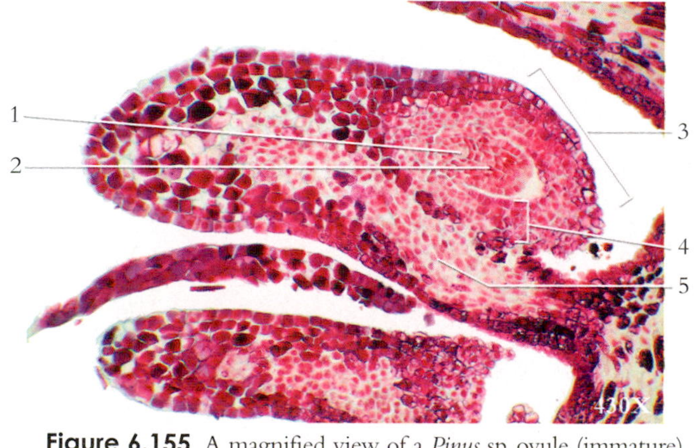

Figure 6.155 A magnified view of a *Pinus* sp. ovule (immature).
1. Megaspore mother cell 2. Nucellus 3. Ovule 4. Integument 5. Cone scale

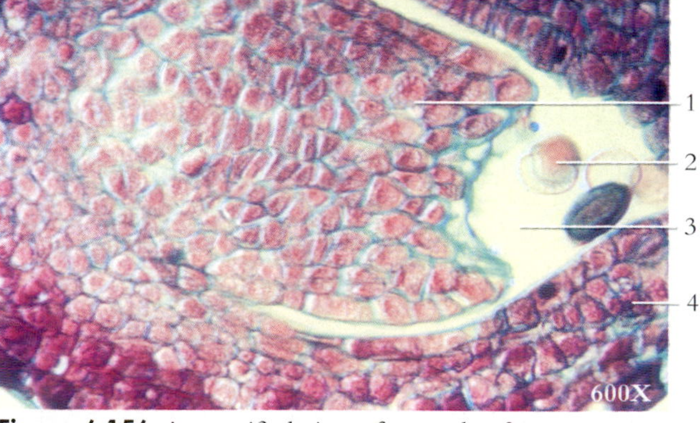

Figure 6.156 A magnified view of an ovule of *Pinus* sp. with pollen grains in the pollen chamber.
1. Nucellus 2. Pollen grain 3. Pollen chamber 4. Integument

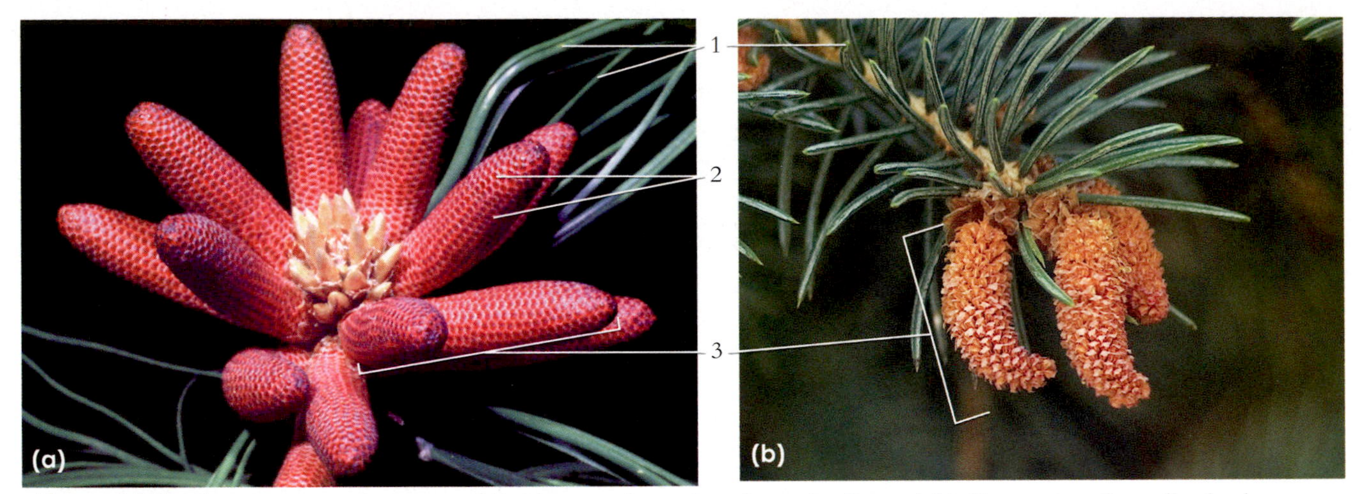

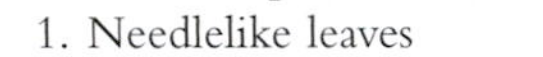

Figure 6.157 The microsporangiate cones of (a) *Pinus* sp. prior to the release of pollen and (b) *Picea pungens* after pollen has been released. The pollen cones are at the end of a branch.
1. Needlelike leaves
2. Microsporophylls
3. Pollen cone

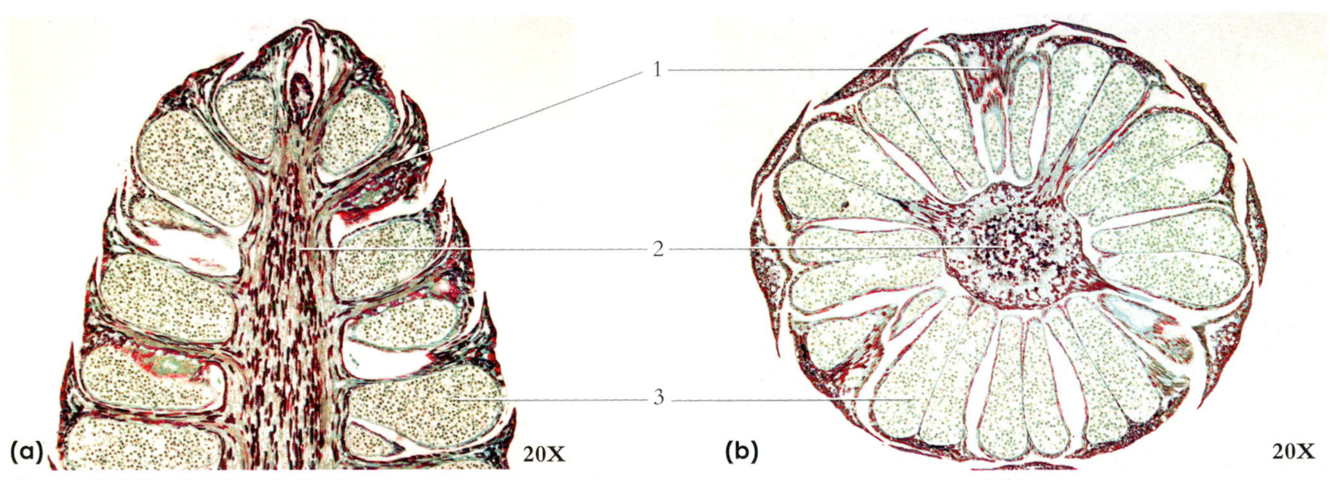

Figure 6.158 (a) A longitudinal section through the tip of a microsporangiate cone of *Pinus* sp. and (b) a transverse section.
1. Sporophyll
2. Cone axis
3. Microsporangium

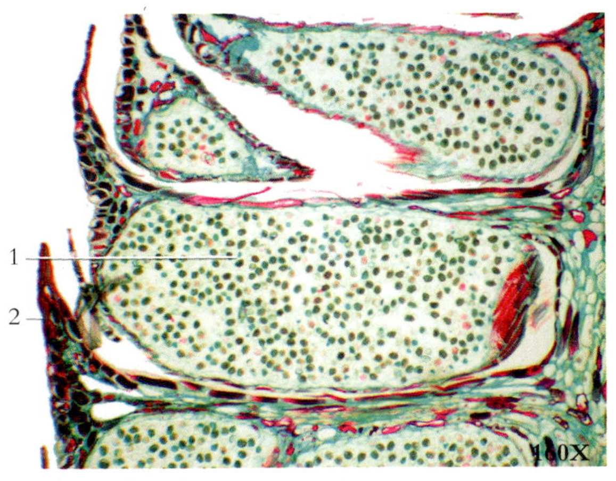

Figure 6.159 A close-up of a microsporangiate cone scale and microsporangium of *Pinus* sp.
1. Microsporangium with pollen grains
2. Microsporophyll

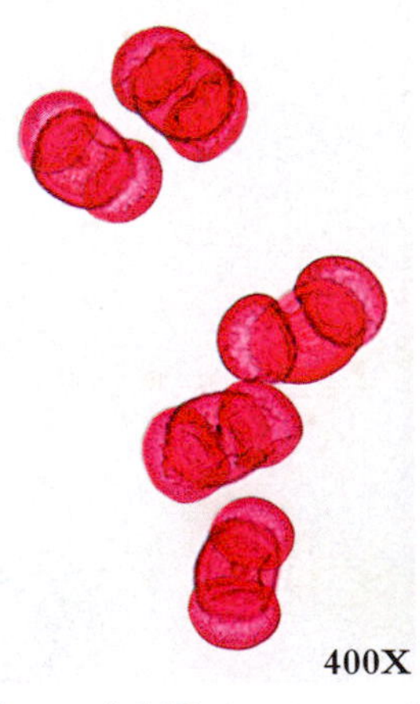

Figure 6.160 A micrograph of stained pollen grains of *Pinus* sp., showing wings.

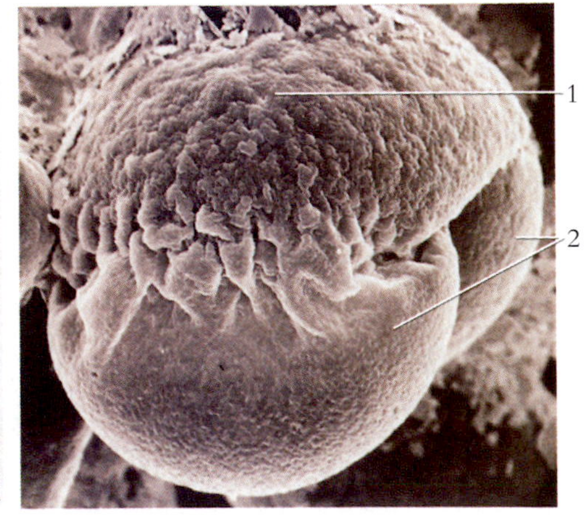

Figure 6.161 A scanning electron micrograph of a *Pinus* sp. pollen grain with inflated bladderlike wings.
1. Pollen body
2. Wings

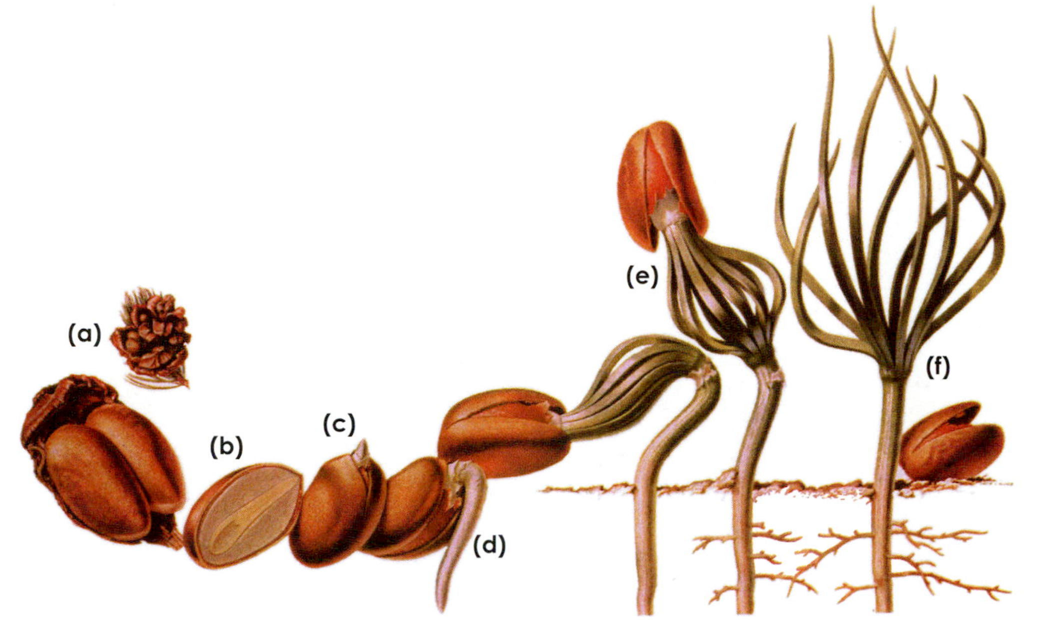

Figure 6.162 A diagram of pinyon pine seed germination producing a young sporophyte. (a) The seeds are protected inside the cone, two seeds formed on each scale. (b) A sectioned seed shows an embryo embedded in the female gametophyte tissue. (c) The growing embryo splits the shell of the seed, enabling the root to grow toward the soil. (d) As soon as the tiny root tip penetrates and anchors into the soil, water and nutrients are absorbed. (e) The cotyledons emerge from the seed coat and create a supply of chlorophyll. Now the sporophyte can manufacture its own food from water and nutrients in the soil and carbon dioxide in the air. (f) Growth occurs at the terminal buds at the base of the leaves.

Figure 6.163 A young sporophyte (seedling) of a pine, *Pinus* sp. (scale in mm).

1. Seedling leaves (needles)
2. Young stem
3. Young roots

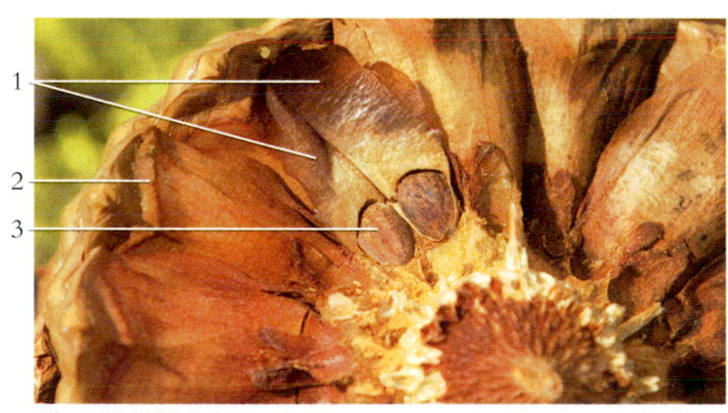

Figure 6.164 A close-up of an ovulate cone scale in *Pinus* sp.

1. Mature seeds (wings)
2. Ovulate cone scale
3. Seed (containing embryo within seed coat)

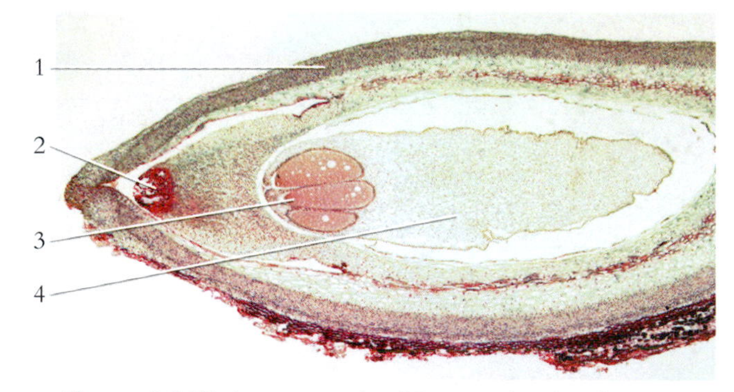

Figure 6.165 A young ovule of *Pinus* sp. showing the megagametophyte.

1. Ovule
2. Micropyle
3. Archegonium
4. Megagametophyte

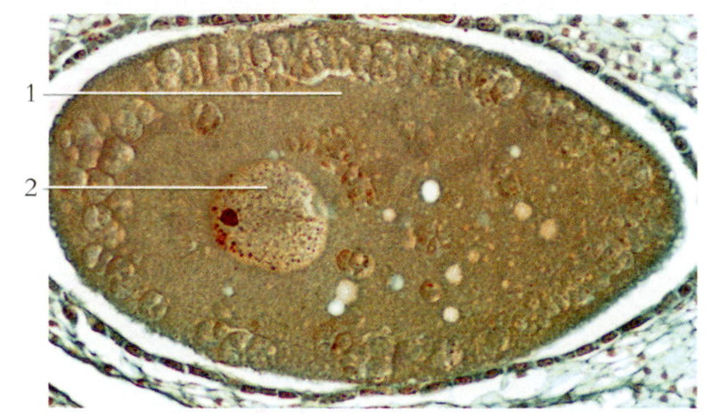

Figure 6.166 A young ovule of *Pinus* sp. showing the egg in archegonium.

1. Egg
2. Nucleus

Figure 6.167 A magnified view of the ovule of *Pinus* sp. showing the embryo.

1. Integument
2. Micropyle
3. Leaf primordium
4. Root primordium

Phylum Magnoliophyta (= Anthophyta) – angiosperms: monocots and dicots

Figure 6.168 A comparison and examples of monocots and dicots.

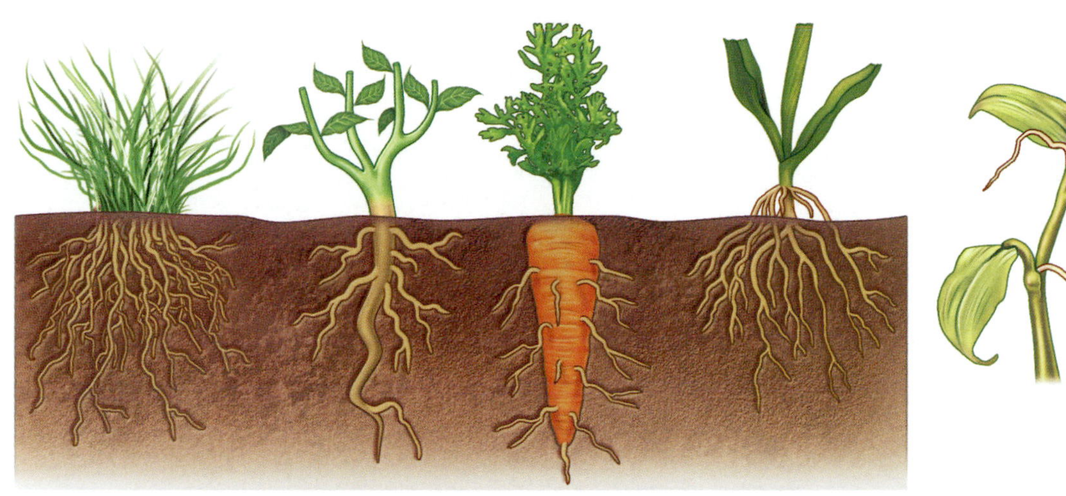

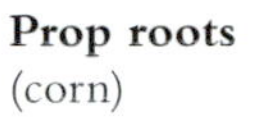

Figure 6.169 The root systems of angiosperms.

Figure 6.170 The root system of an orchid (monocot) (a) showing aerial roots and corn (dicot) (b) showing prop roots. Monocot roots are fibrous, with many roots of more or less equal size. Dicots usually have a taproot system, consisting of a long central root with smaller, secondary roots branching from it.

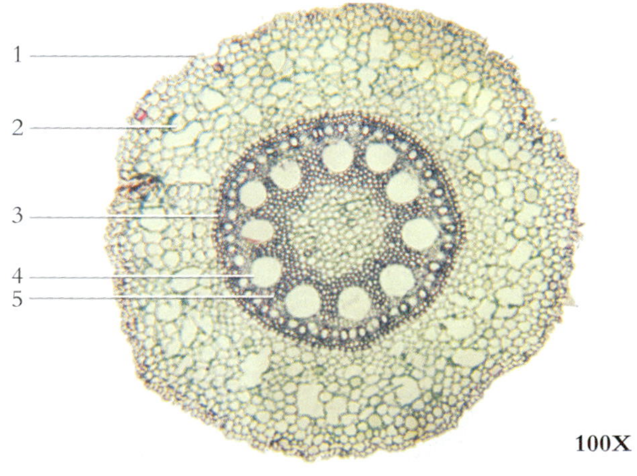

Figure 6.171 A transverse section of the root of the monocot *Smilax* sp.

1. Epidermis
2. Cortex
3. Endodermis
4. Xylem
5. Phloem

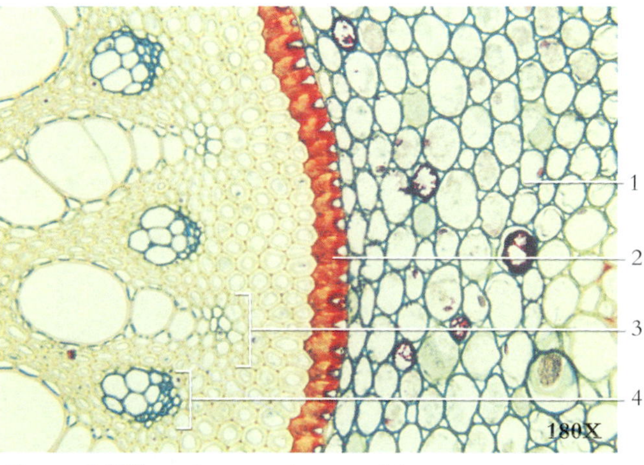

Figure 6.172 A close-up of a root of the monocot *Smilax* sp.

1. Cortex
2. Endodermis
3. Xylem
4. Phloem

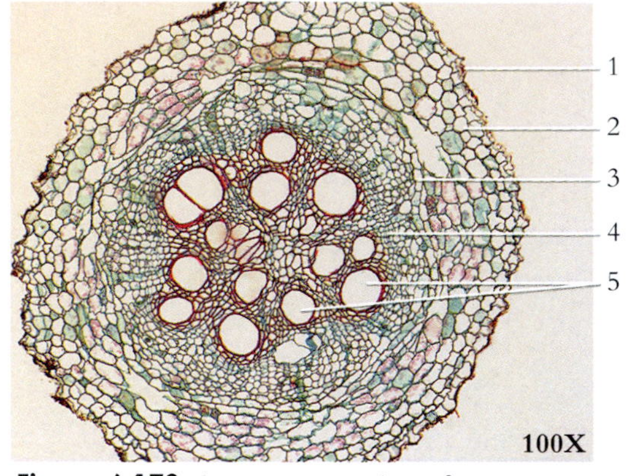

Figure 6.173 A transverse section of a sweet potato root, *Ipomaea* sp.

1. Remnants of epidermis
2. Cortex
3. Endodermis
4. Phloem
5. Xylem

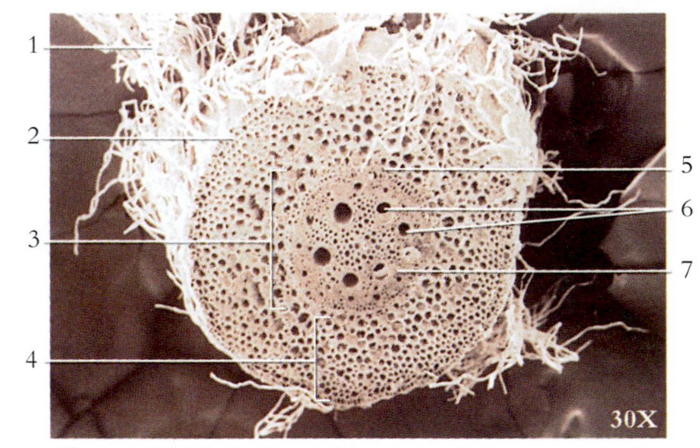

Figure 6.174 A photomicrograph of a young root of wheat, *Triticum* sp., showing root hairs.

1. Root hair
2. Epidermis
3. Stele
4. Cortex
5. Endodermis
6. Primary xylem
7. Primary phloem

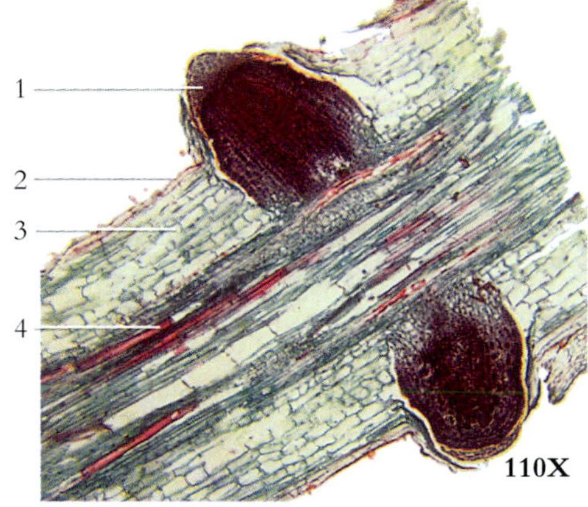

Figure 6.175 A longitudinal section of a willow species showing lateral root formation.

1. Lateral root
2. Epidermis
3. Cortex
4. Vascular tissue

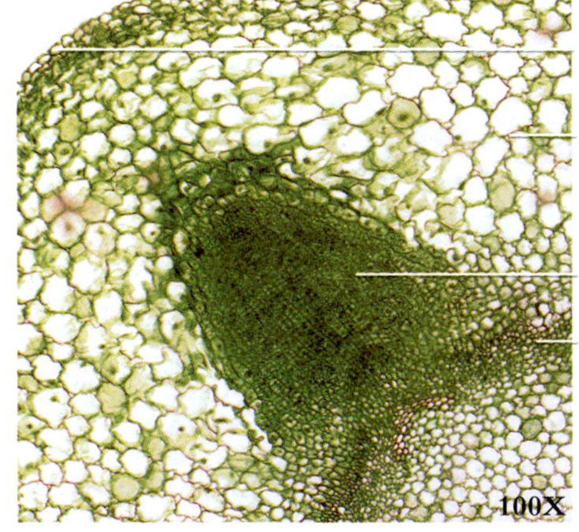

Figure 6.176 A transverse section showing branch root formation of *Phaseolus* sp.

1. Epidermis
2. Cortex
3. Branch root
4. Vascular tissue (stele)

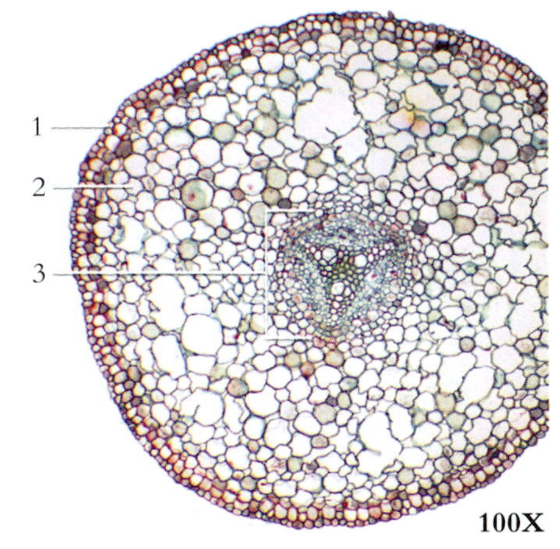

Figure 6.177 A transverse section of a young root of *Salix* sp.

1. Epidermis
2. Cortex
3. Stele

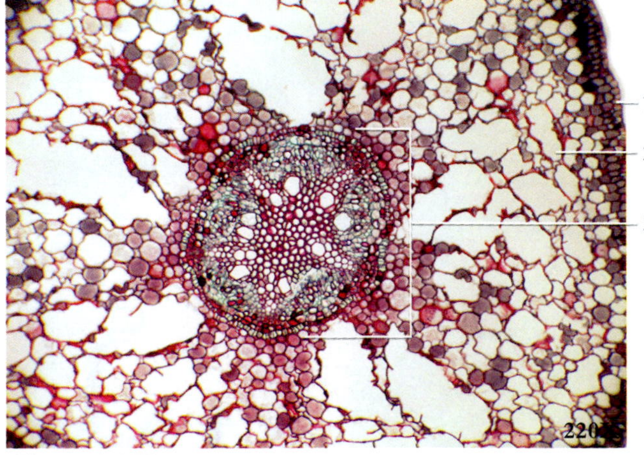

Figure 6.178 A transverse section of an older root of *Salix* sp., showing early secondary growth.

1. Epidermis
2. Cortex
3. Vascular tissue

Stele

Cortex

Root hair

Epidermis

Maturation region
portion where cells are differentiating into epidermal and cortex layers and xylem and phloem tissues

Elongation region
portion where newly added cells increase in size

Meristematic region
portion undergoing mitosis

Root cap
portion protecting the root during growth

Figure 6.179 A diagram of a root tip.

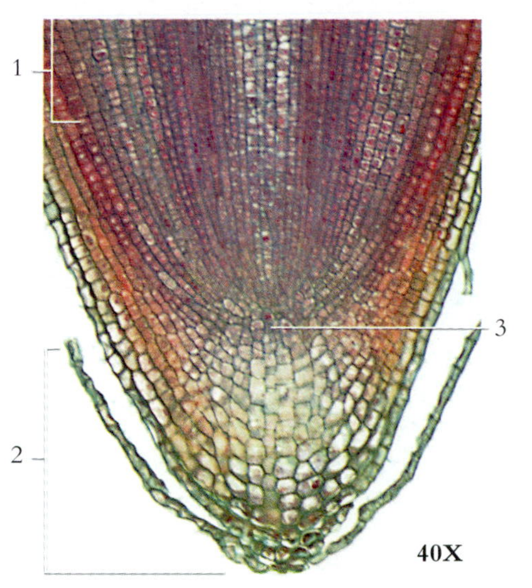

Figure 6.180 A photomicrograph of the root tip of a pear, *Pyrus* sp., seen in longitudinal section.
1. Elongation region
2. Root cap
3. Apical meristem

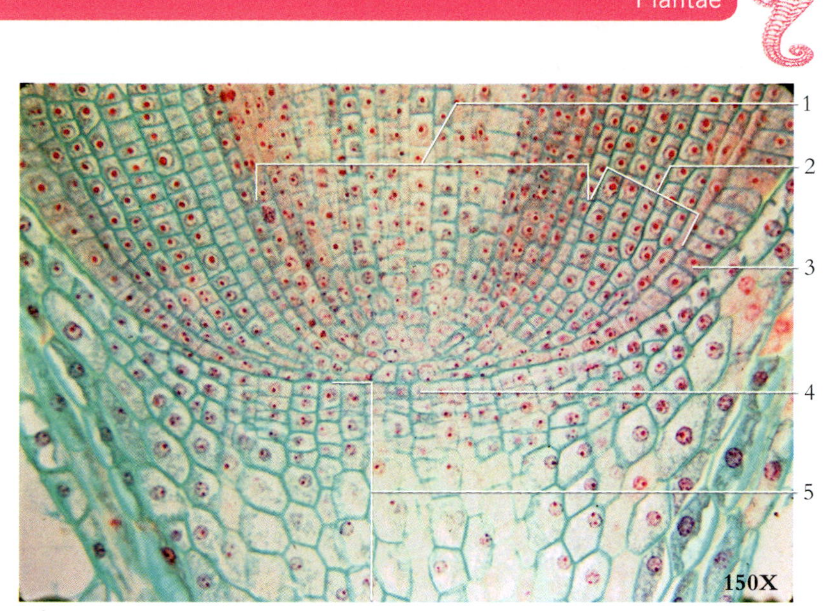

Figure 6.181 A longitudinal section of a root of corn, *Zea mays*, showing primary meristems: protoderm gives rise to the epidermis, ground meristem to cortex, and procambium to primary vascular tissue. The root cap has a separate meristem.
1. Procambium
2. Ground meristem
3. Protoderm
4. Root cap meristem
5. Root cap

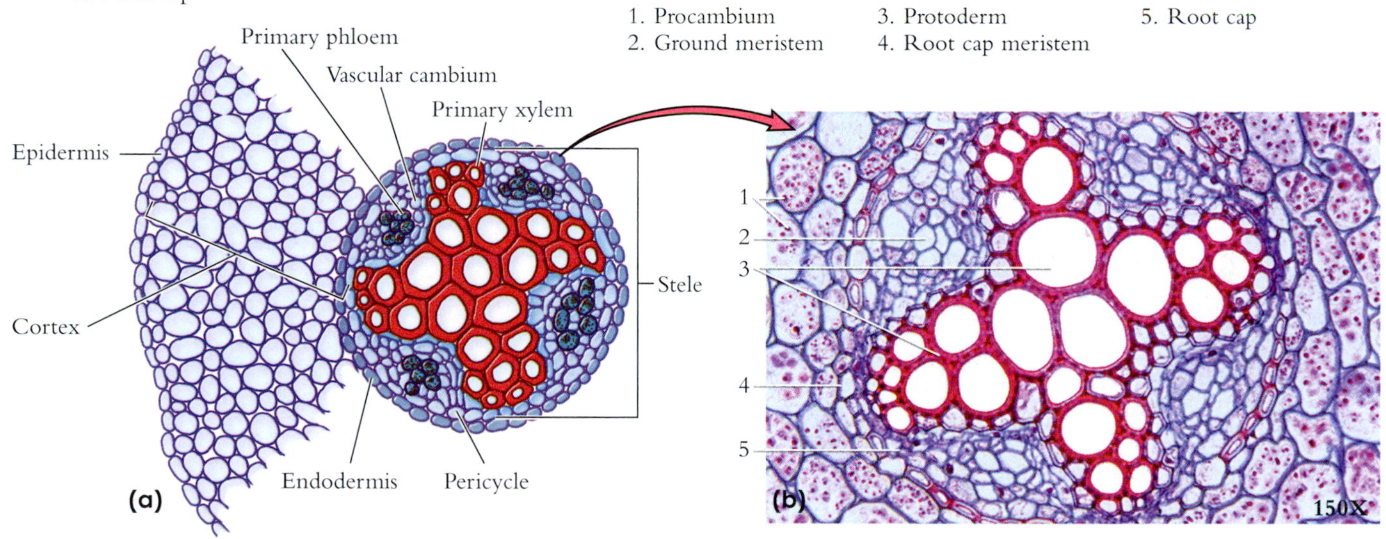

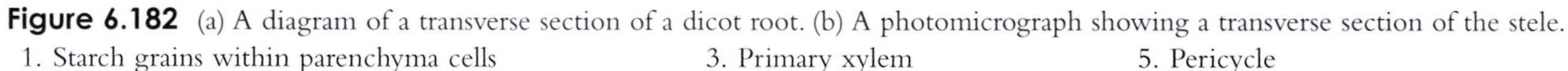

Figure 6.182 (a) A diagram of a transverse section of a dicot root. (b) A photomicrograph showing a transverse section of the stele.
1. Starch grains within parenchyma cells
2. Primary phloem
3. Primary xylem
4. Endodermis
5. Pericycle

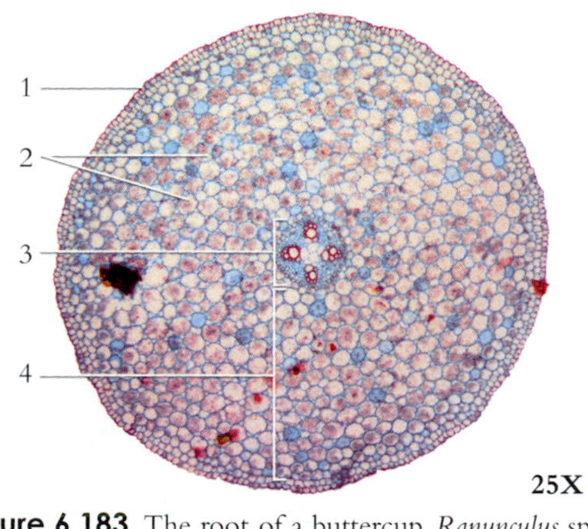

Figure 6.183 The root of a buttercup, *Ranunculus* sp.
1. Epidermis
2. Parenchyma cells of cortex
3. Stele
4. Cortex

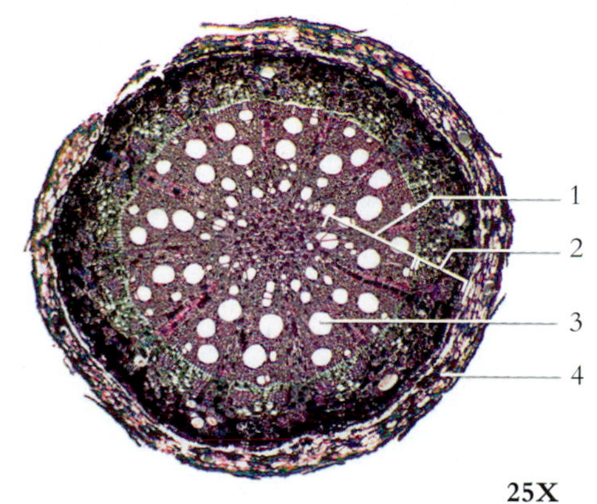

Figure 6.184 A transverse section of the root of basswood, *Tilia* sp., showing secondary growth.
1. Secondary xylem
2. Secondary phloem
3. Vessel element
4. Periderm

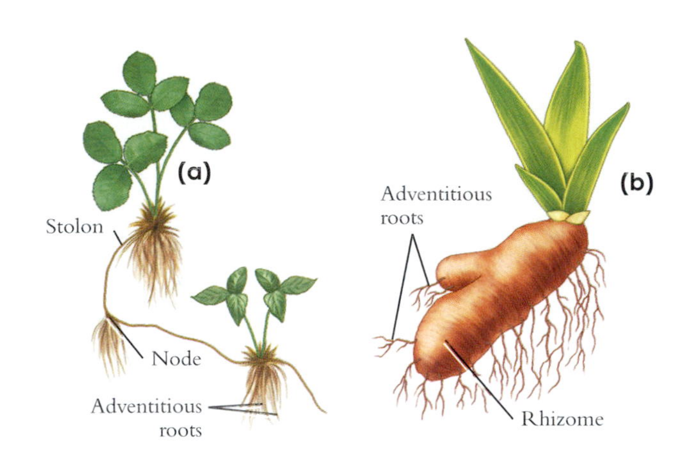

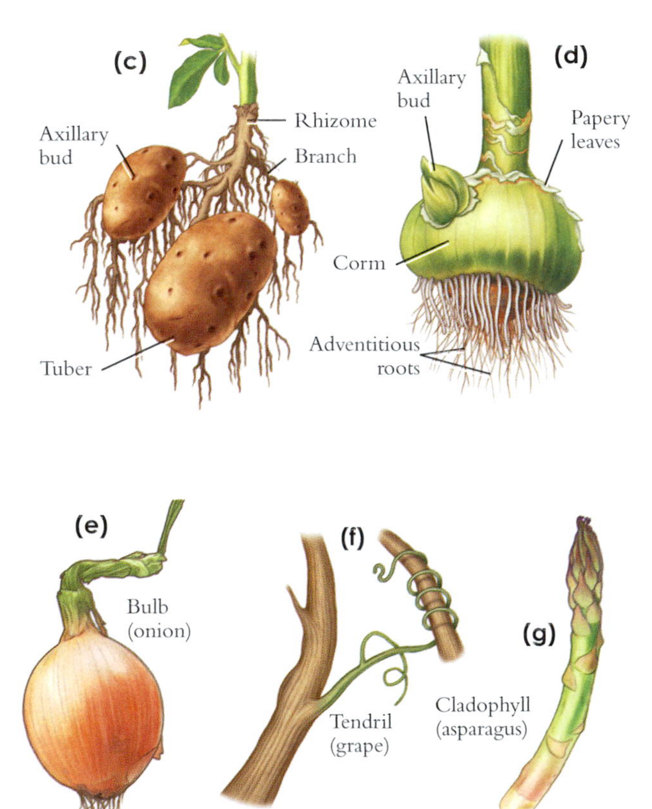

Figure 6.185 Examples of the variety and specialization of angiosperm stems, (a) runners, (b) rhizomes, (c) tubers, (d) corms, (e) bulbs, (f) tendrils, and (g) cladophyll. The stem of an angiosperm is often the ascending portion of the plant specialized to produce and support leaves and flowers, transport and store water and nutrients, and provide growth through cell division. Stems of plants are utilized extensively by humans in products including paper, building materials, furniture, and fuel. In addition, the stems of potatoes, onions, cabbage, and other plants are important food crops.

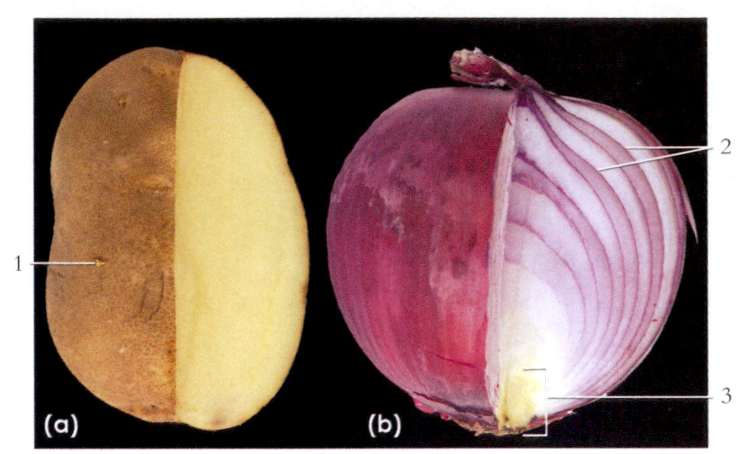

Figure 6.186 Specialized underground stems. (a) A potato (tuber) and (b) an onion (bulb).

1. Node (eye) bearing a minute scale leaf and stem bud
2. Bulb scales (modified leaves)
3. Short stem

Figure 6.187 The woody stem of a dicot seen in early spring just as the buds are beginning to swell. Branches and twigs are small extensions of the stems of angiosperms and often support leaves and flowers.

1. Terminal (apical) bud
2. Internode
3. Terminal bud scale scars
4. Lenticel
5. Lateral (axillary) bud
6. Node
7. Leaf (vascular bundle) scar

Figure 6.188 The terminal bud of a woody stem that has been longitudinally sectioned to show developing leaves.

1. Lateral (axillary) bud
2. Stem
3. Leaf primordia
4. Bud scale

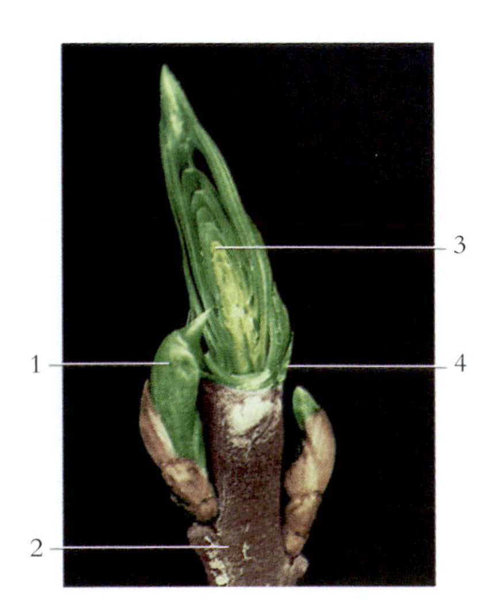

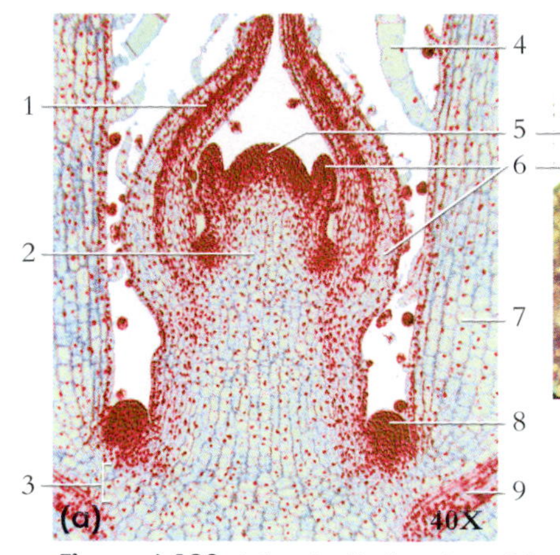

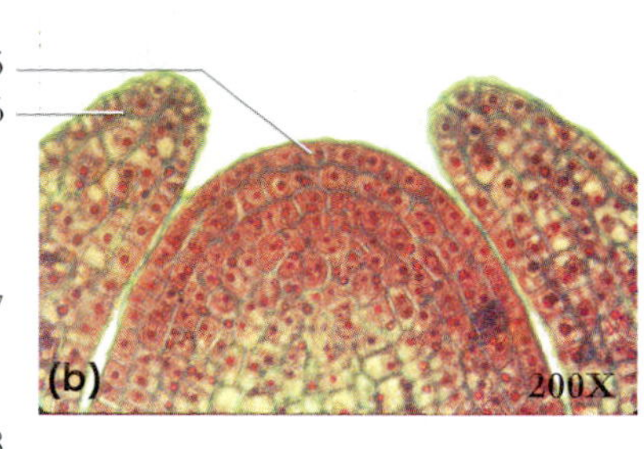

Figure 6.189 A longitudinal section of the stem tip of the common houseplant *Coleus* sp.

1. Procambium
2. Ground meristem
3. Leaf gap
4. Trichome
5. Apical meristem
6. Developing leaf primordia
7. Leaf primordium
8. Axillary bud
9. Developing vascular tissue

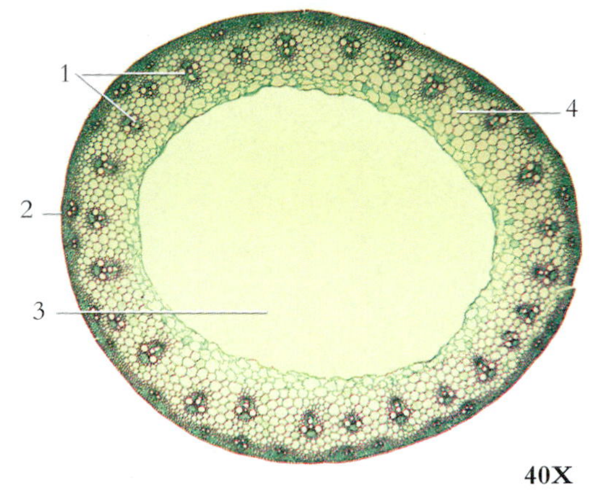

Figure 6.190 A transverse section through the stem of a monocot, *Triticum* sp., wheat.

1. Vascular bundles
2. Epidermis
3. Ground tissue cavity
4. Parenchyma cells

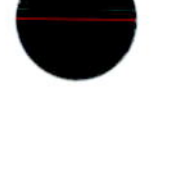

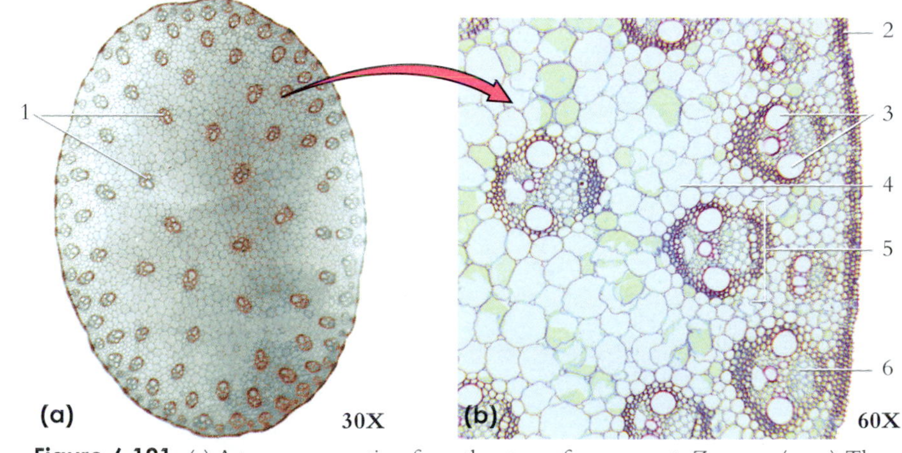

Figure 6.191 (a) A transverse section from the stem of a monocot, *Zea mays* (corn). The pattern of vascular bundles in a monocot is known as an atactostele. (b) a close-up view.

1. Vascular bundles with primary xylem and phloem
2. Epidermis
3. Vessel elements of primary xylem
4. Parenchyma cells
5. Vascular bundle
6. Primary phloem

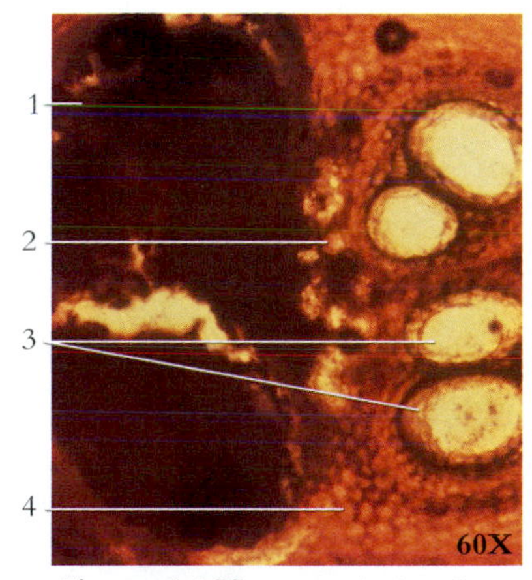

Figure 6.192 A vascular bundle of a fossil palm plant.

1. Bundle cap (fibers)
2. Phloem
3. Vessel elements
4. Ground tissue (parenchyma)

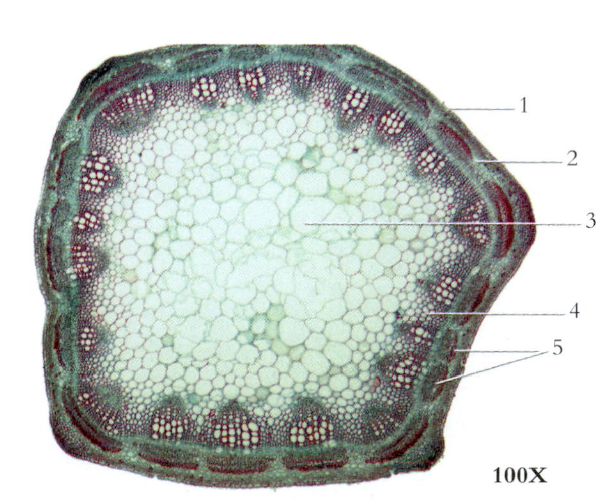

Figure 6.193 A transverse section through a stem of clover, *Trifolium* sp. showing an eustele.

1. Epidermis
2. Cortex
3. Pith
4. Interfascicular region
5. Vascular bundles with caps of phloem fibers

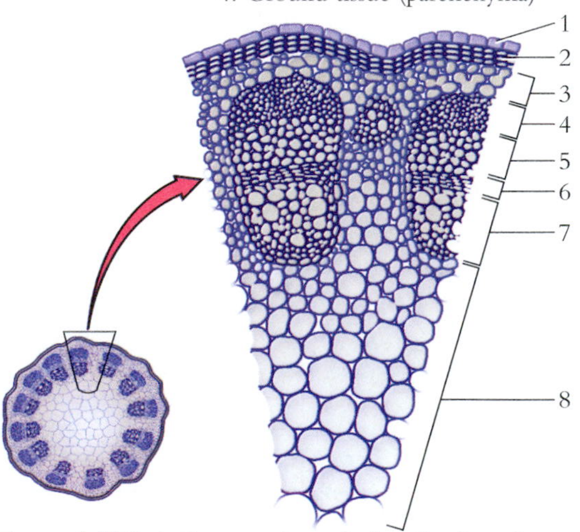

Figure 6.194 A diagram of vascular bundles from the stem of a dicot showing the eustele.

1. Early periderm
2. Cortex
3. Phloem fibers
4. Bundle cap fibers
5. Phloem
6. Vascular cambium
7. Xylem
8. Pith

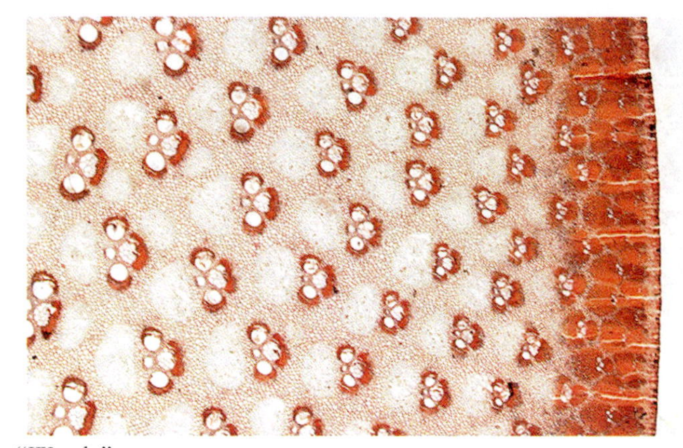

"Woody" monocot

Woody dicot

Figure 6.195 A comparison of the transverse sections of stems of a "woody" monocot (palm tree) and a woody dicot (hickory tree). The stem of the "woody" monocot is rigid because of the fibrous nature of the numerous vascular bundles. The stem of the woody dicot is rigid because of the compact xylem cells impregnated with lignin forming the dense, hardened wood, seen as annual rings.

1. Annual rings
2. Bark

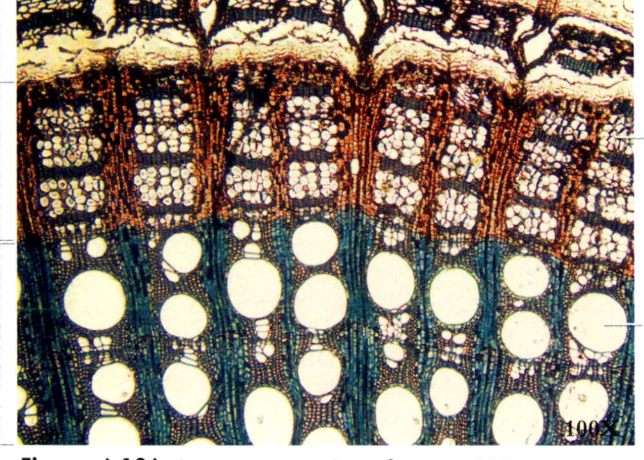

Figure 6.196 A transverse section of a grape, *Vitis* sp., stem showing secondary tissues.

1. Outer bark
2. Secondary phloem
3. Secondary xylem
4. Sieve tube elements
5. Vessel member

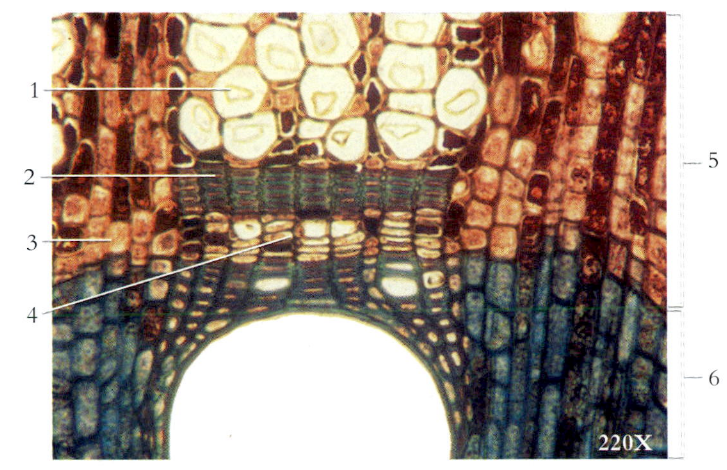

Figure 6.197 A transverse section of a grape, *Vitis* sp., stem.

1. Sieve tube element
2. Phloem fibers
3. Parenchyma cells
4. Vascular cambium
5. Secondary phloem
6. Secondary xylem

1
2
3
4
5
6
100X

Figure 6.198 A transverse section through one-year-old ash, *Fraxinus* sp., stem showing secondary growth.

1. Periderm
2. Cortex
3. Phloem fibers
4. Secondary phloem
5. Secondary xylem
6. Pith

3
1
2
100X

Figure 6.199 A pipevione, *Aristolochia* sp., stem with healing wound.

1. Callus tissue
2. Wound
3. Vascular bundle

Figure 6.200 Samples of bark patterns of representative conifers and angiosperms

(a) **Redwood**—The tough, fibrous bark of a redwood tree may be 30 cm thick. It is highly resistant to fire and insect infestation.
(b) **Ponderosa pine**—The mosaiclike pattern of the bark of mature ponderosa pine is resistant to fire.
(c) **White birch**—The surface texture of bark on the white birch is like white paper. The bark of the white birch was used by Indians in Eastern United States for making canoes.
(d) **Sycamore**—The mottled color of the sycamore bark is due to a tendency for large, thin, brittle plates to peel off, revealing lighter areas beneath. These areas grow darker with exposure, until they, too, peel off.
(e) **Mangrove**—The leathery bark of a mangrove tree is adaptive to brackish water in tropical or semitropical regions.
(f) **Shagbark hickory**—The strips of bark in a mature shagbark hickory tree gives this tree its common name.

Figure 6.201 An angiosperm, *Ruscus aculeatus*, is characterized by stems (a) that resemble leaves in form and function. Note the true leaf (b) arising from the leaflike stem.

1. Stem 2. Leaf 3. Flower bud

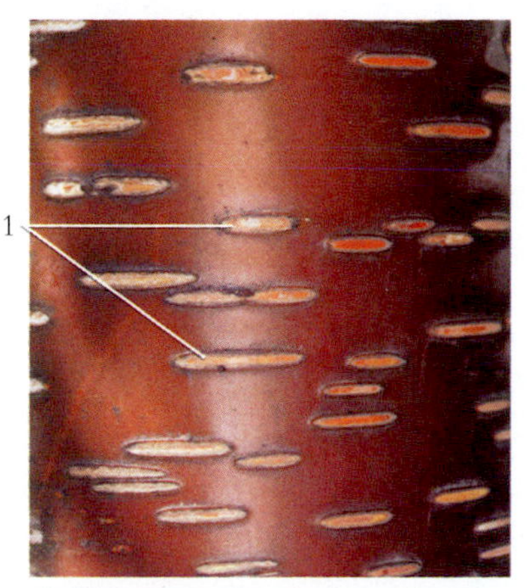

Figure 6.202 The bark of a birch tree, *Betula occidentalis*, showing lenticels. Lenticels are spongy areas in the cork surfaces that permit gas exchange between the internal tissues and the atmosphere.

1. Lenticels

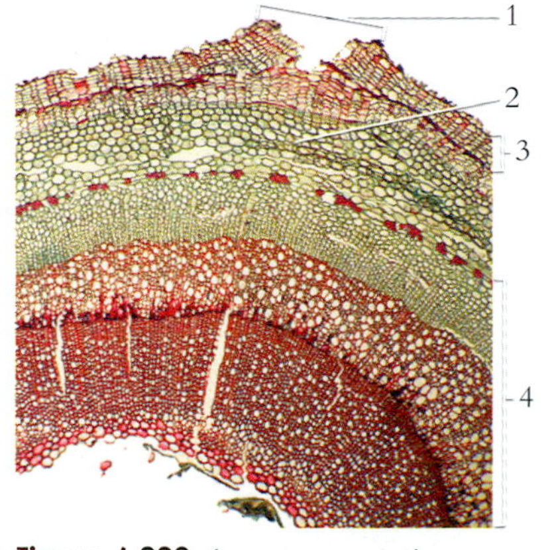

Figure 6.203 A transverse section of a dicot stem showing a lenticel and stem tissues.

1. Lenticel 2. Cortex 3. Periderm 4. Vascular tissue

Figure 6.204 A gall on an oak, *Quercus* sp., stem. The feeding of a gall wasp larva causes abnormal growth and the formation of a gall. The wasp larva feeds upon the gall tissue, pupates within this enclosure, and then chews an exit to emerge.

1. Gall 2. Stem

Venation	Margin	Complexity	Arrangement on Stem
 Pinnate	 Entire	  Palmately compound	 Opposite
 Parallel	 Pinnately lobed	 Simple	 Alternate
 Palmate	 Serrate	 Pinnately compound	 Whorled

Figure 6.205 Several representative angiosperm leaf types. Leaves constitute the foliage of plants, which provides habitat and a food source for many animals including humans. Leaves also provide protective ground cover and are the portion of the plant most responsible for oxygen replenishment into the atmosphere.

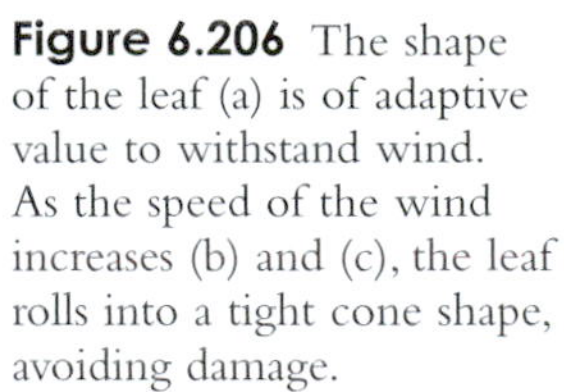

Figure 6.206 The shape of the leaf (a) is of adaptive value to withstand wind. As the speed of the wind increases (b) and (c), the leaf rolls into a tight cone shape, avoiding damage.

Fraxinus sp.

Populus sp.

Allophylus sp.

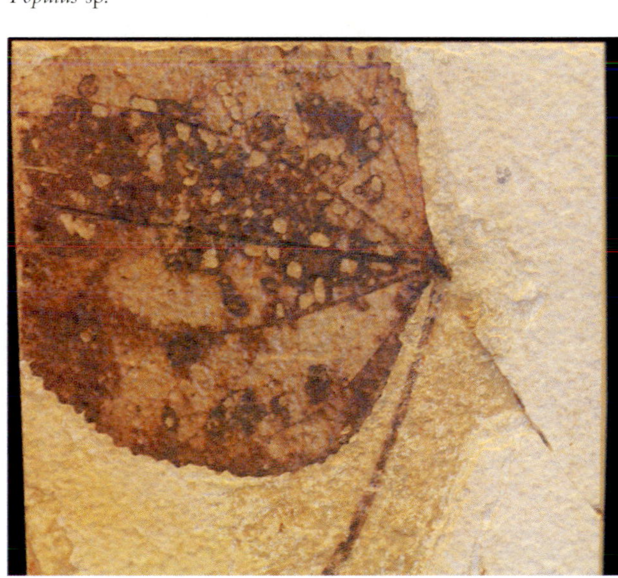

Cercidiphyllum sp.

Figure 6.207 Compression fossils of four angiosperm leaves from the Eocene Epoch, approximately 50 million years old.

Figure 6.208 The brilliant autumn colors of leaves come about when yellow carotenoid pigments are exposed as the chlorophyll breaks down, and colorless flavonoids are converted into anthocyanins.

Figure 6.209 An angiosperm leaf showing characteristic surface features. Leaves are organs modified to carry out photosynthesis. Photosynthesis is the manufacture of food (sugar) from carbon dioxide and water, with sunlight providing energy.

1. Lamina (blade)
2. Serrate margin
3. Midrib
4. Veins
5. Petiole

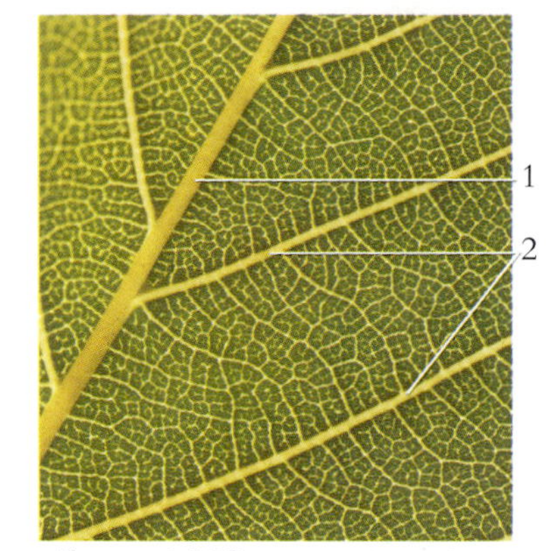

Figure 6.210 The undersurface of an angiosperm leaf showing the vascular tissue lacing through the lamina, or blade, of the leaf.

1. Midrib
2. Secondary veins

Figure 6.211 The organic decomposition of a leaf is a gradual process beginning with the softer tissues of the lamina, leaving only the vascular tissues of the midrib and the veins, as seen in this photograph. With time, these will also decompose.

Figure 6.212 Some examples of specialized leaves for flotation. (a) Leaves from a giant water lily. (b) Water hyacinths, *Eichhornia* sp., have modified leaves that buoy the plants on the water surface. Water hyacinths are common in New World tropical freshwater habitats, where they may become so thick that they choke out bottom-dwelling plants and clog waterways.

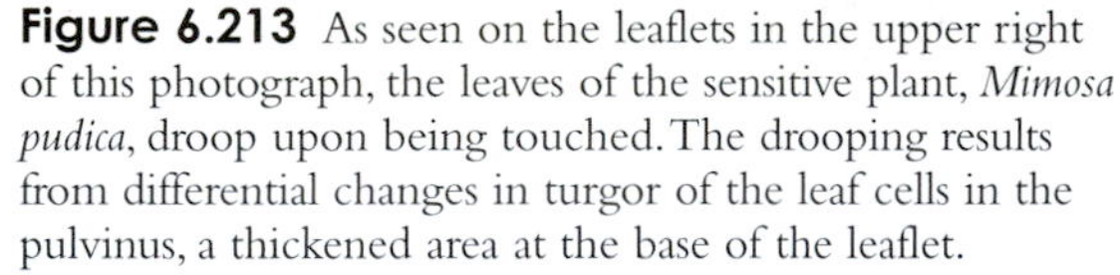

Figure 6.213 As seen on the leaflets in the upper right of this photograph, the leaves of the sensitive plant, *Mimosa pudica*, droop upon being touched. The drooping results from differential changes in turgor of the leaf cells in the pulvinus, a thickened area at the base of the leaflet.

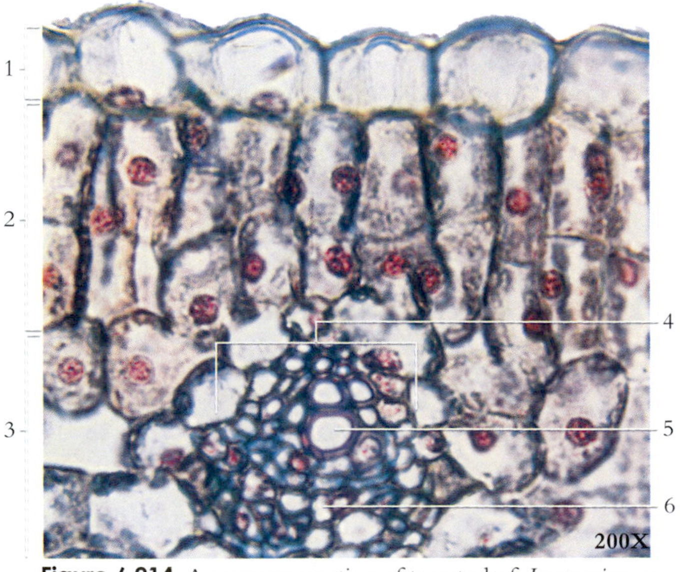

Figure 6.214 A transverse section of tomato leaf, *Lycopersicon* sp.

1. Upper epidermis
2. Palisade mesophyll
3. Spongy mesophyll
4. Leaf vein (vascular bundle)
5. Xylem
6. Phloem

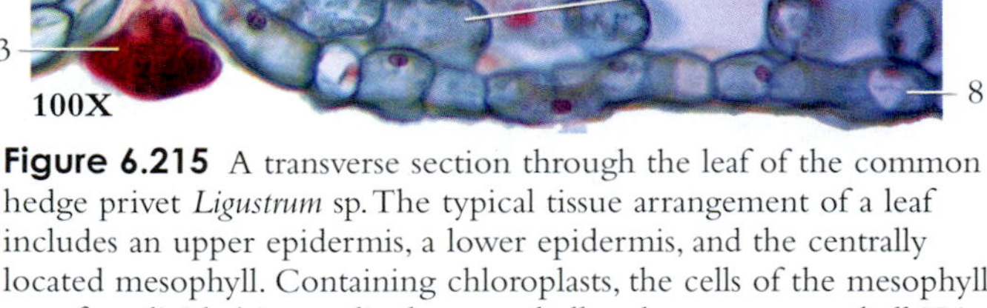

Figure 6.215 A transverse section through the leaf of the common hedge privet *Ligustrum* sp. The typical tissue arrangement of a leaf includes an upper epidermis, a lower epidermis, and the centrally located mesophyll. Containing chloroplasts, the cells of the mesophyll are often divided into palisade mesophyll and spongy mesophyll. Veins within the mesophyll conduct material through the leaf.

1. Upper epidermis
2. Palisade mesophyll
3. Gland
4. Bundle sheath
5. Xylem
6. Phloem
7. Spongy mesophyll
8. Lower epidermis

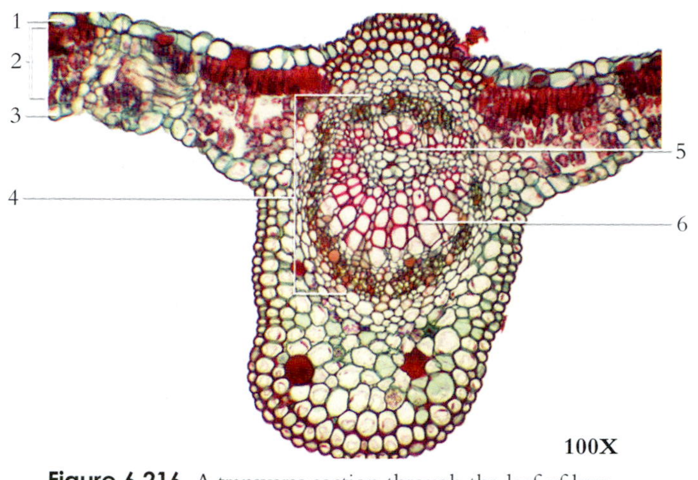

Figure 6.216 A transverse section through the leaf of basswood, *Tilia* sp.

1. Upper epidermis
2. Mesophyll
3. Lower epidermis
4. Leaf vein (midrib)
5. Phloem
6. Xylem

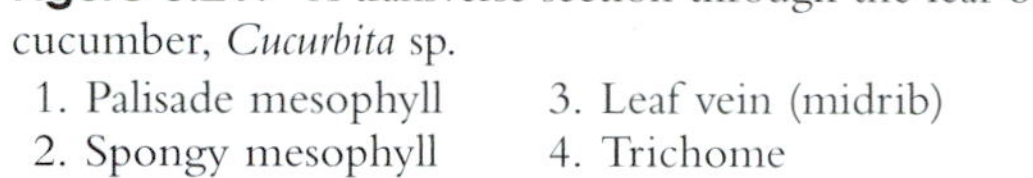

Figure 6.217 A transverse section through the leaf of cucumber, *Cucurbita* sp.

1. Palisade mesophyll
2. Spongy mesophyll
3. Leaf vein (midrib)
4. Trichome

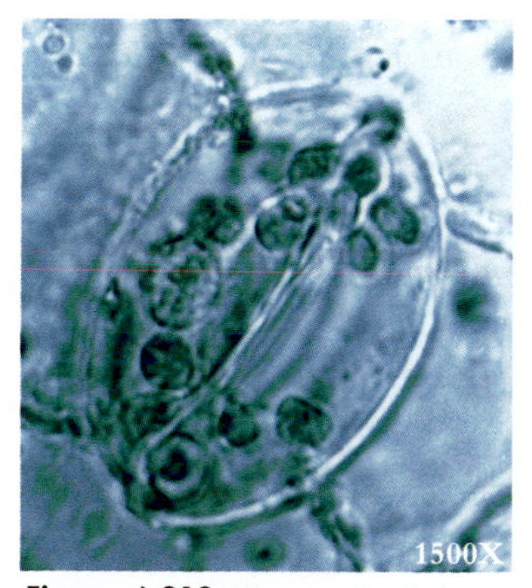

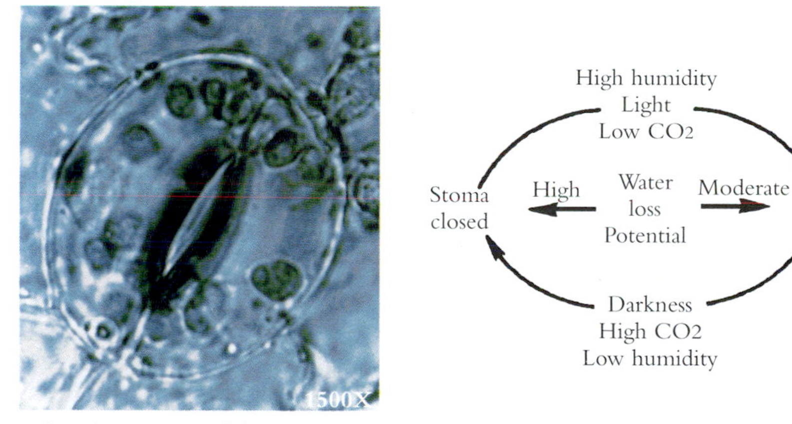

Figure 6.218 The guard cells in many plants regulate the opening of the stomata according to the environmental factors, as indicated in this diagram. (a) Face view of a closed stoma of a geranium, and (b) an open stoma.

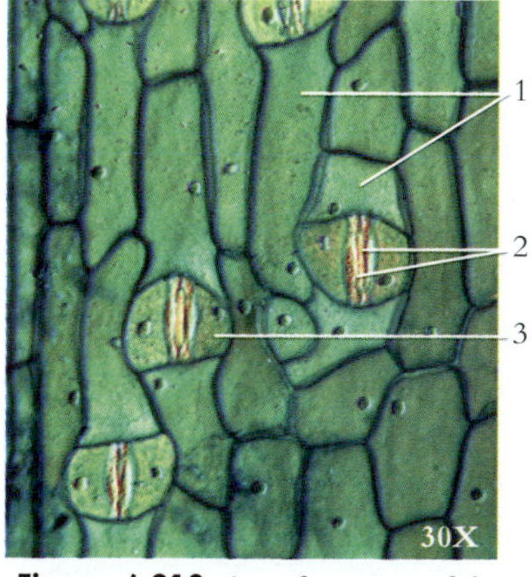

Figure 6.219 A surface view of the leaf epidermis of *Tradescantia* sp.

1. Epidermal cells
2. Guard cells surrounding stomata
3. Subsidiary cells

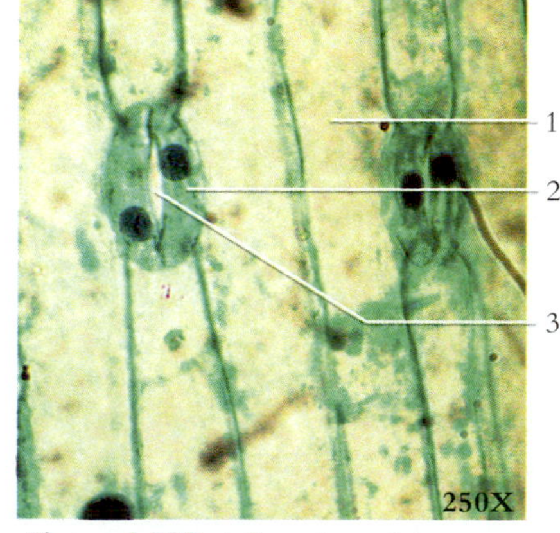

Figure 6.220 A face view of the epidermis of onion, *Allium* sp. Note the twin guard cells with the stoma opened.

1. Lower epidermis
2. Guard cell
3. Stoma

Figure 6.221 The specialized leaves of the carnivorous pitcher plant, *Sarracenia* sp.

Figure 6.222 The leaves of the purple pitcher plant, *Sarracenia purpurea*, are adapted to entrap insects. The leaves are funnel-shaped and have epidermal hairs pointed toward the base of the leaf. Insects are attracted to the funnel where they are entrapped, die, and are digested by the plant.
1. Leaf 2. Epidermal hairs

Figure 6.223 The leaves of the venus flytrap, *Dionaea muscipula*, are adapted to entrap insects. An insect is attracted by nectar secreted on the surface of the leaf. The movement of the insect upon the leaves stimulates the sensitive trichomes on the upper surface of the leaves, triggering the leaves to close, entrapping the insect.

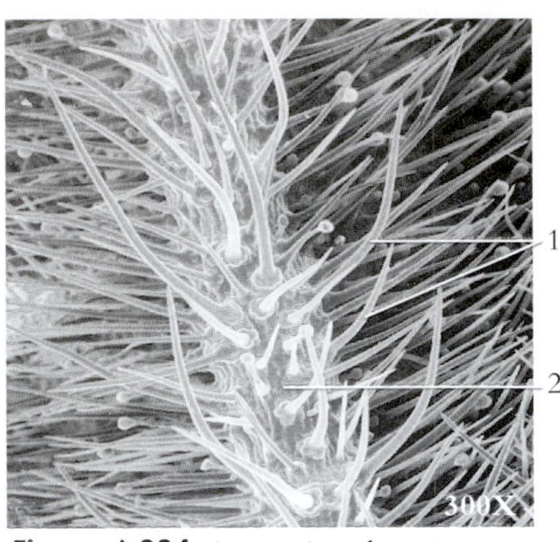

Figure 6.224 A scanning electron micrograph of a geranium leaf showing the prominent and abundant epidermal hairs.
1. Epidermal hairs
2. Epidermis

Figure 6.225 A Joshua tree, *Yucca brevifolia*, is native to the Mojave Desert. Its common name was derived from its resemblance to a bearded kneeling patriarch.

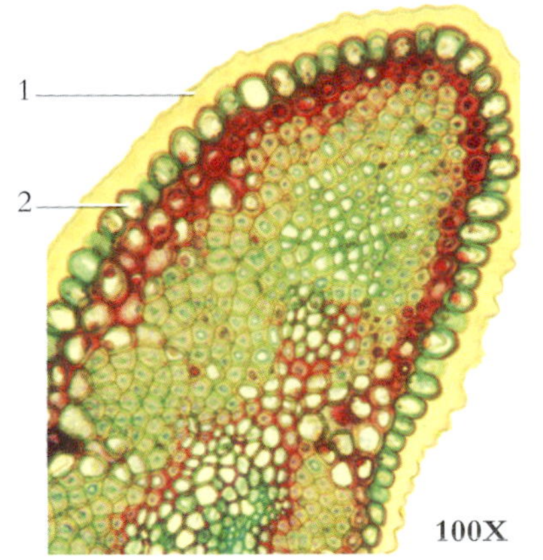

Figure 6.226 The leaf of *Yucca* sp. shows a thick cuticle covering the epidermis of the leaf. The cuticle protects against excessive water loss.
1. Cuticle 2. Epidermis

Figure 6.227 *Euphorbia* sp., is a member of the spurge family, is specialized to survive arid environments in Africa. Euphorbs have undergone convergent evolution to the cacti of the Western Hemisphere.

Figure 6.228 The saguaro cactus, *Carnegiea gigantea*, is the largest of all North American cacti. Arms begin to develop on the saguaro when the plant is about 75 years old. A saguaro cactus may live over 250 years and reach a height of more than 50 feet.

Figure 6.229 The prickly pear, *Opuntia* sp., cacti have several modifications to withstand drought. They have spinelike leaves to prevent water loss through transpiration; they have developed tissue that stores water after rain; and their stems are coated with a waxy substance to aid in water retention.

Figure 6.230 The fruit of the prickly pear, *Opuntia* sp.

Flowers of Angiosperms

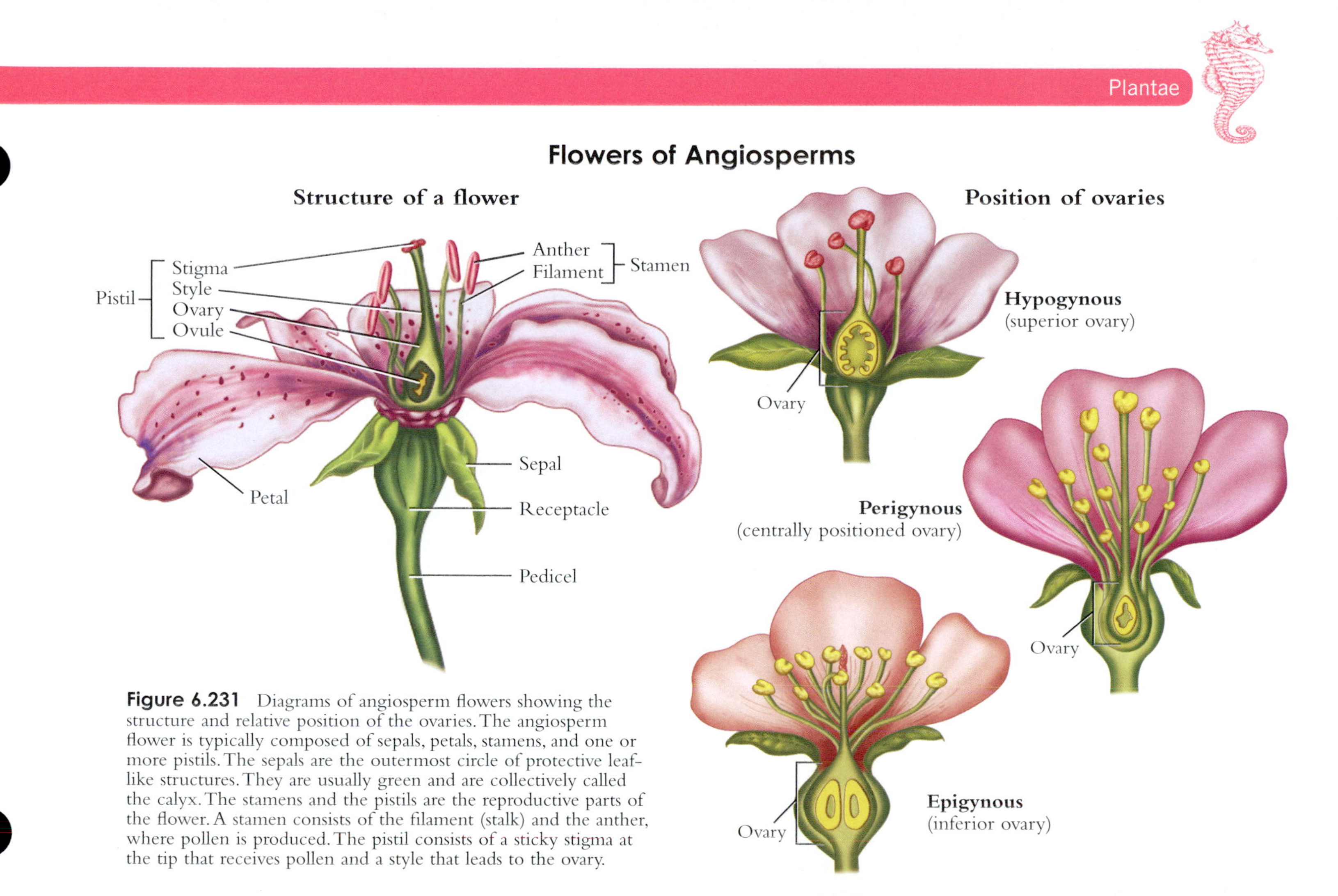

Figure 6.231 Diagrams of angiosperm flowers showing the structure and relative position of the ovaries. The angiosperm flower is typically composed of sepals, petals, stamens, and one or more pistils. The sepals are the outermost circle of protective leaf-like structures. They are usually green and are collectively called the calyx. The stamens and the pistils are the reproductive parts of the flower. A stamen consists of the filament (stalk) and the anther, where pollen is produced. The pistil consists of a sticky stigma at the tip that receives pollen and a style that leads to the ovary.

Sunflower

Dahlia

Passion flower

Figure 6.232 Flowers of angiosperms.

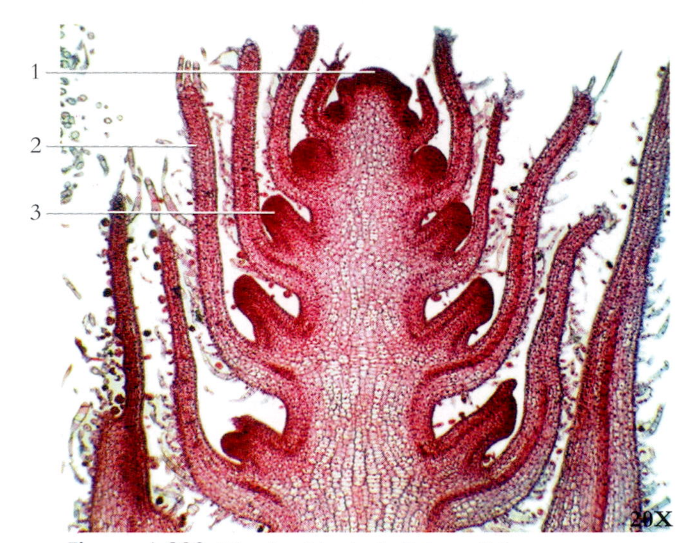

Figure 6.233 The floral bud of Coleus, *Coleus* sp.
1. Apical meristem
2. Bract
3. Floral bud

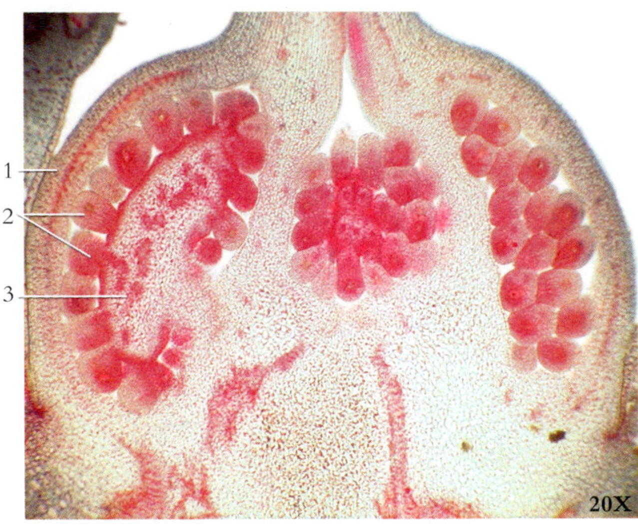

Figure 6.234 The ovary of tomato, *Lycopersicon* sp., with developing ovules
1. Ovary wall
2. Ovules
3. Placenta

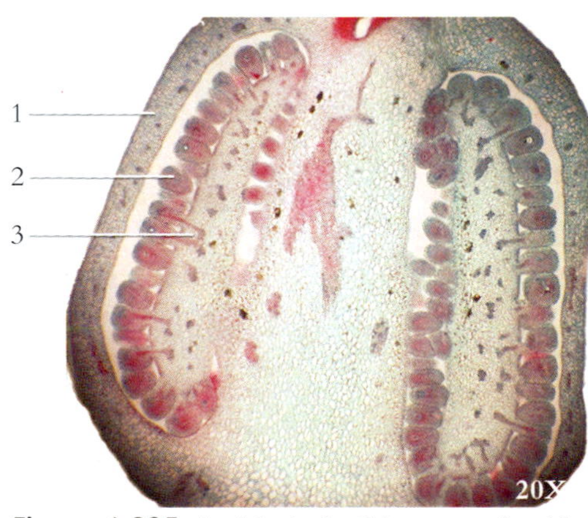

Figure 6.235 A nightshade, *Solanum* sp., floral bud showing ovary with developing ovules.
1. Ovary wall
2. Ovules
3. Placenta

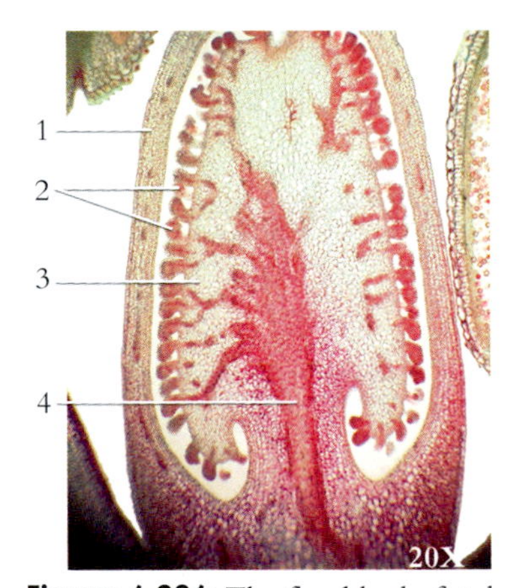

Figure 6.236 The floral bud of tobacco, *Nicotiana* sp., showing the ovary and ovules.
1. Ovary wall
2. Ovules
3. Placenta
4. Vascular tissue

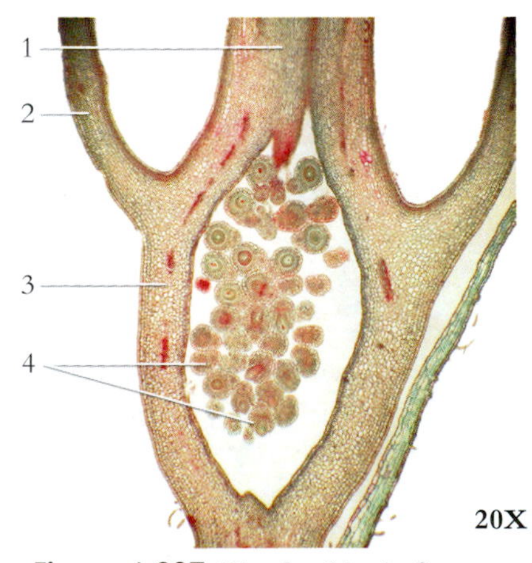

Figure 6.237 The floral bud of a currant, *Ribes* sp., showing an inferior ovary with developing ovules.
1. Style
2. Petal
3. Ovary
4. Ovules

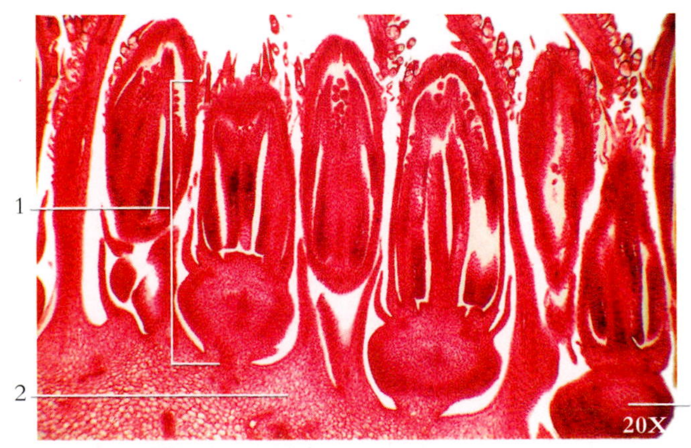

Figure 6.238 The floral bud of sunflower, *Helianthus* sp., with several immature flowers.
1. Individual flower
2. Receptacle
3. Ovary of individual flower

Figure 6.239 The floral structure of a tulip, *Tulipa* sp.

1. Petal
2. Anther
3. Stigma
4. Filament
5. Style

Figure 6.240 The structure of a dissected cherry, *Prunus* sp., showing a perigynous flower.

1. Petal
2. Filaments
3. Sepal
4. Anther
5. Stigma
6. Style
7. Floral tube

Figure 6.241 The structure of a dissected pear, *Pyrus* sp., showing an epigynous flower.

1. Petal
2. Anther
3. Filament
4. Style
5. Sepal
6. Ovary

Figure 6.242 A dissected quince, *Chaenomeles japonica*, showing an epigynous flower.

1. Petal
2. Anther
3. Stigma
4. Filament
5. Style
6. Ovules

Figure 6.243 (a) The floral structure of *Gladiolus* sp. (b) The anthers and stigma and (c) the ovary.

1. Anther
2. Filament
3. Ovules
4. Receptacle
5. Stigma
6. Style
7. Ovary
8. Anther
9. Pollen
10. Stigma
11. Style
12. Filament
13. Ovules (immature seeds)
14. Receptacle
15. Style
16. Ovary

Figure 6.244 A scanning electron micrograph of the stigma of an angiosperm pistil. The stigma is the location where pollen grains adhere and germinate to produce a pollen tube.

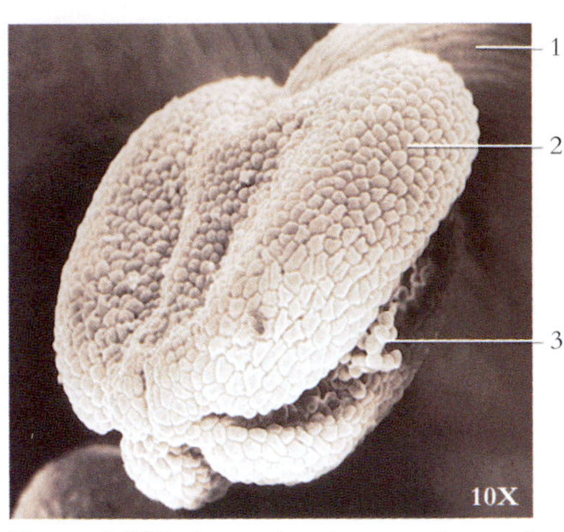

Figure 6.245 A scanning electron micrograph of the anther of candy tuft, *Lobularia* sp. The anther has ruptured, resulting in the release of pollen grains.

1. Filament
2. Anther
3. Pollen grains

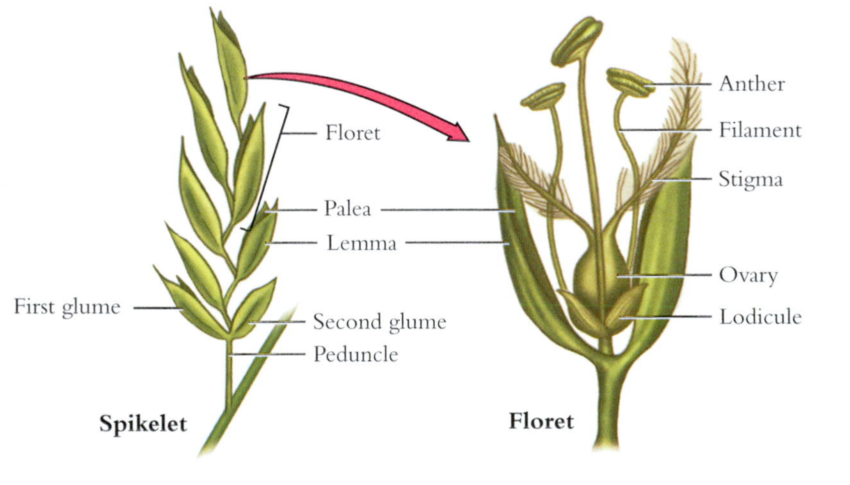

Figure 6.246 The floral structure of grasses.

Figure 6.247 The floral parts of a grass, *Elymus flavescens*, showing spikelets with six florets.

Figure 6.248 Three economically important grasses are: (a) Wheat, *Triticum* sp. is one of the most important human staple foods. (b) Corn, *Zea mays* is a New World native important as human and livestock food. (c) Bamboo is important in commerce and in many natural ecosystems.

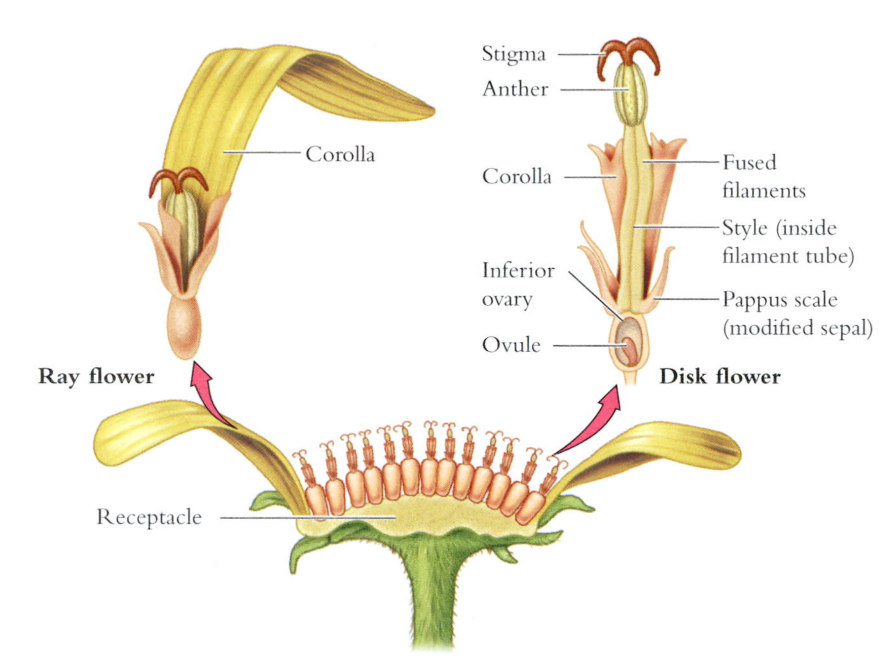

Figure 6.249 The flowers of the family Asteraceae are usually produced in tight heads resembling single large flowers. One of these inflorescences can contain hundreds of individual flowers. Examples of this family include dandelions, sunflowers, asters, and marigolds.

Figure 6.250 A dissected inflorescence of a member of the Asteraceae, *Balsamorhiza sagittata*.

1. Ray flower
2. Disk flower
3. Receptacle

Figure 6.251 A strawberry, *Fragaria* sp., showing (a) the flower, (b) immature aggregate fruits, and (c) a ripening fruit.

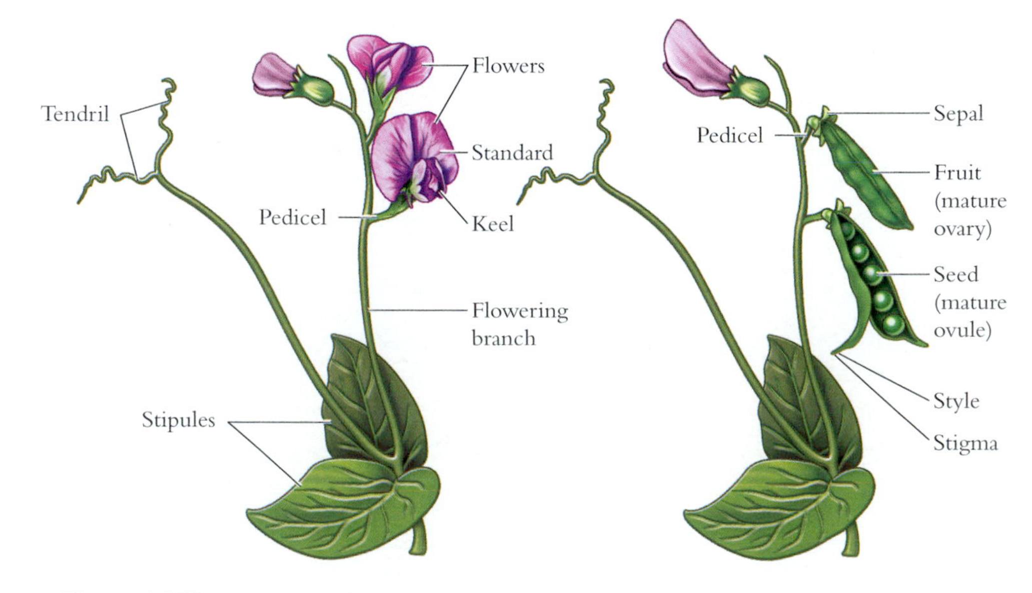

Figure 6.252 The flower and fruit of the pea, *Pisum* sp.

Figure 6.253 The seeds on the receptacle of the giant sunflower.

Figure 6.254 Example flower types: (a) a complete flower, lily and (b) an incomplete flower, orchid. (c) A perfect flower, gerbera daisy, and (d) an imperfect flower, orchid. (e) Actinomorphic symmetry, daffodil, and (f) zygomorphic symmetry, iris. (g) A solitary flower, dahlia, and inflorescent flowers, (h) sunflower, and (i) walnut catkins.

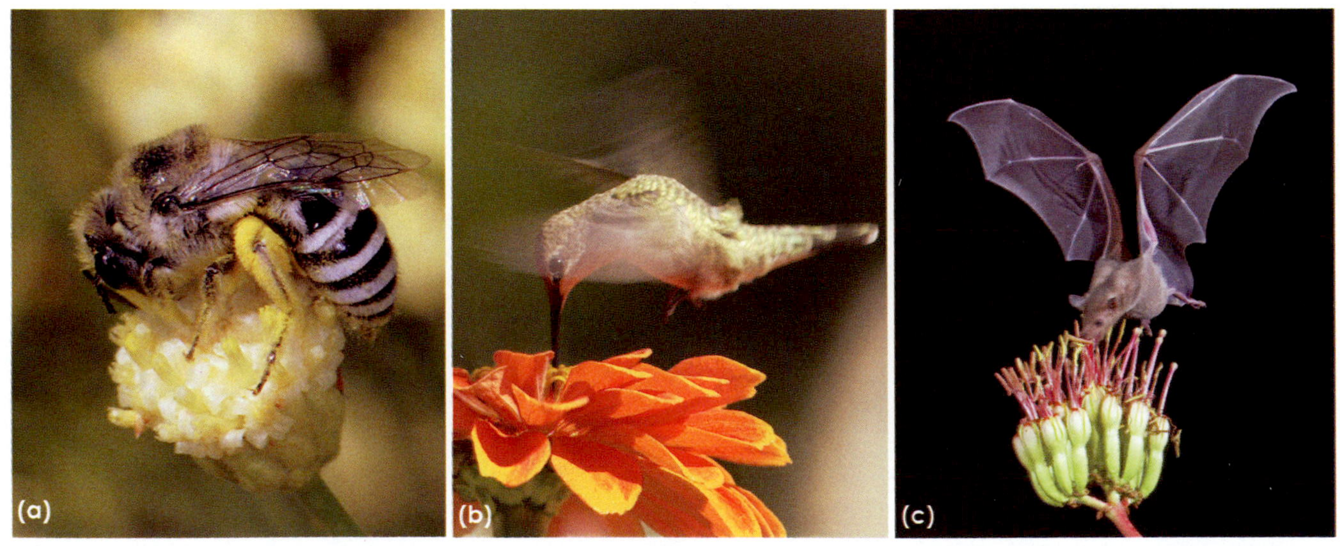

Figure 6.255 Flowers of many angiosperms are uniquely adapted for and rely on specific animals for pollination. Example animal pollinators include: (a) a bee, *Anthophora urbana*, (b) a broad-tailed hummingbird (female), *Selasphorus platycercus*, and (c) a lesser long-nosed bat, *Leptonycteris yerbabuenae.*

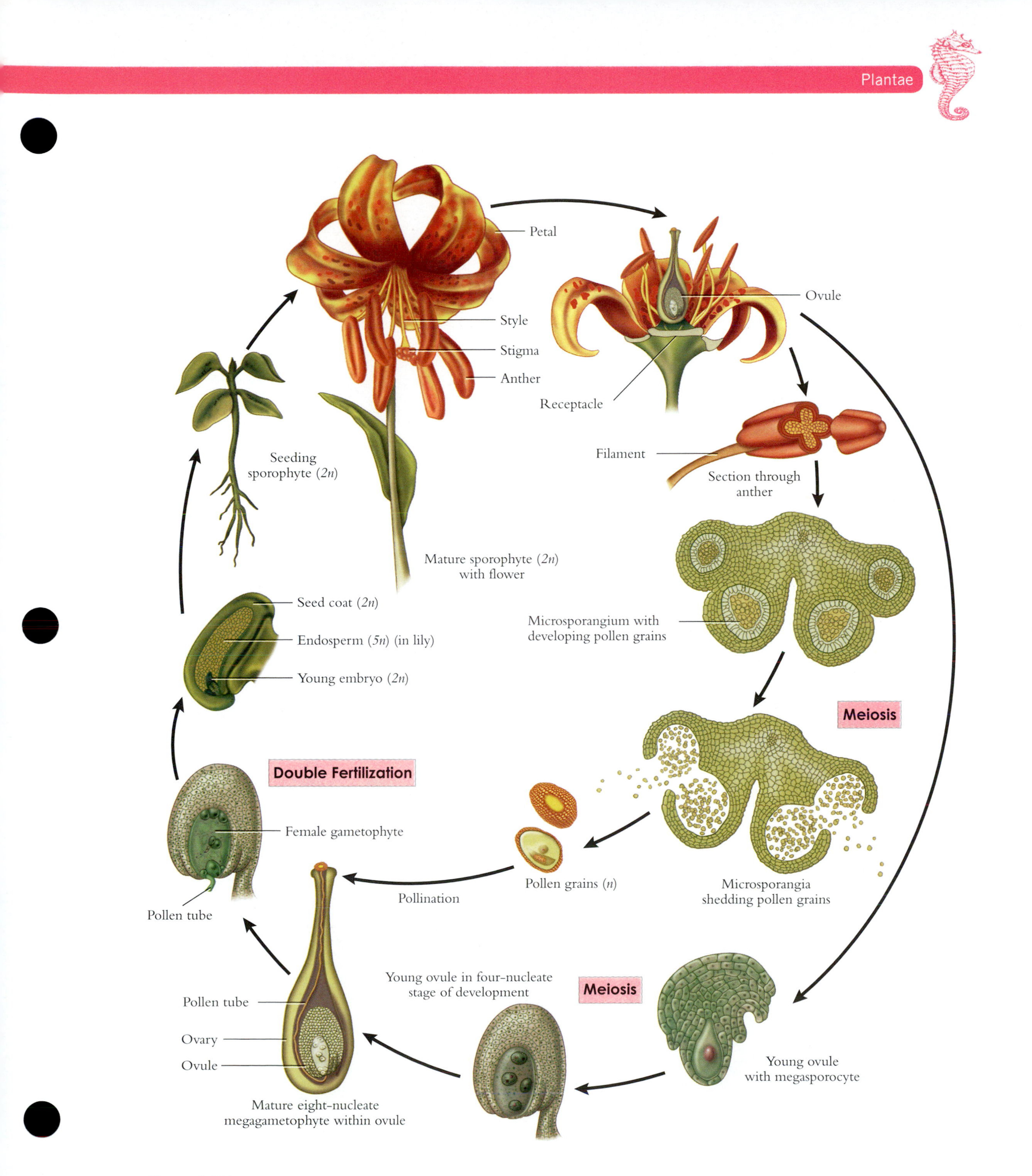

Petal
Ovule
Style
Stigma
Anther
Receptacle
Filament
Section through anther
Seeding sporophyte (2n)
Mature sporophyte (2n) with flower
Seed coat (2n)
Endosperm (5n) (in lily)
Young embryo (2n)
Microsporangium with developing pollen grains
Meiosis
Double Fertilization
Female gametophyte
Pollen grains (n)
Microsporangia shedding pollen grains
Pollination
Pollen tube
Young ovule in four-nucleate stage of development
Meiosis
Pollen tube
Ovary
Ovule
Young ovule with megasporocyte
Mature eight-nucleate megagametophyte within ovule

Figure 6.256 The life cycle of an angiosperm.

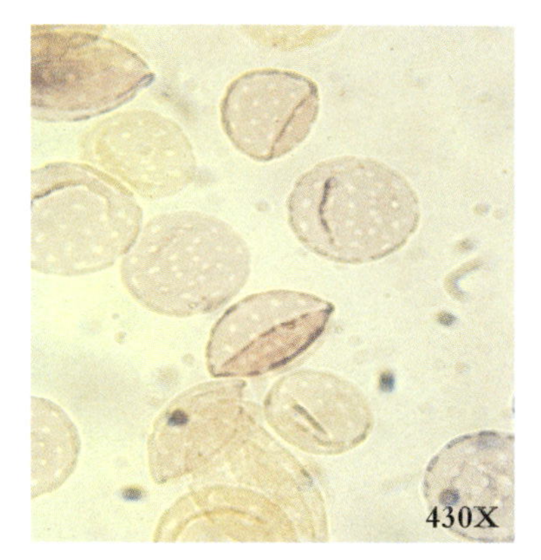

Figure 6.257 The pollen grains of the dicot pigweed, *Amaranthus* sp.

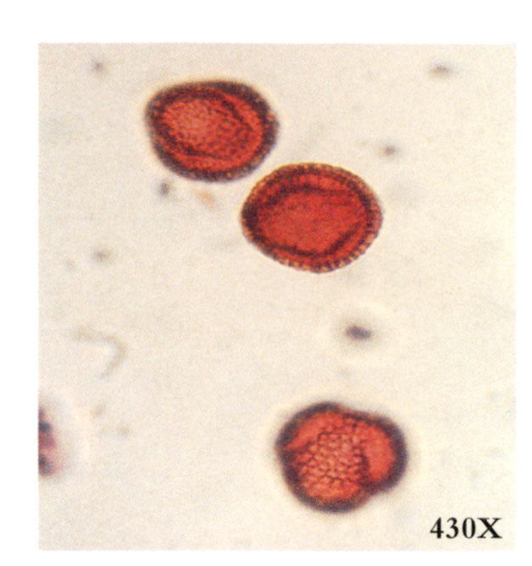

Figure 6.258 The pollen grains of a lilac, *Syringa* sp.

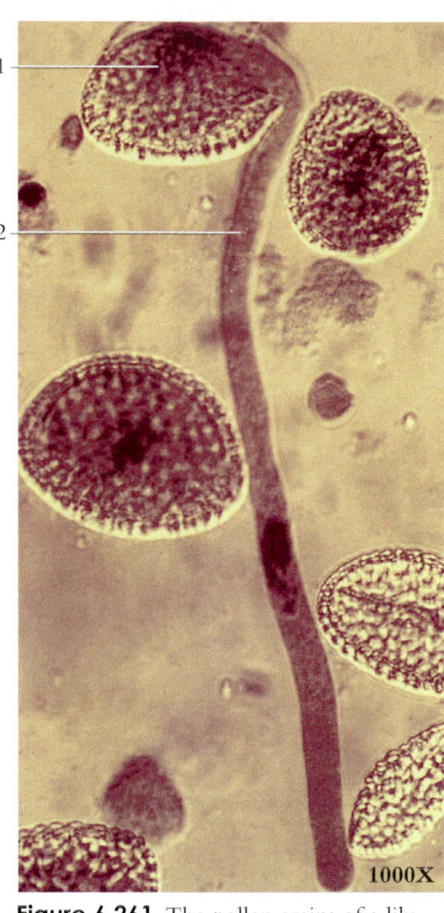

Figure 6.261 The pollen grains of a lily. The pollen grain at the top of the photo has germinated to produce a pollen tube.
1. Pollen grain 2. Pollen tube

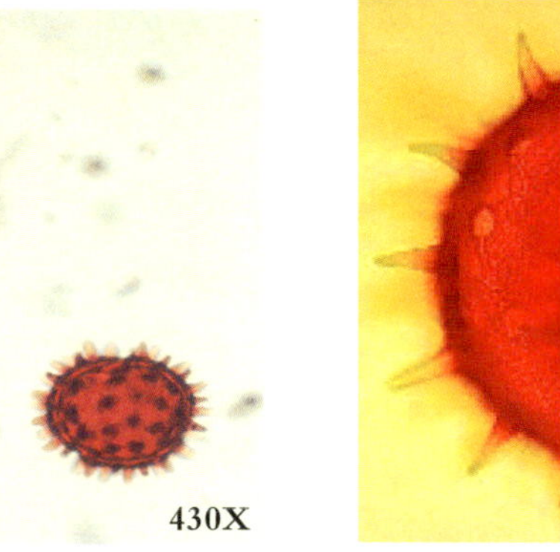

Figure 6.259 The pollen grains of the dicot arrowroot, *Balsamorhiza* sp.

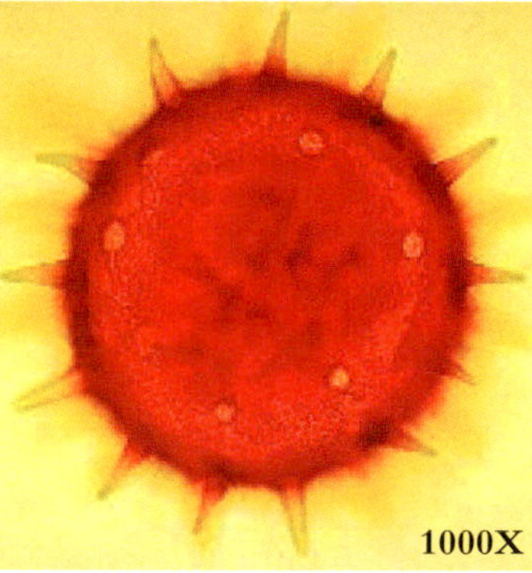

Figure 6.260 The pollen grain of hibiscus, *Hibiscus* sp.

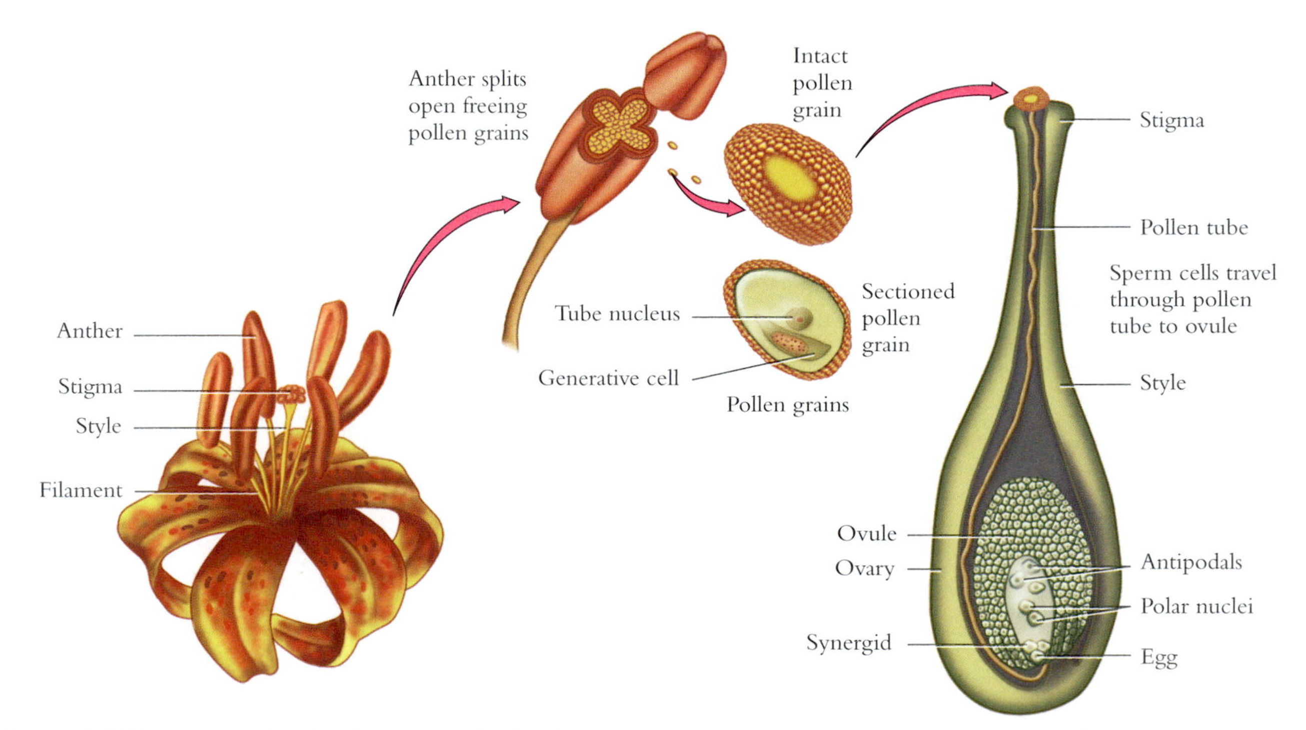

Figure 6.262 A diagram showing the process of pollination.

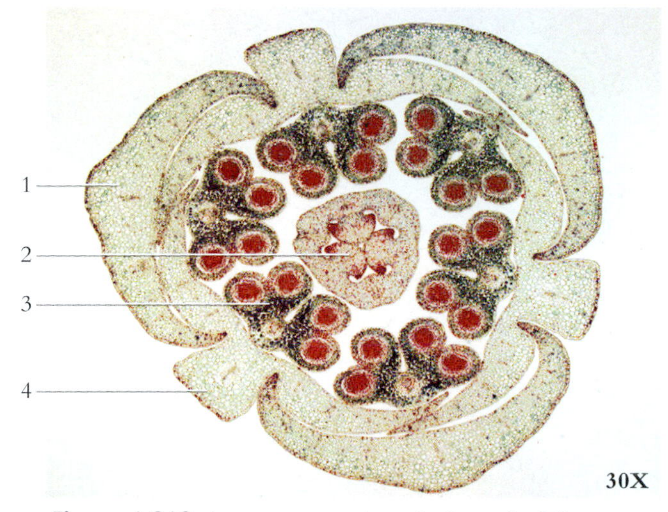

Figure 6.263 A transverse section of a flower bud from a lily, *Lilium* sp.

1. Sepal
2. Ovary
3. Anther
4. Petal

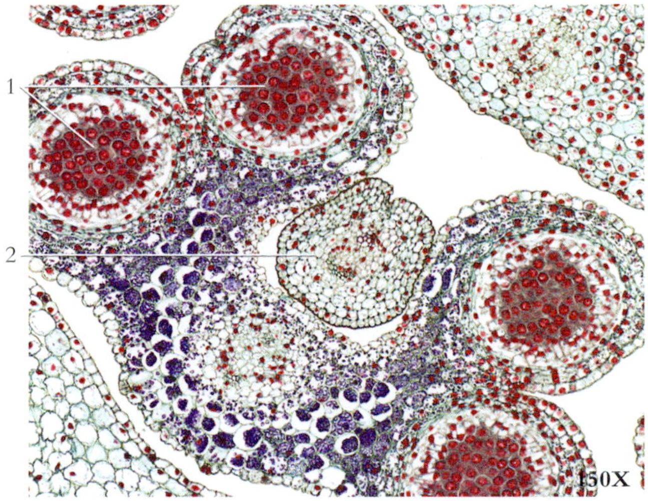

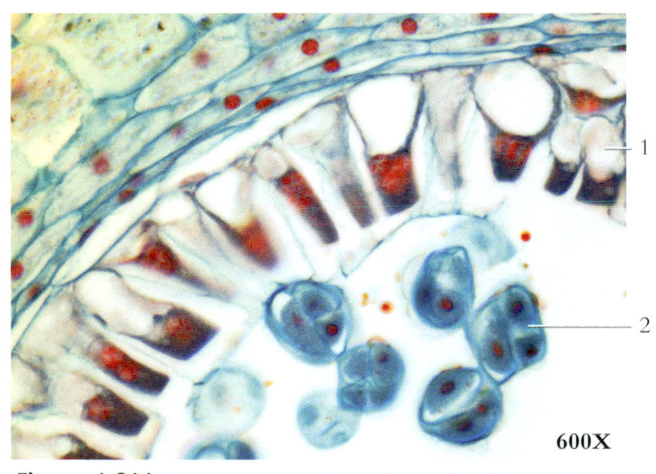

Figure 6.264 A transverse section of an anther from a lily, *Lilium* sp.

1. Sporogenous tissue
2. Filament

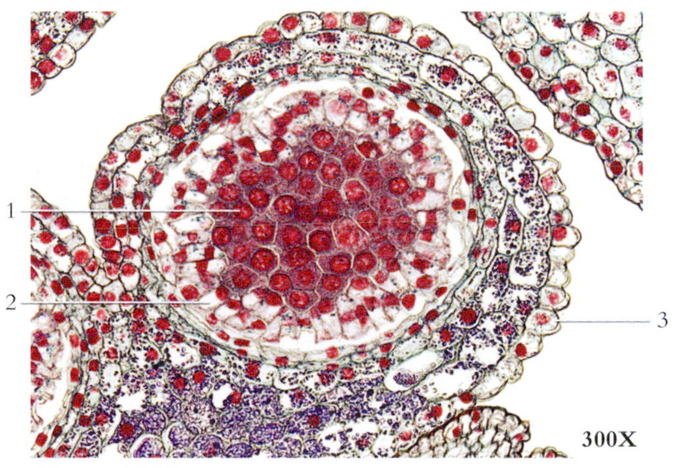

Figure 6.265 A transverse section of an anther from a lily, *Lilium* sp.

1. Young microsporocytes
2. Tapetum
3. Anther wall

Figure 6.266 A transverse section of an anther from a lily, *Lilium* sp., magnified view.

1. Tapetum
2. Tetrad of microspores

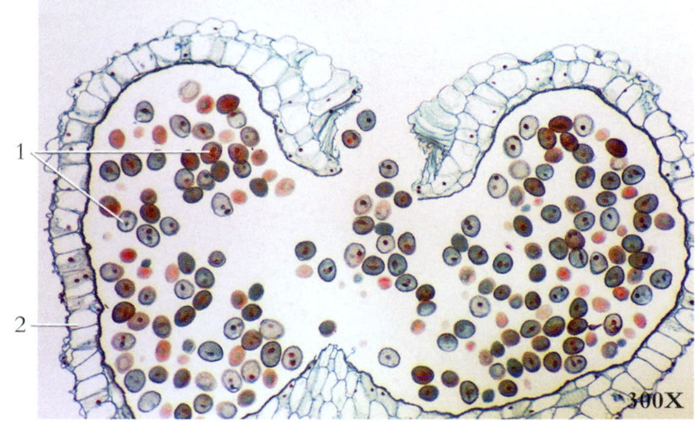

Figure 6.267 A transverse section of an anther from a lily, *Lilium* sp., showing mature pollen.

1. Pollen grains with two cells
2. Anther wall

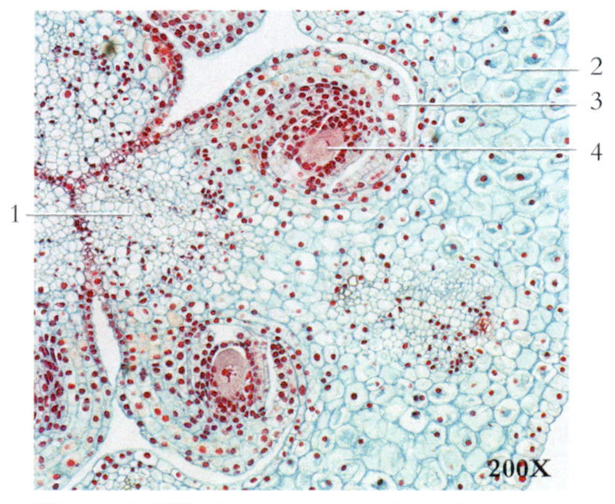

Figure 6.268 A transverse section of a lily, *Lilium* sp., ovary showing ovules.

1. Placenta
2. Ovary wall
3. Ovule
4. Megasporocyte (2*n*)

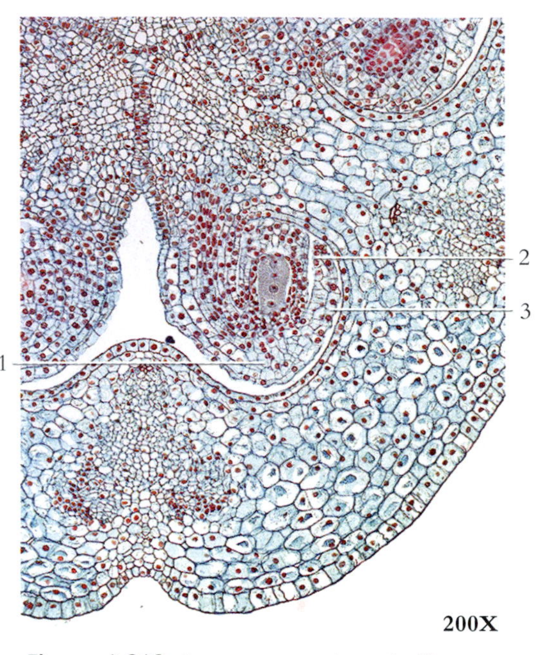

Figure 6.269 A transverse section of a lily, *Lilium* sp., ovary showing megaspore.

1. Ovule
2. Linear tetrad of megaspore
3. Integument

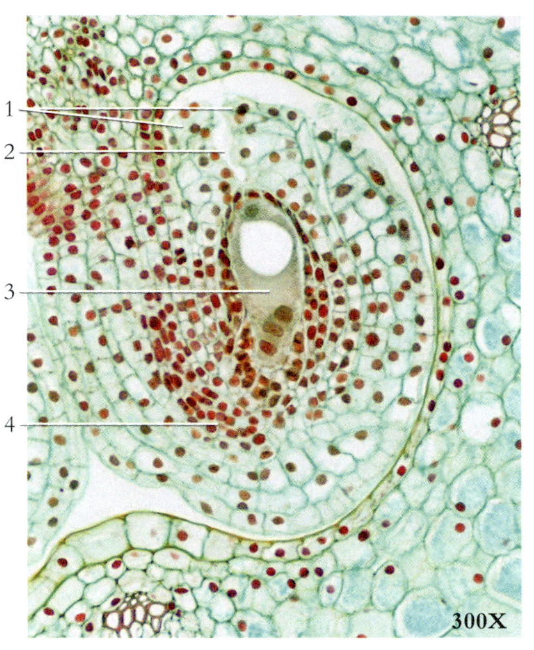

Figure 6.270 A transverse section of a lily, *Lilium* sp., ovary showing ovule with developing embryo sac.

1. Integuments
2. Micropyle
3. Embryo sac
4. Ovule

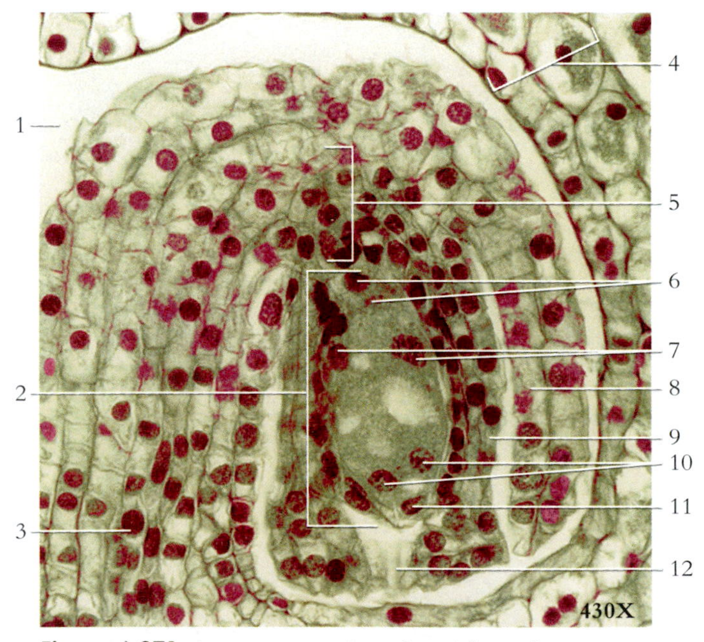

Figure 6.271 A transverse section of an eight-nucleate embryo sac of an ovule from a lily, *Lilium* sp.

1. Locule
2. Megagametophyte
3. Funiculus
4. Wall of ovary
5. Chalaza
6. Antipodal cells (3*n*)
7. Polar nuclei (3*n*)
8. Outer integument (2*n*)
9. Inner integument (2*n*)
10. Synergid cells (*n*)
11. Egg (*n*)
12. Micropyle (pollen tube entrance)

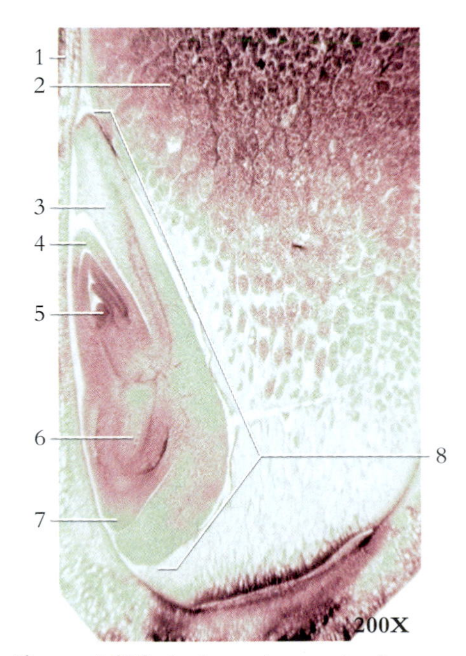

Figure 6.272 A photomicrograph of a mature grain, or kernel, of wheat, *Triticum aestivum*.

1. Pericarp
2. Starchy endosperm
3. Scutellum
4. Coleoptile
5. Shoot apex
6. Radicle
7. Coleorhiza
8. Embryo

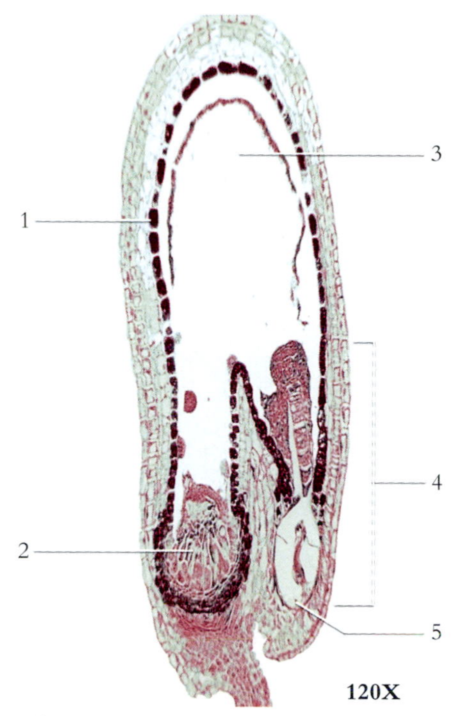

Figure 6.273 A photomicrograph of a developing dicot embryo from a shepherd's purse, *Capsella bursa-pastoris*.

1. Endothelium
2. Cellular endosperm
3. Endosperm
4. Developing embryo
5. Basal cell

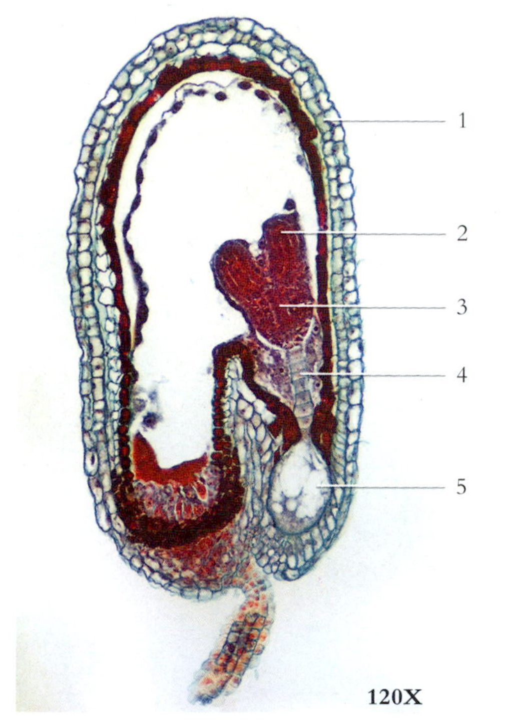

Figure 6.274 A photomicrograph of a developing dicot embryo from a shepherd's purse, *Capsella bursa-pastoris*, showing young embryo.

1. Seed coat
2. Cotyledon
3. Hypocotyl
4. Suspensor
5. Basal cell

Figure 6.275 A photomicrograph of a developing dicot embryo from a shepherd's purse, *Capsella bursa-pastoris*, showing a nearly mature embryo.

1. Endosperm
2. Epicotyl
3. Cotyledon
4. Radicle
5. Seed coat
6. Hypocotyl

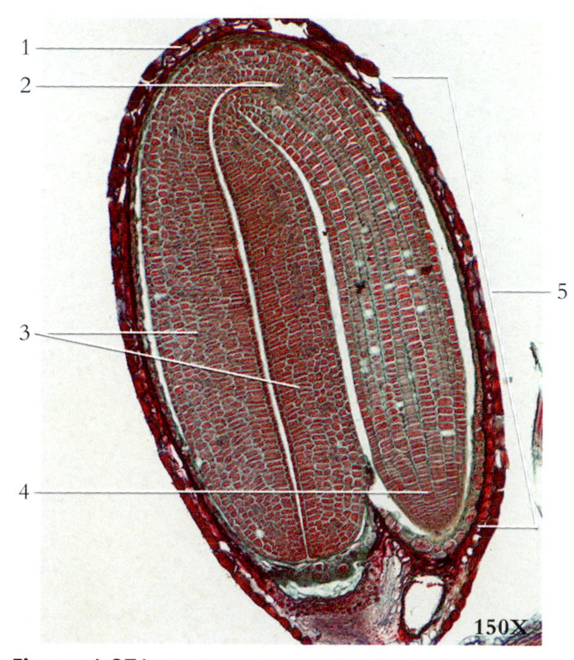

Figure 6.276 A photomicrograph of a developing dicot embryo from a shepherd's purse, *Capsella bursa-pastoris*, showing a mature embryo.

1. Seed coat
2. Epicotyl
3. Cotyledons
4. Radicle
5. Hypocotyl

Figure 6.277 The flower and fruit of the pear *Pyrus* sp. The pear fruit develops from the floral tube (fused perianth) as well as the ovary.

Figure 6.278 The flower (a) and the fruits (b and c) of the dandelion, *Taraxacum* sp. The dandelion has a composite flower. The wind-borne fruit (containing one seed) of a dandelion, and many other members of the family Asteraceae, develop a plumelike pappus that enables the light fruit to float in the air.

1. Pappus
2. Ovary wall, with one seed inside

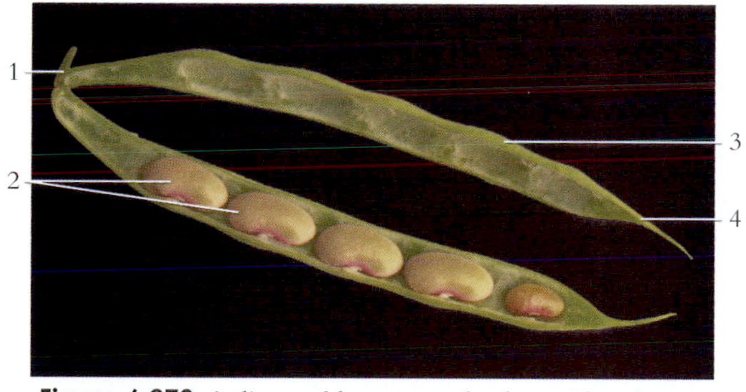

Figure 6.279 A dissected legume, garden bean, *Phaseolus* sp.

1. Pedicel
2. Seeds
3. Fruit
4. Style

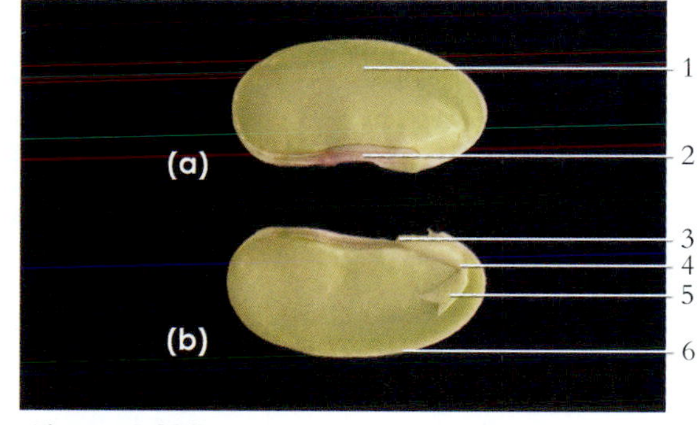

Figure 6.280 A lima bean. (a) The entire bean seed and (b) a longitudinally sectioned seed.

1. Integument (seed coat)
2. Hilum
3. Radicle
4. Hypocotyl
5. Epicotyl (plume)
6. Cotyledon

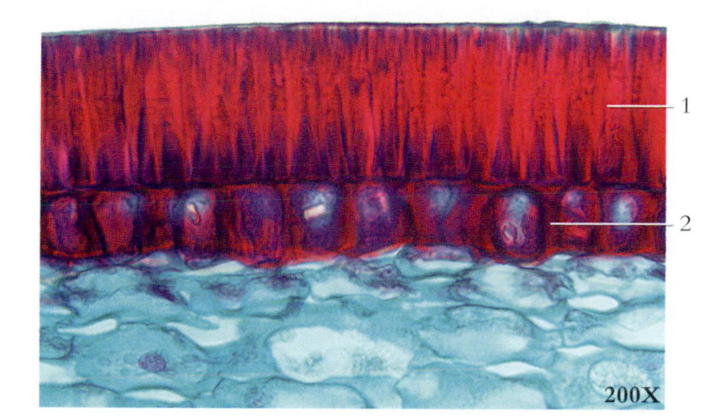

Figure 6.281 A photomicrograph of the seed coat of the garden bean, *Phaseolus* sp., showing the sclerified epidermis

1. Macrosclereids
2. Subepidermal sclereids

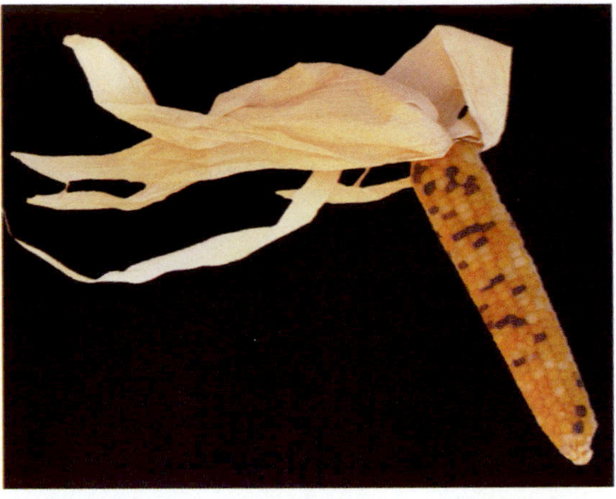

Figure 6.282 A cob of corn, *Zea mays*. Corn was domesticated approximately 7,000 years ago from a Mexican grass, family Poaceae.

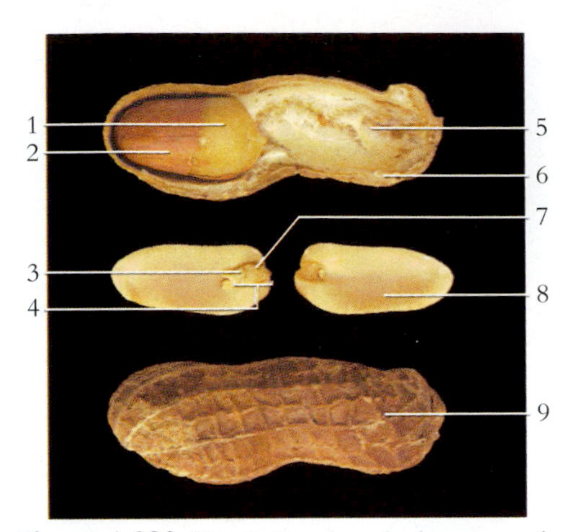

Figure 6.283 The fruit and seed of a peanut plant.

1. Cotyledon
2. Integument (seed coat)
3. Plumule
4. Embryo axis
5. Interior of fruit
6. Mesocarp
7. Radicle
8. Cotyledon
9. Fruit wall (pericarp)

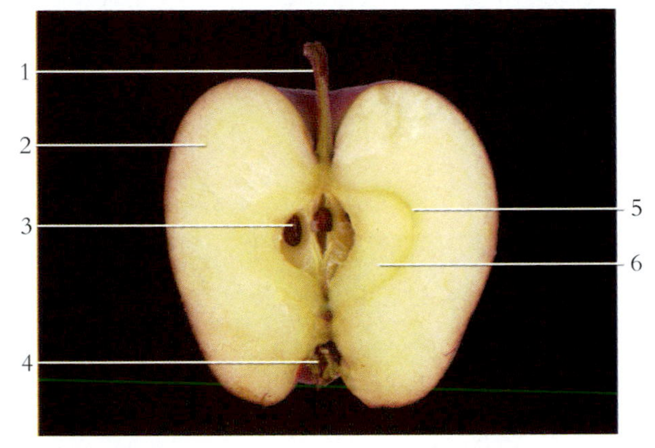

Figure 6.284 A longitudinal section of an apple fruit.

1. Pedicel
2. Mature floral tube
3. Seed (mature ovule)
4. Remnants of floral parts
5. Ovary wall
6. Mature ovary (2 & 6 make up the fruit)

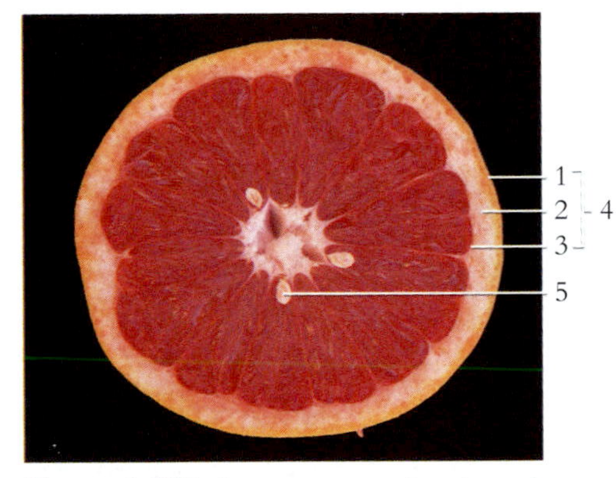

Figure 6.285 A transverse section through a grapefruit fruit.

1. Exocarp
2. Mesocarp
3. Endocarp
4. Pericarp
5. Seed

Figure 6.286 A longitudinal section of a pineapple fruit.

1. Shoot apex
2. Central axis
3. Floral parts

Figure 6.287 A longitudinal section of a tomato fruit (berry).

1. Pedicel
2. Pericarp
3. Locule
4. Placenta
5. Seed
6. Sepals
7. Mature ovary (fruit)

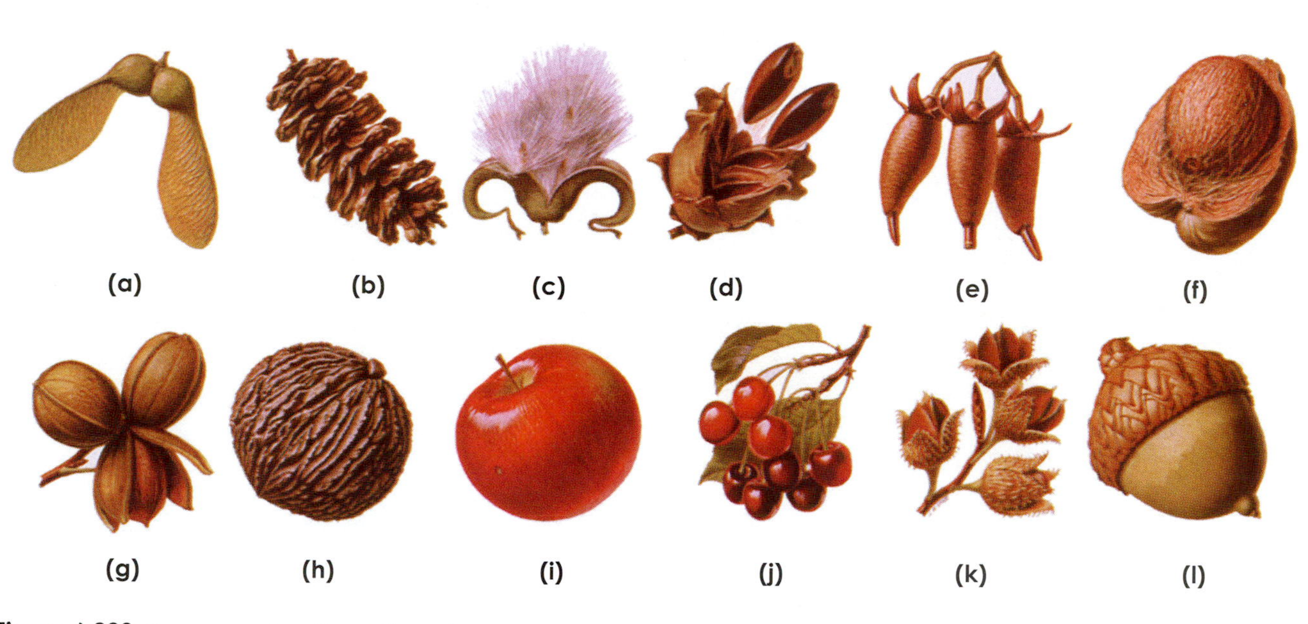

Figure 6.288 Some examples of seed dispersal.

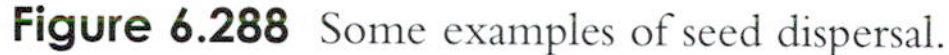

(a) Maple—The winged fruits of a maple fall with a spinning motion that may carry it hundreds of yards from the parent tree.
(b) White pine—The second-year cones of a white pine open to expose the winged seeds to the wind.
(c) Willow—The airborne seeds of a willow may be dispersed over long distances.
(d) Witch hazel—Mature seeds of the witch hazel tree are dispersed up to 10 feet by forceful discharge.
(e) Mangrove—The fruits of this tropical tree begin to germinate while still on the branch, forming pointed roots. When the seeds drop from the tree, they may float to a muddy area where the roots take hold.
(f) Coconut—The buoyant, fibrous husk of a coconut permits dispersal from one island or land mass to another by ocean currents.
(g) Pecan—The fruit husk of a pecan provides buoyancy and protection as it is dispersed by water.
(h) Black walnut—The encapsulated seed of the black walnut is dispersed through burial by a squirrel or floating in a stream.
(i) Apple—The seeds of an apple tree may be dispersed by animals that ingest the fruit and pass the undigested seeds hours later in their feces.
(j) Cherry—Moderate-sized birds, such as robins, may carry a ripe cherry to an eating site where the juicy pulp is eaten and the hard seed is discarded.
(k) Beech—Seeds from a beech tree are dispersed by mammals as the spiny husks adhere to their hair. In addition, many mammals ingest these seeds and disperse them in their feces.
(l) Oak—An oak seed may be dispersed through burial of the acorn fruit by a squirrel or jay.

Figure 6.289 (a) The mature milkweed, *Asclepias* sp.; (b) milkweed pods; and (c) seeds ready for airborne dispersal.

Forcible discharge dispersal

Touch-me-not

Water dispersal

Coconut

Animal dispersal

Cocklebur

Burdock

Blackberries

Wind dispersal

Dandelion

Poppy

Maple

Figure 6.290 Several fruits and seeds to illustrate seed dispersal.

Animalia

Chapter 7

Animals are multicellular, heterotrophic eukaryotes that ingest food materials and store carbohydrate reserves as glycogen or fat. The cells of animals lack cell walls but do contain intercellular connections including desmosomes, gap junctions, and tight junctions. Animal cells are also highly specialized into the specific kinds of tissues described in chapter 1. Most animals are motile through the contraction of muscle fibers containing actin and myosin proteins. The complex body systems of animals include elaborate sensory and neuromotor specializations that accommodate dynamic behavioral mechanisms.

Reproduction in animals is primarily sexual, with the diploid stage generally dominating the life cycle. Primary sex organs, or *gonads*, produce the haploid gametes called *sperm* and *egg*. Propagation begins as a small flagellated sperm fertilizes a larger, nonmotile egg, forming a diploid zygote that has genetic traits of both parents. The zygote then undergoes a succession of mitotic divisions called *cleavage*. In animals, cleavage is followed by the formation of a multicellular stage called a *blastula*. With further development, the *germ layers* form, which eventually give rise to each of the body organs. The developmental cycle of many animals includes *larval forms*, which are still developing, free-living, and sexually immature. Larvae usually have food and habitat requirements different from those of the adults. Larvae eventually undergo metamorphoses that transform them into sexually mature adults.

Animals inhabit nearly all aquatic and terrestrial habitats of the biosphere. The greatest number of animals are marine, where the first animals probably evolved. Depending on the classification scheme, animals may be grouped into as many as 35 phyla. The most commonly known phylum is *Chordata* (table 7.1), which includes the subphylum *Vertebrata*, or the backboned animals. Chordates, however, constitute only about 5% of all the animal species. All other animals are frequently referred to as *invertebrates*, and they account for approximately 95% of the animal species.

Table 7.1 Some Representatives of the Kingdom Animalia

Phylum and Representative Kinds	Characteristics
Porifera — sponges	Multicellular, aquatic animals, with stiff skeletons and bodies perforated by pores
Cnidaria — corals, hydra, and jellyfish	Aquatic animals, radially symmetrical, mouth surrounded by tentacles bearing cnidocytes (stinging cells); body composed of epidermis and gastrodermis, separated by mesoglea
Platyhelminthes — flatworms	Elongated, flattened, and bilaterally symmetrical; distinct head containing ganglia; nerve cords; protonephridia or flame cells
Mollusca — clams, snails, and squids	Bilaterally symmetrical with a true coelom, containing a mantle; many have muscular foot and protective shell
Annelida — segmented worms	Body segmented (except leeches); a series of hearts; hydrostatic skeleton and circular and longitudinal muscles
Nematoda — roundworms	Mostly microscopic; unsegmented wormlike; body enclosed in cuticle; whip-like body movement
Arthropoda — crustaceans, insects, and spiders	Body segmented; paired and jointed appendages; chitinous exoskeleton; hemocoel for blood flow
Echinodermata — sea stars and sea urchins	Larvae have bilateral symmetry; adults have pentaradial symmetry; coelom; most contain a complete digestive tract; regeneration of body parts
Chordata — lancelets, tunicates, and vertebrates	Fibrous notochord, pharyngeal gill slits, dorsal hollow nerve cord, and postanal tail present at some stage in their development

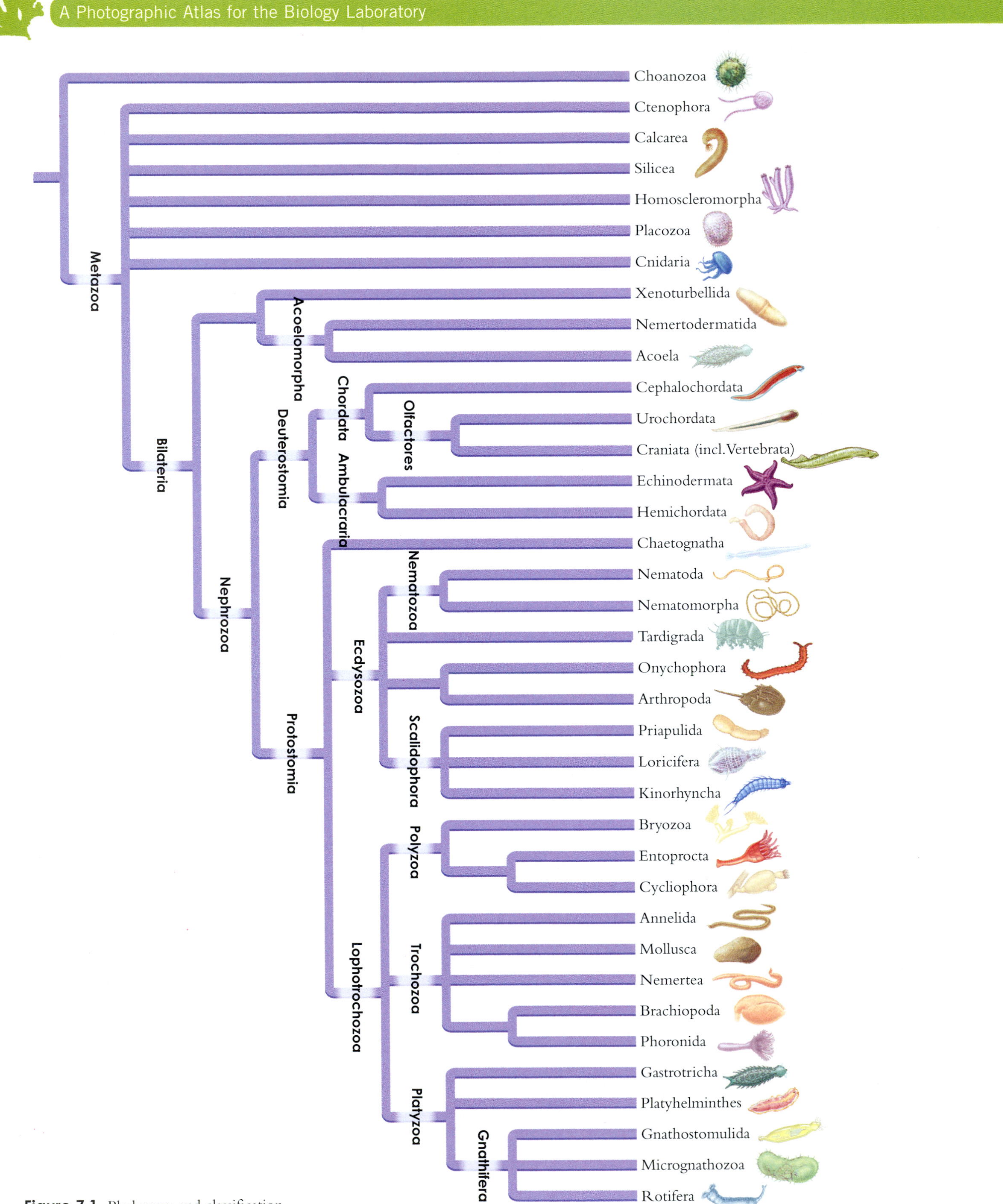

Figure 7.1 Phylogeny and classification of Metazoa (multicellular animals).

Phylum Porifera - sponges

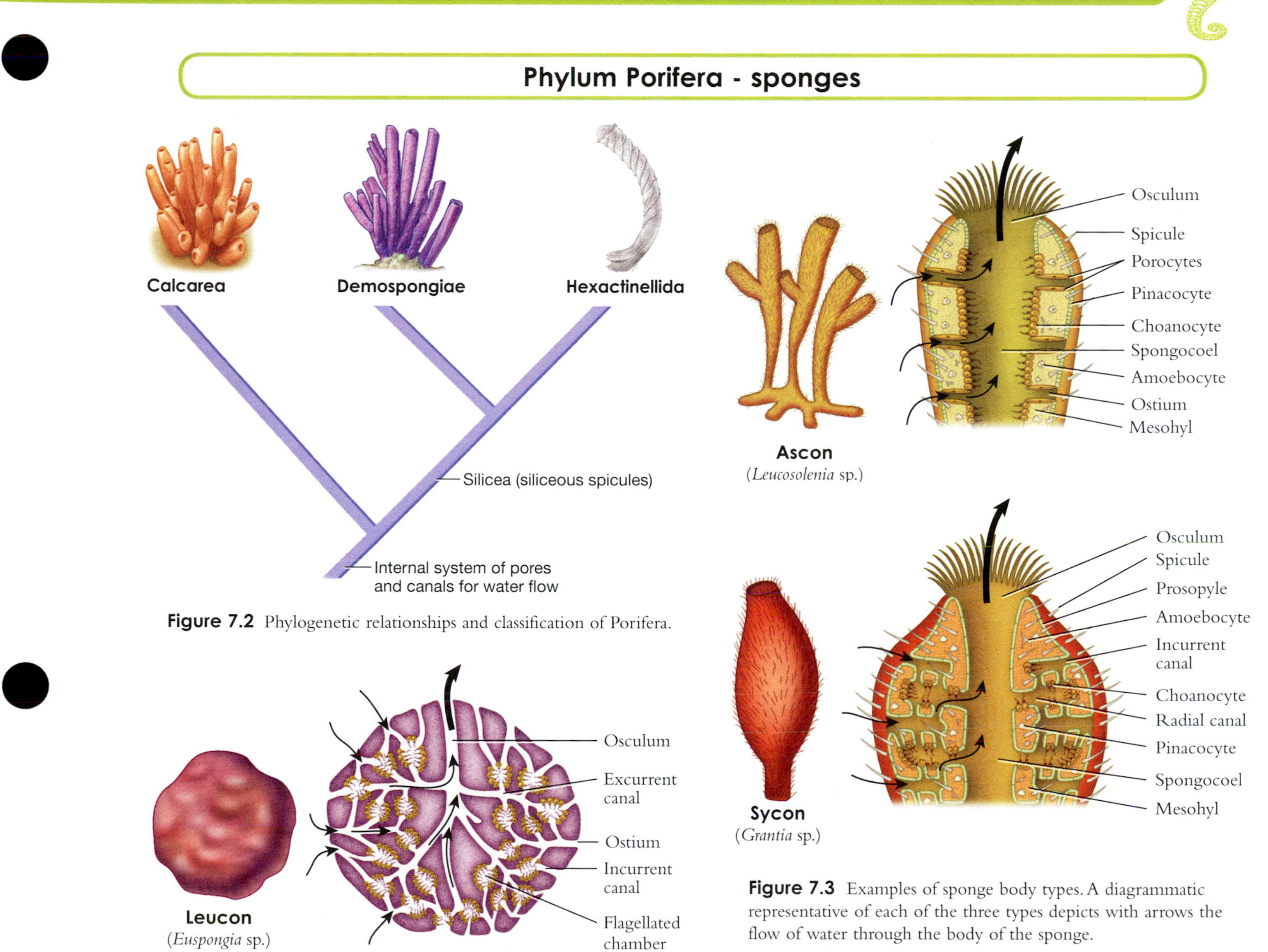

Figure 7.2 Phylogenetic relationships and classification of Porifera.

Figure 7.3 Examples of sponge body types. A diagrammatic representative of each of the three types depicts with arrows the flow of water through the body of the sponge.

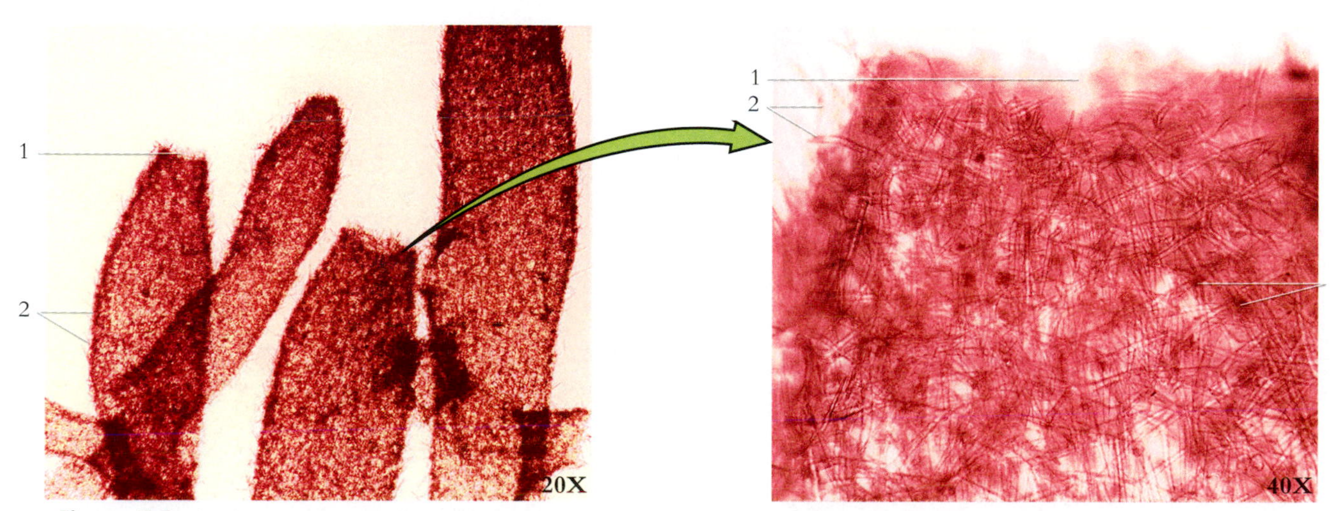

Figure 7.4 (a) The sponge *Leucosolenia* sp. has an ascon body type.
1. Osculum
2. Spicules

Figure 7.5 A higher magnification of the spicules and ostia.
1. Osculum
2. Spicules
3. Ostia

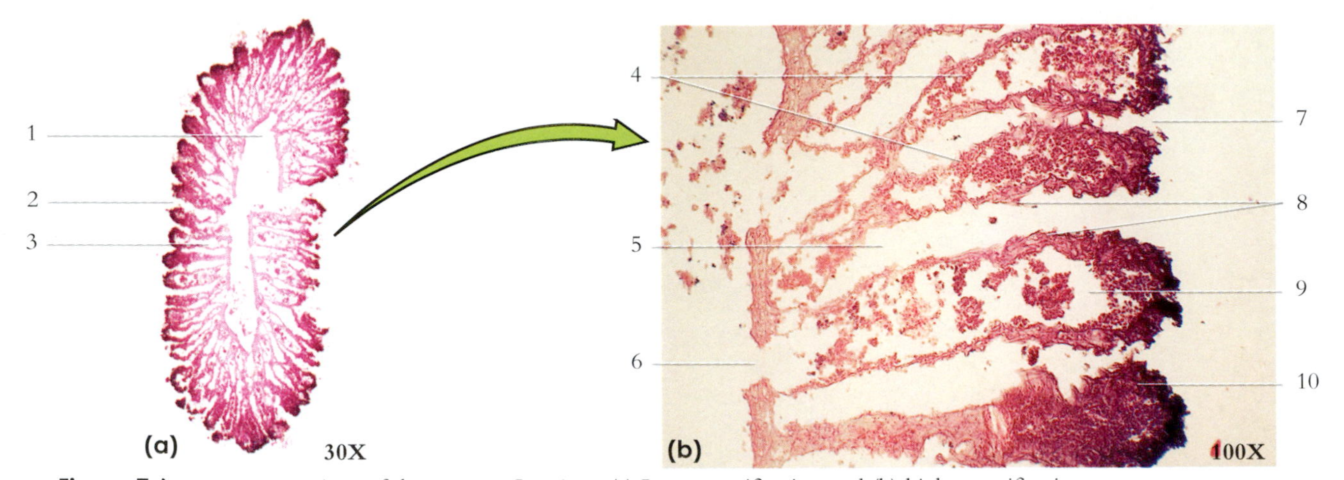

Figure 7.6 Transverse sections of the sponge, *Grantia* sp. (a) Low magnification and (b) high magnification.

1. Spongocoel
2. Incurrent canal
3. Radial canal
4. Choanocytes (collar cells)
5. Incurrent canal
6. Apopyle
7. Ostium
8. Pinacocytes
9. Radial canal
10. Mesohyl

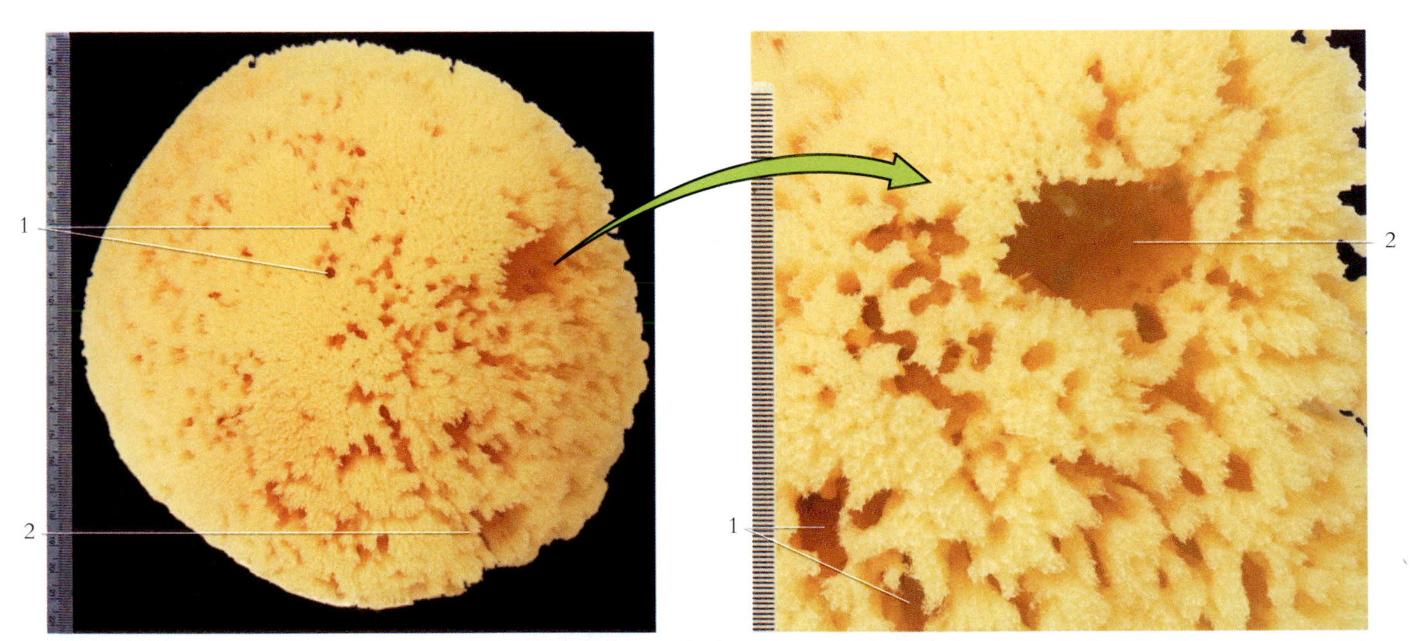

Figure 7.7 A bath sponge, class Demospongiae, has a leuconoid body structure (scale in mm).

1. Ostia
2. Osculum

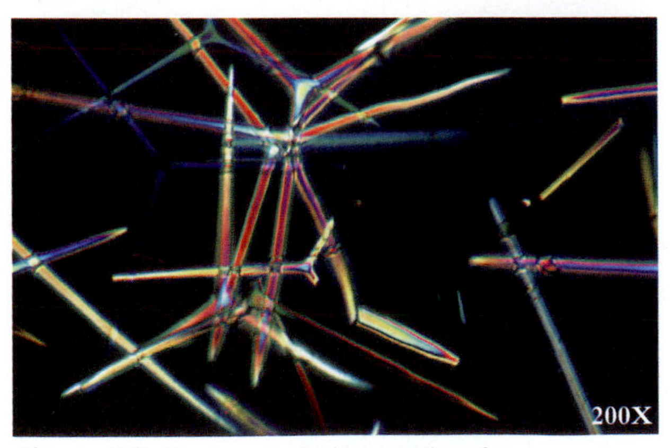

Figure 7.8 The branched silica spicules of a freshwater sponge.

Figure 7.9 An encrusting sponge. Leuconoid sponges display a wide range of color and shape.

Table 7.2 Representatives of the Phylum Ctenophora

Classes and Representative Kinds	Characteristics
Tentaculata – comb jellies	Marine coastal waters; utilize cilia for transportation; most species hermaphroditic; lack stinging cells

Representatives of the Phylum Cnidaria

Classes and Representative Kinds	Characteristics
Hydrozoa – hydra, *Obelia*, and Portuguese man-of-war	Mainly marine; both polyp and medusa stage (polyp form only in hydra); polyp colonies in most
Scyphozoa – jellyfish	Marine coastal waters; polyp stage restricted to small larval forms
Cubozoa – box jellyfish	Marine coastal waters; polyp and medusa stage; square-shaped when viewed from above
Anthozoa – sea anemones, corals, and sea fans	Marine coastal waters; solitary or colonial polyps; no medusa stage; partitioned gastrovascular cavity

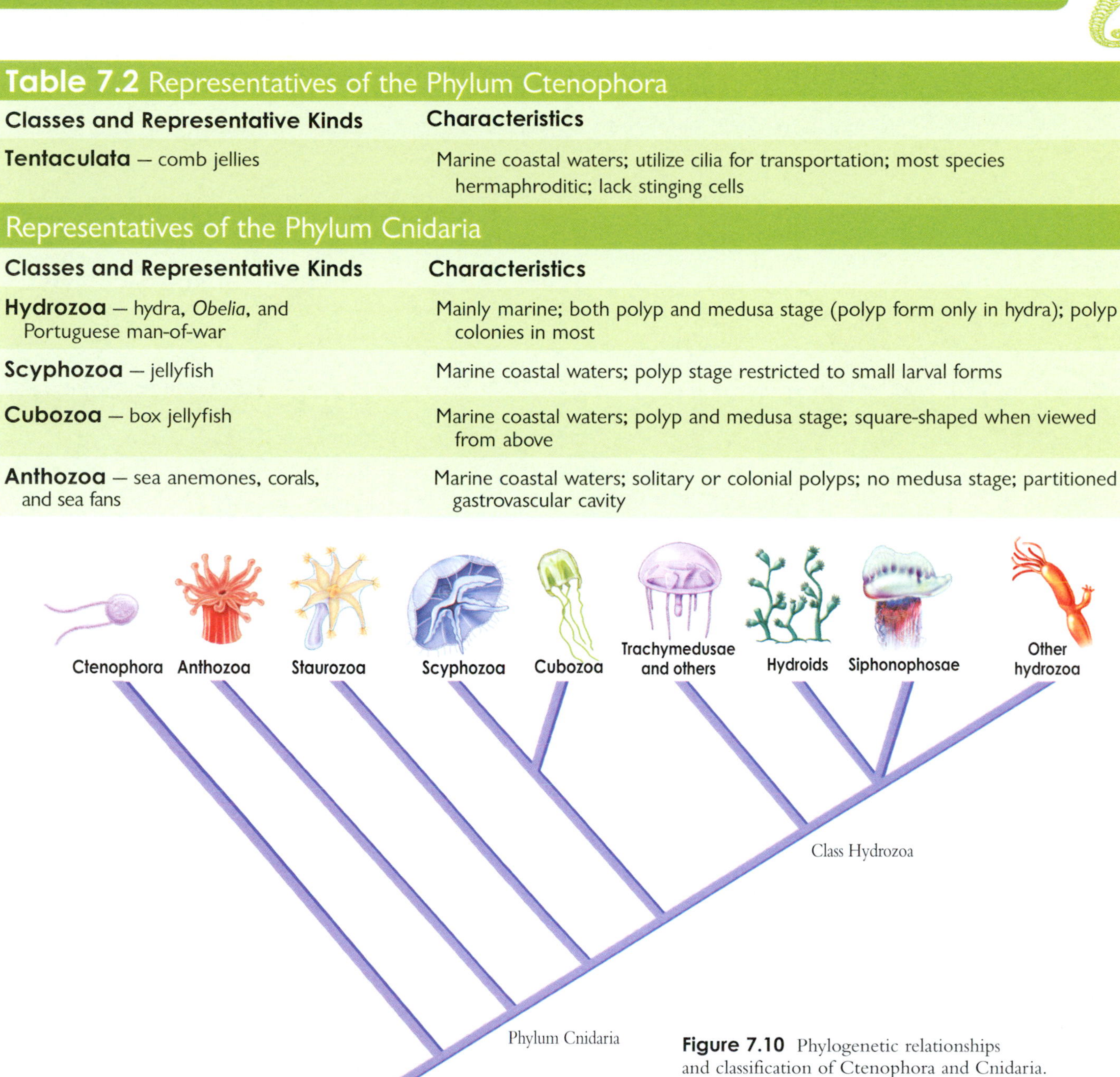

Figure 7.10 Phylogenetic relationships and classification of Ctenophora and Cnidaria.

Phylum Ctenophora - comb jellies

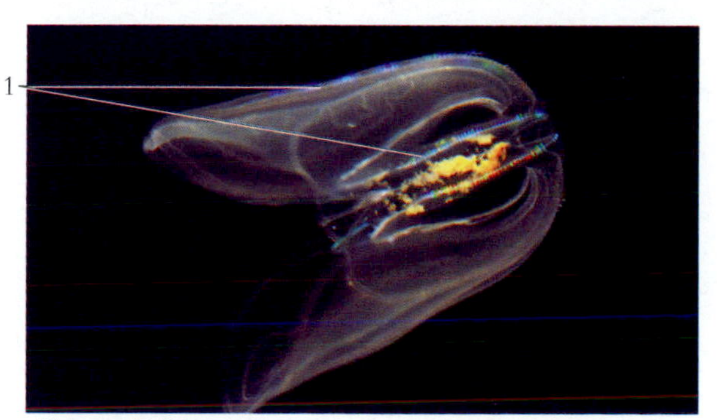

Figure 7.11 The warty comb jelly, *Mnemiopsis leidyi*, is commonly found in the western Atlantic.
1. Rows of cilia

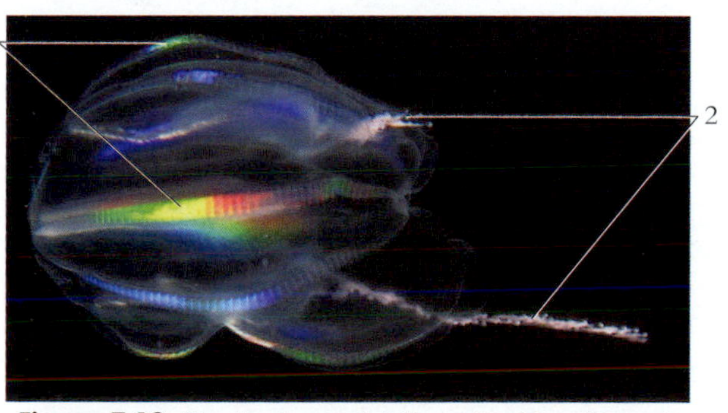

Figure 7.12 The Arctic comb jelly or Sea Nut, *Mertensia ovum*, is found in polar seas.
1. Rows of cilia 2. Tentacles (top tentacle is retracted)

Phylum Cnidaria - hydra, jellyfish, and corals

Class Hydrozoa

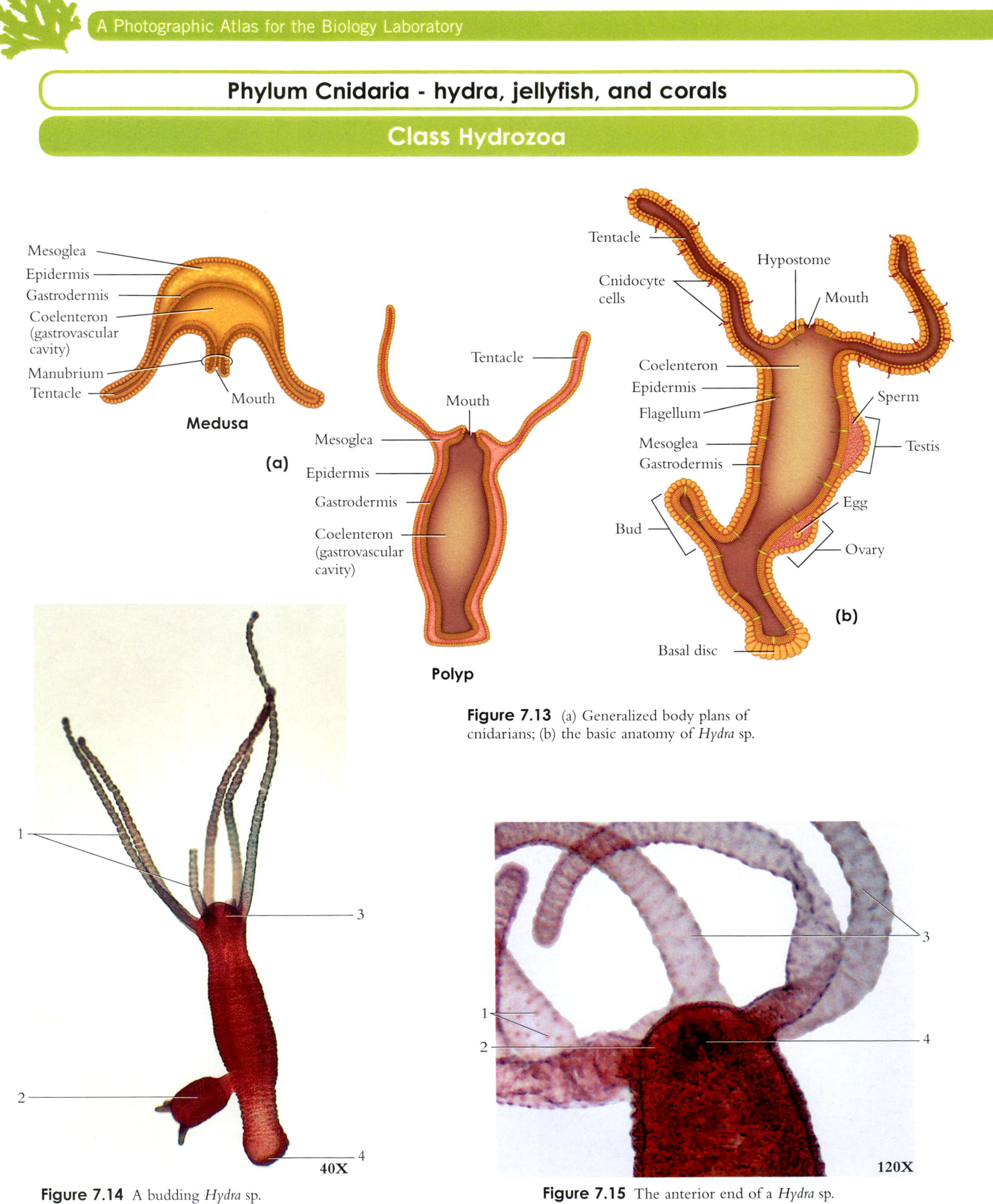

Figure 7.13 (a) Generalized body plans of cnidarians; (b) the basic anatomy of *Hydra* sp.

Figure 7.14 A budding *Hydra* sp.
1. Tentacles
2. Bud
3. Hypostome
4. Basal disc (foot)

Figure 7.15 The anterior end of a *Hydra* sp.
1. Cnidocytes
2. Hypostome
3. Tentacles
4. Mouth

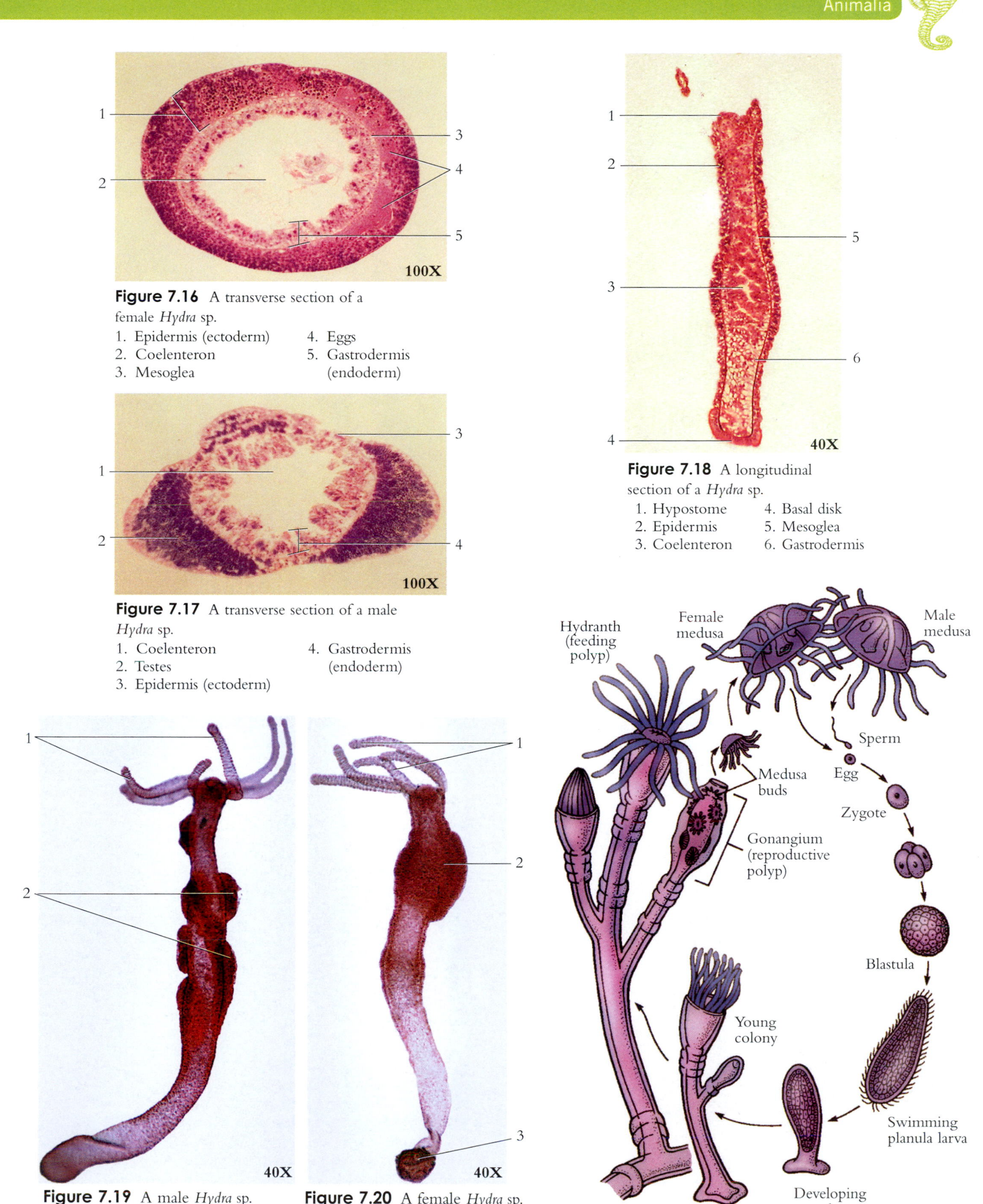

Figure 7.16 A transverse section of a female *Hydra* sp.
1. Epidermis (ectoderm)
2. Coelenteron
3. Mesoglea
4. Eggs
5. Gastrodermis (endoderm)

Figure 7.17 A transverse section of a male *Hydra* sp.
1. Coelenteron
2. Testes
3. Epidermis (ectoderm)
4. Gastrodermis (endoderm)

Figure 7.18 A longitudinal section of a *Hydra* sp.
1. Hypostome
2. Epidermis
3. Coelenteron
4. Basal disk
5. Mesoglea
6. Gastrodermis

Figure 7.19 A male *Hydra* sp.
1. Tentacles
2. Testes

Figure 7.20 A female *Hydra* sp.
1. Tentacles
2. Ovary
3. Basal disk (foot)

Figure 7.21 The life cycle of *Obelia* sp.

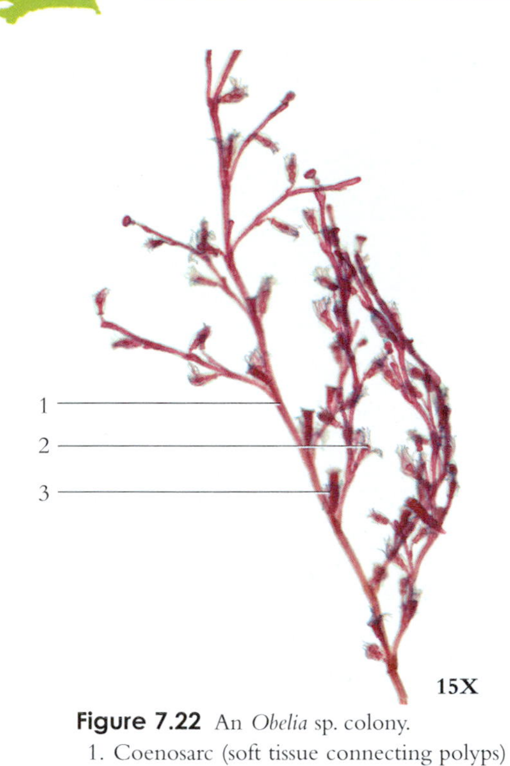

Figure 7.22 An *Obelia* sp. colony.
1. Coenosarc (soft tissue connecting polyps)
2. Hydranth (feeding polyp)
3. Gonangium (reproductive polyp)

Figure 7.23 A detail of an *Obelia* sp. colony.
1. Tentacles
2. Perisarc (horny covering that encloses the polyp)
3. Coenosarc
4. Medusa buds
5. Hydranth (feeding polyp)
6. Gonangium (reproductive polyp)
7. Gonotheca
8. Blastostyle
9. Hypostome

Figure 7.24 An *Obelia* sp. medusa.
1. Tentacles
2. Radial canals
3. Manubrium (seen through the body from above)

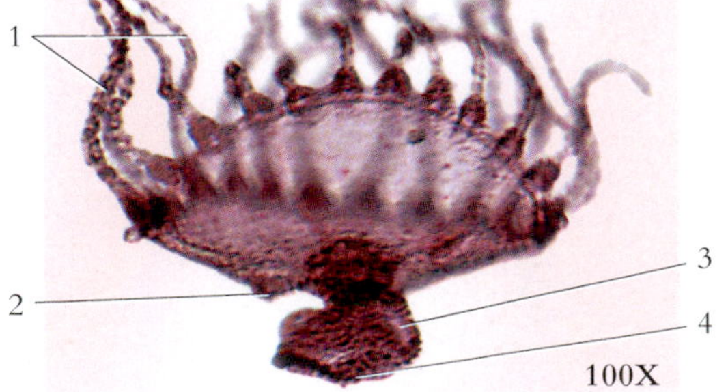

Figure 7.25 An *Obelia* sp. medusa in feeding position.
1. Tentacles
2. Gonad
3. Manubrium
4. Mouth

Figure 7.26 The Portuguese man-of-war, *Physalia* sp. It is a colony of medusae and polyps acting as a single organism. The tentacles are composed of three types of polyps: the gastrozooids (feeding polyps), the dactylozooids (stinging polyps), and the gonozooids (reproductive polyps) (scale in mm).
1. Pneumatophore (float)
2. Tentacles

Class Scyphozoa

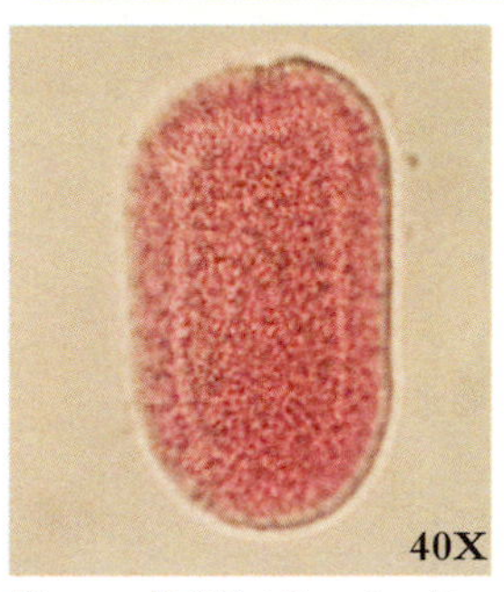

Figure 7.27 The *Aurelia* sp. planula larva develops from a fertilized egg that may be retained on the oral arm of the medusa.

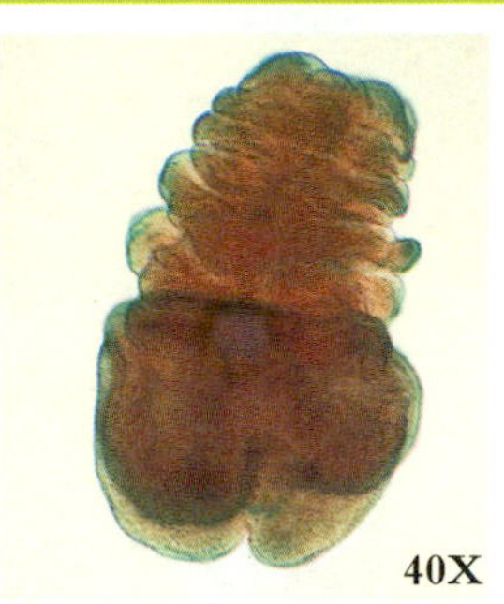

Figure 7.28 An *Aurelia* sp. scyphistoma. The polyp is a developmental stage in the life cycle of the jellyfish.

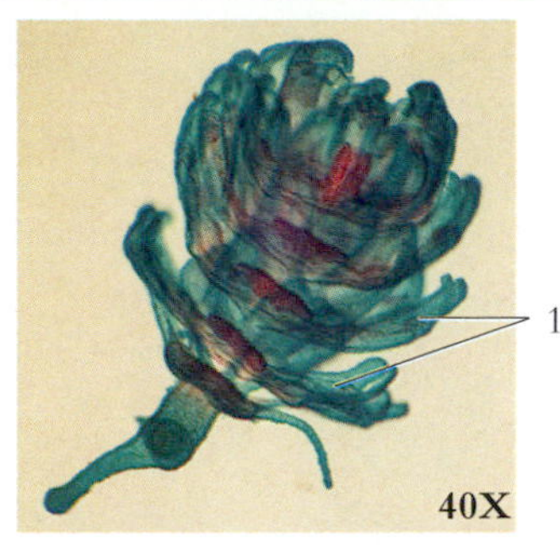

Figure 7.29 An *Aurelia* sp. strobila. Under favorable conditions, the scyphistoma develops into the strobila.
1. Developing ephyrae

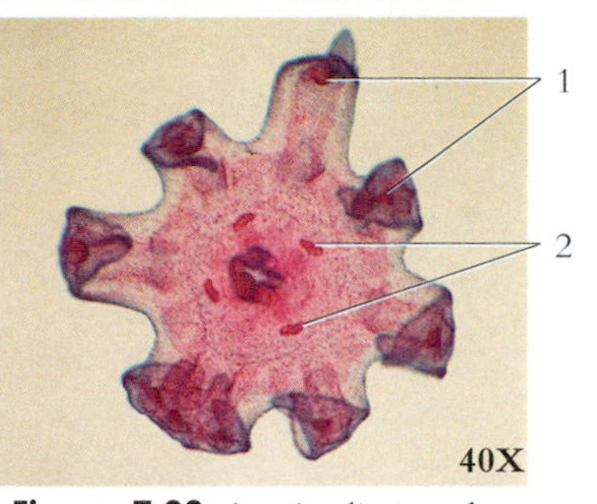

Figure 7.30 An *Aurelia* sp. ephyra larva. It gradually develops into an adult jellyfish.
1. Rhopalia (sense organs)
2. Gonads

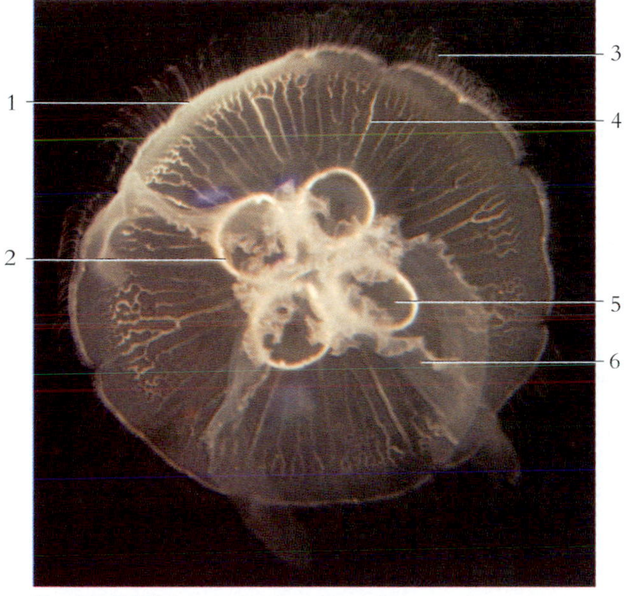

Figure 7.31 An oral view of *Aurelia* sp. medusa.
1. Ring canal
2. Gonad
3. Marginal tentacles
4. Radial canal
5. Subgenital pit
6. Oral arm

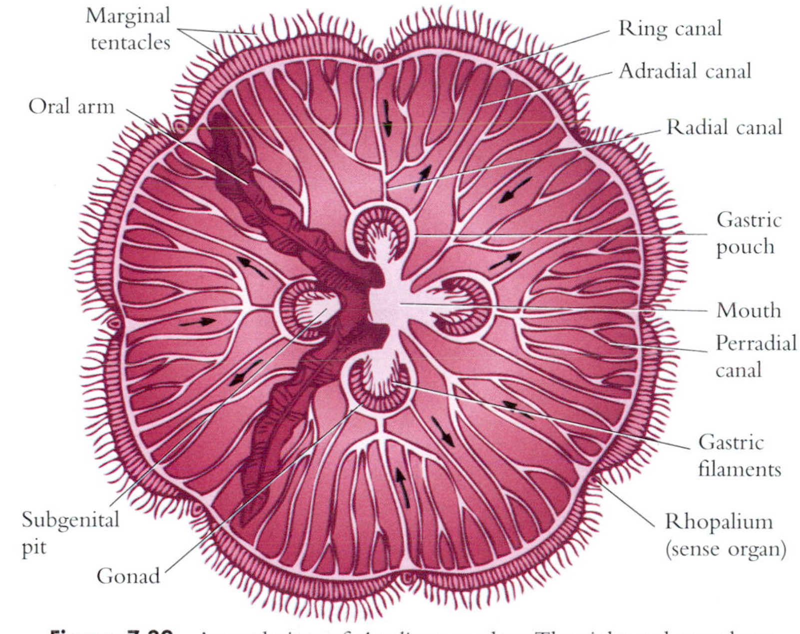

Figure 7.32 An oral view of *Aurelia* sp. medusa. The right oral arms have been removed, and the arrows depict circulation through the canal system.

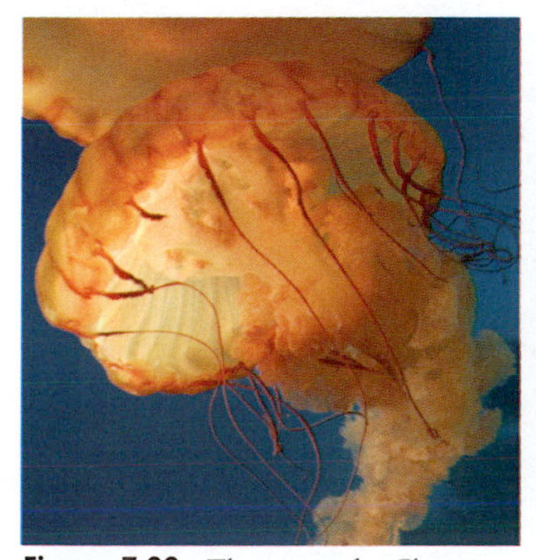

Figure 7.33 The sea nettle, *Chrysaora fuscescens*, They gather in large swarms off the Pacific coast, where they feed on zooplankton.

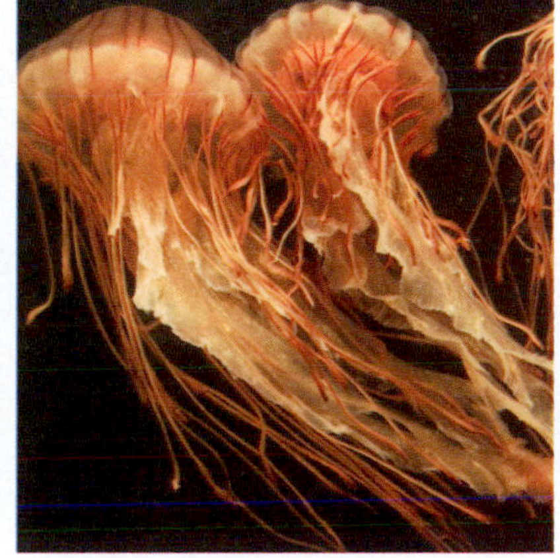

Figure 7.34 The red-striped jellyfish, *Chrysaora melanaster*, is common near the surface of the Bering Sea.

Figure 7.35 The purple-striped jelly, *Chrysaora colorata*, is found off the coast of California and in Monterey Bay.

Class Cubozoa

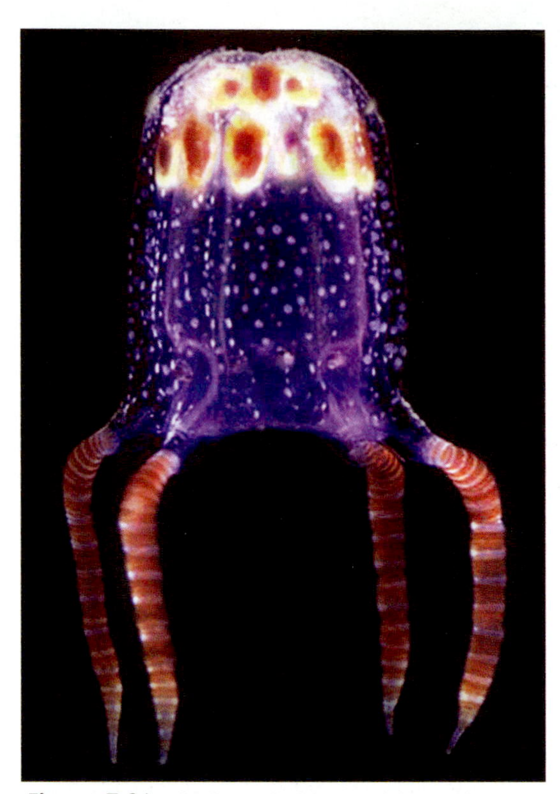

Figure 7.36 The box jellyfish, *Carybdea sivickisi*, is named for its cube-shaped bell. All cubozoans have four tentacles.

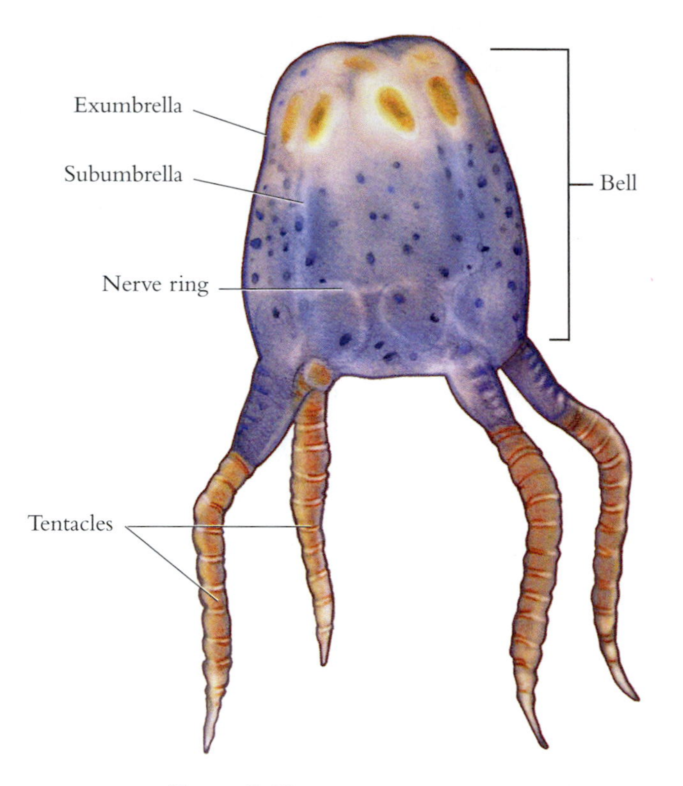

Figure 7.37 An illustration of box jellyfish, *Carybdea sivickisi*, showing basic external structures.

Class Anthozoa

Figure 7.38 The sunburst anemone, *Anthopleura sola*, gets its green coloration from symbiotic algae within it.

Figure 7.39 The firecracker coral, *Dendrophyllia* sp., a filter feeder, actively feeds day and night.

Figure 7.40 The tube anemone, *Pachycerianthus fimbriatus*, makes a leathery tube and sinks it up to two feet into the sand.

Figure 7.41 The sea pen, *Ptilosarcus gurneyi*, is a colony of polyps that may reach two feet in height.

Figure 7.42 The disk anemone, *Actinodiscus* sp. It forms large colonies.

Figure 7.43 A diagram of a partially dissected sea anemone, *Metridium* sp.

Figure 7.44 Brain coral, *Goniastrea* sp.

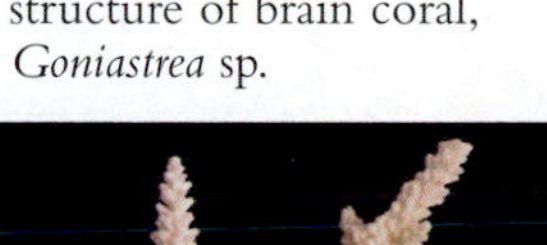

Figure 7.45 The skeletal structure of brain coral, *Goniastrea* sp.

Figure 7.46 Mushroom coral, *Rhodactis* sp.

Figure 7.47 The skeletal structure of mushroom coral, *Rhodactis* sp.

Figure 7.48 Staghorn coral, *Acropora* sp.

Figure 7.49 The skeletal structure of staghorn coral, *Acropora* sp.

Figure 7.50 A detailed view of the polyps of candy cane coral, *Caulastrea furcata*.

Figure 7.51 A detailed view of the polyps of glove xenia, *Xenia umbellata*.

Table 7.3 Some Representatives of the Phylum Platyhelminthes

Classes and Representative Kinds	Characteristics
Turbellaria — planarians	Mostly free-living, carnivorous, aquatic forms; body covered by ciliated epidermis
Trematoda — flukes including schistosomes	Parasitic with wide range of invertebrate and vertebrate hosts; suckers for attachment to host
Cestoda — tapeworms	Parasitic in many vertebrate hosts; complex life cycle with intermediate hosts; suckers or hooks on scolex for attachment to host; eggs are produced and shed within proglottids

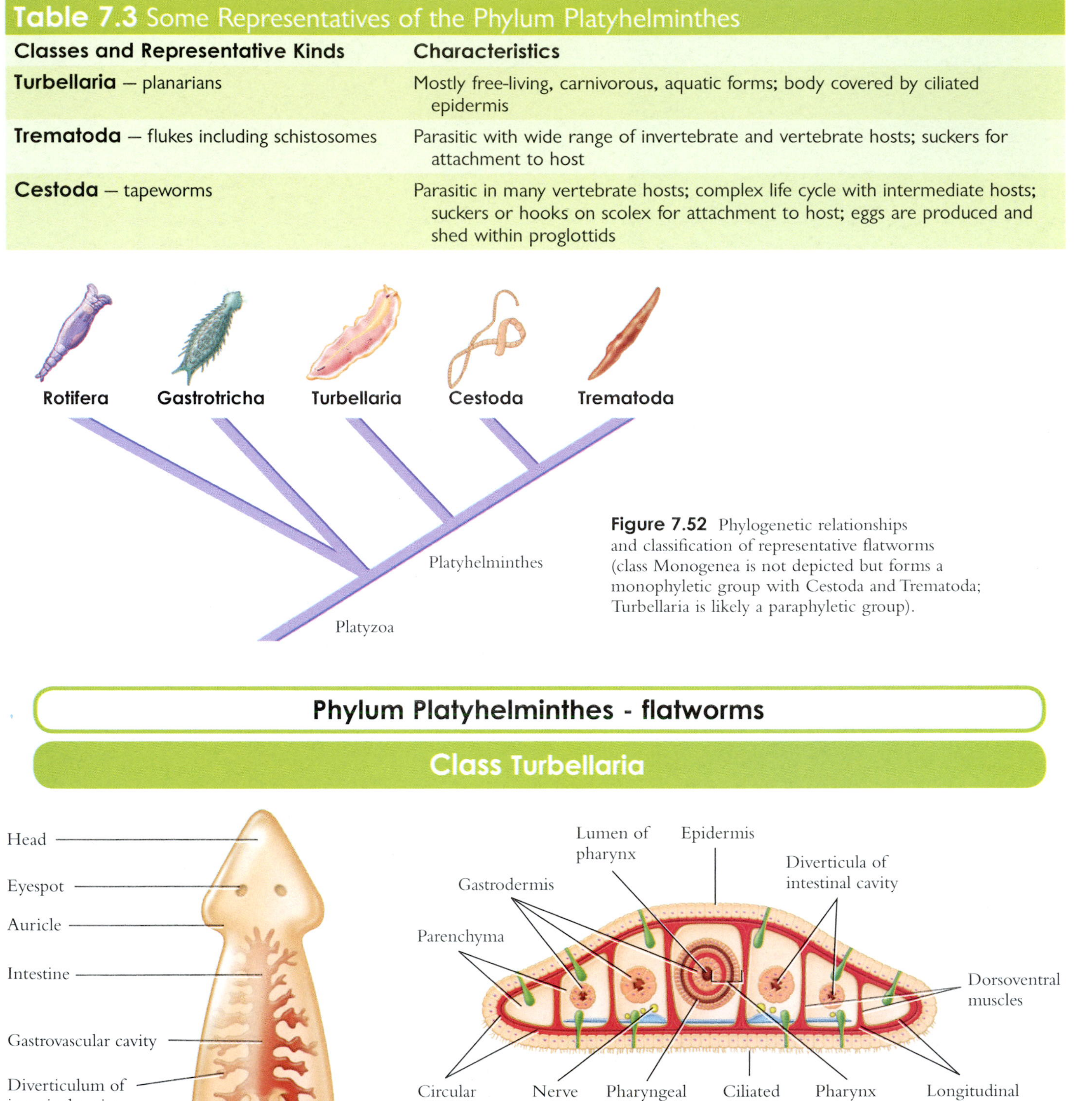

Figure 7.52 Phylogenetic relationships and classification of representative flatworms (class Monogenea is not depicted but forms a monophyletic group with Cestoda and Trematoda; Turbellaria is likely a paraphyletic group).

Phylum Platyhelminthes - flatworms

Class Turbellaria

Figure 7.53 The internal anatomy of planarian. (a) A longitudinal section, and (b) a transverse section through the pharyngeal region.

Figure 7.54 A planarian (a) *Dugesia* sp. is aquatic, while the (b) *Bipalium* sp. is a common inhabitant of gardens.

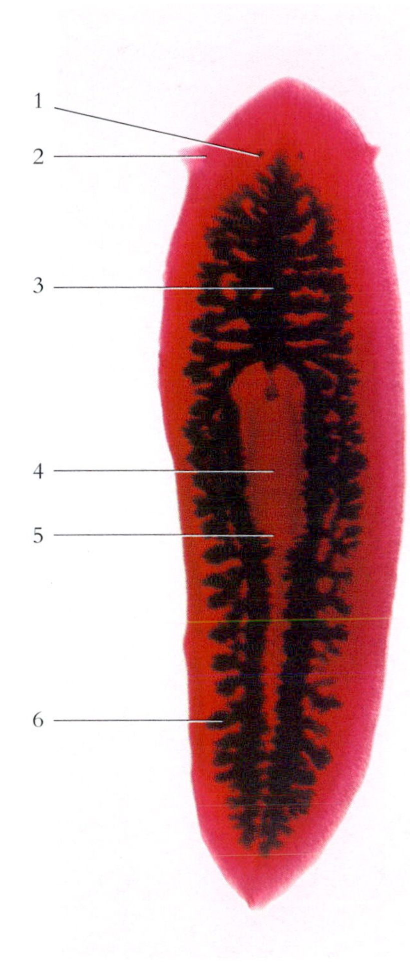

Figure 7.55 *Dugesia* sp.

1. Eyespot
2. Auricle
3. Gastrovascular cavity
4. Pharynx
5. Opening of pharynx (mouth)
6. Diverticulum of intestinal cavity

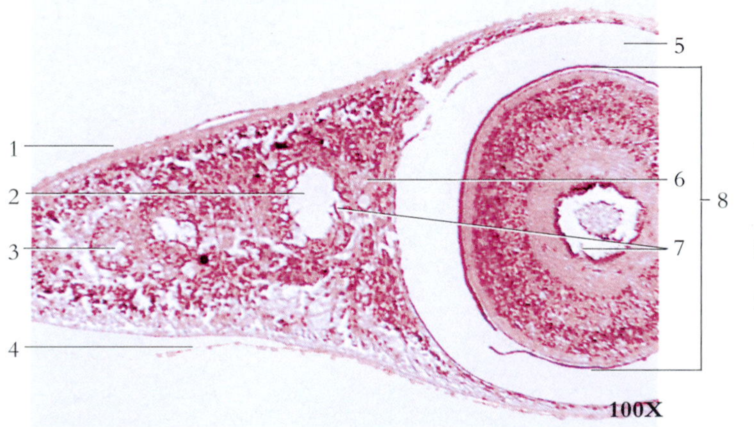

Figure 7.56 A transverse section through the pharyngeal region of *Dugesia* sp.

1. Epidermis
2. Intestinal cavity
3. Testis
4. Cilia
5. Pharyngeal cavity
6. Dorsoventral muscles
7. Gastrodermis
8. Pharynx

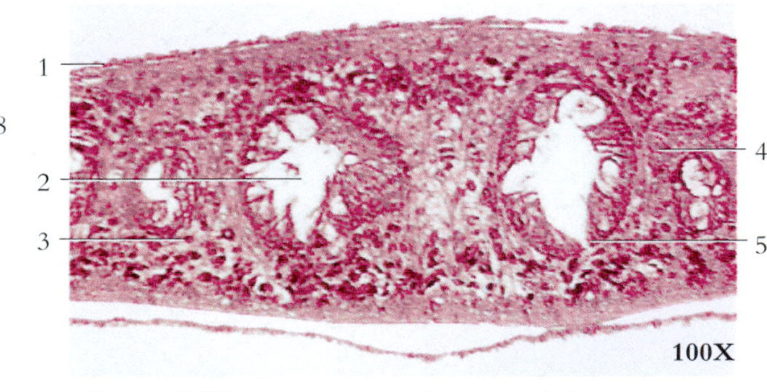

Figure 7.57 A transverse section through the posterior region of *Dugesia* sp.

1. Epidermis
2. Intestinal cavity
3. Mesenchyme
4. Dorsoventral muscles
5. Gastrodermis

Class Trematoda

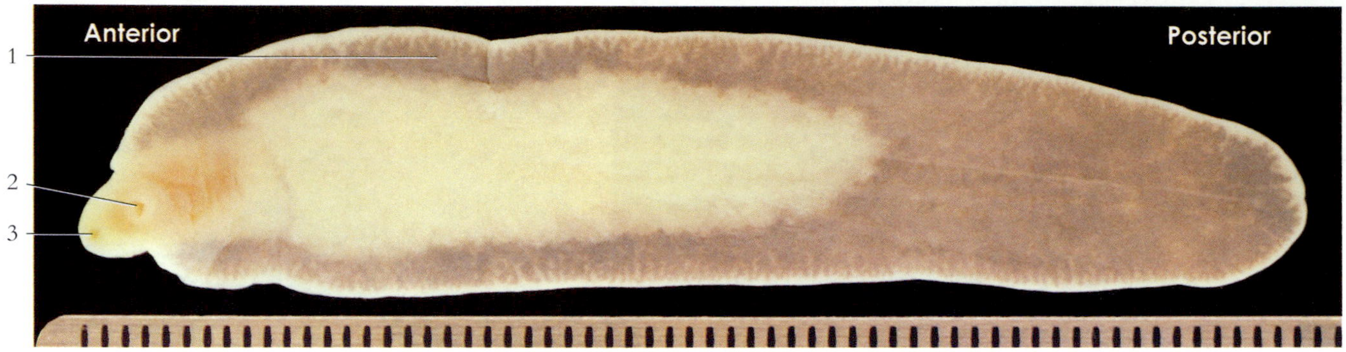

Figure 7.58 The cow liver fluke, *Fasciola magna*, is one of the largest flukes, measuring up to 7.75 cm long (scale in mm).
1. Yolk gland 2. Ventral sucker 3. Oral sucker

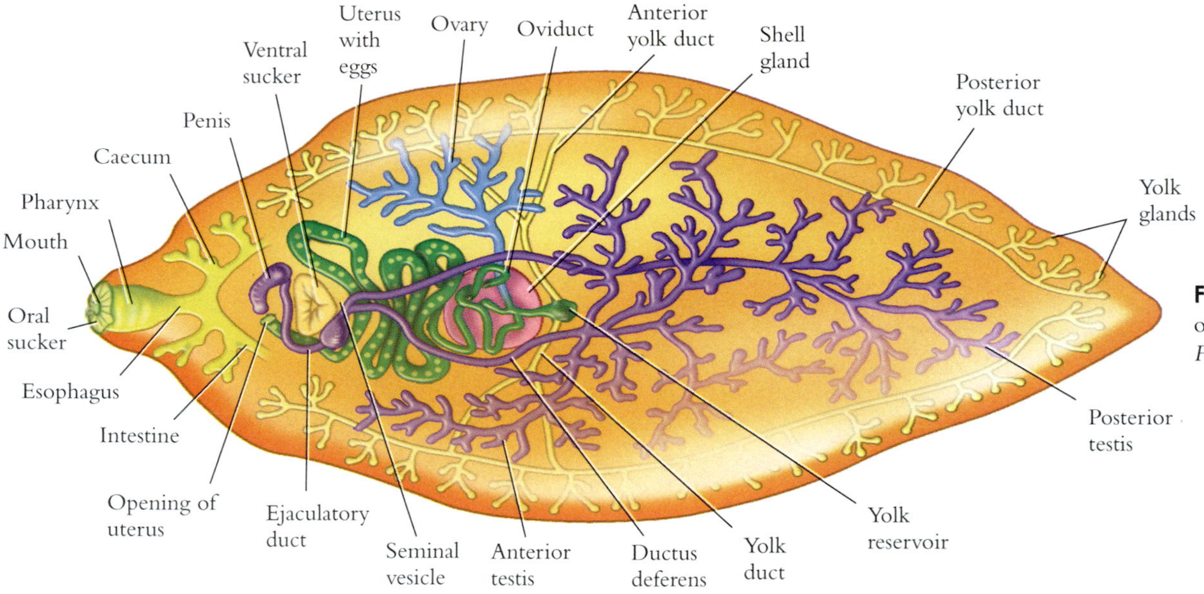

Figure 7.59 A diagram of the sheep liver fluke, *Fasciola hepatica*.

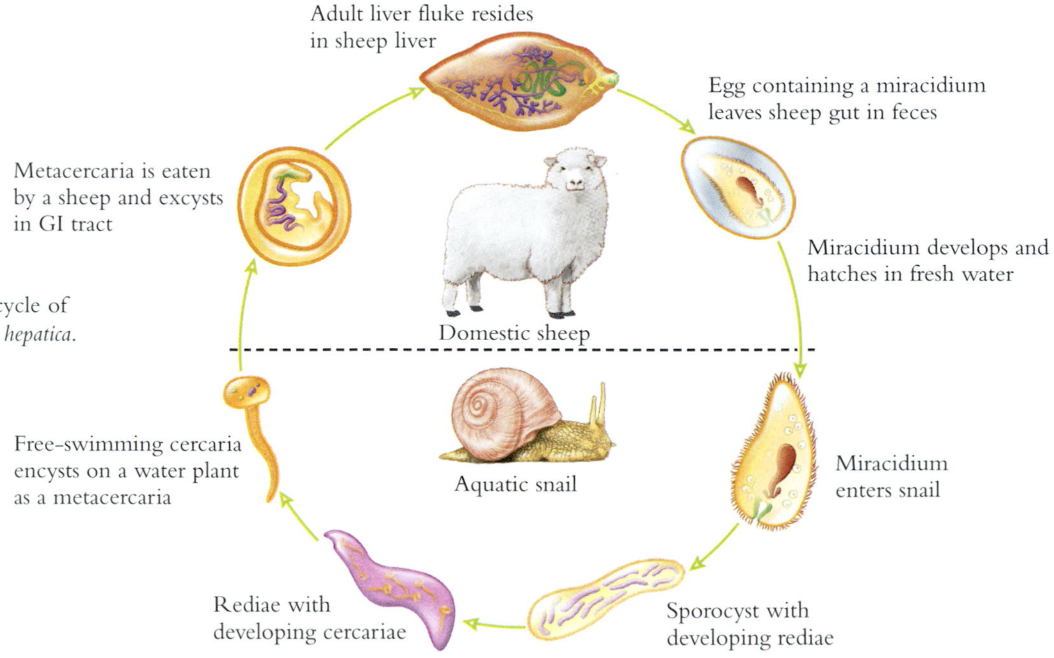

Figure 7.60 The life cycle of sheep liver fluke, *Fasciola hepatica*.

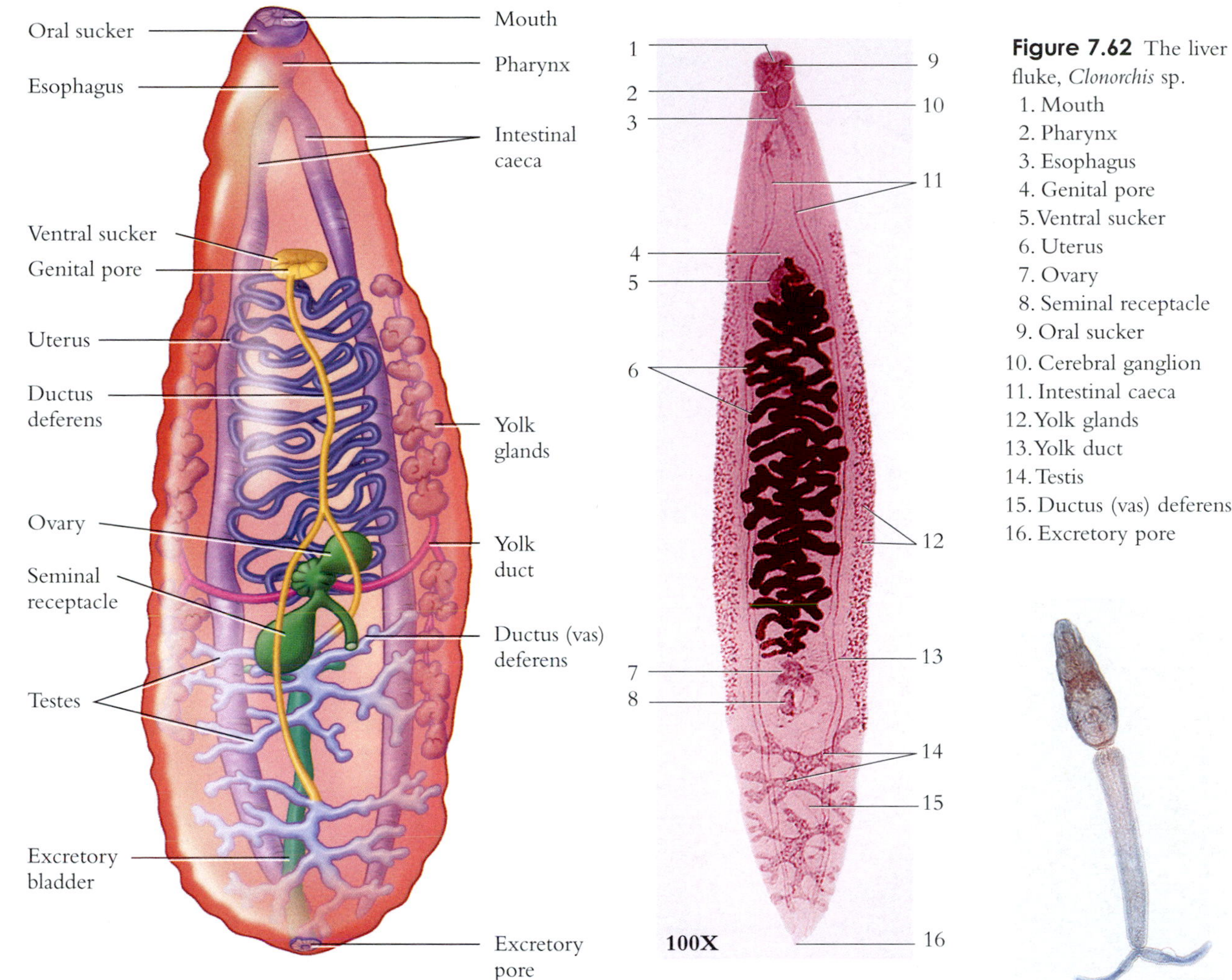

Figure 7.61 A diagram of the human liver fluke, *Clonorchis sinensis*.

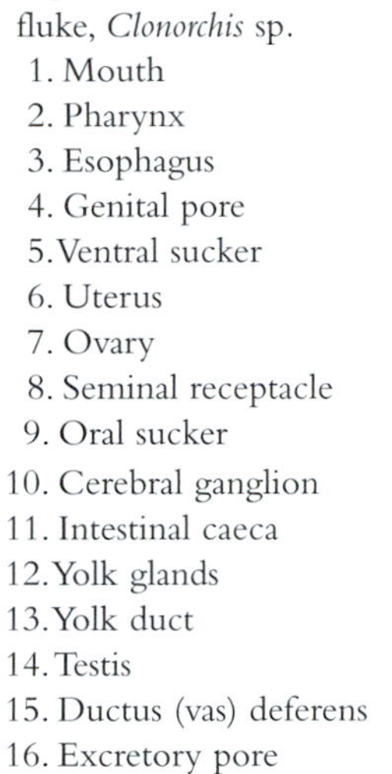

Figure 7.62 The liver fluke, *Clonorchis* sp.

1. Mouth
2. Pharynx
3. Esophagus
4. Genital pore
5. Ventral sucker
6. Uterus
7. Ovary
8. Seminal receptacle
9. Oral sucker
10. Cerebral ganglion
11. Intestinal caeca
12. Yolk glands
13. Yolk duct
14. Testis
15. Ductus (vas) deferens
16. Excretory pore

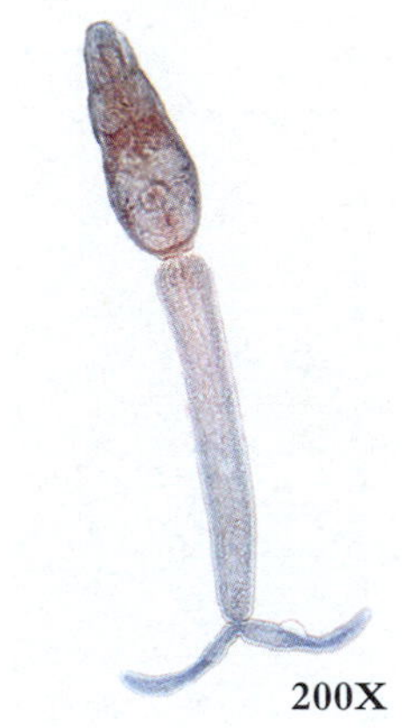

Figure 7.63 The cercaria stage of a trematode species.

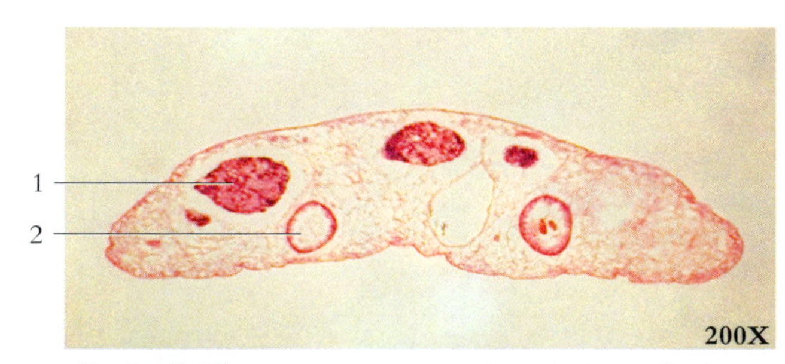

Figure 7.64 A transverse section through the midbody region of *Clonorchis* sp.

1. Uterus
2. Intestine

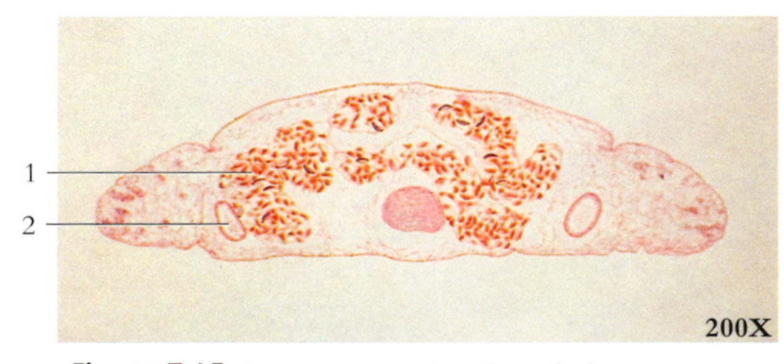

Figure 7.65 A transverse section through the lower body region of *Clonorchis* sp.

1. Testis
2. Intestine

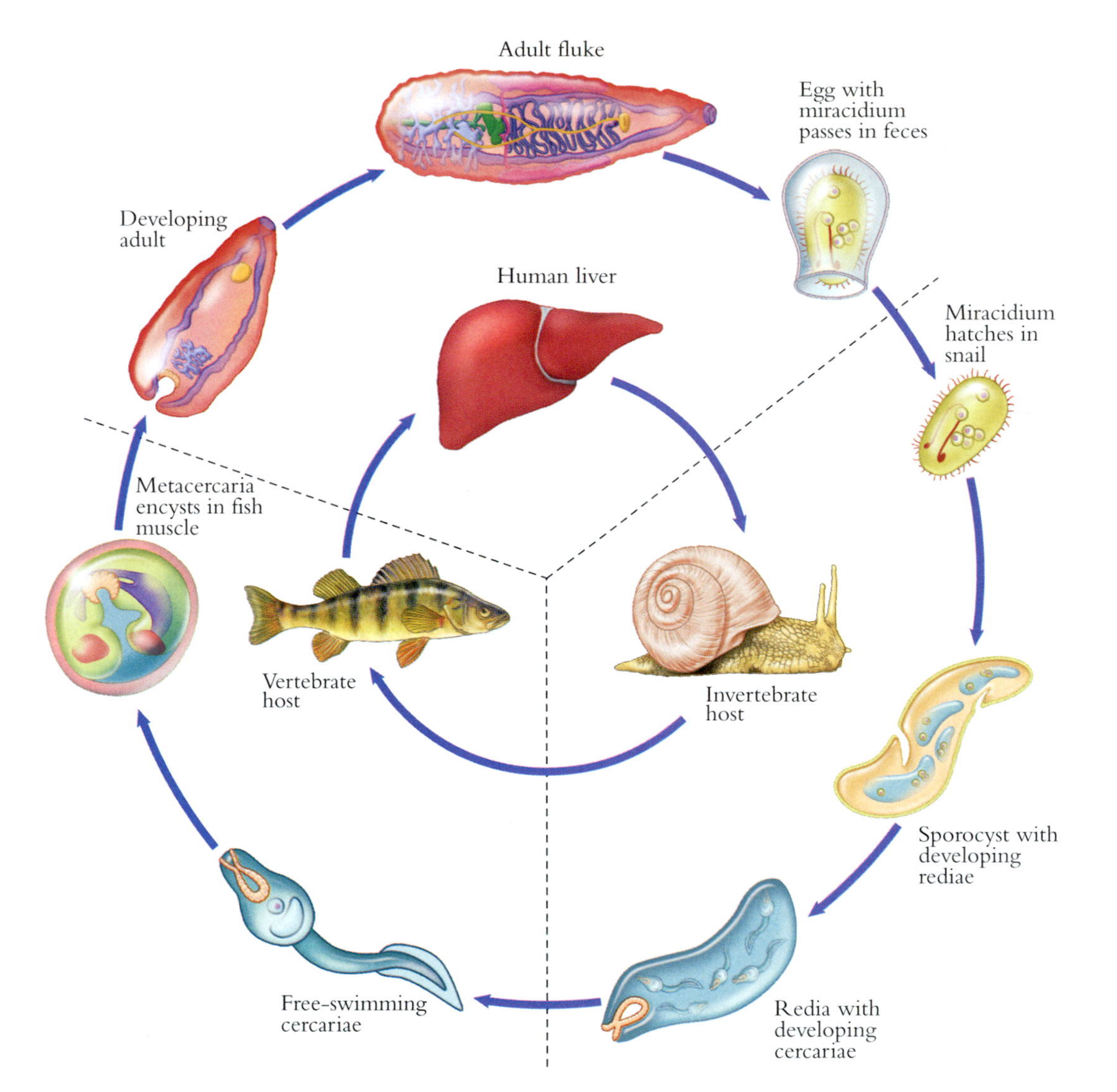

Figure 7.66 The life cycle of the human liver fluke, *Clonorchis sinesis*.

Class Cestoda

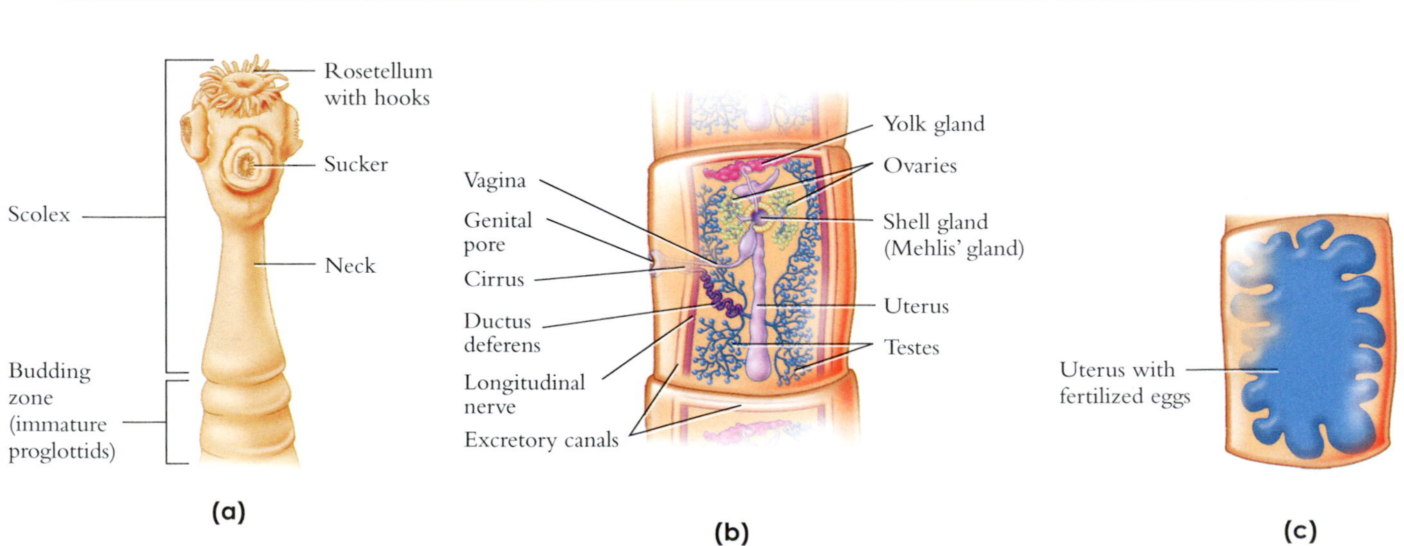

Figure 7.67 The diagrams of a parasitic tapeworm, *Taenia pisiformis*. (a) The anterior end, (b) mature proglottids, and (c) a gravid proglottid.

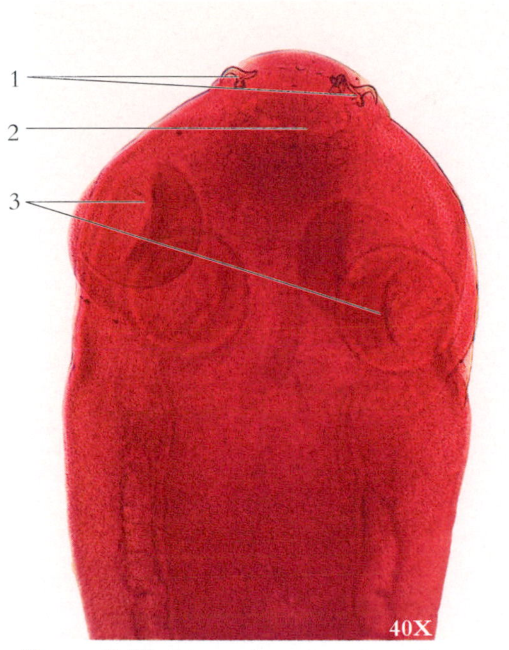

Figure 7.68 The scolex of *Taenia pisiformis*.
1. Hooks
2. Rostellum
3. Suckers

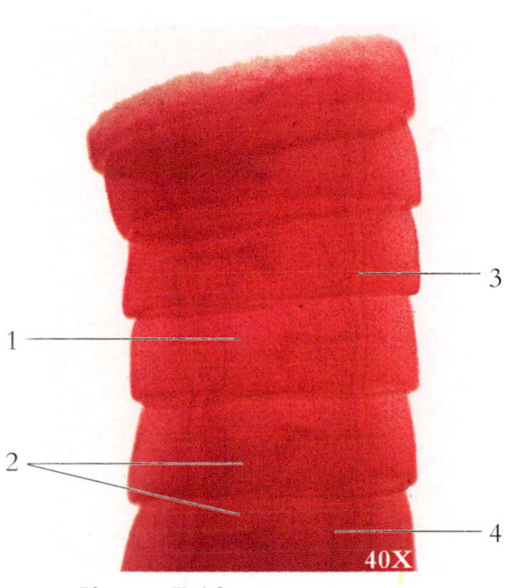

Figure 7.69 The immature proglottids of *Taenia pisiformis*.
1. Early ovary
2. Early testes
3. Excretory canal
4. Immature vagina and ductus deferens

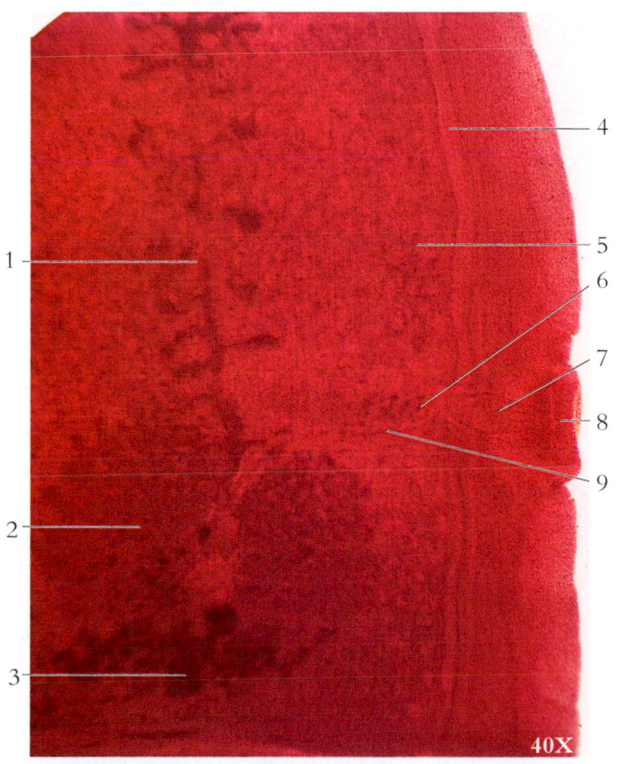

Figure 7.70 A mature proglottid of *Taenia pisiformis*.
1. Uterus
2. Ovary
3. Yolk gland
4. Excretory canal
5. Testes
6. Ductus deferens
7. Cirrus
8. Genital pore
9. Vagina

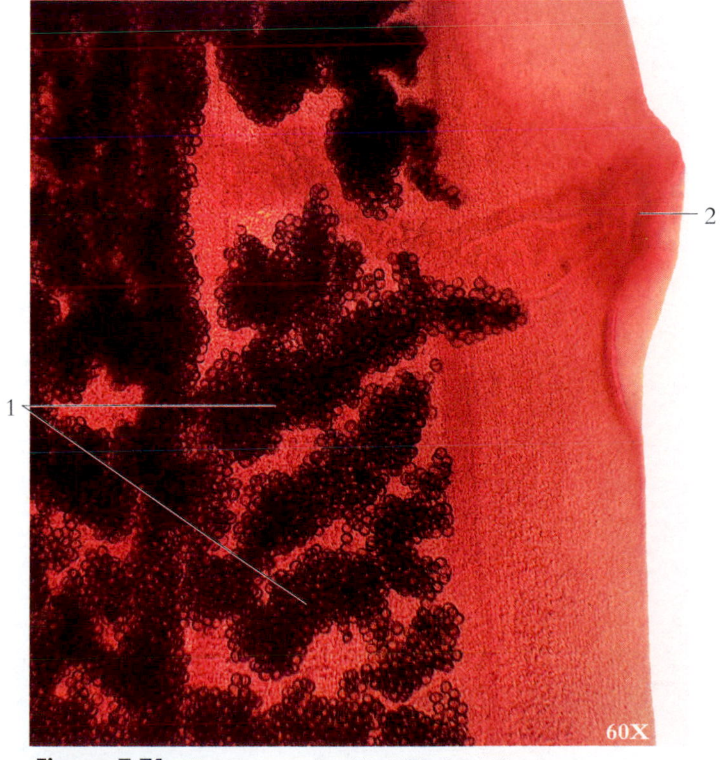

Figure 7.71 The ripe proglottid of *Taenia pisiformis*.
1. Zygotes in branched uterus
2. Genital pore

Table 7.4 Representatives of the Phylum Mollusca

Classes and Representative Kinds	Characteristics
Polyplacophora — chitons	Marine; shell of eight dorsal plates; broad foot
Gastropoda — snails and slugs	Marine, freshwater, and terrestrial; coiled shell; prominent head with tentacles and eyes
Bivalvia — clams, oysters, and mussels	Marine, freshwater; body compressed between two hinged shells in a left and right arrangement; hatchet-shaped foot
Cephalopoda — squids and octopi	Marine; excellent swimmers, predatory; foot separated into arms and tentacles that may contain suckers; well-developed eyes

Representatives of the Phylum Brachiopoda

Classes and Representative Kinds	Characteristics
Lingulata — lamp shells	Marine; body compressed between two hinged shells in a top and bottom arrangement; stalk-like pedicle

Polyplacophora Monoplacophora Gastropoda Cephalopoda Bivalvia Scaphopoda

Figure 7.72 Phylogenetic relationships and classification of Mollusca. Brachiopods, included in the section, are members of a poorly resolved clade containing Mollusca, Annelida, Nemertea, and Phoronida, but share gross morphological affinities with the Mollusca. Brachiopods are sister taxon to the Phoronida, and form a clade with respect to the Annelida, Mollusca, and Nemertea called the Trochozoa (all have a trochophore stage in larval development).

Phylum Mollusca - chitons, snails, clams, and squids

Class Polyplacophora

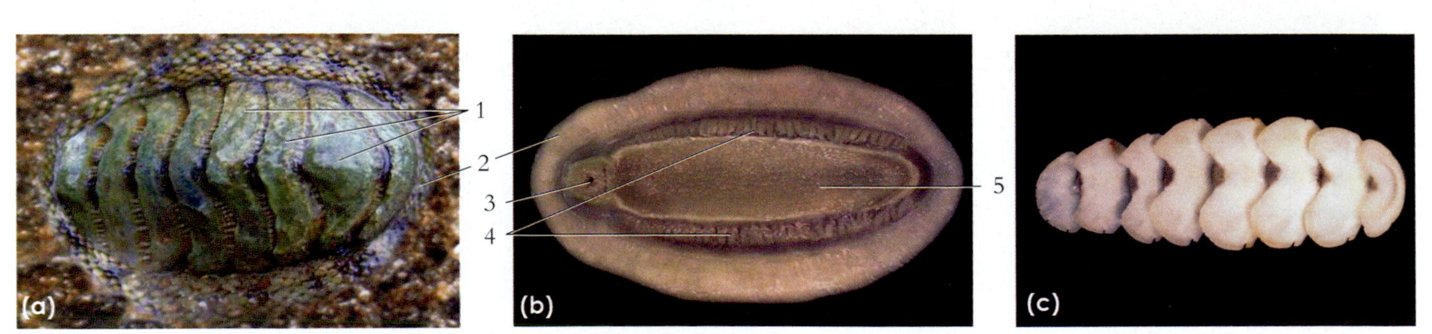

Figure 7.73 Chitons are easily recognized by their eight dorsal plates. (a) A dorsal view and (b) ventral view. (c) A ventral view of a chiton skeleton showing the eight dorsal plates.

1. Dorsal plates 2. Girdle 3. Mouth 4. Gill filaments 5. Ventral foot

Class Gastropoda

Figure 7.74 Many gastropods have ornate shells, such as the Venus comb murex, *Murex pecten* (scale in mm).

Figure 7.75 A keyhole limpet, *Megathura crenulata*.
1. Shell 2. Mantle 3. Foot

Figure 7.76 A snail, *Cornu aspersum*.
1. Shell
2. Foot
3. Occular tentacle
4. Head
5. Sensory tentacle

Figure 7.77 The locomotion of the slug, class Gastropoda, requires the production of mucus. Slugs differ from snails in that a shell is absent.
1. Foot
2. Mucus
3. Mantle
4. Head
5. Occular tentacle
6. Sensory tentacle
7. Pneumostome

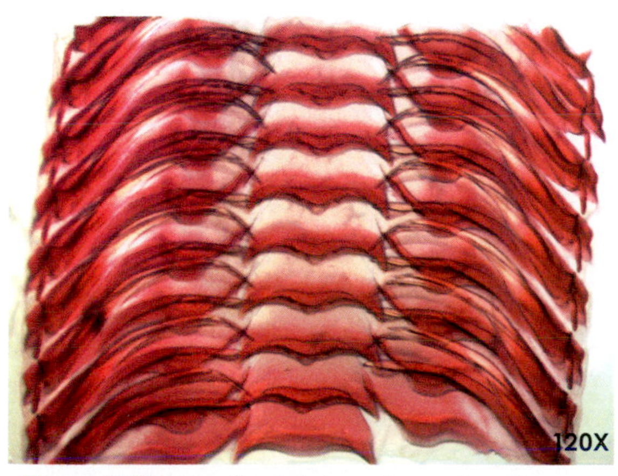

Figure 7.78 A snail radula is made up of small horny teeth made of chitin, called denticles.

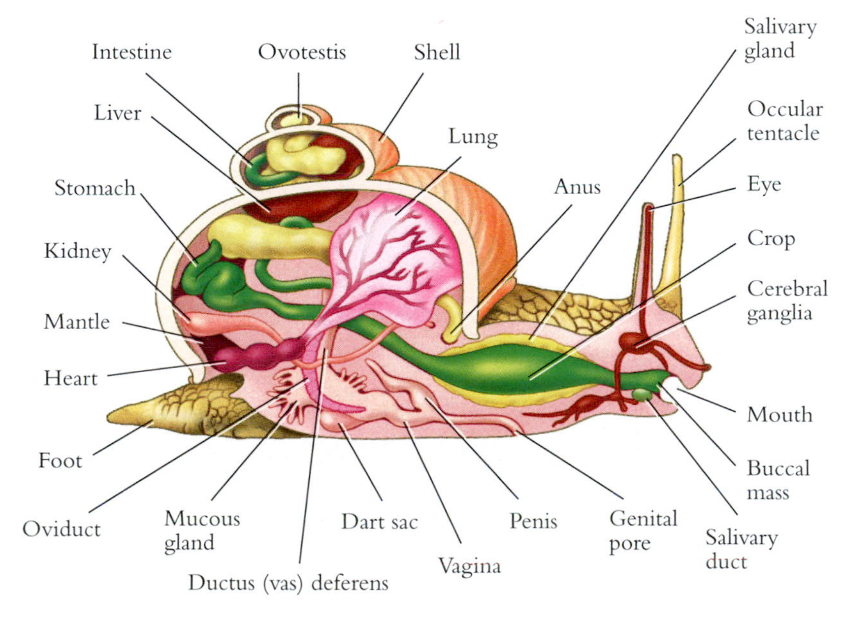

Figure 7.79 A diagram of pulmonate snail anatomy.

Class Bivalvia (= Pelycypoda)

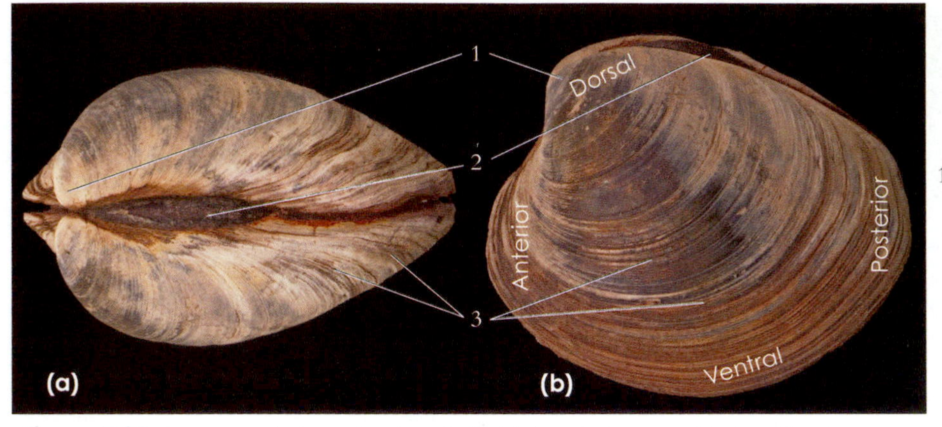

Figure 7.80 An external view of a clam shell: (a) dorsal view and (b) the left valve.
1. Umbo 2. Hinge ligament 3. Growth lines

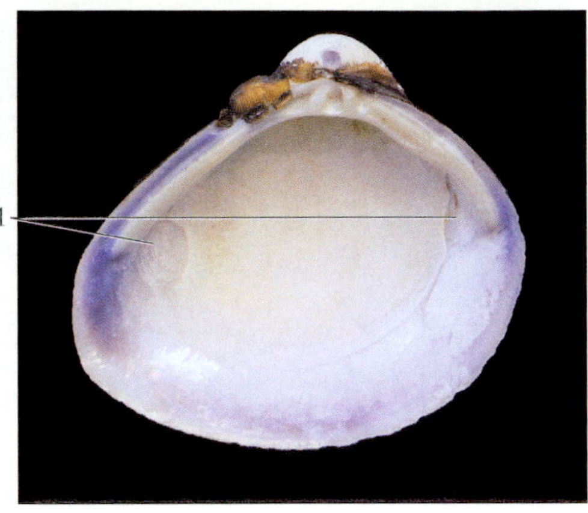

Figure 7.81 Internal view of a clam shell showing the muscle scars where the adductor muscles attached to the shell.
1. Muscle scar

Figure 7.82 A giant clam, *Tridacna derasa*.

Figure 7.83 California mussels, *Mytilus californianus*, form extensive mussel beds.

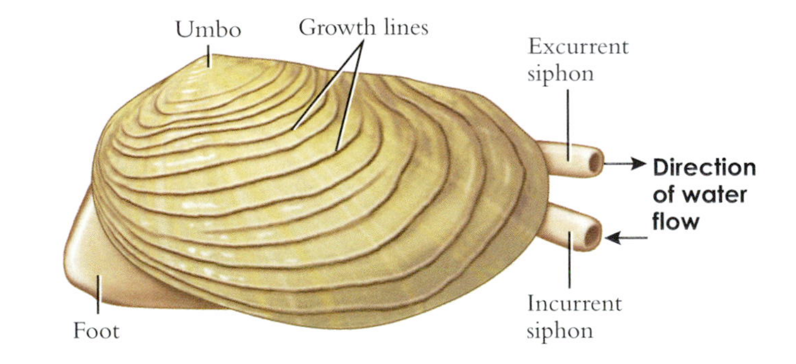

Figure 7.84 The surface anatomy of a freshwater clam, left valve.

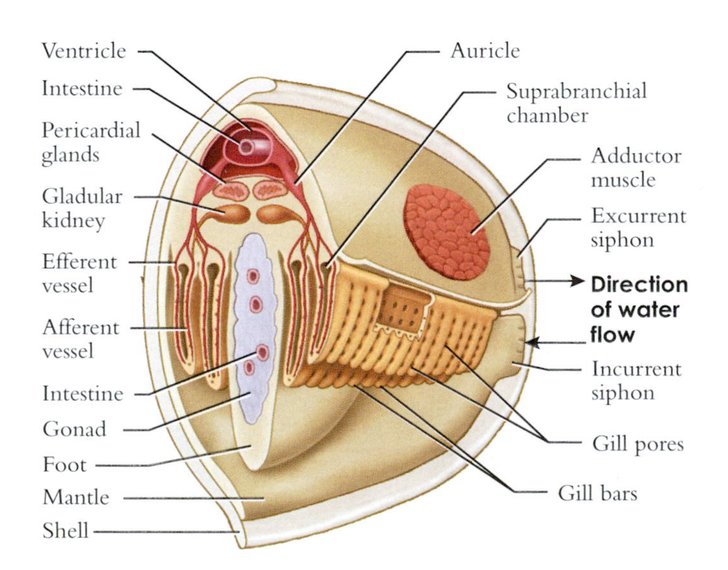

Figure 7.85 A diagram of the circulatory and respiratory systems of a freshwater clam.

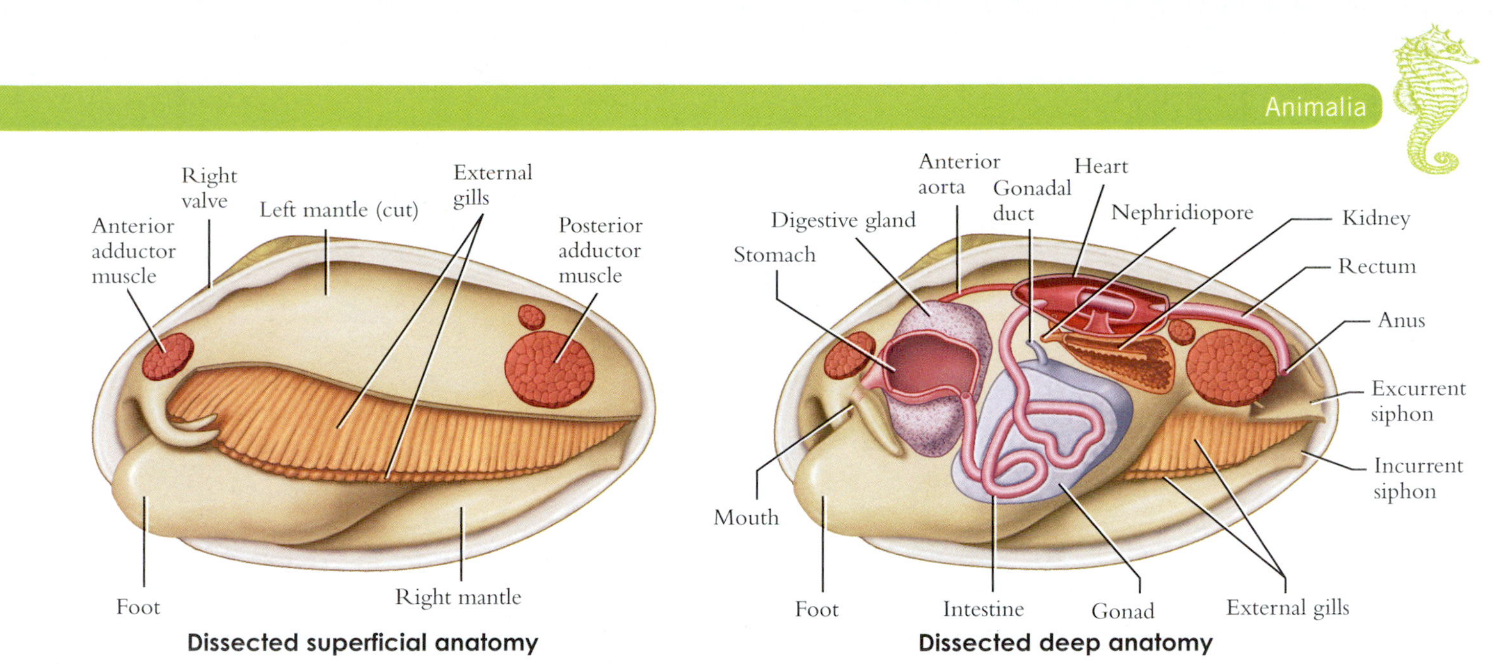

Figure 7.86 The anatomy of a freshwater clam. Bivalves have two shells (valves) that are laterally compressed and dorsally hinged.

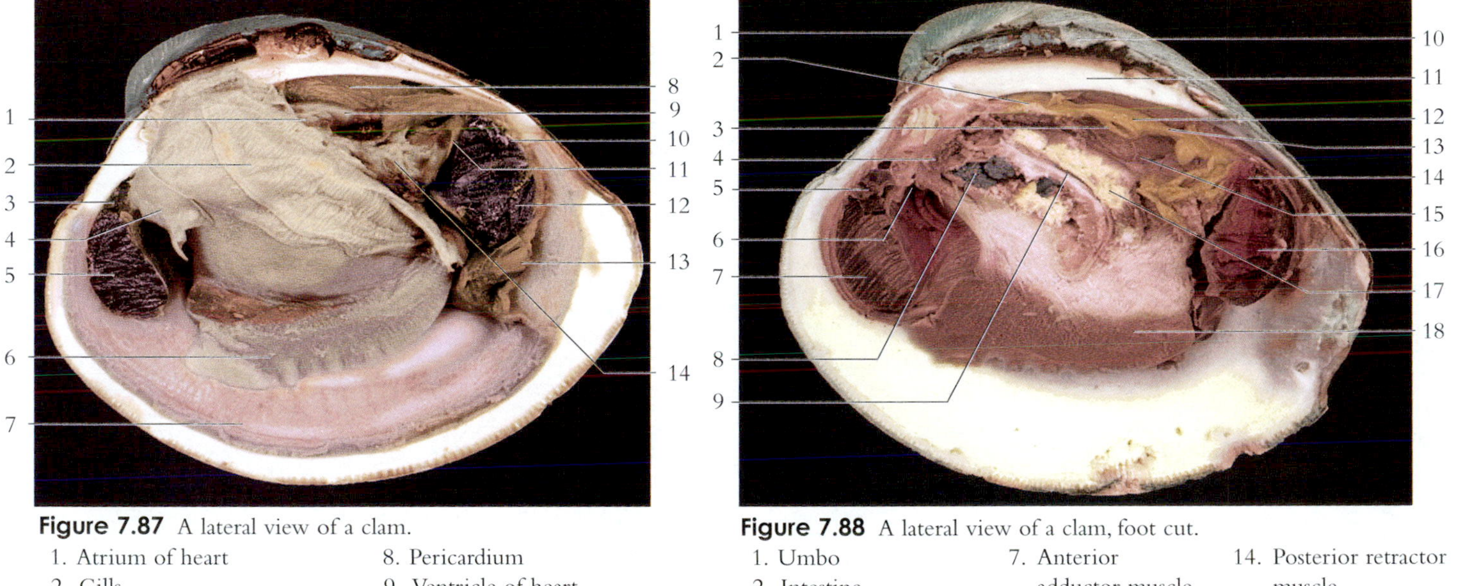

Figure 7.87 A lateral view of a clam.

1. Atrium of heart
2. Gills
3. Anterior retractor muscle
4. Labial palps
5. Anterior adductor muscle
6. Foot
7. Mantle
8. Pericardium
9. Ventricle of heart
10. Anus
11. Posterior retractor muscle
12. Posterior adductor muscle
13. Excurrent siphon
14. Nephridium (kidney)

Figure 7.88 A lateral view of a clam, foot cut.

1. Umbo
2. Intestine
3. Opening between atrium and ventricle
4. Esophagus
5. Anterior retractor muscle
6. Mouth
7. Anterior adductor muscle
8. Digestive gland
9. Intestine
10. Hinge ligament
11. Hinge
12. Ventricle of heart
13. Posterior aorta
14. Posterior retractor muscle
15. Nephridium (kidney)
16. Posterior adductor muscle
17. Gonad
18. Foot

Class Cephalopoda

Figure 7.89 Example cephalopods, (a) the giant octopus, *Enteroctopus* sp., (b) cuttlefish, *Sepiidae* sp., and (c) nautilus, *Nautilus pompilius*.

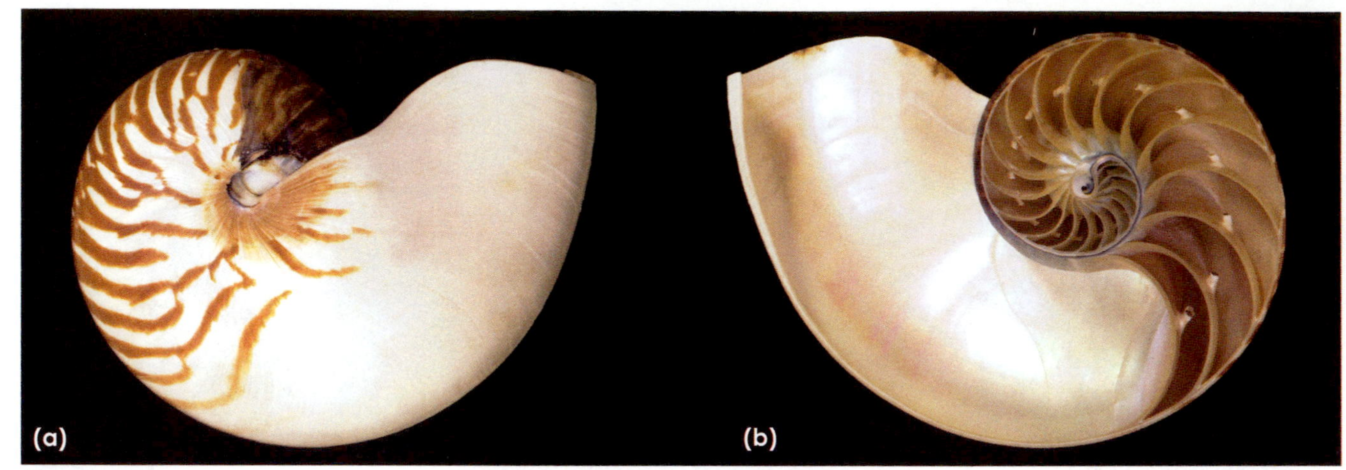

Figure 7.90 The *Nautilus*, a cephalopod, has gas-filled chambers within its shell, as seen in this cross-section of the shell (b). These chambers regulate buoyancy.

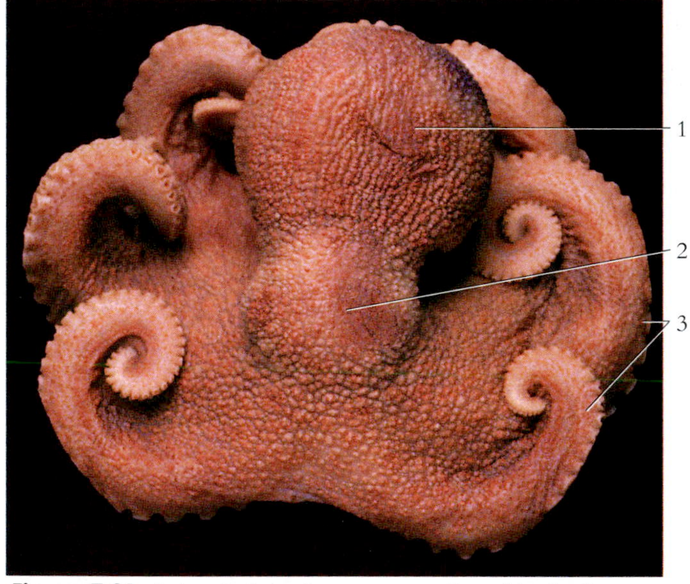

Figure 7.91 A dorsal view of an octopus collected in the Sea of Cortez, San Carlos, Mexico.
1. Mantle 2. Head 3. Arms

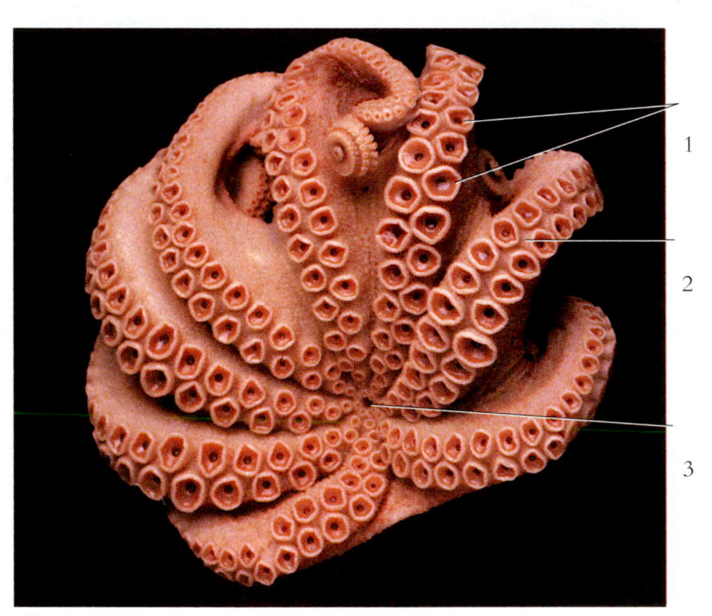

Figure 7.92 A ventral view of an octopus.
1. Suction cups 2. Arm 3. Mouth

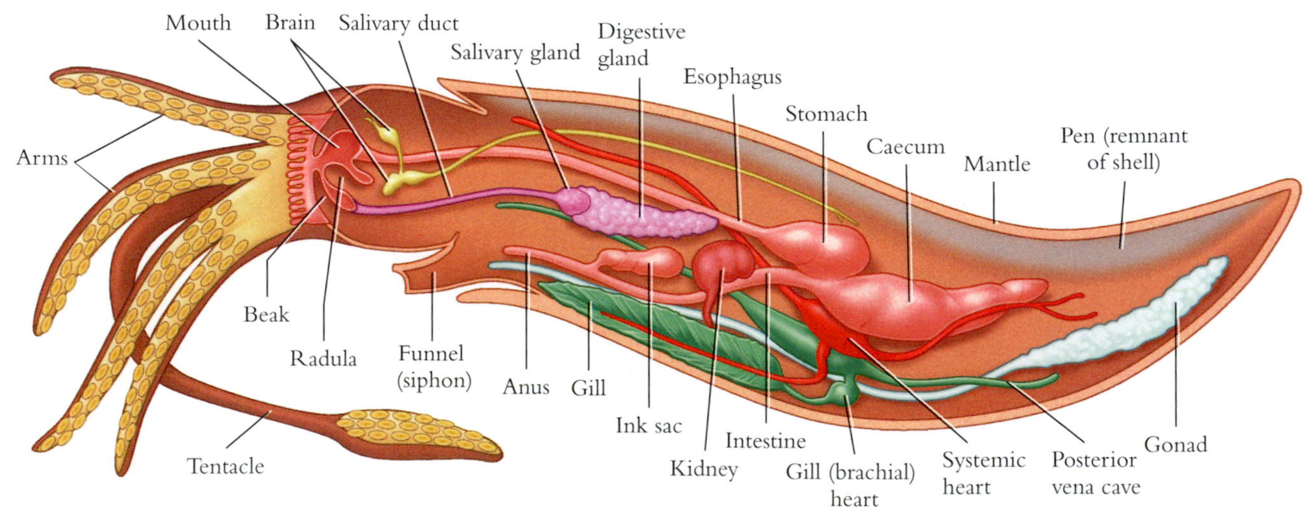

Figure 7.93 The internal anatomy of a squid.

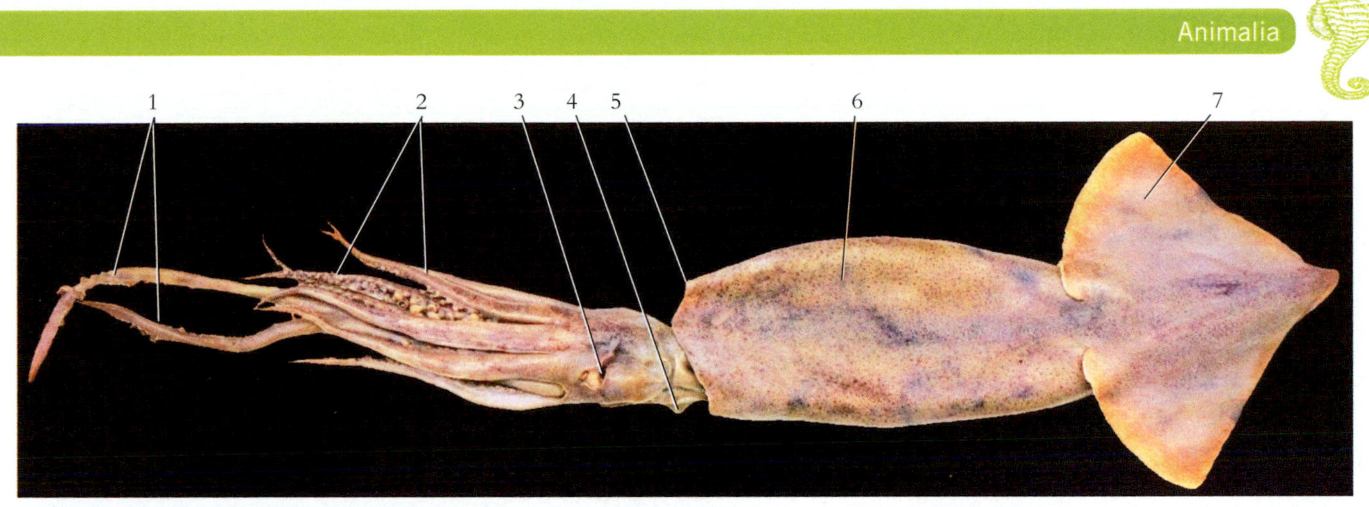

Figure 7.94 The external anatomy of the squid, *Loligo* sp.

1. Tentacles
2. Arms
3. Eye
4. Funnel (siphon)
5. Collar
6. Mantle (body tube)
7. Fin

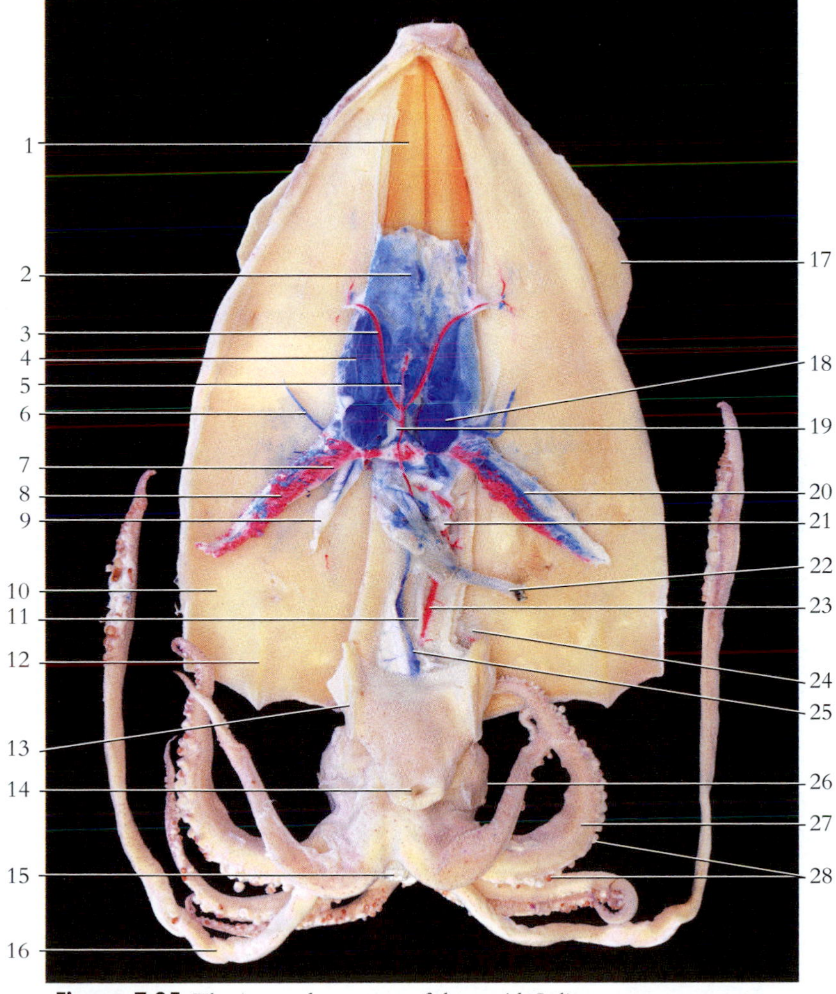

Figure 7.95 The internal anatomy of the squid, *Loligo* sp.

1. Pen (gonad partially resected)
2. Gonad
3. Lateral mantle artery
4. Posterior vena cava
5. Median mantle artery
6. Median mantle vein
7. Afferent branchial artery
8. Gill
9. Genital opening
10. Mantle
11. Esophagus
12. Articulating ridge
13. Articulating cartilage
14. Funnel (siphon)
15. Mouth
16. Tentacle
17. Fin
18. Branchial heart
19. Systemic heart
20. Efferent branchial vein
21. Ink sac
22. Rectum
23. Cephalic aorta
24. Stellate ganglion
25. Cephalic vena cava
26. Eye
27. Arm
28. Suckers

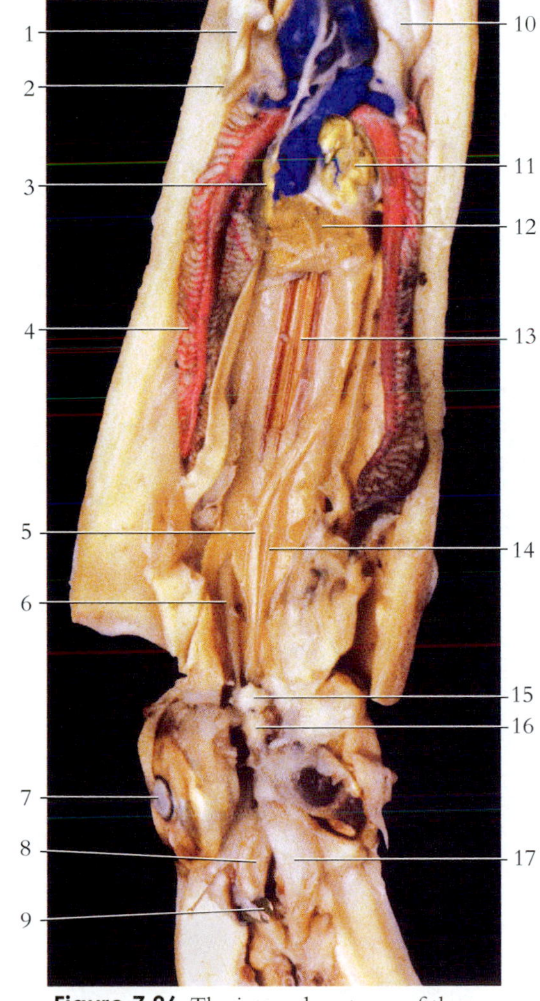

Figure 7.96 The internal anatomy of the squid, *Loligo* sp., including head region.

1. Spermatophoric duct
2. Penis
3. Kidney
4. Gill
5. Esophagus
6. Pleural nerve
7. Eye
8. Radula
9. Beak
10. Stomach
11. Pancreas
12. Digestive gland (cut)
13. Pen
14. Cephalic aorta
15. Visceral ganglion
16. Pedal ganglion
17. Buccal bulb

Phylum Brachiopoda - lamp shells

Figure 7.97 A fossil brachiopod, *Neospirifer* sp., from the Permian period.

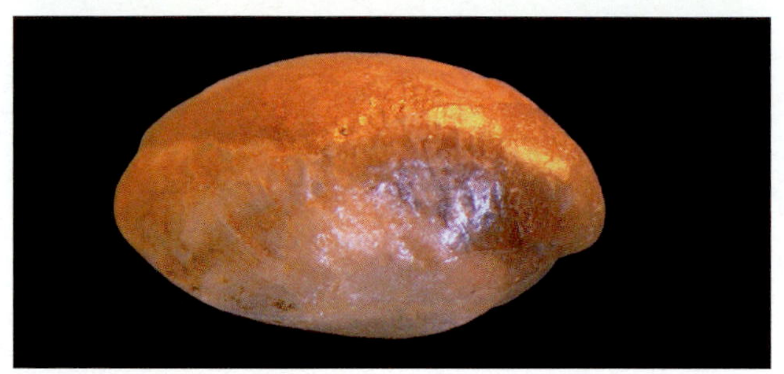

Figure 7.98 A fossil brachiopod, *Kingena* sp., from the Cretaceous period.

Figure 7.99 A living example of a lamp shell, *Lingula* sp.
1. Chaete 2. Pedicle valve 3. Apex 4. Growth lines 5. Pedicle 6. Substrate (sand on pedicle) 7. Muscle 8. Cuticle

Table 7.5 Some Representatives of the Phylum Nemertea

Classes and Representative Kinds	Characteristics
Anopla – proboscis worms	Mostly marine, but some freshwater and terrestrial; use evertible proboscis to catch prey and feed

Some Representatives of the Phylum Annelida

Classes and Representative Kinds	Characteristics
Polychaeta – tubeworms and sandworms	Mostly marine; segments with parapodia
Clitellata (subclass Oligochaeta) – earthworms	Freshwater and burrowing terrestrial forms; small setae; poorly developed head
Clitellata (subclass Hirudinea) – leeches	Freshwater; some are blood-sucking parasites and others are predators; lack setae; prominent muscular suckers

Platyhelminthes Mollusca Polychaeta Hirudinea Oligochaeta

Clitellata

Trochozoa

Spiralia

Figure 7.100 Phylogenetic relationships and classification of Annelida. Nemertea, included in the section, are members of a poorly resolved clade containing Mollusca, Annelida, Nemertea, Brachiopoda, and Phoronida, but share gross morphological affinities with the Annelida (see figure 7.1 on page 150).

Phylum Nemertea - proboscis worms

Class Anopla

Figure 7.101 *Parboriasia corrugatus*, a large nemertean from the Ross Sea, Antarctica. Typical nemertean cuticle is covered with mucus glands, especially at the anterior end, and they use their external cilia and muscular peristaltic undulation to glide on their trails of slime.

Figure 7.102 The milky ribbon worm, *Cerebratulus lacteus*.

Phylum Annelida - segmented worms

Class Polychaeta

Figure 7.103 The sandworm, *Nereis virens*.

1. Parapodia

Figure 7.104 The anterior end of the sandworm, *Nereis virens*. (a) A dorsal view and (b) a ventral view.

1. Palpi
2. Prostomium
3. Peristomial cirri (tentacles)
4. Peristome
5. Parapodia
6. Setae
7. Mouth
8. Everted pharynx

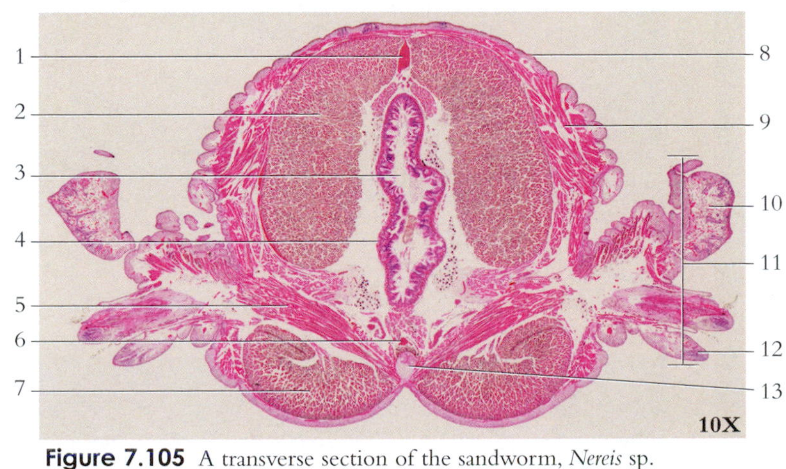

Figure 7.105 A transverse section of the sandworm, *Nereis* sp.

1. Dorsal blood vessel
2. Dorsal longitudinal muscle
3. Lumen of intestine
4. Intestine
5. Oblique muscle
6. Ventral blood vessel
7. Ventral longitudinal muscle
8. Integument
9. Circular muscle
10. Notopodium
11. Parapodium
12. Neuropodium
13. Ventral nerve cord

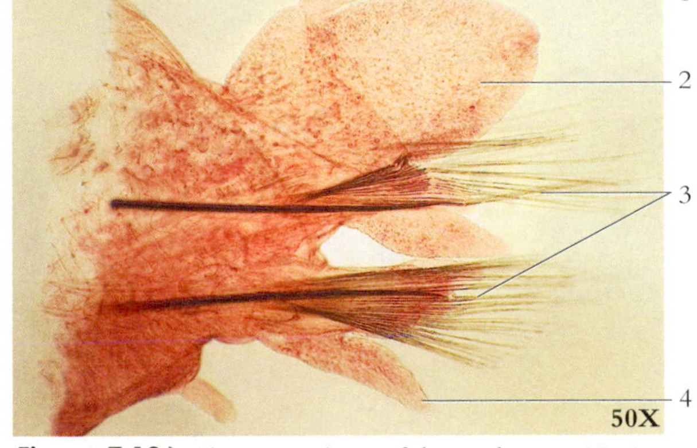

Figure 7.106 The parapodium of the sandworm, *Nereis* sp.

1. Dorsal cirrus
2. Notopodium
3. Setae
4. Neuropodium

Class Clitellata - Subclass Oligochaeta

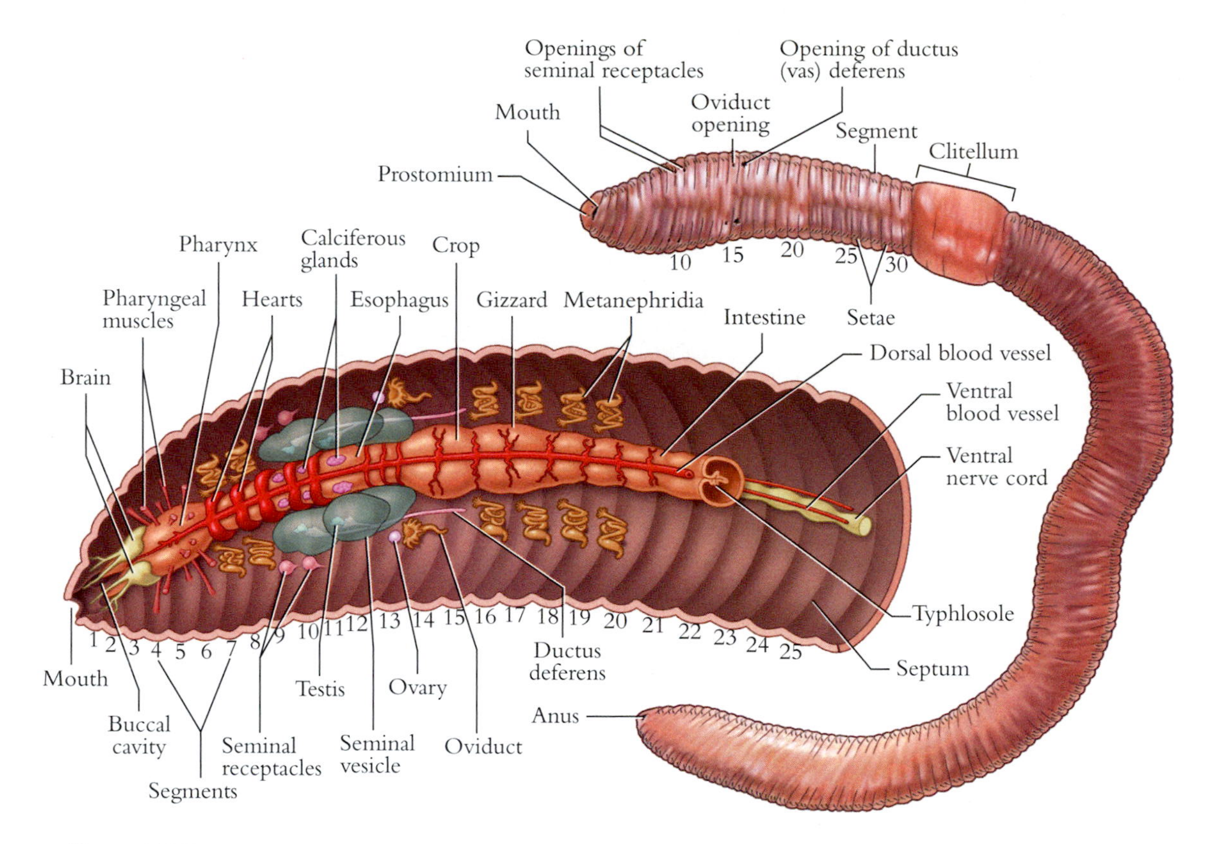

Figure 7.107 A dorsal view of the anterior end of the earthworm, *Lumbricus*.

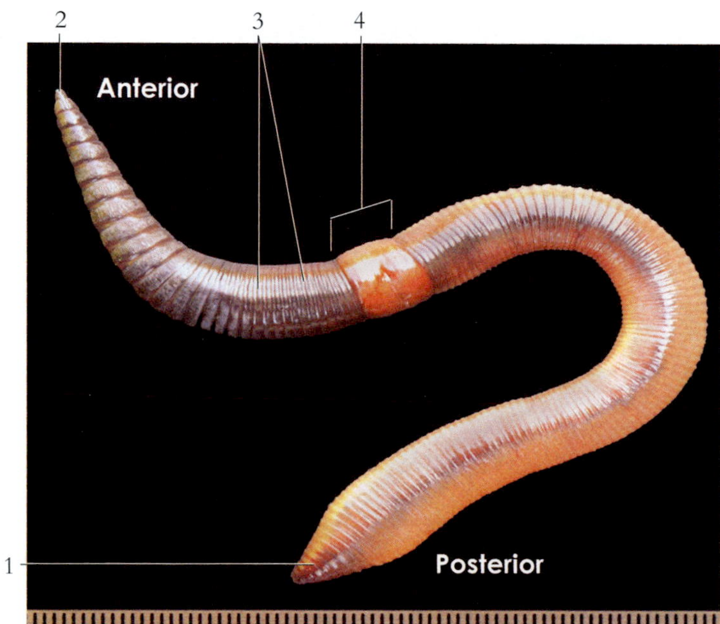

Figure 7.108 A dorsal view of an earthworm, *Lumbricus* sp. (scale in mm).

1. Pygidium
2. Prostomium (located dorsal to mouth)
3. Segments, or metameres
4. Clitellum

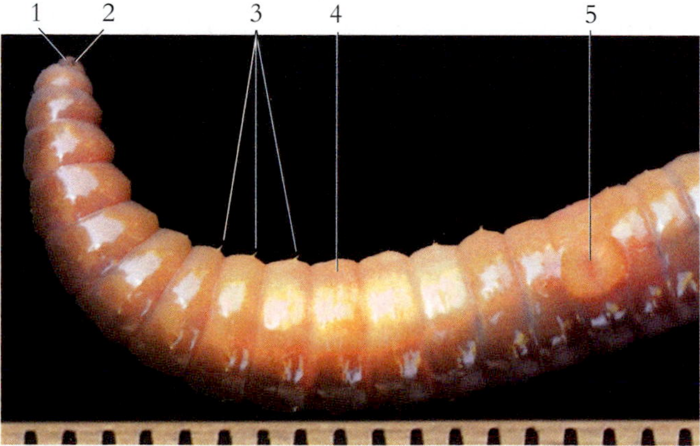

Figure 7.109 An anterior end of an earthworm, *Lumbricus* sp. (scale in mm).

1. Prostomium
2. Mouth
3. Setae
4. Segment 10
5. Opening of ductus (vas) deferens

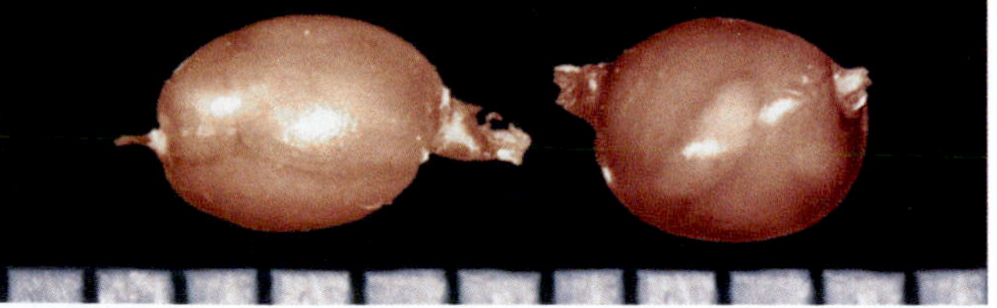

Figure 7.110 Earthworm cocoons (scale in mm).

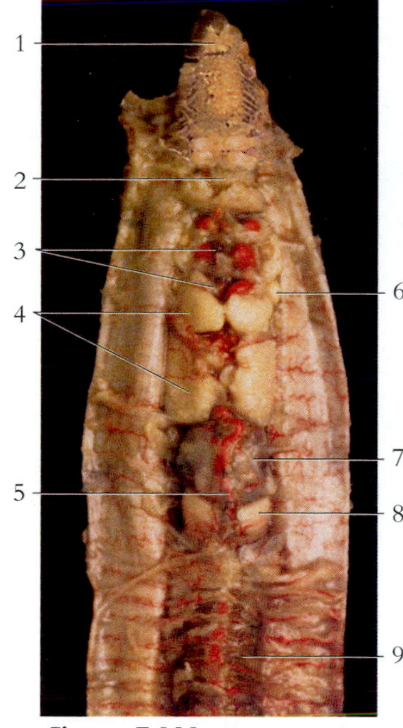

Figure 7.111 The internal anatomy of the anterior end of an earthworm, *Lumbricus* sp.

1. Brain
2. Pharynx
3. Hearts
4. Seminal vesicles
5. Dorsal blood vessel
6. Seminal receptacles
7. Crop
8. Gizzard
9. Intestine

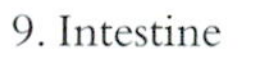

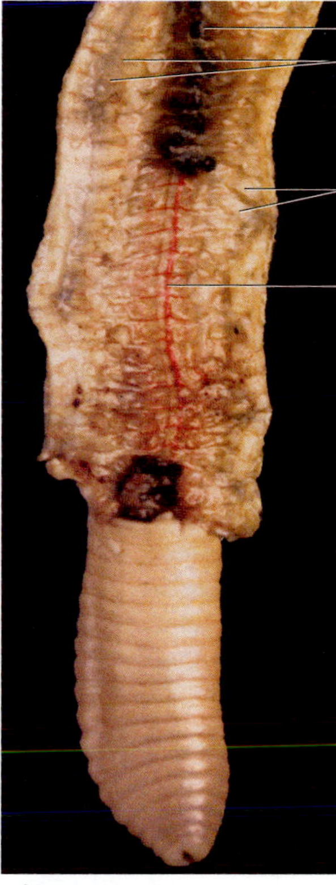

Figure 7.112 The internal anatomy of the posterior end of an earthworm with part of the intestine removed.

1. Intestine
2. Septae
3. Nephridia
4. Ventral blood vessel

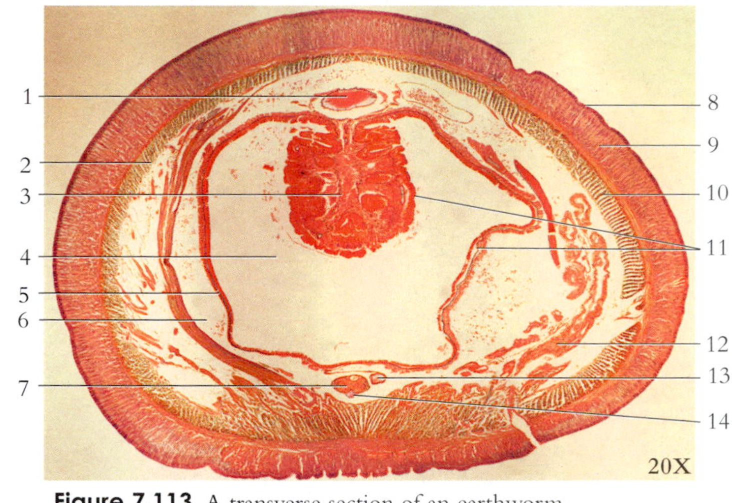

Figure 7.113 A transverse section of an earthworm posterior to the clitellum.

1. Dorsal blood vessel
2. Peritoneum
3. Typhlosole
4. Lumen of intestine
5. Intestine
6. Coelom
7. Ventral nerve cord
8. Epidermis
9. Circular muscles
10. Longitudinal muscles
11. Intestinal epithelium
12. Nephridium
13. Ventral blood vessel
14. Subneural blood vessel

Class Clitellata - Subclass Hirudinea

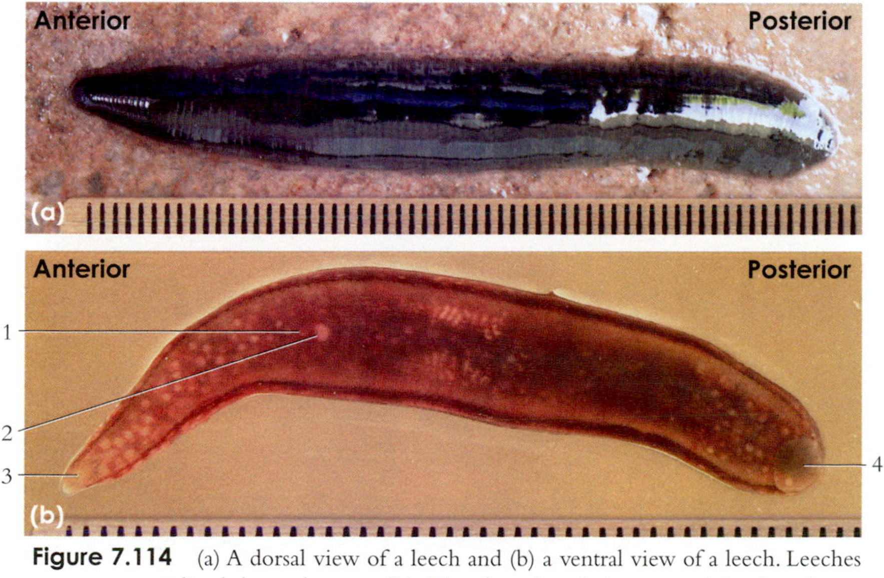

Figure 7.114 (a) A dorsal view of a leech and (b) a ventral view of a leech. Leeches are more specialized than other annelids. They have lost their setae and developed suckers for attachment while sucking blood (scale in mm).

1. Male genital pore
2. Female genital pore
3. Anterior sucker
4. Posterior sucker

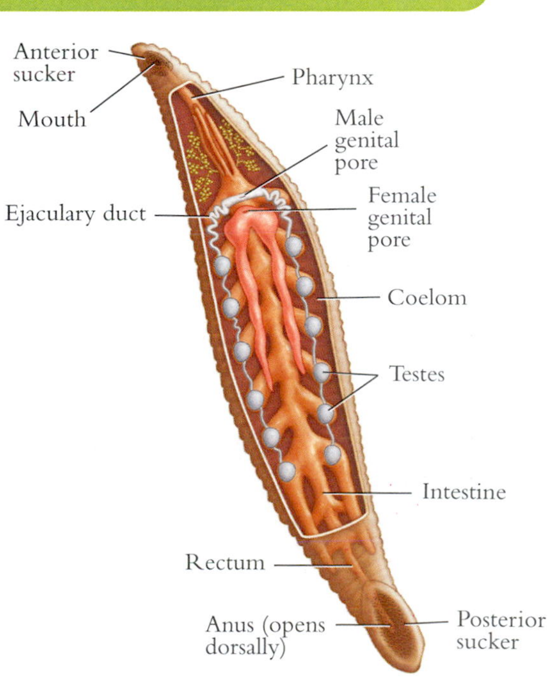

Figure 7.115 A diagram of a leech.

Phylum Nematoda - roundworms and nematodes

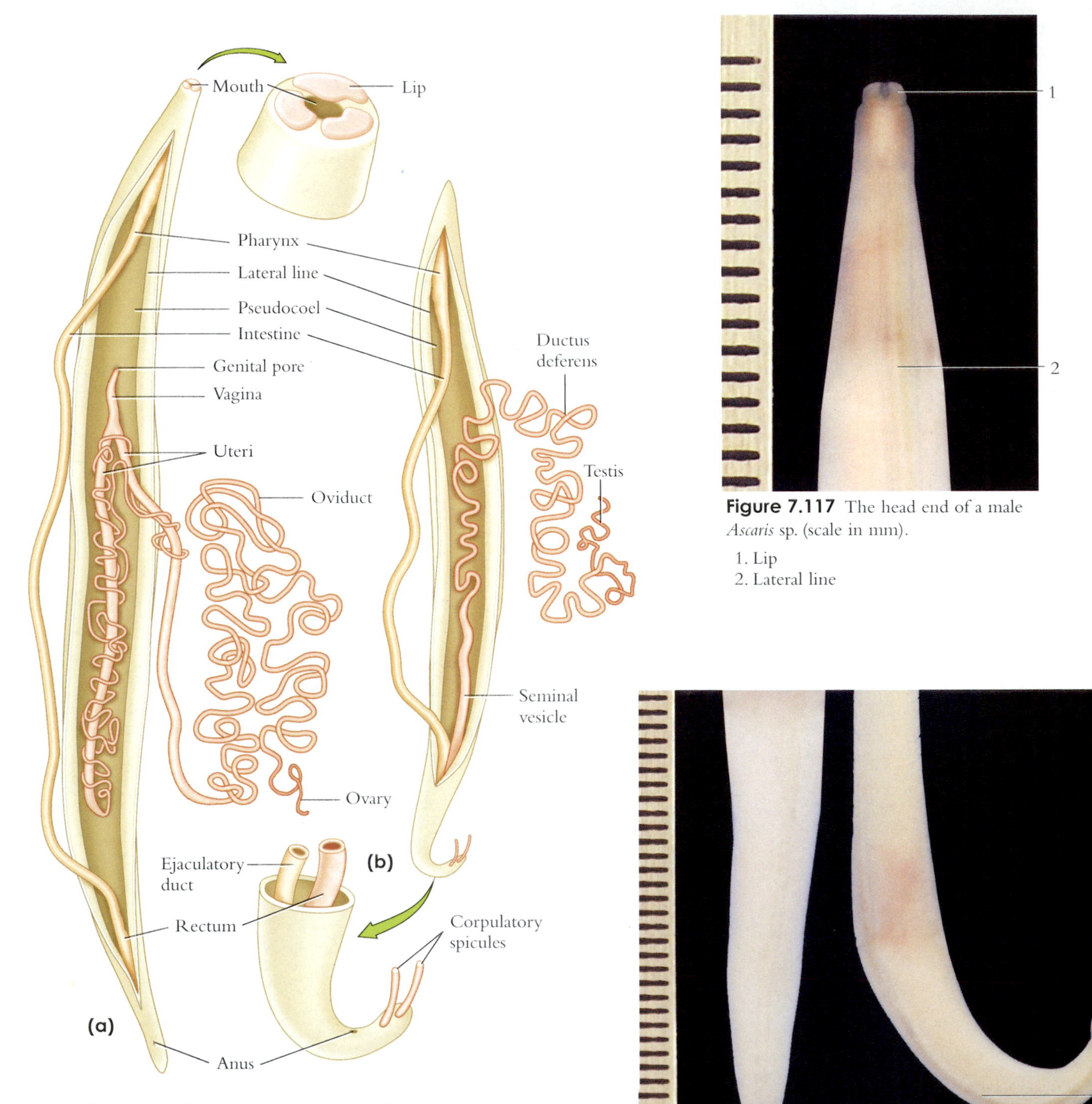

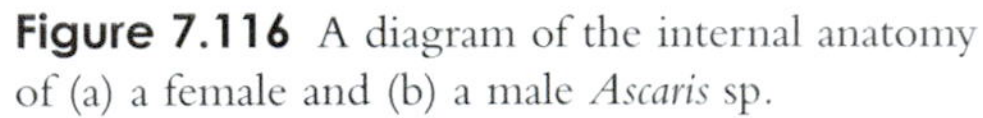

Figure 7.116 A diagram of the internal anatomy of (a) a female and (b) a male *Ascaris* sp.

Figure 7.117 The head end of a male *Ascaris* sp. (scale in mm).

1. Lip
2. Lateral line

Figure 7.118 The posterior end of (a) a female and (b) a male *Ascaris* sp. (scale in mm).

1. Copulatory spicules
2. Ejaculatory duct

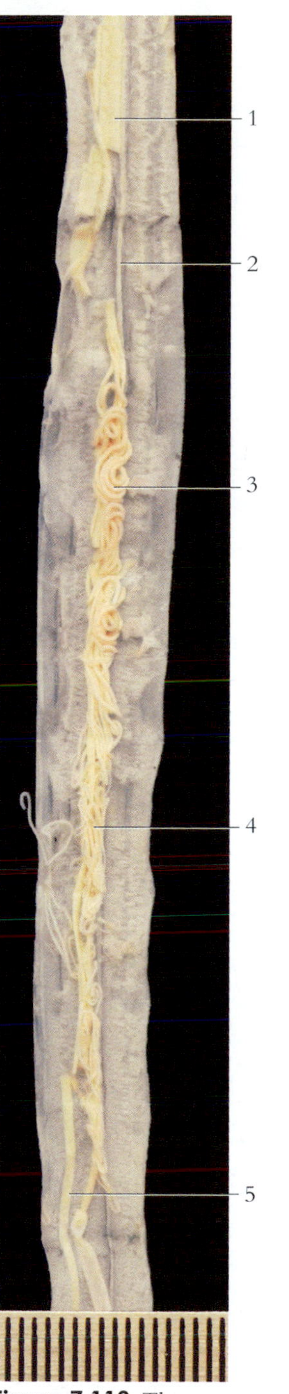

Figure 7.119 The internal anatomy of a male *Ascaris* sp. (scale in mm).

1. Intestine
2. Lateral line
3. Ductus deferens
4. Testes
5. Seminal vesicle

Figure 7.120 The internal anatomy of a female *Ascaris* sp. (scale in mm).

1. Intestine
2. Genital pore
3. Vagina
4. Oviducts
5. Uteri (Y-shaped)
6. Lateral line
7. Ovary

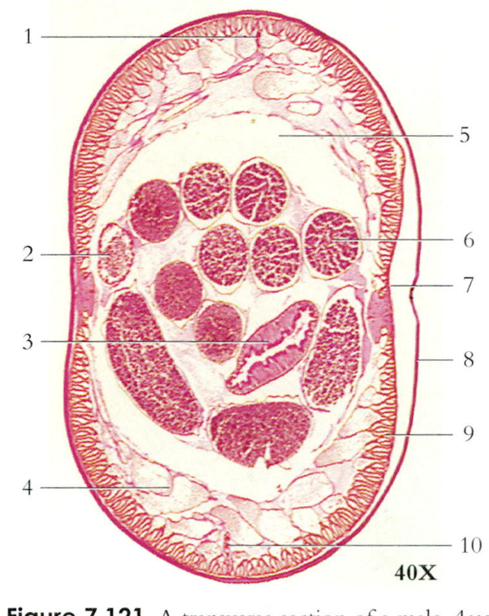

Figure 7.121 A transverse section of a male *Ascaris* sp.

1. Dorsal nerve cord
2. Ductus deferens
3. Intestine
4. Longitudinal muscle cell body
5. Pseudocoel
6. Testis
7. Lateral line
8. Cuticle
9. Contractile sheath of muscle cell
10. Ventral nerve cord

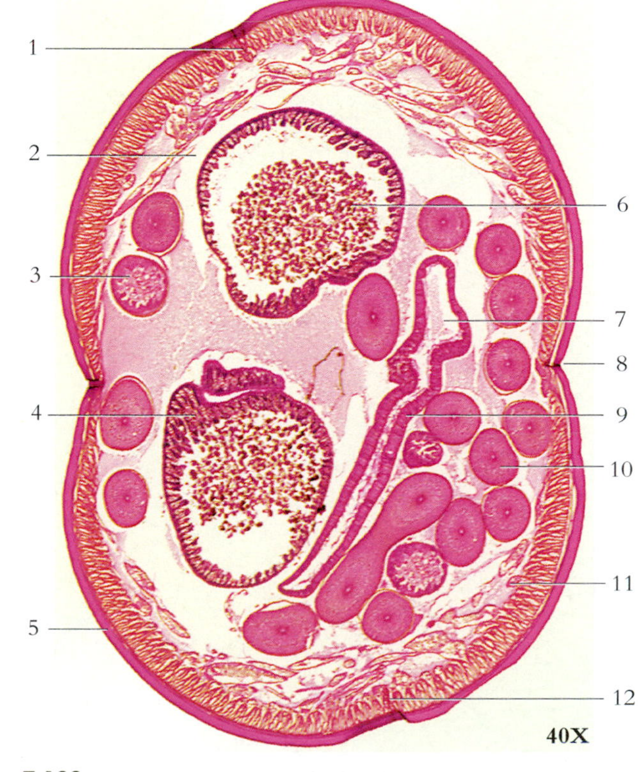

Figure 7.122 A transverse section of a female *Ascaris* sp.

1. Dorsal nerve cord
2. Pseudocoel
3. Oviduct
4. Uterus
5. Cuticle
6. Eggs
7. Lumen of intestine
8. Lateral line
9. Intestine
10. Ovary
11. Longitudinal muscles
12. Ventral nerve cord

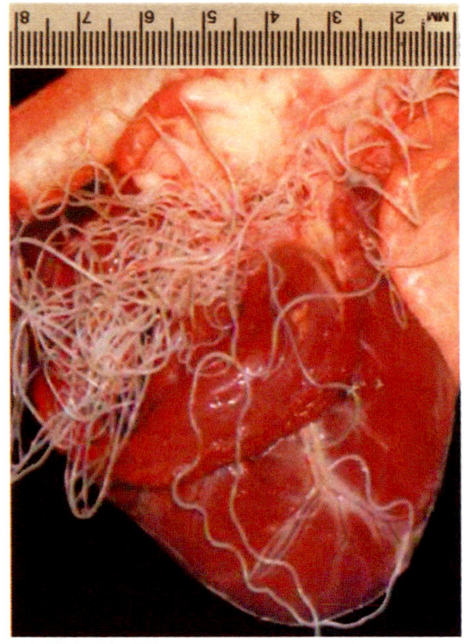

Figure 7.123 A dog heart infested with heartworm, *Dirofilaria immitis* (scale in mm).

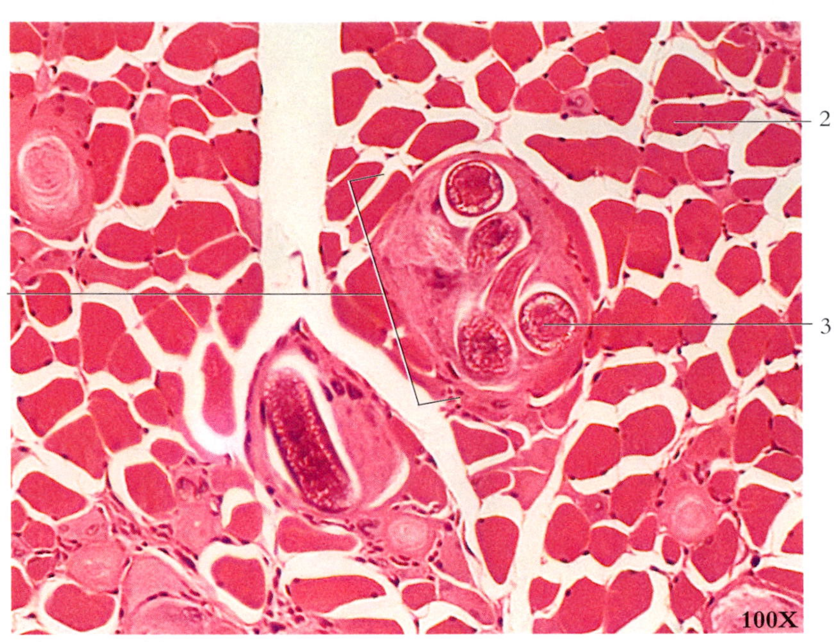

Figure 7.124 A photomicrograph of *Trichinella spiralis* encysted in muscle.
1. Cyst 2. Muscle 3. Larva

Phylum Rotifera - rotifers

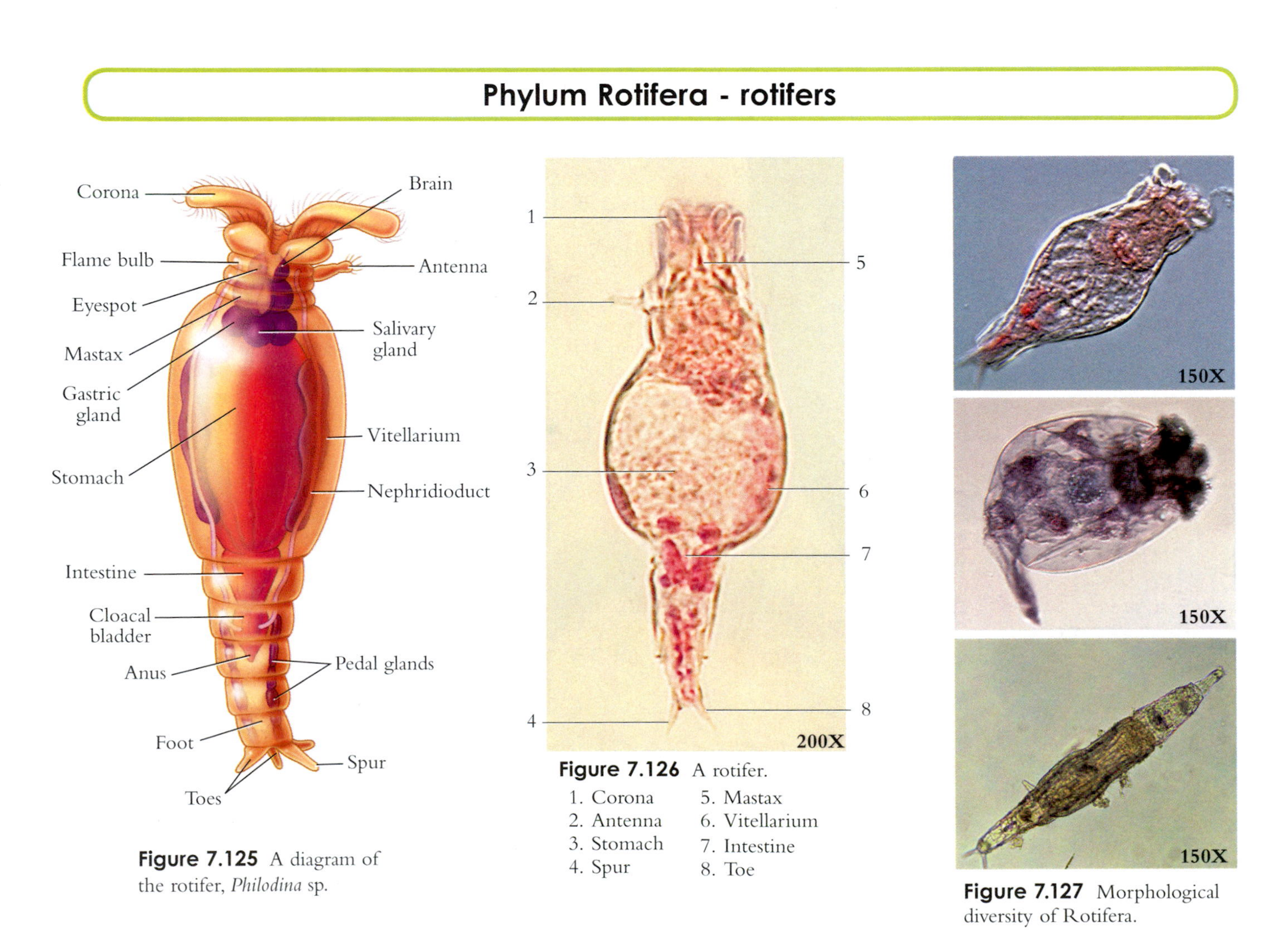

Figure 7.125 A diagram of the rotifer, *Philodina* sp.

Figure 7.126 A rotifer.

1. Corona
2. Antenna
3. Stomach
4. Spur
5. Mastax
6. Vitellarium
7. Intestine
8. Toe

Figure 7.127 Morphological diversity of Rotifera.

Table 7.6 Representatives of the Phylum Arthropoda

Classes and Representative Kinds	Characteristics
Merostomata (Subphylum Chelicerata) – horseshoe crab	Cephalothorax and abdomen; specialized front appendages into chelicerae; lack antennae and mandibles
Arachnida (Subphylum Chelicerata) – spiders, mites, ticks, and scorpions	Cephalothorax and abdomen; chelicerae; four pairs of legs; book lungs or trachea; lack antennae and mandibles
Malacostraca (Subphylum Crustacea) – lobsters, crabs, shrimp, and isopods	Cephalothorax and abdomen; two pairs of antennae; pair of mandibles and two pairs of maxillae; biramous appendages; gills
Maxillopoda (Subphylum Crustacea) – copepods and barnacles	Cephalothorax and abdomen; freshwater and marine; up to six pairs of appendages
Insecta – beetles, butterflies, and ants	Head, thorax, and abdomen; three pairs of legs; well-developed mouth parts; usually two pair of wings; trachea
Chilopoda – centipedes	Head with segmented trunk; one pair of legs per segment; trachea; one pair of antennae
Diplopoda – millipedes	Head with segmented trunk; usually two pair of legs per segment; trachea

Representatives of the Phylum Tardigrada

Phylum	Characteristics
Tardigrada – water bears	Bilaterally symmetrical; four pairs of lobopod legs terminating in claws or sucking disks

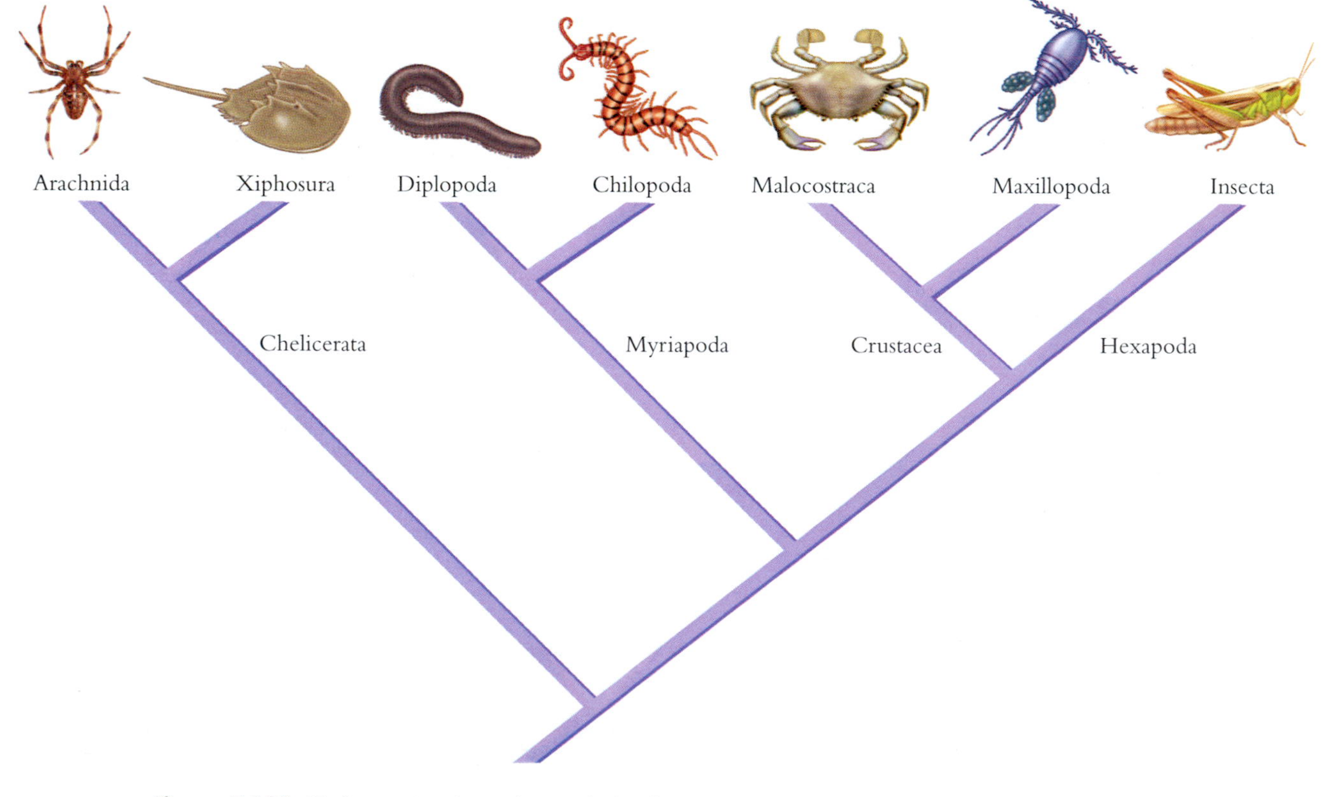

Figure 7.128 Phylogenetic relationships and classification of select arthropods (does not include classes Brachiopoda, Remipedia, Cephalocarida, and Ostracoda. Maxillopoda are likely paraphyletic). Tardigrada probably branch basally to Arthropoda and Onychophora (see fig. 7.1 on page 150).

Phylum Arthropoda - arachnids, crustaceans, and insects

Figure 7.129 Example arthropods include: (a) a flat rock scorpion, *Hadogenes troglodytes*, (b) an American giant millipede, *Narceus americanus*, (c) a brine shrimp, *Artemia salina*, (d) a tiger beetle, *Cicindela fulgida,* (e) a fossil trilobite, *Modicia typicalis* (trilobites are extinct arthropods from the Cambrian and Ordovician periods), (f) a tardigrade, *Macrobiotus* sp., (g) a shieldback katydid, *Neduba carinata*, (h) a water beetle, *Lethocerus medius*, (i) a stripped shore crab, *Pachygrapsus crassipes*, (j) a black widow, *Latrodectus hesperus,* (k) a solpugid, *Eremobates pallipes*, and (l) a painted lady butterfly, *Vanessa annabella*.

Figure 7.130 Harvestmen, *Phalangium opilio,* commonly called daddy long legs, are not really spiders.

Subphylum Chelicerata - Class Merostomata

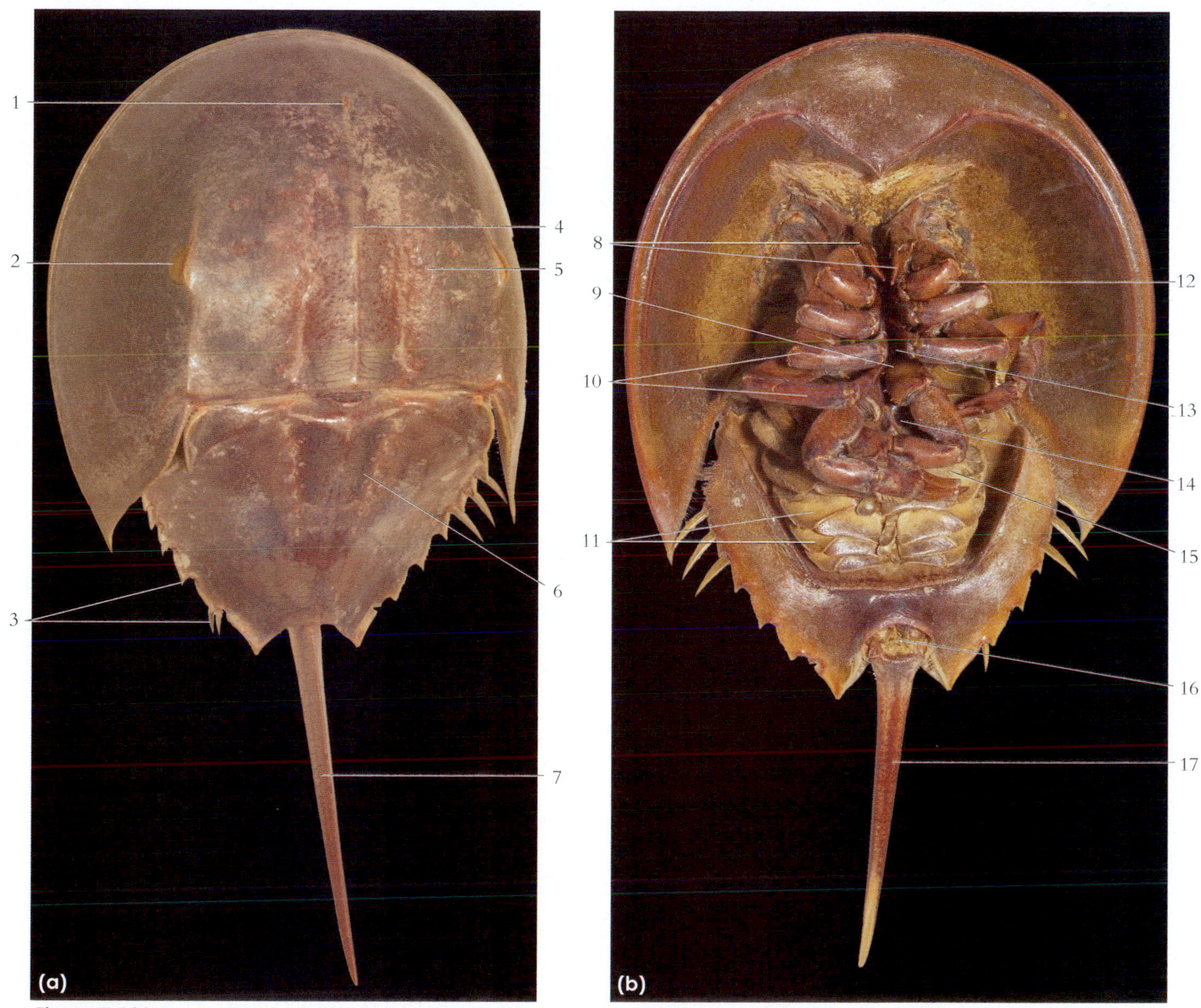

Figure 7.131 (a) A dorsal view and (b) a ventral view of the horseshoe crab, *Limulus* sp. This animal is commonly found in shallow waters along the Atlantic coast from Canada to Mexico.

1. Simple eye
2. Compound eye
3. Abdominal spines
4. Anterior spine
5. Cephalothorax (prosoma)
6. Abdomen (opisthosoma)
7. Telson
8. Chelicerae
9. Gnathobase
10. Chelate legs
11. Book gills
12. Pedipalp
13. Mouth
14. Chilarium
15. Genital operculum
16. Anus
17. Telson

Subphylum Chelicerata - Class Arachnida

Figure 7.132 A garden spider in the process of spinning a web.
1. Spinnerets

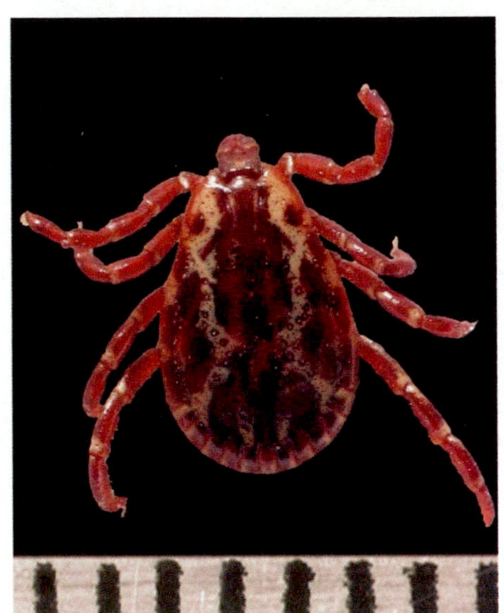

Figure 7.133 A tick, within the family Ixodidae, is a specialized parasitic arthropod (scale in mm).

Figure 7.134 A red mite, *Dermanyssus gallinae*, feeding on a lizard.

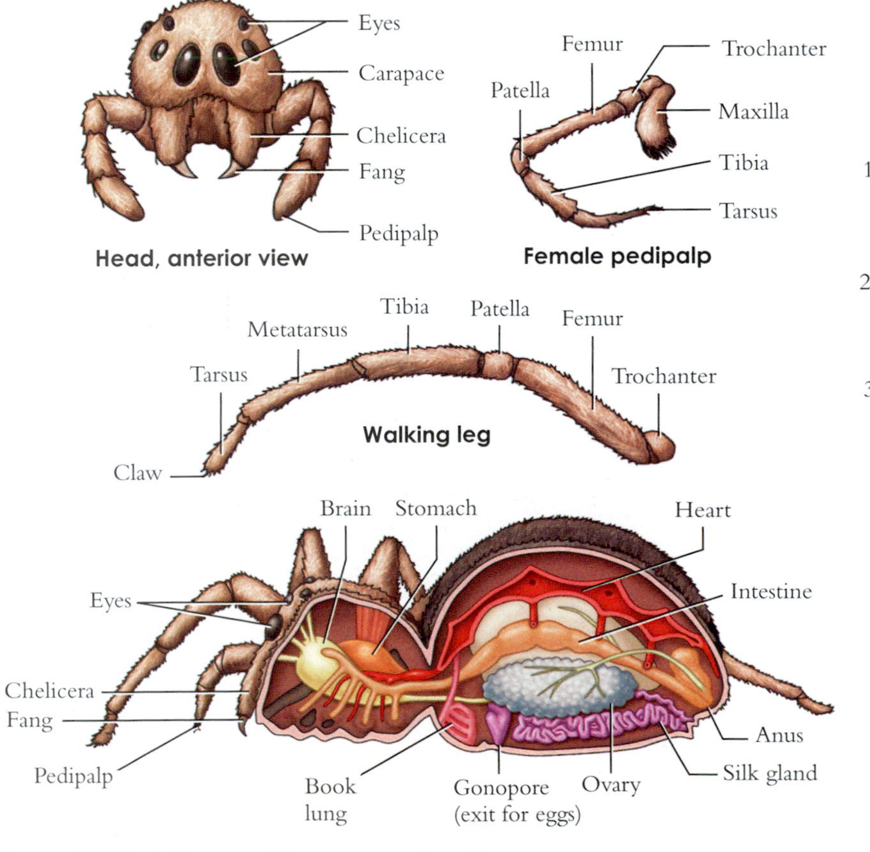

Figure 7.135 A diagram of the anatomy of a spider.

Figure 7.136 A cobalt blue tarantula, *Haplopelma lividum*.
1. Opisthosoma (abdomen)
2. Prosoma (cephalothorax)
3. Pedipalps

Figure 7.137 An Arizona hairy scorpion, *Hadrurus arizonensis.* Scorpions are most commonly found in tropical and subtropical regions, but there are also several species found in arid and temperate zones.

1. Cephalothorax
2. Pedipalp
3. Stinging apparatus
4. Postabdomen (tail)
5. Preabdomen
6. Walking legs

Figure 7.138 Some ticks attached and feeding on a savannah monitor, a large African lizard.

1. Ticks
2. Scales of monitor

Class Malacostraca

Figure 7.139 A pill bug, *Armadillidium* sp.

Figure 7.140 A sea slater, *Ligia italica*.

Figure 7.141 A fire shrimp, *Lysmata debelius*.

Figure 7.142 A Sally Lightfoot crab, *Grapsus grapsus*.

Figure 7.143 A hermit crab, *Coenobita clypeatus*.

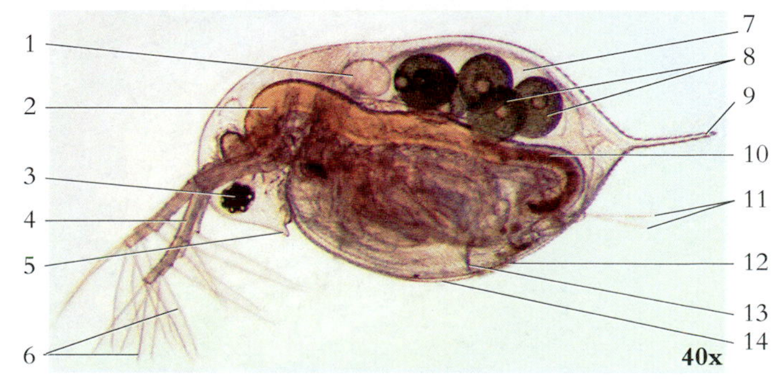

Figure 7.144 The water flea, *Daphnia*, is a common microscopic crustacean.

1. Heart
2. Midgut
3. Compound eye
4. 2nd antenna
5. Rostrum
6. Setae
7. Brood chamber
8. Eggs
9. Apical spine
10. Hindgut
11. Abdominal setae
12. Anus
13. Abdominal claw
14. Carapace

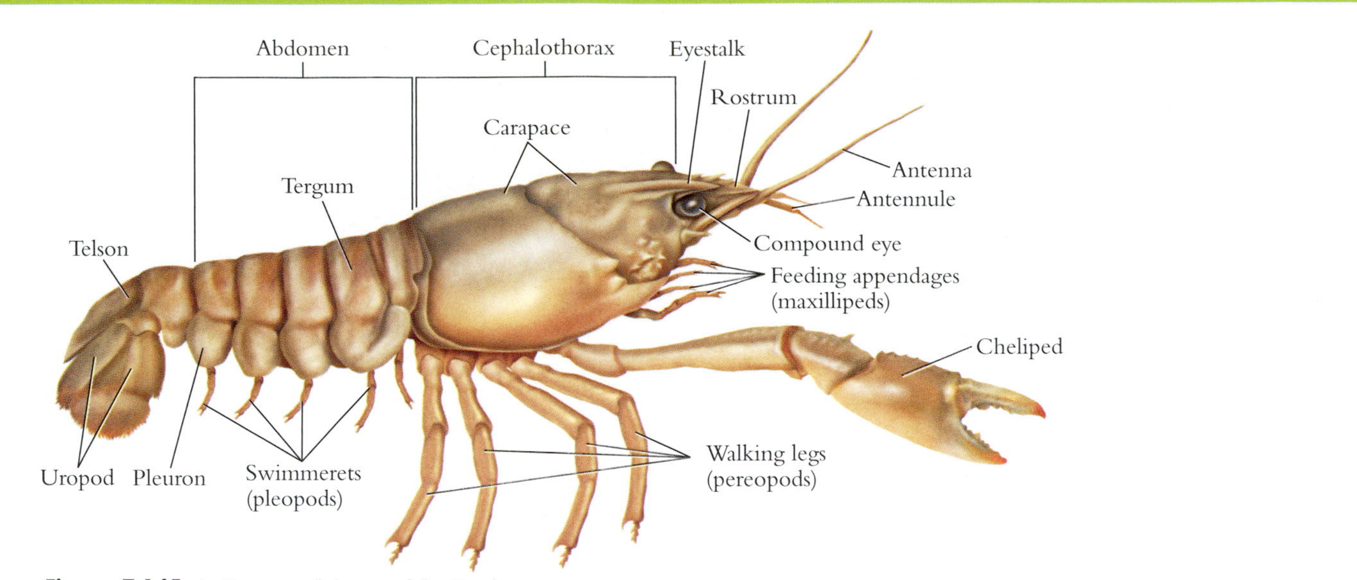

Figure 7.145 A diagram of the crayfish, *Cambarus*.

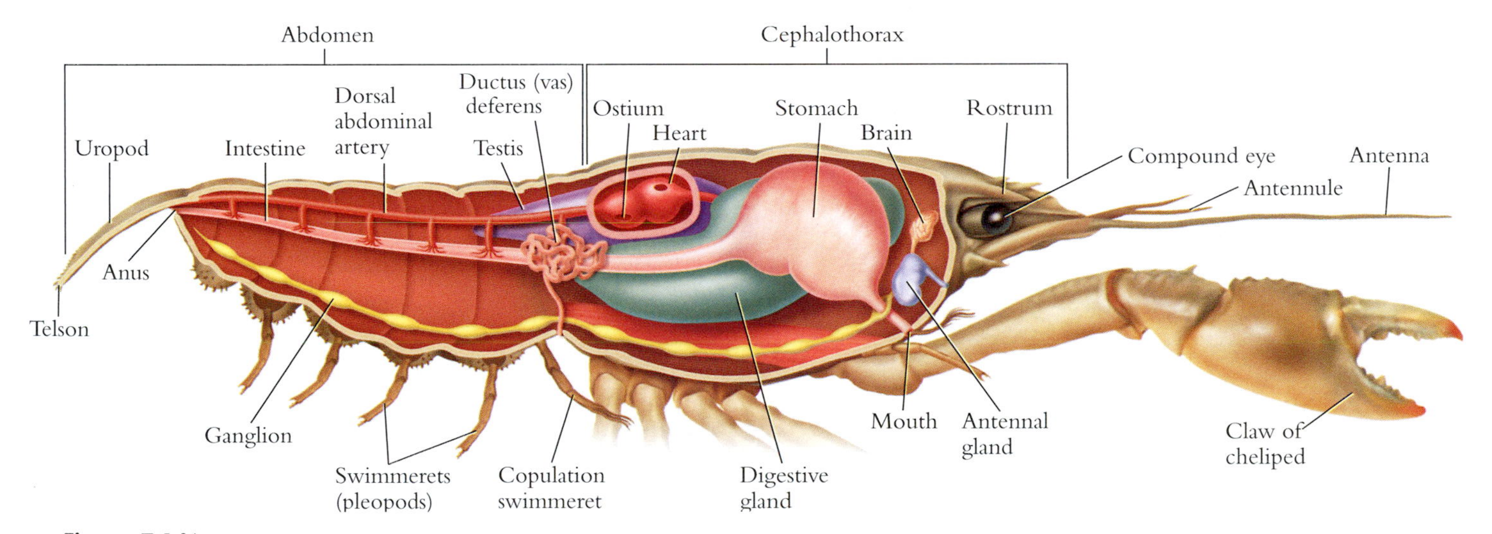

Figure 7.146 The anatomy of a crayfish. A sagittal section of an adult male.

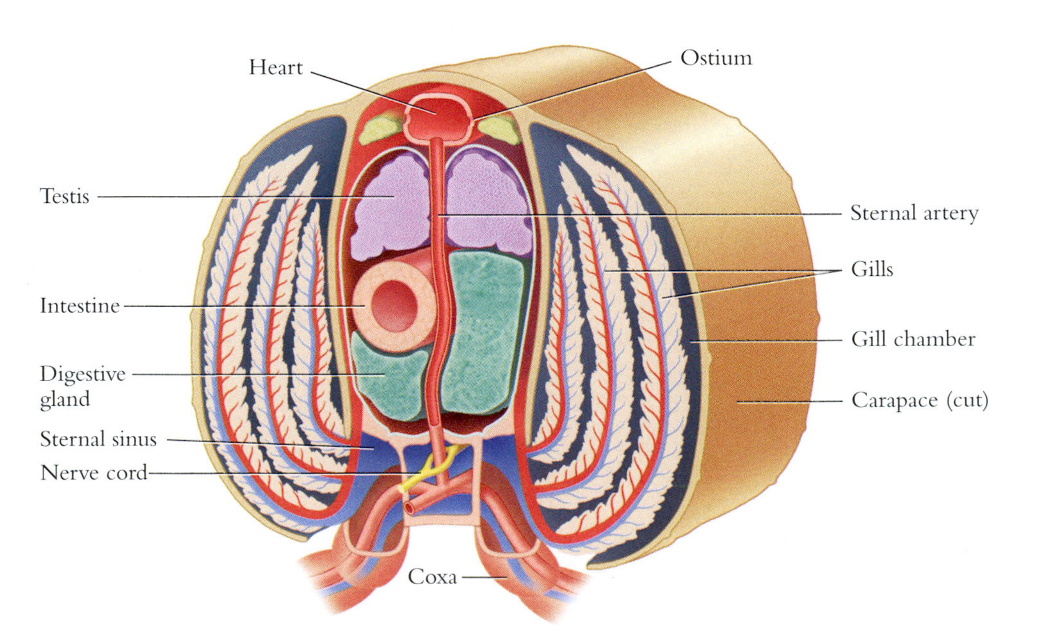

Figure 7.147 The anatomy of a crayfish. A transverse section of an adult male.

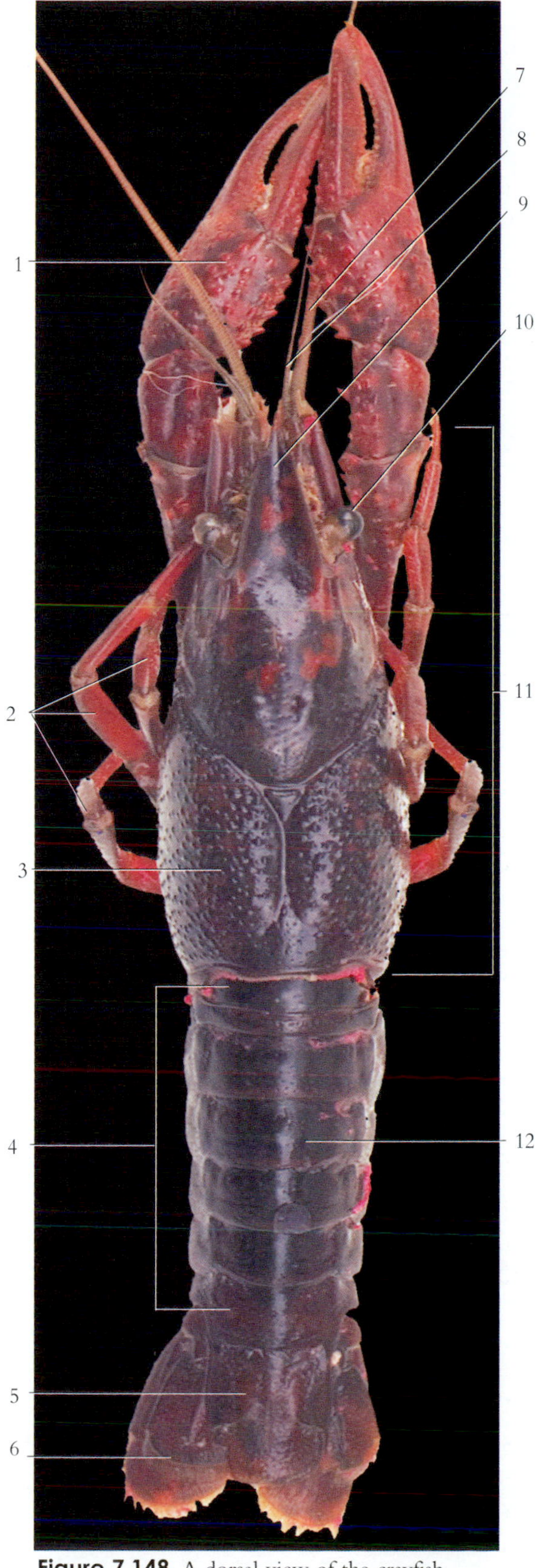

Figure 7.148 A dorsal view of the crayfish.

1. Cheliped
2. Walking legs
3. Carapace
4. Abdomen
5. Telson
6. Uropod
7. Antenna
8. Antennule
9. Rostrum
10. Compound eye
11. Cephalothorax
12. Tergum

Figure 7.149 A lateral view of the crayfish.

1. Carapace
2. Abdomen
3. Uropod
4. Swimmeret (pleopod)
5. Rostrum
6. Compound eye
7. Maxilliped
8. Cheliped
9. Walking legs

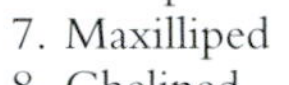

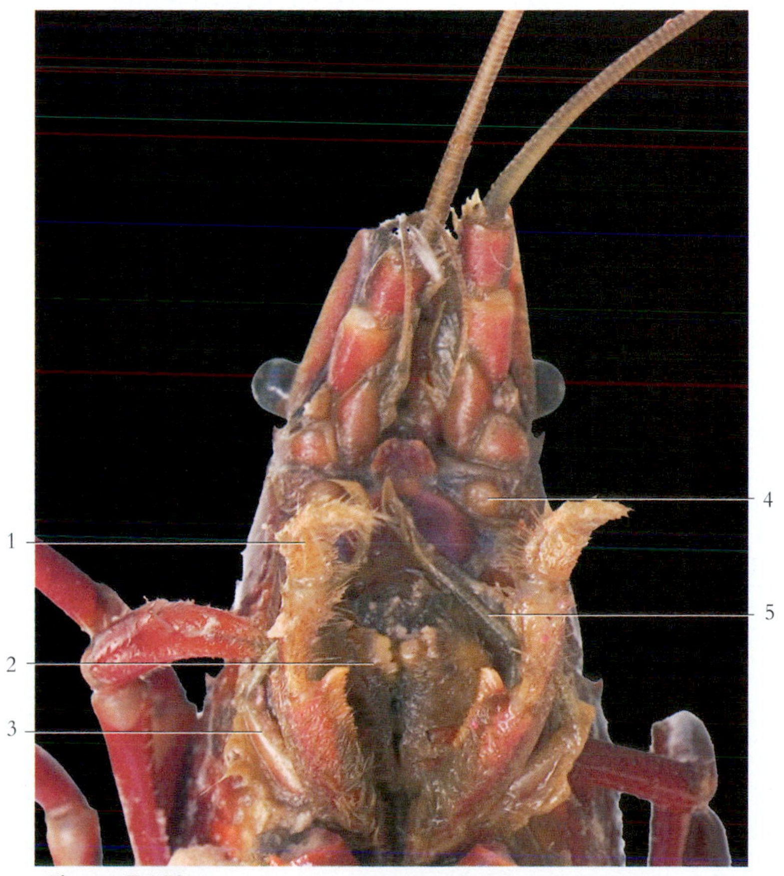

Figure 7.150 A ventral view of the oral region of the crayfish.

1. Third maxilliped
2. Mandible
3. Second maxilla
4. Green gland duct
5. First maxilliped

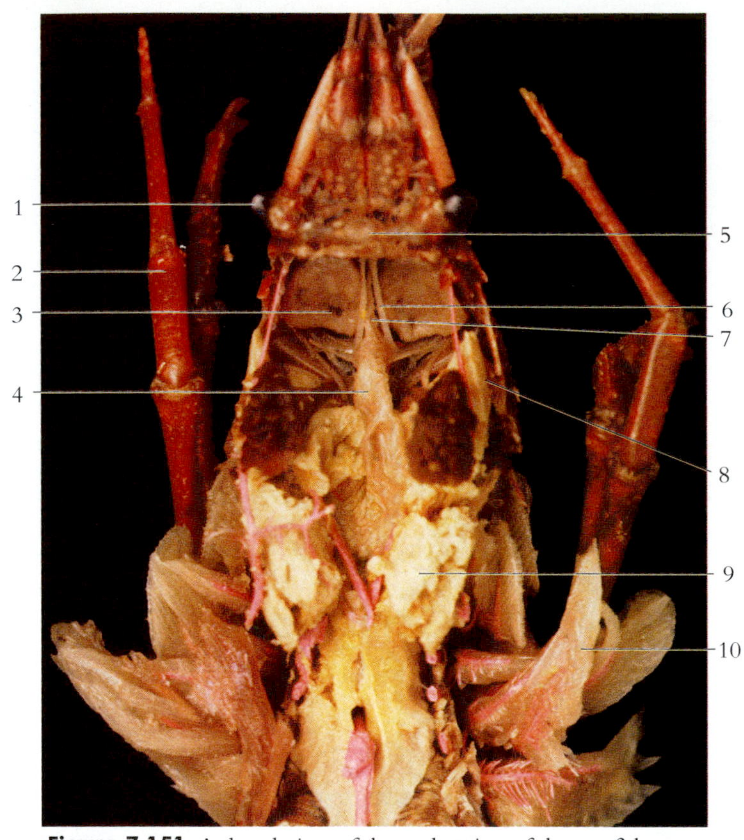

Figure 7.151 A dorsal view of the oral region of the crayfish.

1. Compound eye
2. Walking leg
3. Green gland
4. Cardiac chamber of stomach
5. Brain
6. Circumesophageal connection (of ventral nerve cord)
7. Esophagus
8. Region of gastric mill
9. Digestive gland
10. Gill

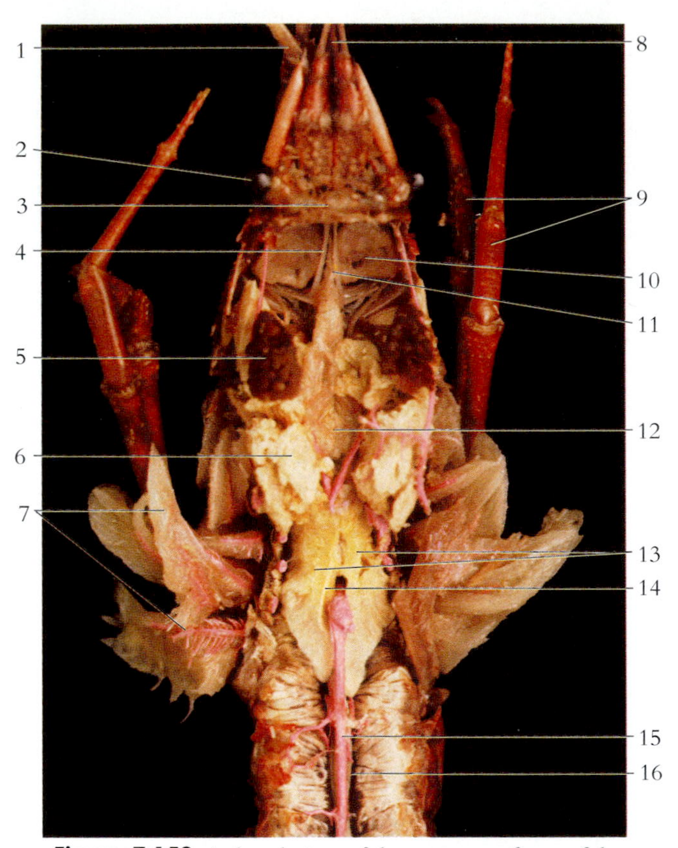

Figure 7.152 A dorsal view of the anatomy of a crayfish.

1. Antenna
2. Compound eye
3. Brain
4. Circumesophageal connection (of ventral nerve cord)
5. Mandibular muscle
6. Digestive gland
7. Gills
8. Antennules
9. Walking legs
10. Green gland
11. Esophagus
12. Pyloric stomach
13. Testis
14. Ductus deferens
15. Aorta
16. Intestine

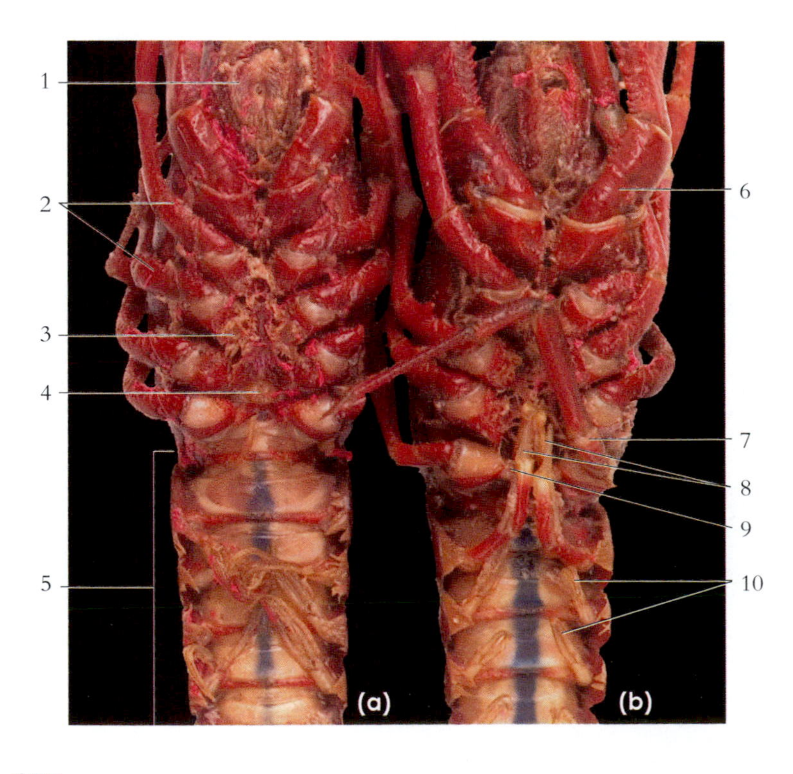

Figure 7.153 A ventral view of (a) a female and (b) a male crayfish. The first pair of swimmerets are greatly enlarged in the male for the depositing of sperm in the female's seminal receptacle.

1. Third maxilliped
2. Walking legs
3. Disk covering oviduct
4. Seminal receptacle
5. Abdomen
6. Base of cheliped
7. Base of last walking leg
8. Copulatory swimmerets (pleopods)
9. Sperm ducts (genital pores)
10. Swimmerets (pleopods)

Class Insecta

(a) (b) (c) (d) (e) (f) (g) (h)

Figure 7.154 Example insects include: (a) a greater arid-land katydid, *Neobarrettia spinosa*, (b) an eastern lubber grasshopper, *Romalea microptera*, (c) a flame skimmer dragonfly, *Libellula saturata*, (d) a cicada, *Diceroprocta apache*, (e) a milkweed beetle, *Tetraopes tetraophthalmus*, (f) a cynthia moth, *Samia cynthia*, (g) a giant cockroach, *Blaberus giganteus*, and (h) a Carolina mantis, *Stagmomantis carolina*.

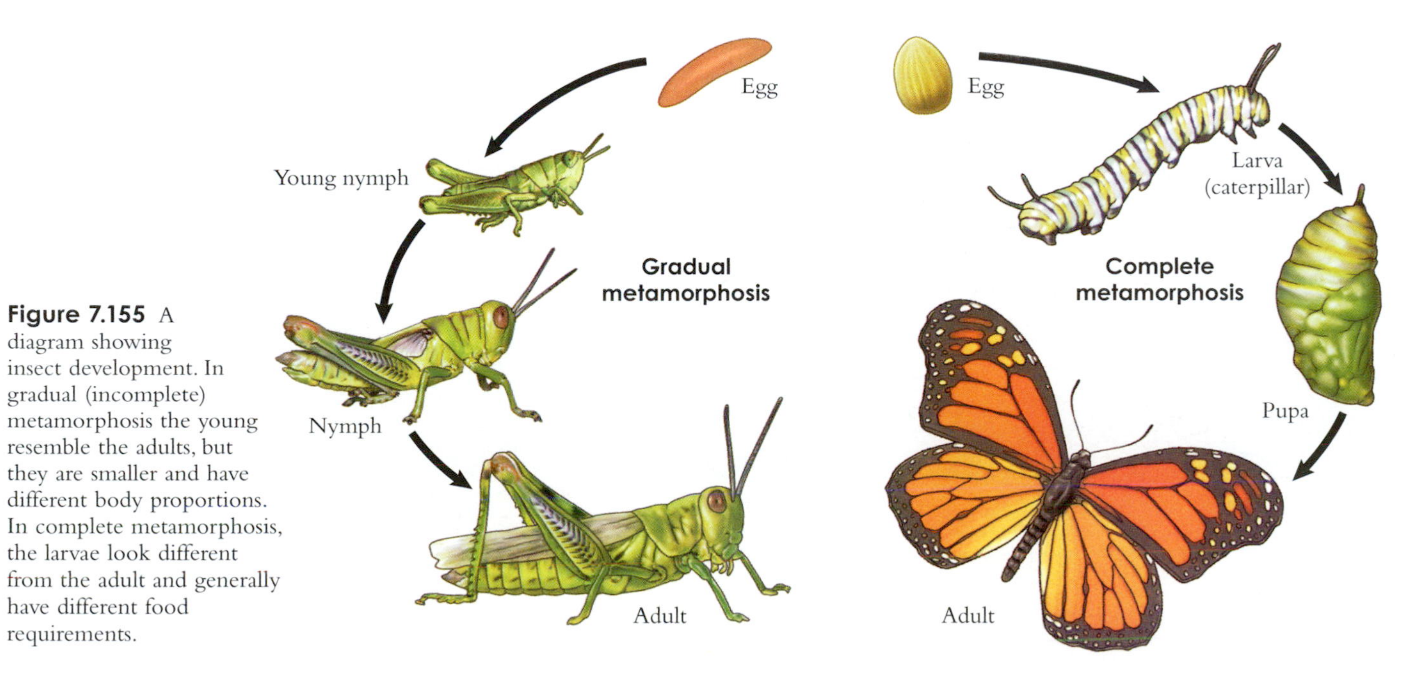

Figure 7.155 A diagram showing insect development. In gradual (incomplete) metamorphosis the young resemble the adults, but they are smaller and have different body proportions. In complete metamorphosis, the larvae look different from the adult and generally have different food requirements.

Figure 7.156 The developmental stages of the monarch butterfly, *Danaus plexippus*, include (a) egg, (b) larval stage, (c) chrysalis, and (d) adult.

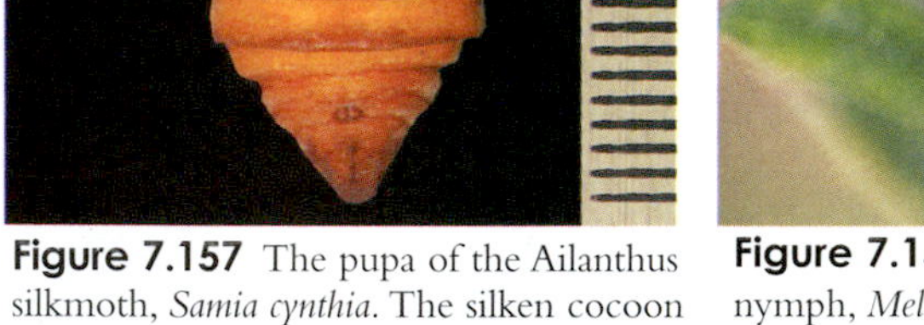

Figure 7.157 The pupa of the Ailanthus silkmoth, *Samia cynthia*. The silken cocoon has been removed (scale in mm).

Figure 7.158 A grasshopper, nymph, *Melanoplus*.

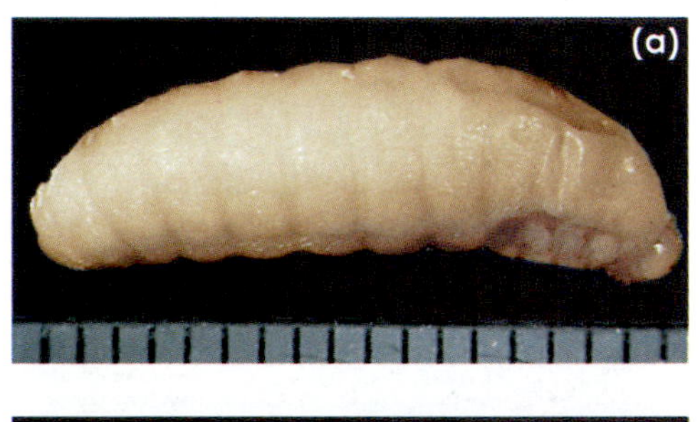

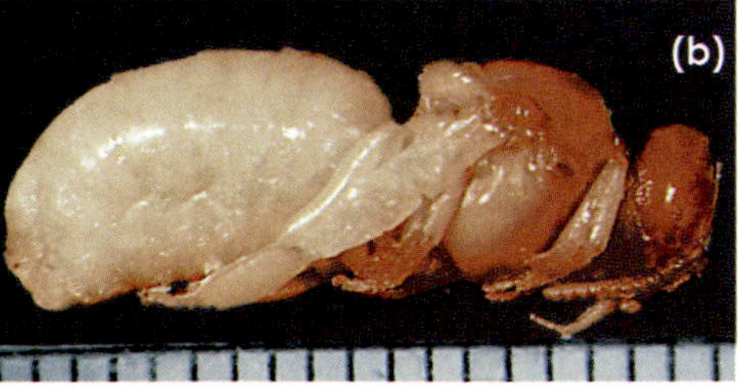

Figure 7.159 A common house cricket, *Acheta domestica*, molting. All arthropods must periodically shed their exoskeleton in order to grow. This process is called molting, or ecdysis.

Figure 7.160 The developmental stages of the common honeybee, *Apis mellifera,* include (a) larval stage, (b) pupa, and (c) adult (scale in mm).

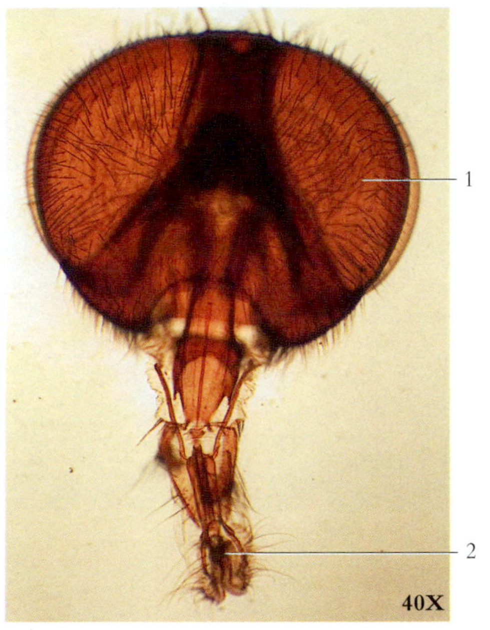

Figure 7.161 The head of a housefly, showing an example of a sponging type mouthpart in insects. Notice the large lobes at the apex of the labium, which function in lapping up liquids.
1. Compound eye
2. Labium

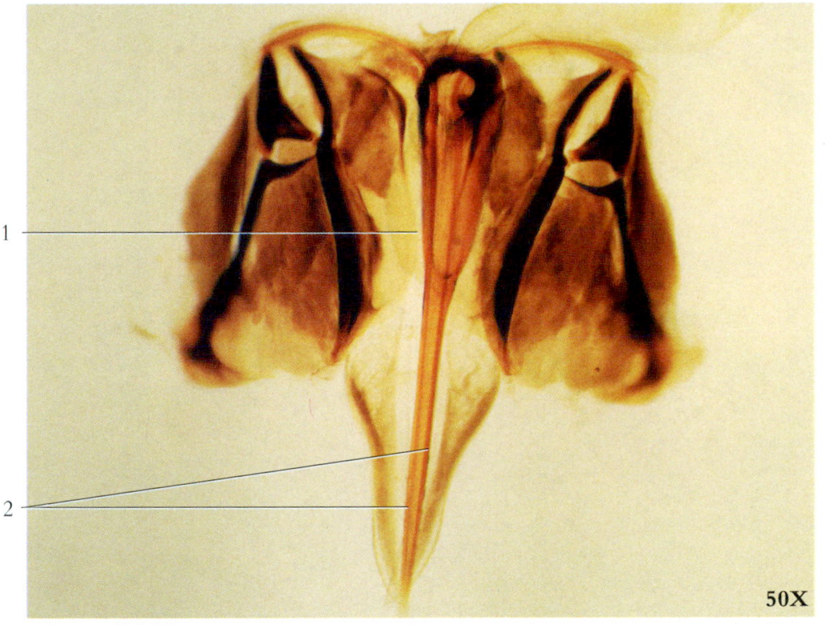

Figure 7.162 A honeybee, *Apis mellifera*, stinger. The two darts contain barbs on the tips that point upward, making it difficult to remove a stinger from a wound.
1. Sheath
2. Darts

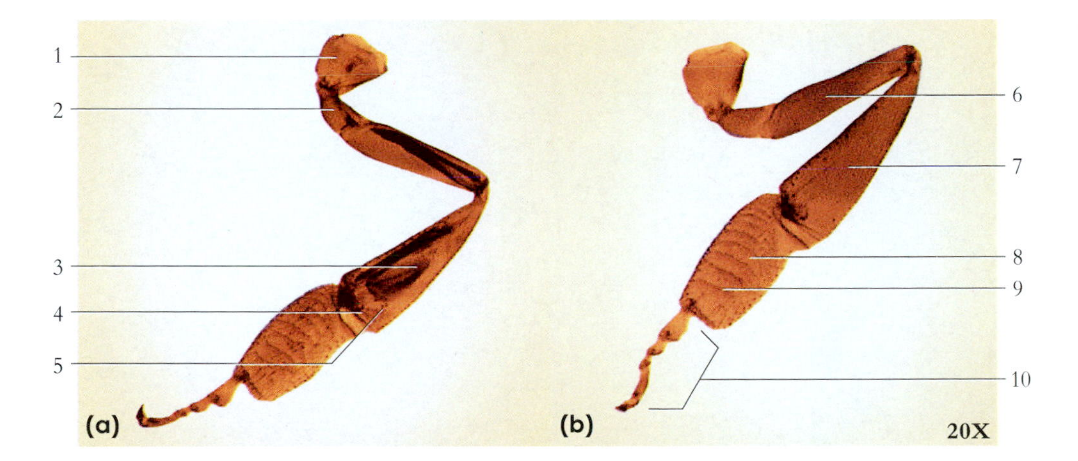

Figure 7.163 The hind legs of a worker honeybee, *Apis mellifera*, (a) outer surface and (b) inner surface.
1. Coxa
2. Trochanter
3. Pollen basket
4. Pollen packer
5. Pecten
6. Femur
7. Tibia
8. Metatarsus
9. Pollen comb
10. Tarsus

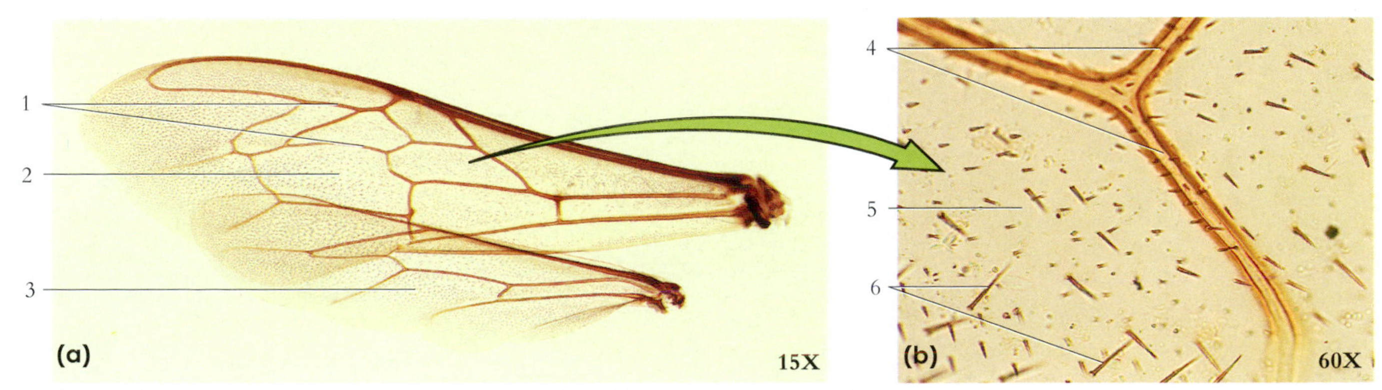

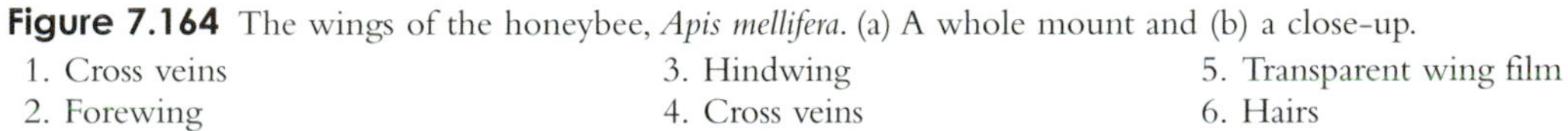

Figure 7.164 The wings of the honeybee, *Apis mellifera*. (a) A whole mount and (b) a close-up.
1. Cross veins
2. Forewing
3. Hindwing
4. Cross veins
5. Transparent wing film
6. Hairs

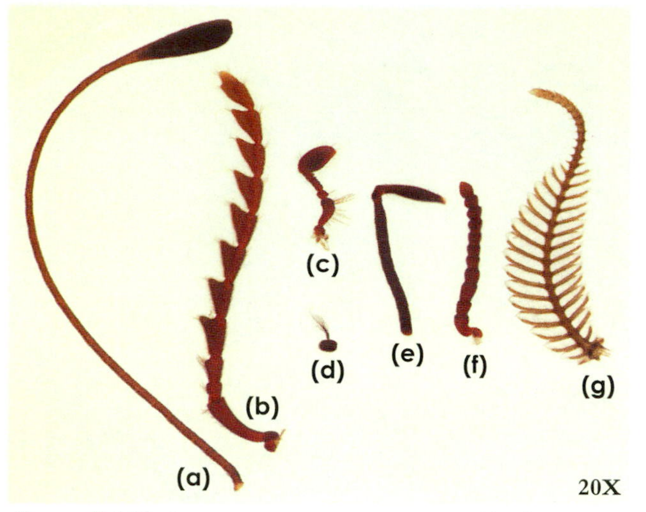

Figure 7.165 Some common insect antennae. (a) Clavate—butterflies, (b) serrate—click beetles, (c) lamellate—scarab beetles, (d) aristate—houseflies, (e) geniculate—weevils (f) moniliform—termites, and (g) plumose—moths.

Figure 7.166 The plumose antennae of the Ailanthus silkmoth, *Samia cynthia*.

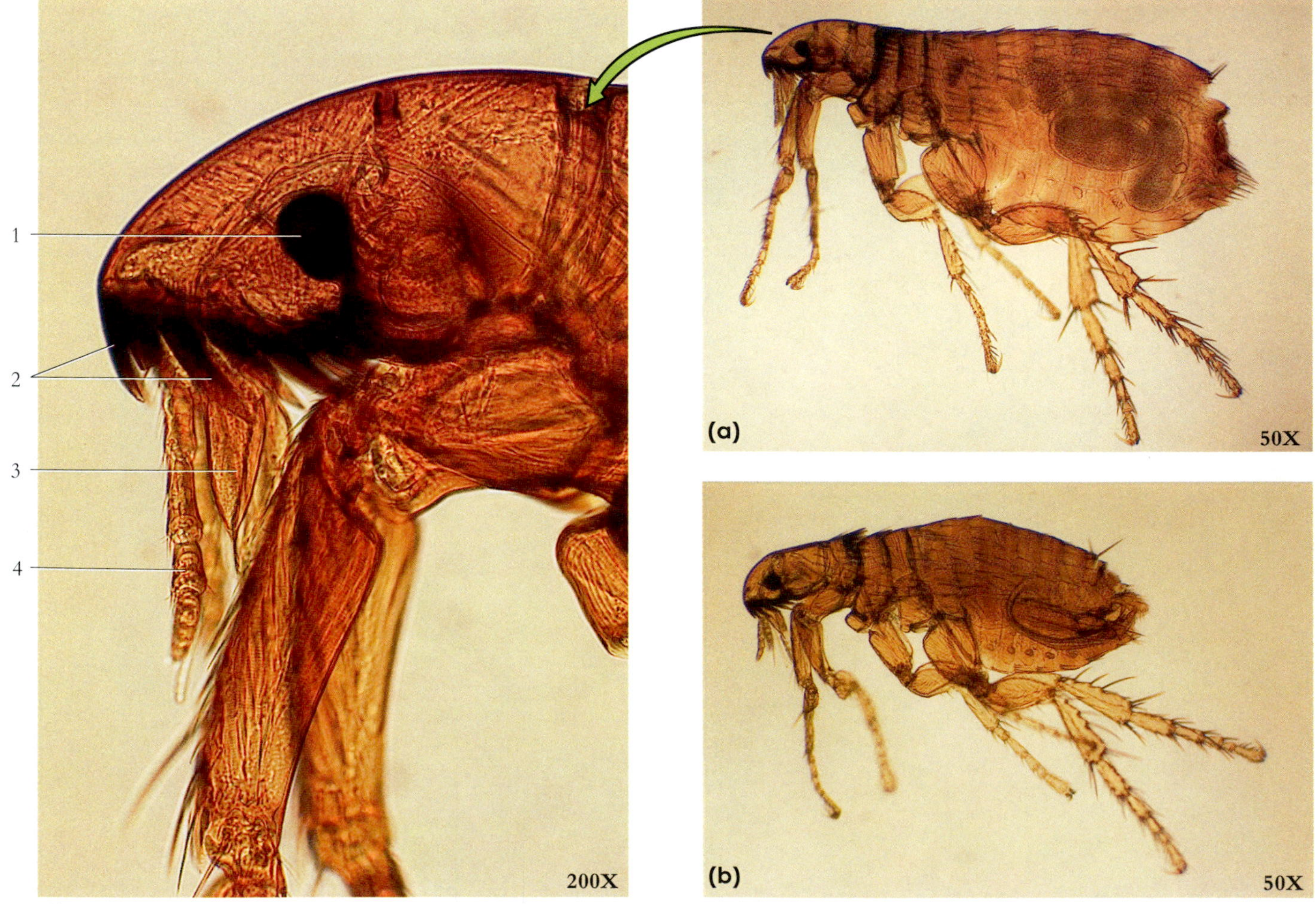

Figure 7.167 The mouthparts of the flea, *Ctenocephalide* sp., which are specialized for parasitism. Notice the oral bristles beneath the mouth that aid the flea in penetrating between hairs to feed on the blood of mammals. (a) Female flea and (b) male flea.

1. Eye
2. Oral bristles
3. Maxilla
4. Maxillary palp

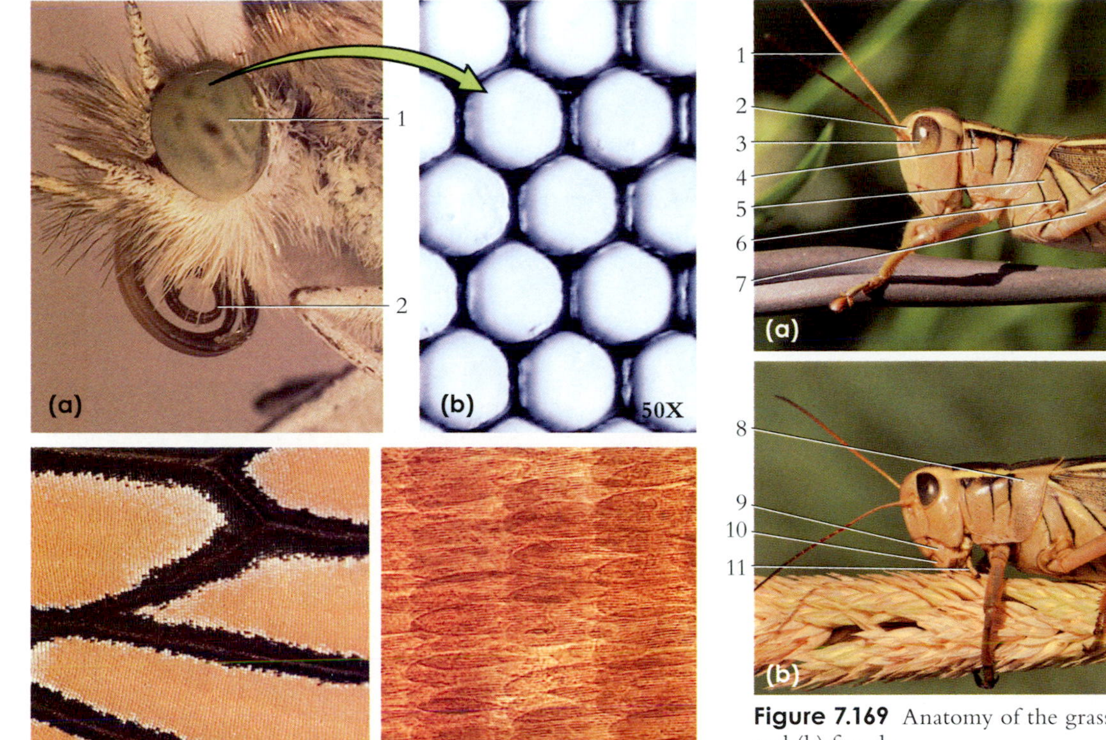

Figure 7.168 (a) A lateral view of the head of a butterfly. The most obvious structures on the head of a butterfly are compound eyes and the curled tongue for siphoning nectar from flowers. (b) A magnified view of the compound eye. (c) A close-up view of the wing scales and (d) a magnified view of the wing scales.

1. Compound eye
2. Tongue

Figure 7.169 Anatomy of the grasshopper. (a) Male and (b) female.

1. Antenna
2. Ocelli
3. Compound eye
4. Prothorax
5. Mesothorax
6. Tympanum
7. Femur
8. Pronotum
9. Mandible
10. Labrum
11. Labial palp
12. Metathorax
13. Tibia
14. Cercus
15. Subgenital plate
16. Spiracle
17. Tarsus
18. Tegmen
19. Wing
20. Abdomen
21. Dorsal valve
22. Ventral valve
23. Ovipositor

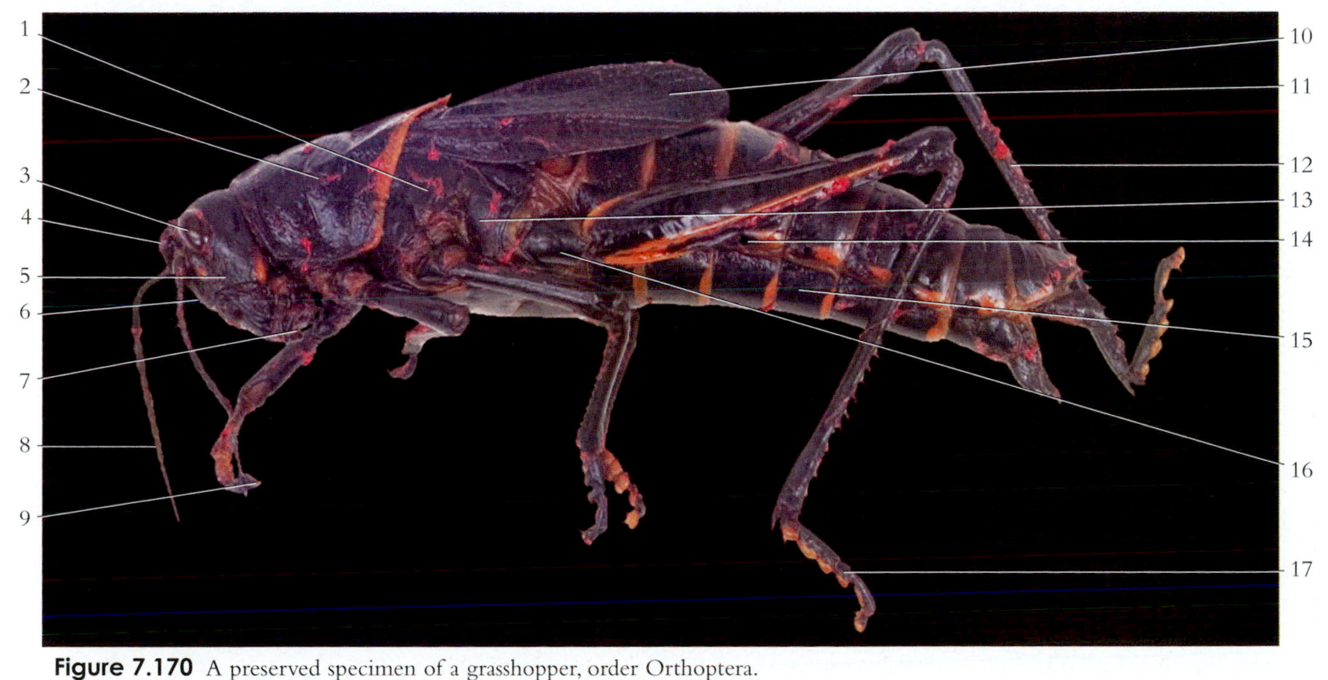

Figure 7.170 A preserved specimen of a grasshopper, order Orthoptera.

1. Mesothorax
2. Pronotum
3. Compound eye
4. Vertex
5. Gena
6. Frons
7. Maxilla
8. Antenna
9. Claw
10. Wing
11. Femur
12. Tibia
13. Mesothorax
14. Spiracle
15. Abdomen
16. Trochanter
17. Tarsus

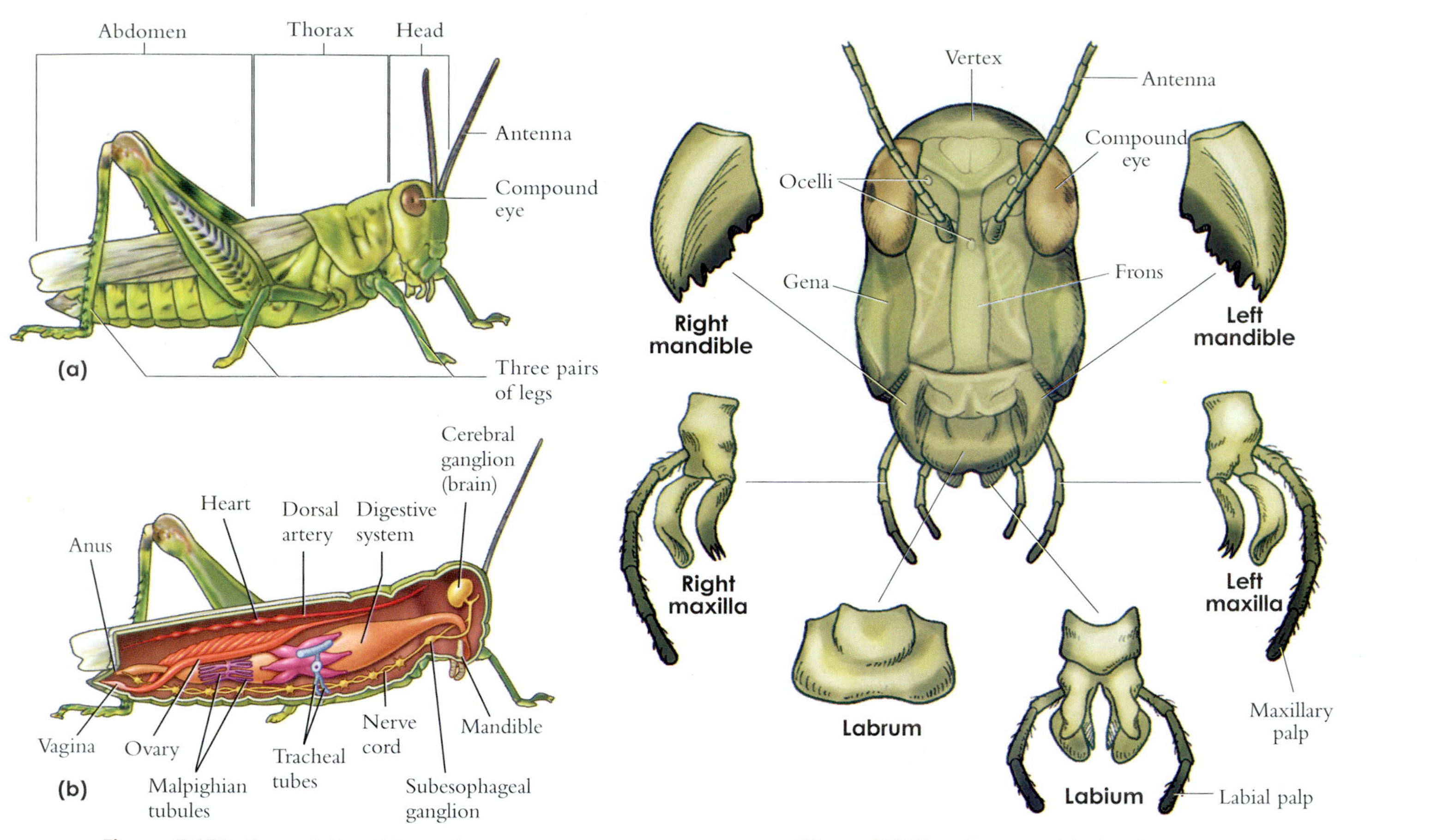

Figure 7.171 External (a) and internal (b) anatomy of a grasshopper.

Figure 7.172 A diagram of the head and mouthparts of a grasshopper.

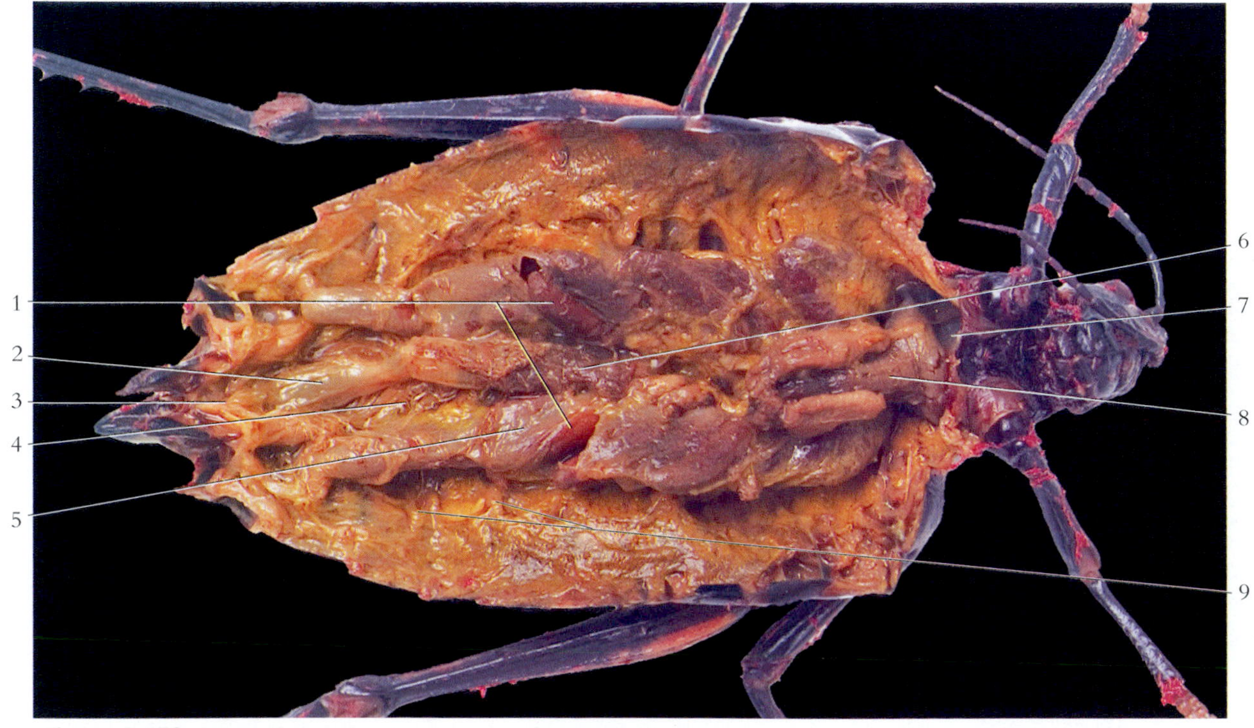

Figure 7.173 A ventral view showing the internal anatomy of a grasshopper.
1. Gastric caecum
2. Hindgut
3. Rectum
4. Malpighian tubules
5. Ovaries
6. Midgut
7. Esophagus
8. Crop
9. Tracheae

Class Chilopoda

Figure 7.174 Examples of centipedes: (a) a giant Sonoran, *Scolopendra heros,* (b) a Florida blue, *Hemiscolopendra marginata*, and (c) a Vietnamese centipede, *Scolopendra subspinipes*.

Class Diplopoda

Figure 7.175 Examples of millipedes: (a) an American giant millipede, *Narceus americanus*, (b) a Sonoran desert, *Orthoporus ornatus*, and (c) an African giant millipede, *Archispirostreptus gigas*.

Phylum Tardigrada - water bears

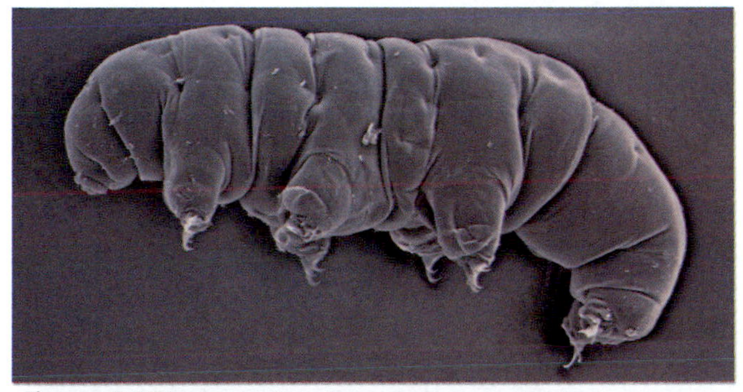

Figure 7.176 A scanning electron micrograph of a eutardigrade. Lateral view, anterior end is to the left.

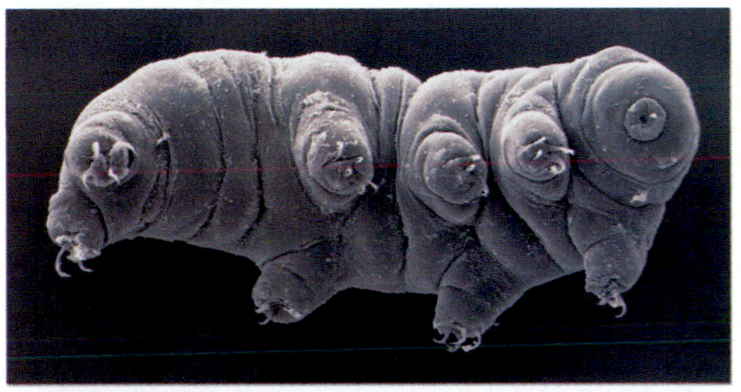

Figure 7.177 A scanning electron micrograph of a eutardigrade. Ventral view, anterior is to the right.

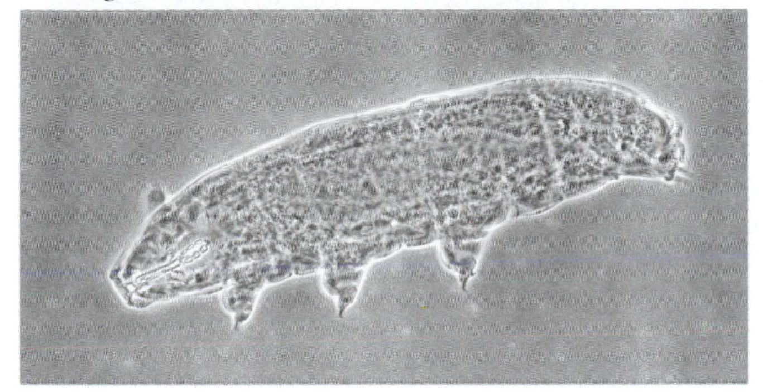

Figure 7.178 A light micrograph of *Macrobiotus polaris*. A lateral view with the anterior end to the left.

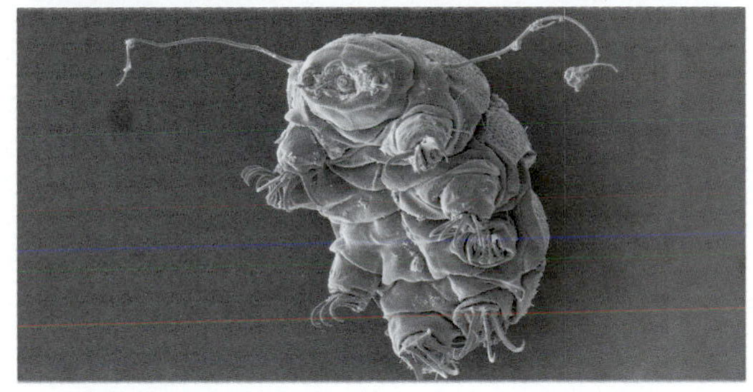

Figure 7.179 A scanning electron micrograph of a heterotardigrade. Ventral view, anterior is to the top.

Table 7.7 Representatives of the Phylum Echinodermata

Classes and Representative Kinds	Characteristics
Crinoidea — sea lilies and feather stars	Sessile during much of life cycle; calyx supported by elongated stalk in some
Asteroidea — sea stars (starfish)	Appendages arranged around a central disk containing the mouth; tube feet with suckers
Echinoidea — sea urchins and sand dollars	Disk-shaped with no arms; skeleton consists of rows of calcium carbonate plates; movable spines; tube feet with suckers
Ophiuroidea — brittle stars	Appendages sharply marked off from central disk; tube feet without suckers
Holothuroidea — sea cucumbers	Cucumber-shaped with no arms; spines absent; tube feet with tentacles and suckers

Representatives of the Phylum Hemichordata

Classes and Representative	Characteristics
Enteropneusta — acorn worm	Vermiform with acorn-shaped proboscis; skin covered with cilia and mucus glands; feed on detritus by swallowing sediment
Pterobranchia — gill-wing worms	Colonial; tentacles with cilia filter food from water

Hemichordata Crinoidea Asteroidea Ophiuroidea Echinoidea Holothuroidea

Echinodermata

Figure 7.180 Phylogenetic relationships and classification of Hemichordata and Echinodermata.

Phylum Echinodermata - sea stars, sea urchins, and sea cucumbers

Figure 7.181 Example echinoderms include: (a) a yellow pyramid sea star, *Pharia Pyramidata*, (b) a group of common sand dollars, *Echinarachnius parma*, and (c) a sea cucumber, *Stichopus fuscus*.

Class Asteroidea

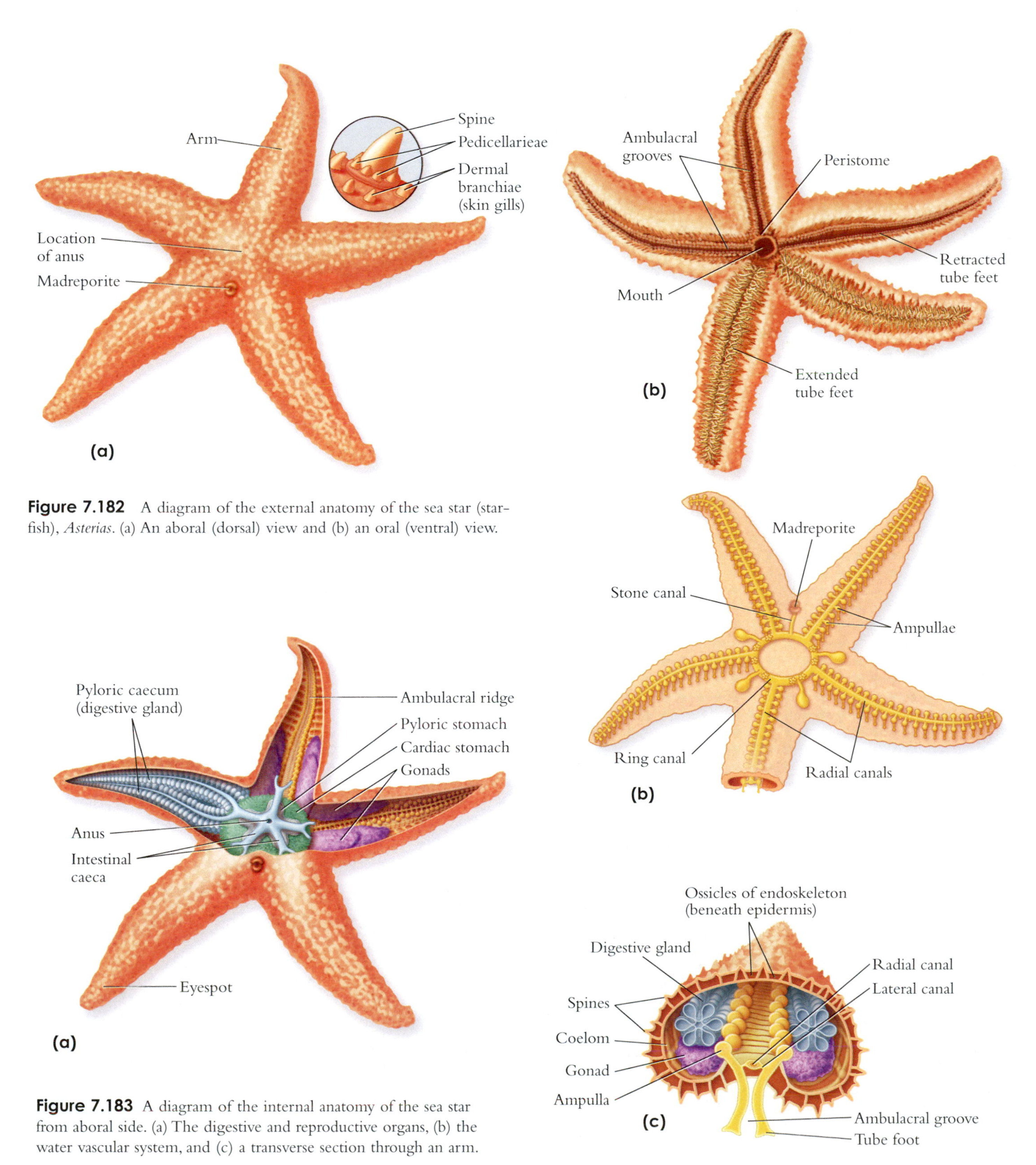

Figure 7.182 A diagram of the external anatomy of the sea star (starfish), *Asterias*. (a) An aboral (dorsal) view and (b) an oral (ventral) view.

Figure 7.183 A diagram of the internal anatomy of the sea star from aboral side. (a) The digestive and reproductive organs, (b) the water vascular system, and (c) a transverse section through an arm.

Figure 7.184 An aboral view of the internal anatomy of a sea star.

1. Ambulacral ridge
2. Gonad
3. Spines
4. Ring canal
5. Pyloric caecum (digestive gland)

Figure 7.185 A magnified aboral view of the internal anatomy of a sea star.

1. Ambulacral ridge
2. Madreporite
3. Stone canal
4. Ampullae
5. Polian vesicle
6. Pyloric duct
7. Pyloric caecum (digestive gland)
8. Gonad
9. Anus
10. Pyloric stomach
11. Spines

Figure 7.186 An oral view of a sea star.

1. Tube feet
2. Peristome
3. Mouth
4. Ambulacral groove
5. Oral spines

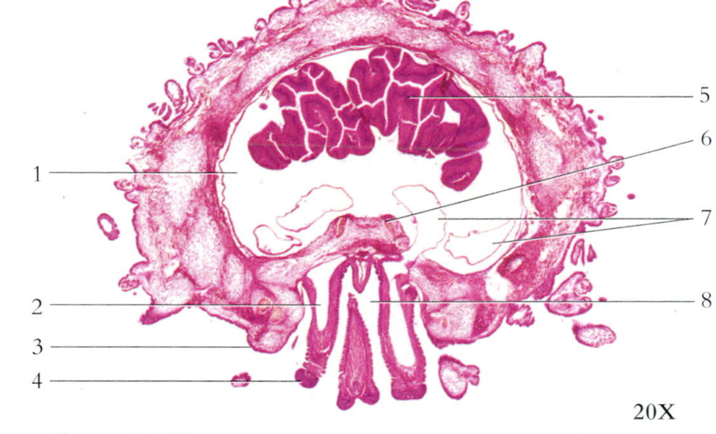

Figure 7.187 A transverse section through the arm of a sea star.

1. Coelom
2. Tube foot
3. Epidermis
4. Sucker
5. Pyloric caecum
6. Ambulacral ridge
7. Ampullae
8. Ambulacral groove

Figure 7.188 An oral view of a sea star (a) showing the cardiac stomach extended through mouth and (b) after retracting stomach.

1. Cardiac stomach

Class Echinoidea

Figure 7.189 A pencil sea urchin, *Heterocentrotus* sp.

Figure 7.190 An oral view of a live sea urchin, *Heterocentrotus* sp.
1. Tips of teeth (of Aristotle's lantern)
2. Mouth
3. Peristome

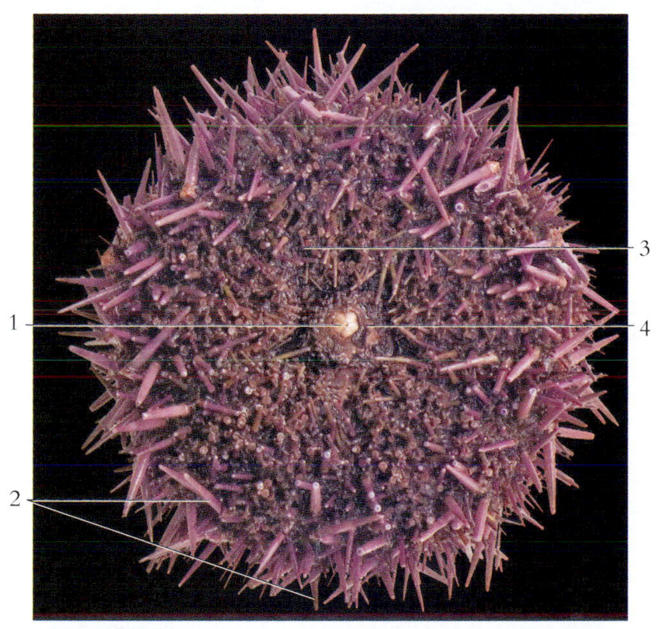

Figure 7.191 An oral view of the sea urchin, *Arbacia* sp.
1. Tip of teeth (of Aristotle's lantern)
2. Spines
3. Pedicellaria
4. Peristome

Figure 7.192 An aboral view of the sea urchin, *Arbacia* sp.
1. Ossicles
2. Madreporite

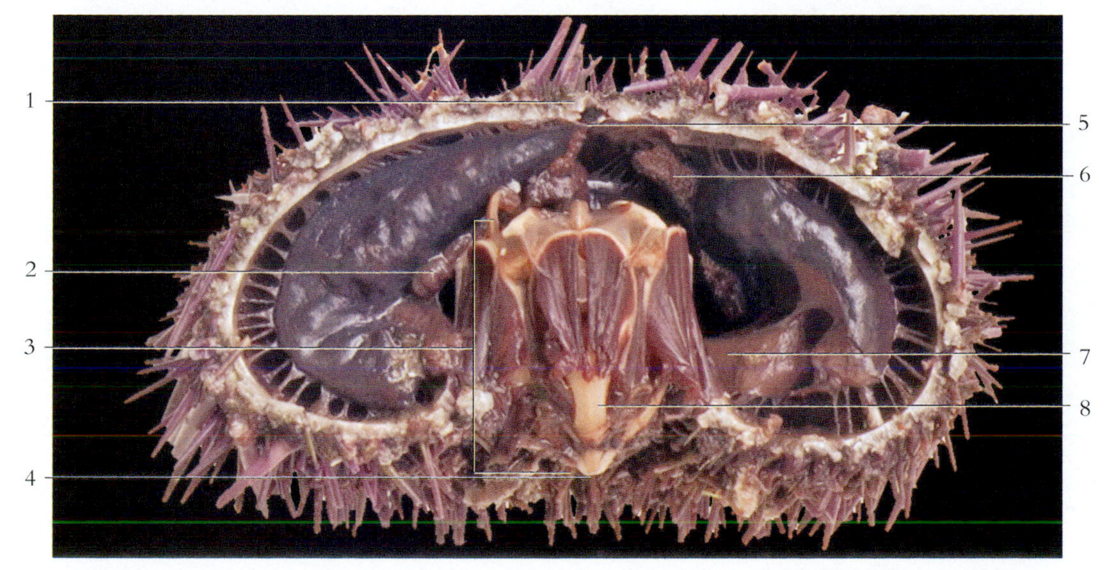

Figure 7.193 The internal anatomy of a sea urchin.
1. Madreporite
2. Intestine
3. Aristotle's lantern
4. Tip of teeth (of Aristotle's lantern)
5. Anus
6. Gonad
7. Stomach
8. Calcareous tooth

Class Holothuroidea

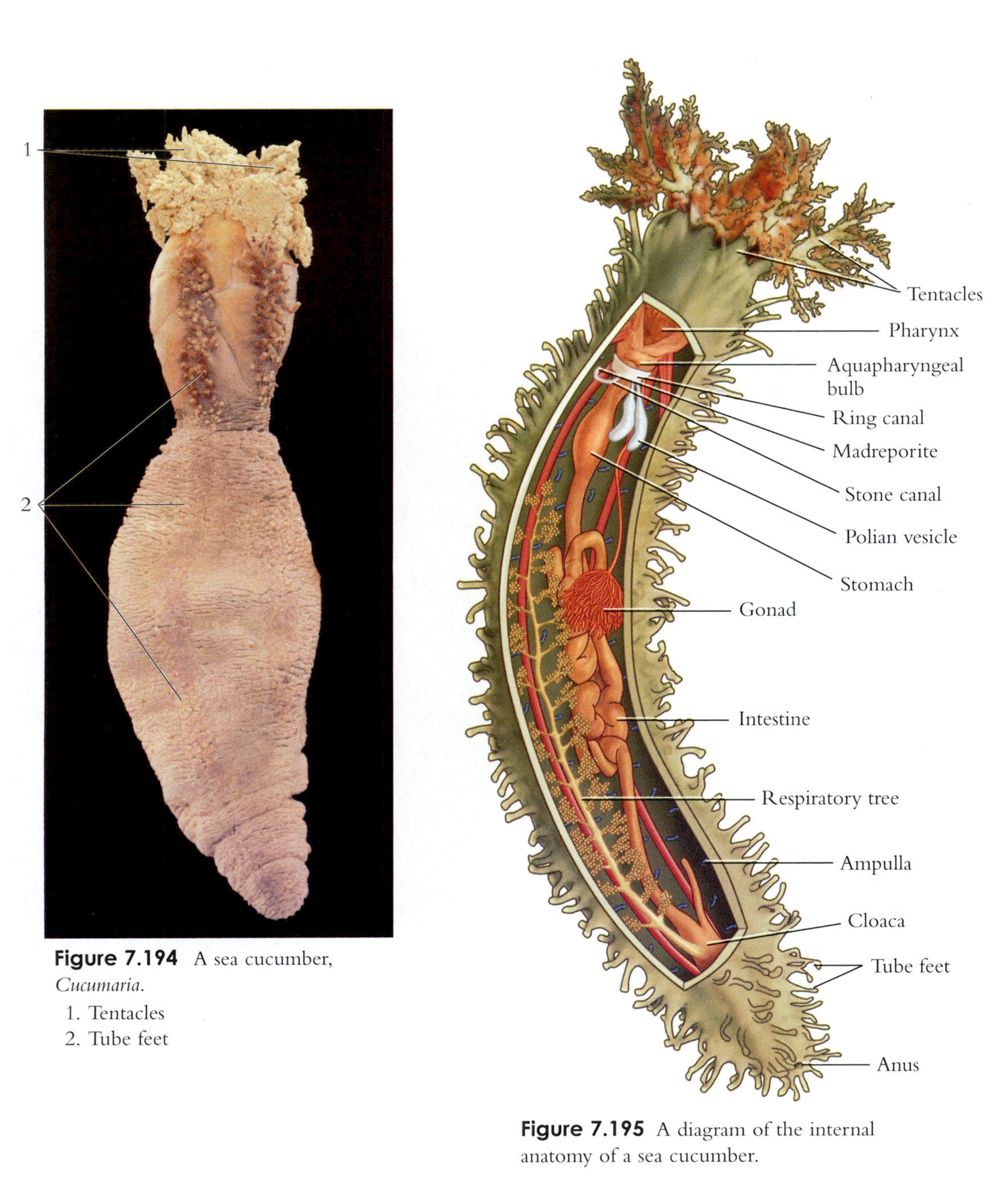

Figure 7.194 A sea cucumber, *Cucumaria.*

1. Tentacles
2. Tube feet

Figure 7.195 A diagram of the internal anatomy of a sea cucumber.

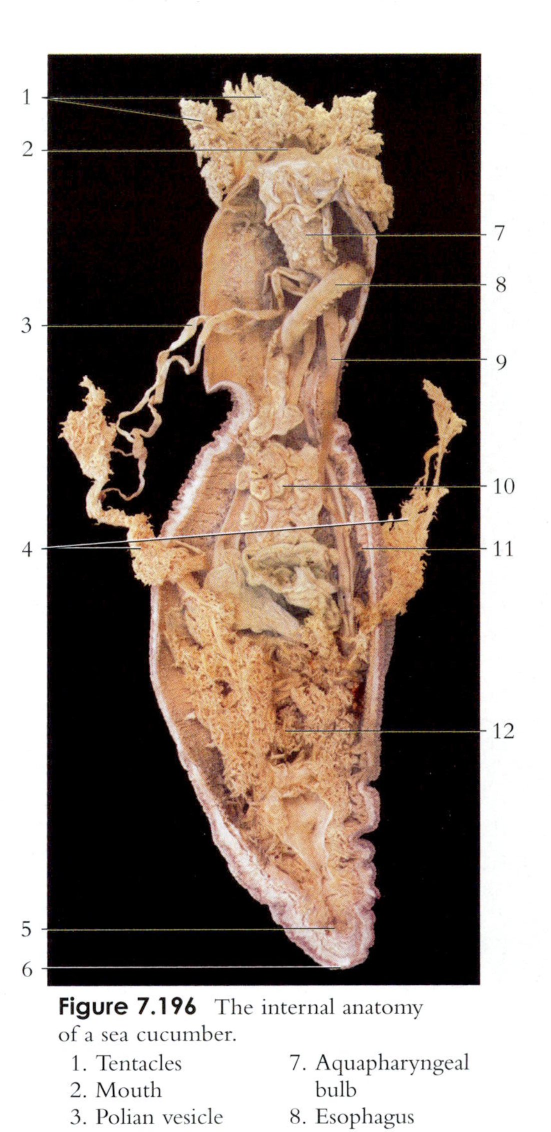

Figure 7.196 The internal anatomy of a sea cucumber.

1. Tentacles
2. Mouth
3. Polian vesicle
4. Respiratory tree
5. Cloaca
6. Anus
7. Aquapharyngeal bulb
8. Esophagus
9. Retractor muscle
10. Intestine
11. Ampulla
12. Gonad

Phylum Hemichordata - acorn worms

Class Enteropneusta

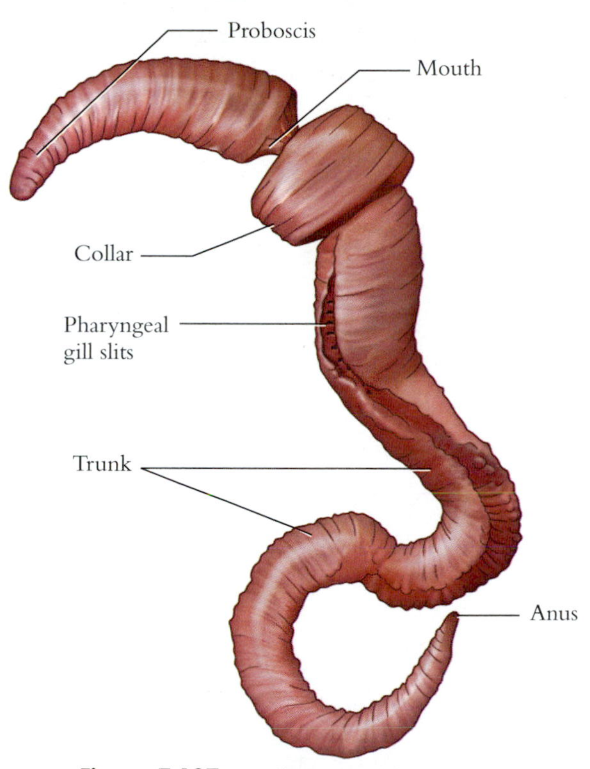

Figure 7.197 An illustration of an acorn worm, *Saccoglossus kowalevskii*.

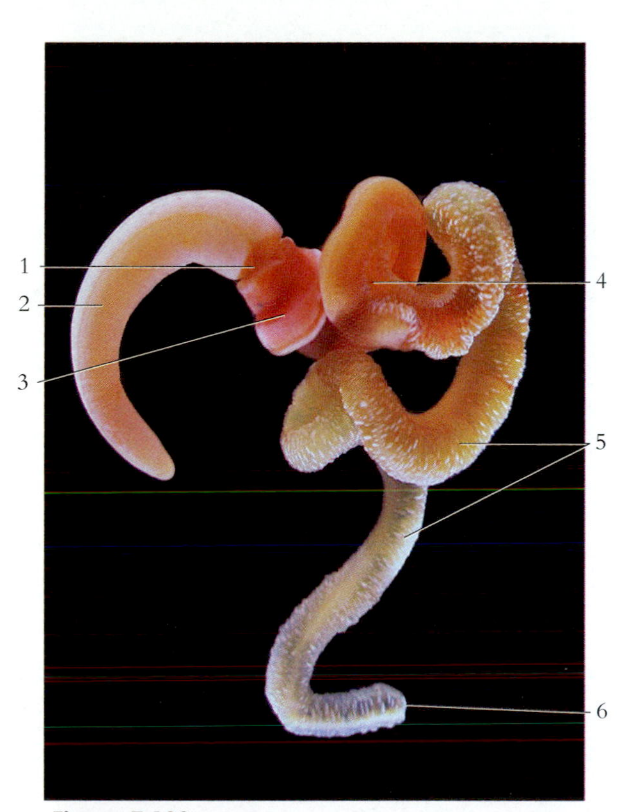

Figure 7.198 An acorn worm, *Saccoglossus kowalevskii*.

1. Mouth
2. Proboscis
3. Collar
4. Location of gills
5. Trunk
6. Anus

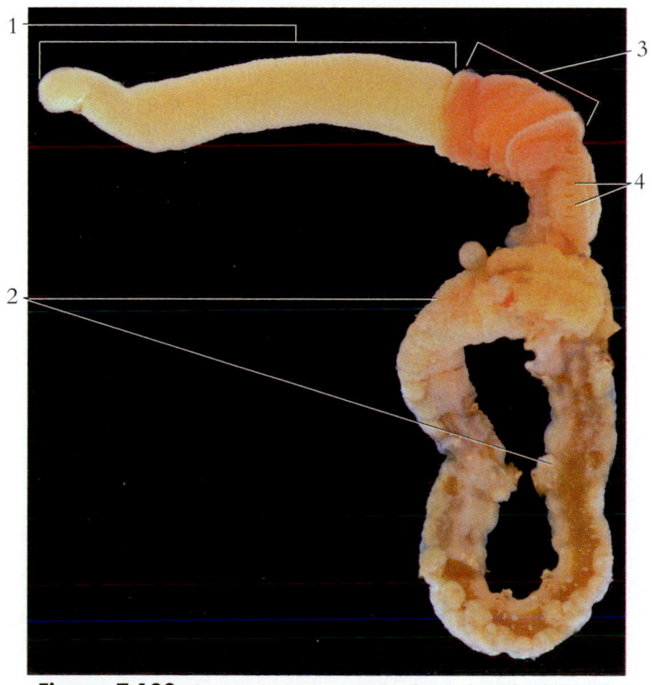

Figure 7.199 An acorn worm, *Saccoglossus kowalevskii*, showing pharyngeal gill slits.

1. Proboscis
2. Trunk
3. Collar
4. Pharyngeal gill slits

Figure 7.200 A deep-sea acorn worm species newly discovered by NOAA (National Oceanic and Atmospheric Administration).

Table 7.8 Representatives of the Phylum Chordata

Subphyla and Representative Kinds	Characteristics
Tunicata — tunicates	Marine, larvae are free-swimming and have notochord, gill slits, and dorsal hollow nerve cord; most adults are sessile (attached), filter-feeders, saclike animals
Cephalochordata — lancelets, amphioxus	Marine, segmented, elongated body with notochord extending the length of the body; cirri surrounding the mouth for obtaining food
Vertebrata — agnathans (lampreys and hagfishes), fishes (cartilaginous and bony), amphibians, reptiles, birds, mammals	Aquatic and terrestrial forms; distinct head and trunk supported by a series of cartilaginous or bony vertebrae in the adult; closed circulatory system and ventral heart; well-developed brain and sensory organs

Table 7.9 Representatives of the Subphylum Vertebrata

Taxa and Representative Kinds	Characteristics
Class Agnatha	Eel-like and aquatic; sucking mouth (some parasitic); lack jaws and paired appendages
Subclass Myxini — hagfishes	Terminal mouth with buccal funnel absent; nasal sac connected to pharynx; four pairs of tentacles; five to ten pairs pharyngeal pouches
Subclass Petromyzontida — lampreys	Suctorial mouth with rasping teeth; nasal sac not connected to buccal cavity; seven pairs of pharyngeal pouches
Infraphylum Gnathostomata	Jawed vertebrates; most with paired appendages
Class Chondrichthyes — sharks, rays, and skates	Cartilaginous skeleton; placoid scales; most have spiracle; spiral valve in digestive tract
Class Osteichthyes	Bony fishes; gills covered by bony operculum; most have swim bladder
Subclass Sarcopterygii	Bony skeleton; lobe-finned; paired pectoral and pelvic fins
Subclass Actinopterygii	Bony skeleton; most have dermal scales; ray-finned
Class Amphibia — salamanders, frogs, and toads	Larvae have gills and adults have lungs; scaleless skin (except apoda); an incomplete double circulation; three–chambered heart
Class Reptilia (= Sauropsida) — reptiles and birds*	Amniotic egg; epidermal scales; three- or four-chambered heart; lungs
Class Aves — birds*	Homeothermous (warm-blooded); feathers; toothless; air sacs; four-chambered heart with right aortic arch
Class Mammalia — mammals	Homeothermous; hair; mammary glands; most have seven cervical vertebrae; muscular diaphragm; three auditory ossicles; four–chambered heart with left aortic arch

* Birds and crocodilians are members of the Archosauria, which include the dinosaurs. For convenience we treat them traditionally as a separate class.

Phylum Chordata - amphioxus, fishes, amphibians, reptiles, birds, and mammals

(a)

(b)

(c)

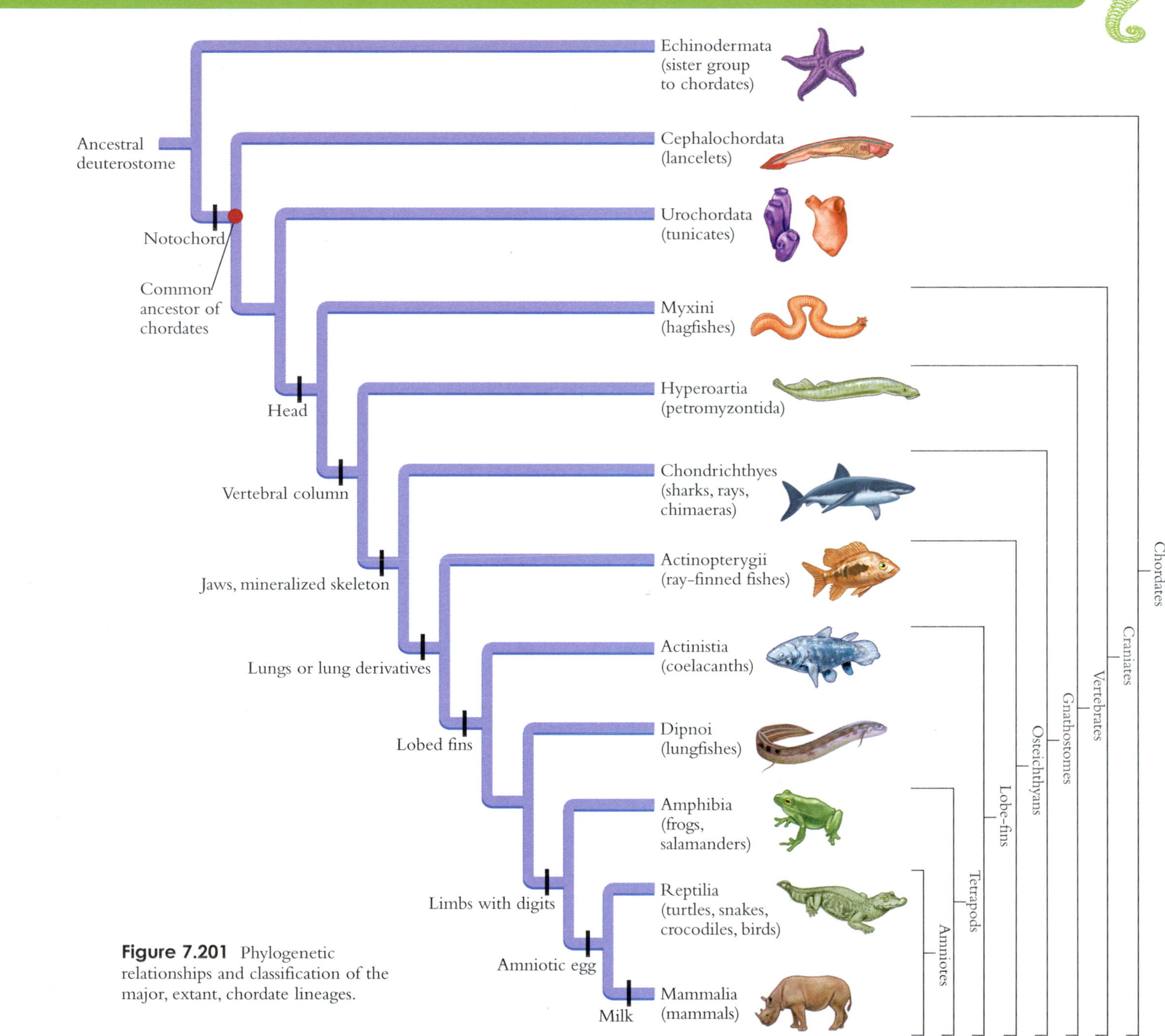

Figure 7.201 Phylogenetic relationships and classification of the major, extant, chordate lineages.

Figure 7.202 Example chordates include (starting on previous page): (a) a lancelet, *Branchiostoma* sp., (b) a giant grouper, *Epinephelus lanceolatus*, (c) a red-eyed tree frog, *Agalychnis callidryas*, (d) a snake-necked turtle, *Chelodina parkeri*, (e) a lazuli bunting, *Passerina amoena*, and (f) a chimpanzee, *Pan troglodytes*.

Subphylum Tunicata

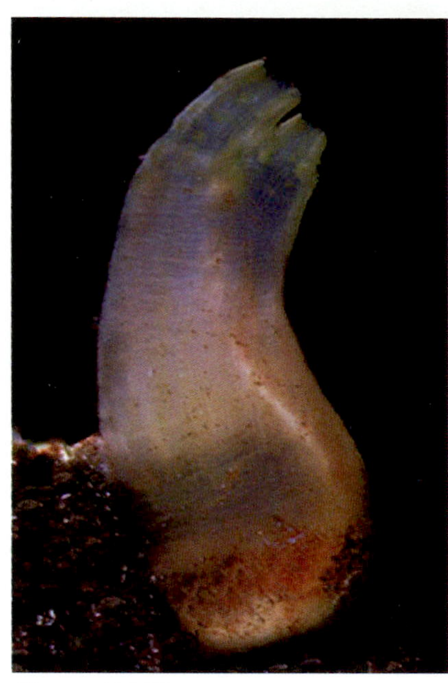

Figure 7.203 An adult tunicate, *Ciona intestinalis*.

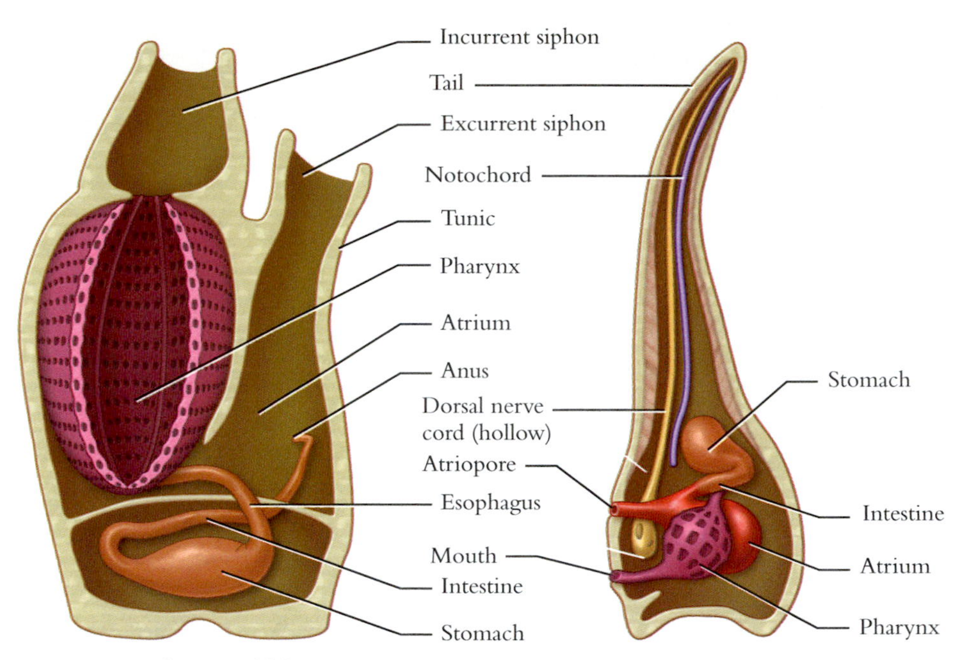

Figure 7.204 A diagram of a tunicate, (a) adult and (b) a larva.

Subphylum Cephalochordata

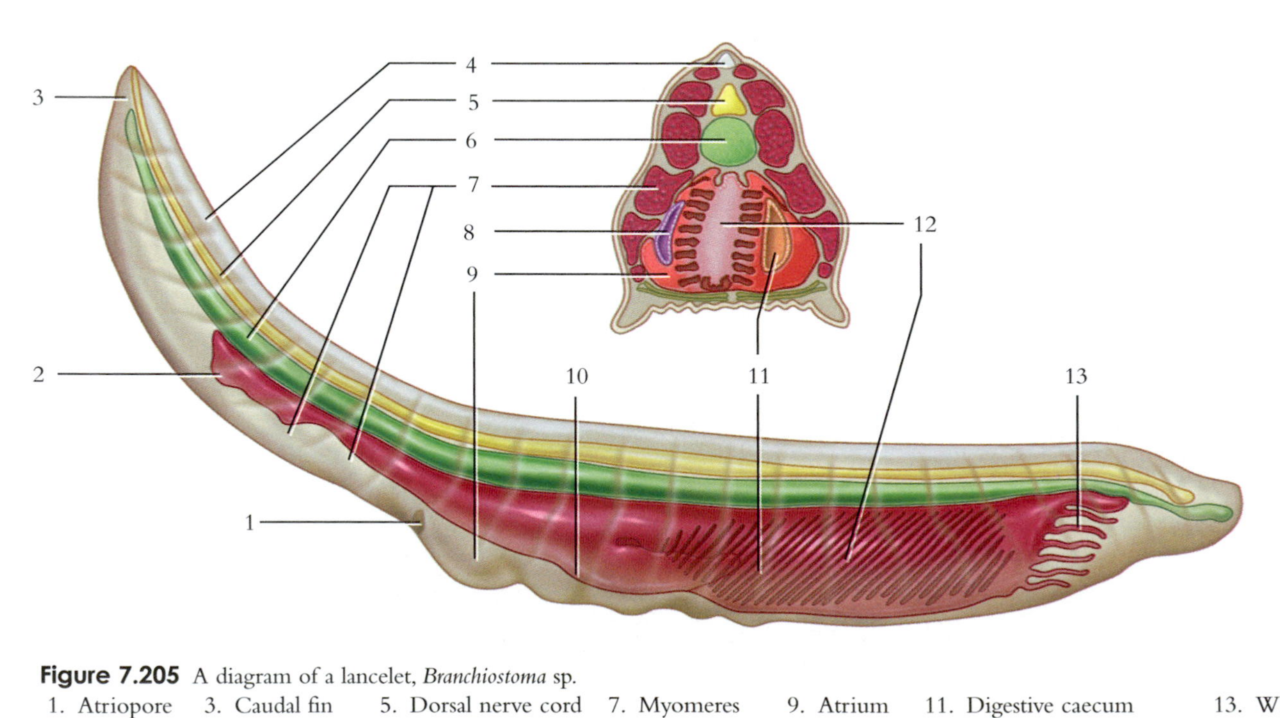

Figure 7.205 A diagram of a lancelet, *Branchiostoma* sp.

1. Atriopore
2. Anus
3. Caudal fin
4. Fin rays
5. Dorsal nerve cord
6. Notochord
7. Myomeres
8. Gonad
9. Atrium
10. Intestine
11. Digestive caecum
12. Pharynx with gill slits
13. Wheel organ

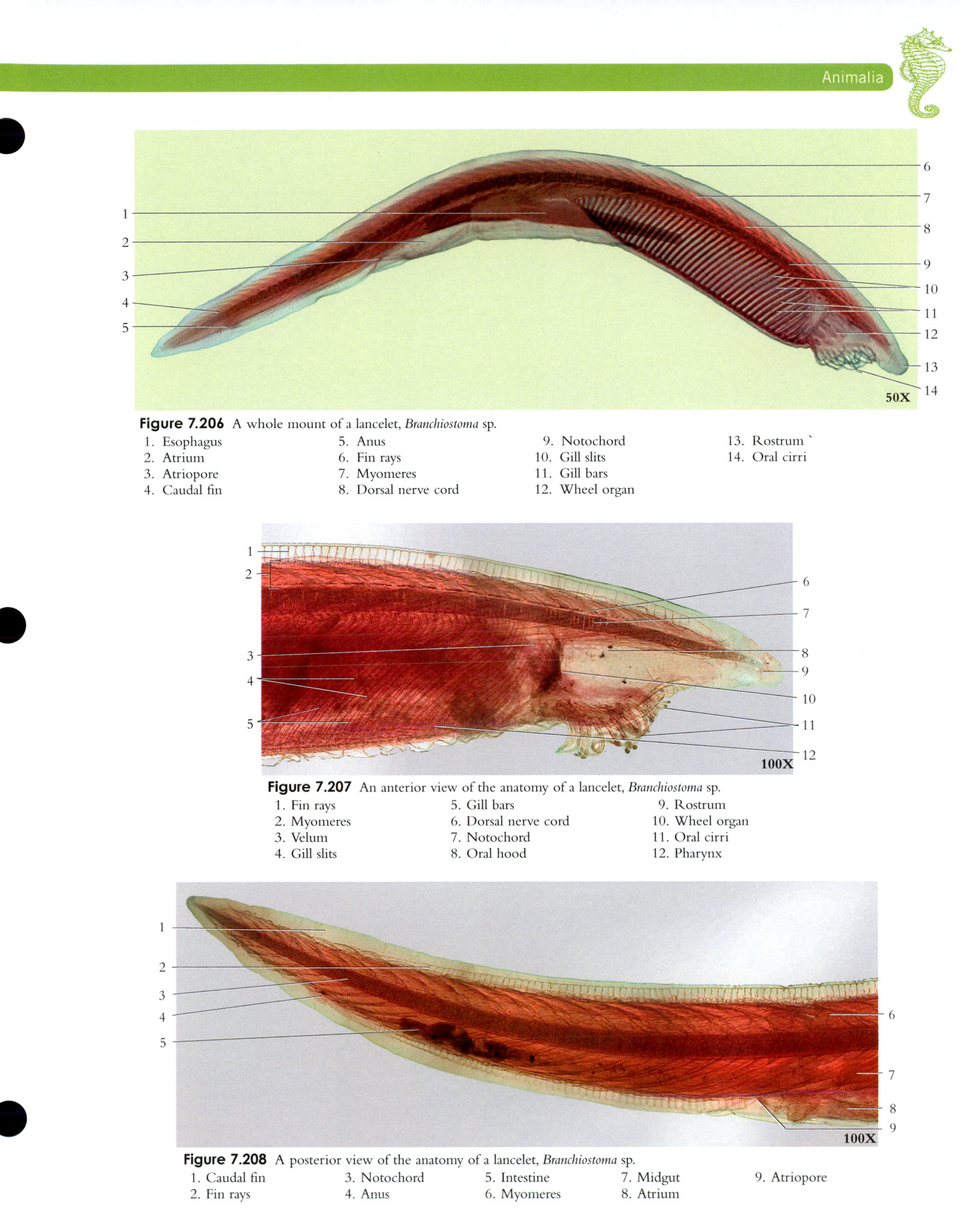

Figure 7.206 A whole mount of a lancelet, *Branchiostoma* sp.

1. Esophagus
2. Atrium
3. Atriopore
4. Caudal fin
5. Anus
6. Fin rays
7. Myomeres
8. Dorsal nerve cord
9. Notochord
10. Gill slits
11. Gill bars
12. Wheel organ
13. Rostrum `
14. Oral cirri

Figure 7.207 An anterior view of the anatomy of a lancelet, *Branchiostoma* sp.

1. Fin rays
2. Myomeres
3. Velum
4. Gill slits
5. Gill bars
6. Dorsal nerve cord
7. Notochord
8. Oral hood
9. Rostrum
10. Wheel organ
11. Oral cirri
12. Pharynx

Figure 7.208 A posterior view of the anatomy of a lancelet, *Branchiostoma* sp.

1. Caudal fin
2. Fin rays
3. Notochord
4. Anus
5. Intestine
6. Myomeres
7. Midgut
8. Atrium
9. Atriopore

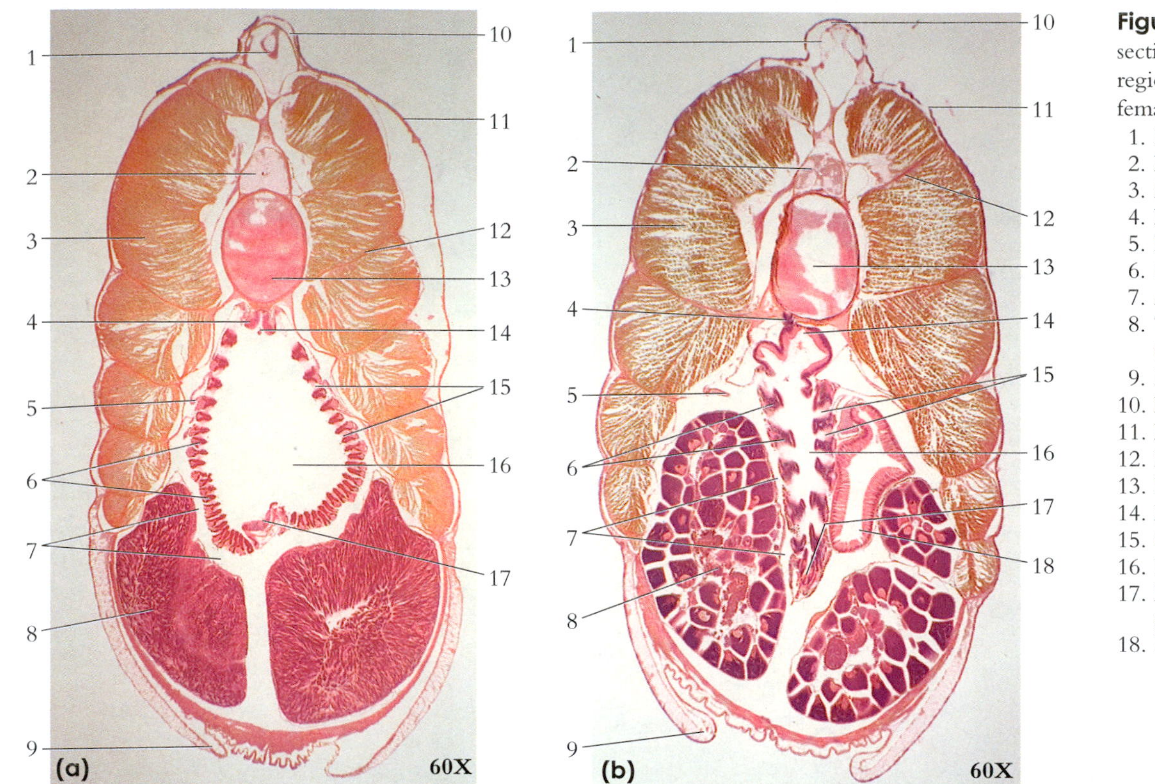

Figure 7.209 A transverse section through the pharyngeal region of (a) a male, and (b) a female lancelet, *Branchiostoma* sp.

1. Fin ray
2. Dorsal nerve cord
3. Myomere
4. Dorsal aorta
5. Nephridium
6. Gill bars
7. Atrium
8. Testis (male)
   Ovary (female)
9. Metapleural fold
10. Dorsal fin
11. Epidermis
12. Myoseptum
13. Notochord
14. Epibranchial groove
15. Gill slits
16. Pharynx
17. Endostyle (hypobranchial groove)
18. Hepatic caecum (liver)

Subclass Petromyzontida

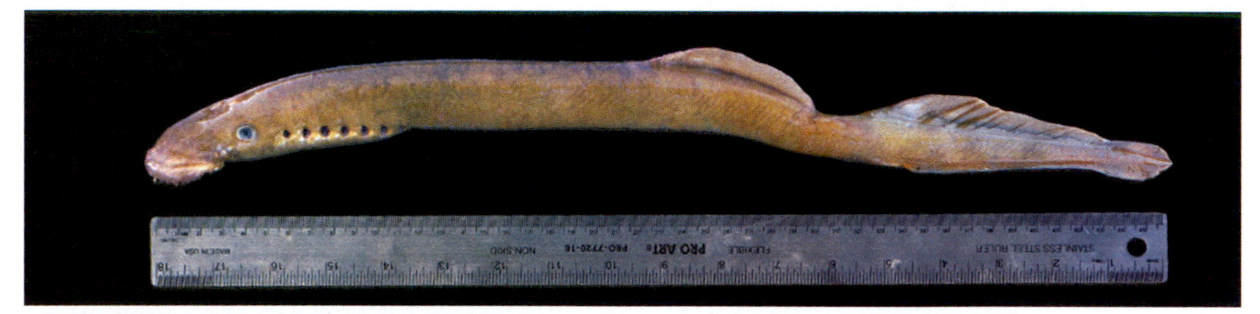

Figure 7.210 A Pacific lamprey, *Lampetra tridentata*.

Figure 7.211 A dorsal view of the external anatomy of a marine lamprey, *Petromyzon marinus*.

1. Head
2. Nostril
3. Pineal body
4. Caudal fin
5. Posterior dorsal fin
6. Trunk
7. Anterior dorsal fin

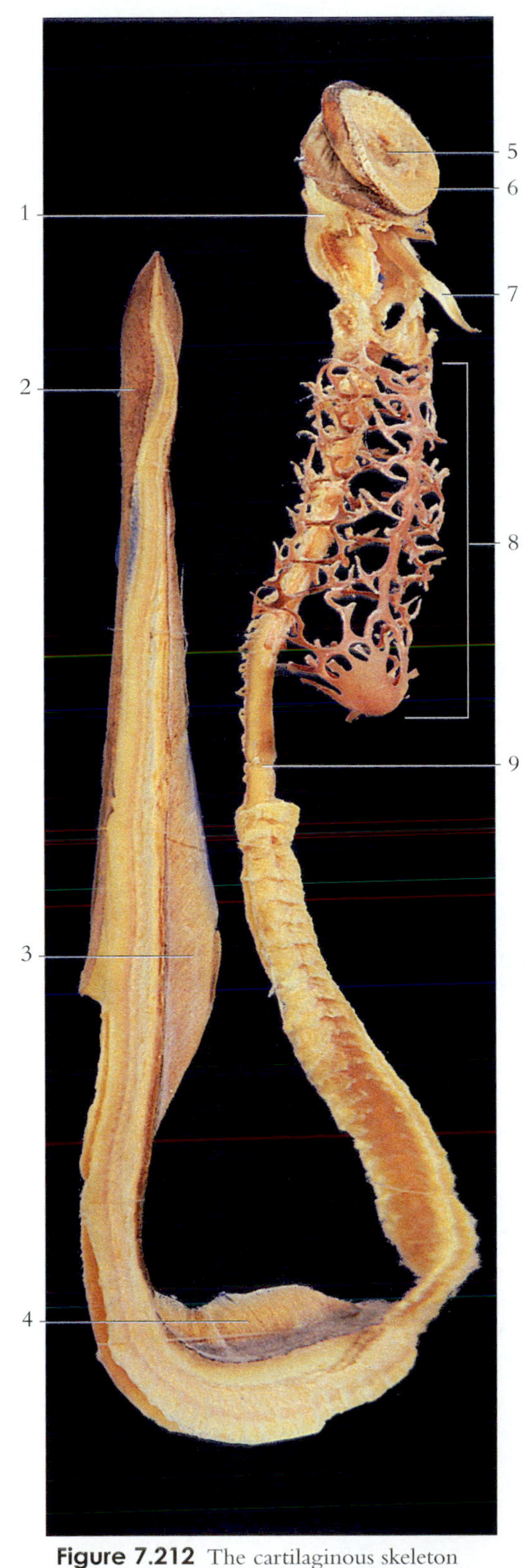

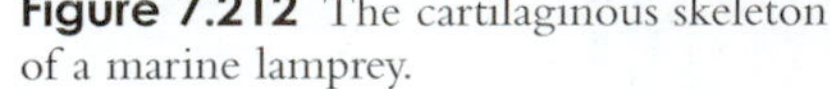

Figure 7.212 The cartilaginous skeleton of a marine lamprey.

1. Cranium
2. Caudal fin
3. Posterior dorsal fin
4. Anterior dorsal fin
5. Buccal cavity
6. Annular cartilage
7. Lingual cartilage
8. Branchial basket
9. Notochord

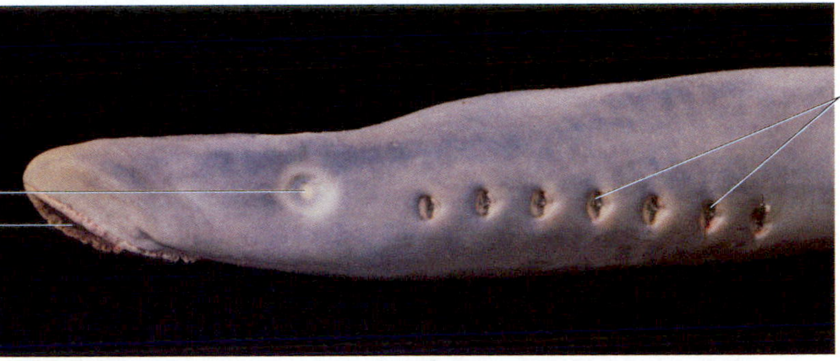

Figure 7.213 A lateral view of the anterior anatomy of a marine lamprey.

1. Eye
2. Buccal funnel
3. External gill slits

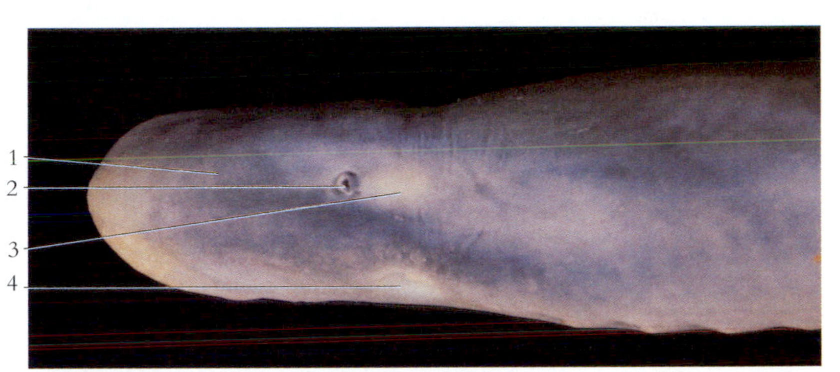

Figure 7.214 A dorsal view of the anterior anatomy of a marine lamprey.

1. Head
2. Nostril
3. Pineal body
2. Eye

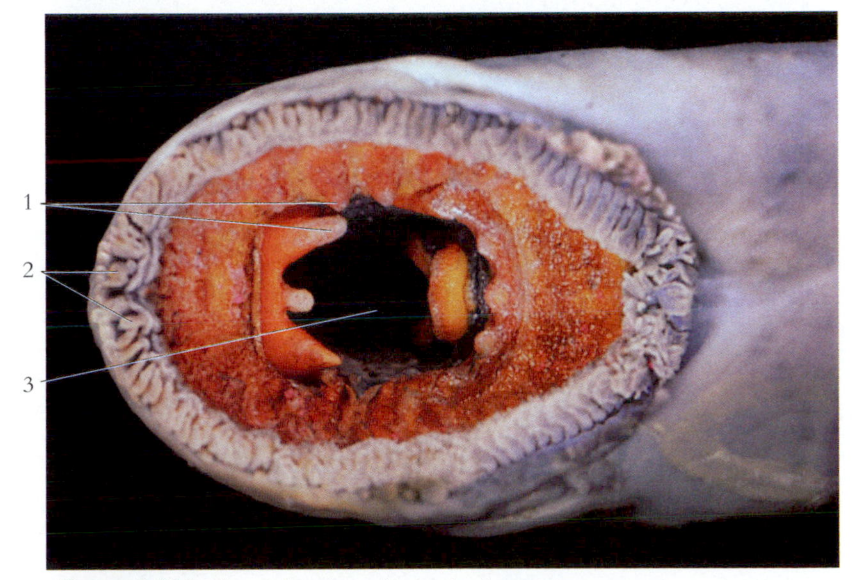

Figure 7.215 The oral region of a marine lamprey.

1. Horny teeth
2. Buccal papillae
3. Mouth

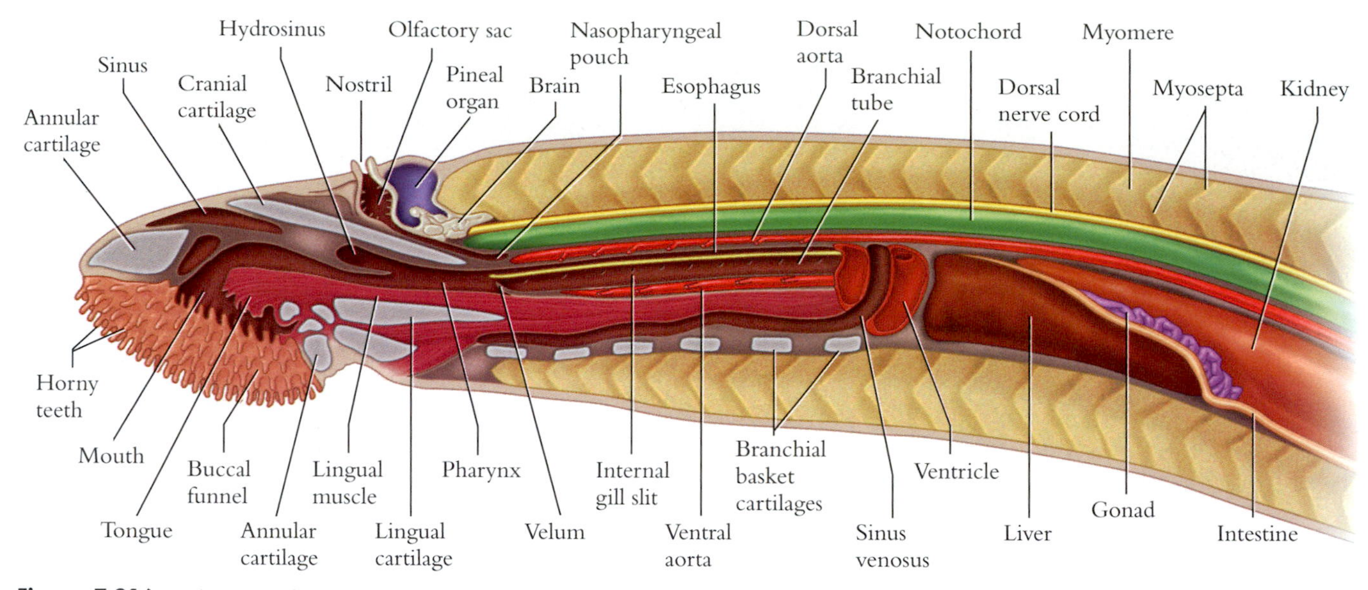

Figure 7.216 A diagram of a sagittal section of a marine lamprey.

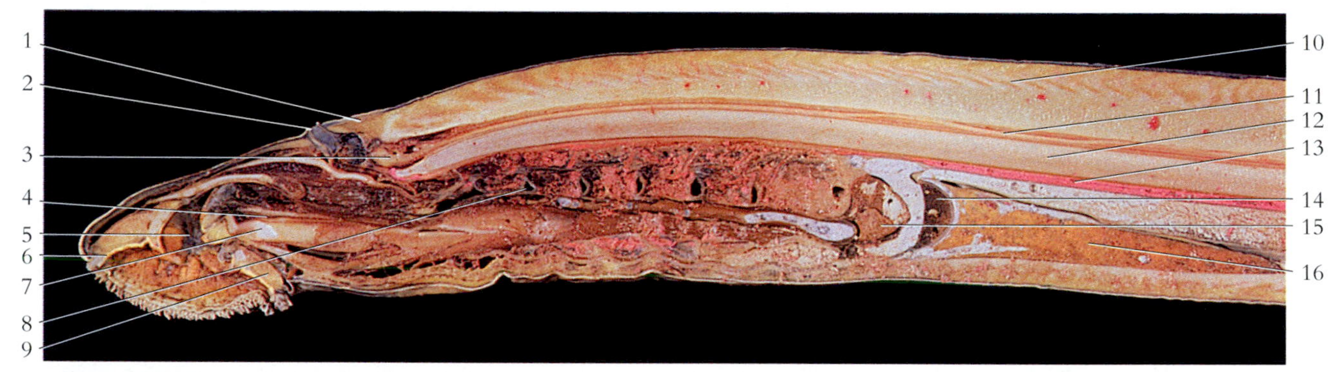

Figure 7.217 A sagittal section through the anterior region of a lamprey.

1. Pineal organ
2. Nostril
3. Brain
4. Pharynx
5. Mouth
6. Annular cartilage
7. Lingual cartilage
8. Internal gill slit
9. Buccal muscle
10. Myomeres
11. Dorsal nerve cord
12. Notochord
13. Dorsal aorta
14. Atrium
15. Ventricle
16. Liver

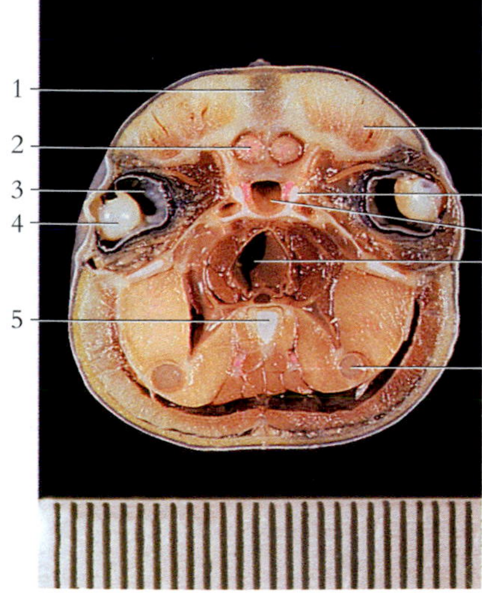

Figure 7.218 A transverse section through the head at the level of the eyes of a lamprey.

1. Pineal organ
2. Brain
3. Retina of eye
4. Lens of eye
5. Lingual cartilage
6. Myomere
7. Cranial cartilage
8. Nasopharyngeal pouch
9. Pharynx
10. Pharyngeal gland

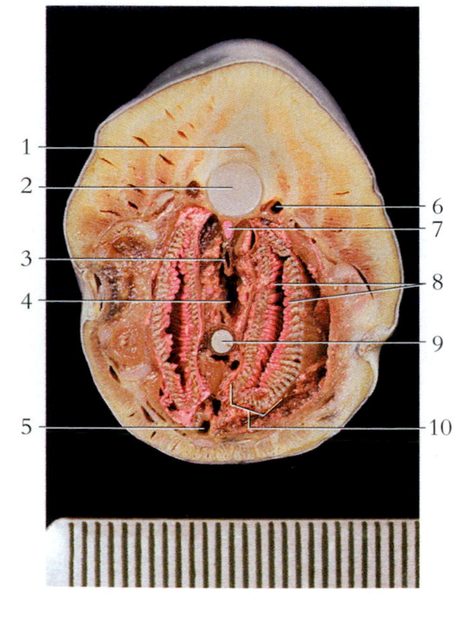

Figure 7.219 A transverse section through the body at the level of the fourth gill slit of a lamprey.

1. Dorsal nerve cord
2. Notochord
3. Esophagus
4. Branchial tube
5. Ventral jugular vein
6. Anterior cardinal vein
7. Dorsal aorta
8. Gill filaments
9. Ventral aorta
10. Branchial pouch

Class Chondrichthyes

Figure 7.220 Examples from class Chondrichthyes include: (a) a black tip reef shark, *Carcharhinus melanopterus*, (b) a gray reef shark, *Carcharhinus amblyrhynchos*, (c) a guitarfish, *Rhina ancylostoma*, (d) a round stingray, *Urobatis halleri*, (e) a blue-spotted stingray, *Taeniura lymma,* and (f) a chimaera, *Hydrolagus colliei*.

Class Osteichthyes - Subclass Sarcopterygii

Figure 7.221 The coelacanth, *Latimeria chalumnae*, a lobe-fin fish, was once thought to be extinct.

Figure 7.222 The African lungfish, *Neoceratodus forsteri*.

Class Osteichthyes - Subclass Actinopterygii

Figure 7.223 Example Actinopterygii include: (a) a tomato clownfish, *Amphiprion melanopus,* (b) chum salmon, *Oncorhynchus keta,* and (c) a lionfish, *Pterois* sp.

Figure 7.224 The external structures of a grouper, *Mycteroperca bonaci.*

Class Amphibia

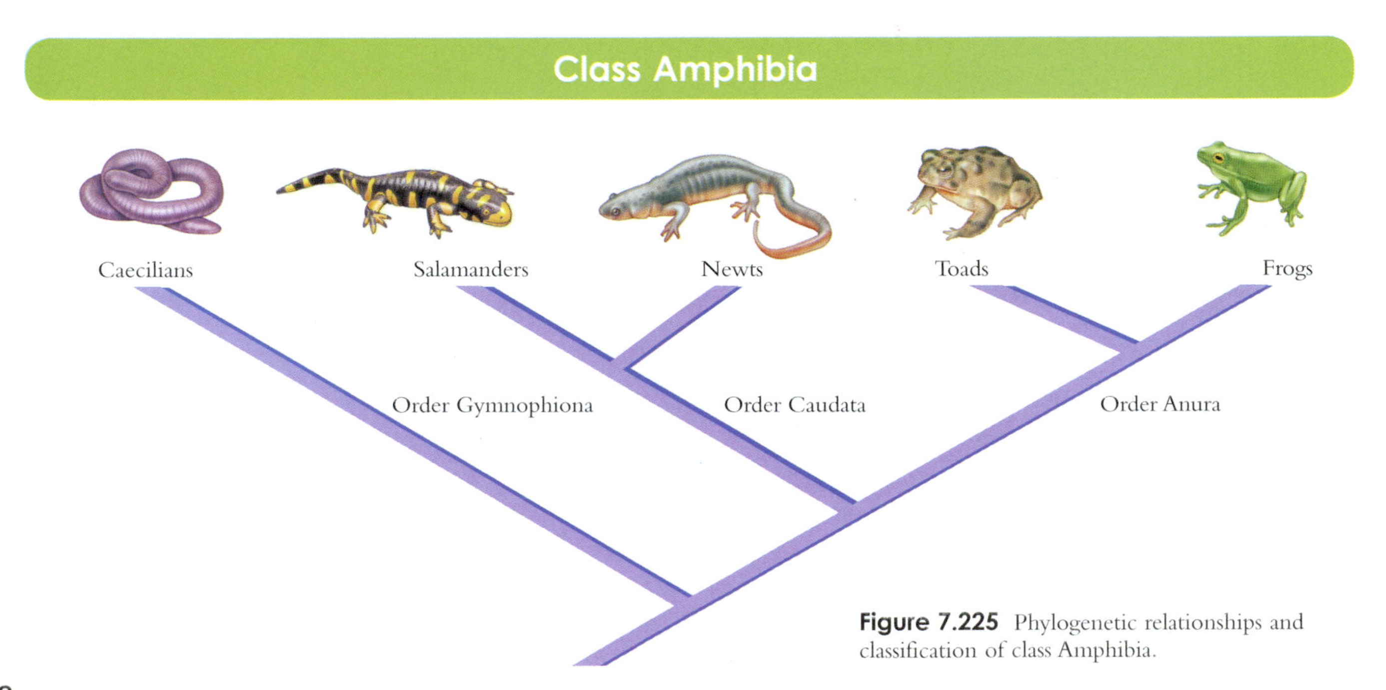

Figure 7.225 Phylogenetic relationships and classification of class Amphibia.

Figure 7.226 Examples of class Amphibia include: (a) a Cameroon caecilian, *Crotaphatrema bornmuelleri*, (b) an amphiuma, *Amphiuma means*, (c) a lesser siren, *Siren intermedia*, (d) an axolotl, *Ambystoma mexicanum*, (e) a tiger salamander, *Ambystoma tigrinum*, (f) a red mud salamander, *Pseudotriton ruber*, (g) an eastern newt, *Notophthalmus viridescens*, (h) a Woodhouse's toad, *Anaxyrus woodhousii*, (i) a Colorado River toad, *Anaxyrus alvarius*, (j) a blue-webbed flying treefrog, *Rhacophorus nigropalmatus*, (k) a canyon tree frog, *Hyla arenicolor*, and (l) a red-eyed tree frog, *Agalychnis callidryas*.

Figure 7.227 Cameroon caecilian, *Crotaphatrema bornmuelleri.* The rings or annuli can clearly be seen. These give caecilians an earthworm-like appearance.

Figure 7.228 An amphiuma, *Amphiuma means.* Note the small vestigial leg. The light colored dots are a lateral line system that aids in hunting.

1. Lateral line system
2. Vestigial limb

Figure 7.229 An axolotl, *Ambystoma mexicanum.* This individual is leucistic.

Figure 7.230 The marine or cane toad, *Bufo marinus*, is an introduced species to Hawaii and has caused many problems for native species.

Figure 7.231 The surface anatomy and body regions of the leopard frog, *Lithobates pipiens.*

1. Ankle
2. Knee
3. Foot
4. Eyes
5. Nostril
6. Tympanic membrane
7. Brachium
8. Antebrachium
9. Digits

Figure 7.232 A white-lipped tree frog, *Litoria infrafrenata.* (a) The frog is crouched on a person's fingers. (b) The adhesive toe disks can be seen in a ventral view.

Figure 7.233 The Vietnamese mossy frog, *Theloderma corticale*, has rough, mottled green skin that resembles moss growing on rock and forms an effective camouflage.

Figure 7.234 An African clawed frog, *Xenopus laevis*, has claws on each of its hind toes.

Figure 7.235 During the day and at rest (a) the red-eyed tree frog, *Agalychnis callidryas*, holds its legs tightly to its body. This not only hides its bright colors, keeping the frog inconspicuous, but keeps the frog from drying out. While it is active, (b) it displays bright colors helping to discourage predators.

Figure 7.236 Color in amphibians, as with other animals, plays a key role. The bumblebee poison-dart frog, *Dendrobates leucomelas*, (a) displays bright colors advertising the fact that it is poisonous. The tiger-legged waxy monkey tree frog, *Phyllomedusa hypochondrialis azurea*, (b) and the fire salamander, *Salamandra salamandra*, (c) lack the poisons of the poison-dart frog but rely on the bright colors to discourage would-be predators.

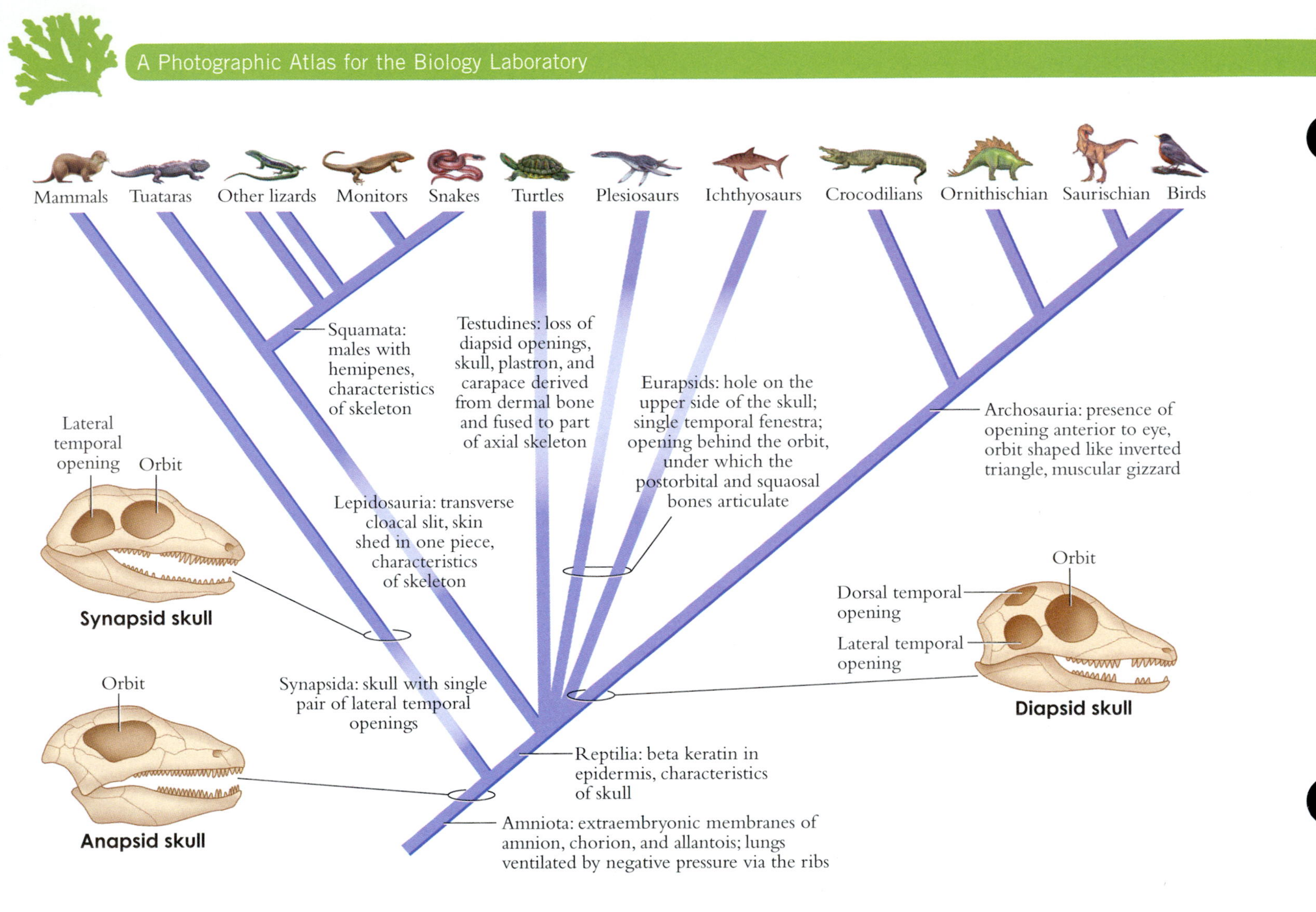
Mammals
Tuataras
Other lizards
Monitors
Snakes
Turtles
Plesiosaurs
Ichthyosaurs
Crocodilians
Ornithischian
Saurischian
Birds
Squamata: males with hemipenes, characteristics of skeleton
Testudines: loss of diapsid openings, skull, plastron, and carapace derived from dermal bone and fused to part of axial skeleton
Eurapsids: hole on the upper side of the skull; single temporal fenestra; opening behind the orbit, under which the postorbital and squaosal bones articulate
Archosauria: presence of opening anterior to eye, orbit shaped like inverted triangle, muscular gizzard
Lateral temporal opening
Orbit
Synapsid skull
Lepidosauria: transverse cloacal slit, skin shed in one piece, characteristics of skeleton
Orbit
Dorsal temporal opening
Lateral temporal opening
Diapsid skull
Orbit
Synapsida: skull with single pair of lateral temporal openings
Anapsid skull
Reptilia: beta keratin in epidermis, characteristics of skull
Amniota: extraembryonic membranes of amnion, chorion, and allantois; lungs ventilated by negative pressure via the ribs

Figure 7.237 The phylogenetic relationships and classification of Reptilia (= Sauropsida).

Class Reptilia (= Sauropsida)

(a)
(b)
(c)

Figure 7.238 Example reptiles include (starting on previous page): (a) a Galapagos green sea turtle, *Chelonia mydas agassisi,* (b) a pair of desert spiny lizards, *Sceloporus magister*, (c) a Jameson's mamba, *Dendroaspis jamesoni*, (d) a green basilisk or plumed basilisk, *Basiliscus plumifrons*, (e) a western coachwhip snake, *Masticophis flagellum*, (f) a spiny soft-shell turtle, *Apalone spinifera*, (g) a gopher tortoise, *Gopherus agassizii*, (h) an American alligator, *Alligator mississippiensis*, (i) a Komodo dragon, *Varanus komodoensis*, (j) a panther chameleon, *Furcifer pardalis*, (k) a Galapagos marine iguana, *Amblyrhynchus cristatus*, and (l) a ring-neck snake, *Diadophis punctatus*.

Figure 7.239 A tuatara, *Sphenodon punctatus.* Endemic to New Zealand, tuataras are the only surviving members of order Rhynchocephalia, which flourished around 200 million years ago.

Figure 7.240 A Galapagos tortoise, *Chelonoidis nigra*, is just one of the many members of order testudines that are threatened or endangered.

Figure 7.241 A red-eared slider, *Trachemys scripta elegans*.

Figure 7.242 An eastern glass lizard, *Ophisaurus ventralis*, is a legless lizard.

Figure 7.243 The gila monster, *Heloderma suspectum*, is a venomous lizard living in the Southwestern United States and Mexico.

Figure 7.244 The Galapagos land iguana, *Conolophus subcristatus*, is threatened due to introduced animals to the Galapagos Archipelago.

Figure 7.245 A garter snake, *Thamnophis* sp., extending its tongue. Reptiles use their tongue in conjunction with the Jacobson's organ or vomeronasal organ, an auxiliary olfactory organ, to aid with smell.

Figure 7.246 A California king snake, *Lampropeltis getula*, in process of ecdysis, or shedding its skin.

Figure 7.247 Color is used by (a) the Arizona coral snake, *Micruroides euryxanthus*, to warn would-be predators that it is venomous. The scarlet kingsnake, *Lampropeltis triangulum elapsoides,* (b) a nonvenomous snake, mimics the colors of venomous snakes to trick would-be predators into leaving it alone. Knowing the pattern, red-yellow-black-yellow versus red-black-yellow-black, can be useful in determining whether it is a venomous coral snake or not.

Figure 7.248 Examples of the four crocodilian types include: (a) American alligator, *Alligator mississippiensis*, (b) Johnston's freshwater crocodile, *Crocodylus johnstoni*, (c) Cuvier's dwarf caiman, *Paleosuchus palpebrosus*, and (d) a gharial, *Gavialis gangeticus*.

Class Aves

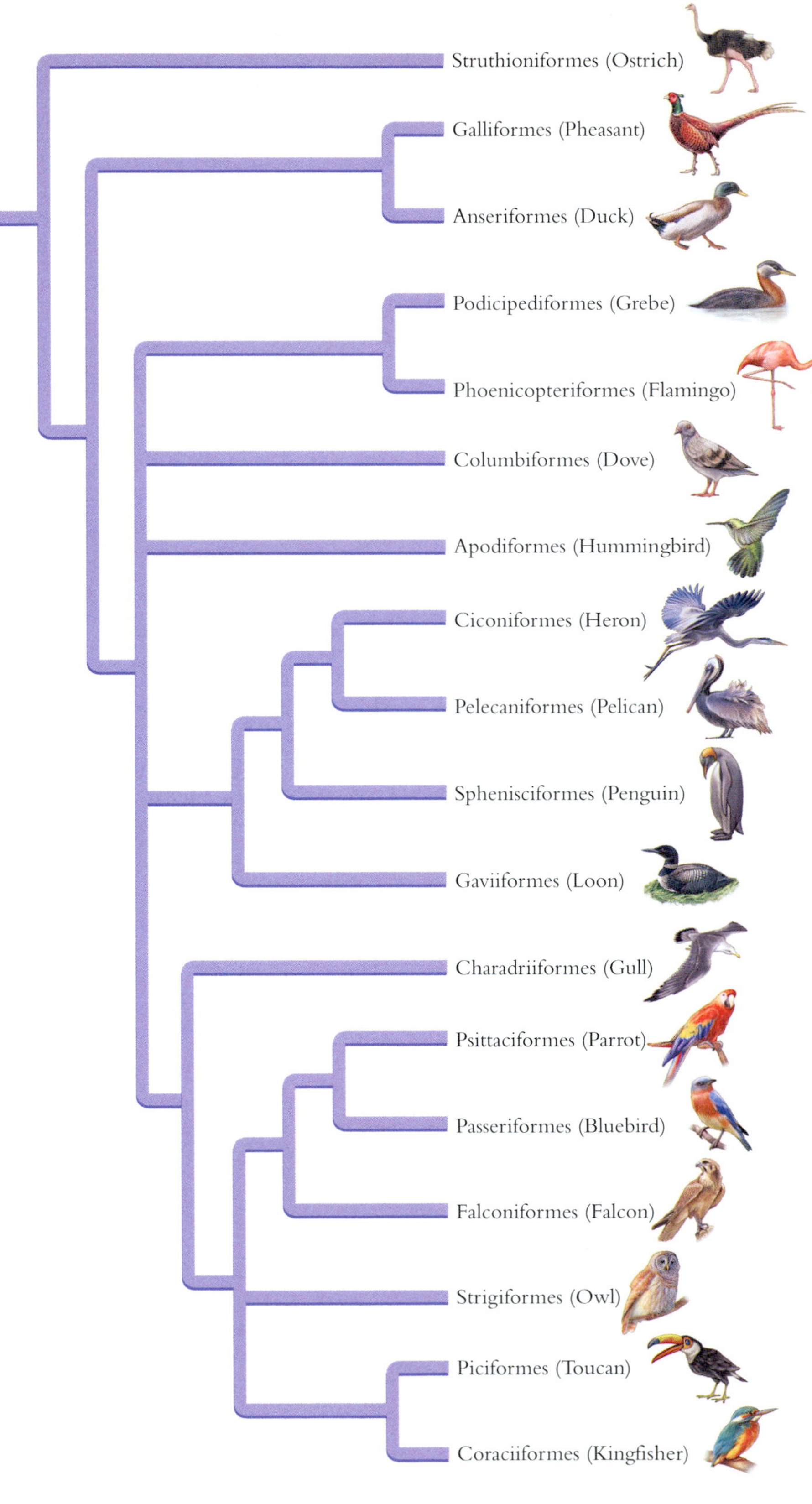

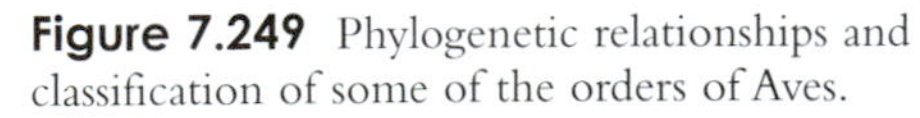

Figure 7.249 Phylogenetic relationships and classification of some of the orders of Aves.

Figure 7.250 Examples of Aves and their associated order include (starting on previous page): (a) an emu, *Dromaius novaehollandiae*, order Struthioniformes, (b) a California quail, *Callipepla californica*, order Galliformes, (c) a redhead duck, *Aythya americana*, order Anseriformes, (d) an eared grebe, *Podiceps nigricollis*, order Podicipediformes, (e) a Chilean flamingo, *Phoenicopterus chilensis*, order Phoenicopteriformes, (f) a mourning dove, *Zenaida macroura*, order Columbiformes, (g) a sparkling violetear hummingbird, *Colibri coruscans*, order Apodiformes, (h) a lava heron, *Butorides sundevalli*, order Ciconiformes, (i) a brown pelican, *Pelecanus occidentalis*, order Pelecaniformes, (j) a Galapagos penguin, *Spheniscus mendiculus*, order Sphenisciformes, (k) a pacific loon, *Gavia pacifica*, order Gaviiformes, (l) a Franklin's gull, *Leucophaeus pipixcan*, order Charadriiformes, (m) a blue and gold macaw, *Ara ararauna,* order Psittaciformes, (n) a Brewer's blackbird, *Euphagus cyanocephalus,* order Passeriformes, (o) a peregrine falcon, *Falco peregrinus*, order Falconiformes, (p) a barn owl, *Tyto alba,* order Strigiformes (q) a red-shafted flicker or northern flicker, *Colaptes auratus cafer,* order Piciformes, and (r) a belted kingfisher, *Megaceryle alcyon*, order Coraciformes.

Figure 7.251 Example beaks and their associated uses include: (a) nectar feeding, a broad-tailed hummingbird, *Selasphorus platycercus*, (b) grain eating, a lazuli bunting, *Passerina amoena*, (c) dip netting, an American white pelican, *Pelecanus erythrorhynchos*, (d) flesh eating, an American bald eagle, *Haliaeetus leucocephalus*, (e) insect eating, a western tanager, *Piranga ludoviciana*, (f) scavenging, a king vulture, *Sarcoramphus papa*, (g) filter feeding, a Chilean flamingo, *Phoenicopterus chilensis*, (h) generalist, a ring-billed gull, *Larus delawarensis*, (i) chiseling, red-shafted flicker or northern flicker, *Colaptes auratus cafer,* (j) probing, an American avocet, *Recurvirostra americana*, (k) spearing, an American darter or anhinga, *Anhinga anhinga*, and (l) nut cracking, a green wing macaw, *Ara chloropterus*.

Figure 7.252 Within the animal kingdom birds are some of the best examples of sexual dimorphism, the morphological difference between males and females of the same species. (a) A pair of wood ducks, *Aix sponsa*, passes a crabapple during courtship. (b) A male Indian peafowl, *Pavo cristatus,* known commonly as a peacock, and (c) a female known as a peahen.

Figure 7.253 (a) A red-tailed hawk, *Buteo jamaicensis*, has long broad wings used for soaring on thermals. Its wing beat is somewhat slow with periods of gliding inbetween several wing beats, while (b) the American kestrel, *Falco sparverius*, a falcon, has narrow, long and pointed wings best suited for direct fast flight. Its wing beats are fast with relatively no breaks for gliding.

Figure 7.254 Plumage plays many roles. For (a) a great horned owl, *Bubo virginianus*, it aids in camouflage, while (b) the Bullock's oriole, *Iterus bullockii*, uses it for territorial display and attracting attention.

Class Mammalia

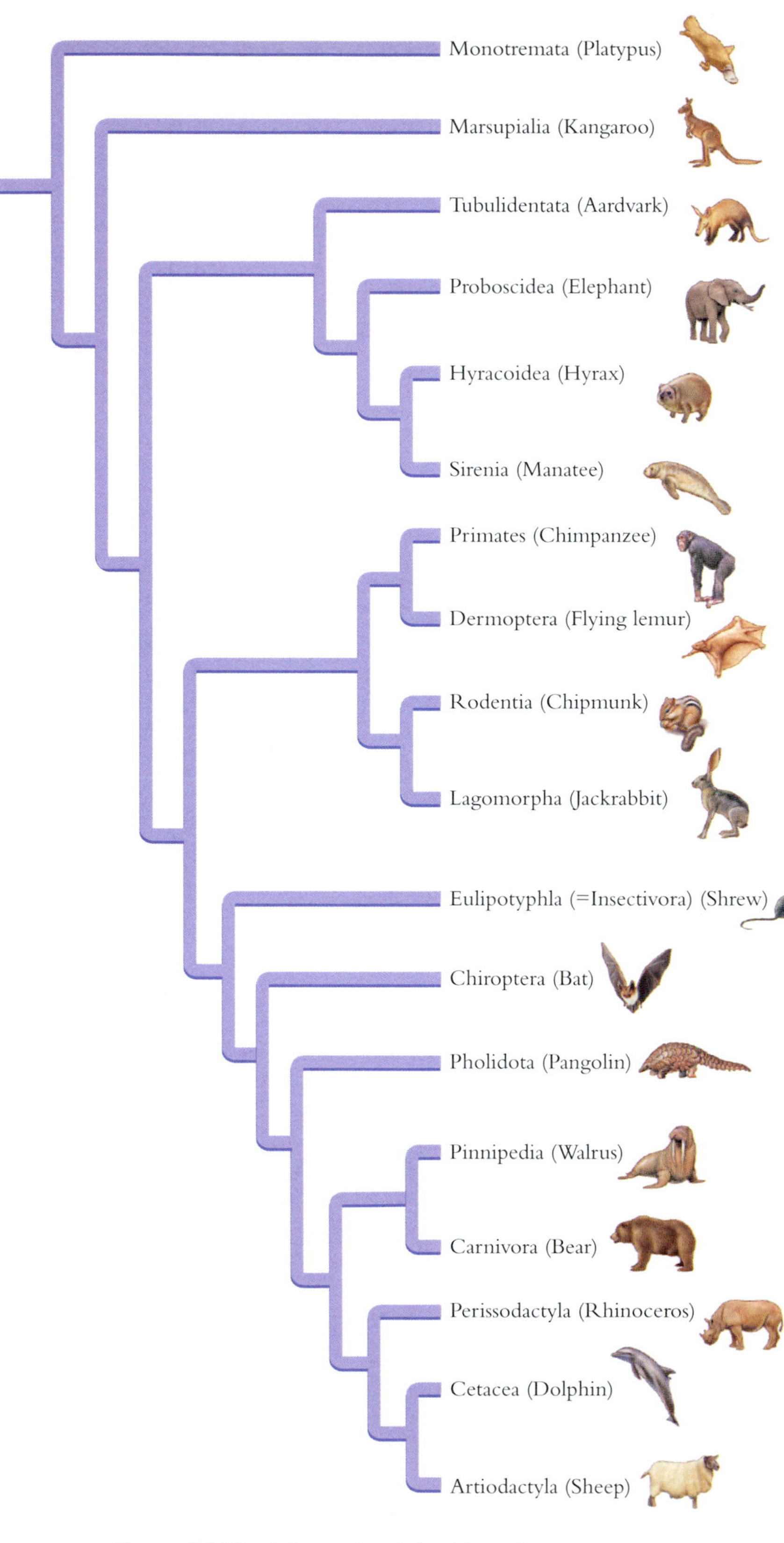

Figure 7.255 Phylogenetic relationships and classification of some of the orders of Mammalia.

Figure 7.256 Examples of Mammalia and their associated order include (starting on previous page): (a) an eastern grey kangaroo, *Macropus giganteus*, order Marsupialia (b) a pika, *Ochotona princeps,* order Lagomorpha, (c) a West Indian manatee, *Trichechus manatus*, order Sirenia, (d) a mandrill, *Mandrillus sphinx*, order Primates, (e) a Utah prairie dog, *Cynomys parvidens*, order Rodentia, (f) a cottontail rabbit, *Sylvilagus audubonii,* order Lagomorpha, (g) Malaysian fruit bats, *Pteropus hypomelanus*, order Chiroptera, (h) a common seal or harbor seal, *Phoca vitulina*, suborder Pinnipedia, (i) a grizzly bear, *Ursus arctos horribilis,* order Carnivora, (j) a black rhinoceros or hook-lipped rhinoceros, *Diceros bicornis*, order Perissodactyla, (k) a bottlenose dolphin, *Tursiops truncatus*, order Cetacea, and (l) a mule deer, *Odocoileus hemionus*, order Artiodactyla.

Figure 7.257 The capybara, *Hydrochoerus hydrochaeris*, is the largest living rodent. It can grow to over 4 feet in length and weigh over 175 pounds.

Figure 7.258 The nine-banded armadillo, *Dasypus novemcinctus,* is the most widespread of the armadillos. It has colonized much of the southeastern United States.

Figure 7.259 The pronghorn antelope, *Antilocapra americana,* is cited as the second fastest land animal behind the cheetah. It is reported that it can reach speeds up to 70 miles per hour.

Figure 7.260 The ringtailed lemur, *Lemur catta,* is a primate. There are approximately 100 species of lemurs, all of which are restricted to the island of Madagascar.

Figure 7.261 The American bison, *Bison bison*, is the largest living land animal in North America.

Figure 7.262 The western lowland gorilla, *Gorilla gorilla gorilla,* is the smallest subspecies of gorilla. An adult male can reach 6 feet tall and weigh as much as 600 pounds. They are critically endangered.

Vertebrate Dissections

Chapter 8

An understanding of the structure of a vertebrate organism is requisite to learning about physiological mechanisms and about how the animal functions in its environment. The selective pressures that determine evolutionary change frequently have an influence on anatomical structures. Studying dissected specimens, therefore, provides phylogenetic information about how groups of organisms are related.

Some biology laboratories have the resources to provide students with opportunities for doing selected vertebrate dissections. For these students, the photographs contained in this chapter will be a valuable source for identification of structures on your specimens as they are dissected and studied. If dissection specimens are not available, the excellent photographs of carefully dissected prepared specimens presented in this chapter will be an adequate substitute. Care has gone into the preparation of these specimens to depict and identify the principal body structures from representative specimens of each of the classes of vertebrates. Selected human cadaver dissections are shown in photographs contained in chapter 9. As the anatomy of vertebrate specimens is studied in this chapter, observe the photographs of human dissections in the next chapter, and note the similarities of body structure, particularly to those of another mammal.

Class Chondrichthyes

Figure 8.1 A lateral view of the leopard shark, *Triakis semifasciata*.

1. Spiracle
2. Lateral line
3. Anterior dorsal fin
4. Posterior dorsal fin
5. Caudal fin (heterocercal tail)
6. Eye
7. Gill slits
8. Pectoral fin
9. Pelvic fin
10. Anal fin

Figure 8.2 A photomicrograph of placoid scales.

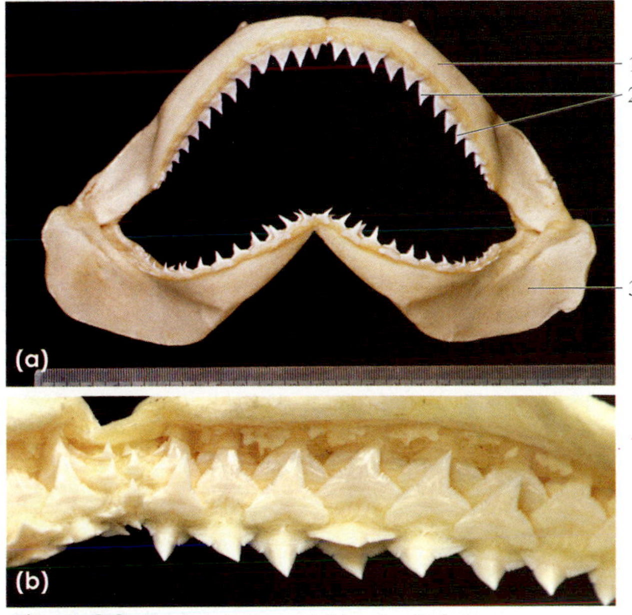

Figure 8.3 Shark jaws (a) and (b) a detailed view showing rows of replacement teeth (scale in mm).

1. Palatopterygoquadrate cartilage (upper jaw)
2. Placoid teeth
3. Meckel's cartilage (lower jaw)

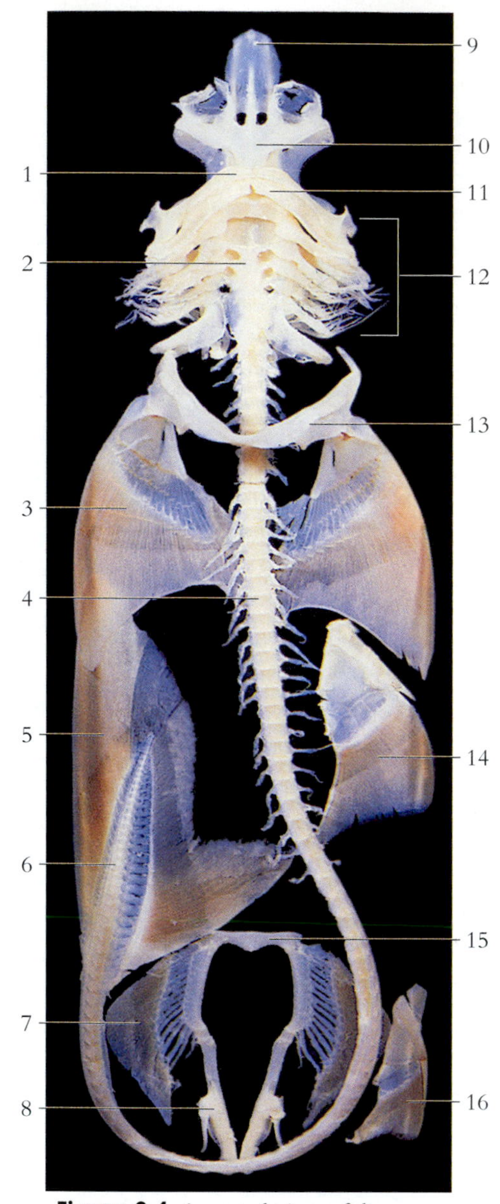

Figure 8.4 A ventral view of the cartilaginous skeleton of a male dogfish shark.

1. Palatopterygoquadrate cartilage (upper jaw)
2. Hypobranchial cartilage
3. Pectoral fin
4. Trunk vertebrae
5. Caudal fin
6. Caudal vertebrae
7. Pelvic fin
8. Clasper
9. Rostrum
10. Chondrocranium
11. Meckel's cartilage (lower jaw)
12. Visceral arches
13. Pectoral girdle
14. Anterior dorsal fin
15. Pelvic girdle
16. Posterior dorsal fin

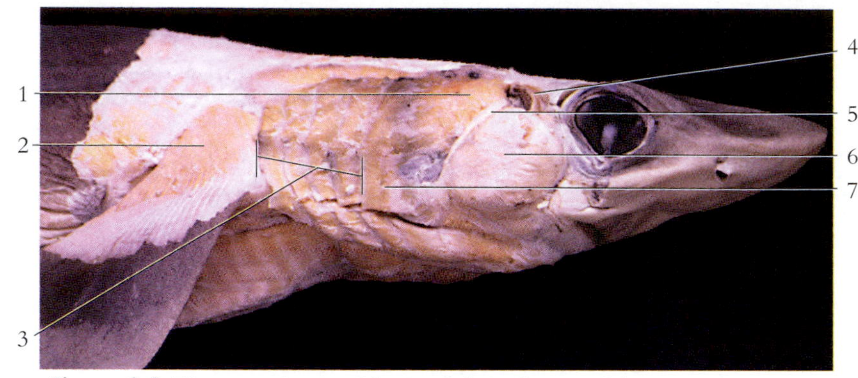

Figure 8.5 The musculature of the jaw, gills, and pectoral fin of a dogfish shark.

1. 2nd dorsal constrictor
2. Levator of pectoral fin
3. 3rd through 6th ventral constrictors
4. Spiracular muscle
5. Facial nerve (hyomandibular branch)
6. Mandibular adductor
7. 2nd ventral constrictor

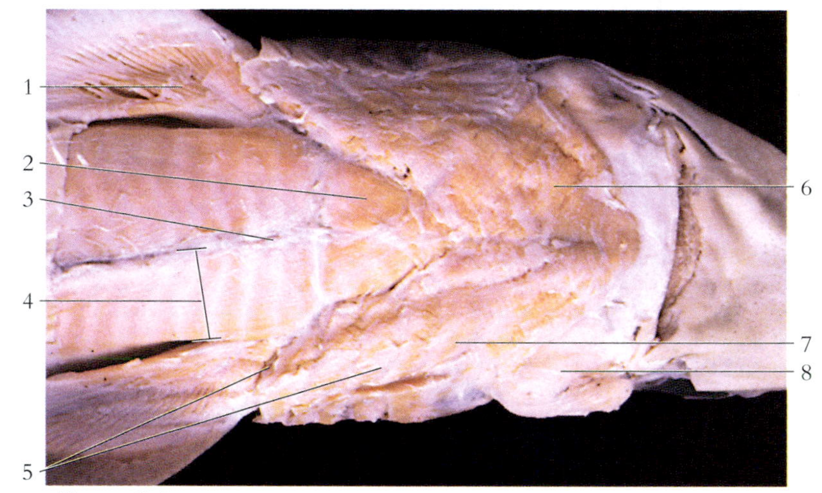

Figure 8.6 A ventral view of the hypobranchial musculature of the dogfish shark.

1. Depressor of pectoral fin
2. Common coracoarcual
3. Linea alba
4. Hypaxial muscle
5. 3rd through 6th ventral constrictors
6. 1st ventral constrictor
7. 2nd ventral constrictor
8. Mandibular adductor

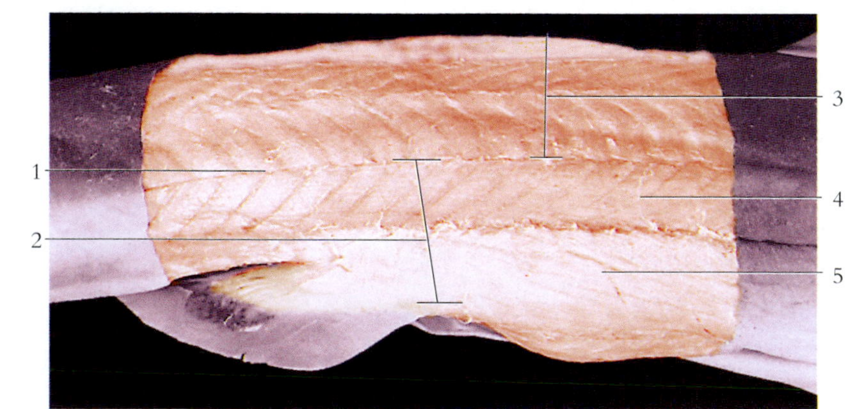

Figure 8.7 A lateral view of the axial musculature of the dogfish shark.

1. Horizontal septum
2. Hypaxial myotome portion
3. Epaxial myotome portion
4. Lateral bundle of myotomes
5. Ventral bundle of myotomes

Figure 8.8 The internal anatomy of a male dogfish shark.

1. Right lobe of liver (reflected)
2. Pyloric sphincter valve
3. Stomach (pyloric region)
4. Spleen
5. Ileum
6. Testis
7. Esophagus
8. Stomach (cardiac region)
9. Kidney
10. Rectal gland
11. Cloaca

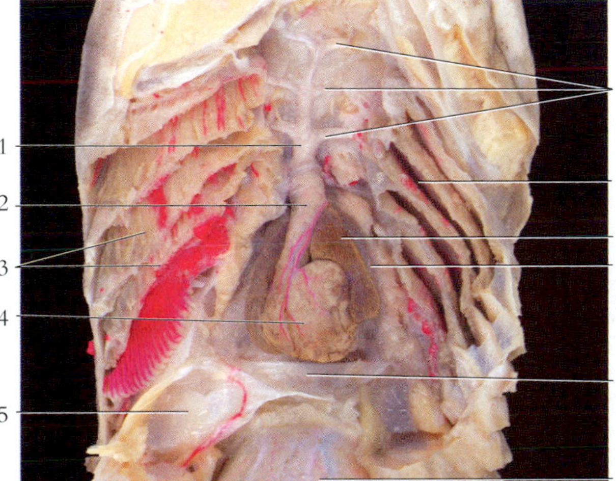

Figure 8.9 The heart, gills, and associated vessels of a dogfish shark.

1. Ventral aorta
2. Conus arteriosus
3. Gills
4. Ventricle
5. Pectoral girdle (cut)
6. Afferent branchial arteries
7. Gill cleft
8. Atrium
9. Pericardial cavity
10. Transverse septum
11. Liver

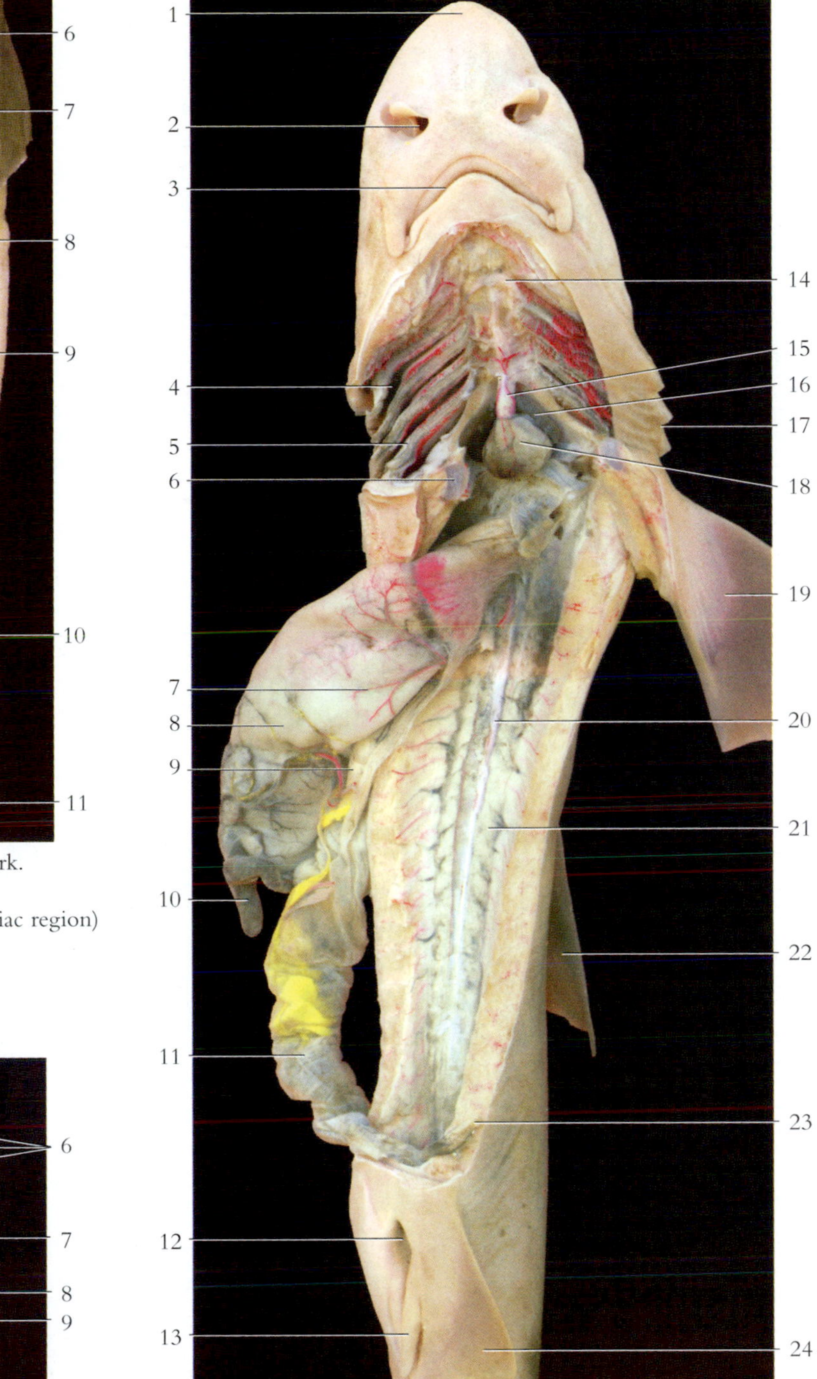

Figure 8.10 The superficial and internal anatomy of a dogfish shark (liver removed).

1. Rostrum
2. Nostril
3. Mouth
4. Gill cleft
5. Gill
6. Pectoral girdle (cut)
7. Gastrosplenic artery
8. Stomach (reflected)
9. Pancreas
10. Spleen (reflected)
11. Intestine (reflected)
12. Cloaca
13. Clasper
14. Afferent branchial artery
15. Ventral aorta
16. Atrium
17. Gill slit
18. Heart
19. Pectoral fin
20. Dorsal aorta
21. Kidney
22. Dorsal fin
23. Rectal gland
24. Pelvic fin

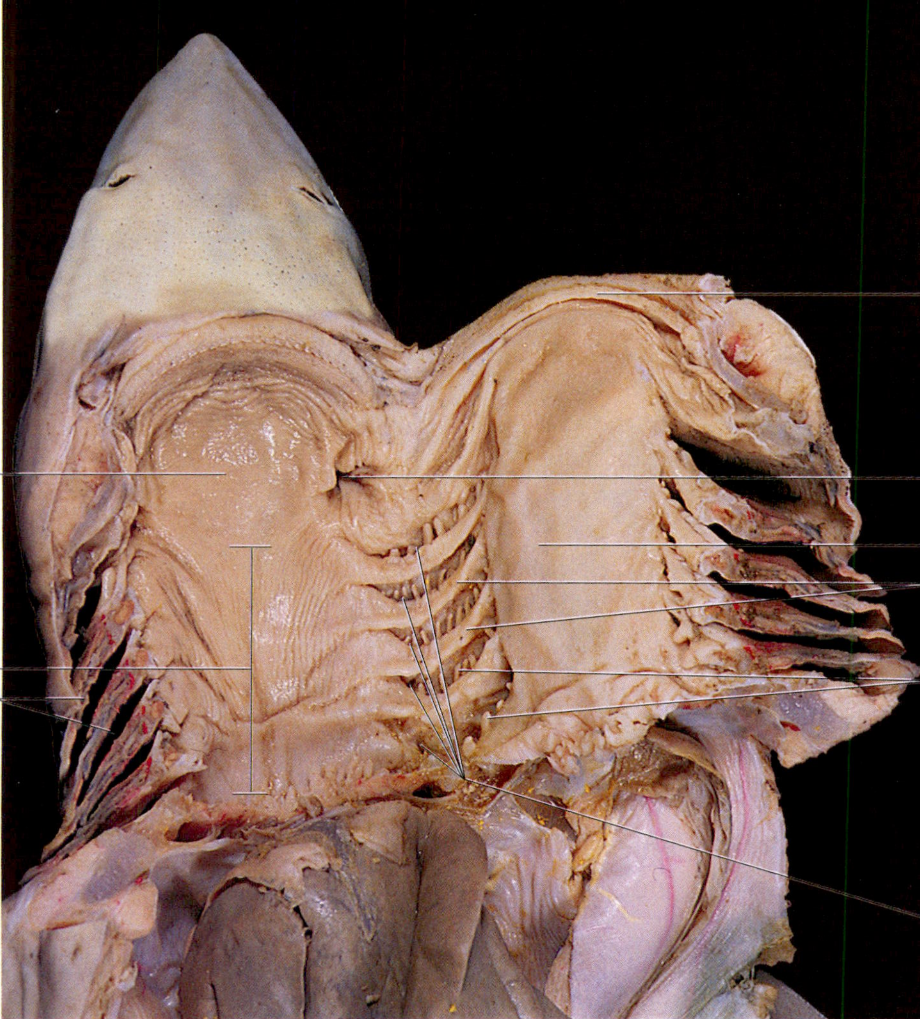

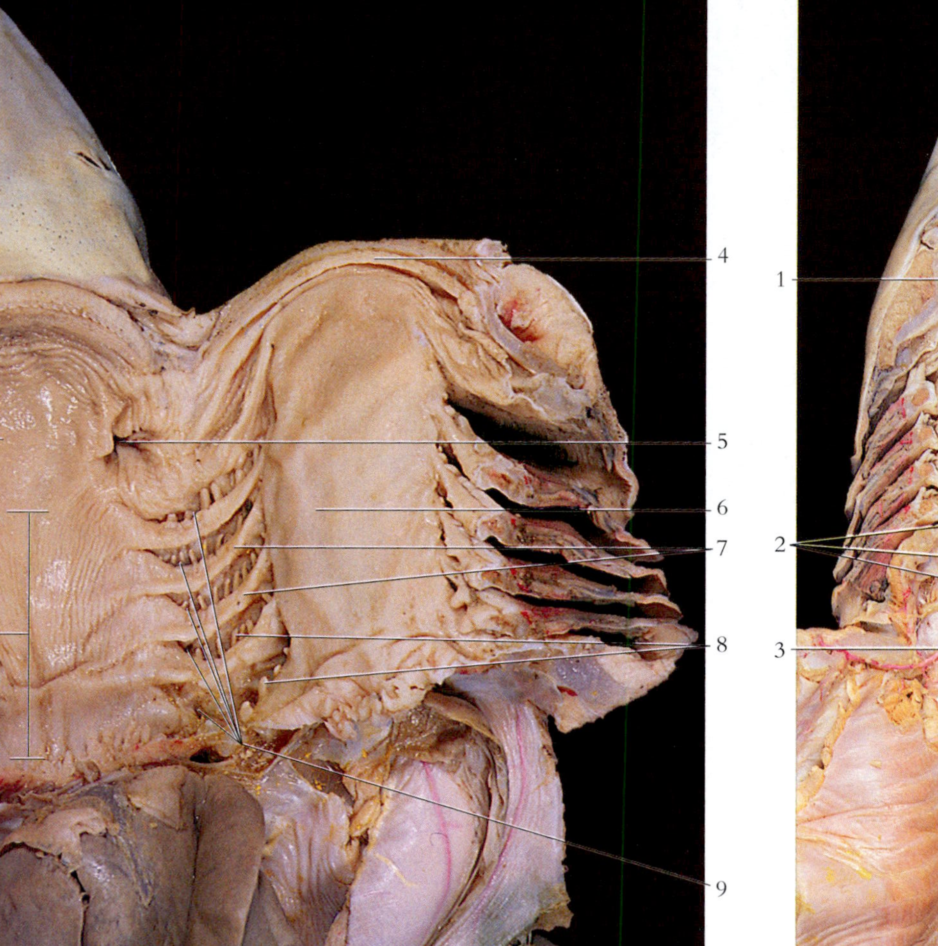

Figure 8.11 A ventral view of the internal respiratory anatomy of the dogfish shark (lower jaw cut and reflected).

1. Oral cavity
2. Pharynx
3. Parabranchial chambers
4. Teeth
5. Spiracle
6. Tongue
7. Gill arches
8. Gill rakers
9. Internal gill slits (5)

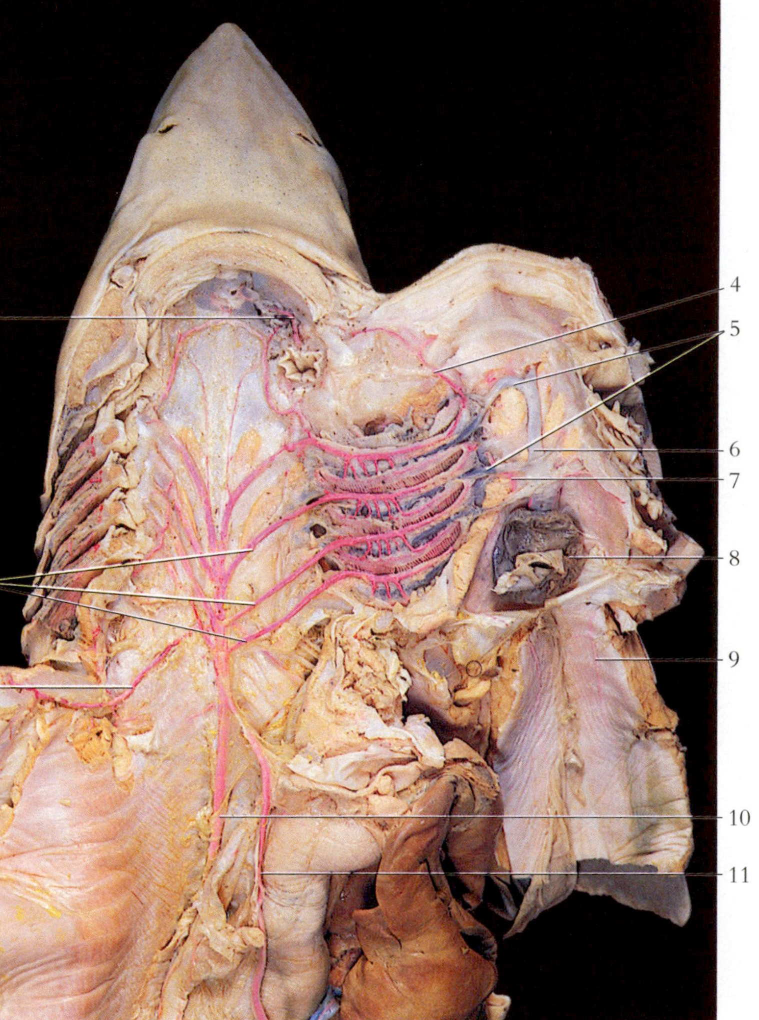

Figure 8.12 A ventral view of the branchial circulation of the dogfish shark (lower jaw cut and reflected).

1. Stapedial artery
2. Efferent branchial arteries
3. Subclavian artery
4. External carotid artery
5. Afferent branchial arteries
6. Ventral aorta
7. Hypobranchial artery
8. Heart
9. Anterior epigastric artery
10. Dorsal aorta
11. Celiac artery

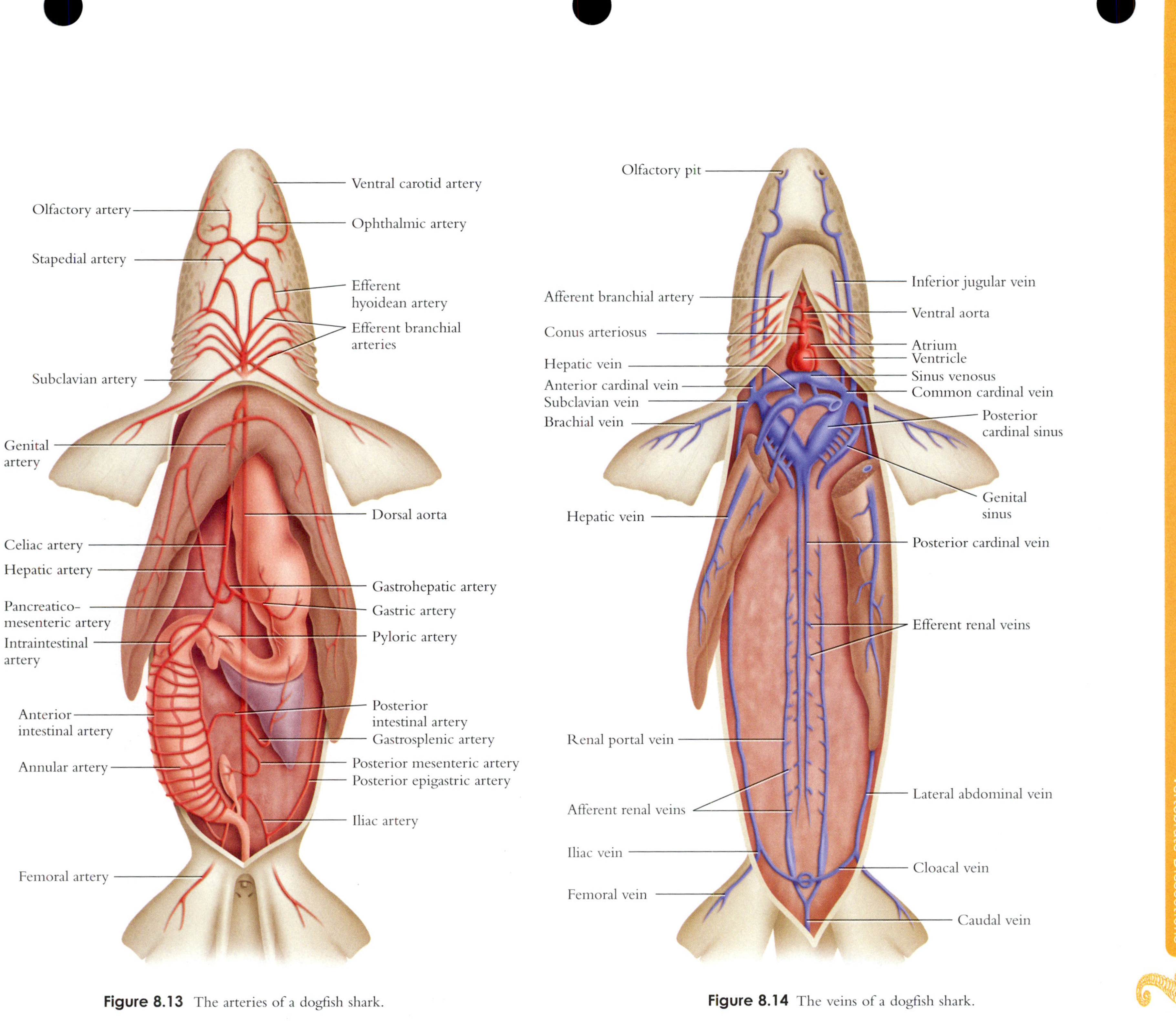
Ventral carotid artery
Olfactory artery
Ophthalmic artery
Stapedial artery
Efferent hyoidean artery
Efferent branchial arteries
Subclavian artery
Genital artery
Dorsal aorta
Celiac artery
Hepatic artery
Gastrohepatic artery
Pancreatico-mesenteric artery
Gastric artery
Intraintestinal artery
Pyloric artery
Anterior intestinal artery
Posterior intestinal artery
Gastrosplenic artery
Annular artery
Posterior mesenteric artery
Posterior epigastric artery
Iliac artery
Femoral artery
Olfactory pit
Inferior jugular vein
Afferent branchial artery
Ventral aorta
Conus arteriosus
Atrium
Ventricle
Hepatic vein
Sinus venosus
Anterior cardinal vein
Common cardinal vein
Subclavian vein
Posterior cardinal sinus
Brachial vein
Genital sinus
Hepatic vein
Posterior cardinal vein
Efferent renal veins
Renal portal vein
Lateral abdominal vein
Afferent renal veins
Iliac vein
Cloacal vein
Femoral vein
Caudal vein

Figure 8.13 The arteries of a dogfish shark.

Figure 8.14 The veins of a dogfish shark.

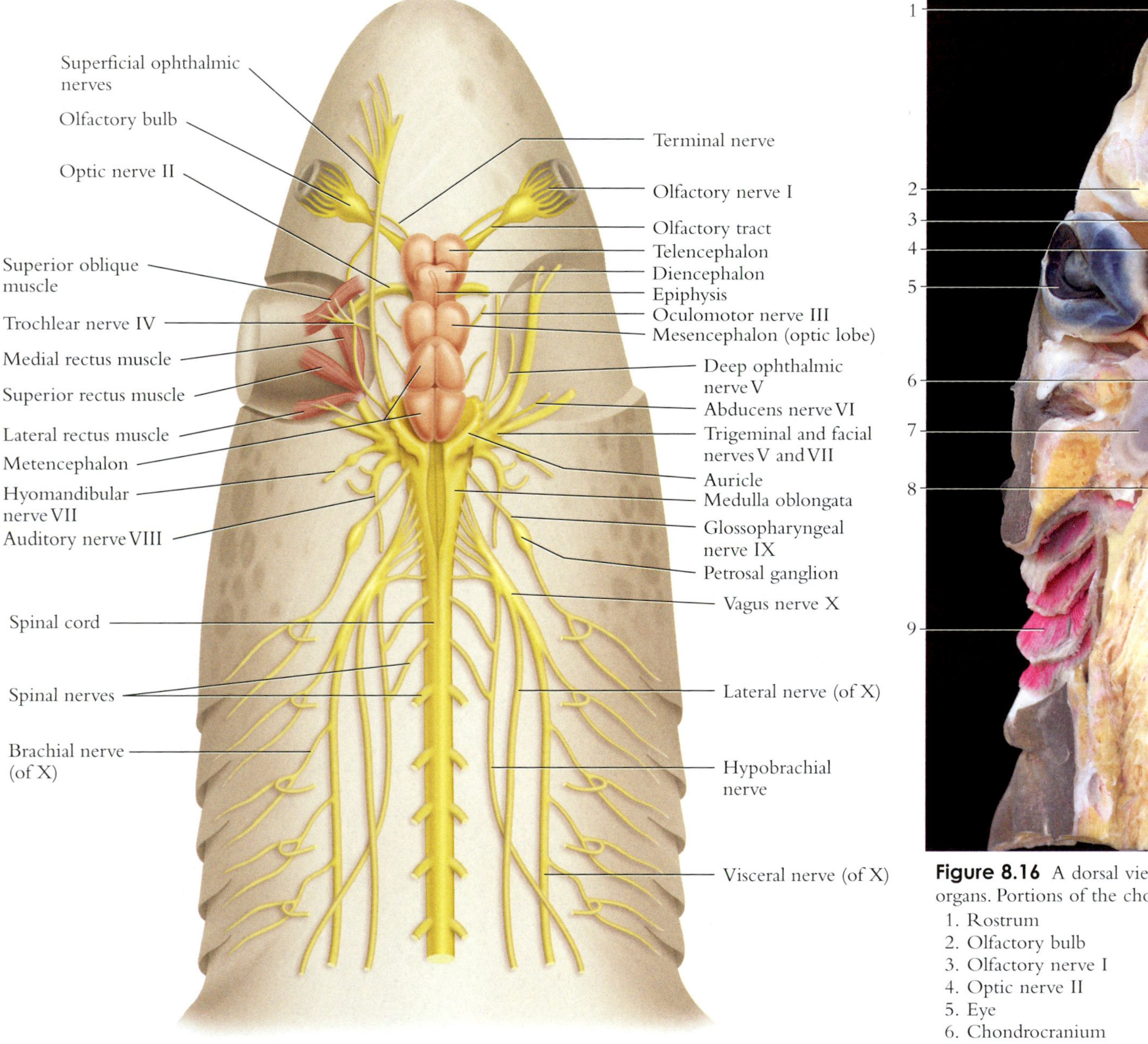

Figure 8.15 A dorsal view of the dogfish brain, cranial nerves, and eye muscles.

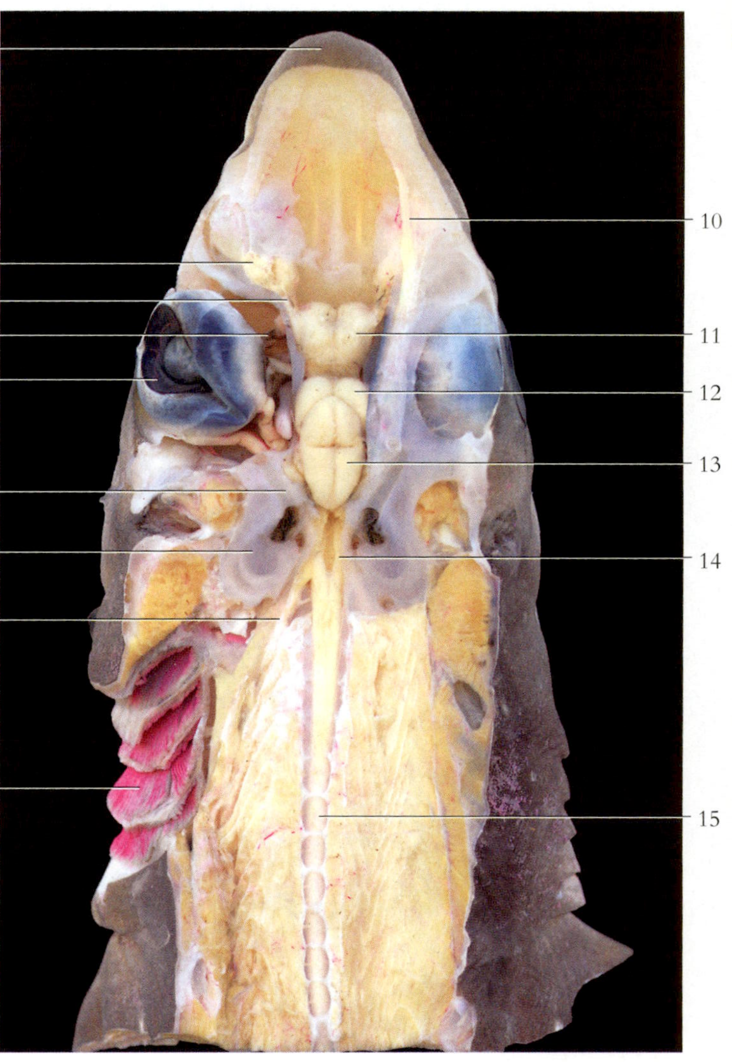

Figure 8.16 A dorsal view of the dogfish brain and sensory organs. Portions of the chondrocranium have been shaved away.

1. Rostrum
2. Olfactory bulb
3. Olfactory nerve I
4. Optic nerve II
5. Eye
6. Chondrocranium
7. Semicircular canal
8. Vagus nerve X
9. Gill
10. Trigeminal nerve V, VII
11. Telencephalon
12. Mesencephalon (optic lobe)
13. Metencephalon
14. Medulla oblongata
15. Spinal cord

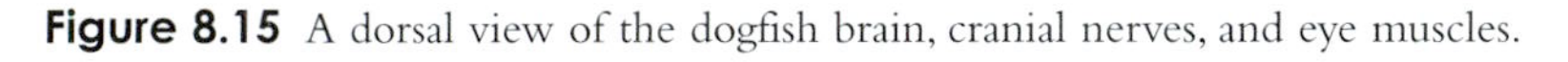

Superclass Osteichthyes - Class Actinopterygii

Figure 8.17 The external anatomy of a perch.

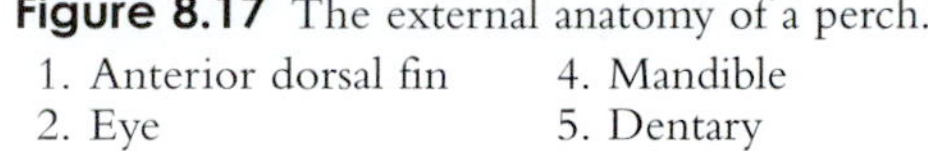

1. Anterior dorsal fin
2. Eye
3. Nostrils
4. Mandible
5. Dentary
6. Operculum
7. Pectoral fin
8. Pelvic fin
9. Lateral line
10. Posterior dorsal fin
11. Caudal fin
12. Anal fin
13. Anus

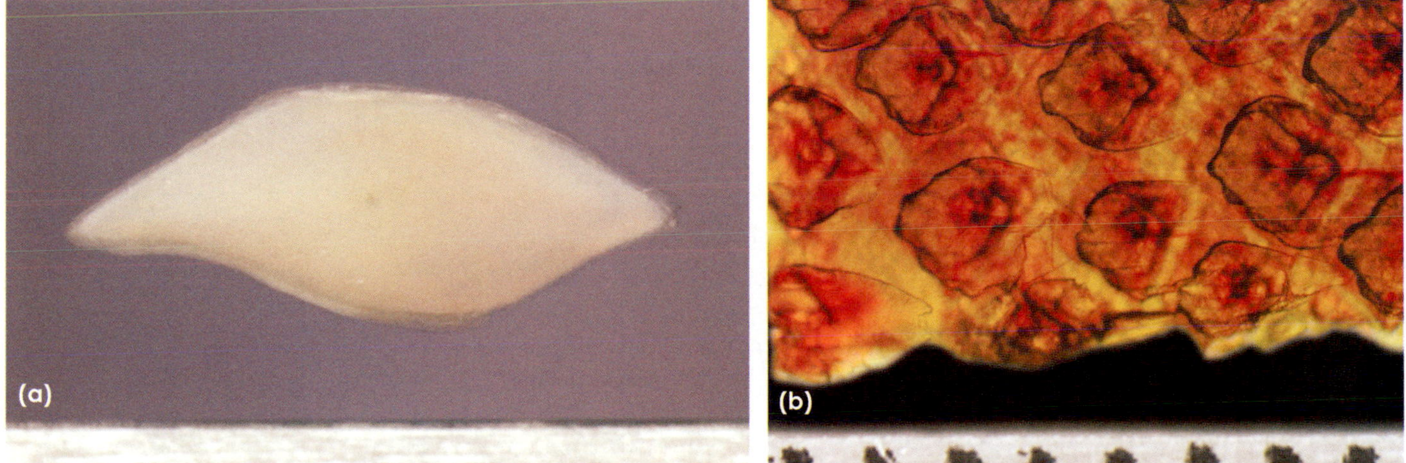

Figure 8.18 Ganoid scales, present in primitive fishes like the gar, are composed of silvery ganoin on the top surface and bone on the bottom. Two different sizes are shown (a) and (b) (scale in mm).

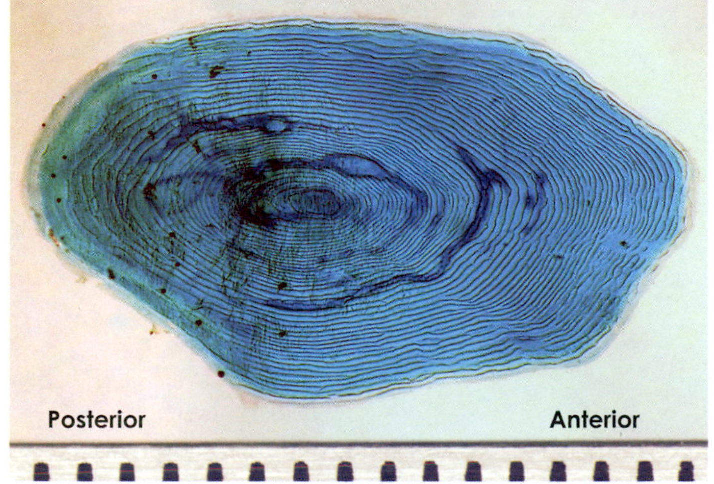

Figure 8.19 Cycloid scales, along with ctenoid scales (fig. 8.20), are found on advanced bony fishes. They are much thinner and more flexible than ganoid scales and overlap each other (scale in mm).

Figure 8.20 Ctenoid scales differ from cycloid scales in that they have comblike ridges on the exposed edge, thought to improve swimming efficiency (scale in mm).

Figure 8.21 The skeleton of a perch.
1. Anterior dorsal fin
2. Fin spines
3. Neurocranium
4. Premaxilla
5. Maxilla
6. Dentary
7. Opercular bones
8. Pectoral fin
9. Pelvic fin
10. Vertebral column
11. Posterior dorsal fin
12. Soft rays
13. Caudal fin
14. Neural spine
15. Haemal spine
16. Anal fin
17. Ribs

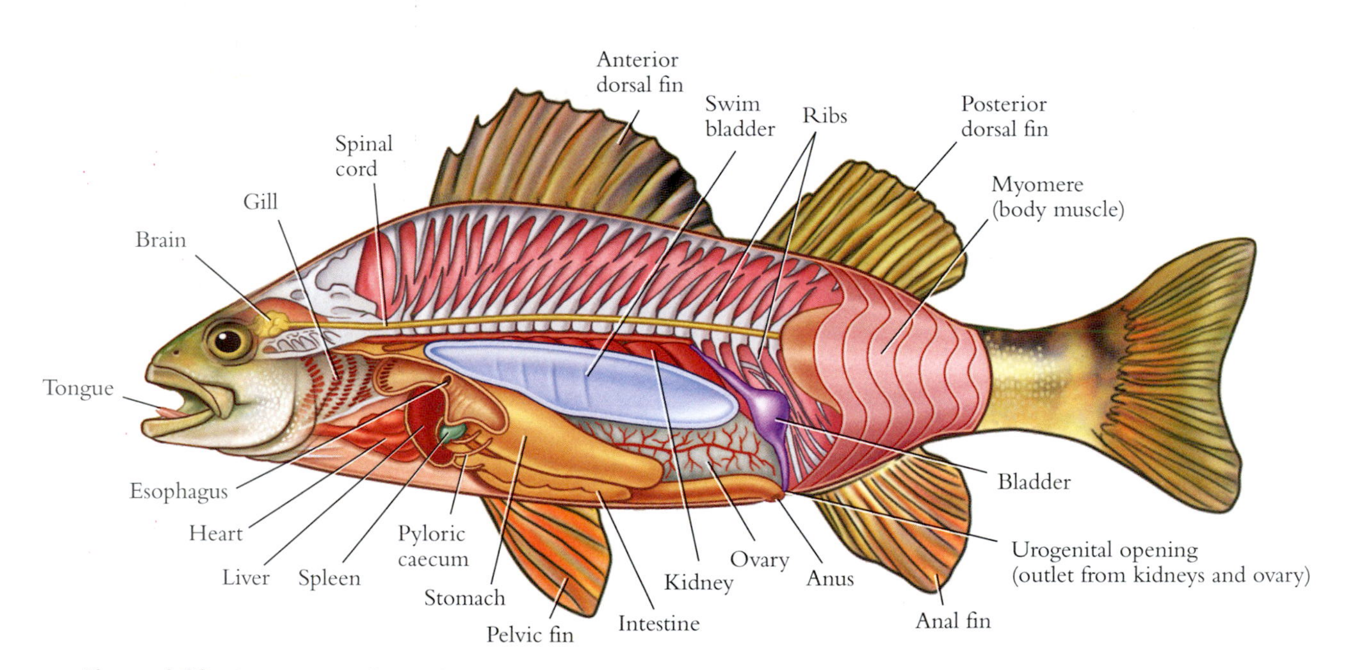

Figure 8.22 The anatomy of a perch.

Figure 8.23 The viscera of a perch.
1. Epaxial muscles
2. Stomach
3. Gill
4. Heart
5. Pyloric caecum
6. Liver (cut)
7. Pancreas
8. Vertebrae
9. Urinary bladder
10. Gonad
11. Anus
12. Intestine

Class Amphibia

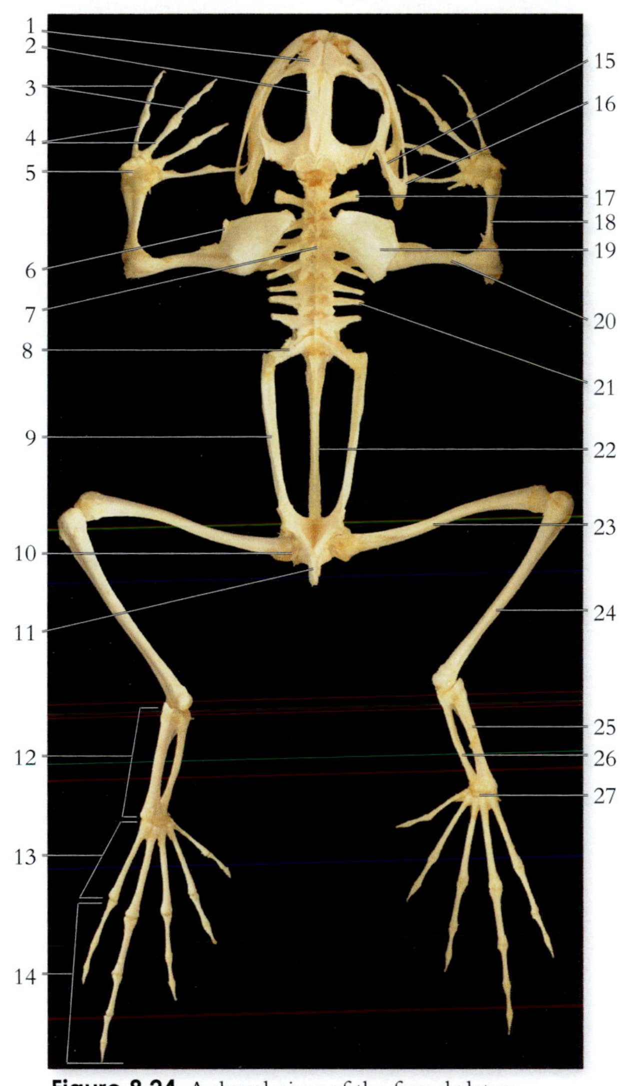

Figure 8.24 A dorsal view of the frog skeleton.

1. Nasal bone
2. Frontoparietal bone
3. Phalanges of digits
4. Metacarpal bones
5. Carpal bones
6. Scapula
7. Vertebra
8. Transverse process of sacral (9th) vertebra
9. Ilium
10. Acetabulum
11. Ischium
12. Tarsal bones
13. Metatarsal bones
14. Phalanges of digits
15. Squamosal bone
16. Quadratojugal bone
17. Transverse process
18. Radioulna
19. Suprascapula
20. Humerus
21. Transverse process
22. Urostyle
23. Femur
24. Tibiofibula
25. Fibulare (calcaneum)
26. Tibiale (astragalus)
27. Distal tarsal bones

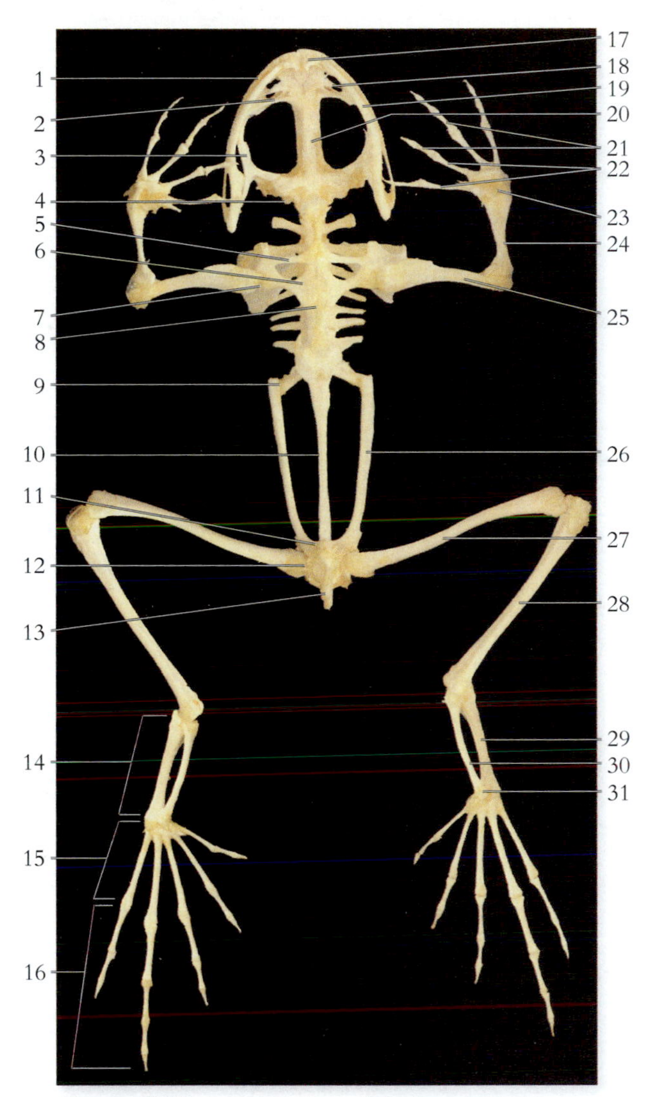

Figure 8.25 A ventral view of the frog skeleton.

1. Maxilla
2. Palatine
3. Pterygoid bone
4. Exoccipital bone
5. Clavicle
6. Coracoid
7. Glenoid fossa
8. Sternum
9. Transverse process of sacral (9th) vertebra
10. Urostyle
11. Pubis
12. Acetabulum
13. Ischium
14. Tarsal bones
15. Metatarsal bones
16. Phalanges of digits
17. Premaxilla
18. Vomer
19. Dentary
20. Parasphenoid bone
21. Phalange of digits
22. Metacarpal bone
23. Carpal bones
24. Radioulna
25. Humerus
26. Ilium
27. Femur
28. Tibiofibula
29. Fibulare (calcaneum)
30. Tibiale (astragalus)
31. Distal tarsal bones

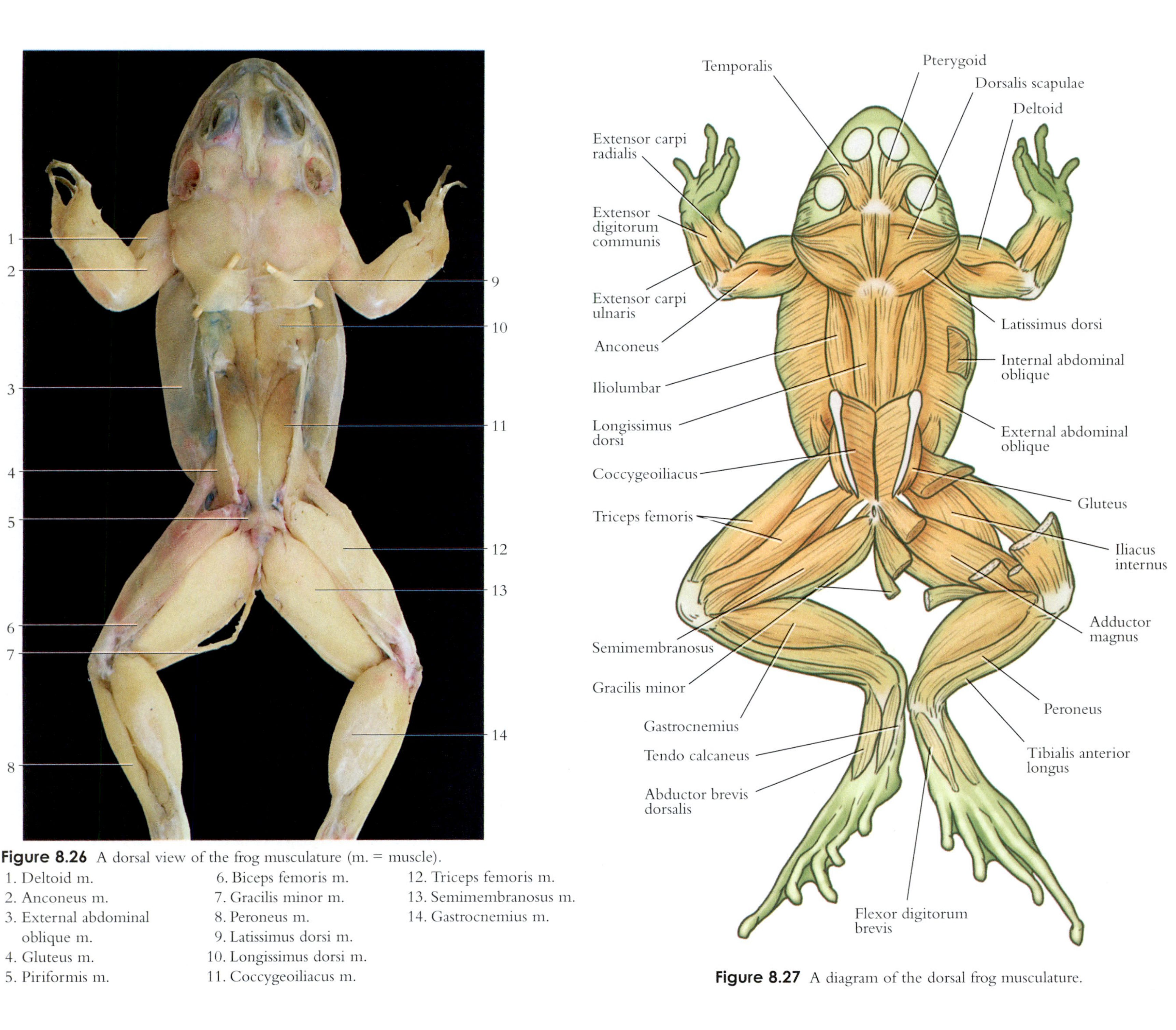

Figure 8.26 A dorsal view of the frog musculature (m. = muscle).

1. Deltoid m.
2. Anconeus m.
3. External abdominal oblique m.
4. Gluteus m.
5. Piriformis m.
6. Biceps femoris m.
7. Gracilis minor m.
8. Peroneus m.
9. Latissimus dorsi m.
10. Longissimus dorsi m.
11. Coccygeoiliacus m.
12. Triceps femoris m.
13. Semimembranosus m.
14. Gastrocnemius m.

Figure 8.27 A diagram of the dorsal frog musculature.

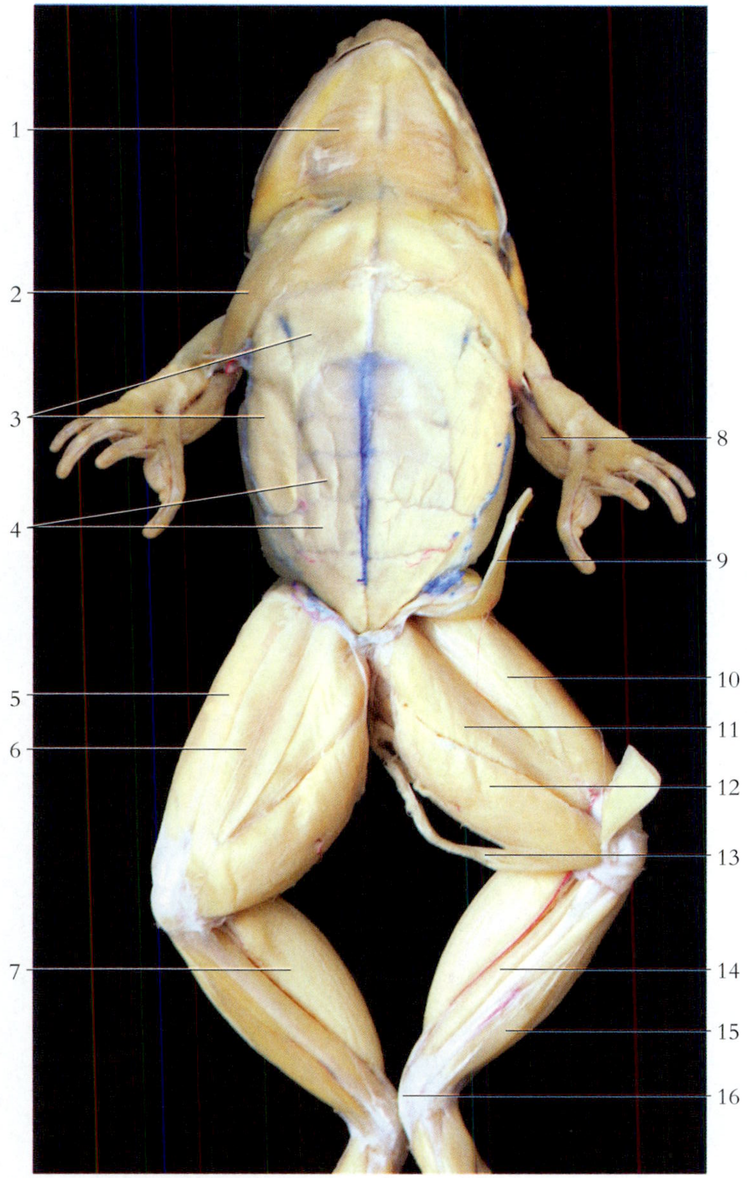

Figure 8.28 A ventral view of the frog musculature (m. = muscle).

1. Mylohyoid m.
2. Deltoid m.
3. Pectoralis m.
4. Rectus abdominis m.
5. Triceps femoris m.
6. Sartorius m.
7. Gastrocnemius m.
8. Palmaris longus m.
9. Sartorius m. (cut and reflected)
10. Adductor longus m.
11. Adductor magnus m.
12. Gracilis major m.
13. Gracilis minor m.
14. Tibialis posterior m.
15. Tibialis anterior m.
16. Tendo calcaneus

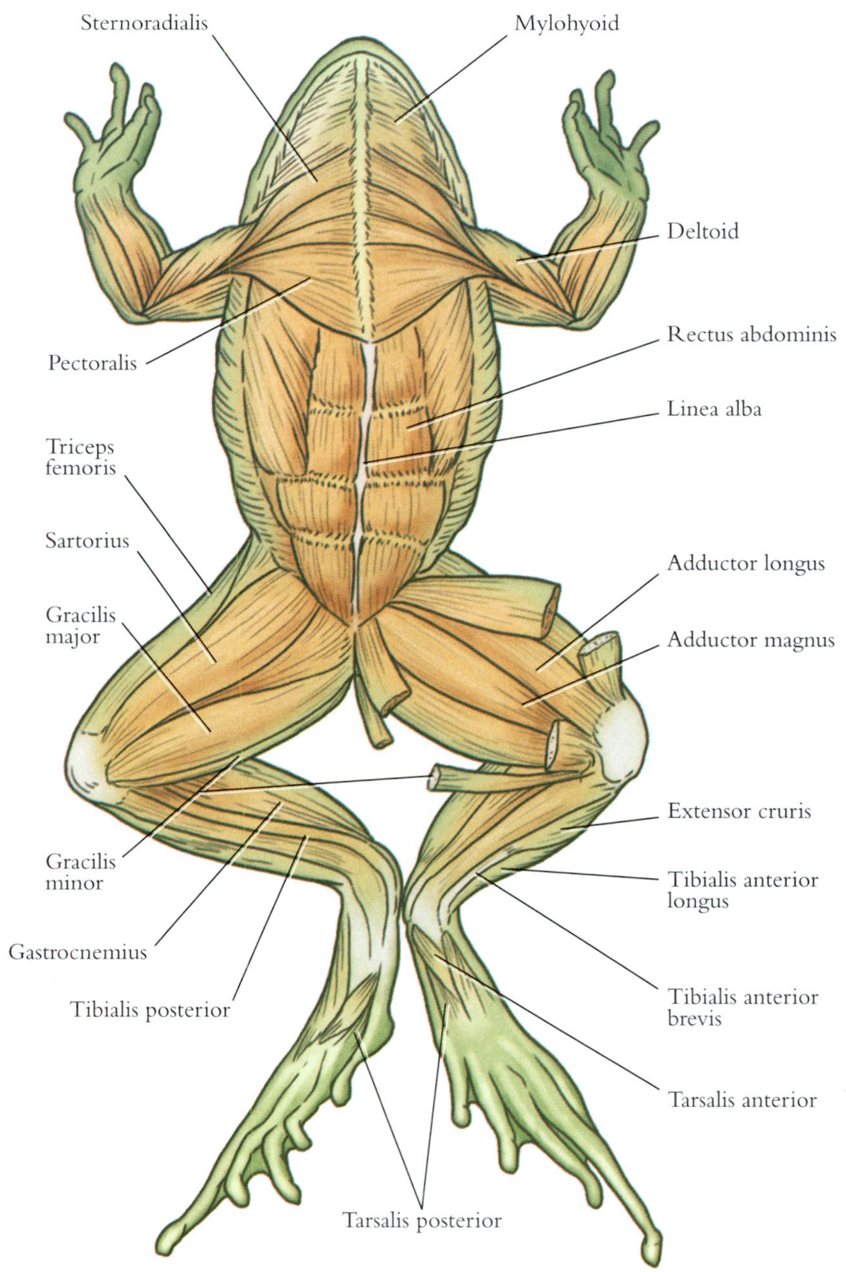

Figure 8.29 A diagram of the ventral frog musculature.

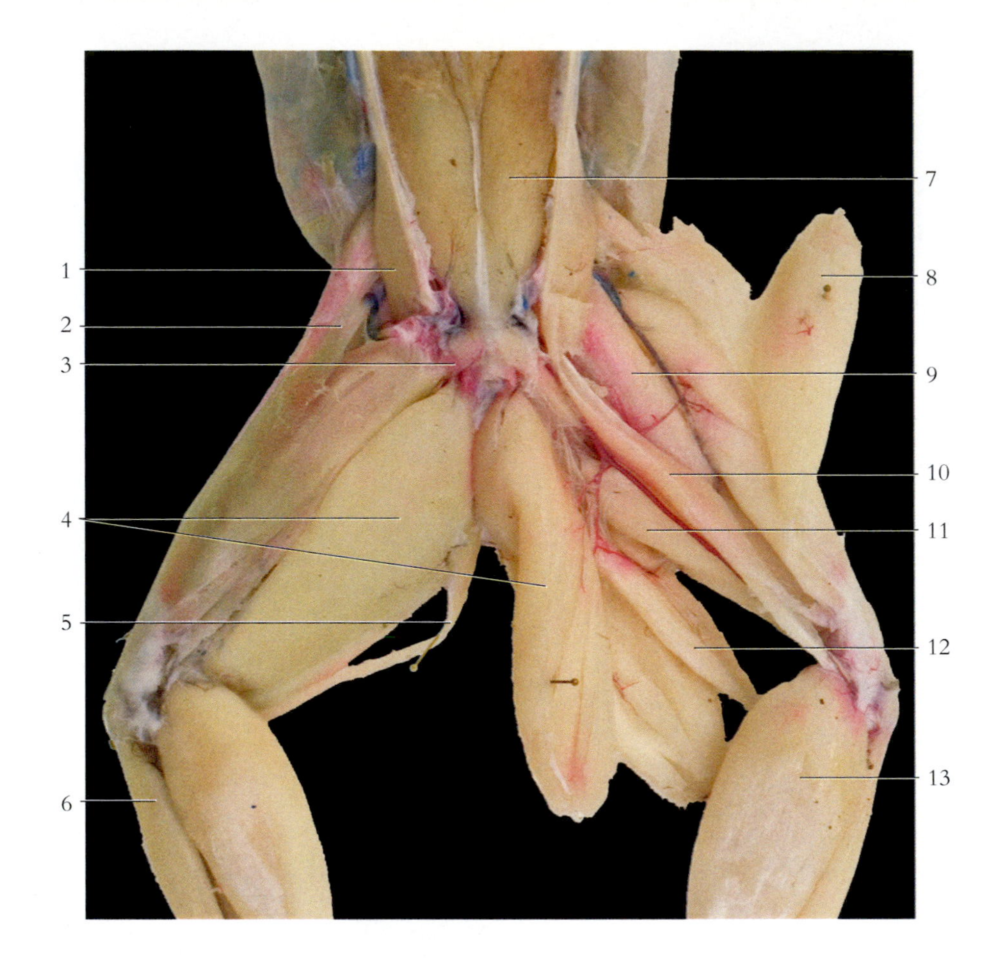

Figure 8.30 A dorsal view of the leg muscles of a frog (m. = muscle).

1. Gluteus m.
2. Cutaneus abdominis m.
3. Piriformis m.
4. Semimembranosus m. (cut and reflected)
5. Gracilis minor m.
6. Peroneus m.
7. Coccygeoiliacus m.
8. Triceps femoris m. (cut and reflected)
9. Iliacus internus m.
10. Biceps femoris m.
11. Adductor magnus m.
12. Semitendinosus m. (cut and reflected)
13. Gastrocnemius m.

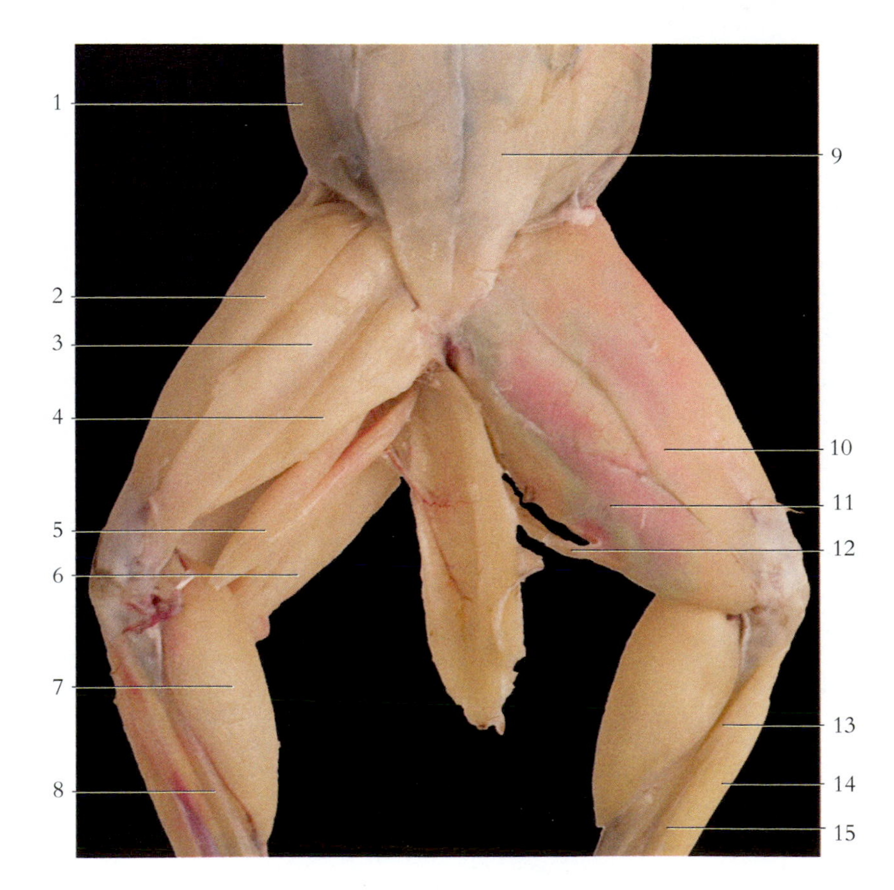

Figure 8.31 A ventral view of the leg muscles of a frog (m. = muscle).

1. External abdominal oblique m.
2. Triceps femoris m.
3. Adductor longus m.
4. Adductor magnus m.
5. Semitendinosus m. (cut)
6. Semimembranosus m.
7. Gastrocnemius m.
8. Tibialis posterior m.
9. Rectus abdominis m.
10. Sartorius m.
11. Gracilis major m.
12. Gracilis minor m.
13. Extensor cruris m.
14. Tibialis anterior longus m.
15. Tibialis anterior brevis m.

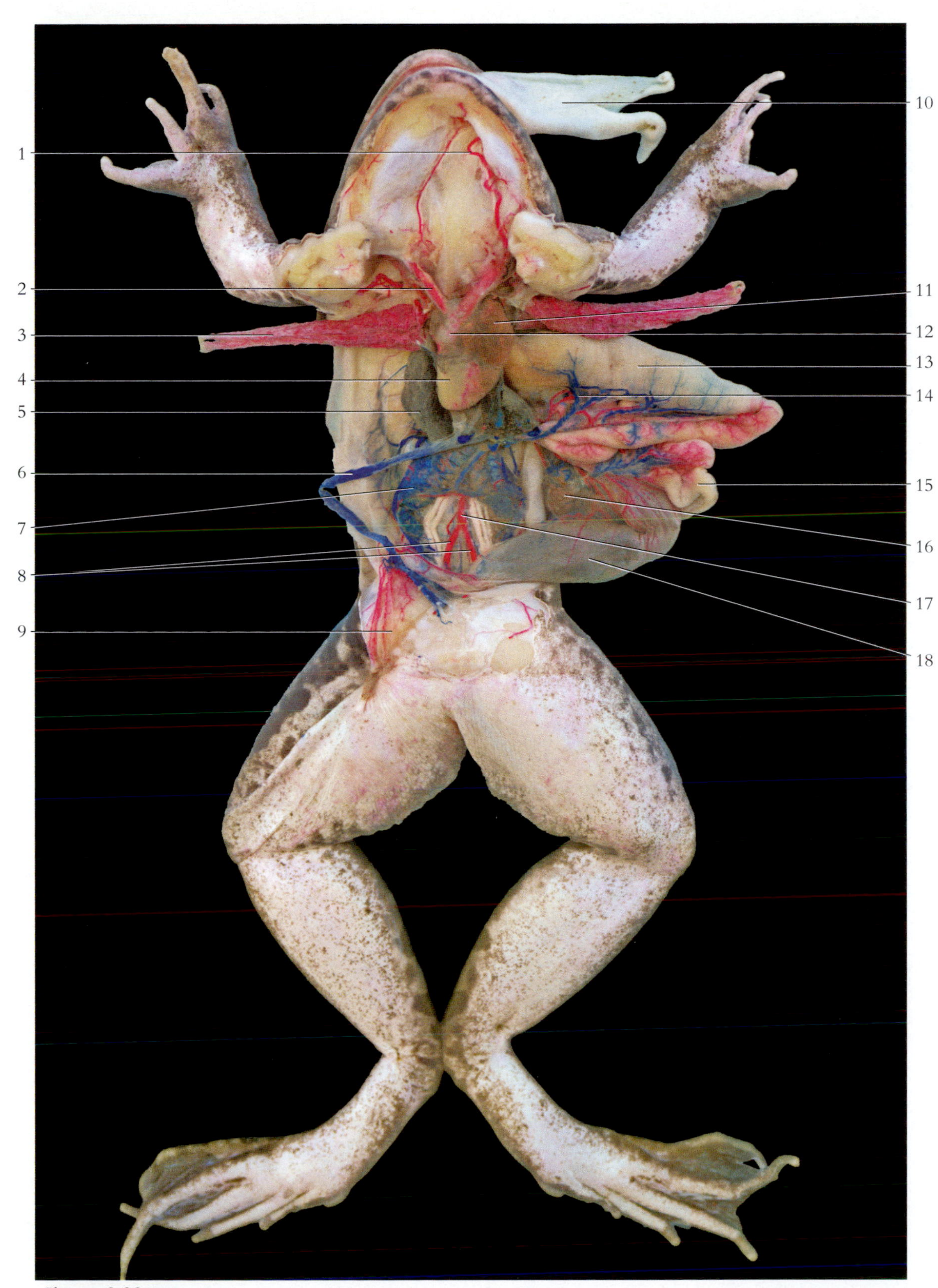

Figure 8.32 The internal anatomy of the frog.

1. External carotid artery
2. Truncus arteriosus
3. Lung (reflected)
4. Ventricle of heart
5. Liver (cut)
6. Ventral abdominal vein
7. Kidney
8. Iliac arteries
9. Bladder (reflected)
10. Tongue
11. Right atrium of heart
12. Conus arteriosus
13. Stomach (reflected)
14. Gastric vein
15. Small intestine
16. Spleen
17. Dorsal aorta
18. Large intestine (reflected)

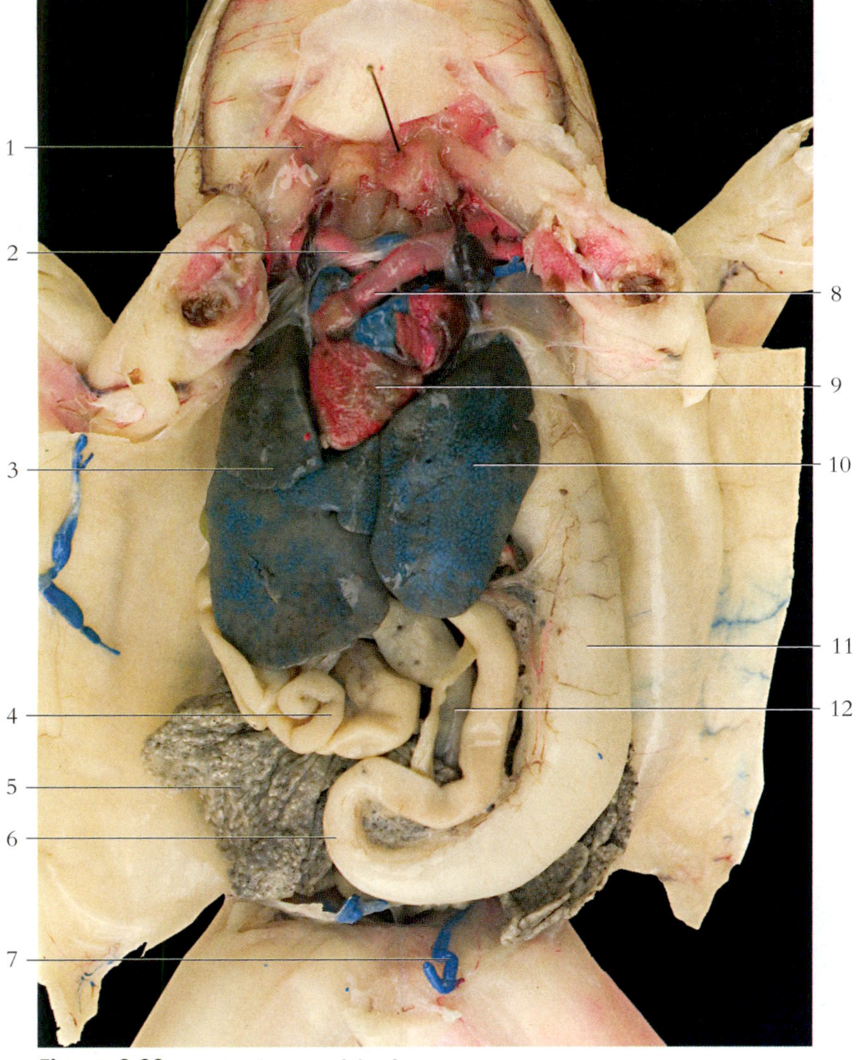

Figure 8.33 A ventral view of the frog viscera.

1. External carotid artery
2. Truncus arteriosus
3. Right lobe of liver
4. Small intestine
5. Ovary
6. Duodenum
7. Ventral abdominal vein (cut)
8. Conus arteriosus
9. Heart
10. Left lobe of liver
11. Stomach
12. Large intestine

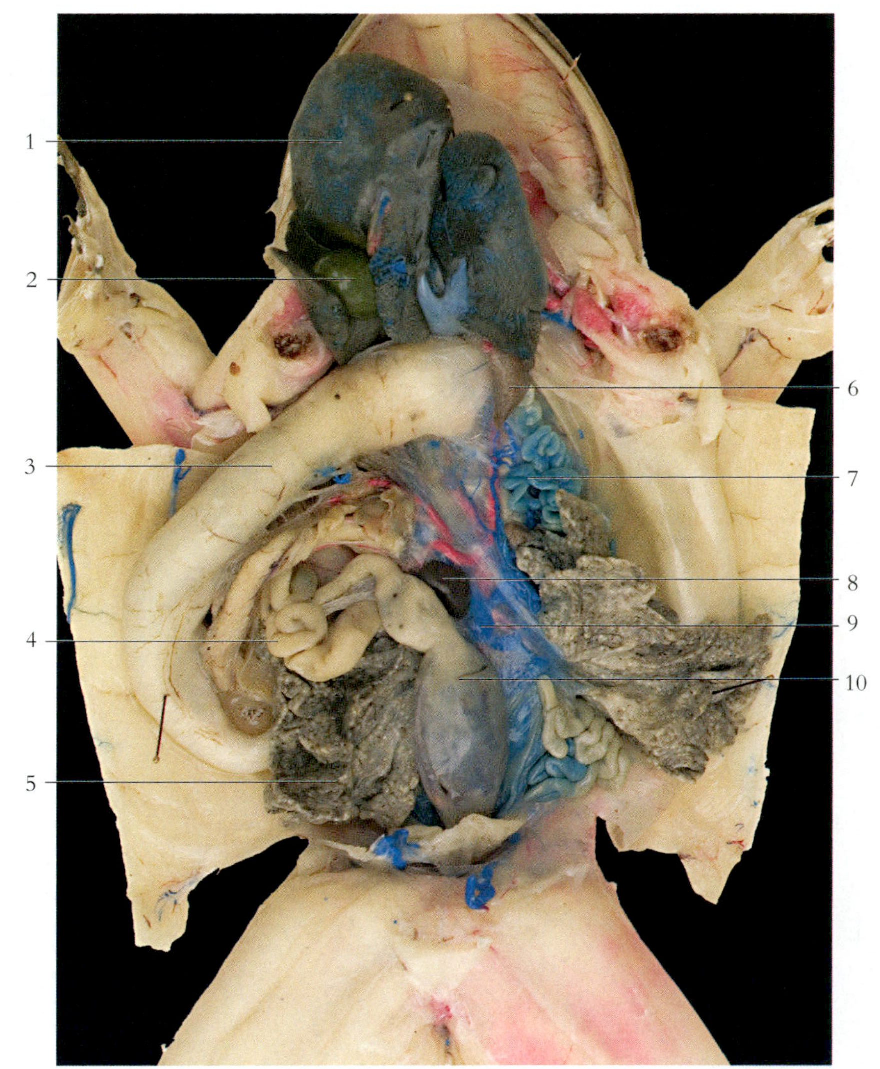

Figure 8.34 A deep view of the frog viscera.

1. Liver (reflected)
2. Gallbladder
3. Stomach
4. Small intestine
5. Ovary
6. Left lung
7. Oviduct
8. Spleen
9. Caudal vena cava
10. Large intestine

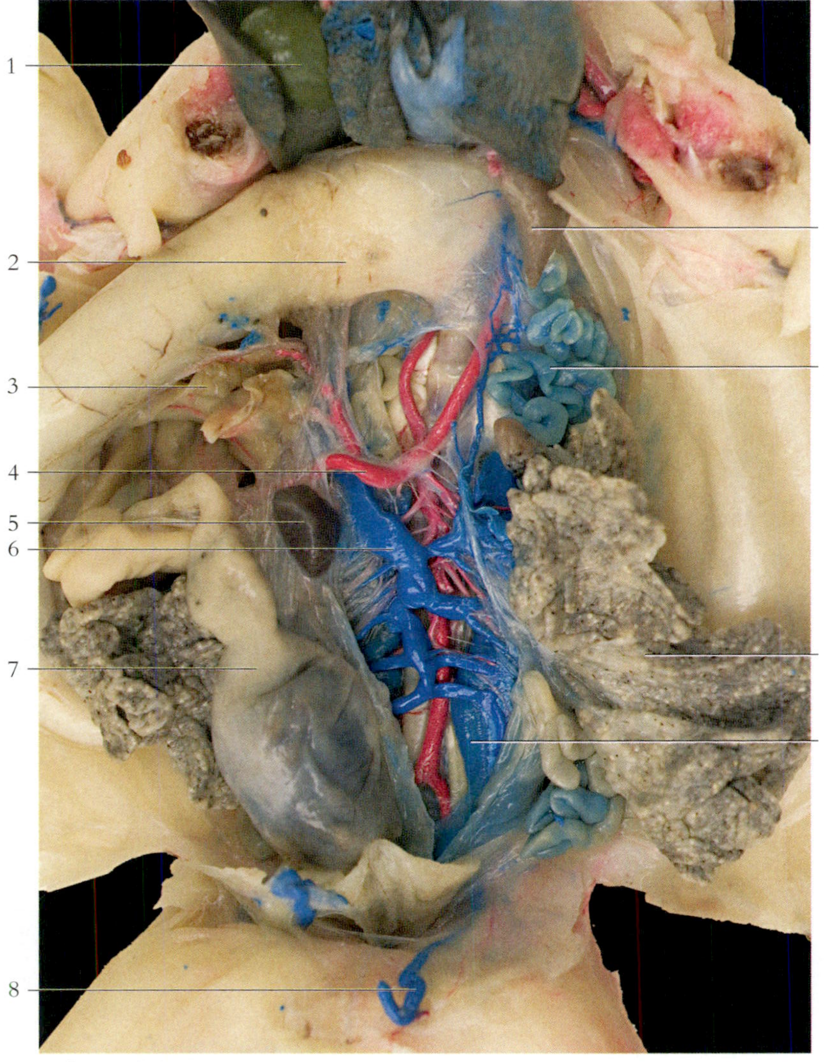

Figure 8.35 A deep view of the frog viscera.

1. Gallbladder
2. Stomach
3. Pancreas
4. Celiacomesenteric trunk
5. Spleen
6. Caudal vena cava
7. Large intestine
8. Ventral abdominal vein (cut)
9. Left lung
10. Oviduct
11. Ovary
12. Left kidney

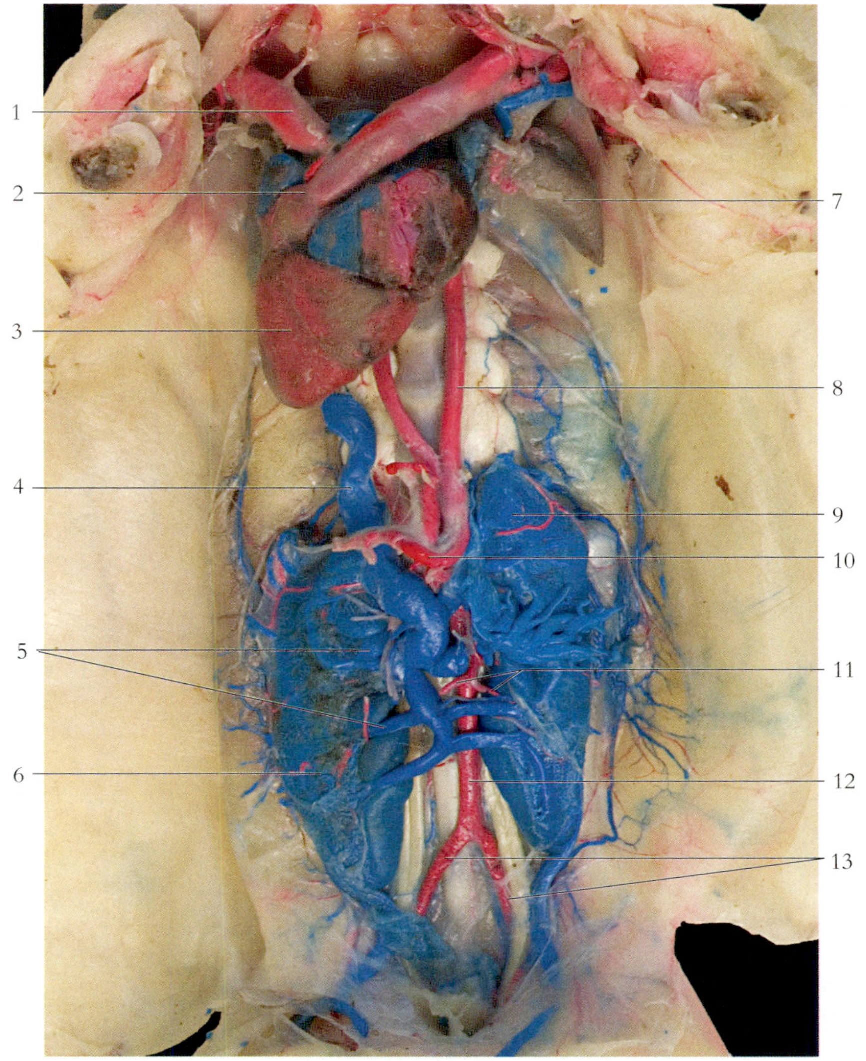

Figure 8.36 The arteries and veins of the frog trunk.

1. Truncus arteriosus
2. Conus arteriosus
3. Ventricle
4. Caudal vena cava
5. Urogenital veins
6. Right kidney
7. Left lung
8. Left systemic arch
9. Left kidney
10. Celiacomesenteric trunk
11. Urogenital arteries
12. Dorsal aorta
13. Iliac arteries

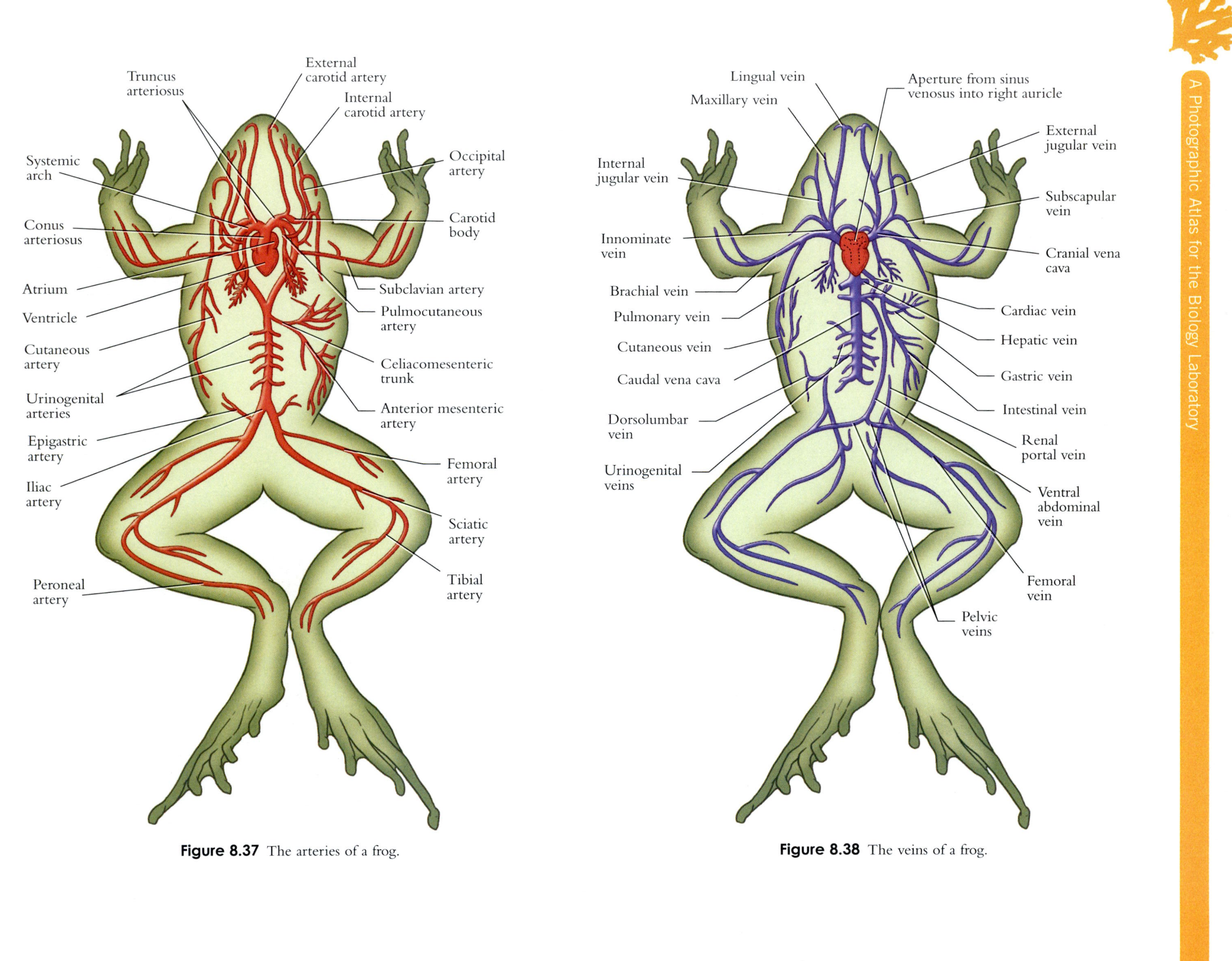

Truncus arteriosus
External carotid artery
Internal carotid artery
Systemic arch
Occipital artery
Conus arteriosus
Carotid body
Atrium
Subclavian artery
Ventricle
Pulmocutaneous artery
Cutaneous artery
Celiacomesenteric trunk
Urinogenital arteries
Anterior mesenteric artery
Epigastric artery
Femoral artery
Iliac artery
Sciatic artery
Peroneal artery
Tibial artery
Lingual vein
Aperture from sinus venosus into right auricle
Maxillary vein
External jugular vein
Internal jugular vein
Subscapular vein
Innominate vein
Cranial vena cava
Brachial vein
Cardiac vein
Pulmonary vein
Hepatic vein
Cutaneous vein
Gastric vein
Caudal vena cava
Intestinal vein
Dorsolumbar vein
Renal portal vein
Urinogenital veins
Ventral abdominal vein
Femoral vein
Pelvic veins

Figure 8.37 The arteries of a frog.

Figure 8.38 The veins of a frog.

Class Sauropsida (= Reptilia)

Figure 8.39 A dorsal view of a turtle.

1. Eye
2. Pentadactyl foot
3. Vertebral scales
4. Marginal scales (encircle the carapace)
5. Nostril
6. Head
7. Nuchal scale
8. Costal scales

Figure 8.40 A ventral view of a turtle.

1. Gular scales
2. Humeral scales
3. Pectoral scales
4. Abdominal scales
5. Femoral scales
6. Anal scales
7. Tail

Figure 8.41 The skull of a turtle.

1. Parietal bone
2. Supraoccipital bone
3. Postorbital bone
4. Jugal bone
5. Quadratojugal bone
6. Exoccipital bone
7. Quadrate bone
8. Supraangular bone
9. Articular bone
10. Angular bone
11. Frontal bone
12. Prefrontal bone
13. Palatine bone
14. Premaxilla
15. Maxilla
16. Beak
17. Dentary

Figure 8.42 The skeleton of a turtle. (The plastron is removed.)

1. Manus (carpal bones, metacarpal bones, phalanges)
2. Radius
3. Ulna
4. Humerus
5. Procoracoid
6. Scapula
7. Vertebra
8. Rib
9. Tibia
10. Fibula
11. Pes (tarsal bones, metatarsal bones, phalanges)
12. Dentary
13. Articular
14. Cervical vertebrae
15. Acromion process
16. Dermal plate of carapace
17. Pubis
18. Femur
19. Ischium
20. Caudal vertebrae

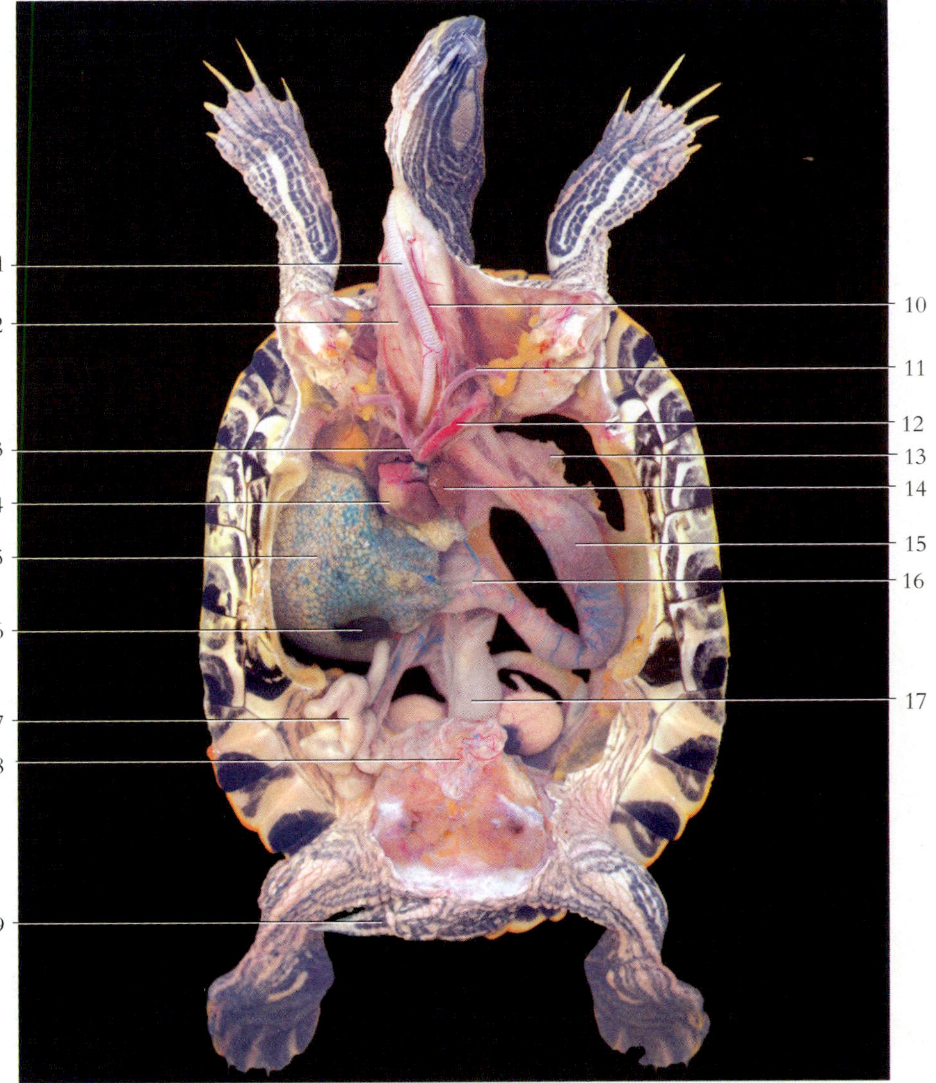

Figure 8.43 A ventral view of the internal anatomy of the turtle.

1. Trachea
2. Esophagus
3. Brachiocephalic trunk
4. Ventricle of heart
5. Liver
6. Gallbladder
7. Small intestine
8. Urinary bladder
9. Anus
10. Common carotid artery
11. Subclavian artery
12. Aortic arch
13. Lung
14. Auricle of heart
15. Stomach
16. Pancreas
17. Colon

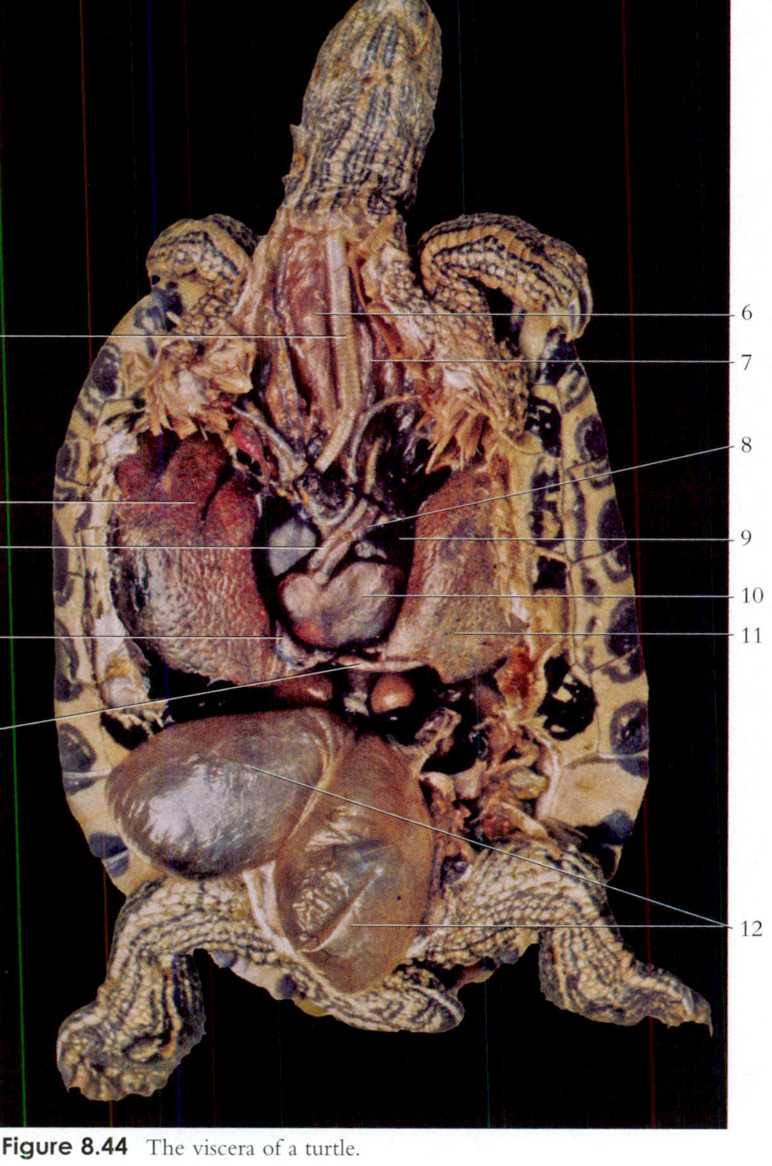

Figure 8.44 The viscera of a turtle.

1. Trachea
2. Liver
3. Pulmonary artery
4. Stomach
5. Pancreas
6. Esophagus
7. Common carotid artery
8. Left aorta
9. Atrium of heart
10. Ventricle of heart
11. Liver
12. Urinary bladder (full)

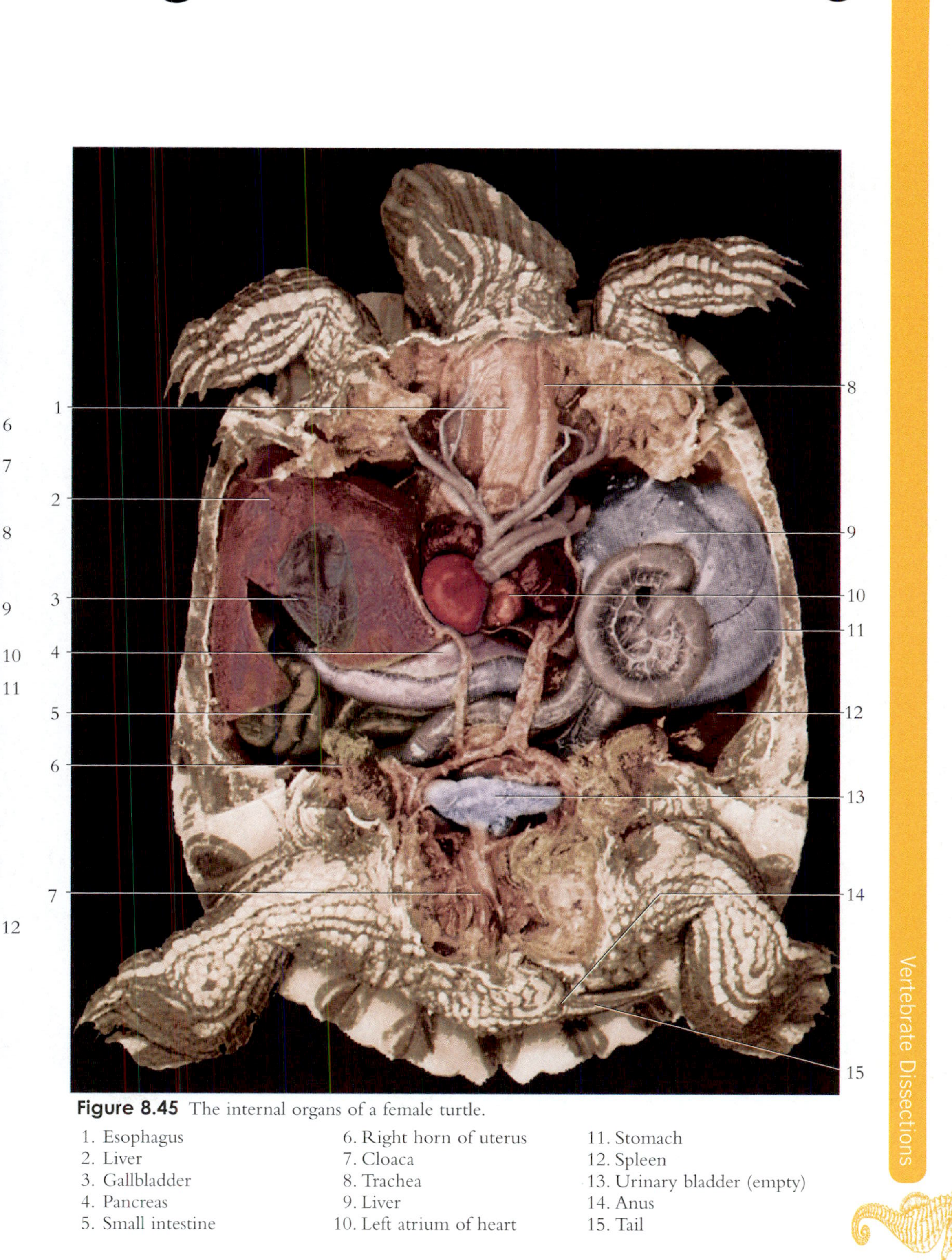

Figure 8.45 The internal organs of a female turtle.

1. Esophagus
2. Liver
3. Gallbladder
4. Pancreas
5. Small intestine
6. Right horn of uterus
7. Cloaca
8. Trachea
9. Liver
10. Left atrium of heart
11. Stomach
12. Spleen
13. Urinary bladder (empty)
14. Anus
15. Tail

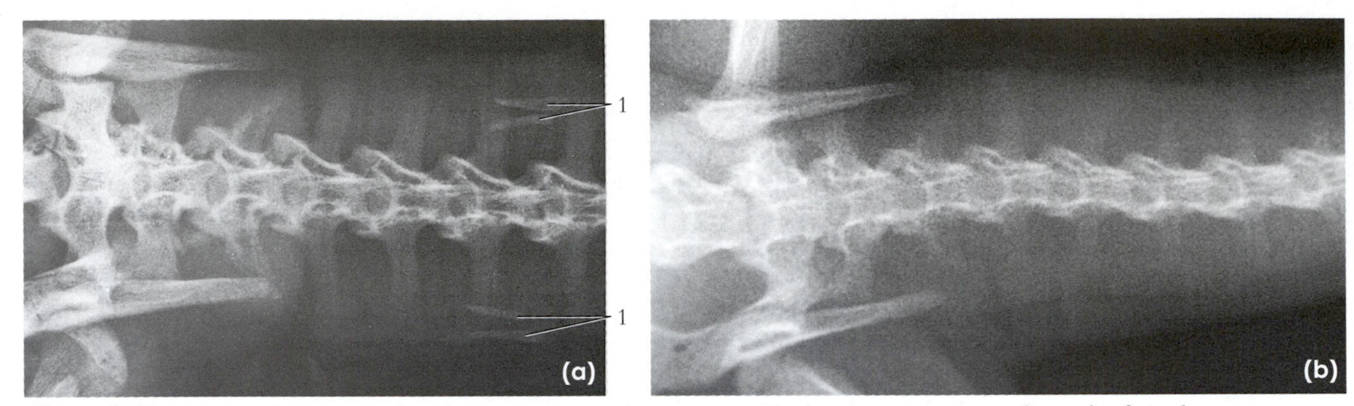

Figure 8.46 Male lizards and snakes have hemipenes as copulatory organs. The hemipene seen in a radiograph of a male (a) crocodile monitor, *Varanus salvadorii*. As seen in a radiograph, a female (b) lacks a hemipenis. The female cloaca is the receptacle of the everted male hemipenis during copulation.
1. Sheaths of hemipenes

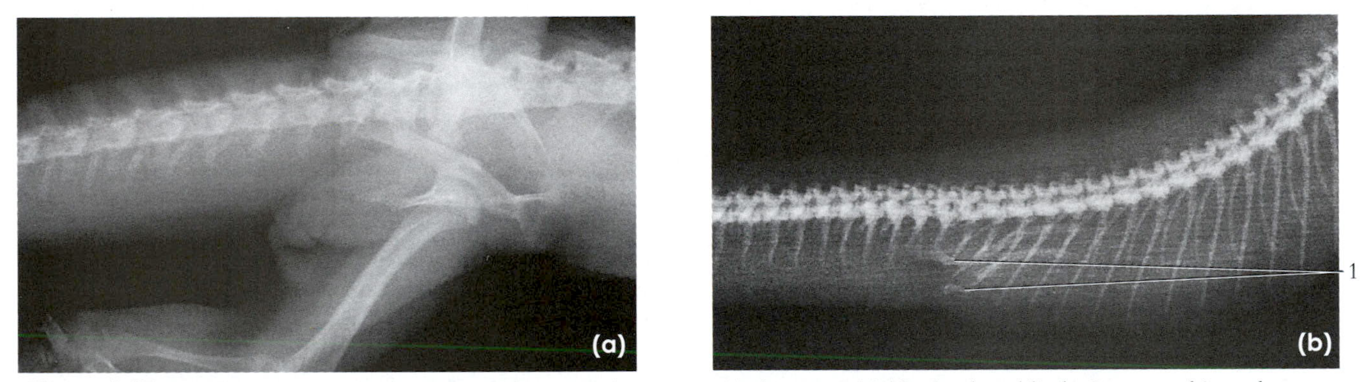

Figure 8.47 A radiograph of the pelvic region of a savannah monitor (a) showing a highly developed limb. Compare this to the radiograph of the pelvic region of a boa (b) showing the vestigeal pelvic girdle.
1. Vestigeal pelvic girdle

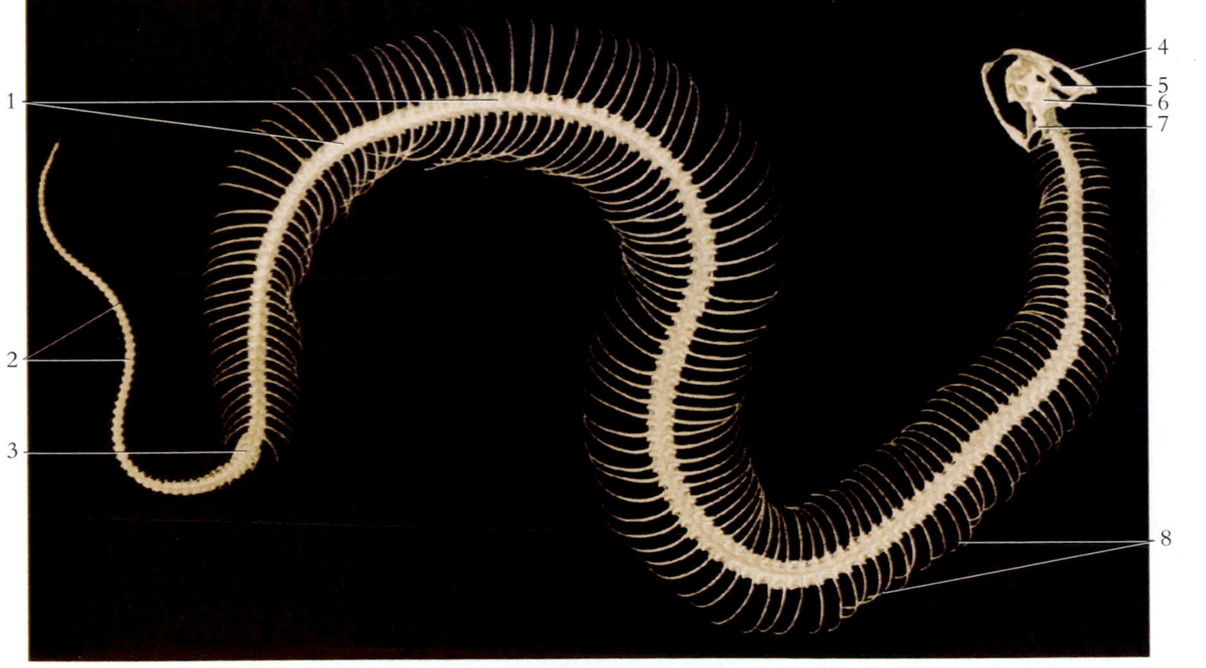

Figure 8.48 The skeleton of a snake (python).
1. Trunk vertebrae
2. Caudal vertebrae
3. Vestigial pelvic girdle
4. Dentary
5. Supratemporal bone
6. Parietal
7. Quadrate bone
8. Ribs

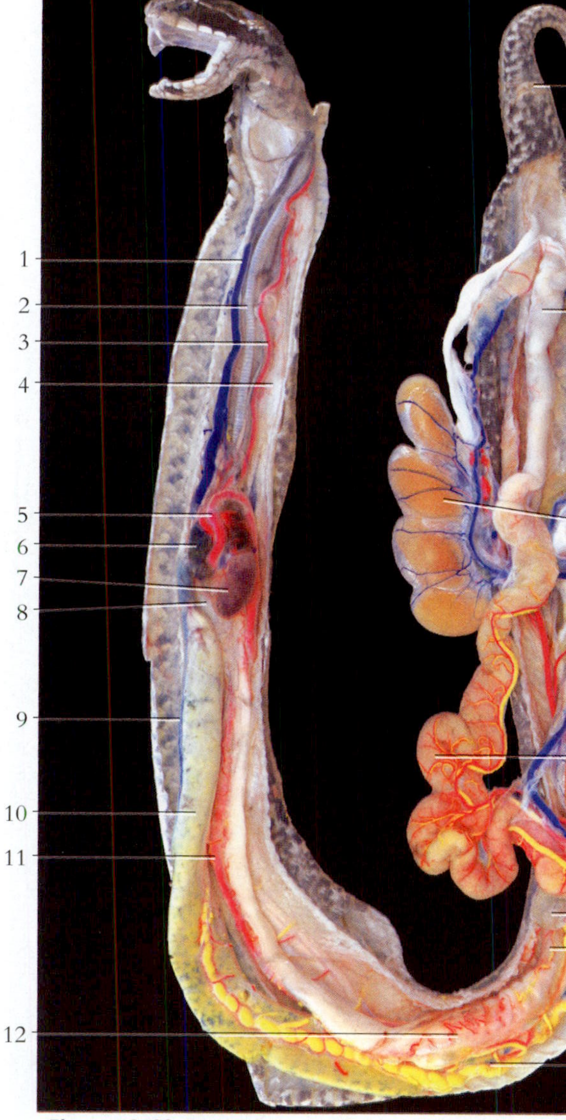

Figure 8.49 A ventral view of the internal anatomy of a female water moccasin, *Agkistrodon piscivorus*.

1. Jugular vein
2. Trachea
3. Common carotid artery
4. Esophagus
5. Aortic arch
6. Auricle of heart
7. Ventricle of heart
8. Lung
9. Hepatic portal vein
10. Liver
11. Dorsal aorta
12. Stomach
13. Anus
14. Colon
15. Kidney
16. Oviduct
17. Eggs
18. Small intestine
19. Pancreas
20. Duodenum
21. Abdominal vein

Figure 8.50 A ventral view of the internal anatomy of a male snake. Note that the testes and kidneys are staggered.

1. Hemipenes
2. Anus
3. Ductus deferens
4. Right testis
5. Intestine
6. Left testis
7. Ductus deferens
8. Right kidney
9. Ureter
10. Left kidney

Class Aves

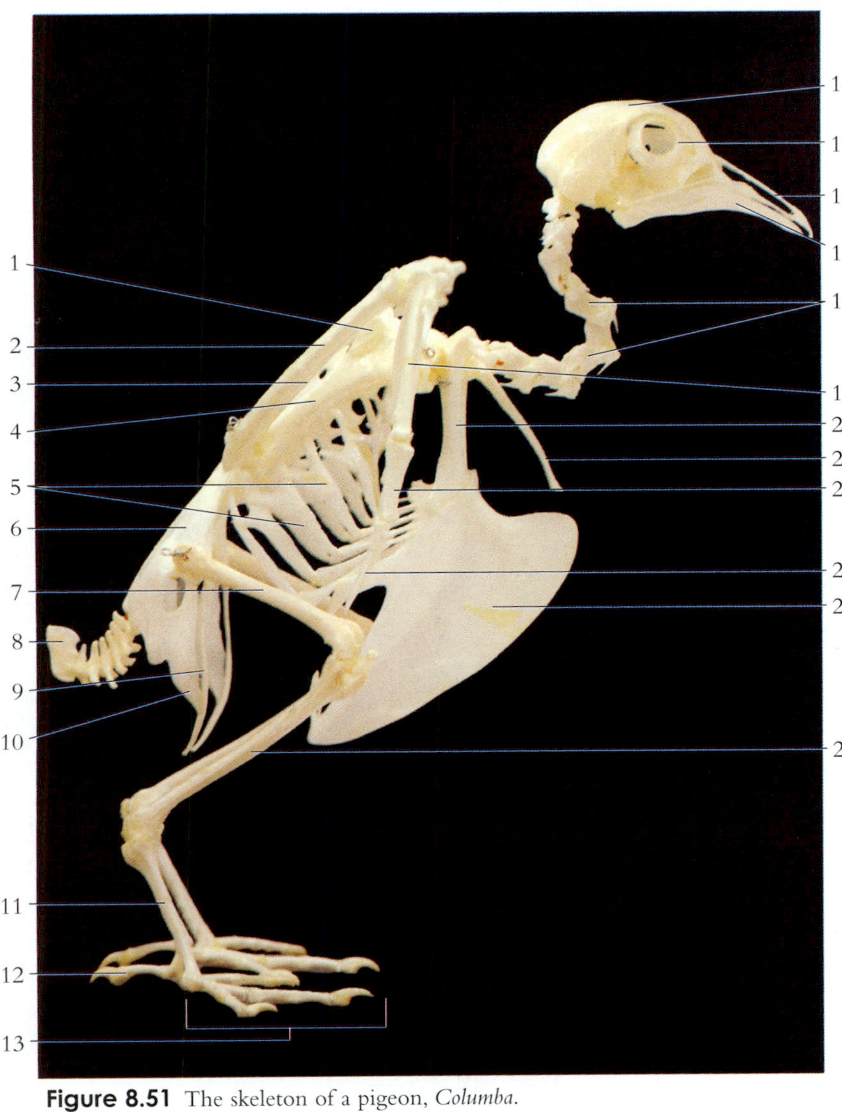

Figure 8.51 The skeleton of a pigeon, *Columba*.

1. Scapula
2. Ulna
3. Radius (behind ulna)
4. Humerus
5. Ribs
6. Ilium
7. Femur
8. Pygostyle
9. Pubis
10. Ischium
11. Tarsometatarsal bone
12. Digit 1
13. Phalanges
14. Cranium
15. Sclerotic bone
16. Premaxilla
17. Dentary
18. Cervical vertebrae
19. Carpometacarpal bones
20. Coracoid bone
21. Furcula
22. Phalanges
23. Phalanx of 3rd digit
24. Keel of sternum
25. Tibiotarsal bone

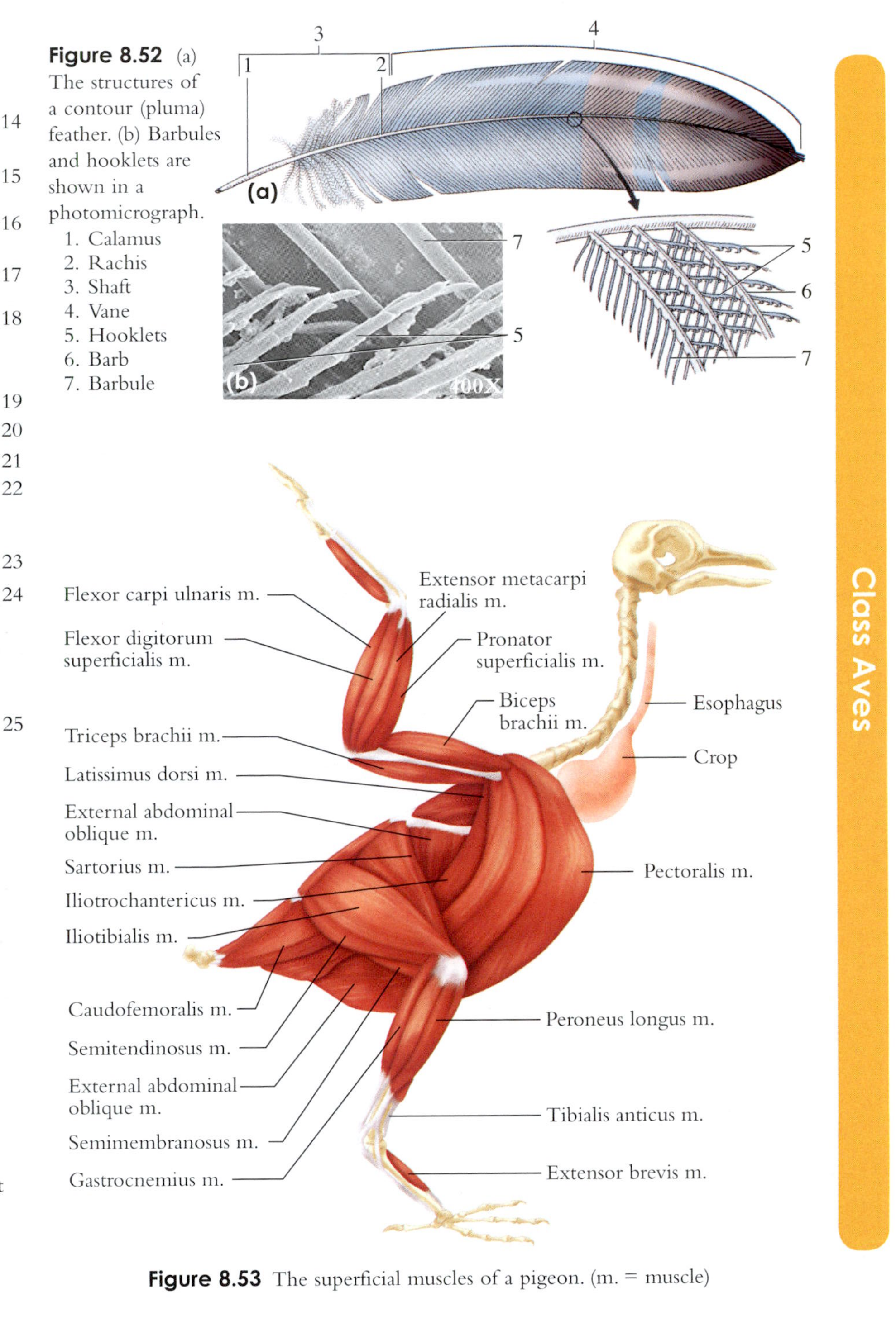

Figure 8.52 (a) The structures of a contour (pluma) feather. (b) Barbules and hooklets are shown in a photomicrograph.

1. Calamus
2. Rachis
3. Shaft
4. Vane
5. Hooklets
6. Barb
7. Barbule

Figure 8.53 The superficial muscles of a pigeon. (m. = muscle)

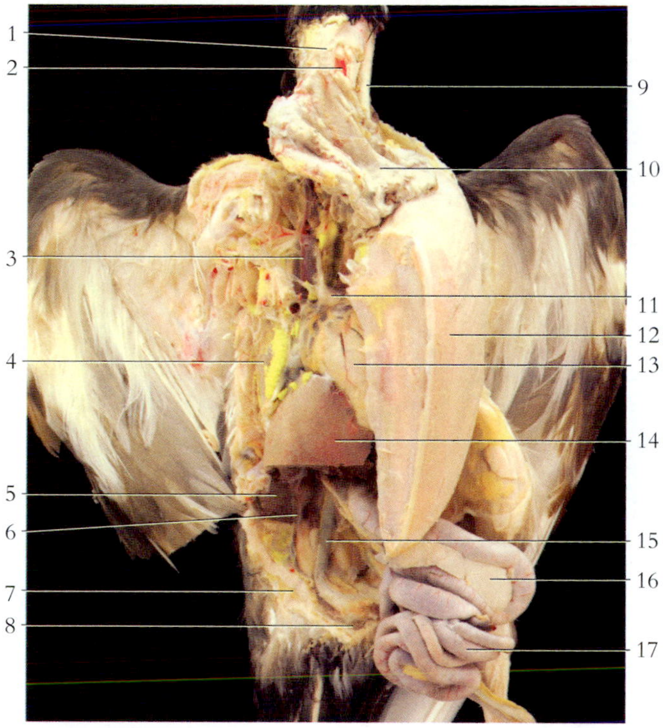

Figure 8.54 A ventral view of the internal anatomy of a pigeon.

1. Esophagus
2. Carotid artery
3. Vena cava
4. Lung
5. Kidney
6. Oviduct
7. Oil gland
8. Cloaca
9. Trachea
10. Crop
11. Aortic arch
12. Pectoralis muscle
13. Heart
14. Liver (cut)
15. Rectum
16. Pancreas
17. Ileum

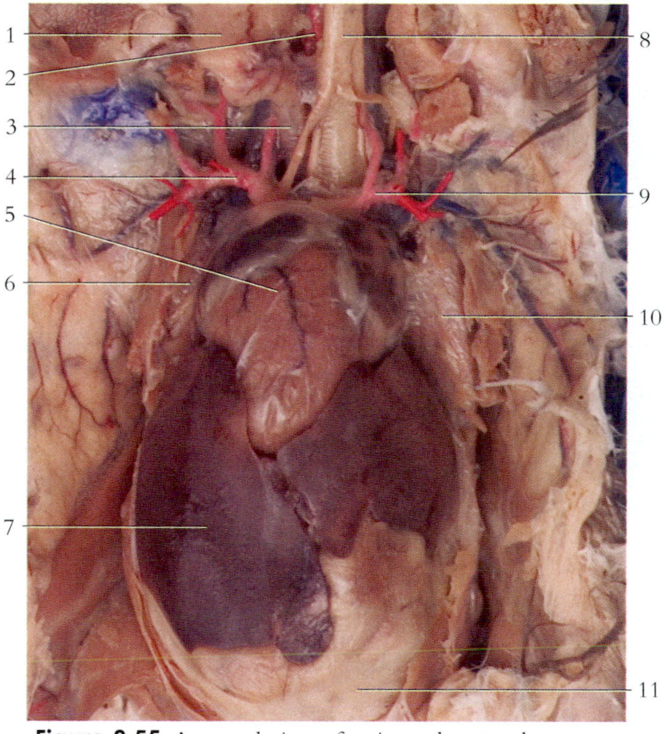

Figure 8.55 A ventral view of a pigeon heart and surrounding organs.

1. Crop
2. Common carotid artery
3. Esophagus
4. Right subclavian artery
5. Heart (within pericardium)
6. Right lung
7. Liver
8. Trachea
9. Left subclavian artery
10. Left lung
11. Greater omentum

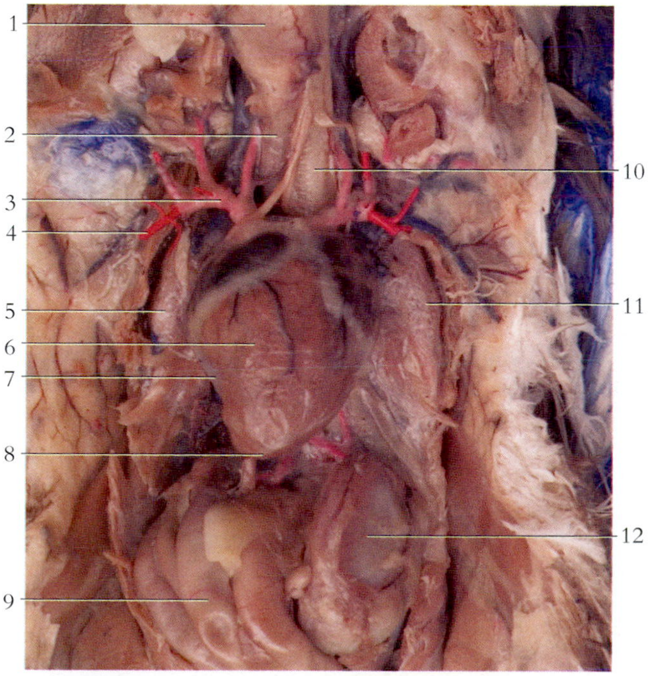

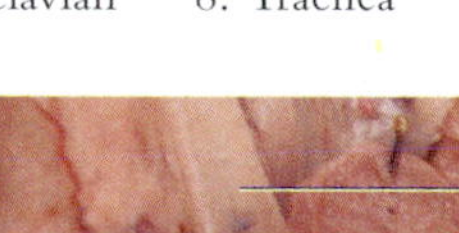

Figure 8.56 A ventral view of the viscera of a pigeon, *Columba*, with the liver removed.

1. Crop
2. Esophagus
3. Right subclavian artery
4. Axillary artery
5. Right lung
6. Heart
7. Pericardium
8. Apex of heart
9. Small intestine
10. Trachea
11. Left lung
12. Gizzard

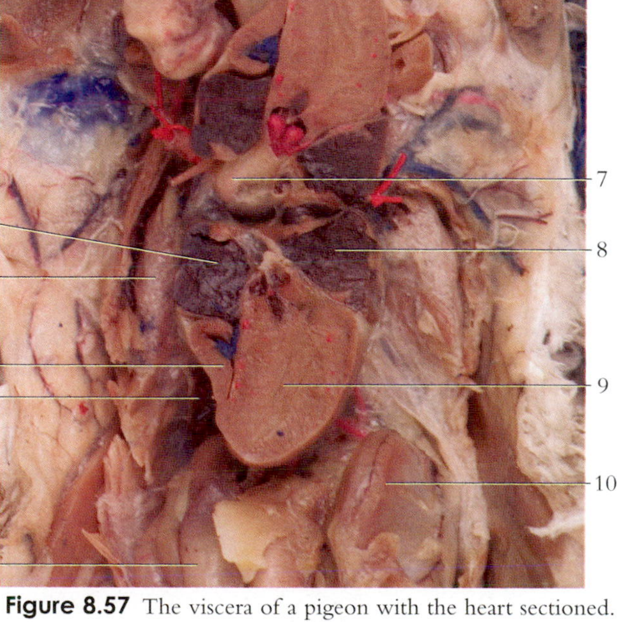

Figure 8.57 The viscera of a pigeon with the heart sectioned.

1. Right atrium
2. Right lung
3. Right ventricle
4. Liver (cut)
5. Small intestine
6. Trachea
7. Aortic arch
8. Left atrium
9. Left ventricle
10. Gizzard

Class Mammalia

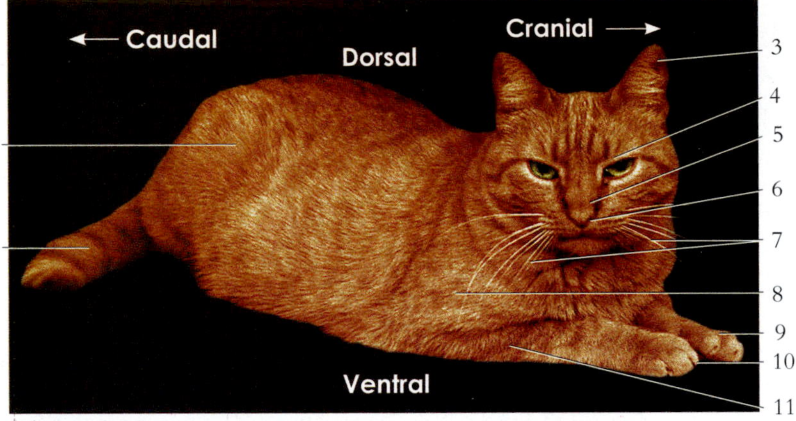

Figure 8.58 Directional terminology and superficial structures in a cat (quadrupedal vertebrate).

1. Thigh
2. Tail
3. Auricle (pinna)
4. Superior palpebra (superior eyelid)
5. Bridge of nose
6. Naris (nostril)
7. Vibrissae
8. Brachium
9. Manus (front foot)
10. Claw
11. Antebrachium

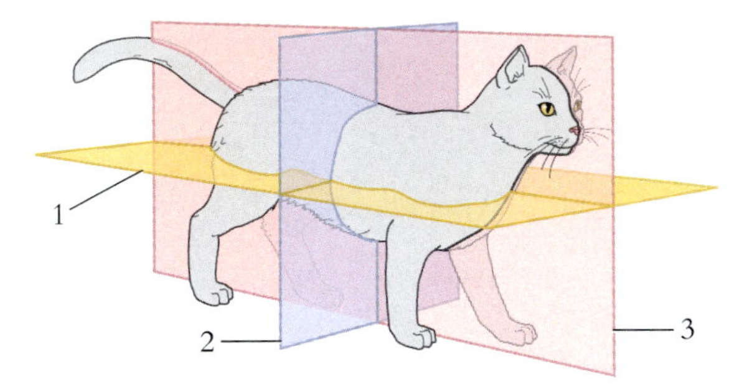

Figure 8.59 The planes of reference in a cat.

1. Coronal plane (frontal plane)
2. Transverse plane (cross-sectional plane)
3. Midsagittal plane (median plane)

Rat Dissection

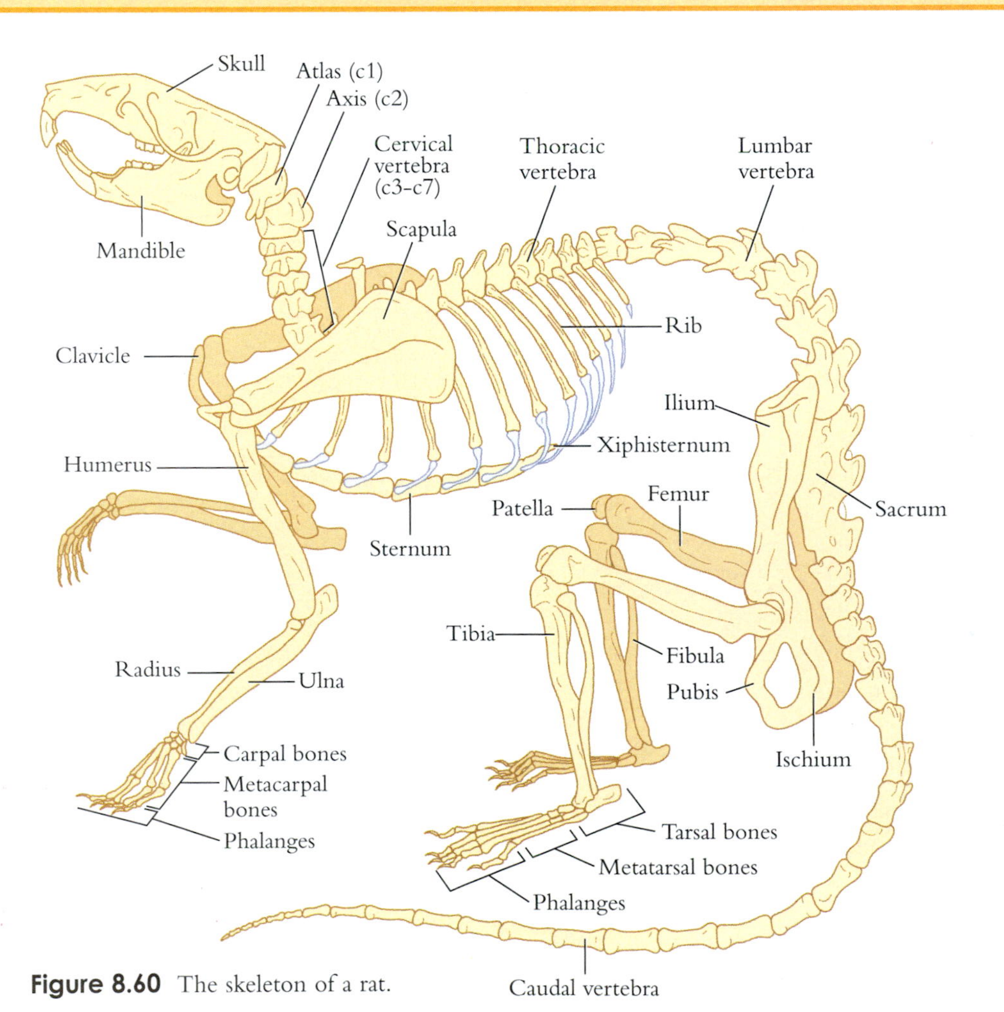

Figure 8.60 The skeleton of a rat.

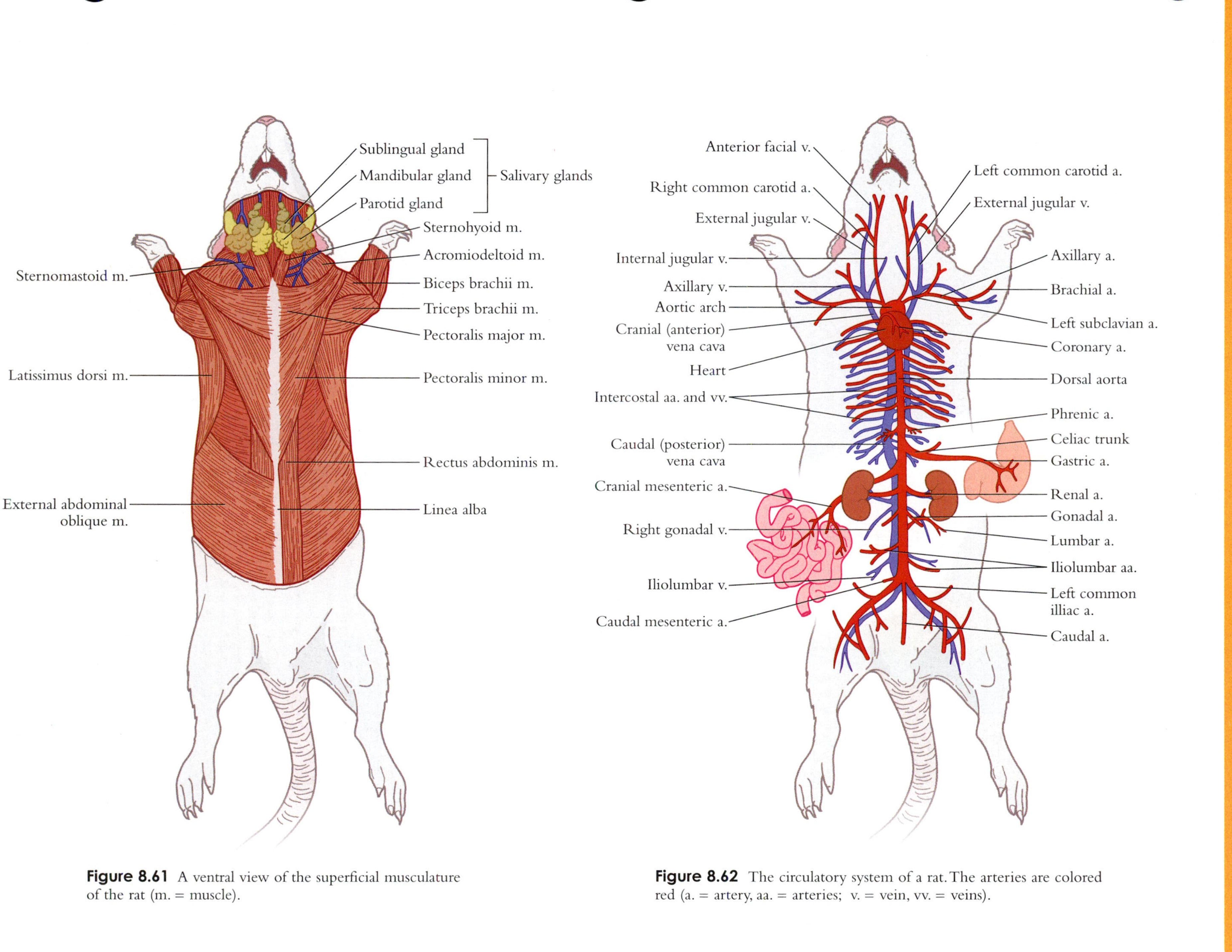

Figure 8.61 A ventral view of the superficial musculature of the rat (m. = muscle).

Figure 8.62 The circulatory system of a rat. The arteries are colored red (a. = artery, aa. = arteries; v. = vein, vv. = veins).

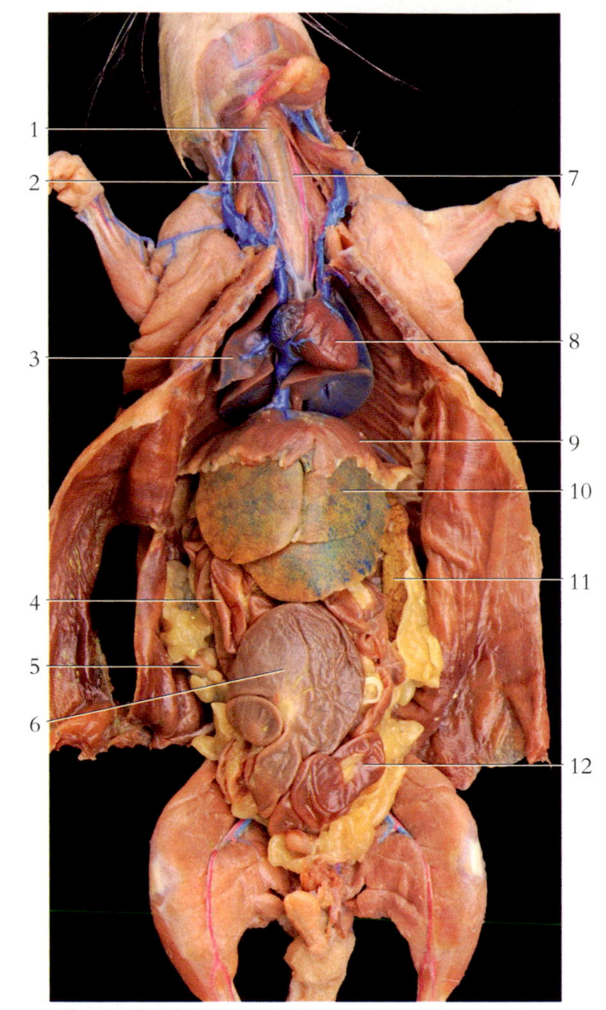

Figure 8.63 A ventral view of the rat viscera.

1. Larynx
2. Trachea
3. Right lung
4. Jejunum
5. Right ovary
6. Caecum
7. Esophagus
8. Heart
9. Diaphragm
10. Liver
11. Spleen
12. Ileum

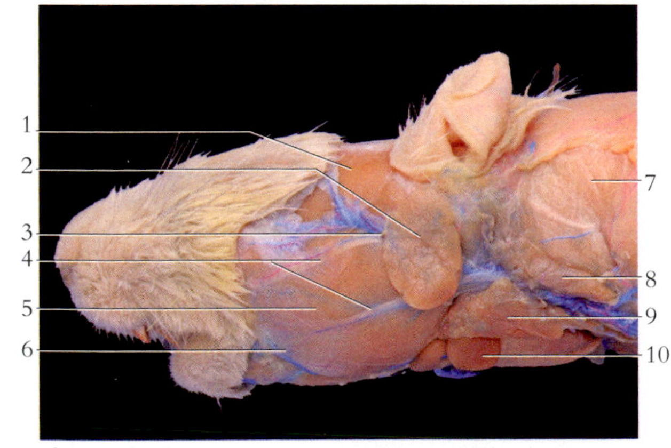

Figure 8.65 The head and neck region of the rat.

1. Temporalis m.
2. Extraorbital lacrimal gland
3. Extraorbital lacrimal duct
4. Facial nerve
5. Masseter m.
6. Parotid duct
7. Cervical trapezius m.
8. Parotid gland
9. Mandibular gland
10. Mandibular lymph node

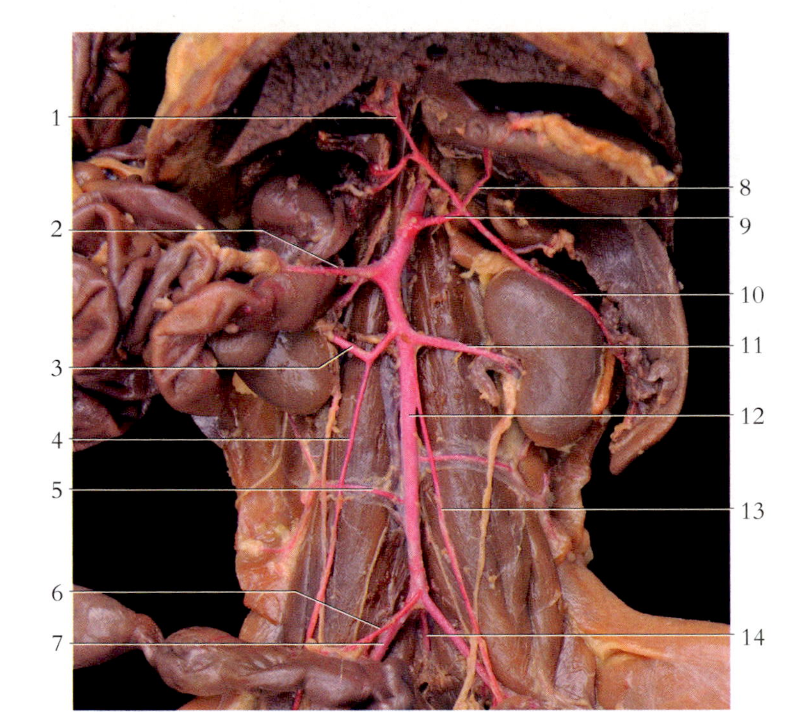

Figure 8.64 The abdominal arteries of the rat.

1. Hepatic artery
2. Right renal artery
3. Cranial mesenteric artery
4. Right testicular artery
5. Right iliolumbar artery
6. Caudal mesenteric artery (cut)
7. Right common iliac artery
8. Gastric artery
9. Celiac trunk
10. Splenic artery
11. Left renal artery
12. Abdominal aorta
13. Left testicular artery
14. Middle sacral artery

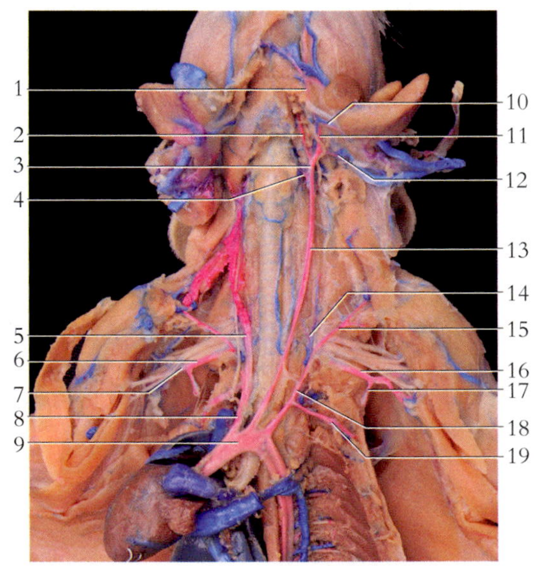

Figure 8.66 The arteries of the thoracic and neck regions of the rat.

1. Facial artery
2. Lingual artery
3. External carotid artery
4. Cranial thyroid artery
5. Right common carotid artery
6. Right axillary artery
7. Right brachial artery
8. Brachiocephalic artery
9. Aortic arch
10. External maxillary artery
11. Internal carotid artery
12. Occipital artery
13. Left common carotid artery
14. Vertebral artery
15. Cervical trunk
16. Lateral thoracic artery
17. Left axillary artery
18. Left subclavian artery
19. Internal thoracic artery

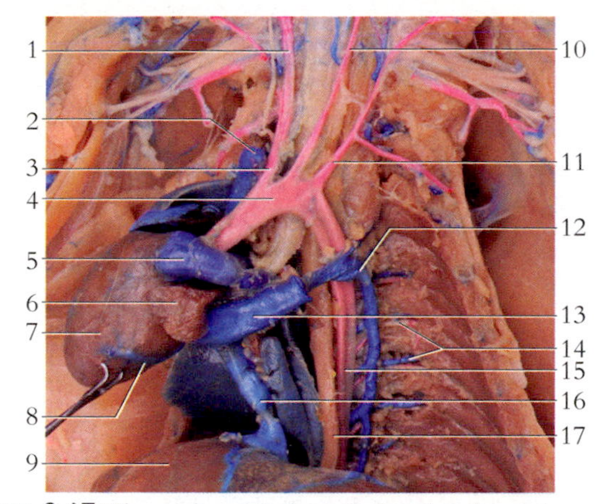

Figure 8.67 The rat heart (reflected) showing the major veins and arteries.

1. Right common carotid artery
2. Right cranial vena cava (cut)
3. Brachiocephalic trunk
4. Aortic arch
5. Pulmonary trunk
6. Left auricle
7. Left ventricle
8. Coronary vein
9. Diaphragm
10. Left common carotid artery
11. Left subclavian artery
12. Azygos vein
13. Coronary sinus (cut)
14. Intercostal artery and vein
15. Aorta
16. Caudal vena cava
17. Esophagus

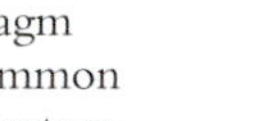

Figure 8.68 The veins of the thoracic and neck regions of the rat.

1. Cephalic vein
2. Right cranial vena cava
3. Caudal vena cava
4. Linguofacial vein
5. Maxillary vein
6. External jugular vein
7. Internal jugular vein
8. Lateral thoracic vein
9. Left cranial vena cava

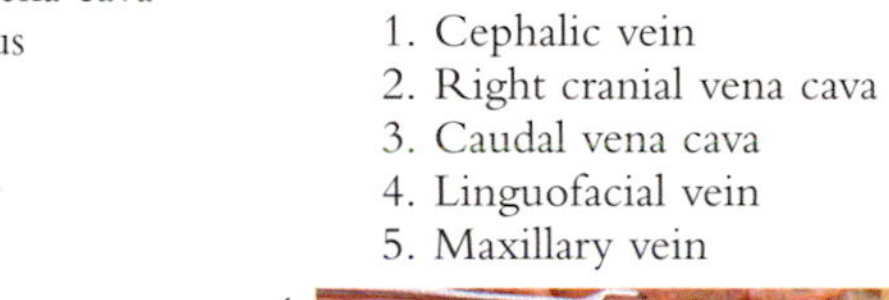

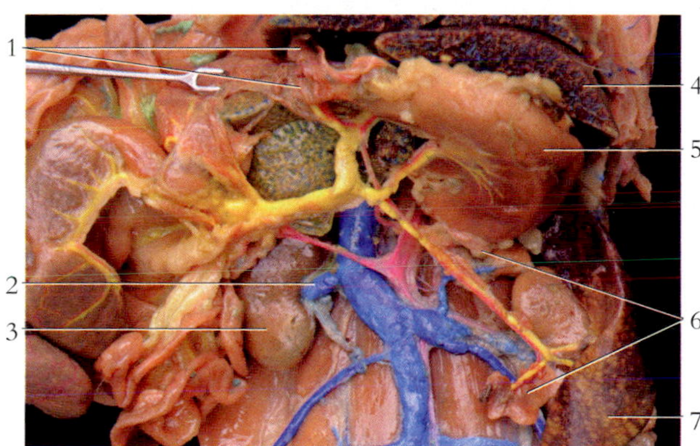

Figure 8.69 The abdominal viscera and vessels of the rat.

1. Biliary and duodenal parts of pancreas
2. Right renal vein
3. Right kidney
4. Liver (cut)
5. Stomach
6. Gastrosplenic part of pancreas
7. Spleen

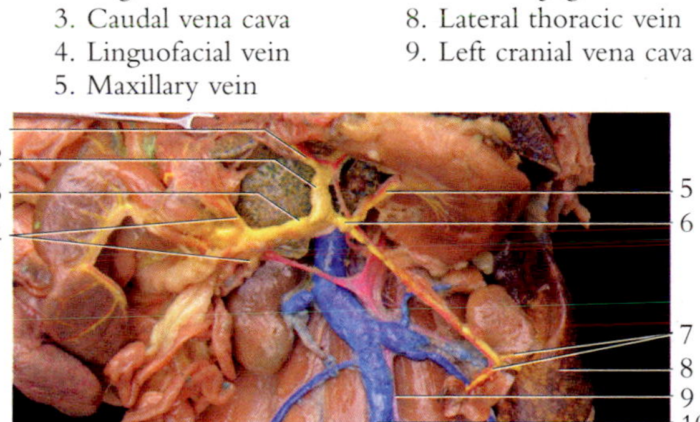

Figure 8.70 The branches of the hepatic portal system.

1. Cranial pancreaticoduodenal vein
2. Hepatic portal vein
3. Cranial mesenteric vein
4. Intestinal branches
5. Gastric vein
6. Gastrosplenic vein
7. Splenic branches
8. Spleen
9. Abdominal aorta
10. Caudal vena cava

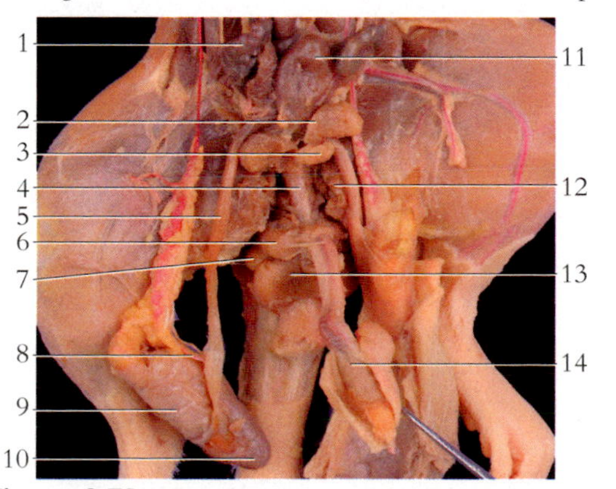

Figure 8.71 The urogenital system of the male rat.

1. Vesicular gland
2. Prostate (ventral part)
3. Prostate (dorsolateral part)
4. Urethra in the pelvic canal
5. Vas (ductus) deferens
6. Crus of penis
7. Bulbourethral gland
8. Head of epididymis
9. Testis
10. Tail of epididymis
11. Urinary bladder
12. Symphysis pubis (cut exposing pelvic canal)
13. Bulbocavernosus muscle
14. Penis

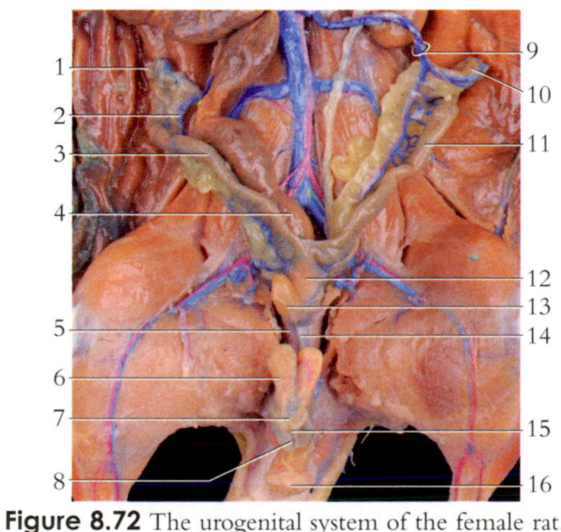

Figure 8.72 The urogenital system of the female rat.

1. Ovary
2. Uterine artery and vein
3. Uterine horn
4. Colon
5. Vagina
6. Vestibular gland
7. Clitoris
8. Vaginal opening
9. Uterine artery and vein
10. Ovary
11. Uterine horn
12. Uterine body
13. Urinary bladder
14. Urethra
15. Urethral opening
16. Anus

Fetal Pig Dissection

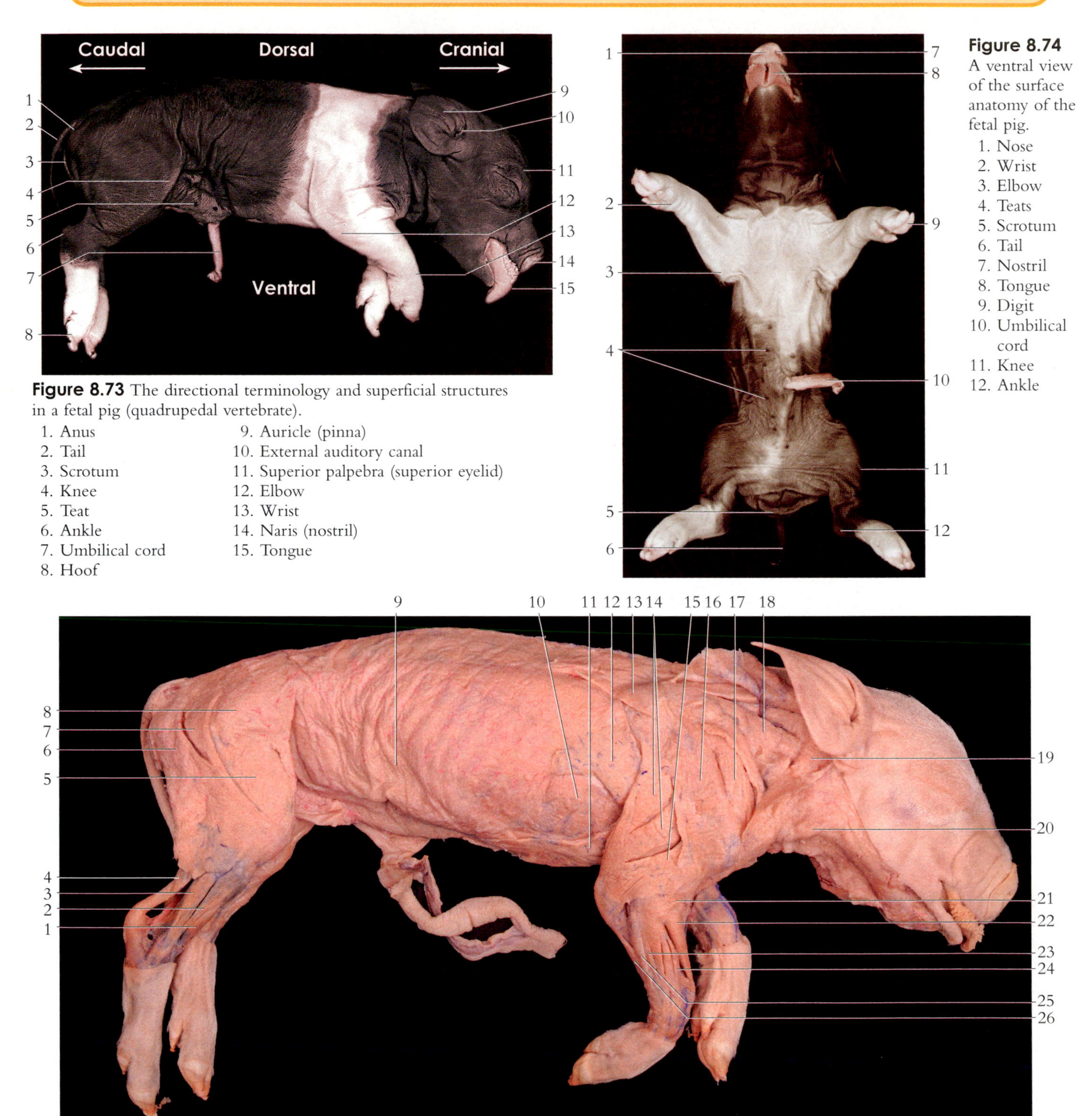

Figure 8.73 The directional terminology and superficial structures in a fetal pig (quadrupedal vertebrate).

1. Anus
2. Tail
3. Scrotum
4. Knee
5. Teat
6. Ankle
7. Umbilical cord
8. Hoof
9. Auricle (pinna)
10. External auditory canal
11. Superior palpebra (superior eyelid)
12. Elbow
13. Wrist
14. Naris (nostril)
15. Tongue

Figure 8.74 A ventral view of the surface anatomy of the fetal pig.

1. Nose
2. Wrist
3. Elbow
4. Teats
5. Scrotum
6. Tail
7. Nostril
8. Tongue
9. Digit
10. Umbilical cord
11. Knee
12. Ankle

Figure 8.75 A lateral view of superficial musculature of the fetal pig.

1. Tibialis anterior m.
2. Peroneus tarius m.
3. Peroneus longus m.
4. Gastrocnemius m.
5. Tensor fasciae latae m.
6. Biceps femoris m.
7. Gluteus superficialis m.
8. Gluteus medius m.
9. External abdominal oblique m.
10. Serratus ventralis m.
11. Pectoralis profundus m.
12. Latissimus dorsi m.
13. Trapezius m.
14. Triceps brachii m. (long head)
15. Triceps brachii m. (lateral head)
16. Deltoid m.
17. Supraspinatus m.
18. Omotransversarius m.
19. Cleidooccipitalis m.
20. Platysma m.
21. Brachialis m.
22. Extensor carpi radialis m.
23. Extensor digiti m.
24. Extensor digitorum communis m.
25. Ulnaris lateralis m.
26. Flexor digitorum profundus m.

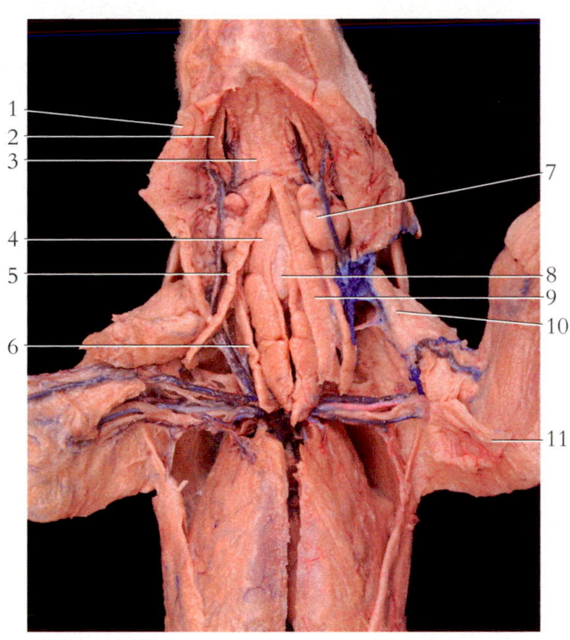

Figure 8.76 A ventral view of superficial muscles of the neck and upper torso.

1. Platysma m. (reflected)
2. Digastric m.
3. Mylohyoid m.
4. Sternohyoid m.
5. Omohyoid m.
6. Sternomastoid m.
7. Mandibular gland
8. Larynx
9. Sternothyroid m.
10. Brachiocephalic m.
11. Pectoralis superficialis m. (cut and reflected)

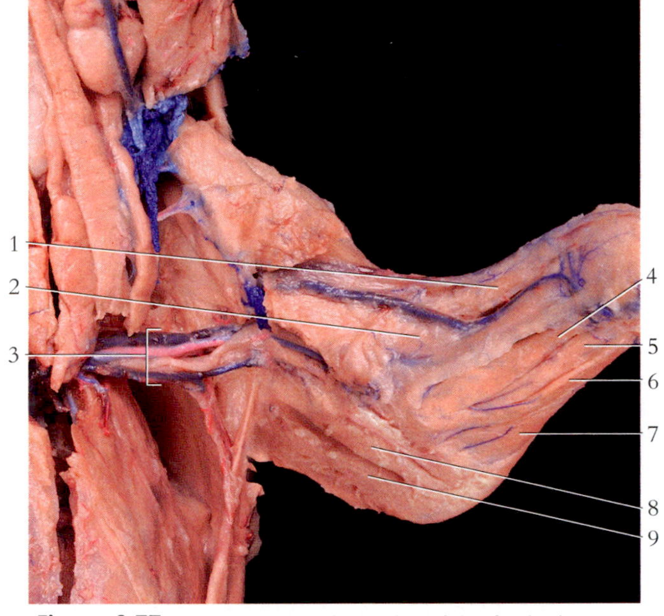

Figure 8.77 Superficial medial muscles of the forelimb.

1. Extensor carpi radialis m.
2. Biceps brachii m.
3. Axillary artery and vein, brachial plexus
4. Flexor carpi radialis m.
5. Flexor digitorum profundus m.
6. Flexor digitorum superficialis m.
7. Flexor carpi ulnaris m.
8. Triceps brachii m. (lateral head)
9. Triceps brachii m. (long head)

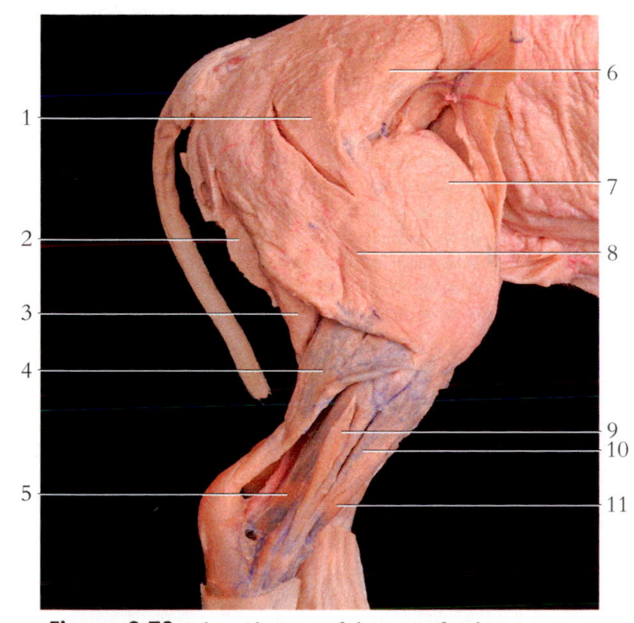

Figure 8.78 A lateral view of the superficial thigh and leg.

1. Gluteus superficialis m.
2. Semitendinosus m.
3. Semimembranosus m.
4. Gastrocnemius m.
5. Extensor digitorum quarti and quinti mm.
6. Gluteus medius m.
7. Tensor fasciae latae m.
8. Biceps femoris m.
9. Fibularis (peroneus) longus m.
10. Fibularis (peroneus) tertius m.
11. Tibialis anterior m.

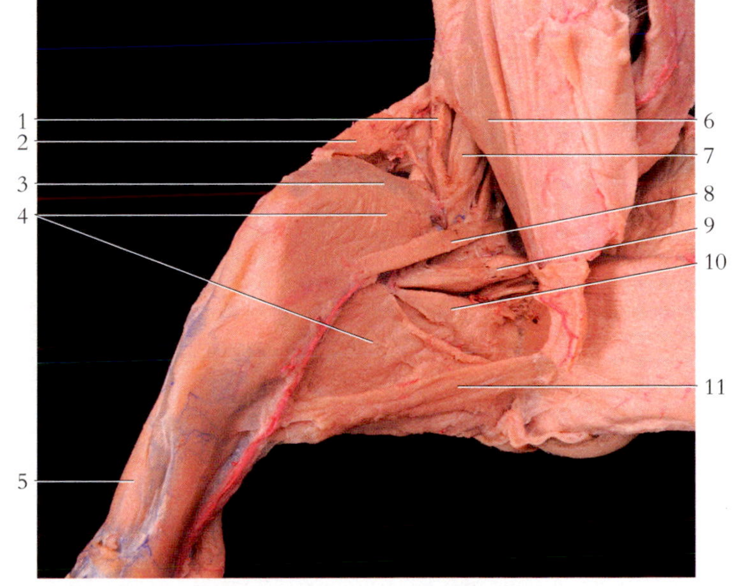

Figure 8.79 Medial muscles of the thigh and leg.

1. Iliacus m.
2. Tensor fasciae latae m.
3. Rectus femoris m.
4. Semimembranosus m.
5. Tibialis anterior m.
6. External abdominal oblique m.
7. Psoas major m.
8. Sartorius m.
9. Pectineus m.
10. Adductor m.
11. Semitendinosus m.

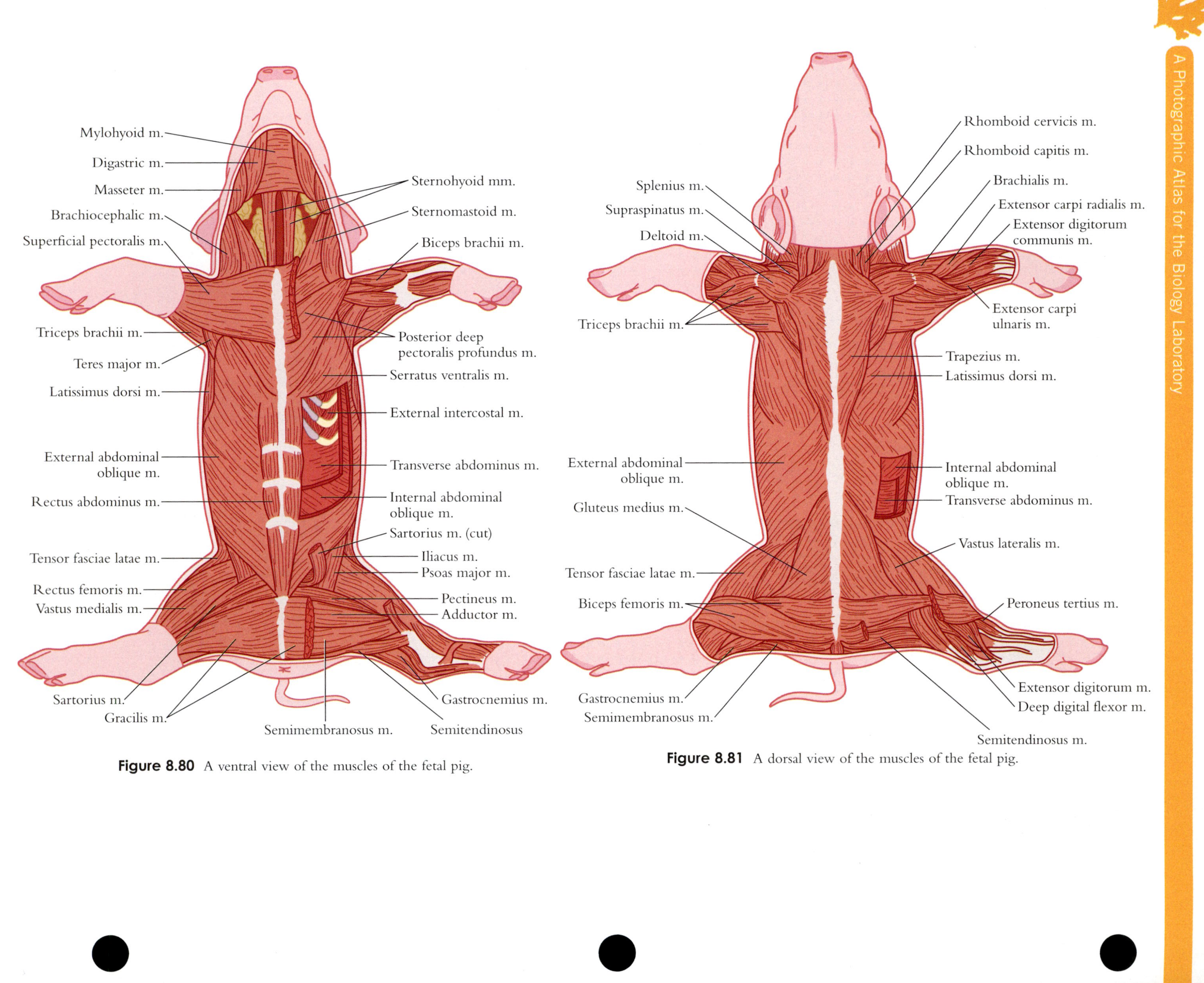
Mylohyoid m.
Digastric m.
Masseter m.
Brachiocephalic m.
Superficial pectoralis m.
Sternohyoid mm.
Sternomastoid m.
Biceps brachii m.
Triceps brachii m.
Teres major m.
Latissimus dorsi m.
Posterior deep pectoralis profundus m.
Serratus ventralis m.
External intercostal m.
External abdominal oblique m.
Rectus abdominus m.
Transverse abdominus m.
Internal abdominal oblique m.
Sartorius m. (cut)
Tensor fasciae latae m.
Iliacus m.
Psoas major m.
Rectus femoris m.
Vastus medialis m.
Pectineus m.
Adductor m.
Sartorius m.
Gracilis m.
Gastrocnemius m.
Semimembranosus m.
Semitendinosus
Rhomboid cervicis m.
Rhomboid capitis m.
Splenius m.
Supraspinatus m.
Deltoid m.
Brachialis m.
Extensor carpi radialis m.
Extensor digitorum communis m.
Extensor carpi ulnaris m.
Triceps brachii m.
Trapezius m.
Latissimus dorsi m.
External abdominal oblique m.
Internal abdominal oblique m.
Transverse abdominus m.
Gluteus medius m.
Vastus lateralis m.
Tensor fasciae latae m.
Biceps femoris m.
Peroneus tertius m.
Gastrocnemius m.
Semimembranosus m.
Extensor digitorum m.
Deep digital flexor m.
Semitendinosus m.

Figure 8.80 A ventral view of the muscles of the fetal pig.

Figure 8.81 A dorsal view of the muscles of the fetal pig.

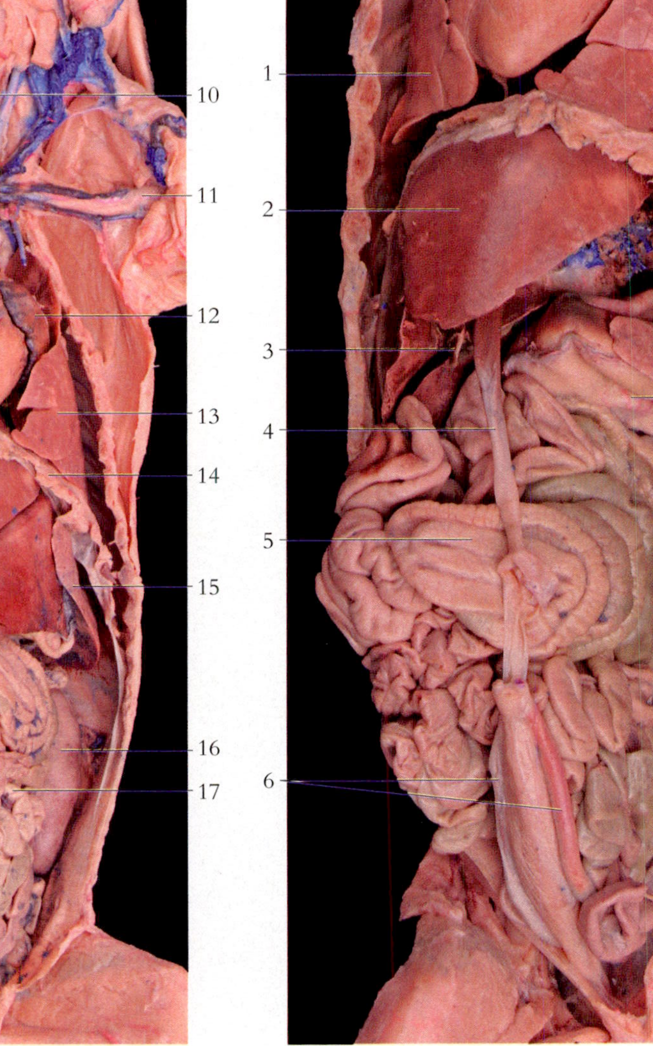

Figure 8.82 The arteries, veins, and viscera of the neck and thoracic region of the fetal pig.

1. Larynx
2. Internal jugular v.
3. External jugular v.
4. Thyroid gland
5. Cranial (superior) vena cava
6. Right auricle
7. Coronary vessels
8. Right lung
9. Liver
10. Trachea
11. Axillary a.
12. Left auricle
13. Left lung
14. Diaphragm
15. Spleen
16. Kidney
17. Small intestine

Figure 8.83 The abdominal organs of the fetal pig.

1. Lung
2. Liver (cut)
3. Gallbladder
4. Umbilical vein
5. Small intestine
6. Umbilical arteries
7. Diaphragm
8. Spleen
9. Stomach
10. Pancreas
11. Kidney

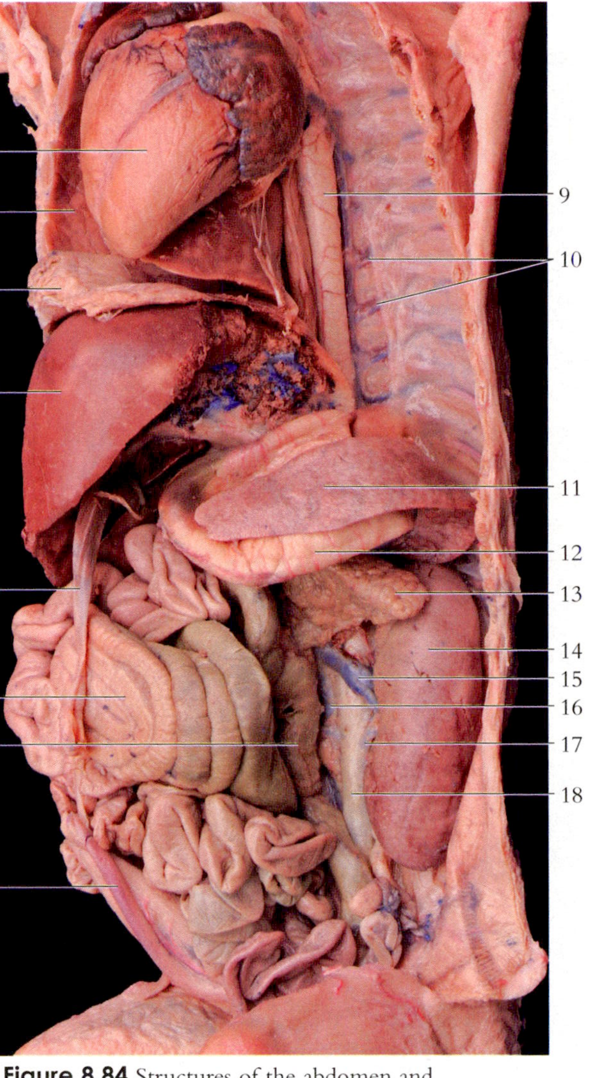

Figure 8.84 Structures of the abdomen and lower extremities.

1. Heart
2. Lung
3. Diaphragm
4. Liver
5. Umbilical vein
6. Small intestine
7. Colon
8. Umbilical artery
9. Thoracic aorta
10. Internal intercostal vessels
11. Spleen
12. Stomach
13. Pancreas
14. Kidney
15. Renal vein
16. Caudal (inferior) vena cava
17. Renal artery
18. Abdominal aorta

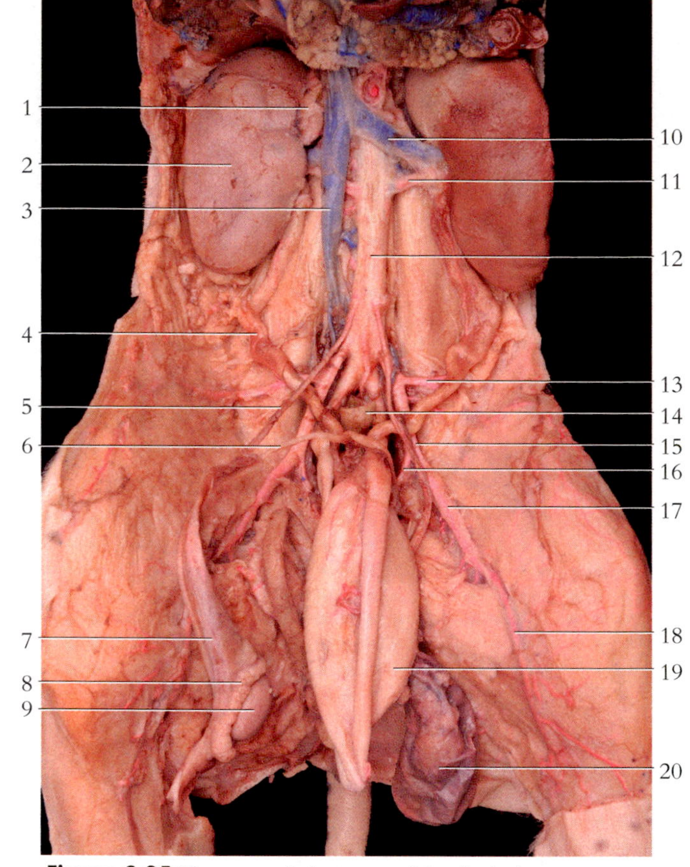

Figure 8.85 The urogenital system of the fetal pig.

1. Adrenal gland
2. Right kidney
3. Caudal (inferior) vena cava
4. Ureter
5. Genital vessels
6. Ductus (vas) deferens
7. Spermatic cord
8. Epididymis
9. Testis
10. Renal vein
11. Renal artery
12. Descending aorta
13. Iliolumbar artery
14. Rectum (cut)
15. Common iliac artery
16. Internal iliac artery
17. External iliac artery
18. Femoral artery
19. Urinary bladder
20. Testis

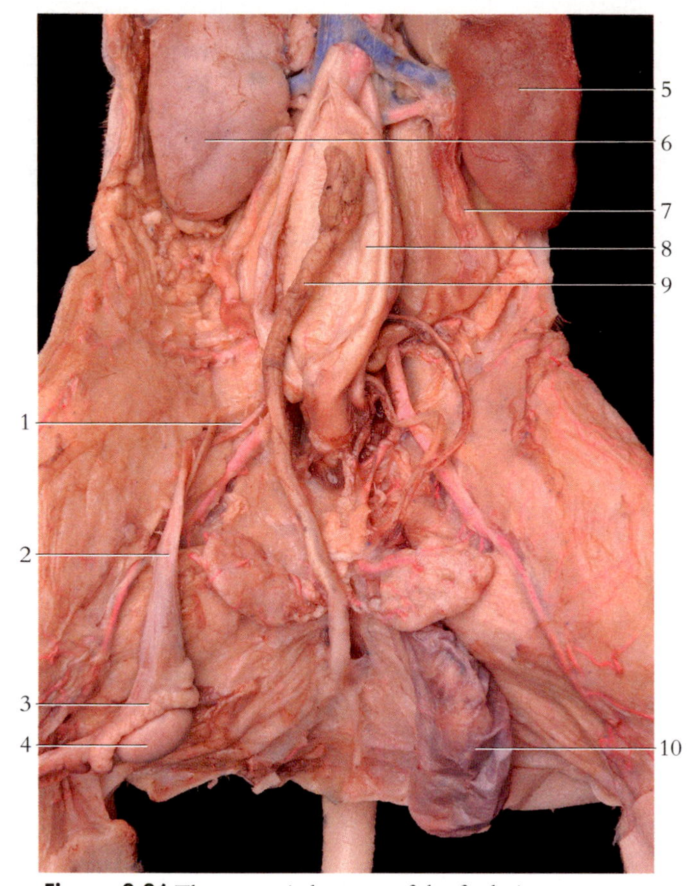

Figure 8.86 The urogenital system of the fetal pig.

1. Ductus (vas) deferens
2. Spermatic cord
3. Epididymis
4. Right testis
5. Left kidney
6. Right kidney
7. Ureter
8. Urinary bladder
9. Penis
10. Left testis

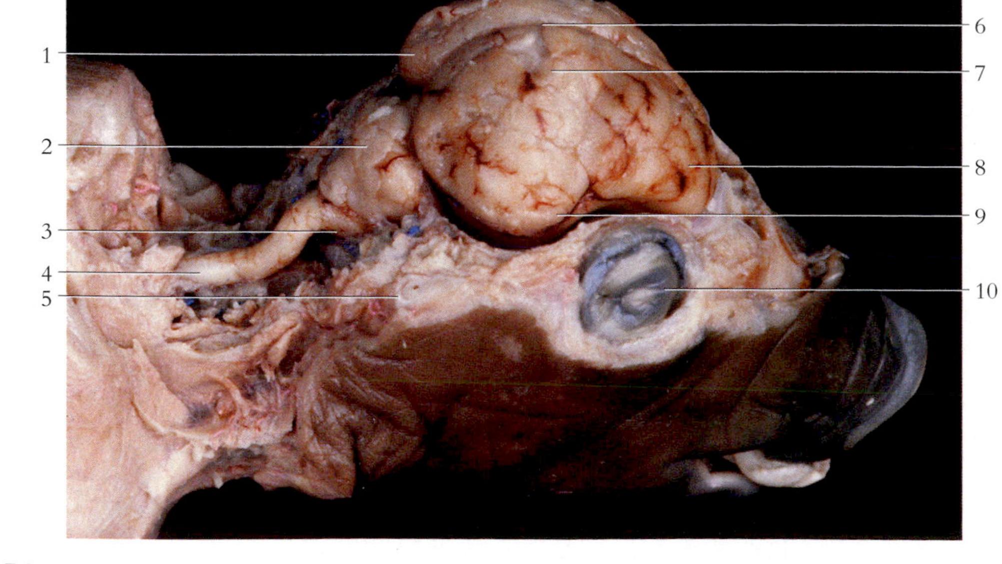

Figure 8.87 The general structures of the fetal pig brain. Because the cerebrum is less defined in pigs, the regions are not known as lobes as they are in humans.

1. Occipital region of cerebrum
2. Cerebellum
3. Medulla oblongata
4. Spinal cord
5. External acoustic meatus
6. Longitudinal fissure
7. Parietal region of cerebrum
8. Frontal region of cerebrum
9. Temporal region of cerebrum
10. Eye

Cat Dissection

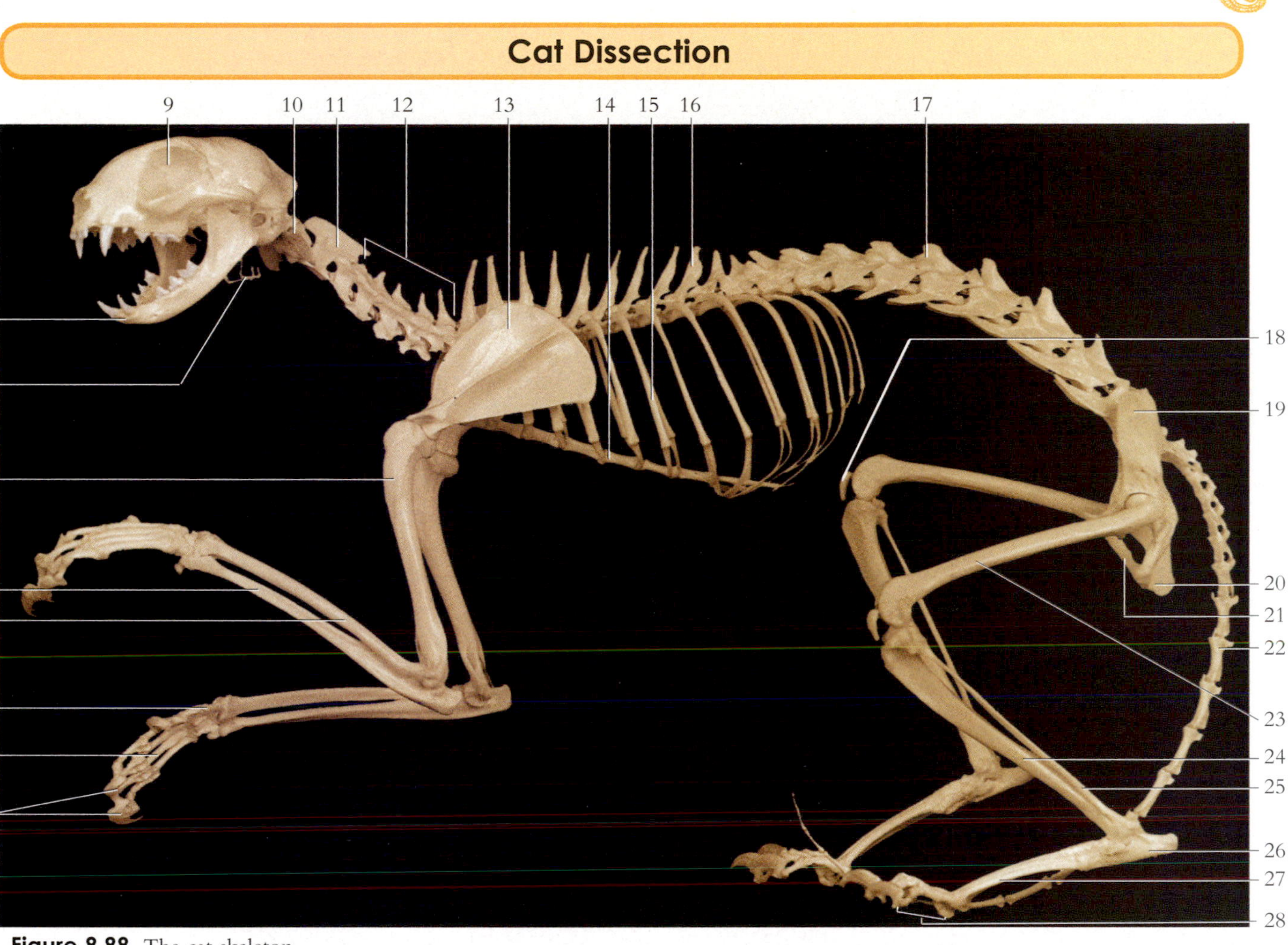

Figure 8.88 The cat skeleton.

1. Mandible
2. Hyoid bone
3. Humerus
4. Ulna
5. Radius
6. Carpal bones
7. Metacarpal bones
8. Phalanges
9. Skull
10. Atlas (c1)
11. Axis (c2)
12. Cervical vertebra (c3-c7)
13. Scapula
14. Sternum
15. Rib
16. Thoracic vertebra
17. Lumbar vertebra
18. Patella
19. Ilium
20. Ischium
21. Pubis
22. Caudal vertebra
23. Femur
24. Tibia
25. Fibula
26. Tarsal bones
27. Metatarsal bones
28. Phalanges

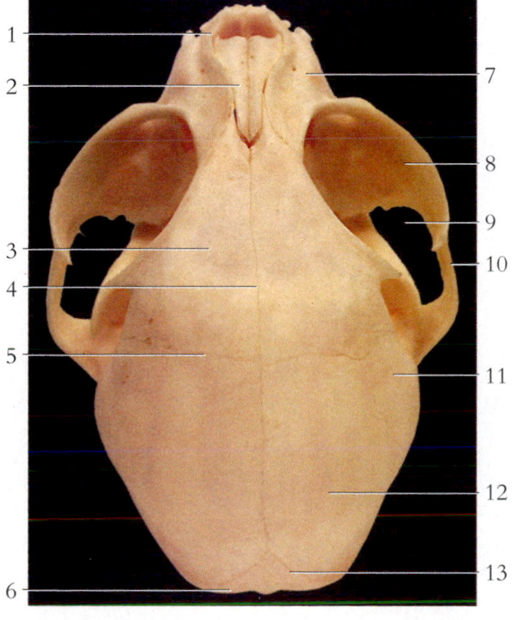

Figure 8.89 A dorsal view of a cat skull.

1. Premaxilla
2. Nasal bone
3. Frontal bone
4. Sagittal suture
5. Coronal suture
6. Nuchal crest
7. Maxilla
8. Zygomatic (malar) bone
9. Orbit
10. Zygomatic arch
11. Temporal bone
12. Parietal bone
13. Interparietal bone

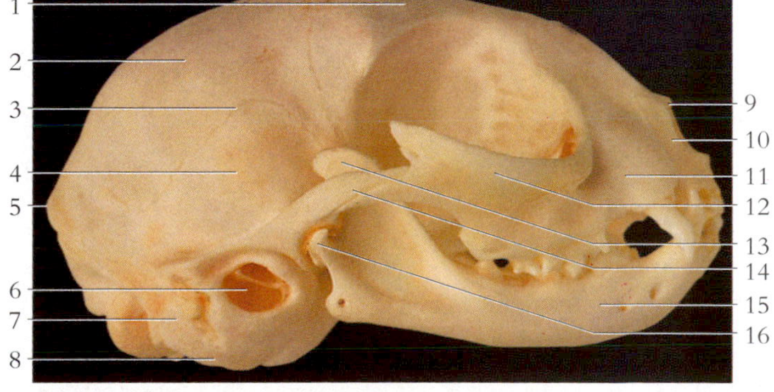

Figure 8.90 A lateral view of a cat skull.

1. Frontal bone
2. Parietal bone
3. Squamosal suture
4. Temporal bone
5. Nuchal crest
6. External acoustic meatus
7. Mastoid process
8. Tympanic bulla
9. Nasal bone
10. Premaxilla bone
11. Maxilla
12. Zygomatic (malar) bone
13. Coronoid process of mandible
14. Zygomatic arch
15. Mandible
16. Condylar process of mandible

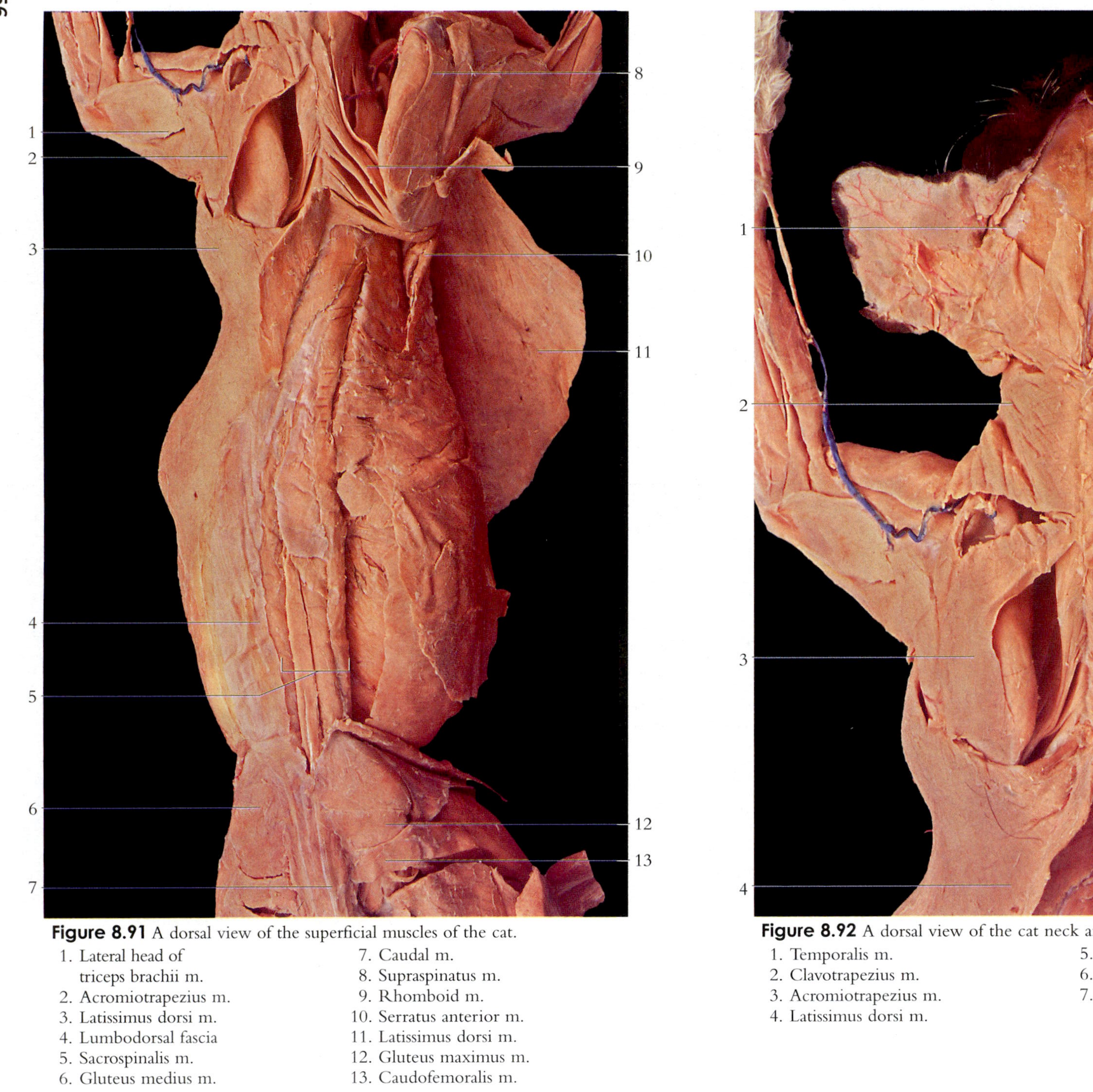

Figure 8.91 A dorsal view of the superficial muscles of the cat.

1. Lateral head of triceps brachii m.
2. Acromiotrapezius m.
3. Latissimus dorsi m.
4. Lumbodorsal fascia
5. Sacrospinalis m.
6. Gluteus medius m.
7. Caudal m.
8. Supraspinatus m.
9. Rhomboid m.
10. Serratus anterior m.
11. Latissimus dorsi m.
12. Gluteus maximus m.
13. Caudofemoralis m.

Figure 8.92 A dorsal view of the cat neck and thorax.

1. Temporalis m.
2. Clavotrapezius m.
3. Acromiotrapezius m.
4. Latissimus dorsi m.
5. Supraspinatus m.
6. Rhomboid m.
7. Serratus anterior m.

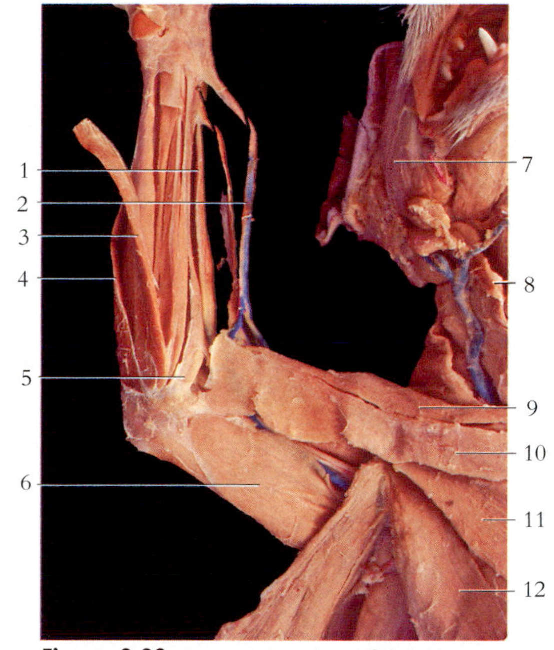

Figure 8.93 An anterior view of the cat brachium and antebrachium.

1. Extensor carpi radialis longus m.
2. Brachioradialis m.
3. Palmaris longus m. (cut)
4. Flexor carpi ulnaris m.
5. Pronator teres m.
6. Epitrochlearis
7. Masseter m.
8. Sternomastoid m.
9. Clavobrachialis m.
10. Pectoantebrachialis m.
11. Pectoralis major m.
12. Pectoralis minor m.

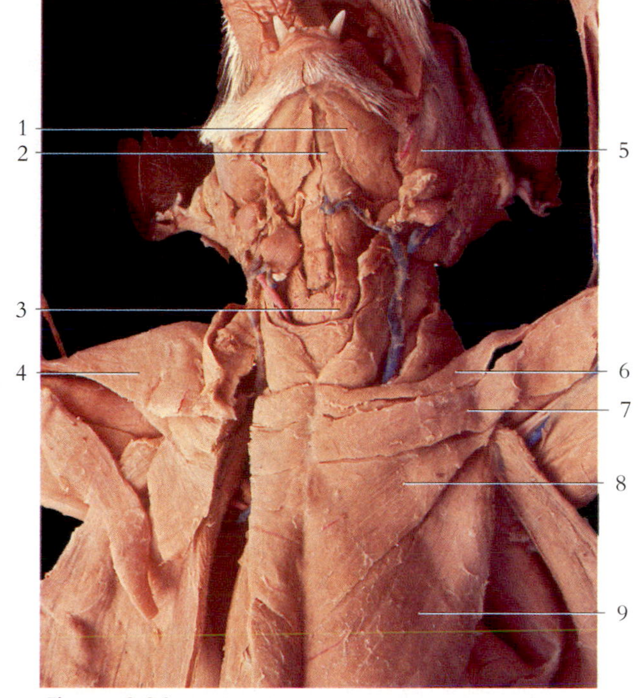

Figure 8.94 A ventral view of the cat neck and thorax.

1. Digastric m.
2. Mylohyoid m.
3. Sternomastoid m.
4. Clavotrapezius m.
5. Masseter m.
6. Clavobrachialis m.
7. Pectoantebrachialis m.
8. Pectoralis major m.
9. Pectoralis minor m.

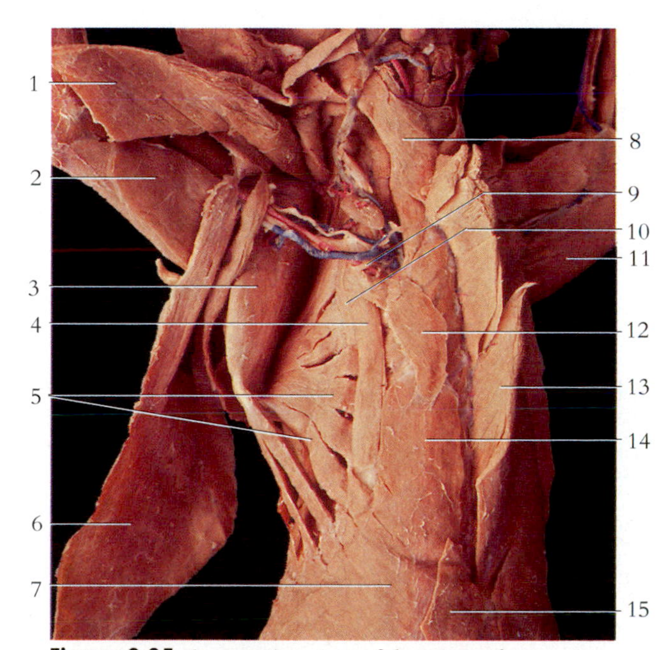

Figure 8.95 An anterior view of the cat trunk.

1. Pectoralis minor (cut)
2. Epitrochlearis m.
3. Subscapularis m.
4. Scalenus medius m.
5. Serratus anterior m.
6. Latissimus dorsi m. (cut)
7. External abdominal oblique m.
8. Sternomastoid m.
9. Scalenus anterior m.
10. Scalenus posterior m.
11. Epitrochlearis m.
12. Transverse costarum m.
13. Pectoralis minor m. (cut)
14. Rectus abdominis m.
15. Xiphihumeralis m. (cut)

Figure 8.96 A lateral view of the cat shoulder and brachium.

1. Acromiotrapezius m.
2. Levator scapulae ventralis m.
3. Spinodeltoid m.
4. Latissimus dorsi m.
5. Long head of triceps brachii m.
6. Clavobrachialis m.
7. Lateral head of triceps brachii m.
8. Clavotrapezius m.
9. Parotid gland
10. Acromiodeltoid m.
11. Brachioradialis m.

Figure 8.97 A lateral view of the cat trunk.
1. Internal abdominal oblique m.
2. Tensor fascia latae
3. Caudofemoris m.
4. Vastus lateralis m.
5. Sartorius m.
6. External abdominal oblique m.
7. Latissimus dorsi m.
8. Spinodeltoid m.
9. Transverse abdominis m.
10. Serratus anterior m.
11. Long head of triceps brachii m.

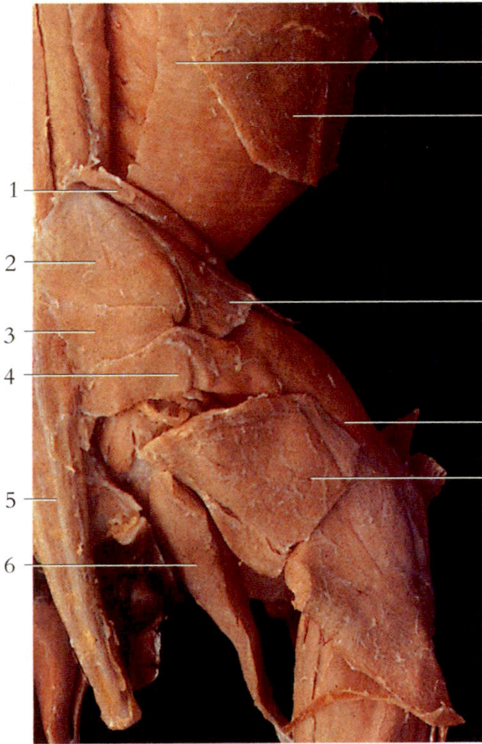

Figure 8.98 A lateral view of the cat superficial thigh.
1. Sartorius m.
2. Gluteus medius m.
3. Gluteus maximus m.
4. Caudofemoris m.
5. Caudal m.
6. Semitendinosus m.
7. Internal abdominal oblique m.
8. External abdominal oblique m.
9. Tensor fascia latae (cut)
10. Vastus lateralis m.
11. Biceps femoris m.

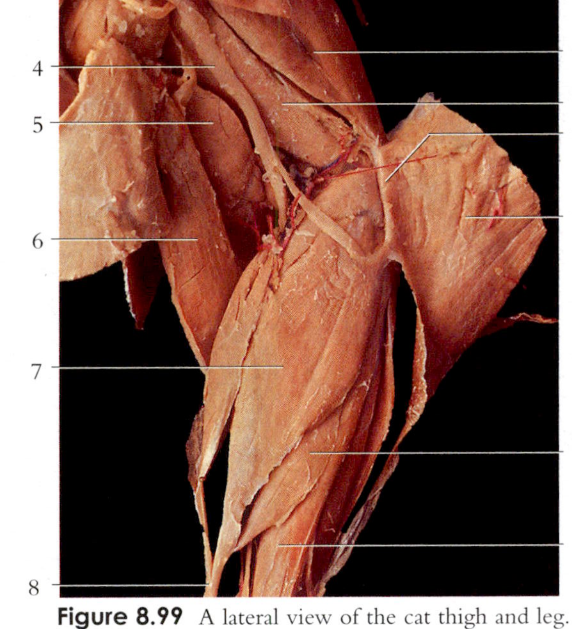

Figure 8.99 A lateral view of the cat thigh and leg.
1. Gluteus medius m.
2. Gluteus maximus m.
3. Caudofemoralis m.
4. Sciatic nerve
5. Semimembranosus m.
6. Semitendinosus m.
7. Gastrocnemius m.
8. Tendo calcaneus
9. Vastus lateralis m.
10. Adductor femoris m.
11. Tenuissimus m.
12. Biceps femoris m. (cut)
13. Soleus m.
14. Peroneal m.

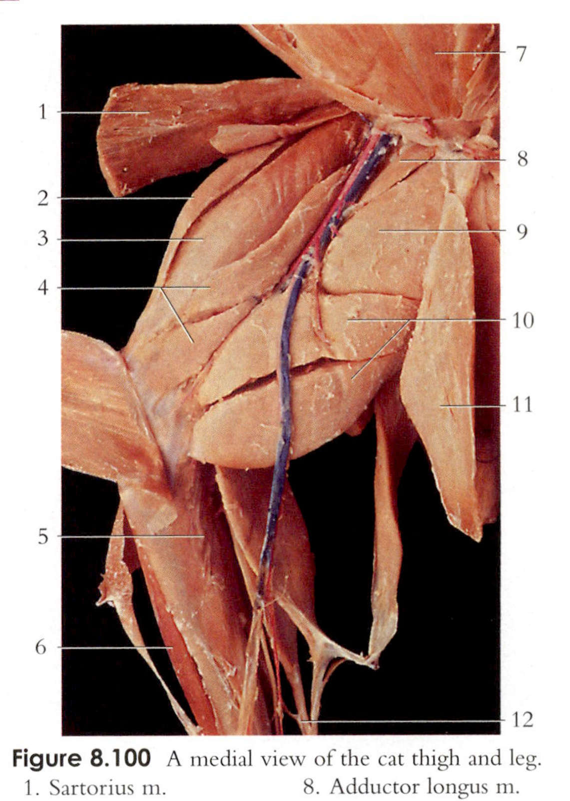

Figure 8.100 A medial view of the cat thigh and leg.
1. Sartorius m.
2. Vastus lateralis m.
3. Rectus femoris m.
4. Vastus medialis m.
5. Flexor digitorum longus m.
6. Tibialis anterior m.
7. Rectus abdominus m.
8. Adductor longus m.
9. Adductor femoris m.
10. Semimembranosus m.
11. Gracilis m. (cut)
12. Tendo calcaneus (Achilles' tendon)

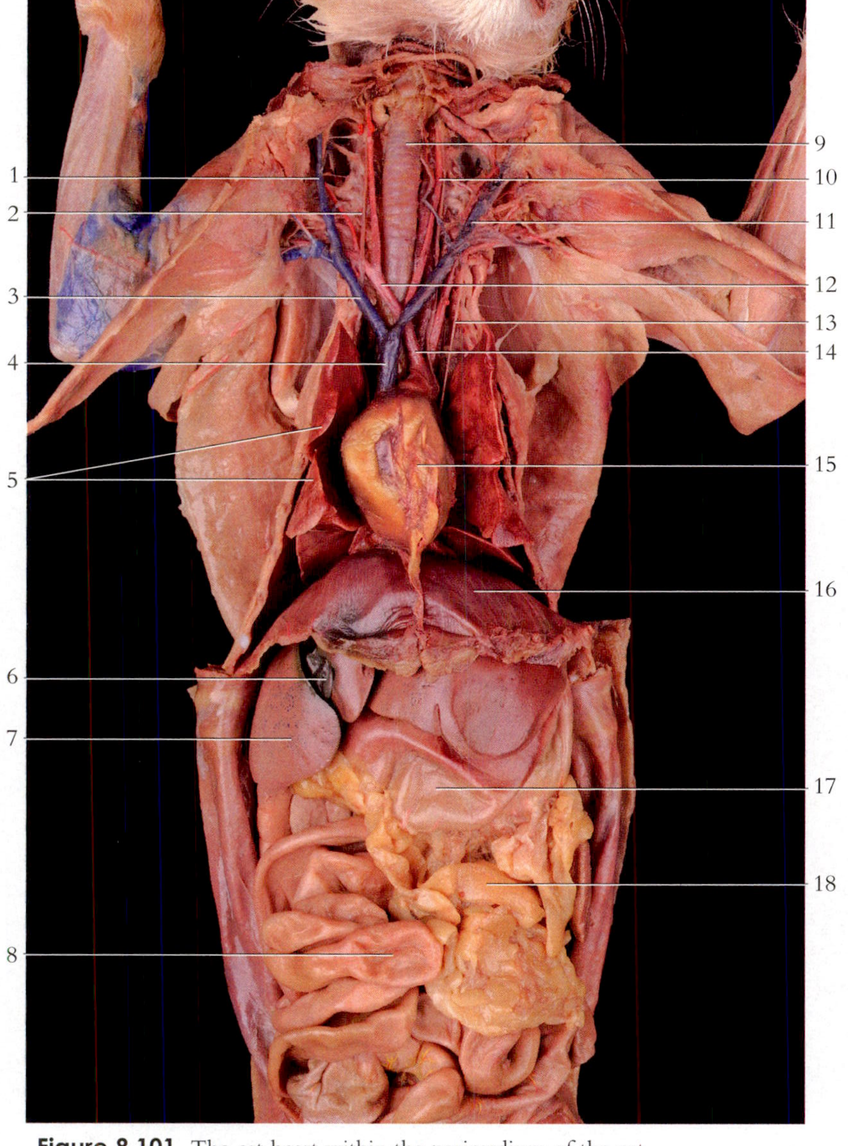

Figure 8.101 The cat heart within the pericardium of the cat.

1. External jugular vein
2. Vagus nerve
3. Right brachiocephalic vein
4. Cranial (superior) vena cava
5. Lung
6. Gallbladder
7. Liver
8. Small intestine
9. Trachea
10. Left carotid artery
11. External jugular vein
12. Common carotid artery
13. Phrenic nerve
14. Brachiocephalic trunk
15. Heart within pericardium
16. Diaphragm
17. Stomach
18. Greater omentum

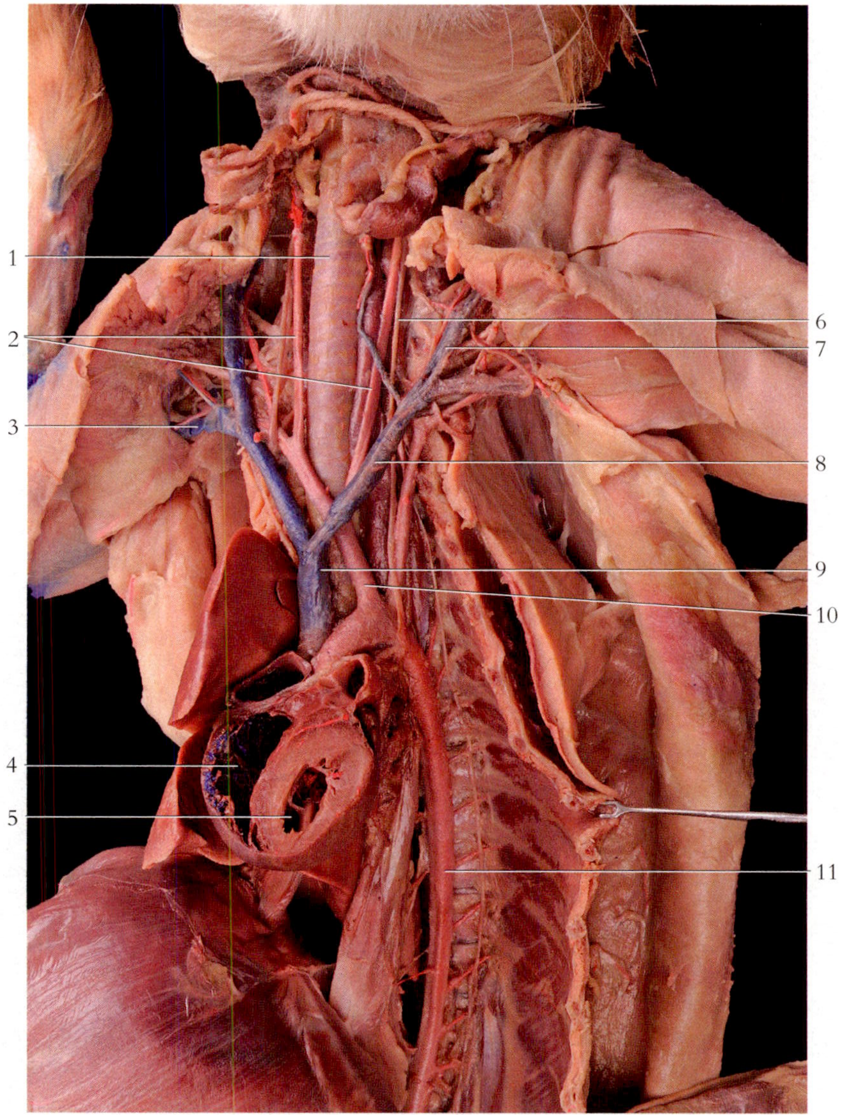

Figure 8.102 The heart and surrounding structures of the cat.

1. Trachea
2. Common carotid arteries
3. Axillary vein
4. Right ventricle
5. Left ventricle
6. Vagus nerve
7. External jugular vein
8. Left brachiocephalic vein
9. Cranial (superior) vena cava
10. Brachiocephalic trunk
11. Dorsal (descending) aorta

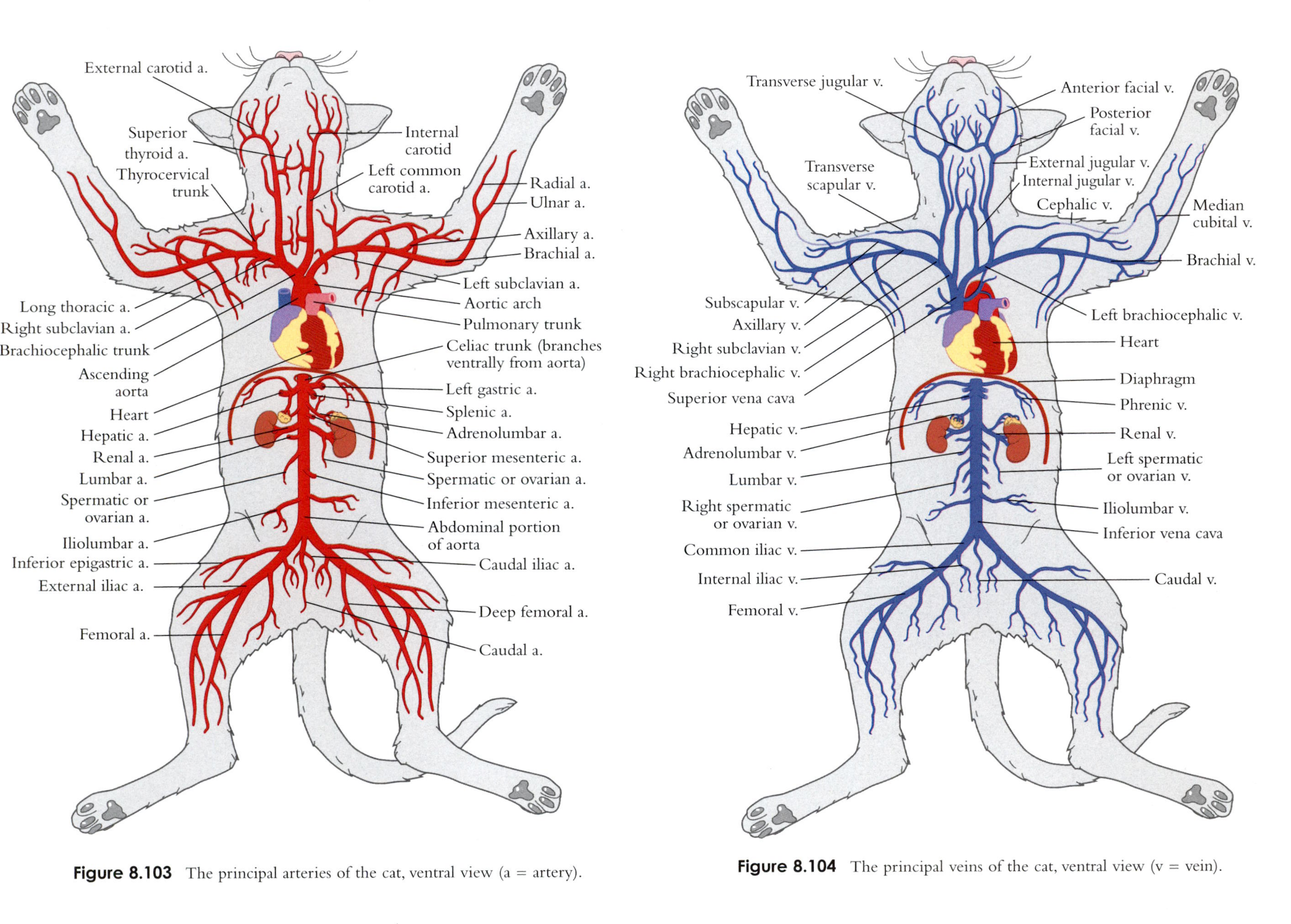

Figure 8.103 The principal arteries of the cat, ventral view (a = artery).

Figure 8.104 The principal veins of the cat, ventral view (v = vein).

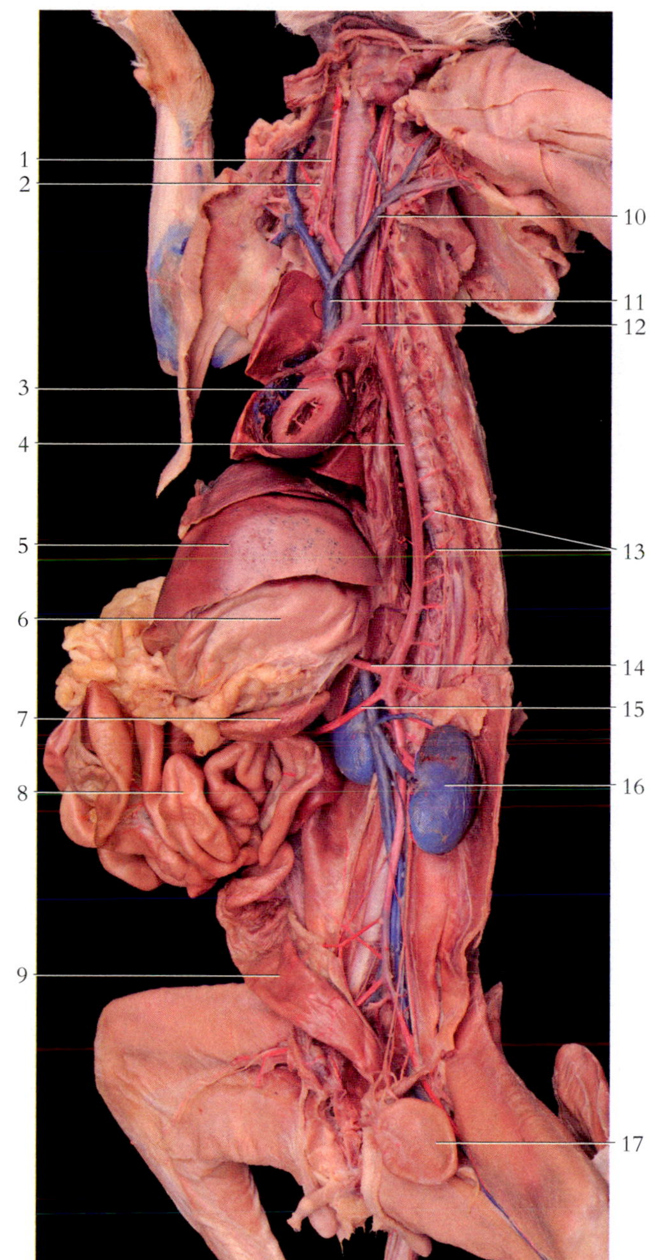

Figure 8.105 An anterior view of the deep structures of the trunk of the cat.

1. Right common carotid artery
2. Vagus nerve
3. Heart (cut)
4. Thoracic aorta
5. Liver
6. Stomach
7. Spleen
8. Small intestine
9. Colon
10. Left brachiocephalic vein
11. Cranial (superior) vena cava
12. Aortic arch
13. Intercostal artery
14. Celiac trunk
15. Superior mesenteric artery
16. Kidney
17. Urinary bladder

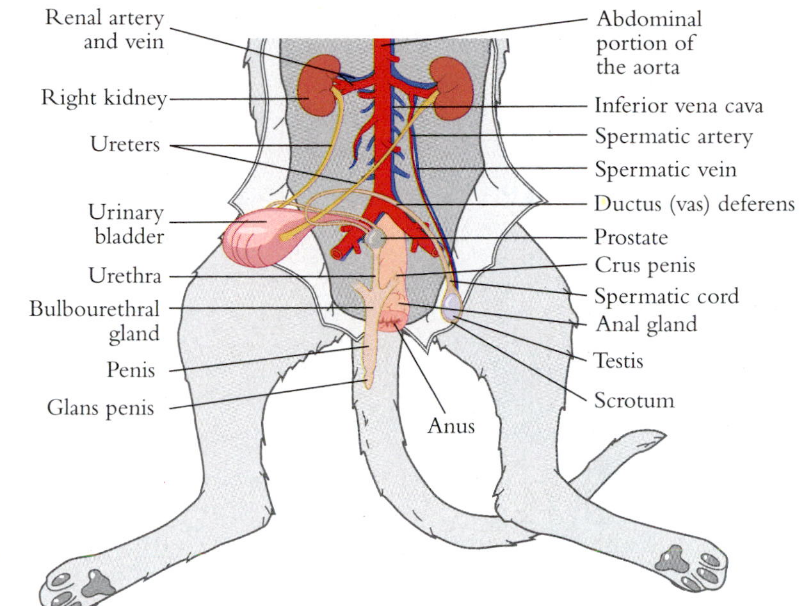

Figure 8.106 A diagram of the urogenital system of a male cat.

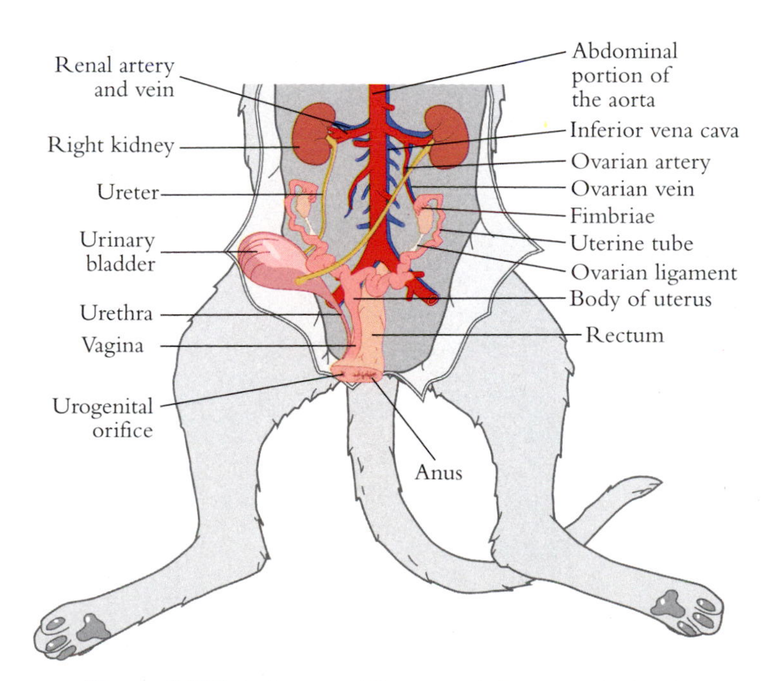

Figure 8.107 A diagram of the urogenital system of a female cat.

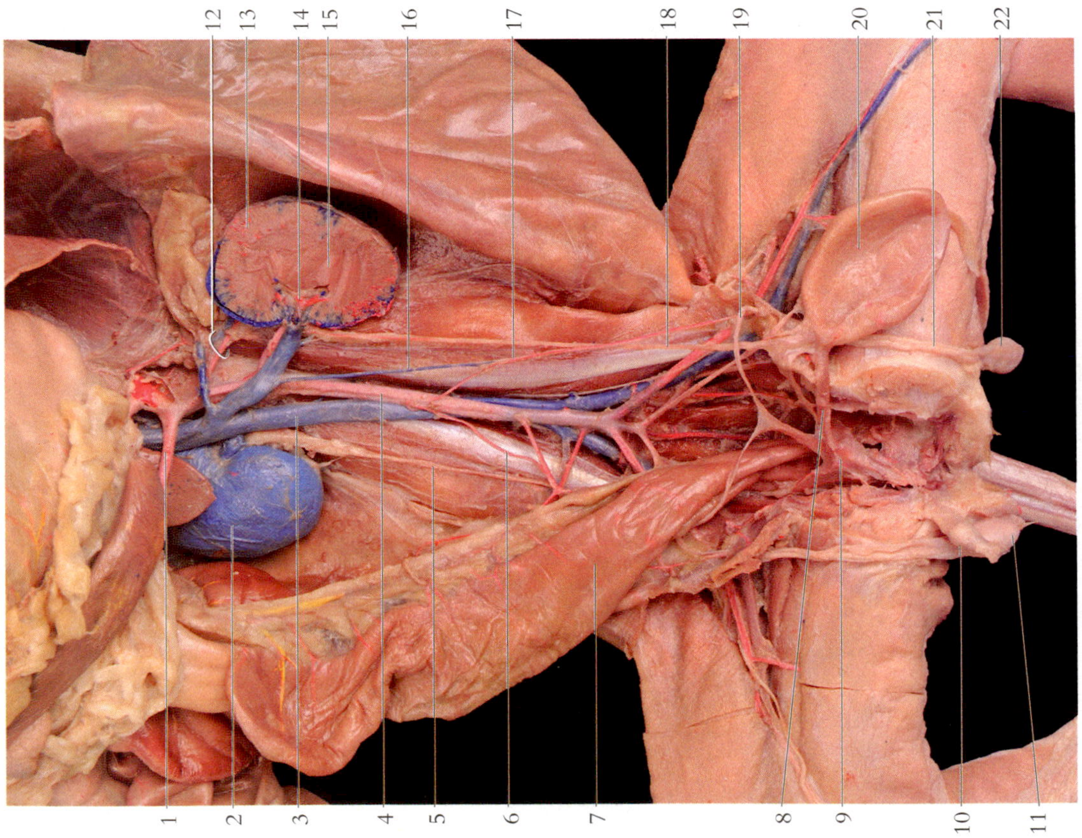

Figure 8.108 The urogenital system of a male cat.

1. Superior mesenteric artery
2. Kidney
3. Caudal (inferior) vena cava
4. Abdominal aorta
5. Right ureter
6. Right testicular artery
7. Colon
8. Urethra
9. Prostate
10. Penis
11. Right testis
12. Renal artery and vein
13. Renal cortex
14. Renal pelvis
15. Renal medulla
16. Left testicular vein
17. Left testicular artery
18. Left ureter
19. Ductus (vas) deferens
20. Urinary bladder
21. Spermatic cord
22. Left testis

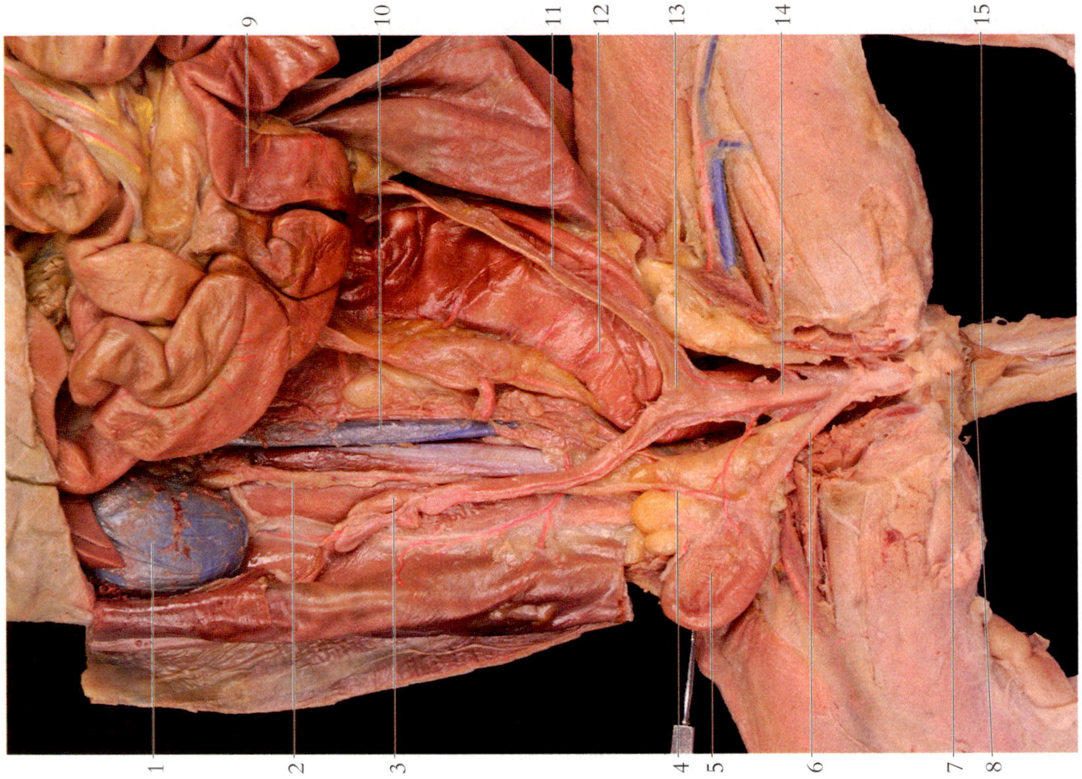

Figure 8.109 The urogenital system of a female cat.

1. Kidney
2. Ureter
3. Ovary
4. Ureter
5. Urinary bladder
6. Urethra
7. Clitoris
8. Vestibule
9. Small intestine
10. Caudal (inferior) vena cava
11. Horn of uterus
12. Colon
13. Body of uterus
14. Vagina
15. Labia

Mammalian Heart and Brain Dissection

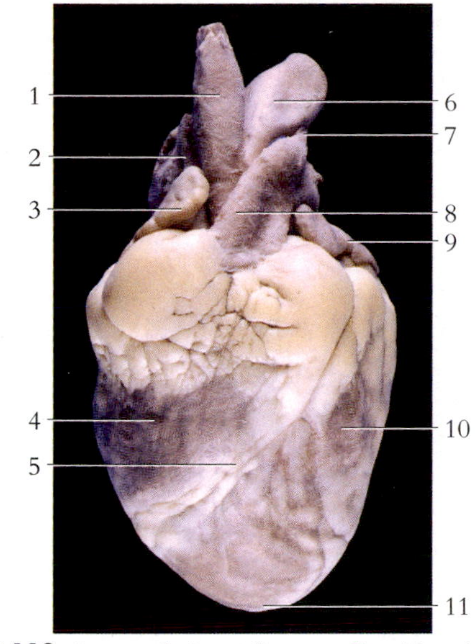

Figure 8.110 A ventral view of mammalian (sheep) heart.
1. Brachiocephalic artery
2. Cranial vena cava
3. Right auricle of right atrium
4. Right ventricle
5. Interventricular groove
6. Aortic arch
7. Ligamentum arteriosum
8. Pulmonary trunk
9. Left auricle of left atrium
10. Left ventricle
11. Apex of heart

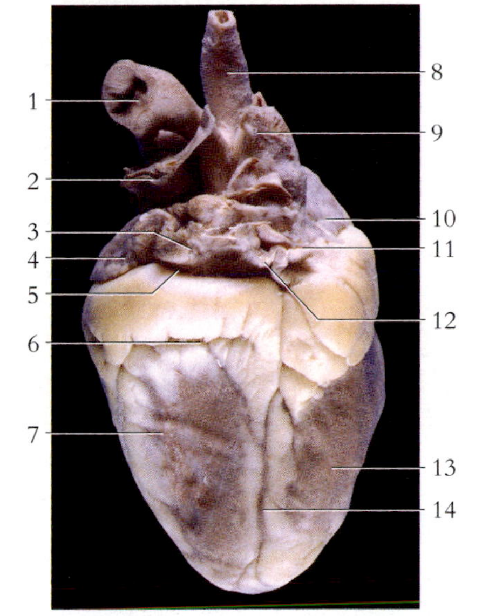

Figure 8.111 A dorsal view of mammalian (sheep) heart.
1. Aorta
2. Pulmonary artery
3. Pulmonary vein
4. Left auricle
5. Left atrium
6. Atrioventricular groove
7. Left ventricle
8. Brachiocephalic artery
9. Cranial vena cava
10. Right auricle
11. Right atrium
12. Pulmonary vein
13. Right ventricle
14. Interventricular groove

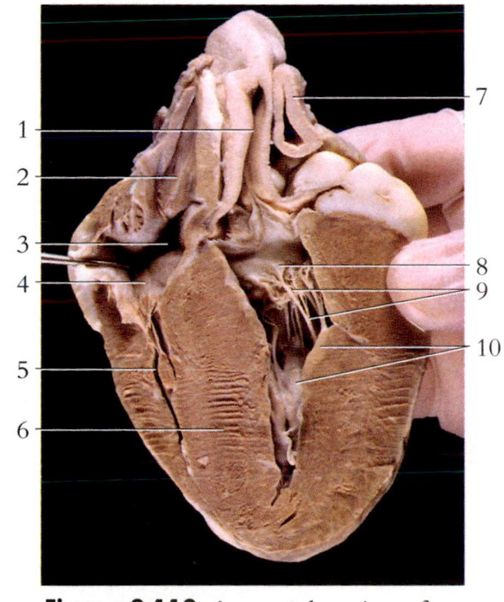

Figure 8.112 A coronal section of the mammalian (sheep) heart.
1. Aorta
2. Cranial vena cava
3. Right atrium
4. Right atrioventricular (tricuspid) valve
5. Right ventricle
6. Interventricular septum
7. Pulmonary artery
8. Left atrioventricular (bicuspid) valve
9. Chordae tendineae
10. Papillary muscles

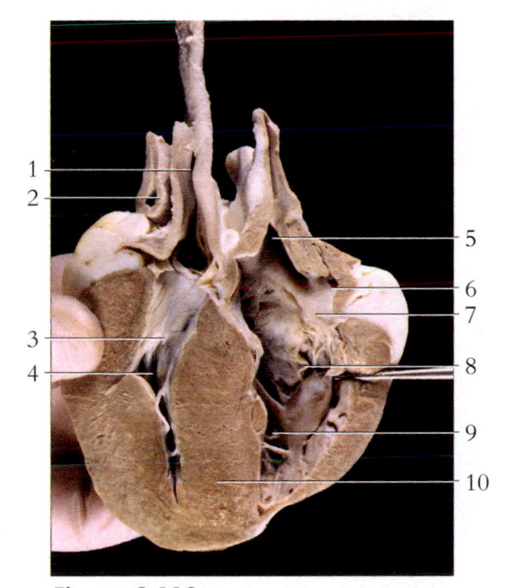

Figure 8.113 A coronal section of the mammalian (sheep) heart showing the valves.
1. Opening of the brachiocephalic artery
2. Pulmonary artery
3. Left atrioventricular (bicuspid) valve
4. Left ventricle
5. Opening of cranial vena cava
6. Opening of coronary sinus
7. Right atrium
8. Right atrioventricular (tricuspid) valve
9. Right ventricle
10. Interventricular septum

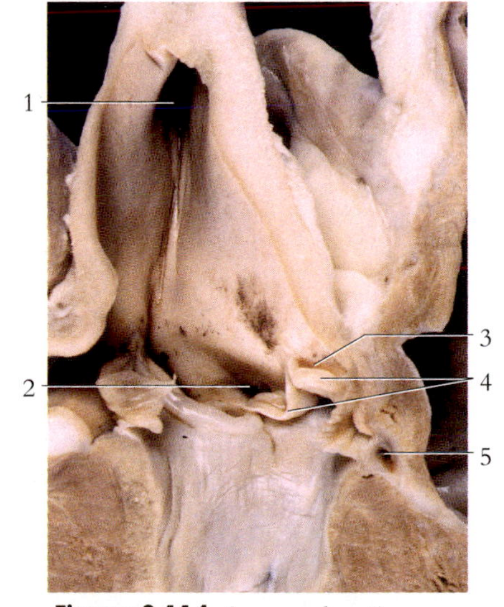

Figure 8.114 A coronal section of the mammalian (sheep) heart showing openings of coronary arteries.
1. Opening of brachiocephalic artery
2. Opening of left coronary artery
3. Opening of right coronary artery
4. Aortic valve
5. Coronary vessel

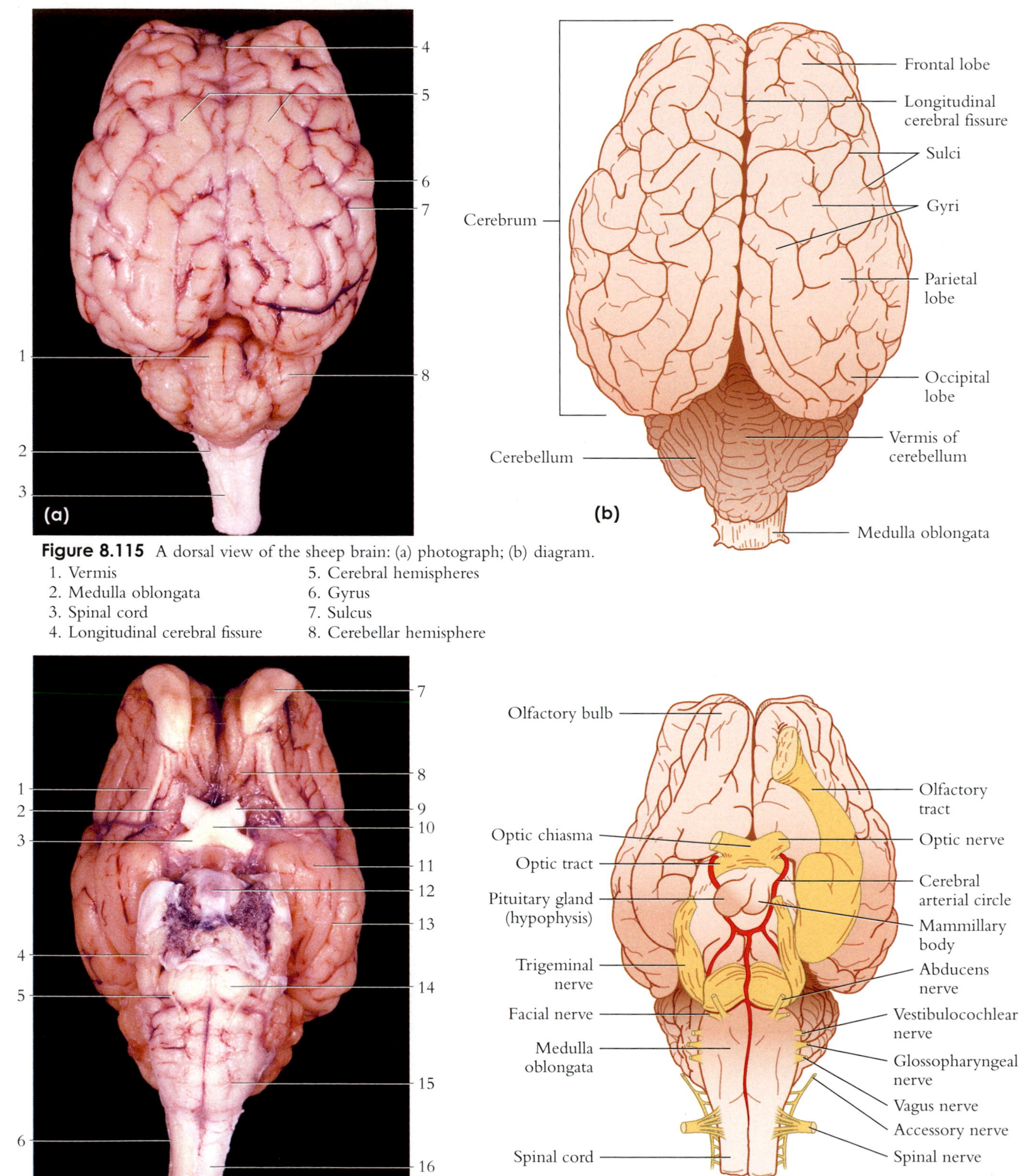

Figure 8.115 A dorsal view of the sheep brain: (a) photograph; (b) diagram.

1. Vermis
2. Medulla oblongata
3. Spinal cord
4. Longitudinal cerebral fissure
5. Cerebral hemispheres
6. Gyrus
7. Sulcus
8. Cerebellar hemisphere

Figure 8.116 A ventral view of sheep brain: (a) photograph; (b) diagram.

1. Lateral olfactory band
2. Olfactory trigone
3. Optic tract
4. Trigeminal nerve
5. Abducens nerve
6. Accessory nerve
7. Olfactory bulb
8. Medial olfactory band
9. Optic nerve
10. Optic chiasma
11. Pyriform lobe
12. Pituitary gland (hypophysis)
13. Rhinal sulcus
14. Pons
15. Medulla oblongata
16. Spinal cord

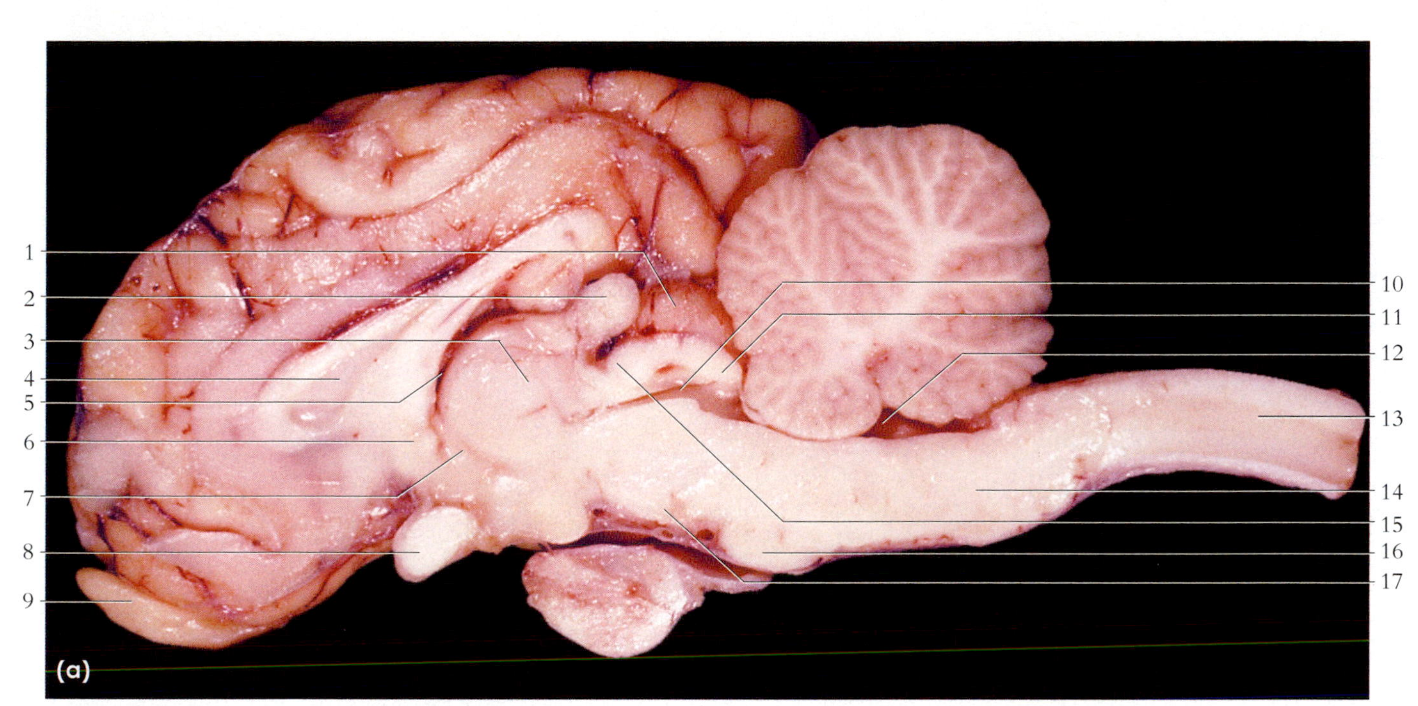

Figure 8.117 A right sagittal view of the sheep brain: (a) photograph; (b) diagram.

1. Superior colliculus
2. Pineal body (gland)
3. Intermediate mass of thalamus
4. Septum pellucidum
5. Interventricular foramen (foramen of Monro)
6. Anterior commissure
7. Third ventricle
8. Optic chiasma
9. Olfactory bulb
10. Mesencephalic (cerebral) aqueduct
11. Inferior colliculus
12. 4th ventricle
13. Spinal cord
14. Medulla oblongata
15. Posterior commissure
16. Pons
17. Cerebral peduncle

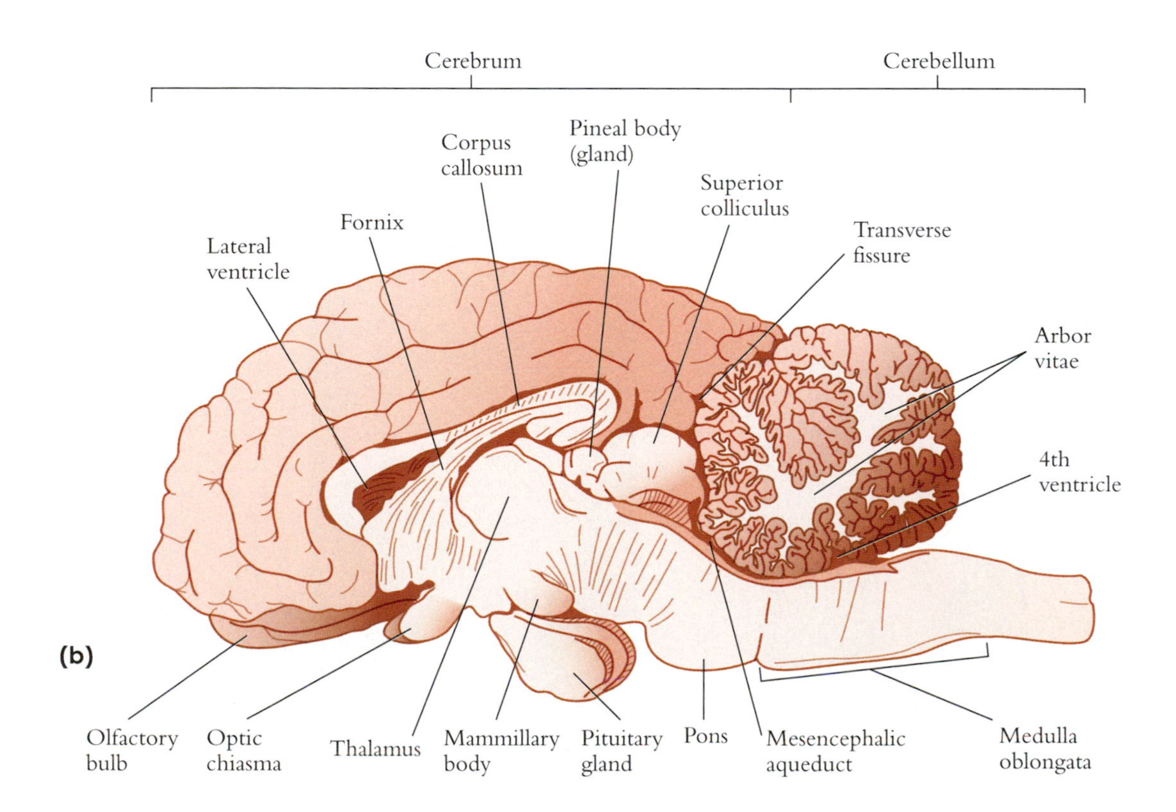

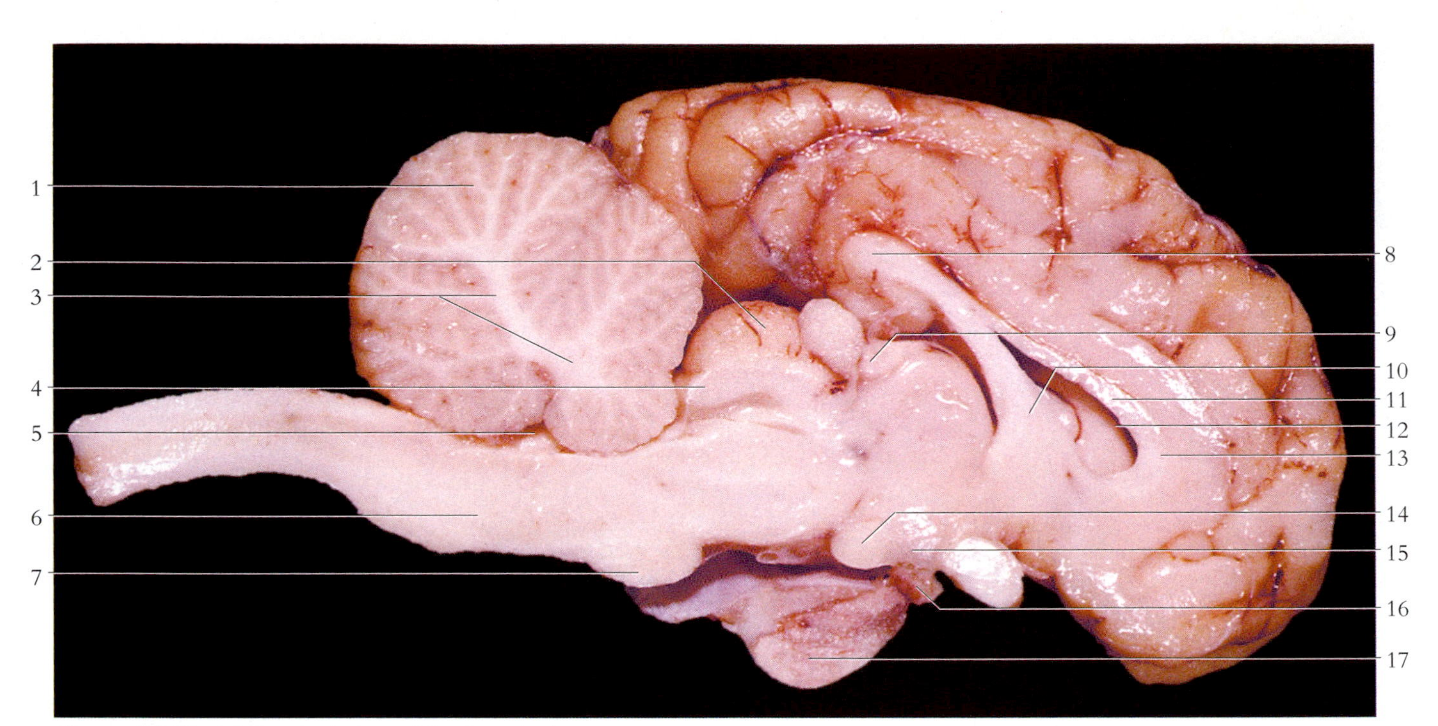

Figure 8.118 A left sagittal view of the sheep brain.

1. Cerebellum
2. Superior colliculus
3. Arbor vitae
4. Inferior colliculus
5. 4th ventricle
6. Medulla oblongata
7. Pons
8. Splenium of corpus callosum
9. Habenular trigone
10. Fornix
11. Body of corpus callosum
12. Lateral ventricle
13. Genu of corpus callosum
14. Mammillary body
15. Tuber cinereum of hypothalamus
16. Pituitary stalk
17. Pituitary gland (hypophysis)

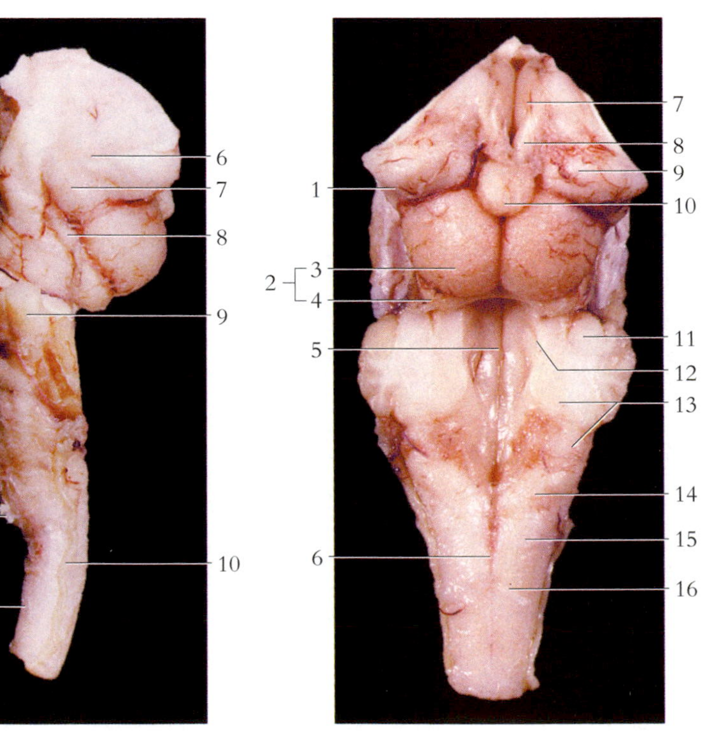

Figure 8.119 A lateral view of the brainstem.

1. Pons
2. Abducens nerve
3. Medulla oblongata
4. Hypoglossal nerve
5. Spinal cord
6. Lateral geniculate body
7. Medial geniculate body
8. Trochlear nerve
9. Trigeminal nerve
10. Accessory nerve

Figure 8.120 A dorsal view of the brainstem.

1. Medial geniculate body
2. Corpora quadrigemina
3. Superior colliculus
4. Inferior colliculus
5. 4th ventricle
6. Dorsal median sulcus
7. Intermediate mass of thalamus
8. Habenular trigone
9. Thalamus
10. Pineal gland
11. Middle cerebellar peduncle
12. Anterior cerebellar penduncle
13. Posterior cerebellar peduncle
14. Tuberculum cuneatum
15. Fasciculus gracilis
16. Fasciculus cuneatus

Human Biology

Chapter 9

Because humans are vertebrate organisms, the study of human biology is appropriate in a general biology course. *Human anatomy* is the scientific discipline that investigates the structure of the body, and *human physiology* is the scientific discipline that investigates how body structures function. The purpose of this chapter is to present a visual overview of the principal anatomical structures of the human body.

Because both the *skeletal system* and the *muscular system* are concerned with body movement, they are frequently discussed together as the *skeletomusculature system*. In a functional sense, the flexible internal framework, or *bones* of the skeleton, support and provide movement at the *joints,* whereas the muscles attached to the bones produce their actions as they are stimulated to contract.

The *nervous system* is anatomically divided into the *central nervous system* (CNS), which includes the *brain* and *spinal cord*, and the *peripheral nervous system* (PNS), which includes the *cranial nerves*, arising from the brain, and the *spinal nerves*, arising from the spinal cord. The *autonomic nervous system* (ANS) is a functional division of the nervous system devoted to regulation of involuntary activities of the body. The brain and spinal cord are the centers for integration and coordination of information. *Nerves*, composed of *neurons*, convey nerve impulses to and from the brain. *Sensory organs*, such as the eyes and ears, respond to impulses in the environment and convey sensations to the CNS. The nervous system functions with the *endocrine system* in coordinating body activities.

The *cardiovascular system* consists of the *heart, vessels* (both blood and lymphatic vessels), *blood*, and the tissues that produce the *blood*. The four-chambered human heart is enclosed by a *pericardial sac* within the thoracic cavity. *Arteries* and *arterioles* transport blood away from the heart, *capillaries* permeate the tissues and are the functional units for product exchange with the cells, and *venules* and *veins* transport blood toward the heart. *Lymphatic vessels* return interstitial fluid back to the circulatory system after first passing it through *lymph nodes* for cleansing. Blood cells are produced in the bone marrow, and once old and worn, they are broken down in the liver.

The *respiratory system* consists of the *conducting division* that transports air to and from the *respiratory division* within the *lungs*. The *alveoli* of the lungs contact the capillaries of the cardiovascular system and are the sites for transport of respiratory gases into and out of the body.

The *digestive system* consists of a *gastrointestinal tract* (GI tract) and *accessory digestive organs*. Food traveling through the GI tract is processed such that it is suitable for absorption through the intestinal wall into the bloodstream. The *pancreas* and *liver* are the principal digestive organs. The pancreas produces hormones and enzymes. The liver processes nutrients, stores glucose as glycogen, and excretes bile.

Because of commonality of prenatal development and dual functions of some of the organs, the *urinary system* and *reproductive system* may be considered together as the *urogenital system*. The urinary system, consisting of the *kidneys, ureters, urinary bladder*, and *urethra*, extracts and processes wastes from the blood in the form of urine. The male and female reproductive systems produce regulatory hormones and gametes (sperm and ova, respectively) within the gonads (testes and ovaries). Sexual reproduction is the mechanism for propagation of offspring that have traits from both parents. The process of prenatal development is made possible by the formation of *extraembryonic membranes* (placenta, umbilical cord, allantois, amnion, chorion, and yolk sac) within the uterus of the pregnant woman.

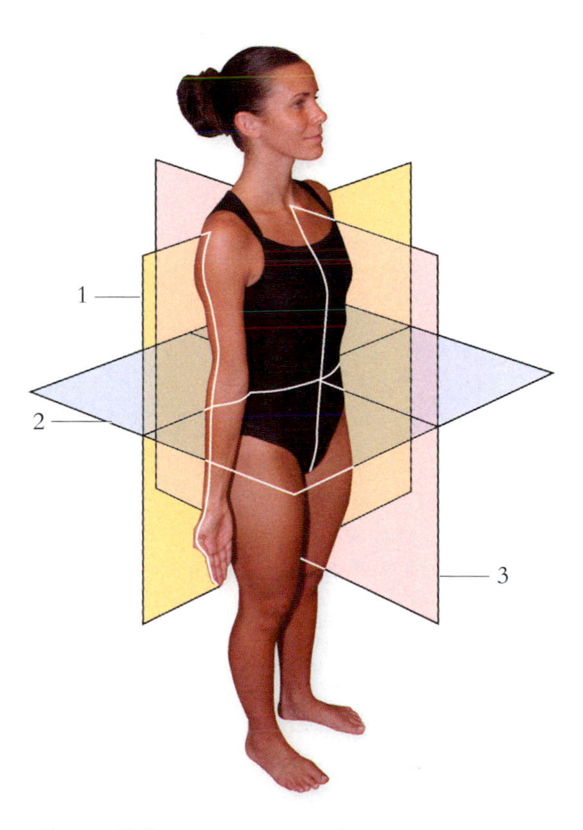

Figure 9.1 The planes of reference in a person while standing in anatomical position. The anatomical position provides a basis of reference for describing the relationship of one body part to another. In the anatomical position, the person is standing, the feet are parallel, the eyes are directed forward, and the arms are to the sides with the palms turned forward and the fingers pointed straight down.

1. Transverse plane (cross-sectional plane)
2. Coronal plane (frontal plane)
3. Sagittal plane

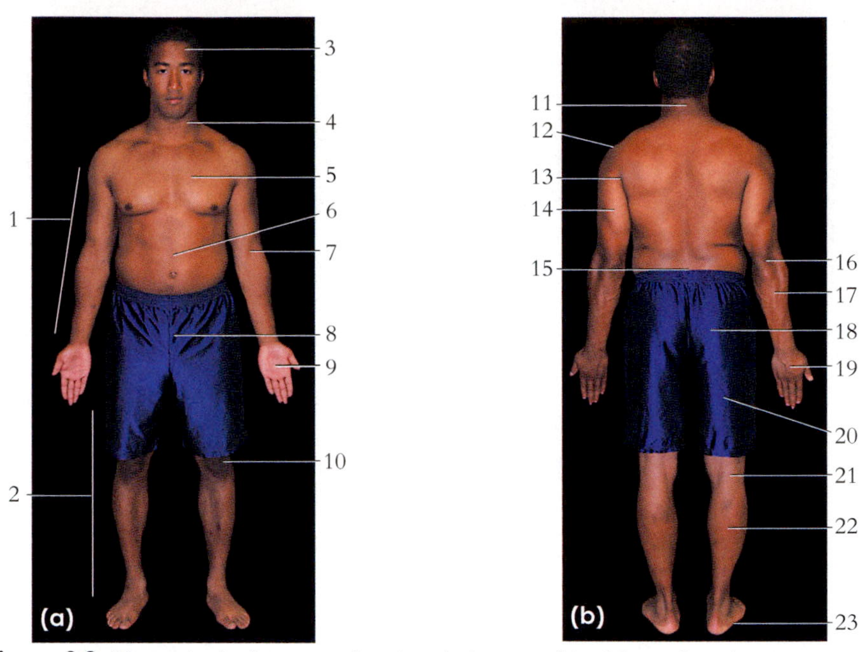

Figure 9.2 The major body parts and regions in humans (bipedal vertebrate). (a) An anterior view and (b) a posterior view.

1. Upper extremity
2. Lower extremity
3. Head
4. Neck, anterior aspect
5. Thorax (chest)
6. Abdomen
7. Cubital fossa
8. Pubic region
9. Palmar region (palm)
10. Patellar region (patella)
11. Cervical region
12. Deltoid region (shoulder)
13. Axilla (armpit)
14. Brachium (upper arm)
15. Lumbar region
16. Elbow
17. Antebrachium (forearm)
18. Gluteal region (buttock)
19. Dorsum of hand
20. Femoral region (thigh)
21. Popliteal fossa
22. Calf
23. Plantar surface (sole)

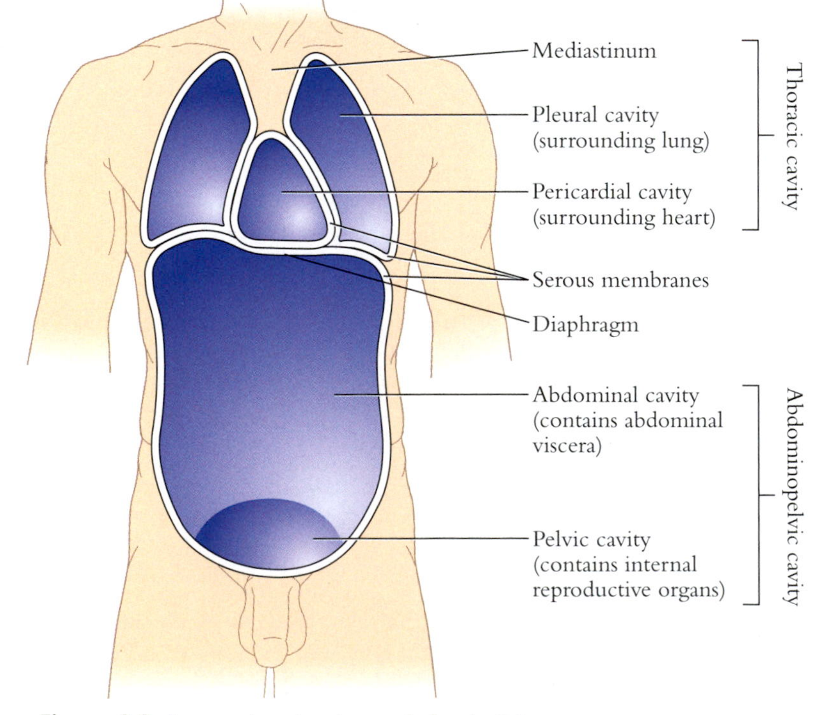

Figure 9.3 An anterior view (coronal plane) of the body cavities of the trunk.

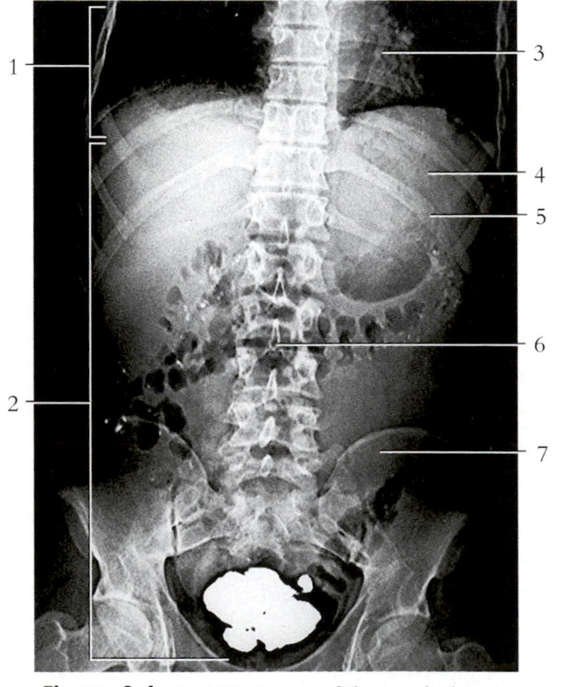

Figure 9.4 An MR image of the trunk showing the body cavities and their contents.

1. Thoracic cavity
2. Abdominopelvic cavity
3. Image of heart
4. Image of diaphragm
5. Image of rib
6. Image of lumbar vertebra
7. Image of ilium

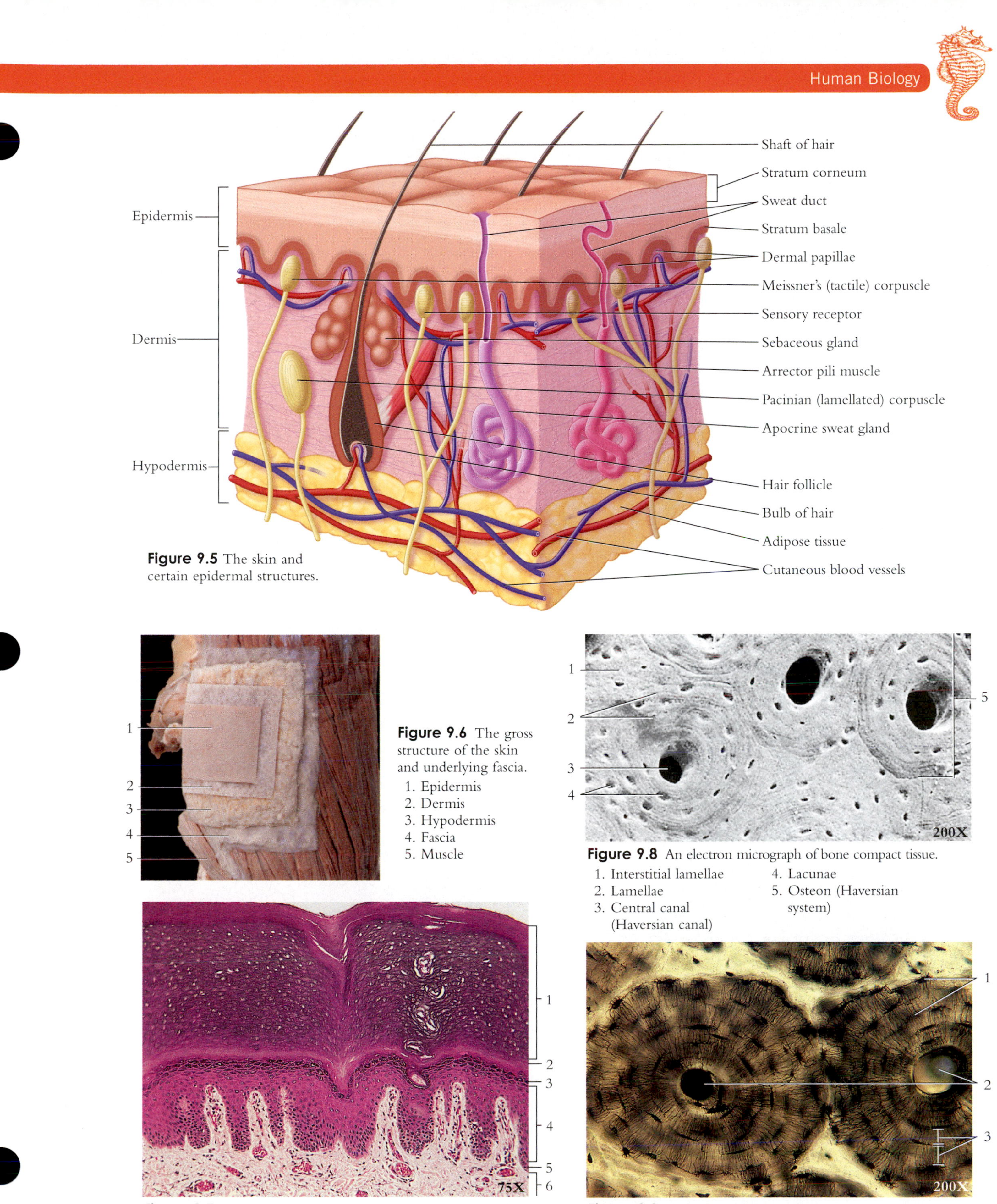

Figure 9.5 The skin and certain epidermal structures.

Figure 9.6 The gross structure of the skin and underlying fascia.
1. Epidermis
2. Dermis
3. Hypodermis
4. Fascia
5. Muscle

Figure 9.8 An electron micrograph of bone compact tissue.
1. Interstitial lamellae
2. Lamellae
3. Central canal (Haversian canal)
4. Lacunae
5. Osteon (Haversian system)

Figure 9.7 The epidermis and dermis of thick skin.
1. Stratum corneum
2. Stratum lucidum
3. Stratum granulosum
4. Stratum spinosum
5. Stratum basale
6. Dermis

Figure 9.9 A transverse section of two osteons.
1. Lacunae with contained osteocytes
2. Central (Haversian) canals
3. Lamellae

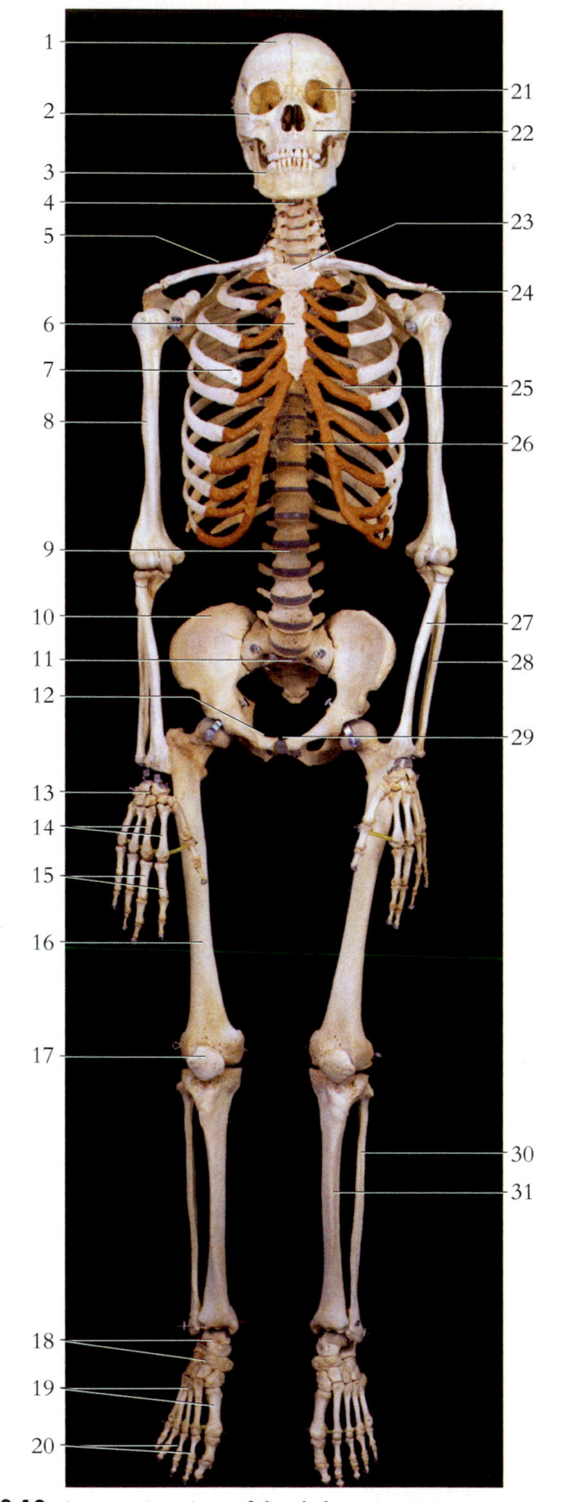

Figure 9.10 An anterior view of the skeleton.

1. Frontal bone
2. Zygomatic bone
3. Mandible
4. Cervical vertebra
5. Clavicle
6. Body of sternum
7. Rib
8. Humerus
9. Lumbar vertebra
10. Ilium
11. Sacrum
12. Pubis
13. Carpal bones
14. Metacarpal bones
15. Phalanges
16. Femur
17. Patella
18. Tarsal bones
19. Metatarsal bones
20. Phalanges
21. Orbit
22. Maxilla
23. Scapula
24. Manubrium
25. Costal cartilage
26. Thoracic vertebra
27. Radius
28. Ulna
29. Symphysis pubis
30. Fibula
31. Tibia

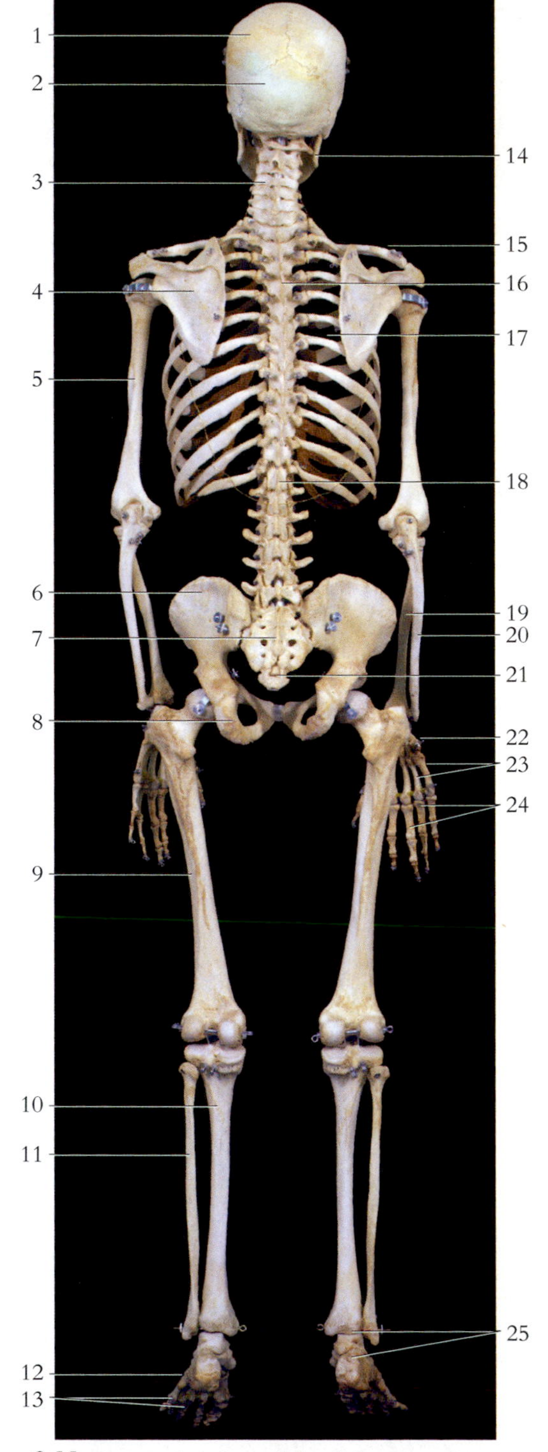

Figure 9.11 A posterior view of the skeleton.

1. Parietal bone
2. Occipital bone
3. Cervical vertebra
4. Scapula
5. Humerus
6. Ilium
7. Sacrum
8. Ischium
9. Femur
10. Tibia
11. Fibula
12. Metatarsal bones
13. Phalanges
14. Mandible
15. Clavicle
16. Thoracic vertebra
17. Rib
18. Lumbar vertebra
19. Radius
20. Ulna
21. Coccyx
22. Carpal bones
23. Metacarpal bones
24. Phalanges
25. Tarsal bones

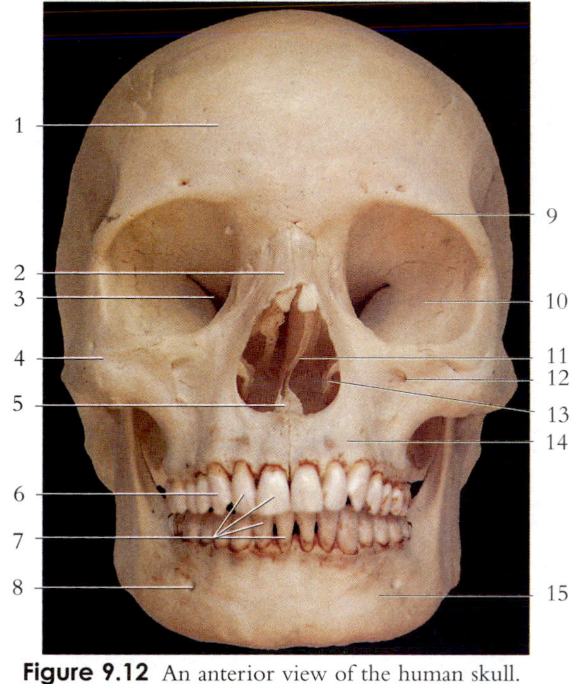

Figure 9.12 An anterior view of the human skull.

1. Frontal bone
2. Nasal bone
3. Superior orbital fissure
4. Zygomatic bone
5. Vomer
6. Canine
7. Incisors
8. Mental foramen
9. Supraorbital margin
10. Sphenoid bone
11. Perpendicular plate of ethmoid bone
12. Infraorbital foramen
13. Inferior nasal concha
14. Maxilla
15. Mandible

Figure 9.13 A lateral view of the human skull.

1. Coronal suture
2. Frontal bone
3. Lacrimal bone
4. Nasal bone
5. Zygomatic bone
6. Maxilla
7. Premolars
8. Molars
9. Mandible
10. Parietal bone
11. Squamosal suture
12. Temporal bone
13. Lambdoidal suture
14. External acoustic meatus
15. Occipital bone
16. Condylar process of mandible
17. Mandibular notch
18. Mastoid process of temporal bone
19. Coronoid process of mandible
20. Angle of mandible

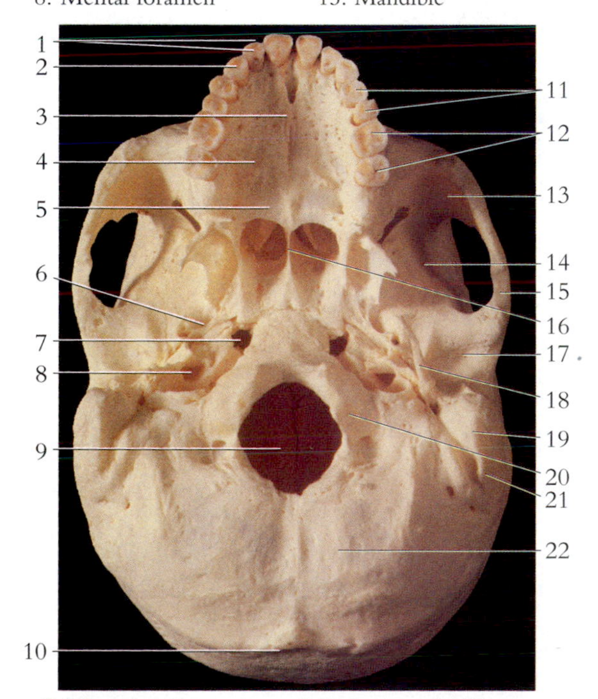

Figure 9.14 An inferior view of the human skull.

1. Incisors
2. Canine
3. Intermaxillary suture
4. Maxilla
5. Palatine bone
6. Foramen ovale
7. Foramen lacerum
8. Carotid canal
9. Foramen magnum
10. Superior nuchal line
11. Premolars
12. Molars
13. Zygomatic bone
14. Sphenoid bone
15. Zygomatic arch
16. Vomer
17. Mandibular fossa
18. Styloid process of temporal bone
19. Mastoid process of temporal bone
20. Occipital condyle
21. Temporal bone
22. Occipital bone

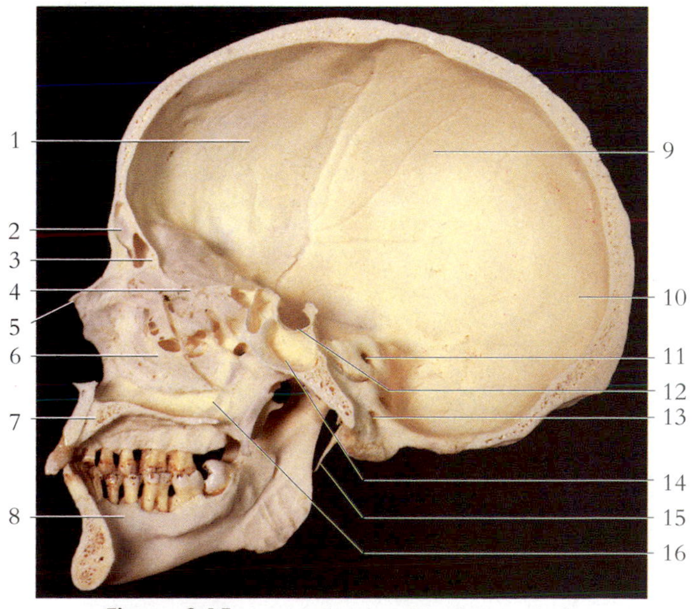

Figure 9.15 A sagittal view of the human skull.

1. Frontal bone
2. Frontal sinus
3. Crista galli of ethmoid bone
4. Cribriform plate of ethmoid bone
5. Nasal bone
6. Nasal concha
7. Maxilla
8. Mandible
9. Parietal bone
10. Occipital bone
11. Internal acoustic meatus
12. Sella turcica
13. Hypoglossal canal
14. Sphenoidal sinus
15. Styloid process of temporal bone
16. Vomer

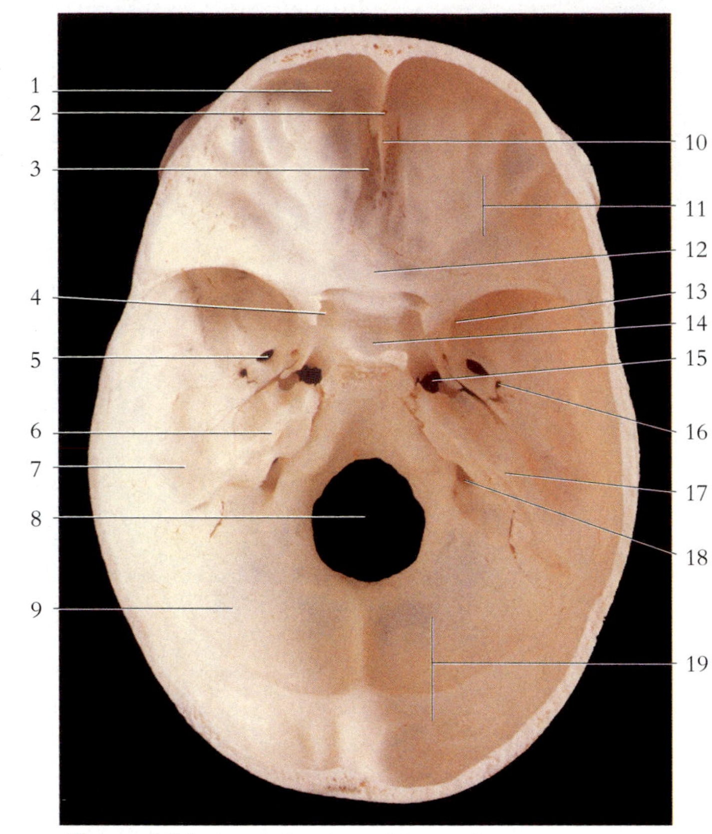

Figure 9.16 A superior view of the cranium.

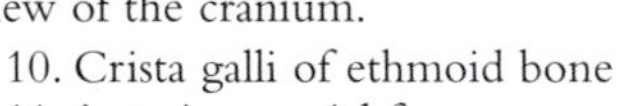

1. Frontal bone
2. Foramen caecum
3. Cribriform plate of ethmoid bone
4. Optic canal
5. Foramen ovale
6. Petrous part of temporal bone
7. Temporal bone
8. Foramen magnum
9. Occipital bone
10. Crista galli of ethmoid bone
11. Anterior cranial fossa
12. Sphenoid bone
13. Foramen rotundum
14. Sella turcica of sphenoid bone
15. Foramen lacerum
16. Foramen spinosum
17. Internal acoustic meatus
18. Jugular foramen
19. Posterior cranial fossa

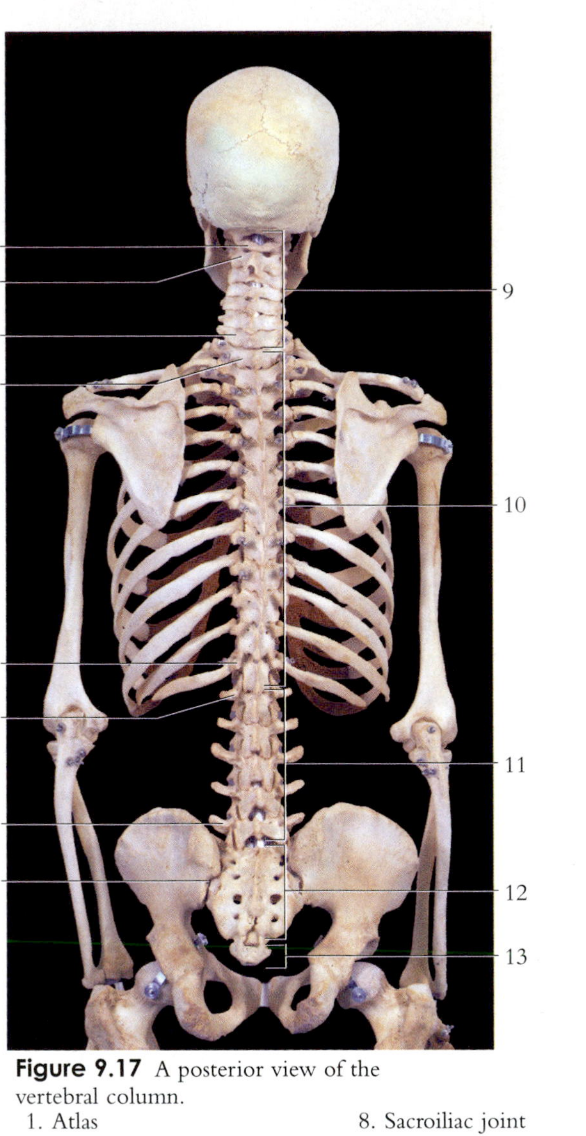
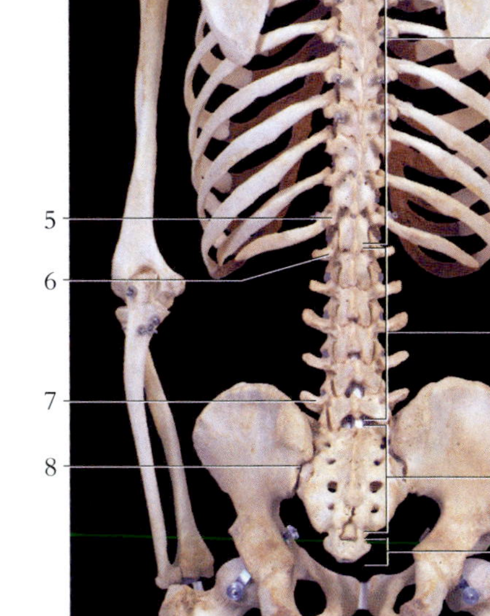

Figure 9.17 A posterior view of the vertebral column.

1. Atlas
2. Axis
3. 7th cervical vertebra
4. 1st thoracic vertebra
5. 12th thoracic vertebra
6. 1st lumbar vertebra
7. 5th lumbar vertebra
8. Sacroiliac joint
9. Cervical vertebrae
10. Thoracic vertebrae
11. Lumbar vertebrae
12. Sacrum
13. Coccyx

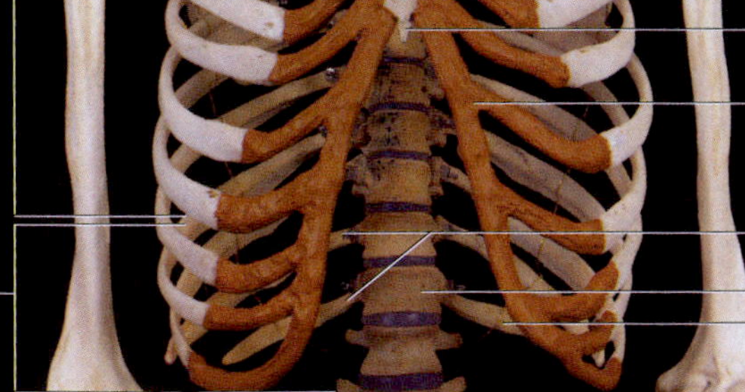

Figure 9.18 An anterior view of the rib cage.

1. True ribs (seven pairs)
2. False ribs (five pairs)
3. Jugular notch
4. Manubrium
5. Body of sternum
6. Xiphoid process
7. Costal cartilage
8. Floating ribs (inferior two pairs of false ribs)
9. 12th thoracic vertebra
10. 12th rib

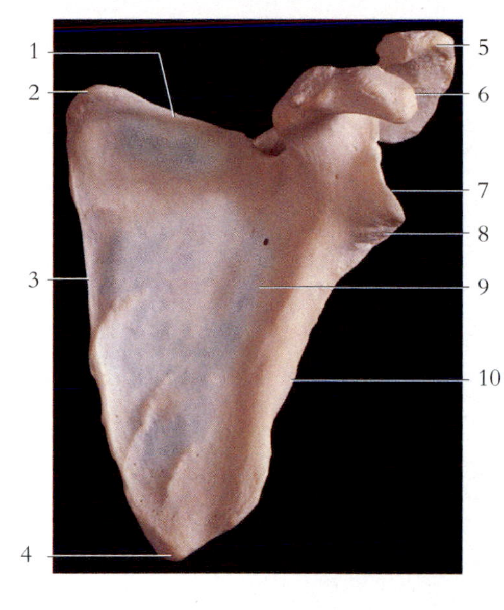

Figure 9.19 An anterior view of the left scapula.
1. Superior border
2. Superior angle
3. Medial (vertebral) border
4. Inferior angle
5. Acromion
6. Coracoid process
7. Glenoid fossa
8. Infraglenoid tubercle
9. Subscapular fossa
10. Lateral (axillary) border

Figure 9.20 A posterior view of the left scapula.
1. Acromion
2. Glenoid fossa
3. Lateral (axillary) border
4. Superior angle
5. Supraspinous fossa
6. Spine
7. Infraspinous fossa
8. Medial (vertebral) border
9. Inferior angle

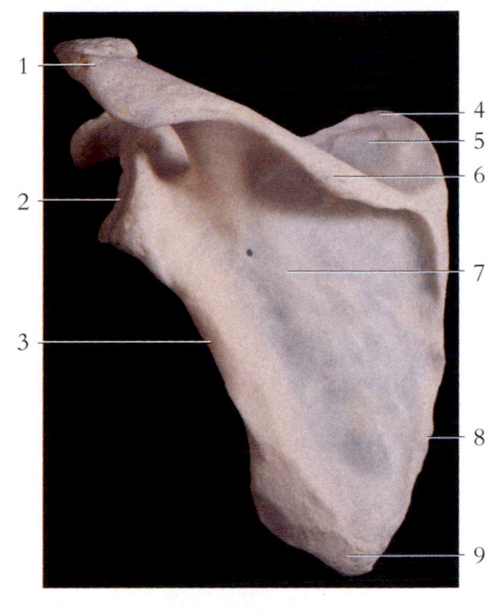

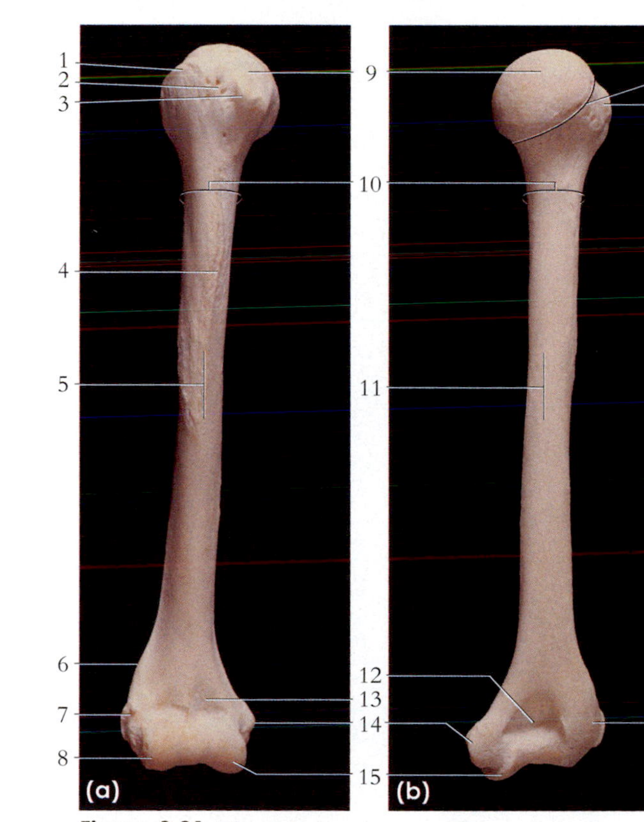

Figure 9.21 The right humerus. (a) Anterior view and (b) posterior view.
1. Greater tubercle
2. Intertubercular groove
3. Lesser tubercle
4. Deltoid tuberosity
5. Anterior body (shaft) of humerus
6. Lateral supracondylar ridge
7. Lateral epicondyle
8. Capitulum
9. Head of humerus
10. Surgical neck
11. Posterior body (shaft) of humerus
12. Olecranon fossa
13. Coronoid fossa
14. Medial epicondyle
15. Trochlea
16. Anatomical neck
17. Greater tubercle
18. Lateral epicondyle

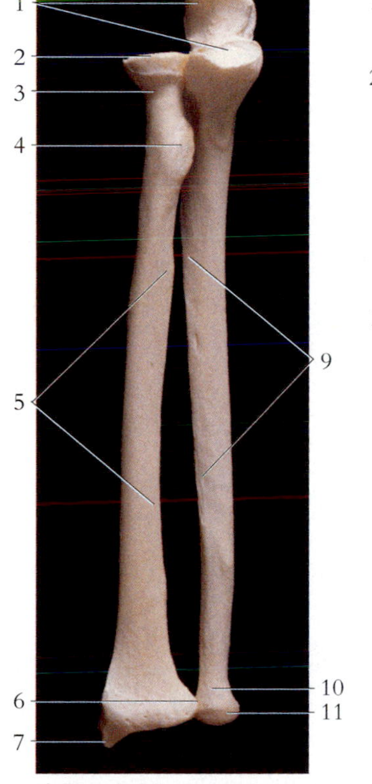

Figure 9.22 An anterior view of the right ulna and radius.
1. Trochlear notch
2. Head of radius
3. Neck of radius
4. Radial tuberosity
5. Interosseous margin
6. Location of ulnar notch of radius
7. Styloid process of radius
8. Olecranon
9. Interosseous margin
10. Neck of ulna
11. Head of ulna

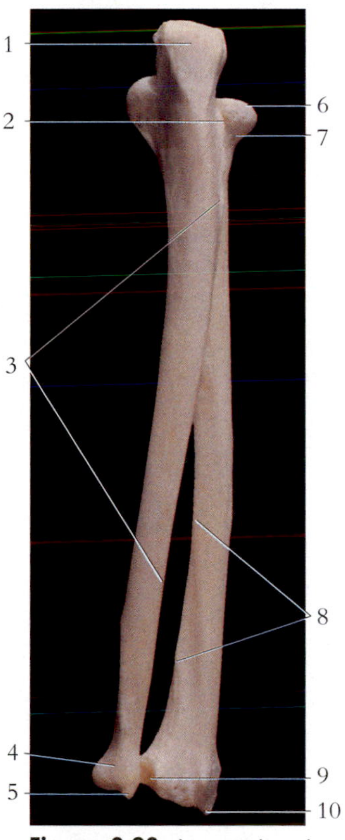

Figure 9.23 A posterior view of the right ulna and radius.
1. Olecranon
2. Location of radial notch of ulna
3. Interosseous margin
4. Head of ulna
5. Styloid process of ulna
6. Head of radius
7. Neck of radius
8. Interosseous margin
9. Ulnar notch
10. Styloid process of radius

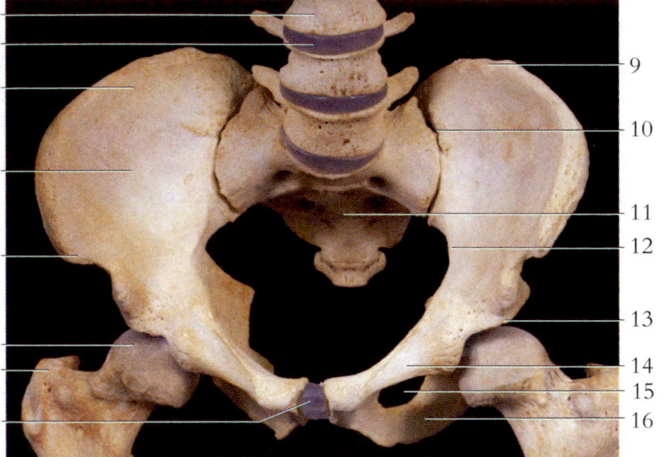

Figure 9.24 An anterior view of the articulated pelvic girdle showing the two coxal bones, the sacrum, and the two femora.

1. Lumbar vertebra
2. Intervertebral disk
3. Ilium
4. Iliac fossa
5. Anterior superior iliac spine
6. Head of femur
7. Greater trochanter
8. Symphysis pubis
9. Crest of the ilium
10. Sacroiliac joint
11. Sacrum
12. Pelvic brim
13. Acetabulum
14. Pubic crest
15. Obturator foramen
16. Ischium

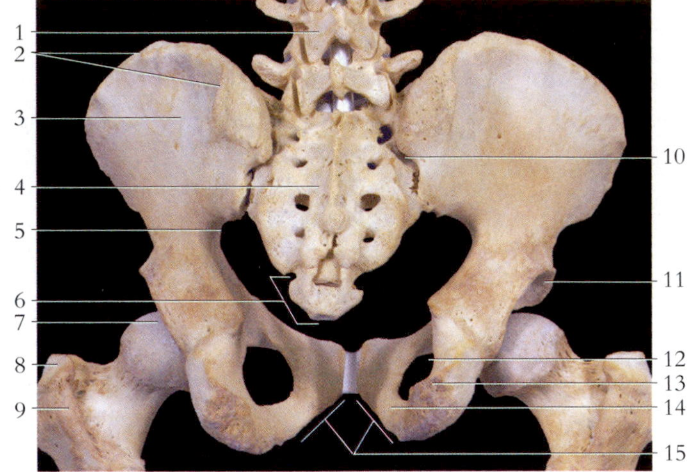

Figure 9.25 A posterior view of the articulated pelvic girdle showing the two coxal bones, the sacrum, and the two femora.

1. Lumbar vertebra
2. Crest of ilium
3. Ilium
4. Sacrum
5. Greater sciatic notch
6. Coccyx
7. Head of femur
8. Greater trochanter
9. Intertrochanteric crest
10. Sacroiliac joint
11. Acetabulum
12. Obturator foramen
13. Ischium
14. Pubis
15. Pubic angle

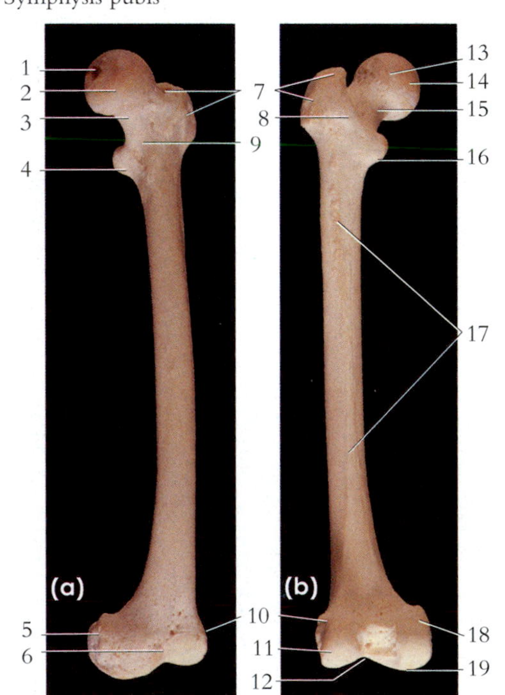

Figure 9.26 The left femur. (a) An anterior view and (b) a posterior view.

1. Fovea capitis femoris
2. Head
3. Neck
4. Lesser trochanter
5. Medial epicondyle
6. Patellar surface
7. Greater trochanter
8. Intertrochanteric crest
9. Intertrochanteric line
10. Lateral epicondyle
11. Lateral condyle
12. Intercondylar fossa
13. Head
14. Fovea capitis femoris
15. Neck
16. Lesser trochanter
17. Linea aspera on shaft (body) of femur
18. Medial epicondyle
19. Medial condyle

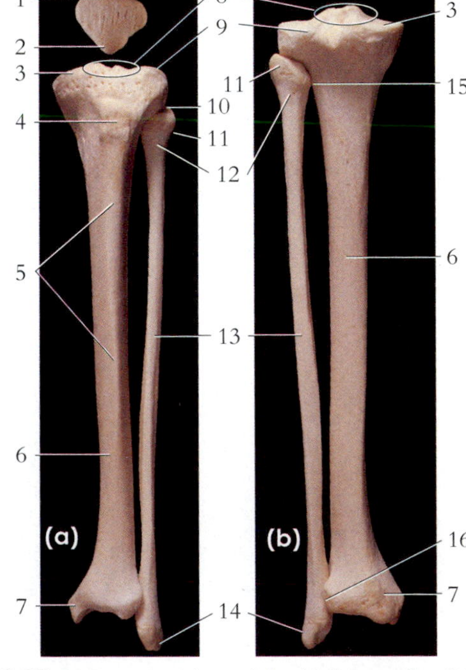
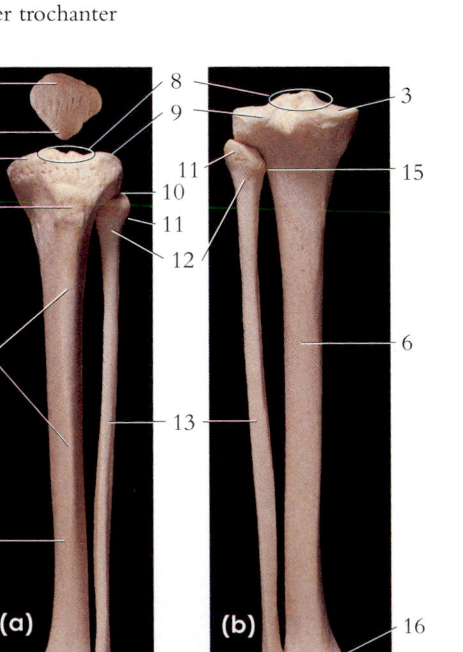

Figure 9.27 An anterior view of the (a) left patella, tibia, and fibula. (b) A posterior view of the left tibia and fibula.

1. Base of patella
2. Apex of patella
3. Medial condyle
4. Tibial tuberosity
5. Anterior crest of tibia
6. Body (shaft) of tibia
7. Medial malleolus
8. Intercondylar tubercles
9. Lateral condyle
10. Tibial articular facet of fibula
11. Head of fibula
12. Neck of fibula
13. Body (shaft) of fibula
14. Lateral malleolus
15. Fibular articular facet of tibia
16. Fibular notch of tibia

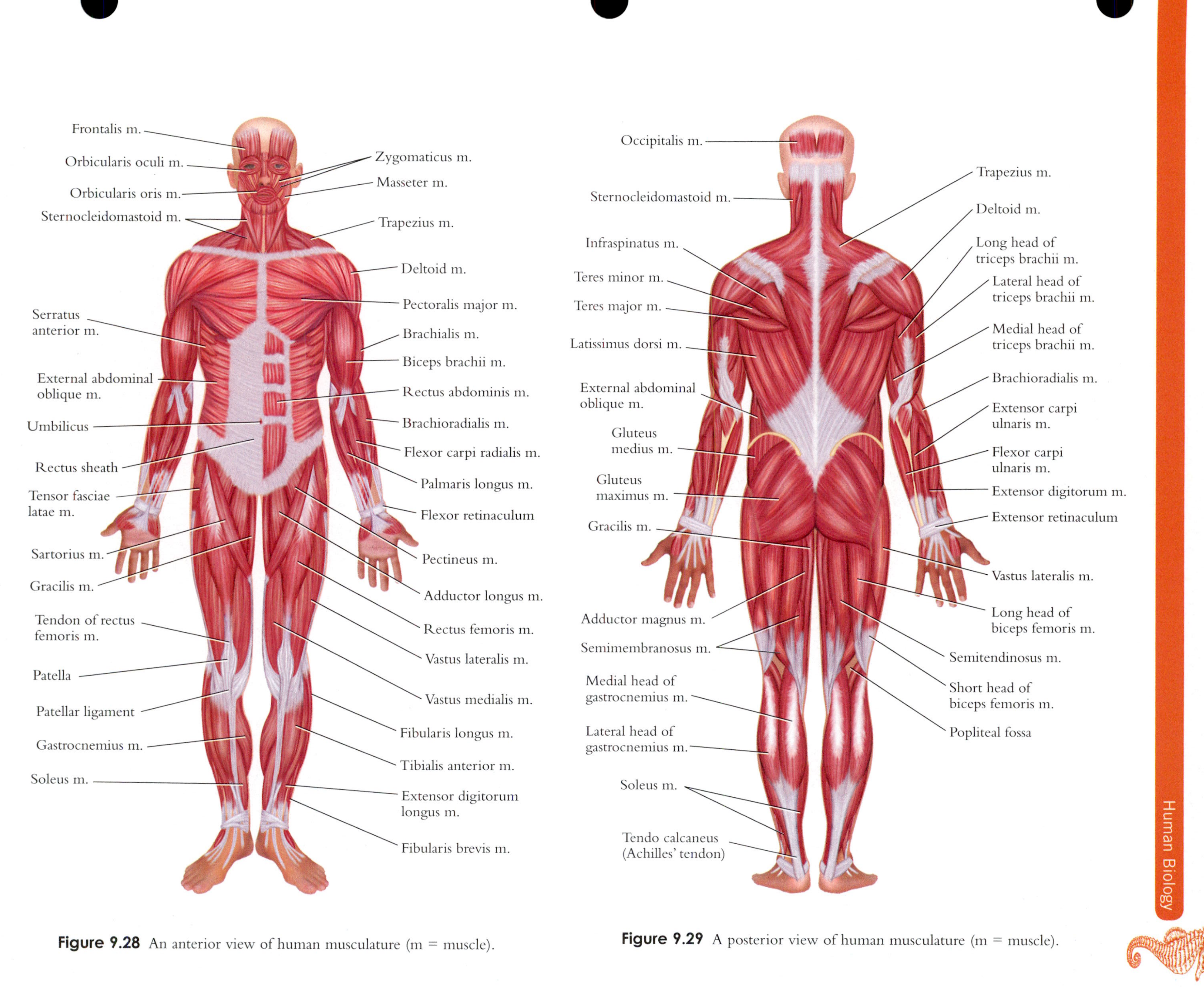
Frontalis m.
Orbicularis oculi m.
Orbicularis oris m.
Sternocleidomastoid m.
Zygomaticus m.
Masseter m.
Trapezius m.
Deltoid m.
Pectoralis major m.
Serratus anterior m.
Brachialis m.
Biceps brachii m.
External abdominal oblique m.
Rectus abdominis m.
Umbilicus
Brachioradialis m.
Flexor carpi radialis m.
Rectus sheath
Palmaris longus m.
Tensor fasciae latae m.
Flexor retinaculum
Sartorius m.
Pectineus m.
Gracilis m.
Adductor longus m.
Rectus femoris m.
Tendon of rectus femoris m.
Vastus lateralis m.
Patella
Vastus medialis m.
Patellar ligament
Fibularis longus m.
Gastrocnemius m.
Tibialis anterior m.
Soleus m.
Extensor digitorum longus m.
Fibularis brevis m.
Occipitalis m.
Sternocleidomastoid m.
Trapezius m.
Deltoid m.
Infraspinatus m.
Long head of triceps brachii m.
Teres minor m.
Lateral head of triceps brachii m.
Teres major m.
Latissimus dorsi m.
Medial head of triceps brachii m.
Brachioradialis m.
External abdominal oblique m.
Extensor carpi ulnaris m.
Gluteus medius m.
Flexor carpi ulnaris m.
Gluteus maximus m.
Extensor digitorum m.
Extensor retinaculum
Gracilis m.
Vastus lateralis m.
Long head of biceps femoris m.
Adductor magnus m.
Semimembranosus m.
Semitendinosus m.
Short head of biceps femoris m.
Medial head of gastrocnemius m.
Popliteal fossa
Lateral head of gastrocnemius m.
Soleus m.
Tendo calcaneus (Achilles' tendon)

Figure 9.28 An anterior view of human musculature (m = muscle).

Figure 9.29 A posterior view of human musculature (m = muscle).

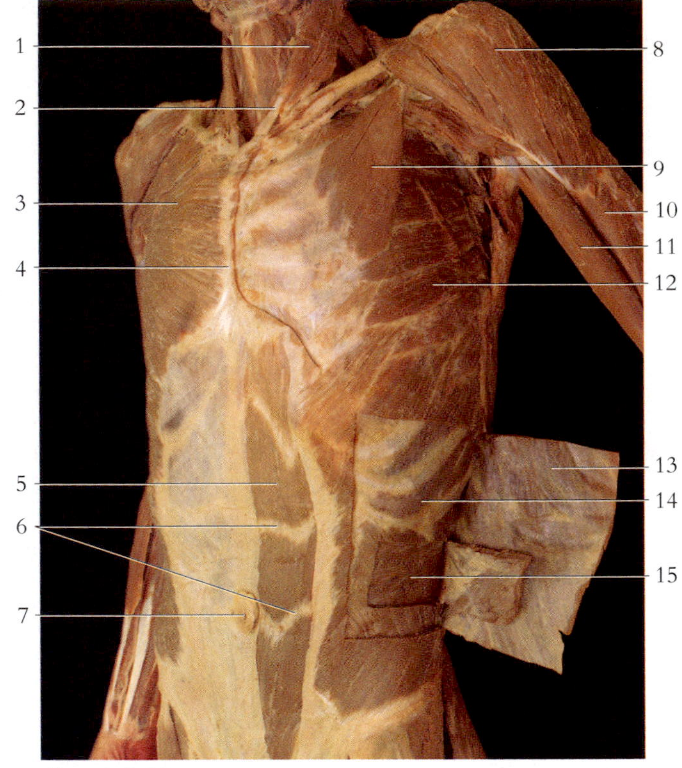

Figure 9.30 An anterolateral view of the trunk and upper arm.

1. Sternocleidomastoid m.
2. Tendon of sternocleidomastoid m.
3. Pectoralis major m.
4. Sternum
5. Rectus abdominis m.
6. Tendinous inscriptions of rectus abdominis m.
7. Umbilicus
8. Deltoid m.
9. Pectoralis minor m.
10. Brachialis m.
11. Biceps brachii m. (long head)
12. Serratus anterior m.
13. External abdominal oblique m. (reflected)
14. External intercostal m.
15. Transverse abdominis m.

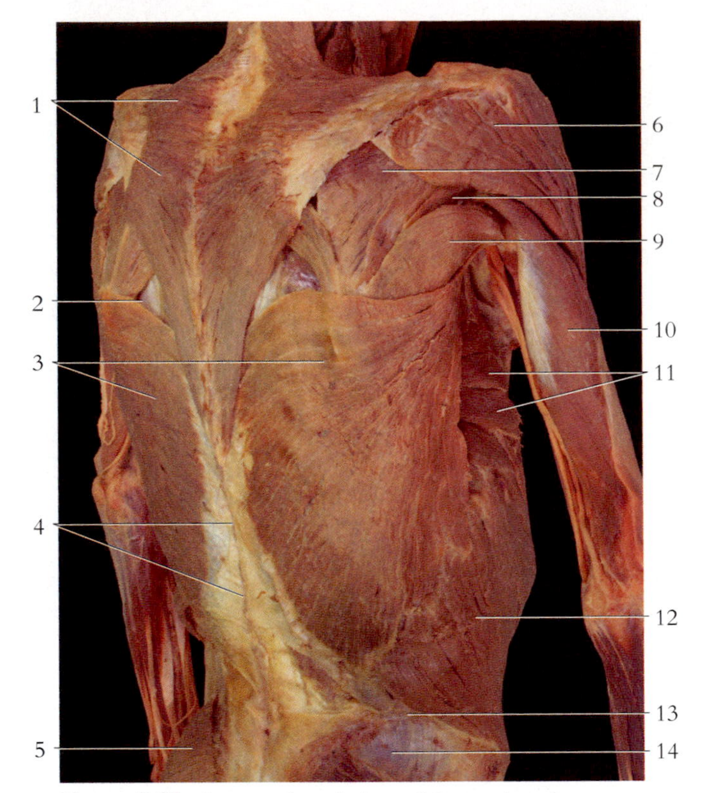

Figure 9.31 A posterolateral view of the trunk and upper arm.

1. Trapezius m.
2. Triangle of ausculation
3. Latissimus dorsi mm.
4. Vertebral column (spinous processes)
5. Gluteus maximus m.
6. Deltoid m.
7. Infraspinatus m.
8. Teres minor m.
9. Teres major m.
10. Triceps brachii m. (long head)
11. Serratus anterior mm.
12. External abdominal oblique m.
13. Iliac crest
14. Gluteus medius m.

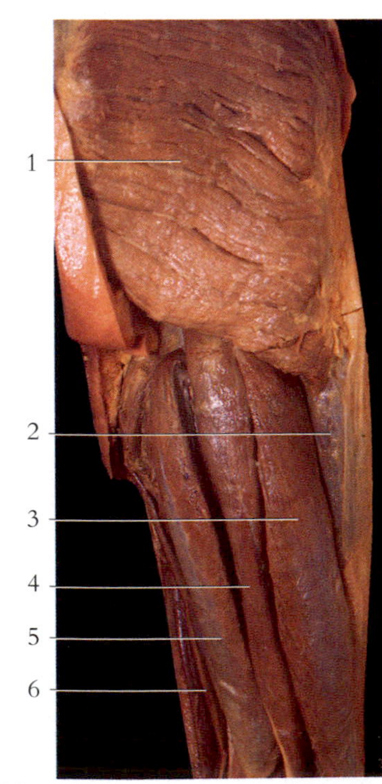

Figure 9.32 The superficial muscles of gluteal and thigh regions.

1. Gluteus maximus m.
2. Vastus lateralis m.
3. Biceps femoris m.
4. Semitendinosus m.
5. Semimembranosus m.
6. Gracilis m.

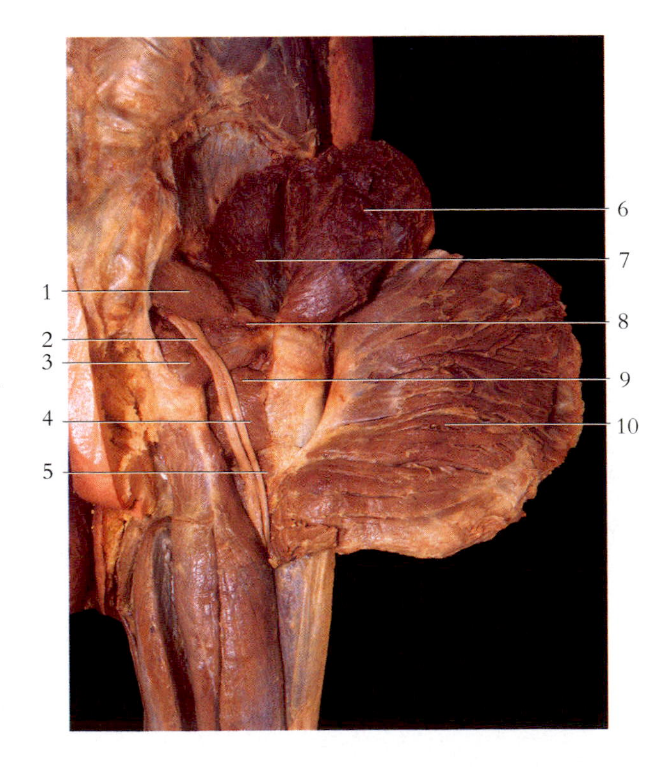

Figure 9.33 The deep structures of gluteal region.

1. Piriformis m.
2. Sciatic n.
3. Obturator internus m.
4. Quadratus femoris m.
5. Adductor minimus m.
6. Gluteus medius m. (reflected)
7. Gluteus minimus m.
8. Superior gemellus m.
9. Inferior gemellus m.
10. Gluteus maximus m. (reflected)

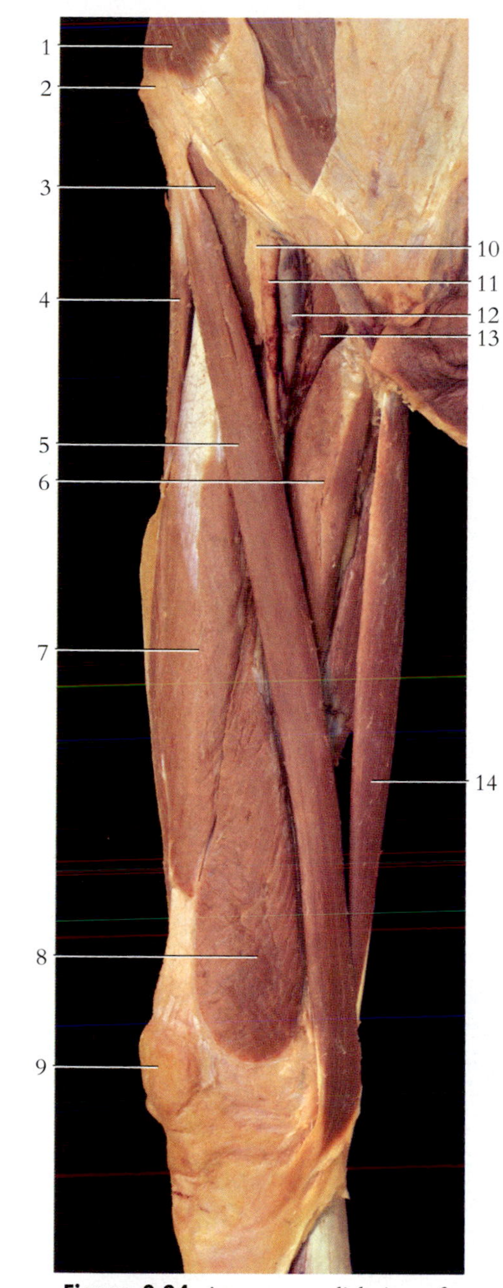

Figure 9.34 An anteromedial view of the right thigh.

1. External oblique m.
2. Anterior superior iliac spine
3. Iliopsoas m.
4. Tensor fascia lata m.
5. Sartorius m.
6. Adductor longus m.
7. Rectus femoris m.
8. Vastus medialis m.
9. Patella
10. Femoral nerve
11. Femoral artery
12. Femoral vein
13. Pectineus m.
14. Gracilis m.

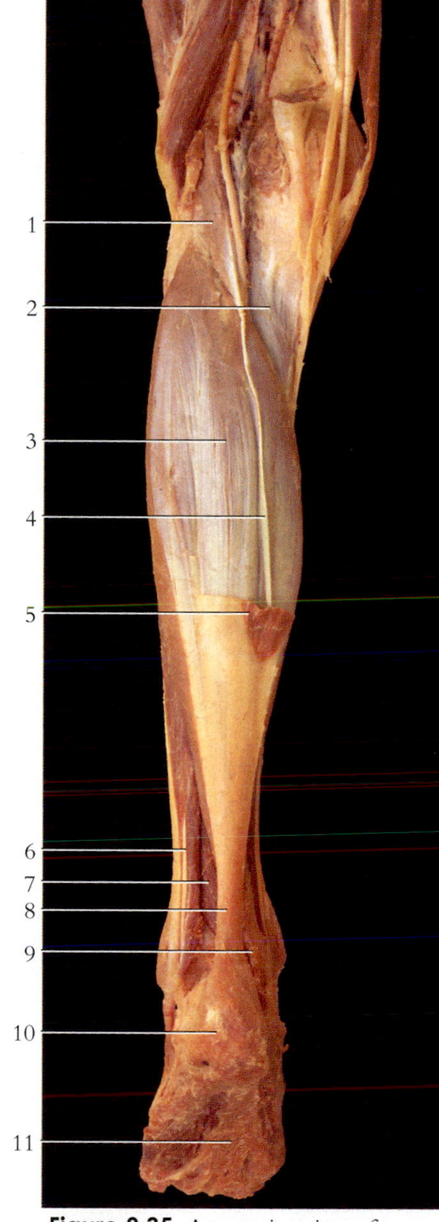

Figure 9.35 A posterior view of lower leg.

1. Plantaris m.
2. Popliteus m.
3. Soleus m.
4. Plantaris tendon
5. Gastrocnemius m. (cut)
6. Fibularis longus and brevis mm.
7. Flexor hallucis longus m.
8. Calcaneal tendon (Achilles' tendon)
9. Tendon of flexor hallucis longus m.
10. Calcaneus
11. Plantar aponeurosis

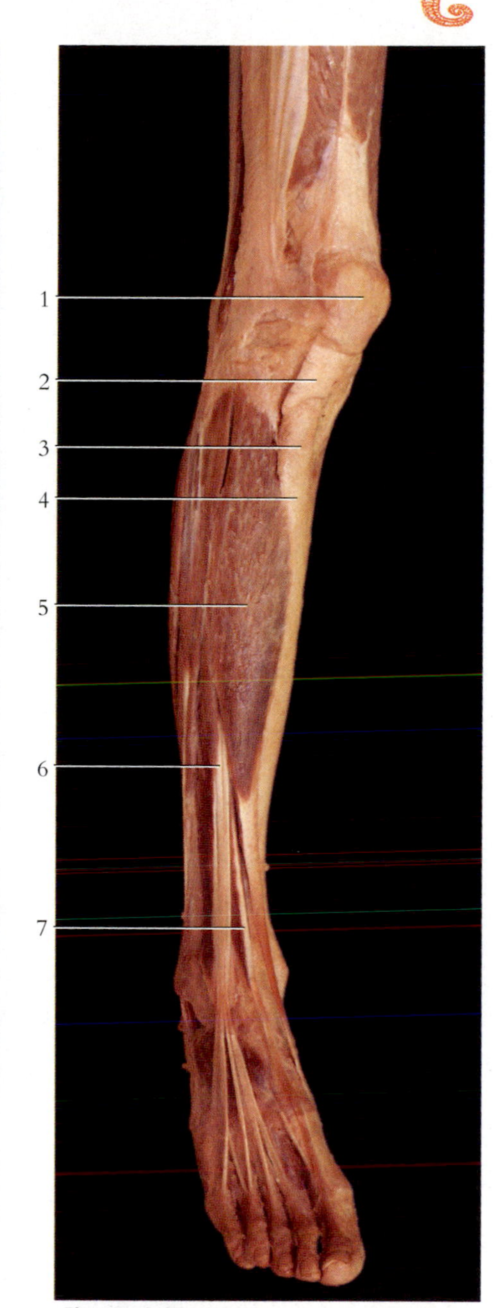

Figure 9.36 An anterior view of lower leg.

1. Patella
2. Patellar ligament
3. Tibial tuberosity
4. Tibia
5. Tibialis anterior m.
6. Tendon of extensor digitorum longus m.
7. Tendon of extensor hallucis longus m.

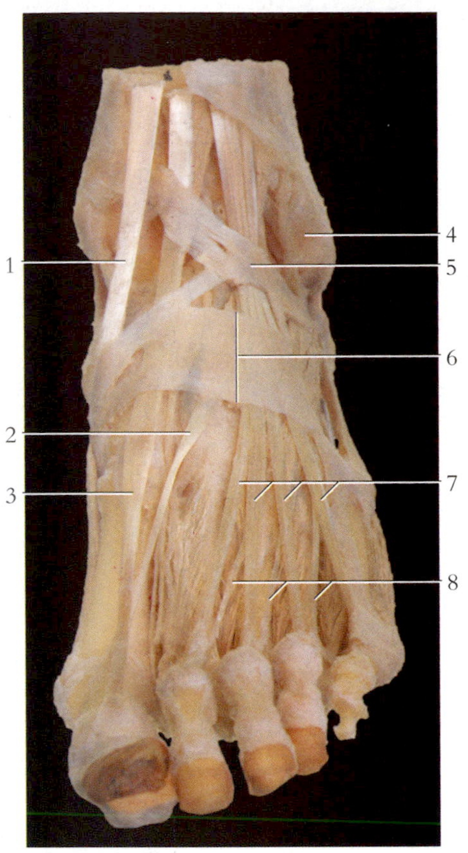

Figure 9.37 An anterior view of dorsum of foot.

1. Tendon of tibialis anterior m.
2. Tendon of extensor hallucis brevis m.
3. Tendon of extensor hallucis longus m.
4. Lateral malleolus
5. Superior extensor retinaculum
6. Inferior extensor retinaculum
7. Tendon of extensor digitorum longus m.
8. Tendon of extensor digitorum brevis m.

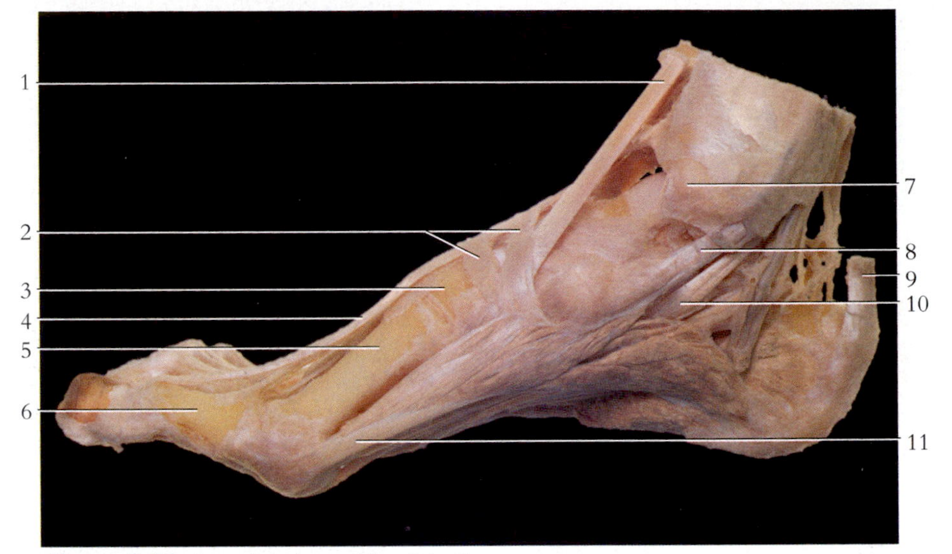

Figure 9.38 A medial view of the right foot

1. Tendon of tibialis anterior m.
2. Extensor retinaculum
3. Medial cuneiform
4. Tendon of extensor hallucis longus m.
5. First metatarsal bone
6. Proximal phalanx of hallux
7. Medial malleolus of tibia
8. Tendon of tibialis posterior m.
9. Tendo calcaneus
10. Tendon of flexor digitorum longus m.
11. Abductor hallucis m.

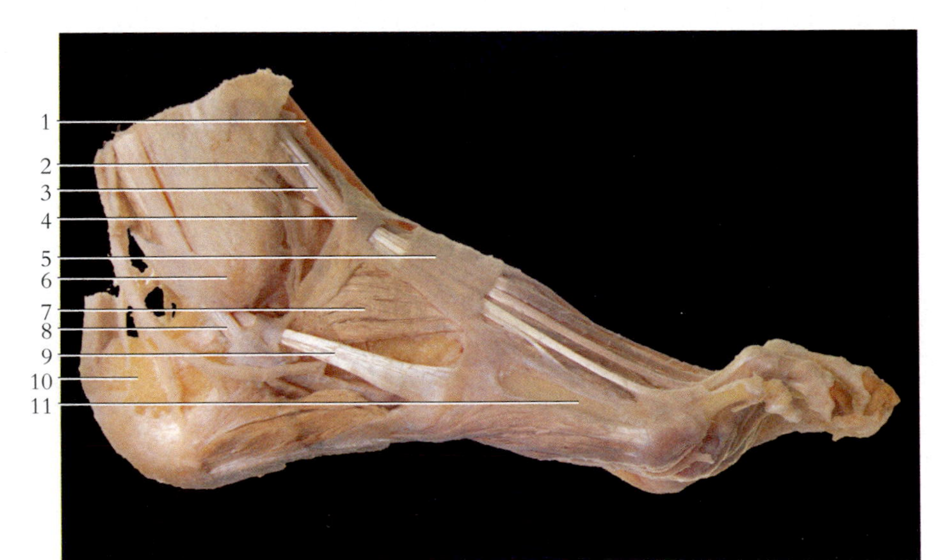

Figure 9.39 A lateral view of the right foot.

1. Tendon of tibialis anterior
2. Tendon of extensor digitorum longus m.
3. Tendon of fibularis tertius m.
4. Superior extensor retinaculum
5. Inferior extensor retinaculum
6. Lateral malleolus of fibula
7. Extensor digitorum brevis m.
8. Tendon of fibularis longus m.
9. Tendon of fibularis brevis m.
10. Calcaneus
11. 5th metatarsal bone

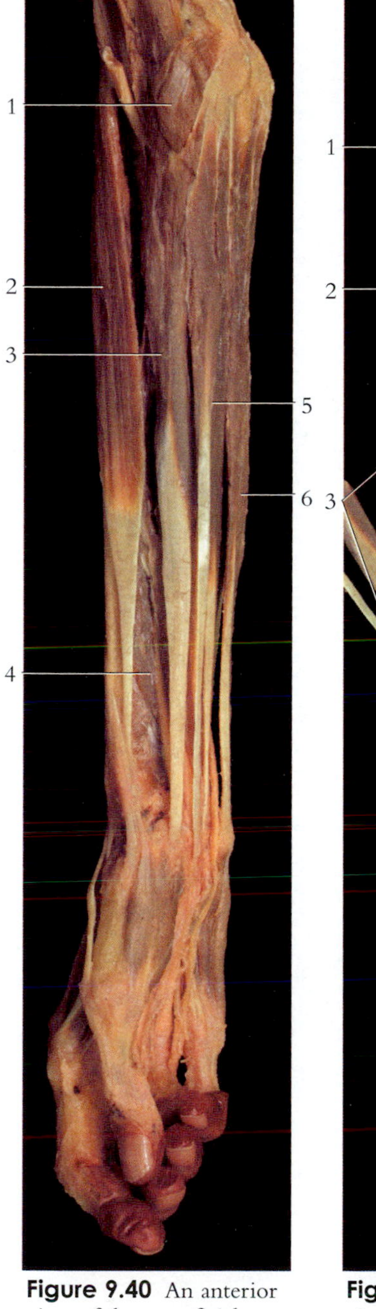

Figure 9.40 An anterior view of the superficial muscles of the right forearm.

1. Pronator teres m.
2. Brachioradialis m.
3. Flexor carpi radialis m.
4. Flexor pollicis longus m.
5. Palmaris longus m.
6. Flexor carpi ulnaris m.

Figure 9.41 An anterior view of the muscles of the right forearm.

1. Pronator teres m.
2. Brachioradialis m. (cut and reflected)
3. Palmaris longus m. (cut and reflected)
4. Flexor carpi ulnaris m. (cut)
5. Flexor carpi radialis m. (cut and reflected)
6. Flexor digitorum superficialis m.
7. Flexor pollicis longus m.
8. Pronator quadratus m.

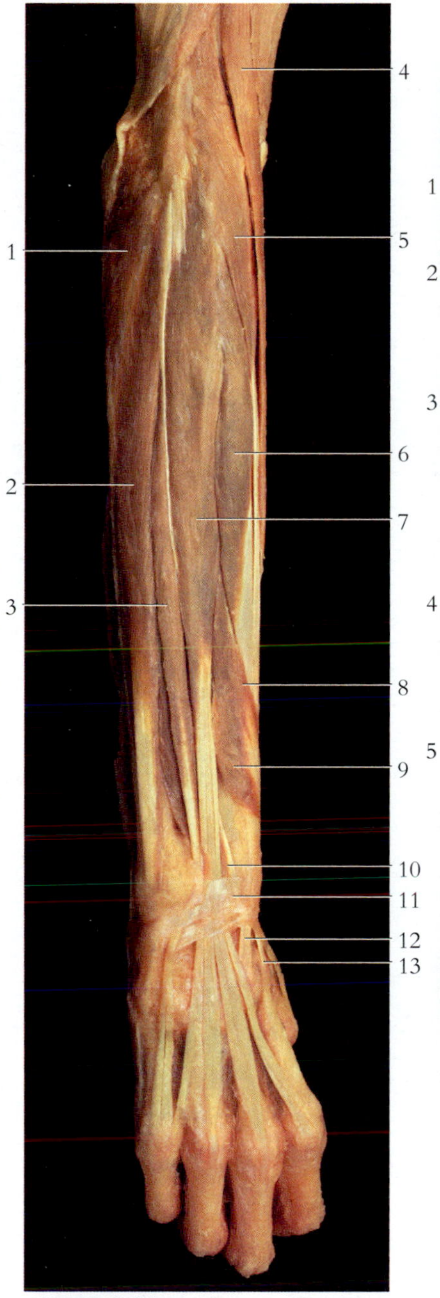

Figure 9.42 A posterior view of the superficial muscles of the right forearm.

1. Anconeus m.
2. Extensor carpi ulnaris m.
3. Extensor digiti minimi m.
4. Brachioradialis m.
5. Extensor carpi radialis longus m.
6. Extensor carpi radialis brevis m.
7. Extensor digitorum m.
8. Abductor pollicis longus m.
9. Extensor pollicis brevis m.
10. Extensor pollicis longus m.
11. Extensor retinaculum
12. Tendon of extensor carpi radialis brevis
13. Tendon of extensor carpi radialis longus

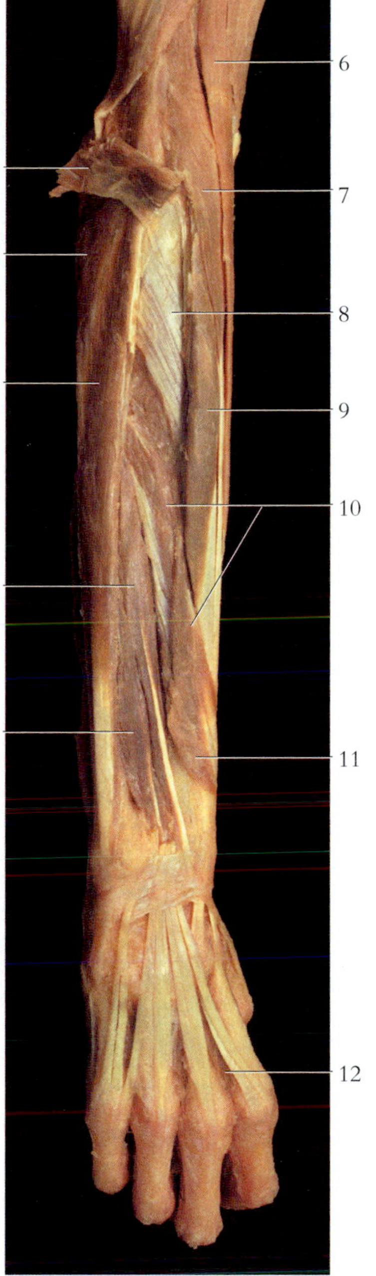

Figure 9.43 A posterior view of the deep muscles of the right forearm.

1. Extensor digitorum m. (cut and reflected)
2. Anconeus m.
3. Extensor carpi ulnaris m.
4. Extensor pollicis longus m.
5. Extensor indicis m.
6. Brachioradialis m.
7. Extensor carpi radialis longus m.
8. Supinator m.
9. Extensor carpi radialis brevis m.
10. Abductor pollicis longus m.
11. Extensor pollicis brevis m.
12. Dorsal interosseous m.

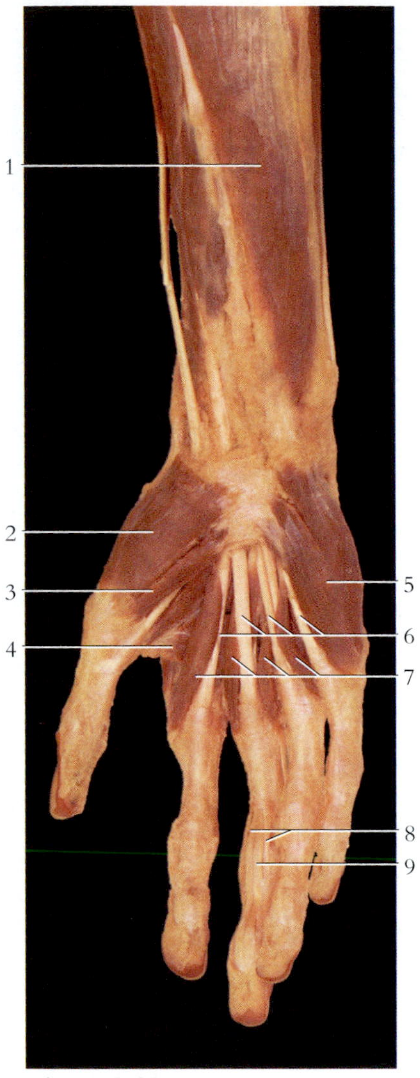

Figure 9.44 An anterior view of right hand

1. Flexor carpi ulnaris m.
2. Abductor pollicis brevis m.
3. Flexor pollicis brevis m.
4. Adductor pollicis m.
5. Hypothenar mm.
6. Tendons of flexor digitorum superficialis
7. Lumbrical mm.
8. Flexor digitorum superficialis tendon (bifurcated for insertion)
9. Tendon of flexor digitorum profundus

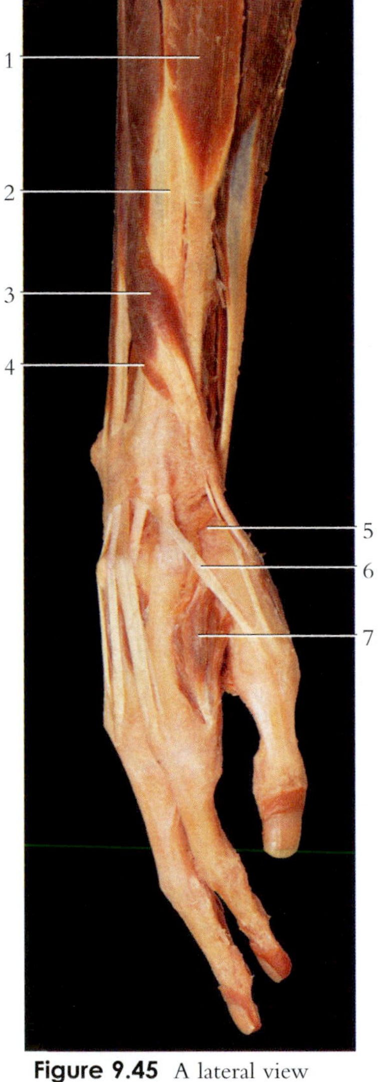

Figure 9.45 A lateral view of right hand

1. Extensor carpi radialis longus m.
2. Tendon of extensor carpi radialis brevis m.
3. Abductor pollicis longus m.
4. Extensor pollicis brevis m.
5. Anatomical snuff box
6. Tendon of extensor pollicis longus m.
7. First dorsal interosseus m.

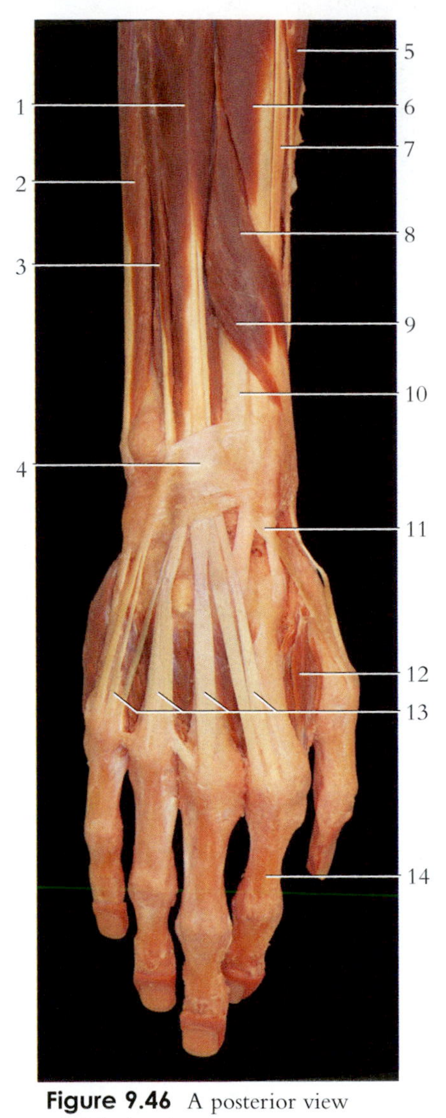

Figure 9.46 A posterior view of right hand

1. Extensor digitorum m.
2. Extensor carpi ulnaris m.
3. Extensor digiti minimi m.
4. Extensor retinaculum
5. Brachioradialis m.
6. Extensor carpi radialis brevis m.
7. Tendon of extensor carpi radialis longus m.
8. Abductor pollicis longus m.
9. Extensor pollicis brevis m.
10. Radius
11. Tendon of extensor pollicis longus m.
12. First dorsal interosseus m.
13. Extensor digitorum tendons
14. Extensor expansion

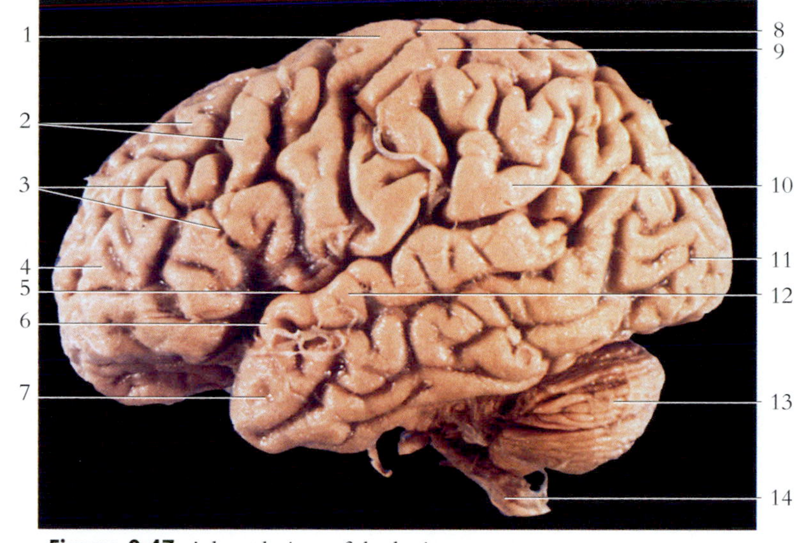

Figure 9.47 A lateral view of the brain.

1. Primary motor cerebral cortex
2. Gyri
3. Sulci
4. Frontal lobe of cerebrum
5. Lateral sulcus
6. Olfactory cerebral cortex
7. Temporal lobe of cerebrum
8. Central sulcus
9. Primary sensory cerebral cortex
10. Parietal lobe of cerebrum
11. Occipital lobe of cerebrum
12. Auditory cerebral cortex
13. Cerebellum
14. Medulla oblongata

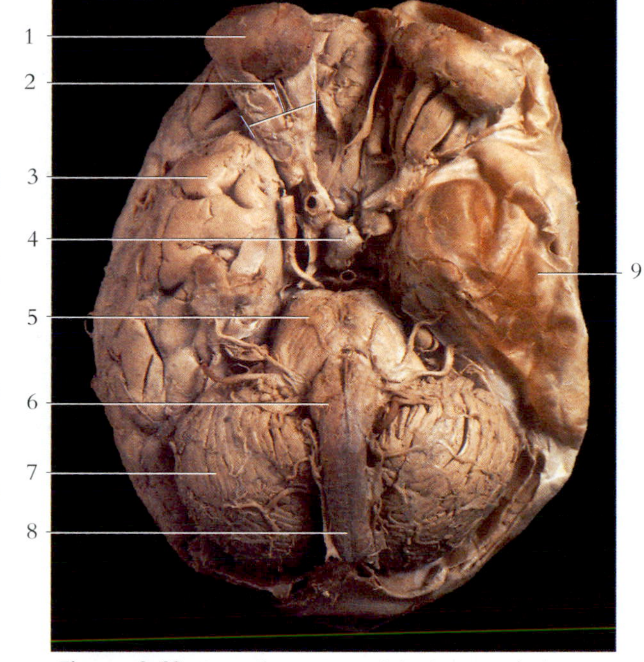

Figure 9.48 An inferior view of the brain with the eyes and part of the meninges still intact.

1. Eyeball
2. Muscles of the eye
3. Temporal lobe of cerebrum
4. Pituitary gland
5. Pons
6. Medulla oblongata
7. Cerebellum
8. Spinal cord
9. Dura mater

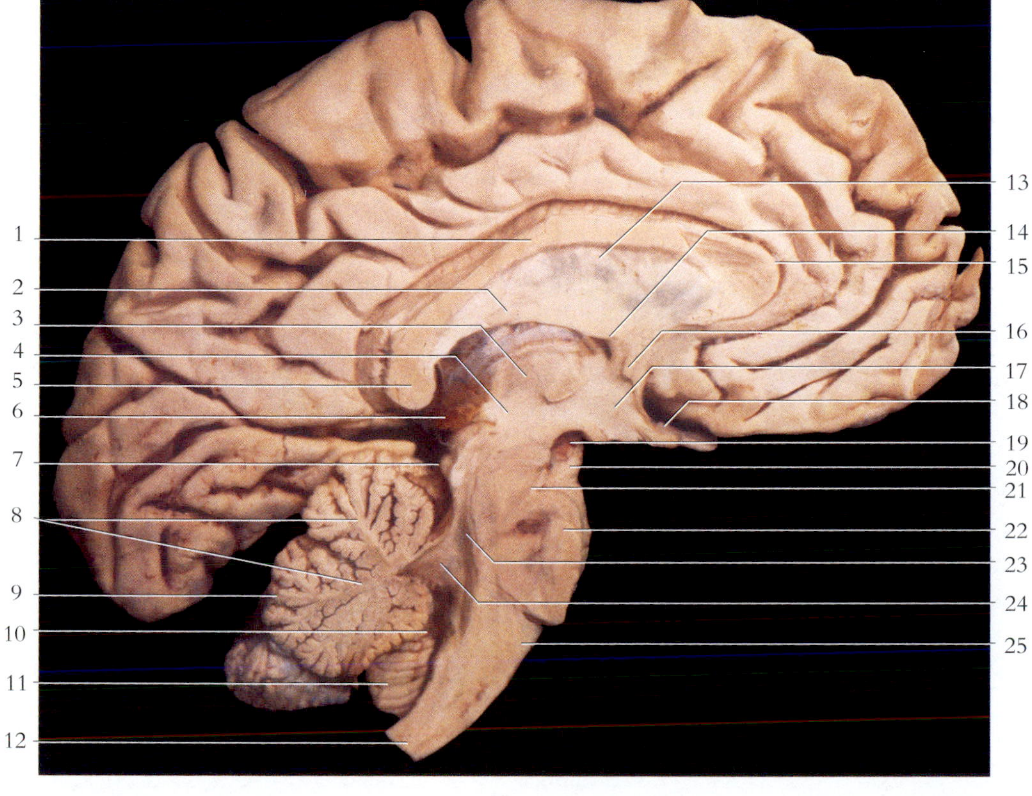

Figure 9.49 A sagittal view of the brain.

1. Body of corpus callosum
2. Crus of fornix
3. 3rd ventricle
4. Posterior commissure
5. Splenium of corpus callosum
6. Pineal body
7. Inferior colliculus
8. Arbor vitae of cerebellum
9. Vermis of cerebellum
10. Choroid plexus of 4th ventricle
11. Tonsilla of cerebellum
12. Medulla oblongata
13. Septum pellucidum
14. Intraventricular foramen
15. Genu of corpus callosum
16. Anterior commissure
17. Hypothalamus
18. Optic chiasma
19. Oculomotor nerve
20. Cerebral peduncle
21. Midbrain
22. Pons
23. Mesencephalic (cerebral) aqueduct
24. 4th ventricle
25. Pyramid of medulla oblongata

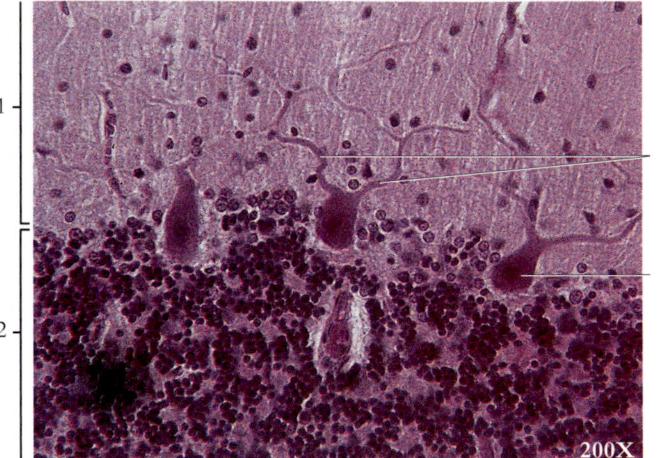

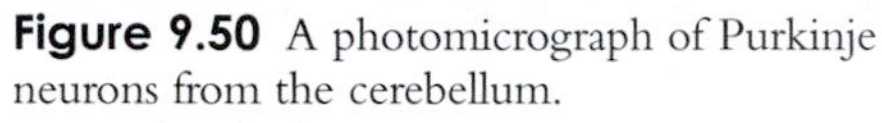

Figure 9.50 A photomicrograph of Purkinje neurons from the cerebellum.
1. Molecular layer of cerebellar cortex
2. Granular layer of cerebellar cortex
3. Dendrites of Purkinje cell
4. Purkinje cell body

Figure 9.51 A transverse section of the spinal cord.
1. Posterior (dorsal) root of spinal nerve
2. Posterior (dorsal) horn (gray matter)
3. Spinal cord tract (white matter)
4. Anterior (ventral) horn (gray matter)

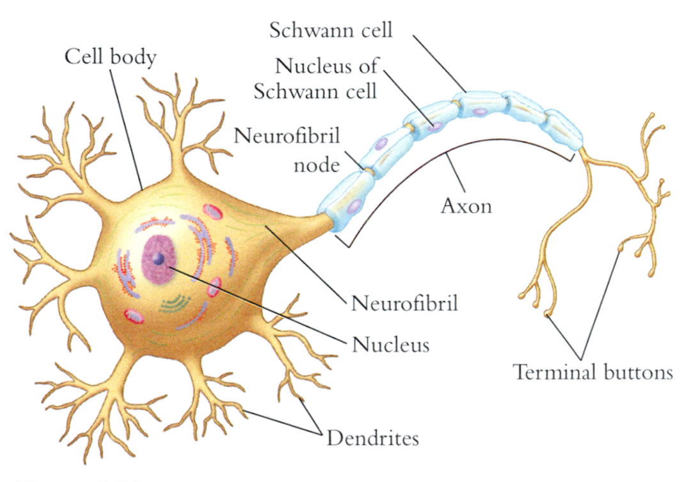

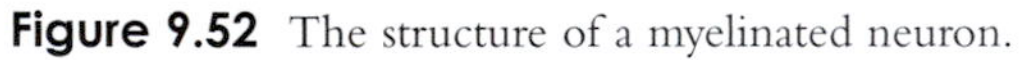

Figure 9.52 The structure of a myelinated neuron.

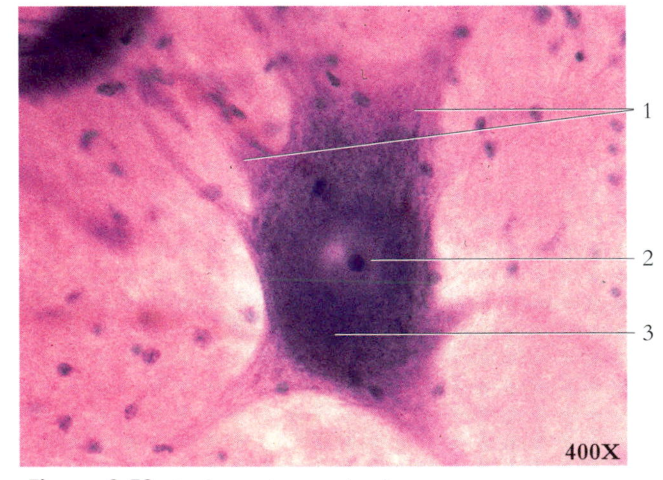

Figure 9.53 A photomicrograph of a neuron.
1. Cytoplasmic extensions
2. Nucleolus
3. Cell body of neuron

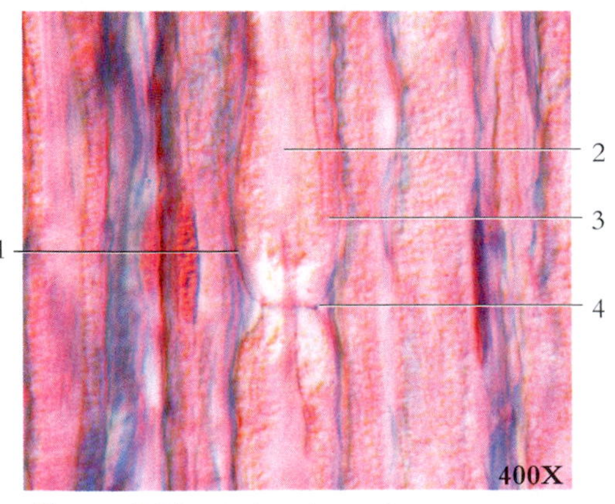

Figure 9.54 The histology of a myelinated nerve.
1. Endoneurium
2. Axon
3. Myelin layer
4. Neurofibril node (node of Ranvier)

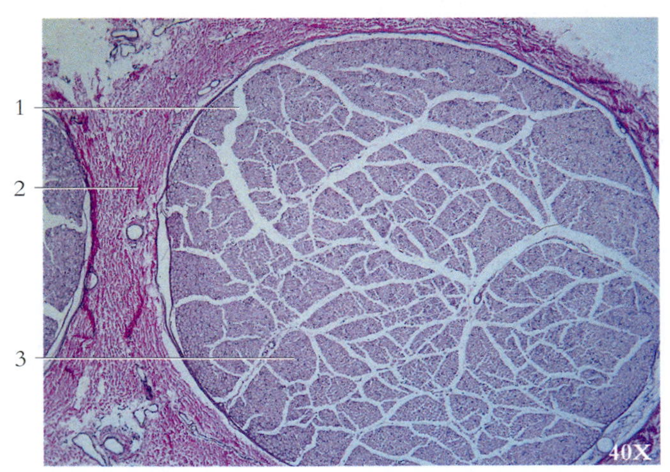

Figure 9.55 A transverse section of a nerve.
1. Perineurium
2. Epineurium
3. Bundle of axons

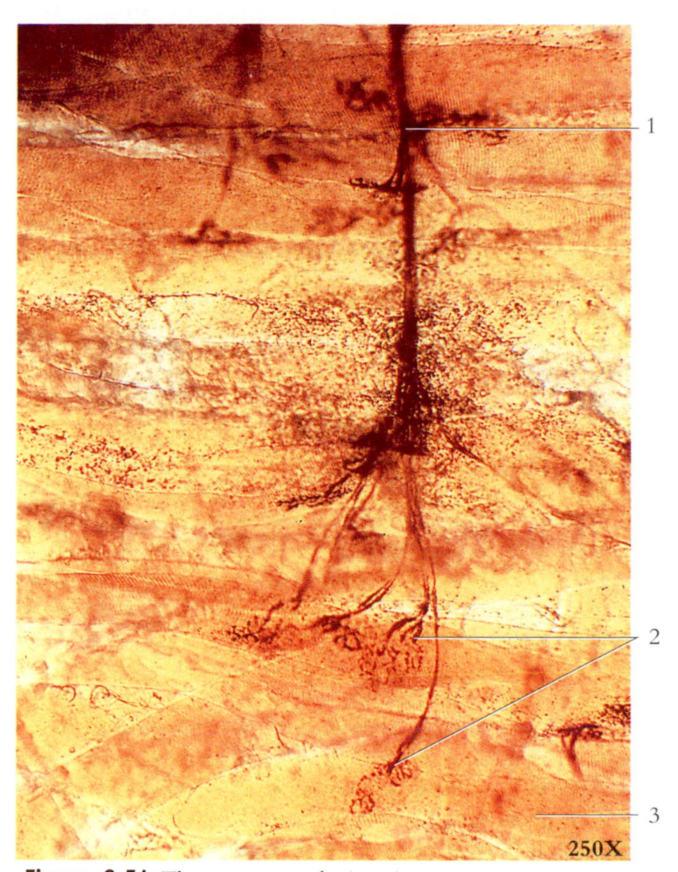

Figure 9.56 The neuromuscular junction.
1. Motor nerve
2. Motor end plates
3. Skeletal muscle fiber

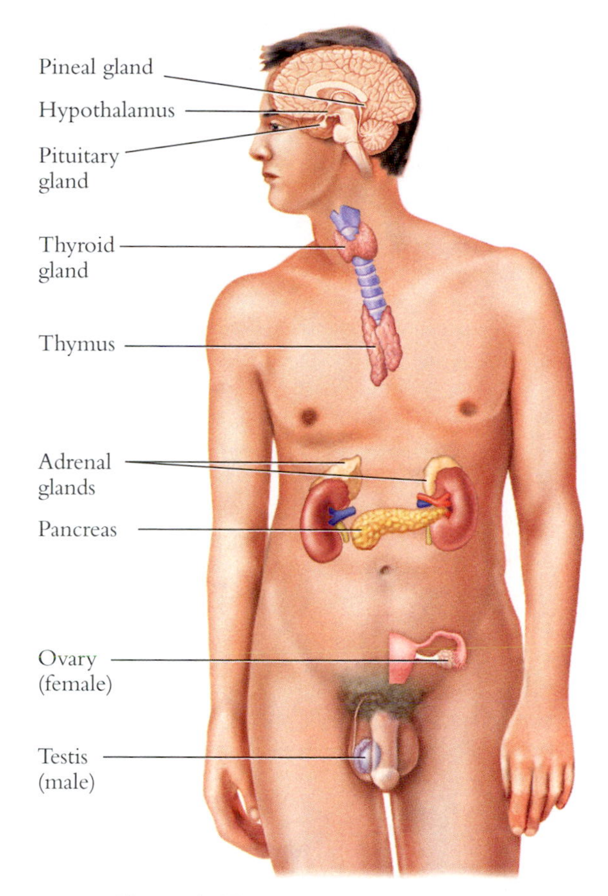

Figure 9.57 The principal endocrine glands.

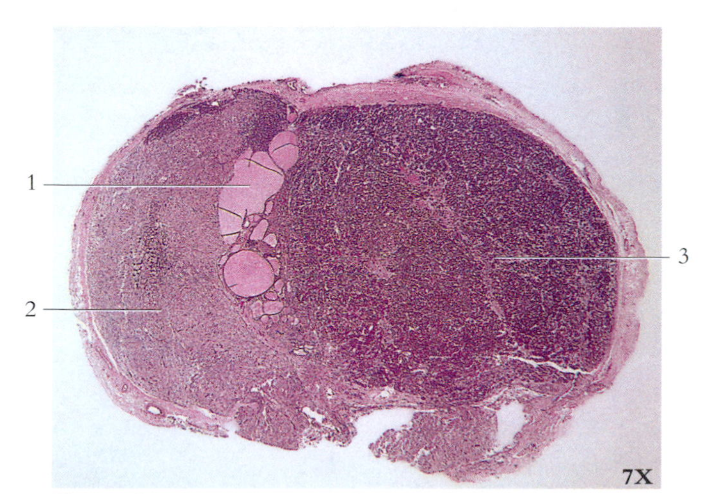

Figure 9.58 The pituitary gland.
1. Pars intermedia (adenohypophysis)
2. Pars nervosa (neurohypophysis)
3. Pars distalis (adenohypophysis)

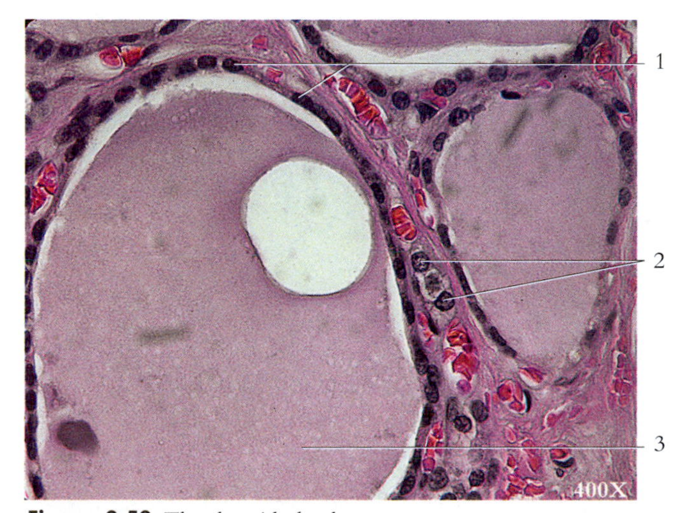

Figure 9.59 The thyroid gland.
1. Follicle cells
2. C cells
3. Colloid within follicle

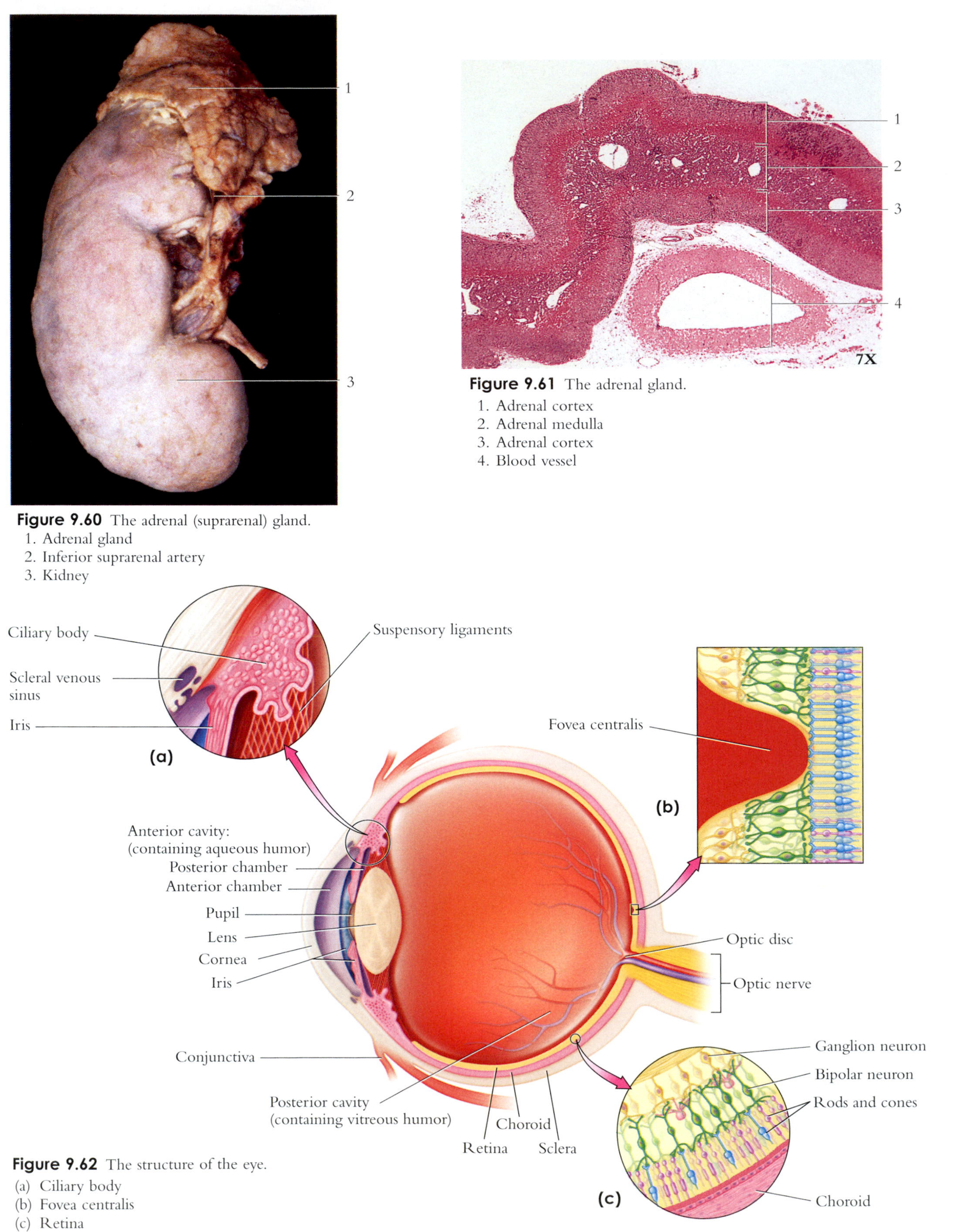

Figure 9.60 The adrenal (suprarenal) gland.
1. Adrenal gland
2. Inferior suprarenal artery
3. Kidney

Figure 9.61 The adrenal gland.
1. Adrenal cortex
2. Adrenal medulla
3. Adrenal cortex
4. Blood vessel

Figure 9.62 The structure of the eye.
(a) Ciliary body
(b) Fovea centralis
(c) Retina

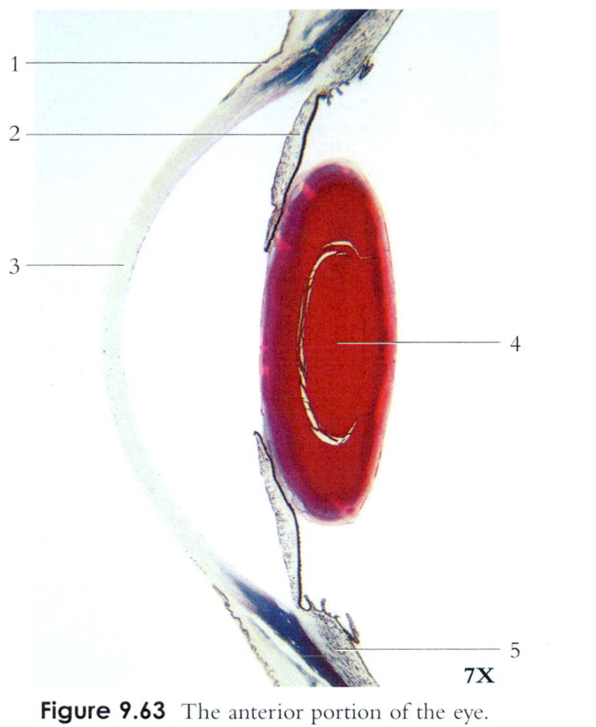

Figure 9.63 The anterior portion of the eye.
1. Conjunctiva
2. Iris
3. Cornea
4. Lens
5. Ciliary body

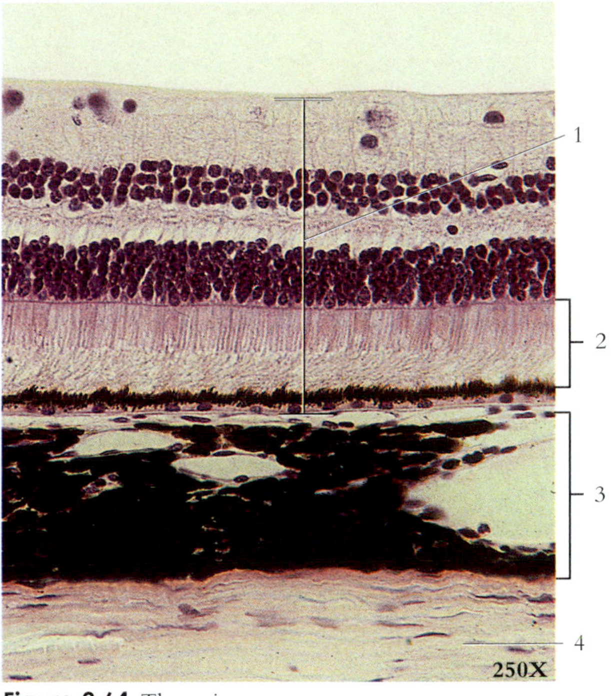

Figure 9.64 The retina.
1. Retina
2. Rods and cones
3. Choroid
4. Sclera

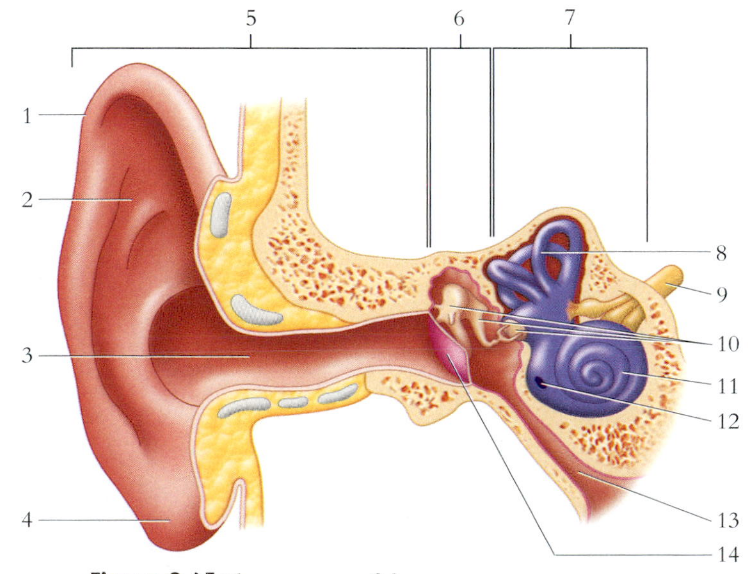

Figure 9.65 The structure of the ear.
1. Helix
2. Auricle
3. External auditory canal
4. Earlobe
5. Outer ear
6. Middle ear
7. Inner ear
8. Semicircular canals
9. Vestibulocochlear nerve
10. Auditory ossicles
11. Cochlea
12. Vestibular (oval) window
13. Auditory tube
14. Tympanic membrane

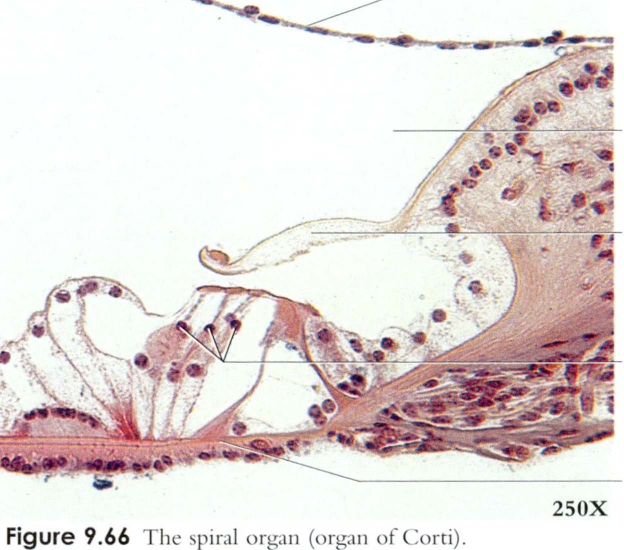

Figure 9.66 The spiral organ (organ of Corti).
1. Vestibular membrane
2. Cochlear duct
3. Tectorial membrane
4. Hair cells
5. Basilar membrane

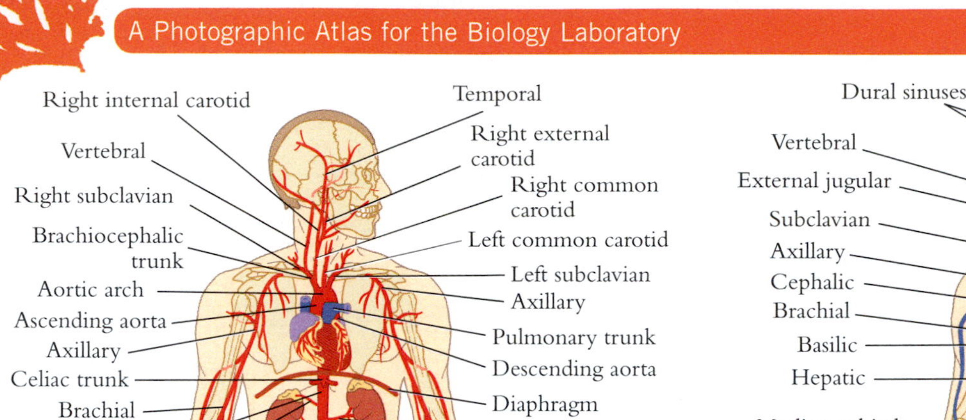
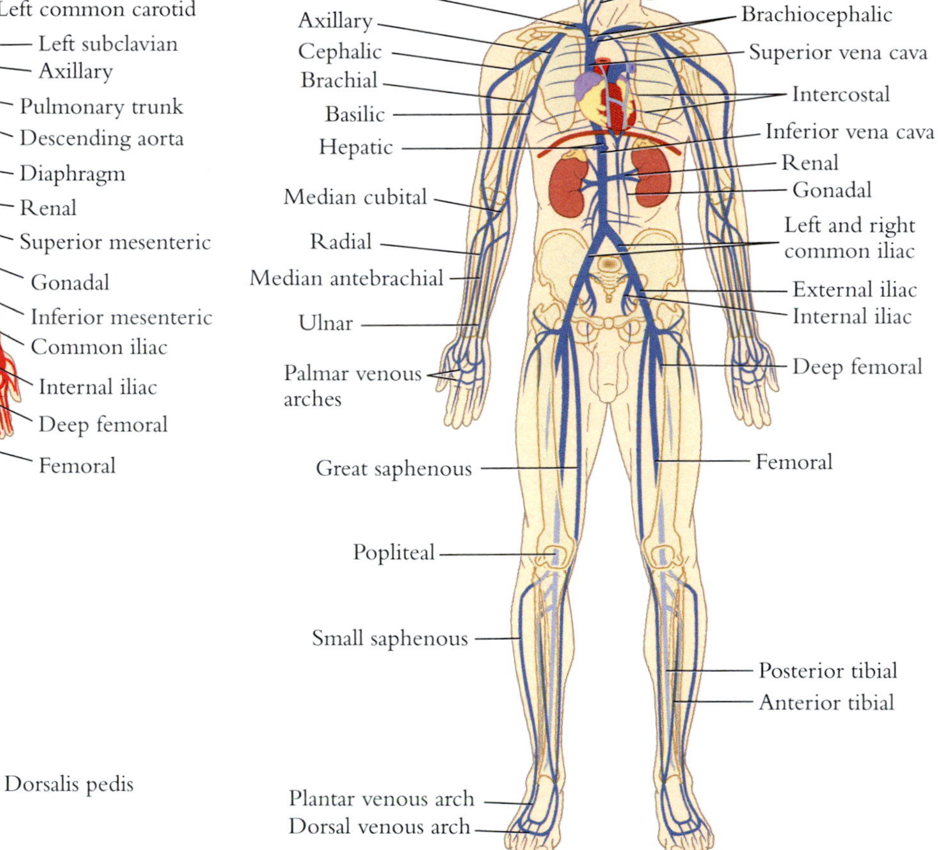

Figure 9.67 The principal arteries of the body.

Figure 9.68 The principal veins of the body.

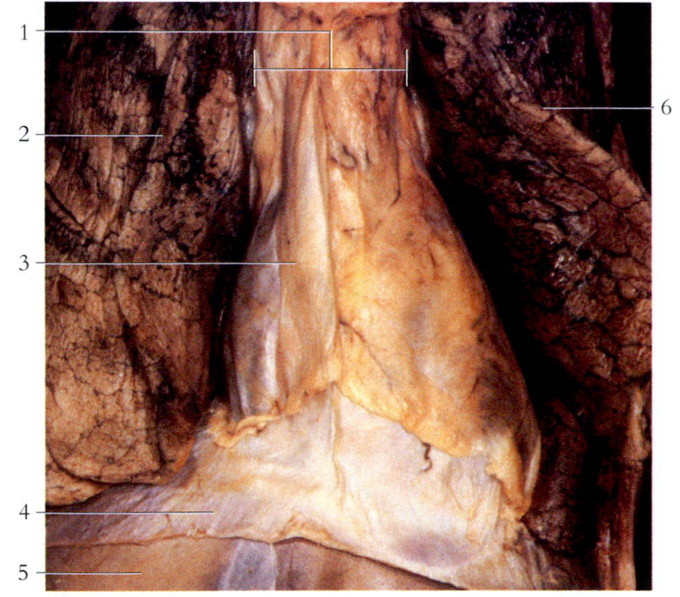

Figure 9.69 The position of the heart within the pericardium.

1. Mediastinum
2. Right lung
3. Pericardium
4. Diaphragm
5. Liver
6. Left lung

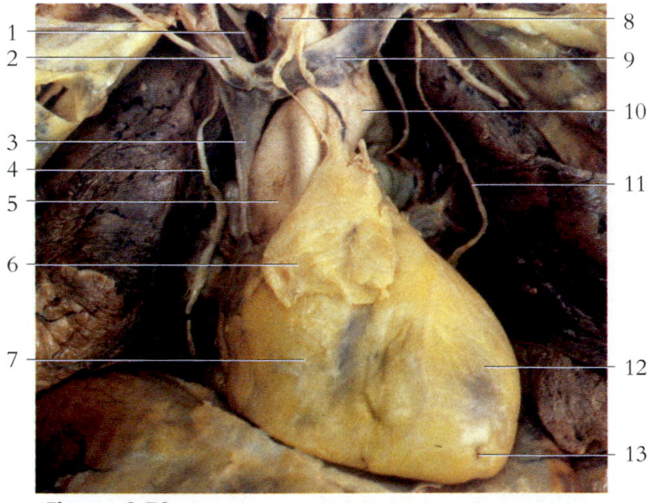

Figure 9.70 Anterior view of the heart and associated structures.

1. Right vagus nerve
2. Right brachiocephalic vein
3. Superior vena cava
4. Right phrenic nerve
5. Ascending aorta
6. Pericardium (cut)
7. Right ventricle of heart
8. Brachiocephalic artery
9. Left brachiocephalic vein
10. Aortic arch
11. Left phrenic nerve
12. Left ventricle of heart
13. Apex of heart

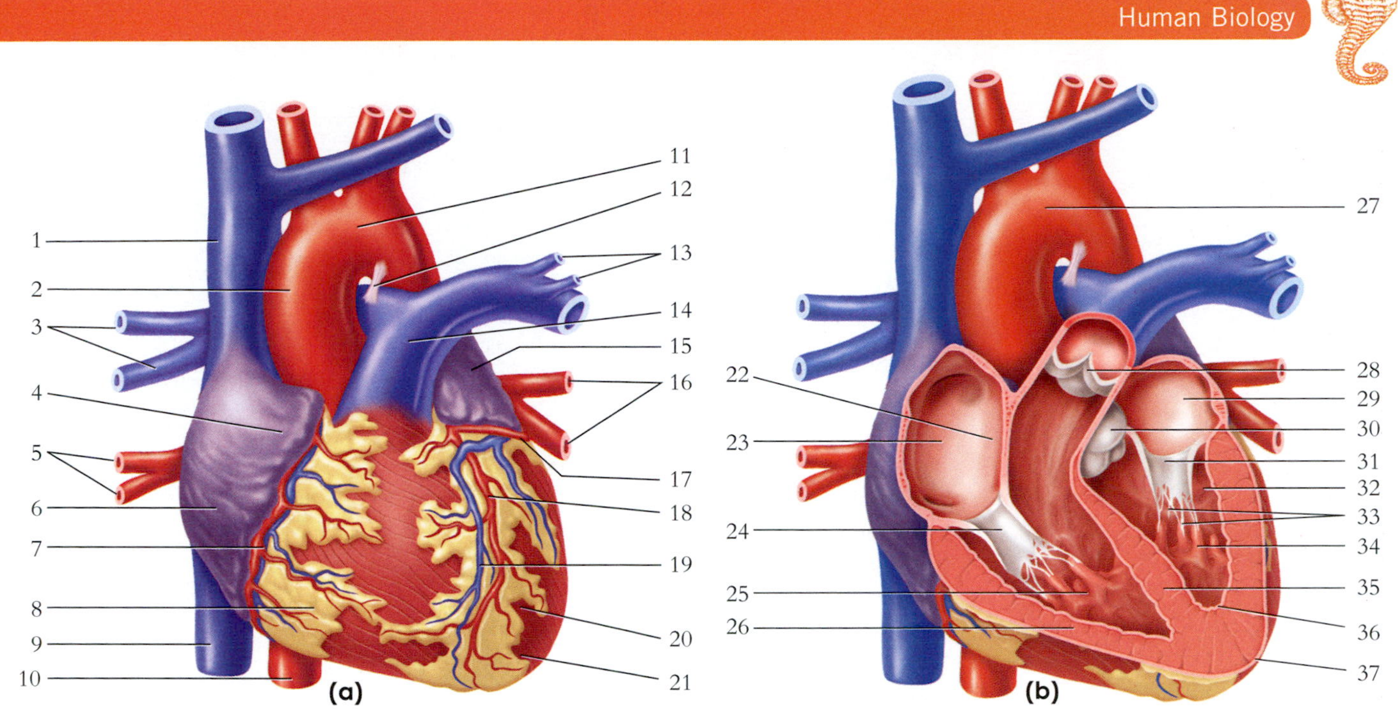

Figure 9.71 (a) An anterior view of the structure of the heart and (b) the internal view of the structure of the heart.

1. Superior vena cava
2. Ascending aorta
3. Branches of right pulmonary artery
4. Auricle of right atrium
5. Right pulmonary veins
6. Right atrium
7. Right coronary artery and vein
8. Right ventricle
9. Inferior vena cava
10. Thoracic aorta
11. Aortic arch
12. Ligamentum arteriosum
13. Branches of left pulmonary artery
14. Pulmonary trunk
15. Left atrium
16. Left pulmonary veins
17. Circumflex artery
18. Anterior interventricular artery
19. Anterior interventricular vein
20. Left ventricle
21. Apex of heart
22. Interatrial septum
23. Right atrium
24. Tricuspid valve
25. Right ventricle
26. Myocardium
27. Aortic arch
28. Pulmonary valve
29. Left atrium
30. Aortic valve
31. Bicuspid valve
32. Left ventricle
33. Chordae tendinae
34. Papillary muscle
35. Interventricular septum
36. Endocardium
37. Visceral pericardium

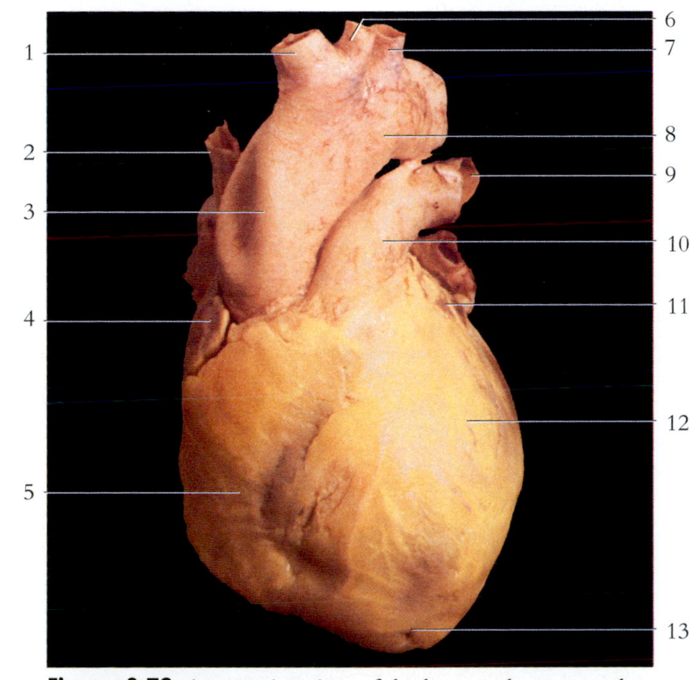

Figure 9.72 An anterior view of the heart and great vessels.

1. Brachiocephalic trunk
2. Superior vena cava
3. Ascending aorta
4. Right atrium
5. Right ventricle
6. Left common carotid artery
7. Left subclavian artery
8. Aortic arch
9. Pulmonary artery
10. Pulmonary trunk
11. Left atrium
12. Left ventricle
13. Apex of heart

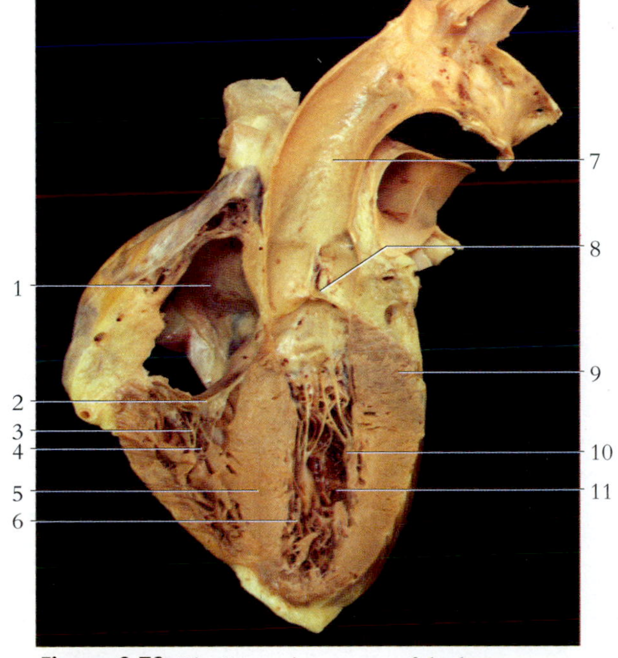

Figure 9.73 The internal structure of the heart.

1. Right atrium
2. Right atrioventricular valve
3. Chordae tendinae
4. Right ventricle
5. Interventricular septum
6. Trabeculae carneae
7. Ascending aorta
8. Aortic valve
9. Myocardium
10. Papillary muscle
11. Left ventricle

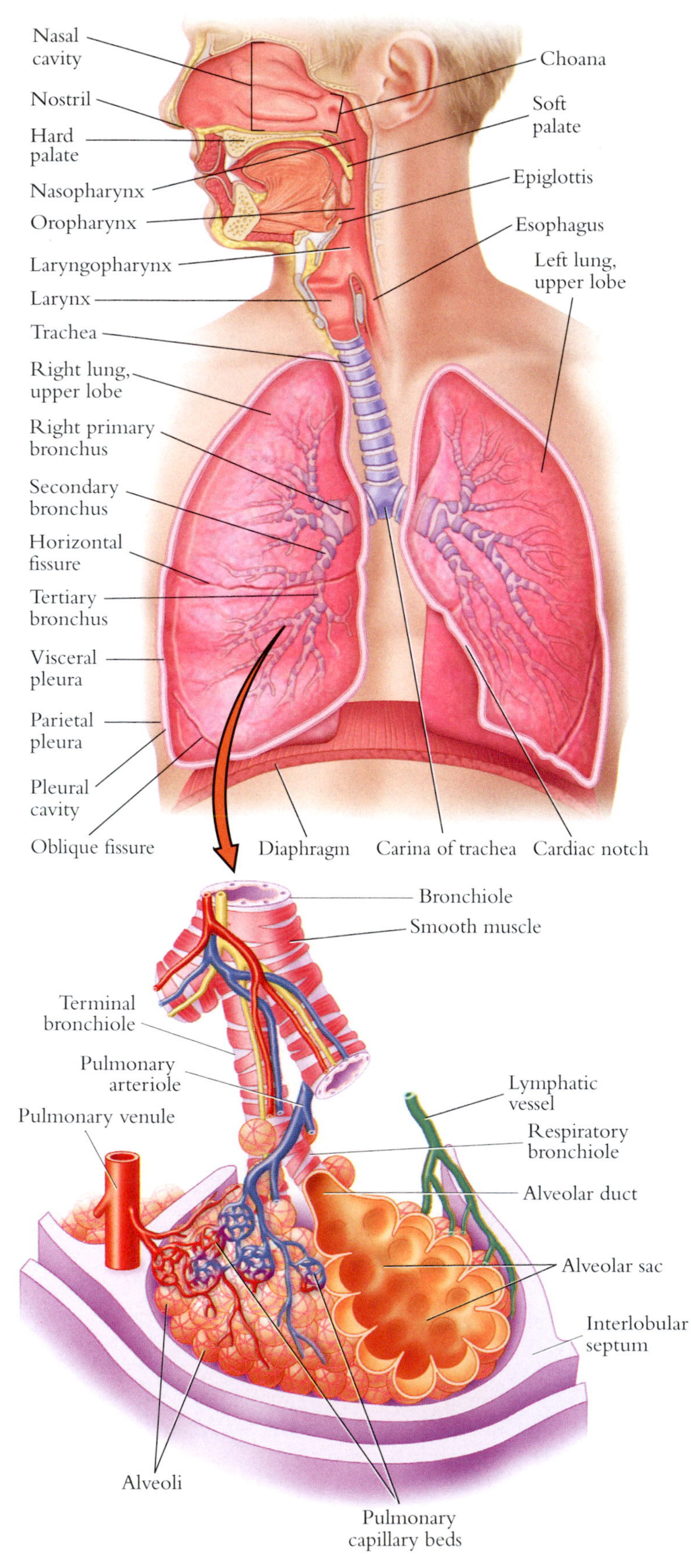

Figure 9.74 The structure of the respiratory system.

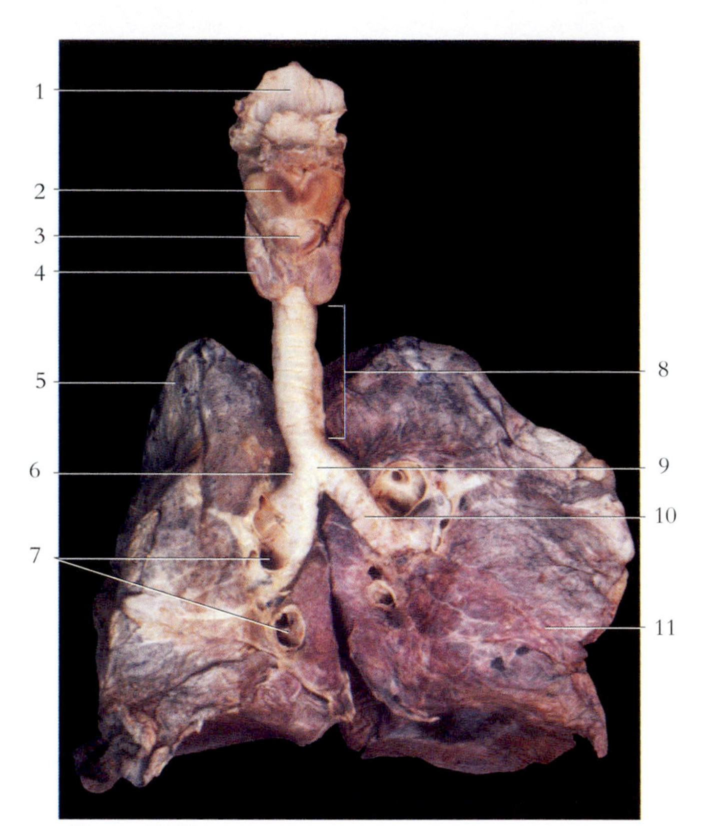

Figure 9.75 An anterior view of the larynx, trachea, and lungs.

1. Epiglottis
2. Thyroid cartilage
3. Cricoid cartilage
4. Thyroid gland
5. Right lung
6. Right principal (primary) bronchus
7. Pulmonary vessels
8. Trachea
9. Carina
10. Left principal (primary) bronchus
11. Left lung

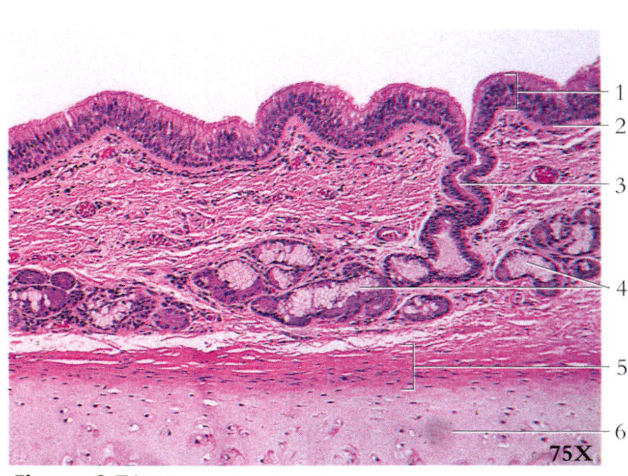

Figure 9.76 The tracheal wall.

1. Respiratory epithelium
2. Basement membrane
3. Duct of seromucous gland
4. Seromucous glands
5. Perichondrium
6. Hyaline cartilage

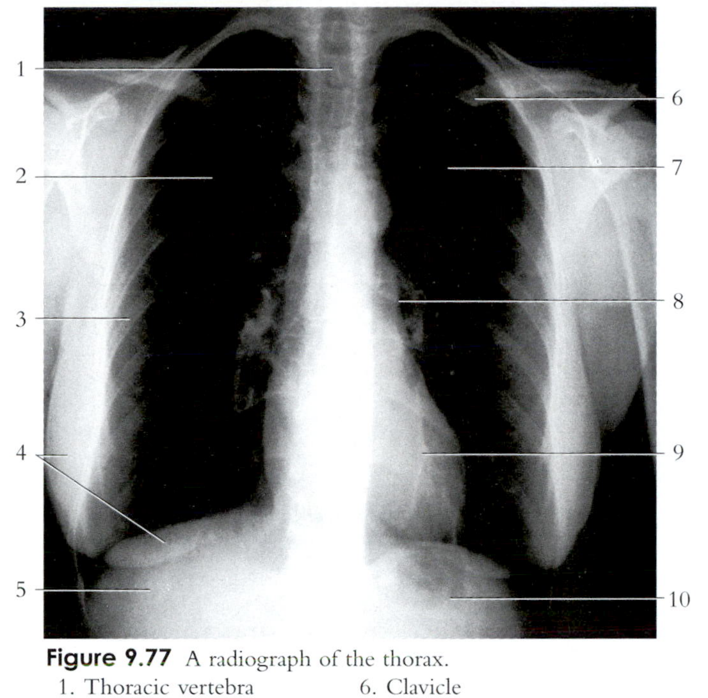

Figure 9.77 A radiograph of the thorax.

1. Thoracic vertebra
2. Right lung
3. Rib
4. Image of right breast
5. Diaphragm/liver
6. Clavicle
7. Left lung
8. Mediastinum
9. Heart
10. Diaphragm/stomach

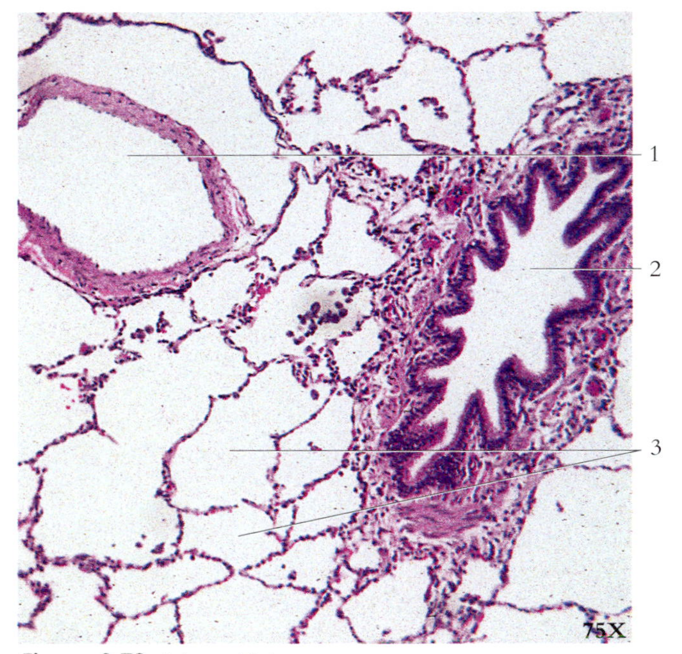

Figure 9.78 A bronchiole.

1. Pulmonary arteriole
2. Bronchiole
3. Pulmonary alveoli

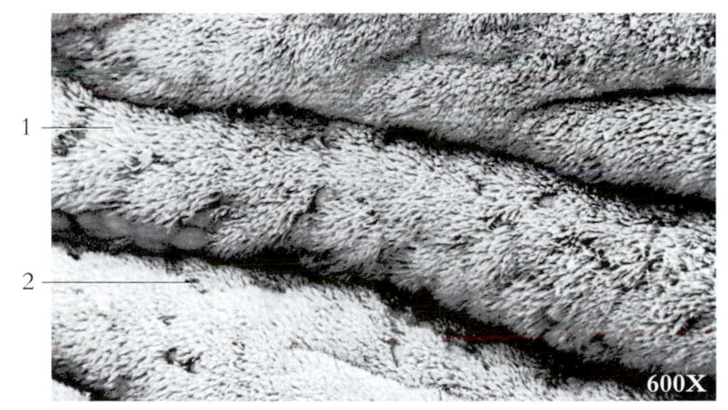

Figure 9.79 An electron micrograph of the lining of the trachea.

1. Cilia
2. Goblet cell

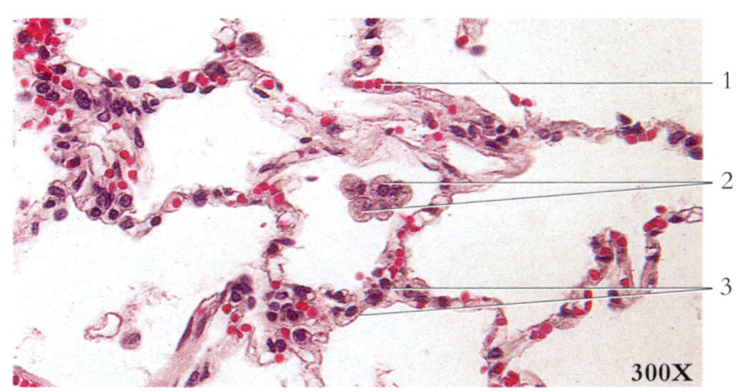

Figure 9.80 The pulmonary alveoli.

1. Capillary in alveolar wall
2. Macrophages
3. Type II pneumocytes

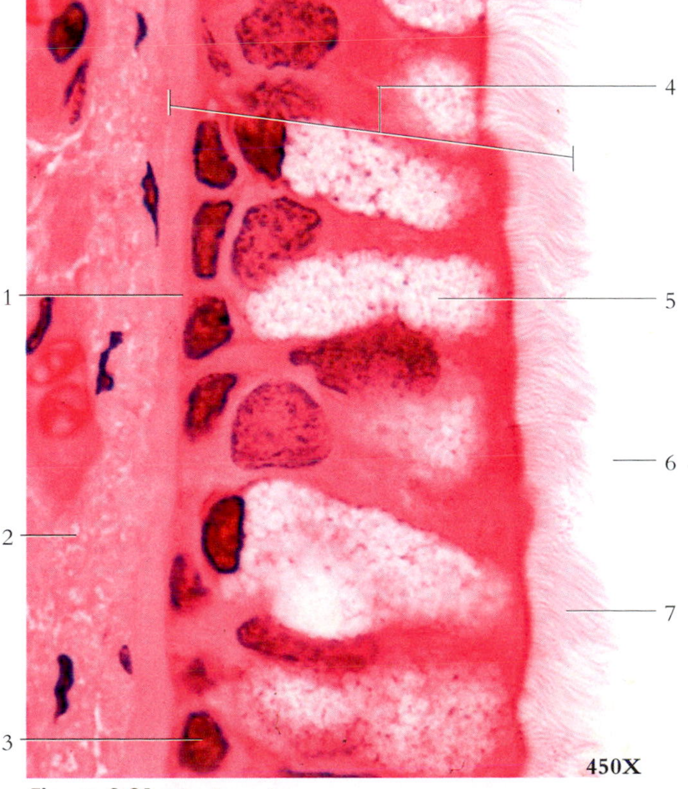

Figure 9.81 The bronchus.

1. Basement membrane
2. Lamina propria
3. Nucleus of epithelial cell
4. Pseudostratified squamous epithelium
5. Goblet cell
6. Lumen of bronchus
7. Cilia

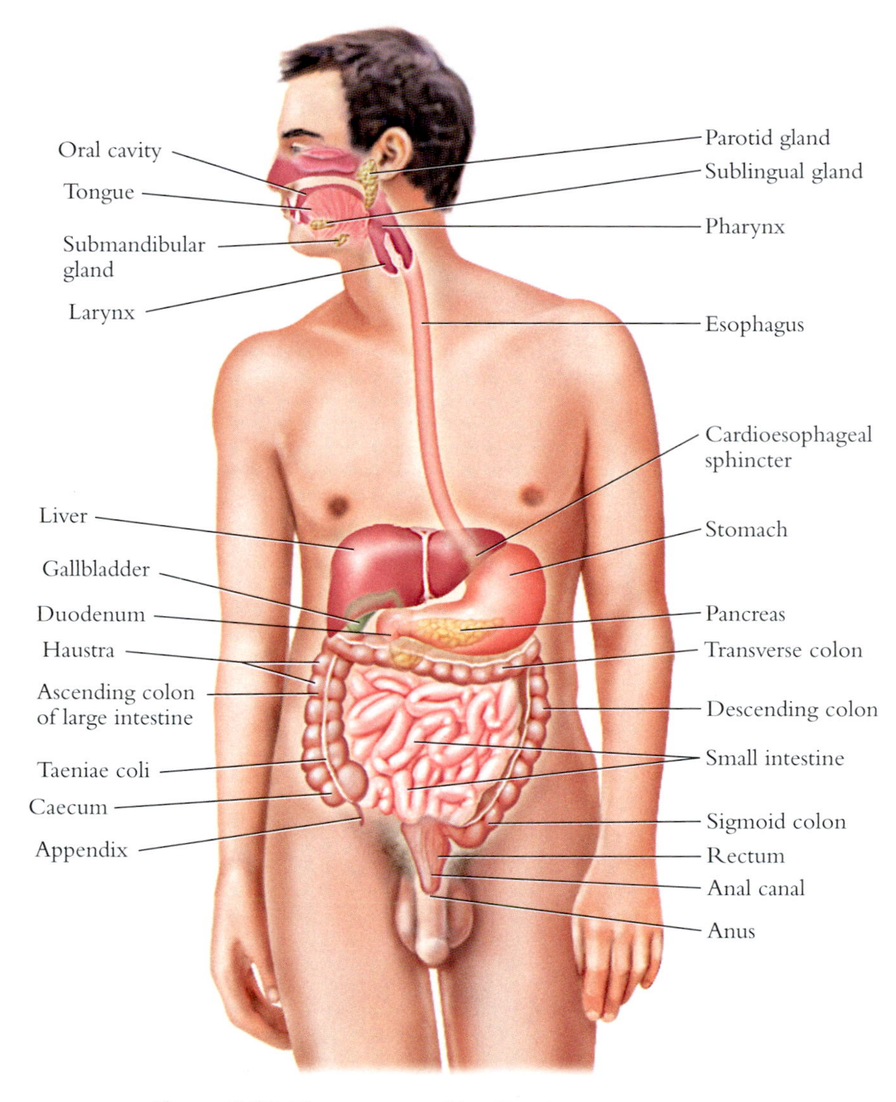

Figure 9.82 The structure of the digestive system.

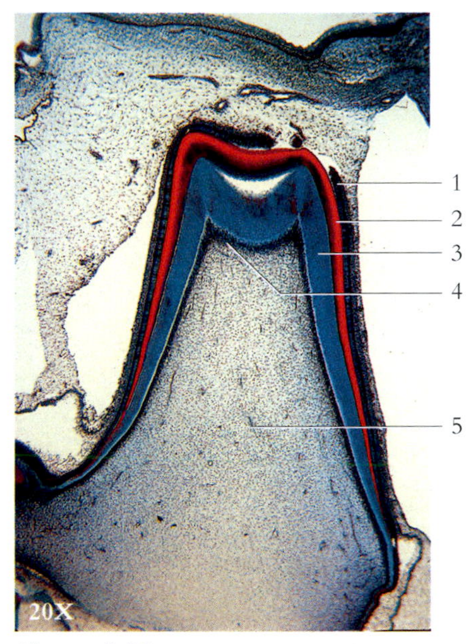

Figure 9.83 A developing tooth.

1. Ameloblasts
2. Enamel
3. Dentin
4. Odontoblasts
5. Pulp

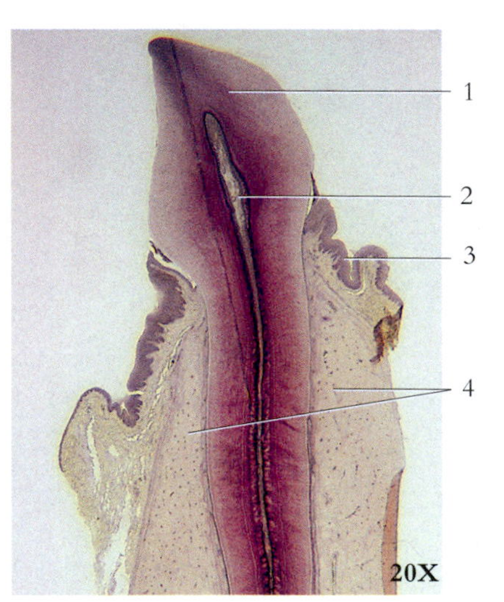

Figure 9.84 A mature tooth.

1. Dentin (enamel has been dissolved away)
2. Pulp
3. Gingiva
4. Alveolar bone

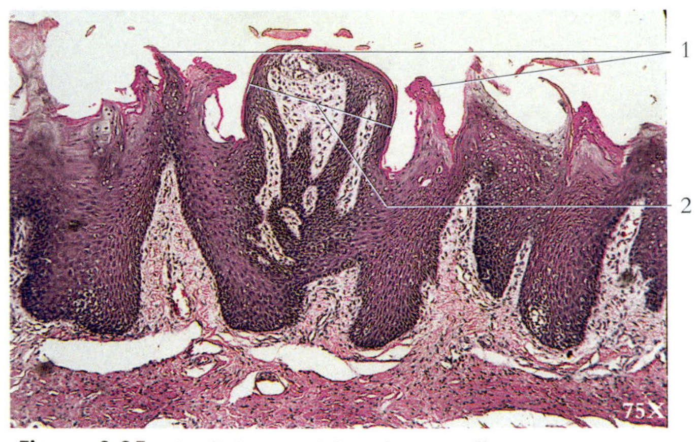

Figure 9.85 The filiform and fungiform papillae.

1. Filiform papillae
2. Fungiform papilla

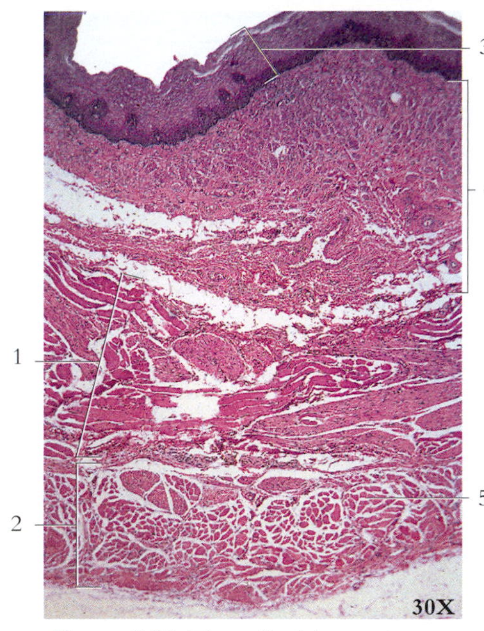

Figure 9.86 The wall of the esophagus.
1. Inner circular layer (muscularis externa)
2. Outer longitudinal layer (muscularis externa)
3. Mucosa
4. Submucosa
5. Muscularis externa

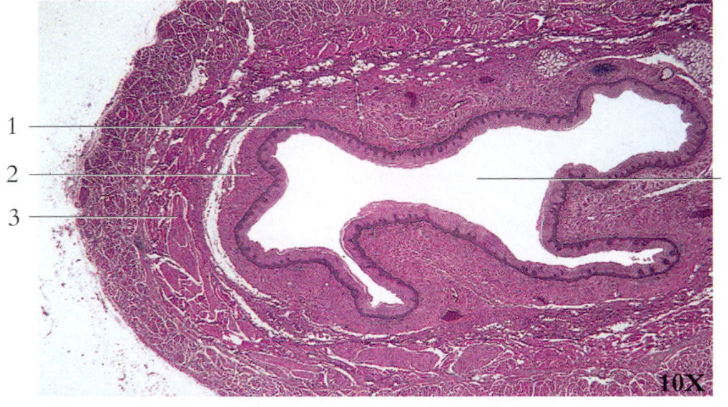

Figure 9.87 A transverse section of esophagus.
1. Mucosa
2. Submucosa
3. Muscularis
4. Lumen

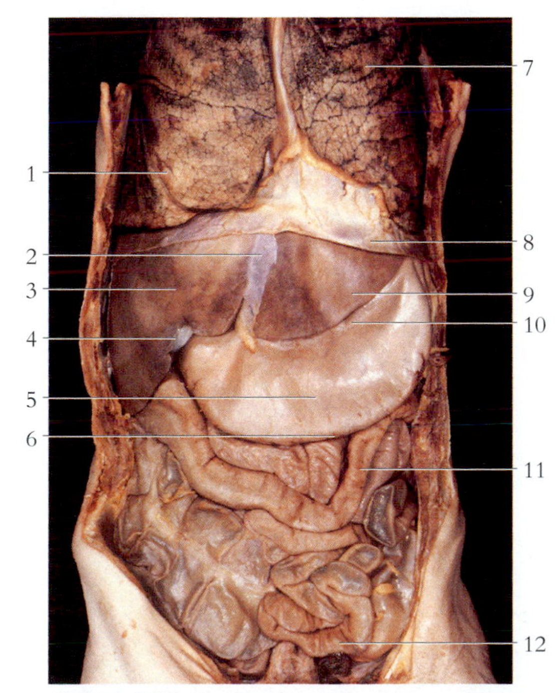

Figure 9.88 An anterior view of the trunk.
1. Right lung
2. Falciform ligament
3. Right lobe of liver
4. Gallbladder
5. Body of stomach
6. Greater curvature of stomach
7. Left lung
8. Diaphragm
9. Left lobe of liver
10. Lesser curvature of stomach
11. Transverse colon
12. Small intestine

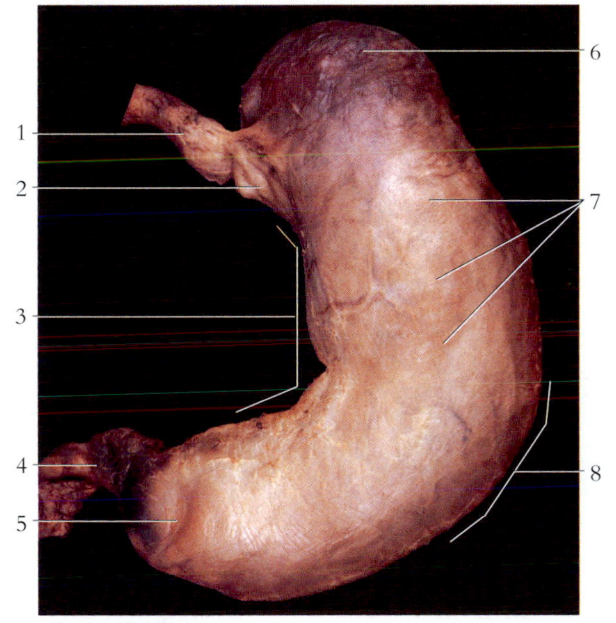

Figure 9.89 The major regions and structures of the stomach.
1. Esophagus
2. Cardiac portion of stomach
3. Lesser curvature of stomach
4. Duodenum
5. Pylorus of stomach
6. Fundus of stomach
7. Body of stomach
8. Greater curvature of stomach

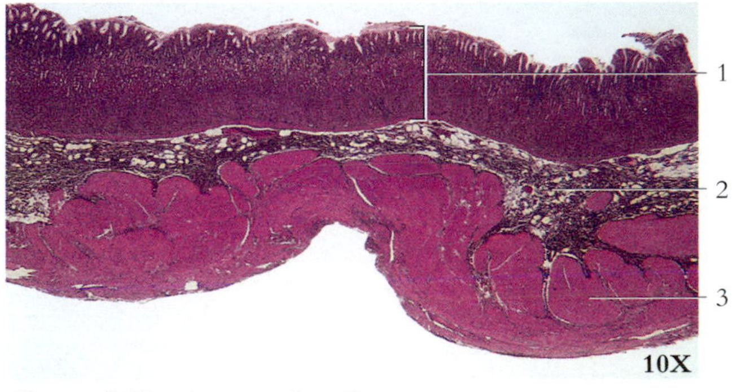

Figure 9.90 The stomach wall.
1. Mucosa
2. Submucosa
3. Muscularis externa

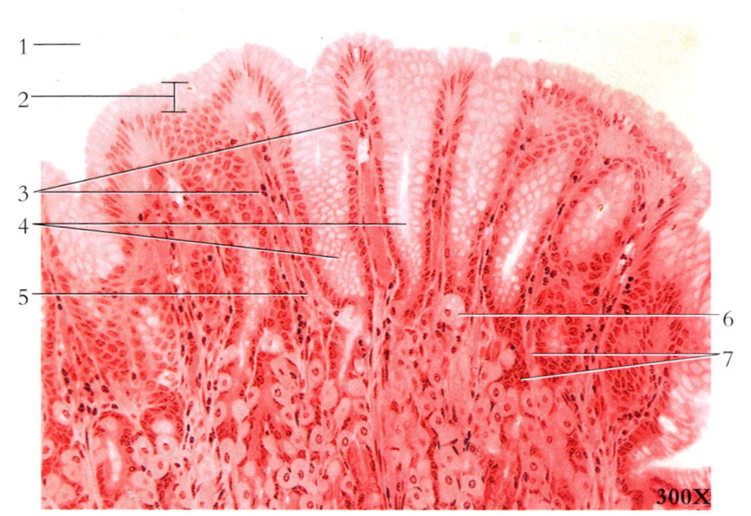

Figure 9.91 The histology of the cardiac region of the stomach.

1. Lumen of stomach
2. Surface epithelium
3. Mucosal ridges
4. Gastric pits
5. Lamina propria
6. Parietal cells
7. Chief (zymogenic) cells

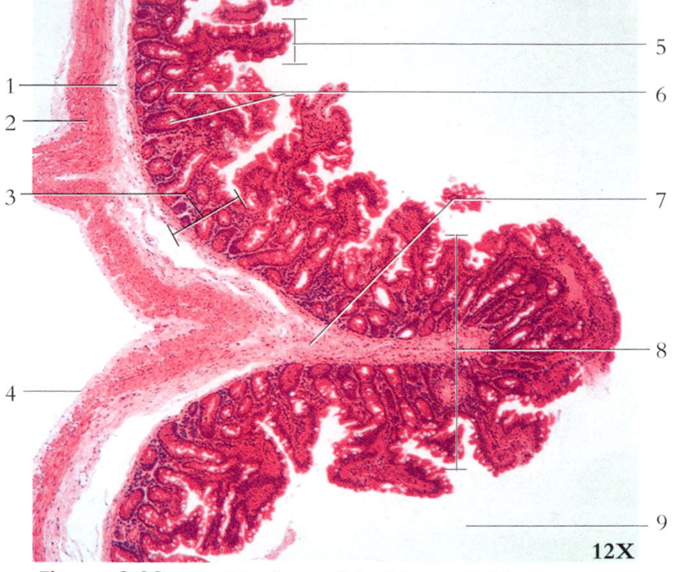

Figure 9.92 The histology of the jejunum of the small intestine.

1. Submucosa
2. Circular and longitudinal muscles
3. Mucosa
4. Serosa
5. Villus
6. Intestinal glands
7. Submucosa
8. Plica circulares
9. Lumen of small intestine

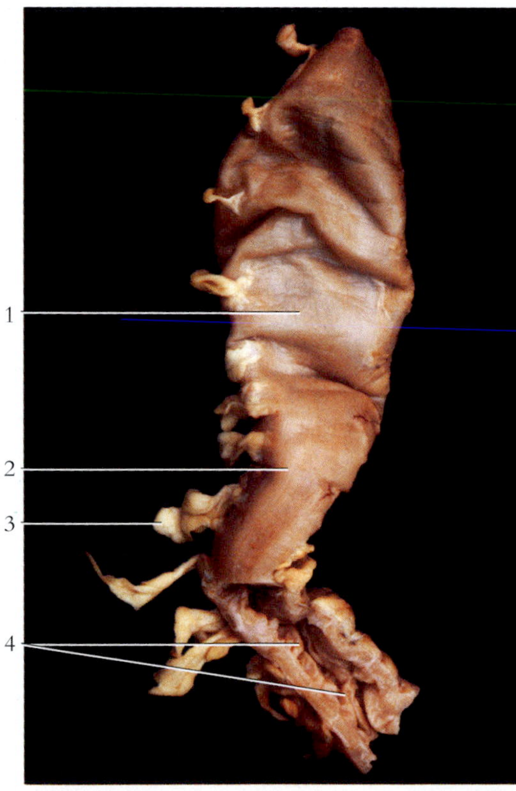

Figure 9.93 A section of the large intestine (colon).

1. Haustrum
2. Taeniae coli
3. Epiploic appendage
4. Semilunar folds (plicae) of colon

Figure 9.94 A radiograph of the large intestine.

1. Right colon (hepatic) flexure
2. Ascending colon
3. Sigmoid colon
4. Caecum
5. Left colic (splenic) flexure
6. 12th rib
7. Transverse colon
8. Lumbar vertebra
9. Descending colon
10. Rectum
11. Hip joint

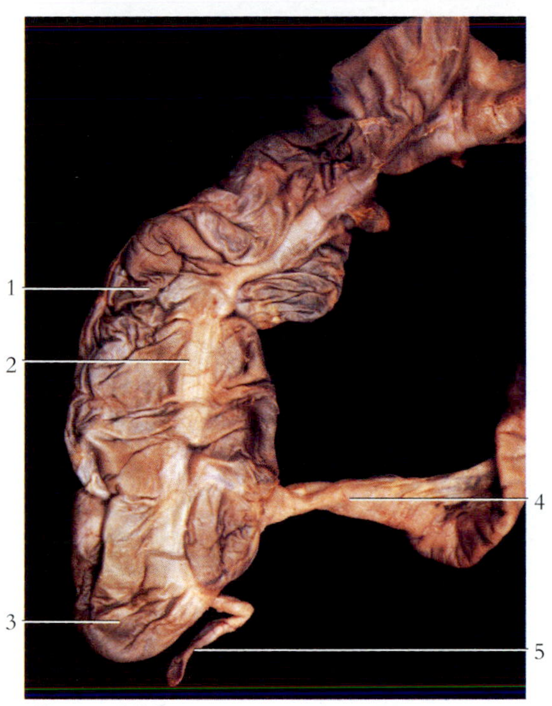

Figure 9.95 The caecum and appendix.
1. Ascending colon
2. Taeniae coli
3. Caecum
4. Ileum
5. Appendix

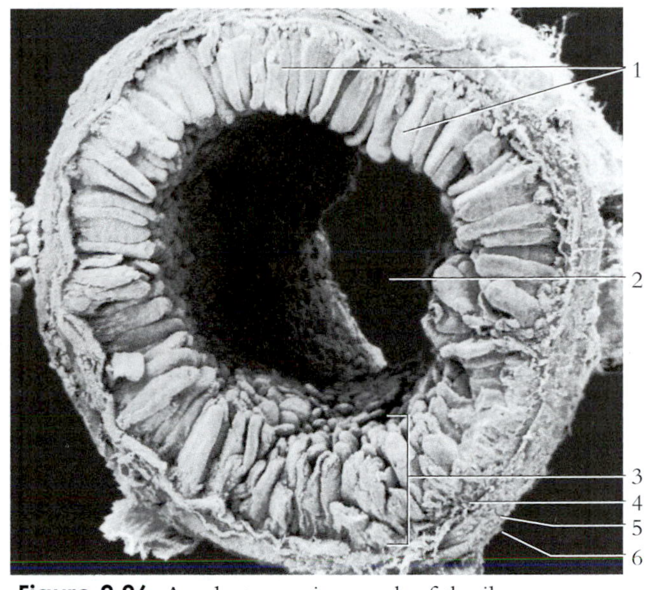

Figure 9.96 An electron micrograph of the ileum, shown in cross section.
1. Intestinal villi
2. Lumen
3. Mucosa
4. Submucosa
5. Tunica muscularis
6. Adventitia

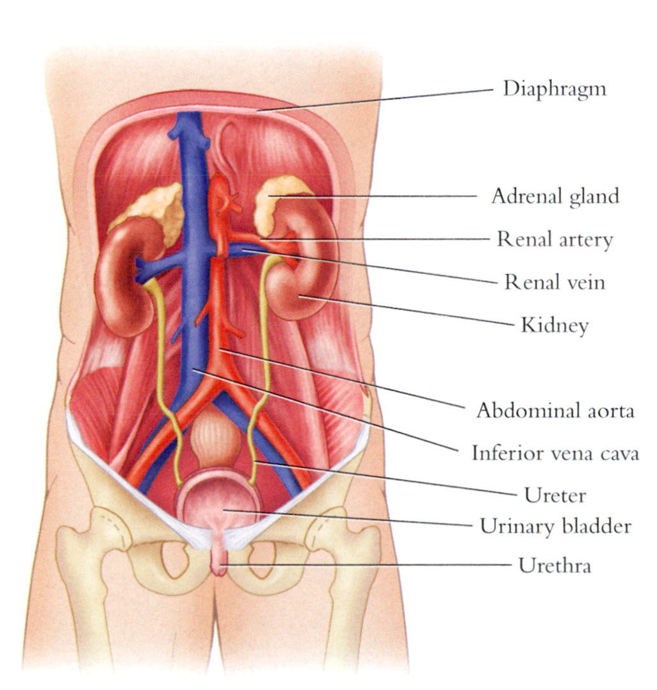

Figure 9.97 The organs of the urinary system.

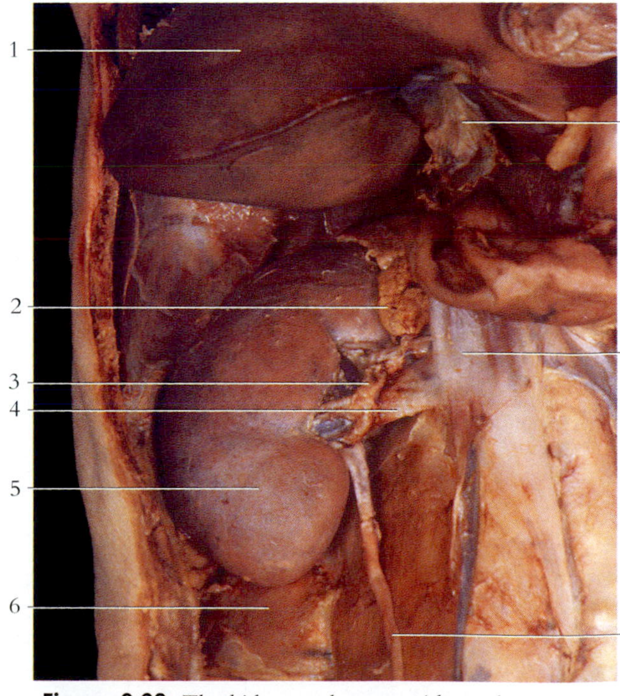

Figure 9.98 The kidney and ureter with overlying viscera removed.
1. Liver
2. Adrenal gland
3. Renal artery
4. Renal vein
5. Right kidney
6. Quadratus lumborum muscle
7. Gallbladder
8. Inferior vena cava
9. Ureter

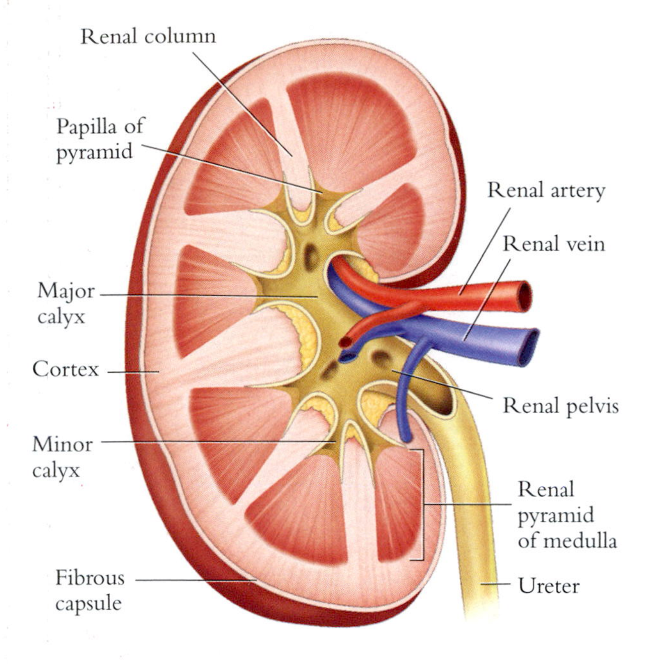

Figure 9.99 The structure of the kidney.

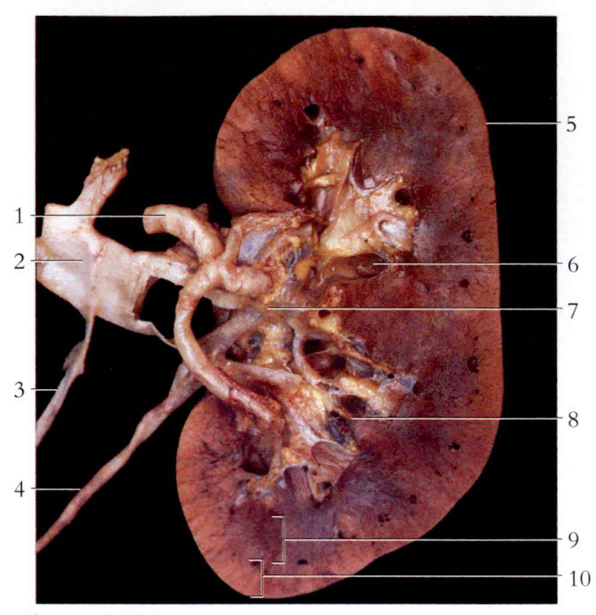

Figure 9.100 A coronal section of the left kidney.

1. Renal artery
2. Renal vein
3. Left testicular vein
4. Ureter
5. Renal capsule
6. Major calyx
7. Renal pelvis
8. Renal papilla
9. Renal medulla
10. Renal cortex

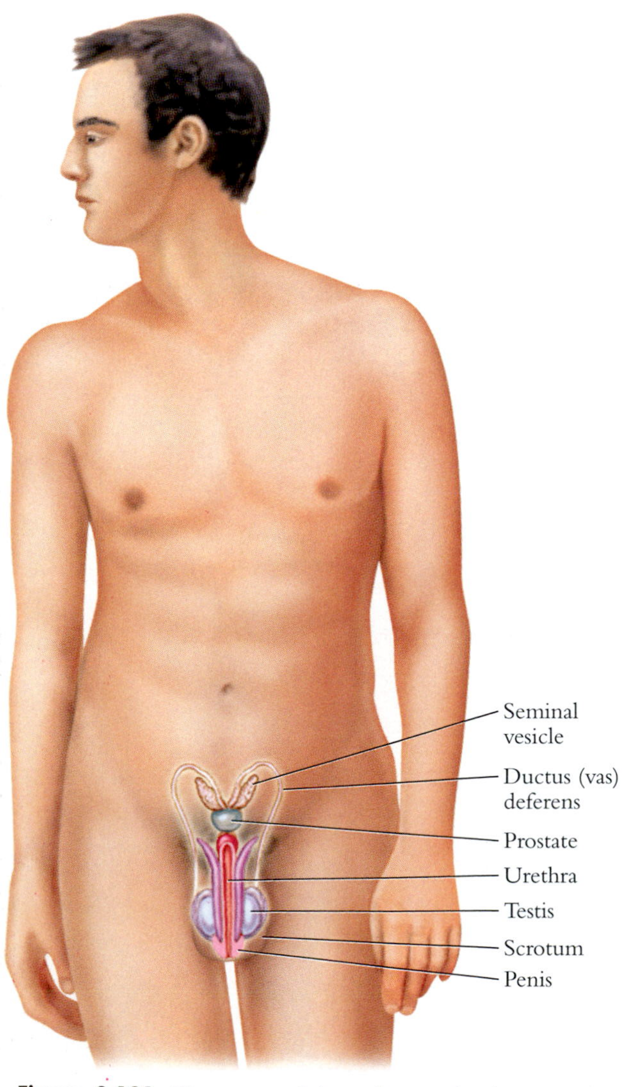

Figure 9.101 The organs of the male reproductive system.

Nipple
Breast
Uterine tube
Ovary
Uterus
Cervix
Vagina

Figure 9.102 The organs of the female reproductive system.

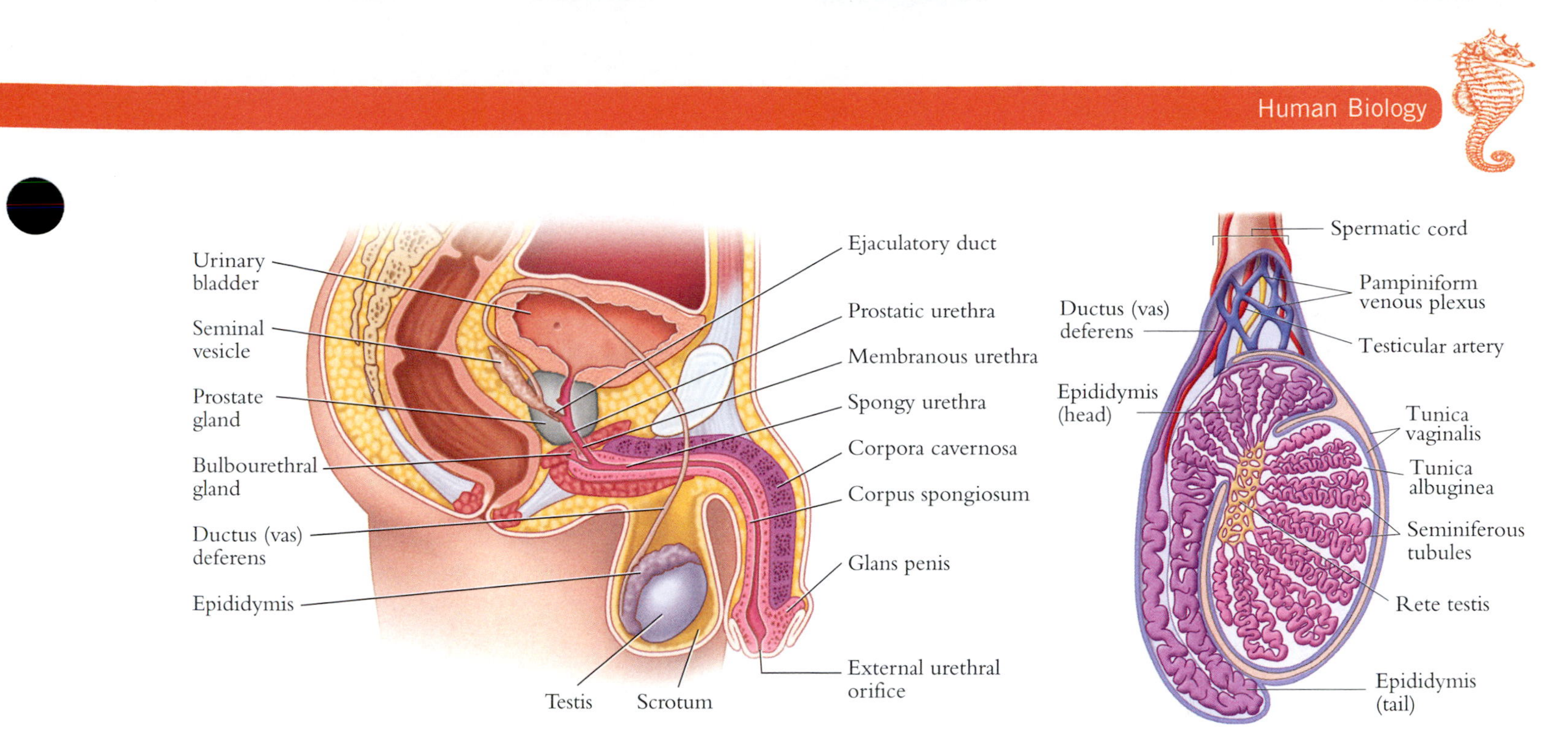

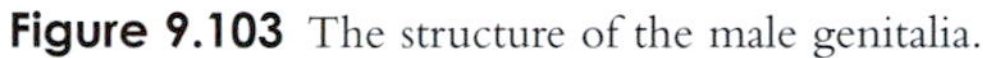

Figure 9.103 The structure of the male genitalia.

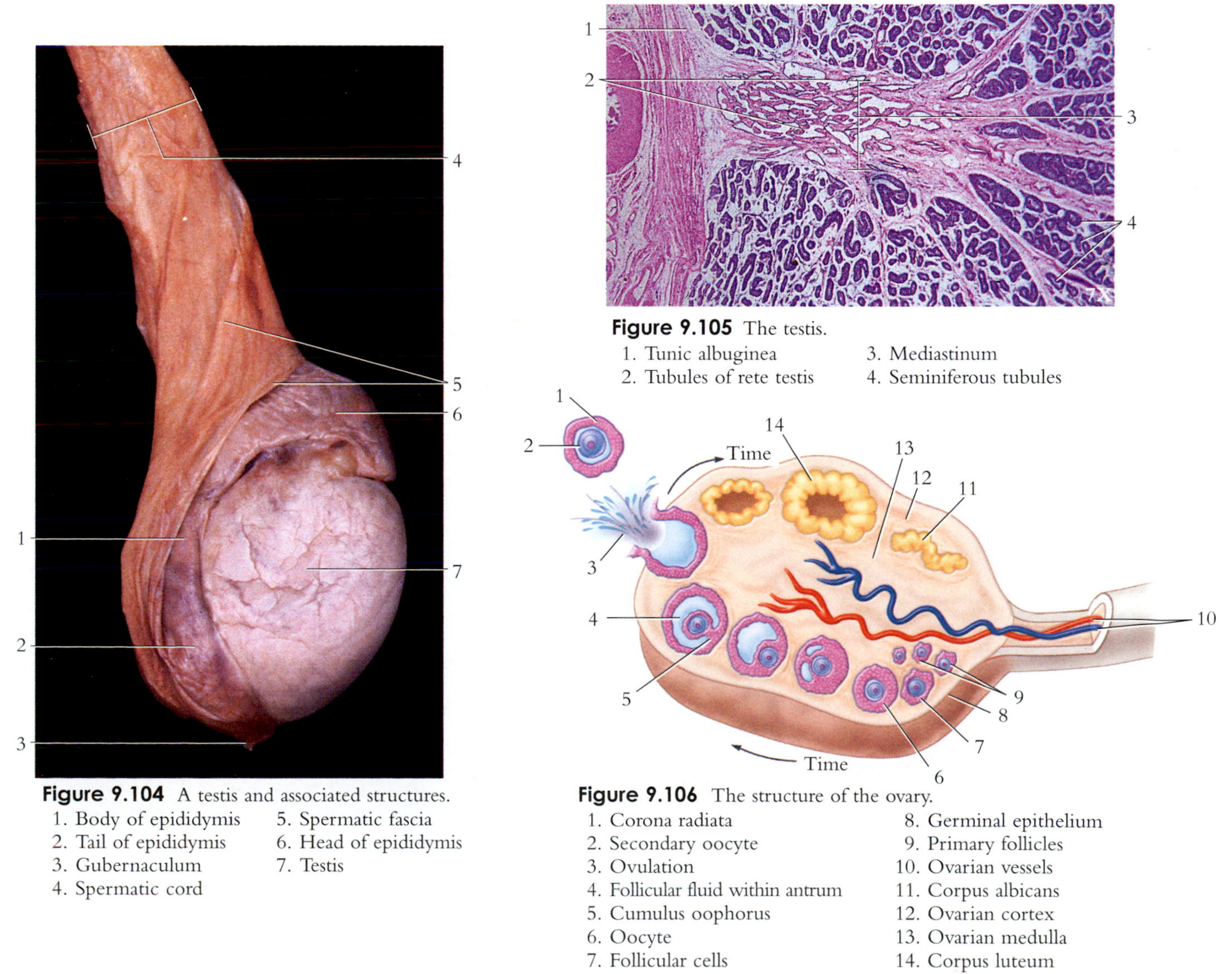

Figure 9.104 A testis and associated structures.

1. Body of epididymis
2. Tail of epididymis
3. Gubernaculum
4. Spermatic cord
5. Spermatic fascia
6. Head of epididymis
7. Testis

Figure 9.105 The testis.

1. Tunic albuginea
2. Tubules of rete testis
3. Mediastinum
4. Seminiferous tubules

Figure 9.106 The structure of the ovary.

1. Corona radiata
2. Secondary oocyte
3. Ovulation
4. Follicular fluid within antrum
5. Cumulus oophorus
6. Oocyte
7. Follicular cells
8. Germinal epithelium
9. Primary follicles
10. Ovarian vessels
11. Corpus albicans
12. Ovarian cortex
13. Ovarian medulla
14. Corpus luteum

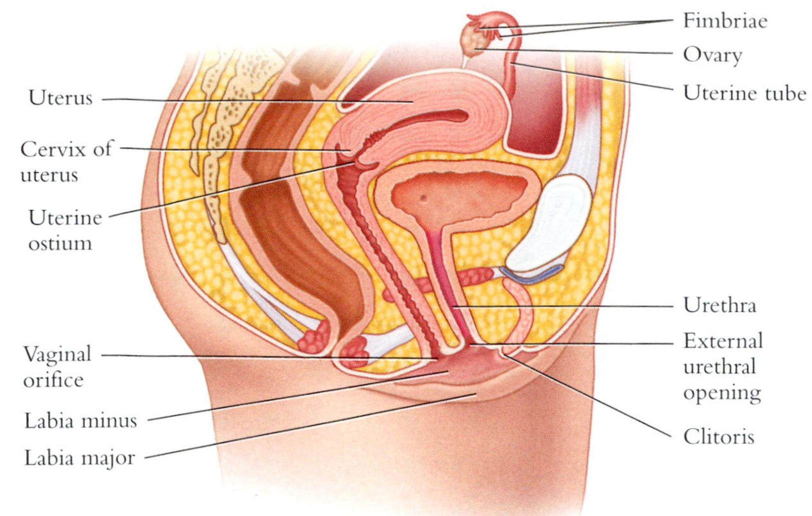

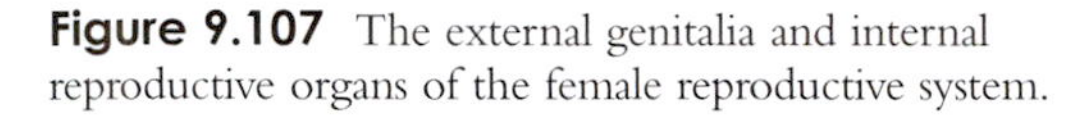

Figure 9.107 The external genitalia and internal reproductive organs of the female reproductive system.

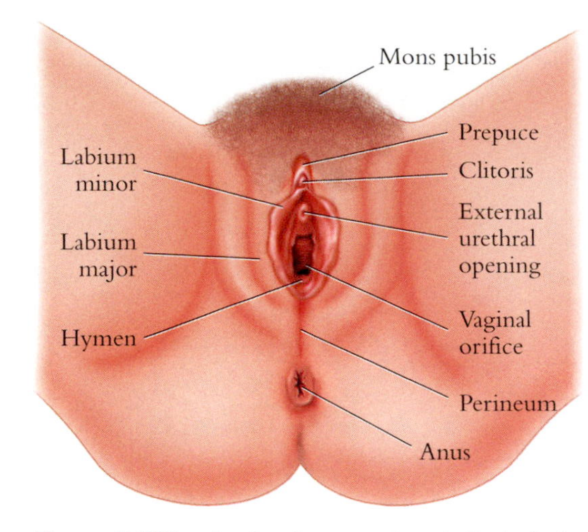

Figure 9.108 The female external genitalia (vulva).

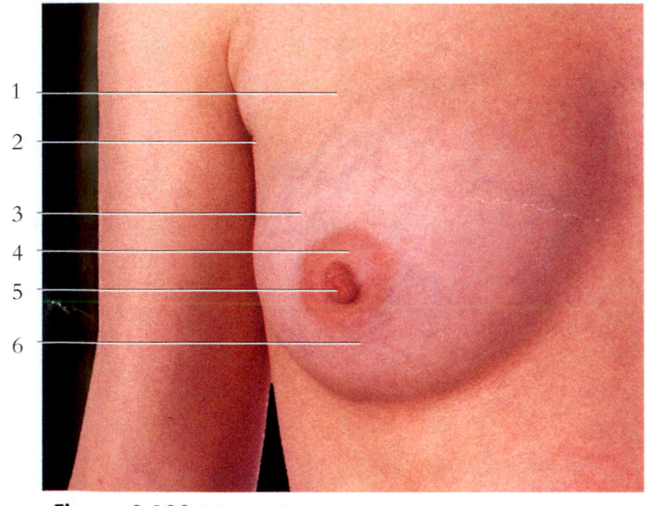

Figure 9.109 The surface anatomy of the female breast.
1. Pectoralis major muscle
2. Axilla
3. Lateral process of breast
4. Areola
5. Nipple
6. Breast (containing mammary glands)

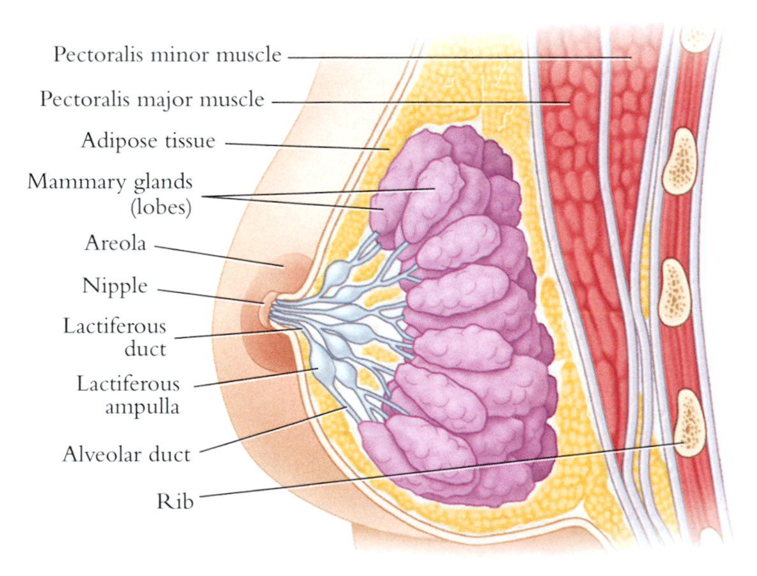

Figure 9.110 Internal structure of the female breast.

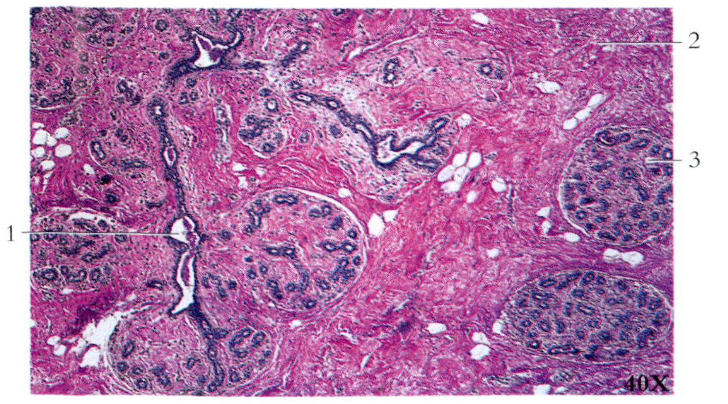

Figure 9.111 Mammary glands (nonlactating glands).
1. Interlobular duct
2. Interlobular connective tissue
3. Lobule of glandular tissue

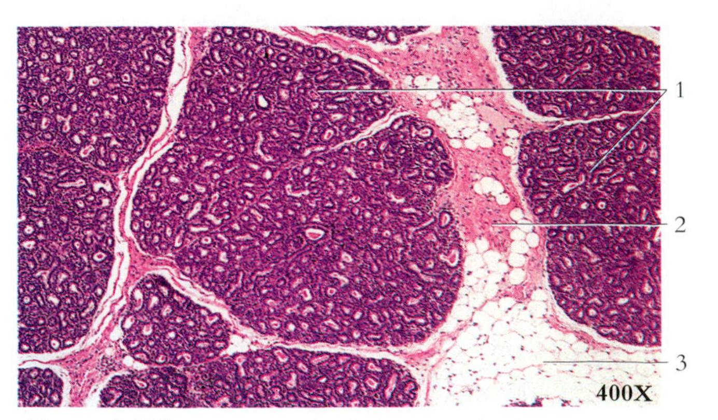

Figure 9.112 Mammary glands (lactating glands).
1. Lobules of glandular tissue
2. Intralobular connective tissue
3. Adipose cells

Glossary

abdomen: the portion of the trunk of the mammalian body located between the diaphragm and the pelvis, which contains the abdominal cavity and its visceral organs; one of the three principal body regions (head, thorax, and abdomen) of many animals.
abduction: a movement away from the axis or midline of the body; opposite of adduction, a movement of a digit away from the axis of a limb.
abiotic: without living organisms; nonliving portions of the environment.
abscission: the shedding of leaves, flowers, fruits, or other plant parts, usually following the formation of an abscission zone.
absorption: movement of a substance into a cell or an organism, or through a surface within an organism.
acapnia: a decrease in normal amount of CO_2 in the blood.
accommodation: a change in the shape of the lens of the eye so that vision is more acute; focusing for various distances.
acetone: an organic compound that may be present in the urine of diabetics; also called *ketone body.*
acetylcholine: a neurotransmitter chemical secreted at the terminal ends of many neurons, responsible for postsynaptic transmission; also called *ACh.*
acetylcholinesterase: an enzyme that breaks down acetylcholine; also called AChE.
Achilles' tendon: see *tendo calcaneous.*
acid: a substance that releases hydrogen ions (H^+) in a solution.
acidosis: a disorder of body chemistry in which the alkaline substances of the blood are reduced in an amount below normal.
acoelomate: without a coelomic cavity, as in flatworms.
acoustic: referring to sound or the sense of hearing.
actin: a protein in muscle fibers that together with myosin is responsible for contraction.
action potential: the change in ionic charge propagated along the membrane of a neuron; the *nerve impulse.*
active transport: movement of a substance into or out of a cell from a lesser to a greater concentration, requiring a carrier molecule and expenditure of energy.
adaptation: structural, physiological, or behavioral traits of an organism that promote its survival and contribute to its ability to reproduce under environmental conditions.

adduction: a movement toward the axis or midline of the body; opposite of abduction, a movement of a digit toward the axis of a limb.
adenohypophysis: anterior pituitary.
adenoid: paired lymphoid structures in the nasopharynx; also called *pharyngeal tonsils.*
adenosine triphosphate (ATP): a chemical compound that provides energy for cellular use.
adhesion: an attraction between unlike substances.
adipose: fat, or fat-containing, such as adipose tissue.
adrenal glands: endocrine glands; one superior to each kidney; also called *suprarenal glands.*
adventitious root: Supportive root developing from the stem of a plant.
aerobic: requiring free O_2 for growth and metabolism as in the case of certain bacteria also called *aerobes.*
agglutination: clumping of cells; particular reference to red blood cells in an antigen-antibody reaction.
aggregate fruit: ripened ovaries from a single flower with several separate carpels.
aggression: provoking, domineering behavior.
alga (pl. algae): any of a diverse group of aquatic photosynthesizing organisms that are either unicellular or multicellular; algae comprise the phytoplankton and seaweeds of the Earth.
alkaline: a substance having a pH greater than 7.0; *basic.*
allantois: an extraembryonic membranous sac that forms blood cells and gives rise to the fetal umbilical arteries and vein. It also contributes to the formation of the urinary bladder.
allele: an alternative form of a gene occurring at a given chromosome site, or *locus.*
all-or-none response: functioning completely when exposed to a stimulus of threshold strength; applies to action potentials through neurons and muscle fiber contraction.
alpha helix: right-handed spiral typical in proteins and DNA.
alternation of generations: two-phased life cycle characteristic of many plants in which there are sporophyte and gametophyte generations.
altruism: behavior benefiting other organisms without regard to its possible advantage or detrimental effect on the performer.
alveolus: An individual air capsule within the lung. Pulmonary alveoli are the basic functional units of respiration. Also, the socket that secures a tooth (dental alveolus).
amino acid: a unit of protein that contains an amino group (NH_2) and an acid group ($COOH$).
amnion: a membrane that surrounds the fetus to contain the amniotic fluid.
amniote: an animal that has an amnion during embryonic development; reptiles, birds, and mammals.
amoeba: protozoans that move by means of pseudopodia.
amphiarthrosis: a slightly moveable joint in a functional classification of joints.
anaerobic respiration: metabolizing and growing in the absence of oxygen.
analogous: similar in function regardless of developmental origin; generally in reference to similar adaptations.
anatomical position: the position in human anatomy in which there is an erect body stance with the eyes directed forward, the arms at the sides, and the palms of the hands facing forward.
anatomy: the branch of science concerned with the structure of the body and the relationship of its organs.
angiosperm: flowering plant, having double fertilization resulting in development of specialized seeds within fruits.
annual: a flowering plant that completes its entire life cycle in a single year or growing season.
annual ring: yearly growth demarcation in woody plants formed by buildup of secondary xylem.
annulus: a ring-like segment, such as body rings on an earthworm.
antebrachium: the forearm.
antenna: a sensory appendage on many species of invertebrate animals.
anterior (ventral): toward the front; the opposite of *posterior (dorsal).*
anther: the position of a plant stamen in which pollen is produced.
antheridium: male reproductive organ in certain nonseed plants and algae where motile sperm are produced.
anticodon: three ("a triplet") nucleotides sequence in transfer RNA that pairs with a complementary codon (triplet) in messenger RNA.
antigen: a foreign material, usually a protein, that triggers the immune system to produce antibodies.
anus: the terminal end of the gastrointestinal tract, opening of the anal canal.
aorta: the major systemic vessel of the arterial portion of the circulatory system, emerging from the left ventricle.
apical meristem: embryonic plant tissue in the tip of a root, bud, or shoot where continual cell divisions cause growth in length.
apocrine gland: a type of sweat gland that functions in evaporative cooling.
apopyle: opening of the radial canal into the spongocoel of sponges.
appeasement: submission behavior, usually soliciting an end to aggression.
appendix: a short pouch that attaches to the caecum.
aqueous humor: the watery fluid that fills the anterior and posterior chambers of the eye.
arachnoid: the weblike middle covering (meninx) of the central nervous system.
arbor vitae: the branching arrangement of white matter within the cerebellum.
archaebacteria: organisms within the kingdom Monera that represent an early group of simple life forms.
archegonium: female reproductive organ in certain nonseed plants; a gametangium where eggs are produced.
archenteron: the principal cavity of an embryo during the gastrula stage. Lined with endoderm, the archenteron develops into the digestive tract.
areola: the pigmented ring around the nipple.
arteries: blood vessels that carry blood away from the heart.
articular cartilage: a hyaline cartilaginous covering over the articulating surface of bones of synovial joints.
ascending colon: the portion of the large intestine between the caecum and the hepatic (right colic) flexure.
asexual: lacking distinct sexual organs and lacking the ability to produce gametes.
aster: minute rays of microtubules at the ends of the spindle apparatus in animal cells during cell division.
asymmetry: not symmetrical.
atom: the smallest unit of an element that can exist and still have the properties of the element; collectively, atoms form molecules in a compound.
atomic number: the weight of the atom of a particular element.

atomic weight: the number of protons together with the number of neutrons within the nucleus of an atom.
ATP (adenosine triphosphate): a compound of adenine, ribose, and three phosphates; it is the energy source for most cellular processes.
atrium: either of the two superior chambers of the heart that receive venous blood.
atrophy: wasting away or decrease in size of a cell or organ.
auditory tube: a narrow canal that connects the middle ear chamber to the pharynx; also called the *eustachian canal.*
autonomic: self-governing; pertaining to the division of the nervous system that controls involuntary activities.
autosome: a chromosome other than a sex chromosome.
autotroph: an organism capable of synthesizing its own organic molecules (food) from inorganic molecules.
axilla: the depressed hollow under the arm; the armpit.
axillary bud: a group of meristematic cells at the junction of a leaf and stem that develops branches or flowers; also called *lateral bud.*
axon: The elongated process of a neuron (nerve cell) that transmits an impulse away from the cell body.

bacillus (pl. bacilli): a rod-shaped bacterium.
bacteria: prokaryotes within the kingdom Monera, lacking the organelles of eukaryotic cells.
bark: outer tissue layers of a tree consisting of cork, cork cambium, cortex, and phloem.
basal: at or near the base or point of attachment, as of a plant shoot.
base: a substance that contributes or liberates hydroxide ions in a solution.
basement membrane: a thin sheet of extracellular substance to which the basal surfaces of membranous epithelial cells are attached.
basidia: club-shaped reproductive structures of club fungi that produce basidiospores during sexual reproduction.
basophil: a granular leukocyte that readily stains with basophilic dye.
belly: the thickest circumference of a skeletal muscle.
benign: nonmalignant; a confined tumor.
berry: a simple fleshy fruit.
biennial: a plant that lives through two growing seasons; generally, these plants have vegetative growth during the first season and flower and set seed during the second season.
bilateral symmetry: the morphologic condition of having similar right and left halves.
binary fission: a process of sexual reproduction that does not involve a mitotic spindle.
binomial nomenclature: assignment of two names to an organism, the first of which is the genus and the second the specific epithet; together constituting the scientific name.
biome: a major climax community characterized by a particular group of plants and animals.
biosphere: the portion of the Earth's atmosphere and surface where living organisms exist.
biotic: pertaining to aspects of life, especially to characteristics of ecosystems.
bisexual flower: a flower that contains both male and female sexual structures.
blade: the broad expanded portion of a leaf.
blastocoel: the cavity of any blastula.
blastula: stage of embryonic development of animals near the end of cleavage but before gastrulation.
blood: the fluid connective tissue that circulates through the cardiovascular system to transport substances throughout the body.
bolus: a moistened mass of food that is swallowed from the oral cavity into the pharynx.
bone: an organ composed of solid, rigid connective tissue, forming a component of the skeletal system.
Bowman's capsule: see *glomerular capsule.*
brain: the enlarged superior portion of the central nervous system, located in the cranial cavity of the skull.
brainstem: the portion of the brain consisting of the medulla oblongata, pons, and midbrain.
bronchial tree: the bronchi and their branching bronchioles.
bronchiole: a small division of a bronchus within the lung.
bronchus: a branch of the trachea that leads to a lung.
buccal cavity: the mouth, or oral cavity.
budding: a type of asexual reproduction in which outgrowths from the parent plant pinch off to live independently or may remain attached to form colonies.
buffer: a compound or substance that prevents large changes in the pH of a solution.
bulb: a thickened underground stem often enclosed by enlarged, fleshy leaves containing stored food.
bursa: a saclike structure filled with synovial fluid that occurs around joints.
buttock: the rump or fleshy mass on the posterior aspect of the lower trunk, formed primarily by the gluteal muscles.

caecum: the pouch-like portion of the large intestine to which the ileum of the small intestine is attached.
calorie: the heat required to raise one kilogram of water one degree centigrade.
calyx: a cup-shaped portion of the renal pelvis that encircles renal papillae; the collective term for the sepals of a flower.
cambium: the layer of meristematic tissue in roots and stems of many vascular plants that continues to produce tissue.
cancellous bone: spongy bone; bone tissue with a lattice-like structure.
capillary: a microscopic blood vessel that connects an arteriole and a venule; the functional unit of the circulatory system.
carapace: protective covering over the dorsal part of the body of certain crustaceans and turtles.
carcinogenic: stimulating or causing the growth of a malignant tumor, or cancer.
carnivore: any animal that feeds upon another; especially, flesh-eating mammal.
carpus: the proximal portion of the hand that contains the carpal bones.
carrying capacity: the maximum number of organisms of a species that can be maintained indefinitely in an ecosystem.
cartilage: a type of connective tissue with a solid elastic matrix.
catalyst: a chemical, such as an enzyme, that accelerates the rate of a reaction of a chemical process but is not used up in the process.
caudal: referring to a position more toward the tail.
cell: the structural and functional unit of an organism; the smallest structure capable of performing all the functions necessary for life.
cell wall: a rigid protective structure of a plant cell surrounding the cell (plasma) membrane; often composed of cellulose fibers embedded in a polysaccharide/protein matrix.
cellular respiration: the reactions of glycolysis, Krebs cycle, and electron transport system that provide cellular energy and accompanying reactions to produce ATP.
cellulose: a polysaccharide produced as fibers that forms a major part of the rigid cell wall around a plant cell.
central nervous system (CNS): the brain and the spinal cord.
centromere: a portion of the chromosome to which a spindle fiber attaches during mitosis or meiosis.
centrosome: a dense body near the nucleus of a cell that contains a pair of centrioles.
cephalothorax: fusion of the head and thoracic regions, characteristic of certain arthropods.
cercaria: larva of trematodes (flukes).
cerebellum: the portion of the brain concerned with the coordination of movements and equilibrium.
cerebrospinal fluid (CSF): a liquid that buoys and cushions the central nervous system.
cerebrum: the largest portion of the brain, composed of the right and left hemispheres.
cervical: pertaining to the neck or a neck-like portion of an organ.
chelipeds: front pair of pincer-like legs in most decapod crustaceans, adapted for seizing and crushing.
chitin: strong, flexible polysaccharide forming the exoskeleton of arthropods.
chlorophyll: green pigment in photosynthesizing organisms that absorbs energy from the sun.
chloroplast: a membrane-enclosed organelle that contains chlorophyll and is the site of photosynthesis.
choanae: the two posterior openings from the nasal cavity into the nasopharynx.
cholesterol: a lipid used in the synthesis of steroid hormones.
chondrocyte: a cartilage cell.
chorion: an extraembryonic membrane that participates in the formation of the placenta.
choroid: the vascular, pigmented middle layer of the wall of the eye.
chromatin: threadlike network of DNA and proteins within the nucleus.
chromosome: structure in the nucleus that contains the genes for genetic expression.
chyme: the mass of partially digested food that passes from the stomach into the duodenum of the small intestine.
cilia: microscopic, hairlike processes that move in a wavelike manner on the exposed surfaces of certain epithelial cells.
ciliary body: a portion of the choroid layer of the eye that secretes aqueous humor and contains the ciliary muscle.
ciliates: protozoans that move by means of cilia.
circadian rhythm: a daily physiological or behavioral event occurring on an approximate 24-hour cycle.
circumduction: a cone-like movement of a body part, such that the distal end moves in a circle while the proximal portion remains relatively stable.
clitoris: a small, erectile structure in the vulva of the female.
cochlea: the spiral portion of the inner ear that contains the spiral organ (organ of Corti).
climax community: the final stable stage in succession.
clone: asexually produced organisms having a consistent genetic constitution.
cnidarian: small aquatic organisms having radial symmetry and stinging cells with nematocysts.
cocoon: protective, or resting, stage of development in certain invertebrate animals.
codon: a "triplet" of three nucleotides in RNA that directs the placement of an amino acid into a polypeptide chain.
coelom: body cavity of higher animals, containing visceral organs.

collar cells: flagella-supporting cells in the inner layer of the wall of sponges.
colon: the first portion of the large intestine.
common bile duct: a tube that is formed by the union of the hepatic duct and cystic duct; transports bile to the duodenum.
compact bone: tightly packed bone that is superficial to spongy bone; also called *dense bone*.
competition: interaction between individuals of the same or different species for a mutually necessary resource.
complete flower: a flower that has four whorls of floral components including sepals, petals, stamens, and carpels.
compound eye: arthropod eye consisting of multiple lenses.
compound leaf: a leaf blade divided into distinct leaflets.
condyle: a rounded process at the end of a long bone that forms an articulation.
conidia: spores produced by fungi during asexual reproduction.
conifer: a cone-bearing woody seed plant, such as pine, fir, and spruce
conjugation: sexual union in which the nuclear material of one cell enters another.
connective tissue: one of the four basic tissue types within an animal's body. It is a binding and supportive tissue with abundant matrix.
consumer: an organism that derives nutrients by feeding upon another.
control: a sample in an experiment that undergoes all the steps in the experiment except the one being investigated.
convergent evolution: the evolution of similar structures in different groups of organisms exposed to similar environments.
coral: a cnidarian that has a calcium carbonate skeleton whose remains contribute to form reefs.
cork: the protective outer layer of bark of many trees, composed of dead cells that may be sloughed off.
cornea: the transparent, convex, anterior portion of the outer layer of the vertebrate eye.
cortex: the outer layer of an organ such as the convoluted cerebrum, adrenal gland, or kidney.
costal cartilage: the cartilage that connects the ribs to the sternum.
cranial: pertaining to the cranium.
cranial nerve: one of 12 pairs of nerves that arise from the inferior surface of the brain.
cranium: the bones of the skull that enclose the brain and support the organs of sight, hearing, and balance.
crossing over: the exchange of corresponding chromatid segments of genetic material of homologous chromosomes during synapsid of meiosis I.
cuticle: waxlike covering on the epidermis of nonwoody plants to prevent water loss.
cyanobacteria: photosynthetic prokaryotes that have chlorophyll and release oxygen.
cytokinesis: division of the cellular cytoplasm.
cytology: the science dealing with the study of cells.
cytoplasm: the protoplasm of a cell located outside of the nucleus.
cytoskeleton: protein filaments throughout the cytoplasm of certain cells that help maintain the cell shape.

deciduous: plants that seasonally shed their leaves.
dendrite: a nerve cell process that transmits impulses toward a neuron cell body.
denitrifying bacteria: single-cellular organisms that convert nitrate to atmospheric nitrogen.
dentin: the principal substance of a tooth, covered by enamel over the crown and by cementum on the root.
dermis: the second, or deep, layer of skin beneath the epidermis.
descending colon: the segment of the large intestine that descends on the left side from the level of the spleen to the level of the left iliac crest.
detritus: nonliving organic matter important in the nutrient cycle in soil formation.
diaphragm: a flat dome of skeletal muscle and connective tissue that separates the thoracic and abdominal cavities in mammals.
diaphysis: the body, or shaft, of a long bone.
diarthrosis: a freely movable joint in a functional classification of joints.
diastole: the portion of the cardiac cycle during which the ventricular heart chamber wall is relaxed.
diatoms: aquatic unicellular algae characterized by a cell wall composed of two silica-impregnated valves.
dicot: a kind of angiosperm characterized by the presence of two cotyledons in the seed; also called *dicotyledon*.
diffusion: movement of molecules from an area of greater concentration to an area of lesser concentration.
dihybrid cross: a breeding experiment in which parental varieties differing in two traits are mated.
dimorphism: two distinct forms within a species, with regard to size, color, organ structure, and so on.
diphyodent: two sets of teeth, deciduous and permanent.
diploid: having two copies of each different chromosome, pairs of homologous chromosomes (2N).
distal: away from the midline or origin; the opposite of *proximal*.
dominant: a hereditary characteristic that expresses itself even when the genotype is heterozygous.
dormancy: a period of suspended activity and growth.
dorsal: pertaining to the back or posterior portion of a body part; the opposite of *ventral*.
double helix: a double spiral used to describe the three-dimensional shape of DNA.
ductus deferens: a tube that carries spermatozoa from the epididymis to the ejaculatory duct; also called the *vas deferens* or *seminal duct*.
duodenum: the first portion of the small intestine.
dura mater: the outermost meninx covering the central nervous system.

eccrine gland: a sweat gland that functions in body cooling.
ecology: the study of the relationship of organisms and the physical environment and their interactions.
ecosystem: a biological community and its associated abiotic environment.
ectoderm: the outermost of the three primary embryonic germ layers.
edema: an excessive retention of fluid in the body tissues.
effector: an organ such as a gland or muscle that responds to motor stimulation.
efferent: conveying away from the center of an organ or structure.
ejaculation: the discharge of semen from the male urethra during climax.
electrocardiogram: a recording of the electrical activity that accompanies the cardiac cycle; also called *ECG* or *EKG*.
electroencephalogram: a recording of the brain wave pattern; also called *EEG*.
electromyogram: a recording of the activity of a muscle during contraction; also called *EMG*.
electrolyte: a solution that conducts electricity by means of charged ions.
electron: the unit of negative electricity.
element: a structure composed of only one type of atom (e.g., carbon, hydrogen, oxygen).
embryo: a plant or an animal at an early stage of development.
emulsification: the process of dispersing one liquid in another.
enamel: the outer, dense substance covering the crown of a tooth.
endocardium: the fibrous lining of the heart chambers and valves.
endochondral bone: bone that forms as hyaline cartilage models first and then is ossified.
endocrine gland: a hormone-producing gland that secretes directly into the blood or body fluids.
endoderm: the innermost of the three primary germ layers of an embryo.
endodermis: a plant tissue composed of a single layer of cells that surrounds and regulates the passage of materials into the vascular cylinder of roots.
endometrium: the inner lining of the uterus.
endoskeleton: hardened, supportive internal tissue of echinoderms and vertebrates.
endosperm: a plant tissue of angiosperm seeds that stores nutrients.
endothelium: the layer of epithelial tissue that forms the thin inner lining of blood vessels and heart chambers.
enzyme: a protein catalyst that activates a specific reaction.
eosinophil: a type of white blood cell that becomes stained by acidic eosin dye; constitutes about 2%–4% of the human white blood cells.
epicardium: the thin, outer layer of the heart; also called the *visceral pericardium*.
epicotyl: part of the plant embryo that contributes to stem development.
epidermis: the outermost layer of the skin, composed of stratified squamous epithelium; also the outer part of plants.
epididymis: a coiled tube located along the posterior border of the testis; stores spermatozoa and discharges them during ejaculation.
epidural space: a space between the spinal dura mater and the bone of the vertebral canal.
epiglottis: a cartilaginous leaflike structure positioned on top of the larynx that covers the glottis during swallowing.
epinephrine: a hormone secreted from the adrenal medulla resulting in actions similar to those from sympathetic nervous system stimulation; also called *adrenaline*.
epiphyseal plate: a cartilaginous layer located between the epiphysis and diaphysis of a long bone that functions in longitudinal bone growth.
epiphysis: the end segment of a long bone, distinct in early life but later becoming part of the larger bone.
epiphyte: nonparasitic plant, such as orchid and Spanish moss, that grows on the surface of other plants.
epithelial tissue: one of the four basic tissue types; the type of tissue that covers or lines all exposed body surfaces.
erection: a response within an organ, such as the penis, when it becomes turgid and erect as opposed to being flaccid.
erythrocyte: a red blood cell.
esophagus: a tubular organ of the gastrointestinal tract that leads from the pharynx to the stomach.
estrogen: female sex hormone secreted from the ovarian (Graafian) follicle.
estuary: a zone of mixing between freshwater and sea water.
eukaryotic: possessing the membranous organelles characteristic of complex cells.
eustachian canal: see *auditory tube*.
evolution: genetic and phenotypic changes occurring in populations of organisms through time,

generally resulting in increased adaptation for continued survival.
excretion: discharging waste material.
exocrine gland: a gland that secretes its product to an epithelial surface, directly or through ducts.
exoskeleton: an outer, hardened supporting structure secreted by ectoderm or epidermis.
expiration: the process of expelling air from the lungs through breathing out; also called *exhalation*.
extension: a movement that increases the angle between two bones of a joint.
external ear: the outer portion of the ear, consisting of the auricle (pinna), external auditory canal, and tympanum.
extracellular: outside a cell or cells.
extraembryonic membranes: membranes that are not a part of the embryo but are essential for the health and development of the organism.
extrinsic: pertaining to an outside or external origin.

facet: a small, smooth surface of a bone where articulation occurs.
facilitated transport: transfer of a particle into or out of a cell along a concentration gradient by a process requiring a carrier.
fallopian tube: see *uterine tube*.
fascia: a tough sheet of fibrous connective tissue binding the skin to underlying muscles or supporting and separating muscle.
fasciculus: a bundle of muscle or nerve fibers.
feces: waste material expelled from the gastrointestinal tract during defecation, composed of food residue, bacteria, and secretions; also called *stool*.
fertilization: the fusion of two haploid gamete nuclei to form a diploid zygote nucleus.
fetus: the unborn offspring during the last stage of prenatal development.
fibrous root: an intertwining mass of roots of about equal size.
filament: a long chain of cells.
filter feeder: an animal that obtains food by straining it from the water.
filtration: the passage of a liquid through a filter or a membrane.
fimbriae: fringe-like extensions from the borders of the open end of the uterine tube.
fissure: a groove or narrow cleft that separates two parts of an organ.
flagella: long, slender locomotor processes characteristic of flagellate protozoans, certain bacteria, and sperm.
flexion: a movement that decreases the angle between two bones of a joint; opposite of extension.
flora: a general term for the plant life of a region or area.
flower: the blossom of an angiosperm that contains the reproductive organs.
fluke: a parasitic flatworm within the class Trematoda.
follicle: the portion of the ovary that produces the egg and the female sex hormone estrogen; the depression that supports and develops a feather or hair.
fontanel: a membranous-covered region on the skull of a fetus or baby where ossification has not yet occurred; also called a *soft spot*.
food web: the food links between populations in a community.
foot: the terminal portion of the lower extremity, consisting of the tarsus, metatarsus, and digits.
foramen: an opening in a bone for the passage of a blood vessel or a nerve.
foramen ovale: the opening through the interatrial septum of the fetal heart.
fossa: a depressed area, usually on a bone.
fossil: any preserved remains or impressions of an organism.
fourth ventricle: a cavity within the brain containing cerebrospinal fluid.
fovea centralis: a depression on the macula lutea of the eye where only cones are located, which is the area of keenest vision.
frond: a leaf of a fern usually containing many leaflets.
fruit: a mature ovary enclosing a seed or seeds.

gallbladder: a pouch-like organ attached to the inferior side of the liver that stores and concentrates bile.
gamete: a haploid sex cell, sperm or egg.
gametophyte: the haploid, gamete-producing generation in the life cycle of a plant.
gamma globulins: protein substances that act as antibodies, often found in immune serums.
ganglion: an aggregation of nerve cell bodies outside the central nervous system.
gastrointestinal tract: the tubular portion of the digestive system that includes the stomach and the small and large intestines; *(GI tract)*.
gene: part of the DNA molecule located in a definite position on a certain chromosome and coding for a specific product.
gene pool: the total of all the genes of the individuals in a population.
genetic drift: evolution by chance process.
genetics: the study of heredity.
genotype: the genetic makeup of an organism.
genus: the taxonomic category above species and below family.
geotropism: plant growth oriented with respect to gravity; stems grow upward, roots grow downward.
germ cells: gametes or cells that give rise to gametes or other cells.
germination: the process by which a spore or seed ends dormancy and resumes normal metabolism, development, and growth.
gill: a gas-exchange organ characteristic of fishes and other aquatic or semiaquatic animals.
gingiva: the fleshy covering over the mandible and maxilla through which the teeth protrude within the mouth; also called the *gum*.
girdling: removal of a strip of bark from around a tree down to the wood layer.
gland: an organ that produces a specific substance or secretion.
glans penis: the enlarged, distal end of the penis.
glomerular capsule: the double-walled proximal portion of a renal tubule that encloses the glomerulus of a *nephron;* also called *Bowman's capsule*.
glomerulus: a coiled tuft of capillaries that is surrounded by the glomerular capsule and filters urine from the blood.
glottis: a slit-like opening into the larynx, positioned between the true vocal folds.
glycogen: the principal storage carbohydrate in animals. It is stored primarily in the liver and is made available as glucose when needed by the body cells.
goblet cell: a unicellular gland within columnar epithelia that secretes mucus.
gonad: a reproductive organ, testis or ovary, that produces gametes and sex hormones.
granum: a "stack" of membrane flattened disks within the chloroplast that contain chlorophyll.
gray matter: the portion of the central nervous system that is composed of nonmyelinated nervous tissue.
grazer: animal that feeds on low-growing vegetation, such as grasses.
growth ring: a growth layer of secondary xylem (wood) or secondary phloem in gymnosperms or angiosperms.
guard cell: an epidermal cell to the side of a leaf stoma that helps to control the stoma size.
gut: pertaining to the intestine; generally a developmental term.
gymnosperm: a vascular seed-producing plant that does not produce flowers.
gyrus: a convoluted elevation or ridge.

habitat: the ecological abode of a plant or animal species.
hair: an epidermal structure consisting of keratinized dead cells that have been pushed up from a dividing basal layer.
hair cells: specialized receptor nerve endings for responding to sensations, such as in the spiral organ of the inner ear.
hair follicle: a tubular depression in the skin in which a hair develops.
hand: the terminal portion of the upper extremity, consisting of the carpus, metacarpus, and digits.
haploid: having one copy of each different chromosome.
hard palate: the bony partition between the oral and nasal cavities formed by the maxillae and palatine bones.
Haversian system: see *osteon*.
heart: a muscular, pumping organ positioned in the thoracic cavity.
hematocrit: the volume percentage of red blood cells in whole blood.
hemoglobin: the pigment of red blood cells that transports O_2 and CO_2.
hemopoiesis: production of red blood cells.
hepatic portal circulation: the return of venous blood from the digestive organs and spleen through a capillary network within the liver before draining into the heart.
herbaceous: a nonwoody plant.
herbaceous stem: stem of a nonwoody plant.
herbivore: an organism that feeds exclusively on plants.
heredity: the transmission of certain characteristics, or traits, from parents to offspring, via the genes.
heterodont: having teeth differentiated into incisors, canines, premolars, and molars for specific functions.
heterotroph: an organism that utilizes preformed food.
heterozygous: having two different alleles (i.e., *Bb*) for a given trait.
hiatus: an opening or fissure.
hilum: a concave or depressed area where vessels or nerves enter or exit an organ.
histology: microscopic anatomy of the structure and function of tissues.
holdfast: basal extension of a multicellular alga that attaches it to a solid object.
homeostasis: a consistency and uniformity of the internal body environment that maintains normal body function.
homologous: similar in developmental origin and sharing a common ancestry.
homothallic: species in which individuals produce both male and female reproductive structures and are self-fertile.
hormone: a chemical substance that is produced in an endocrine gland and secreted into the bloodstream to cause an effect in a specific target organ.
host: an organism on or in which another organism lives.

ileum: the terminal portion of the small intestine, between the jejunum and caecum.
imprinting: a type of learned behavior during a limited critical period.
indigenous: organisms that are native to a specific region; not introduced.
inguinal: pertaining to the groin region.
insertion: the more movable attachment of a muscle, usually more distal in location.
inspiration: the act of breathing air into the pulmonary alveoli of the lungs; also called *inhalation*.

instar: stage of insect or other arthropod development between molts.
integument: pertaining to the skin.
internal ear: the innermost portion or chamber of the ear, containing the cochlea and the vestibular organs.
internode: region between stem nodes.
interstitial: pertaining to spaces or structures between the functioning active tissue of any organ.
intracellular: within the cell itself.
intervertebral disk: a pad of fibrocartilage between the bodies of adjacent vertebrae.
intestinal gland: a simple tubular digestive gland that opens onto the surface of the intestinal mucosa and secretes digestive enzymes; also called *crypt of Lieberkühn*.
intrinsic: situated or pertaining to internal origin.
invertebrate: an animal that lacks a vertebral column.
iris: the pigmented vascular tunic portion of the eye that surounds the pupil and regulates its diameter.
islets of Langerhans: see *pancreatic islets*.
isotope: a chemical element that has the same atomic number as another but a different atomic weight.

jejunum: the middle portion of the small intestine, located between the duodenum and the ileum.
joint capsule: a fibrous tissue cuff surrounding a movable joint.
jugular: pertaining to the veins of the neck that drain the areas supplied by the carotid arteries.

karyotype: the arrangement of chromosomes that is characteristic of the species or of a certain individual.
keratin: an insoluble protein present in the epidermis and in epidermal derivatives such as scales, feathers, hair, and nails.
kidney: one of the paired organs of the urinary system that contain nephrons and filter wastes from the blood in the formation of the urine.
kingdom: a taxonomic category grouping related phyla.

labia major: a portion of the external genitalia of a female, consisting of two longitudinal folds of skin extending downward and backward from the mons pubis.
labia minora: two small folds of skin, devoid of hair and sweat glands, lying between the labia majora of the external genitalia of a female.
lacrimal gland: a tear-secreting gland located on the superior lateral portion of the eyeball underneath the upper eyelid.
lacteal: a small lymphatic duct within a villus of the small intestine.
lacuna: a hollow chamber that houses an osteocyte in mature bone tissue or a chondrocyte in cartilage tissue.
lamella: a concentric ring of matrix surrounding the central canal in an osteon of mature bone tissue.
large intestine: the last major portion of the gastrointestinal tract, consisting of the caecum, colon, rectum, and anal canal.
larva: an immature, developmental stage that is quite different from the adult.
larynx: the structure located between the pharynx and trachea that houses the vocal cords; commonly called the *voice box*.
lateral root: a secondary root that arises by branching from an older root.
leaf veins: plant structures that contain the vascular tissues in a leaf.
legume: a member of the pea, or bean, family.
lens: a transparent refractive structure of the eye, derived from ectoderm and positioned posterior to the pupil and iris.
lenticel: spongy area in the bark of a stem or root that permits interchange of gases between internal tissues and the atmosphere.
leukocyte: a white blood cell; also spelled *leucocyte*.
lichen: alga or bacteria and fungi coexisting in a mutualistic relationship.
ligament: a fibrous band of connective tissue that binds bone to bone to strengthen and provide support to the joint; also may support viscera.
limbic system: portion of the brain concerned with emotions and autonomic activity.
locus: the specific location or site of a gene within the chromosome.
lumbar: pertaining to the region of the loins.
lumen: the space within a tubular structure through which a substance passes.
lung: one of the two major organs of respiration within the thoracic cavity.
lymph: a clear fluid that flows through lymphatic vessels.
lymph node: a small, oval mass located along the course of lymph vessels.
lymphocyte: a type of white blood cell characterized by a granular cytoplasm.
macula lutea: a depression in the retina that contains the fovea centralis, the area of keenest vision.
malnutrition: any abnormal assimilation of food; receiving insufficient nutrients.
mammary gland: the gland of the mammalian female breast responsible for lactation and nourishment of the young.
mantle: fleshy fold of the body wall of a mollusk, typically involved in shell formation.
marine: pertaining to the sea or ocean.
marrow: the soft vascular tissue that occupies the inner cavity of certain bones and produces blood cells.
matrix: the intercellular substance of a tissue.
mediastinum: the partition in the center of the thorax between the two pleural cavities.
medulla: the center portion of an organ.
medulla oblongata: a portion of the brainstem between the pons and the spinal cord.
medullary cavity: the hollow center of the diaphysis of a long bone, occupied by red bone marrow.
megaspore: a plant spore that will germinate to become a female gametophyte.
meiosis: cell division by which gametes, or haploid sex cells, are formed. In plants, meiosis yields spores.
melanocyte: a pigment-producing cell in the deepest epidermal layer of the skin.
membranous bone: bone that forms from membranous connective tissue rather than from cartilage.
menarche: the first menstrual discharge.
meninges: a group of three fibrous membranes that covers the central nervous system.
meniscus: wedge-shaped cartilage in certain synovial joints.
menopause: the cessation of menstrual periods in the human female.
menses: the monthly flow of blood from the human female genital tract.
menstrual cycle: the rhythmic female reproductive cycle, characterized by changes in hormone levels and physical changes in the uterine lining.
menstruation: the discharge of blood and tissue from the uterus at the end of the menstrual cycle.
meristem tissue: undifferentiated plant tissue that is capable of dividing and producing new cells.
mesentery: a fold of peritoneal membrane that attaches an abdominal organ to the abdominal wall.
mesoderm: the middle one of the three primary germ layers.
mesophyll: the middle tissue layer of a leaf containing cells that are active in photosynthesis.
mesothelium: a simple squamous epithelial tissue that lines body cavities and covers visceral organs; also called *serosa*.
metabolism: the chemical changes that occur within a cell.
metacarpus: the region of the hand between the wrist and the phalanges, including the five bones that constitute the palm of the hand.
metamorphosis: change in morphologic form, such as when an insect larva develops into the adult or a tadpole develops into an adult frog.
metatarsus: the region of the foot between the ankle and the phalanges, consisting of five bones.
microbiology: the science dealing with microscopic organisms, including bacteria, fungi, protozoa, and viruses.
microspore: a spore in seed plants that develops into a pollen grain, the male gametophyte.
microvilli: microscopic, hairlike projections of cell membranes on certain epithelial cells.
midbrain: the portion of the brain between the pons and the forebrain.
middle ear: the middle of the three ear chambers, containing the three auditory ossicles.
migration: movement of organisms from one geographical site to another.
mimicry: a protective resemblance of an organism to another.
mitosis: the process of cell division, in which the two daughter cells are identical and contain the same number of chromosomes.
mitral valve: the left atrioventricular heart valve; also called the *bicuspid valve*.
mixed nerve: a nerve containing both motor and sensory nerve fibers.
molecule: a minute mass of matter composed of a combination of atoms that form a given chemical substance or compound.
molting: periodic shedding of an epidermal-derived structure.
monocot: a type of angiosperm in which the seed has only a single cotyledon; also called *monocotyledon*.
motor neuron: a nerve cell that conducts action potential away from the central nervous system and innervates effector organs (muscles and glands); also called *efferent neuron*.
motor unit: a single motor neuron and the muscle fibers it innervates.
mucosa: a mucous membrane that lines cavities and tracts opening to the exterior.
muscle: an organ adapted to contract; three types of muscle tissue are cardiac, smooth, and skeletal.
mutation: a variation in an inheritable characteristic, a permanent transmissible change in which the offspring differ from the parents.
mutualism: a beneficial relationship between two organisms of different species.
myelin: a lipoprotein material that forms a sheath-like covering around nerve fibers.
myocardium: the cardiac muscle layer of the heart.

myofibril: a bundle of contractile fibers within muscle cells.
myoneural junction: the site of contact between an axon of a motor neuron and a muscle fiber.
myosin: a thick filament protein that, together with actin, causes muscle contraction.

nail: a hardened, keratinized plate that develops from the epidermis and forms a protective covering on the surfaces of the digits.
nares: the openings into the nasal cavity; also called *nostrils*.
nasal cavity: a mucosa-lined space above the oral cavity that is divided by a nasal septum and is the first chamber of the respiratory system.
nasal septum: a bony and cartilaginous partition that separates the nasal cavity into two portions.
natural selection: the evolutionary mechanism by which better adapted organisms are favored to reproduce and pass on their genes to the next generation.
nephron: the functional unit of the kidney, consisting of a glomerulus, glomerular capsule, convoluted tubules, and the nephron loop.
nerve: a bundle of nerve fibers outside the central nervous system.
neurofibril node: a gap in the myelin sheath of a nerve fiber; also called the *node of Ranvier*.
neuroglia: specialized supportive cells of the central nervous system.
neurolemmocyte: a specialized neuroglial cell that surrounds an axon fiber of a peripheral nerve and forms the neurilemmal sheath; also called the *Schwann cell*.
neuron: the structural and functional unit of the nervous system, composed of a cell body, dendrites, and an axon; also called a *nerve cell*.
neutron: a subatomic particle in the nucleus of an atom that has a weight of one atomic mass unit and carries no charge.
neutrophil: a type of phagocytic white blood cell.
niche: the position and functional role of an organism in its ecosystem.
nipple: a dark pigmented, rounded projection at the tip of the breast.
nitrogen fixation: a process carried out by certain organisms, such as by soil bacteria, whereby free atmospheric nitrogen is converted into ammonia or nitrate compounds.
node: location on a stem where a leaf is attached.
node of Ranvier: see *neurofibril node*.
notochord: a flexible rod of tissue that extends the length of the back of an embryo and in some adults.
nucleic acid: an organic molecule composed of joined nucleotides, such as RNA and DNA.
nucleus: a spheroid body within a eukaryotic cell that contains the chromosomes of the cell.
nut: a hardened and dry single-seeded fruit.

olfactory: pertaining to the sense of smell.
oocyte: a developing egg cell.
oogenesis: the process of female gamete formation.
oogonium: a unicellular female reproductive organ of various protists that contains a single or several eggs.
optic: pertaining to the eye and the sense of vision.
optic chiasma: an X-shaped structure on the inferior aspect of the brain where there is a partial crossing over of fibers in the optic nerves.
optic disk: a small region of the retina where the fibers of the ganglion neurons exit from the eyeball to form the optic nerve; also called the *blind spot*.
oral: pertaining to the mouth; also called *buccal*.
organ: a structure consisting of two or more tissues that performs a specific function.
organelle: a minute structure of the eukaryotic cell that performs a specific function.
organism: an individual living creature.
orifice: an opening into a body cavity or tube.
origin: the place of muscle attachment onto the more stationary point or proximal bone; opposite the insertion.
osmosis: the diffusion of water from a solution of lesser concentration to one of greater concentration through a semipermeable membrane.
ossicle: one of the three bones of the middle ear.
osteocyte: a mature bone cell.
osteon: a group of osteocytes and concentric lamellae surrounding a central canal within bone tissue; also called a *Haversian system*.
oval window: see *vestibular window*
ovarian follicle: a developing ovum and its surrounding epithelial cells.
ovary: the female gonad, in which ova and certain sexual hormones are produced.
oviduct: the tube that transports ova from the ovary to the uterus; also called the *uterine tube* or *fallopian tube*.
ovipositor: a structure at the posterior end of the abdomen in many female insects for laying eggs.
ovulation: the rupture of an ovarian follicle with the release of an ovum.
ovule: the female reproductive organ in a seed plant that contains megasporangium where meiosis occurs and the female gametophyte is produced.
ovum: a secondary oocyte after ovulation but before fertilization.
palisade layer: the upper layer of the mesophyll of a leaf, which carries out photosynthesis.
pancreas: organ in the abdominal cavity that secretes gastric juices into the gastrointestinal tract and insulin and glucagon into the blood.
pancreatic islets: a cluster of cells within the pancreas that forms the endocrine portion of the pancreas; also called *islets of Langerhans*.
papillae: small nipple-like projections.
paranasal sinus: a mucous-lined air chamber that communicates with the nasal cavity.
parasite: an organism that resides in or on another from which it derives sustenance.
parasympathetic: pertaining to the division of the autonomic nervous system concerned with activities that are antagonistic to the sympathetic division of the autonomic nervous system.
parathyroids: small endocrine glands that are embedded on the posterior surface of the thyroid glands and are concerned with calcium metabolism.
parenchyma: the principal structural cells of plants.
parietal: pertaining to a wall of an organ or cavity.
parotid gland: one of the paired salivary glands on the sides of the face over the masseter muscle.
parturition: the process of childbirth.
pathogen: any disease-producing organism.
pectin: an organic compound in the intercellular layer and primary wall of plant cell walls; the basis of fruit jellies.
pedicel: the stalk of a flower in an inflorescence.
pectoral girdle: the portion of the skeleton that supports the upper extremities.
pelvic: pertaining to the pelvis.
pelvic girdle: the portion of the skeleton to which the lower extremities are attached.
penis: the external male genital organ, through which urine passes during urination and that transports semen to the female during coitus.
perennial: a plant that lives throughout several to many growing seasons.
pericardium: a protective serous membrane that surrounds the heart.
pericarp: the fruit wall that forms from the wall of a mature ovary.
perineum: the floor of the pelvis.
periosteum: a fibrous connective tissue covering the surface of bone.
peripheral nervous system: the nerves and ganglia of the nervous system that lie outside of the brain and spinal cord.
peristalsis: rhythmic contractions of smooth muscle in the walls of various tubular organs that move the contents along.
peritoneum: the serous membrane that lines the abdominal cavity and covers the abdominal viscera.
petal: the leaf of a flower, which is generally colored.
petiole: structure of a leaf that connects the blade to the stem.
phagocyte: any cell that engulfs other cells, including bacteria, or small foreign particles.
phalanx: a bone of the finger or toe.
pharynx: the region of the gastrointestinal tract and respiratory system located at the back of the oral and nasal cavities and extending to the larynx anteriorly and the esophagus posteriorly; also called the *throat*.
phenotype: the appearance of an organism caused by the genotype and environmental influences.
pheromone: a chemical secreted by one organism that influences the behavior of another.
phloem: vascular tissue in plants that transports nutrients.
photoperiodism: the response of an organism to periods of light and dark.
photosynthesis: the process of using the energy of the sun to make carbohydrate from carbon dioxide and water.
phototropism: plant growth or movement in response to a directional light source.
physiology: the science that deals with the study of body functions.
phytoplankton: microscopic, free-floating, photosynthetic organisms that are the major primary producers in freshwater and marine ecosystems.

pia mater: the innermost meninx, which is in direct contact with the brain and spinal cord.
pineal gland: a small cone-shaped gland located in the roof of the third ventricle of the brain.
pistil: a reproductive structure of a flower composed of the stigma, style, and ovary.
pith: a centrally located tissue within a dicot stem.
pituitary gland: a small, pea-shaped endocrine gland situated on the inferior surface of the brain that secretes a number of hormones; also called the *hypophysis*.
placenta: the organ of metabolic exchange between the mother and the fetus.
plankton: aquatic, free-floating microscopic organisms.
plasma: the fluid, extracellular portion of circulating blood.
plastid: an organelle in the cell of certain plants that is the site for food manufacture and storage.
platelets: fragments of specific bone marrow cells that function in blood coagulation; also called *thrombocytes*.
pleural membranes: serous membranes that surround the lungs and line the thoracic cavity.
plexus: a network of interlaced nerves or vessels.
pollen grain: immature male gametophyte generation of seed plants.
pollination: the delivery by wind, water, or animals of pollen to the ovule of a seed plant, leading to fertilization.

polypeptide: a molecule of many amino acids linked by peptide bonds.
pons: the portion of the brainstem just above the medulla oblongata and anterior to the cerebellum.
population: all the organisms of the same species in a particular location.
posterior (*dorsal*): toward the back.
predation: the consumption of one organism by another.
pregnancy: a condition in which a female has a developing offspring in the uterus.
prenatal: the period of offspring development during pregnancy; before birth.
prey: organisms that are food for a predator.
producer: organisms within an ecosystem that synthesize organic compounds from inorganic constituents.
prokaryote: organism, such as a bacterium, that lacks the specialized organelles characteristic of complex cells.
proprioceptor: a sensory nerve ending that responds to changes in tension in a muscle or tendon.
prostate: a walnut-shaped gland surrounding the male urethra just below the urinary bladder that secretes an additive to seminal fluid during ejaculation.
protein: a macromolecule composed of one or several polypeptides.
prothallus: a heart-shaped structure that is the gametophyte generation of a fern.
proton: a subatomic particle of the atom nucleus that has a weight of one atomic mass unit and carries a positive charge; also called a *hydrogen ion*.
proximal: closer to the midline of the body or origin of an appendage; opposite of distal.
puberty: the period of human development in which the reproductive organs become functional.
pulmonary: pertaining to the lungs.
pupil: the opening through the iris that permits light to pass through the lens and enter the posterior cavity.
radial symmetry: symmetry around a central axis so that any half of an organism is identical to the other.
receptacle: the tip of the axis of a flower stalk that bears the floral organs.
receptor: a sense organ or a specialized end of a sensory neuron that receives stimuli from the environment.
rectum: the terminal portion of the gastrointestinal tract, between the sigmoid colon and the anal canal.
reflex arc: the basic conduction pathway through the nervous system, consisting of a sensory neuron, an interneuron (association), and a motor neuron.
regeneration: regrowth of tissue or the formation of a complete organism from a portion.
renal: pertaining to the kidney.
renal corpuscle: the portion of the nephron consisting of the glomerulus and a glomerular capsule.
renal pelvis: the inner cavity of the kidney formed by the expanded ureter, into which the calyces open.
renewable resource: a commodity that is not used up because it is continually produced in the environment.
replication: the process of producing a duplicate; a copying or duplication, such as DNA replication.
respiration: the exchange of gases between the external environment and the cells of an organism; the metabolic activity of cells resulting in the production of ATP.
retina: the inner layer of the eye, which contains the rods and cones.
rhizome: an underground stem in some plants that stores photosynthetic products and gives rise to roots and above-ground stems and leaves.
rod: a photoreceptor in the retina of the eye that is specialized for colorless, dim-light vision.
root: the anchoring subterranean portion of a plant that permits absorption and conduction of water and minerals.
root cap: end mass of parenchyma cells that protects the apical meristem of a root.
root hair: unicellular epidermal projection from the root of a plant that functions in absorption.
rugae: the folds or ridges of the mucosa of an organ.

sagittal: a vertical plane through the body that divides it into right and left portions.
salinity: saltiness in water or soil; a measure of the concentration of dissolved salts.
salivary gland: an accessory digestive gland that secretes saliva into the oral cavity.
sarcolemma: the cell membrane of a muscle fiber.
sarcomere: the portion of a skeletal muscle fiber between the two adjacent Z lines that is considered the functional unit of a myofibril.
savanna: open grassland with scattered trees.
Schwann cell: see *neurolemmocyte*.
sclera: the outer white layer of connective tissue that forms the protective covering of the eye.
sclerenchyma: supporting tissue in plants composed of hollow cells with thickened walls.
scolex: head region of a tapeworm.
scrotum: a pouch of skin that contains the testes and their accessory organs.
sebaceous gland: an exocrine gland of the skin that secretes *sebum*, an oily protective product.
secondary growth: plant growth in girth from secondary or lateral meristems.
seed: a plant embryo with a food reserve that is enclosed in a protective seed coat; seeds develop from matured ovules.
semen: the secretion of the reproductive organs of the male, consisting of spermatozoa and additives.
semilunar valve: crescent-shaped heart valves positioned at the entrances to the aorta and the pulmonary trunk.
sensory neuron: a nerve cell that conducts an impulse from a receptor organ to the central nervous system; also called *afferent neuron*.
sepal: outermost whorl of flower structures beneath the petals; collectively called the *calyx*.
serous membrane: an epithelial and connective tissue membrane that lines body cavities and covers viscera; also called *serosa*.
sesamoid bone: a membranous bone formed in a tendon in response to joint stress.
sessile: organisms that lack locomotion and remain stationary, such as sponges and plants.
shoot: portion of a vascular plant that includes a stem with its branches and leaves.
sinoatrial node (SA node): a mass of cardiac tissue in the wall of the right atrium that initiates the cardiac cycle; also called the *pacemaker*.
sinus: a cavity or hollow space within a body organ such as a bone.
skeletal muscle: a type of muscle tissue that is multinucleated, occurs in bundles, has crossbands of proteins, and contracts either in a voluntary or involuntary fashion.
small intestine: the portion of the gastrointestinal tract between the stomach and the caecum; functions in absorption of food nutrients.
smooth muscle: a type of muscle tissue that is nonstriated, composed of fusiform and single-nucleated fibers, and contracts in an involuntary, rhythmic fashion within the walls of visceral organs.
solute: a substance dissolved in a solvent to form a solution.
solvent: a fluid such as water that dissolves solutes.
somatic: pertaining to the nonvisceral parts of the body.
somatic cells: all the cells of the body of an organism except the germ cells.
sorus: a cluster of sporangia on the underside of fern pinnae.
species: a group of morphologically similar (common gene pool) organisms that are reproductively isolated and capable of interbreeding and producing fertile offspring.
spermatic cord: the structure of the male reproductive system, composed of the ductus (vas) deferens, spermatic vessels, nerve, cremasteric muscle, and connective tissue.
spermatogenesis: the production of male sex gametes, or spermatozoa.
spermatozoan: a sperm cell, or gamete.
sphincter: a circular muscle that constricts a body opening or the lumen of a tubular structure.
spinal cord: the portion of the central nervous system that extends from the brainstem through the vertebral canal.
spinal nerve: one of the 31 pairs of nerves that arise from the spinal cord.
spiracle: a respiratory opening in certain animals such as arthropods and sharks.
spirillum (pl. spirilla): a spiral-shaped bacterium.
spleen: a large, blood-filled organ located in the upper left of the abdomen and attached by the mesenteries to the stomach.
spongy bone: a type of bone that contains many porous spaces; also called *cancellous bone*.
sporangium: any structure within which spores are produced.
spore: a reproductive cell capable of developing into an adult organism without fusion with another cell.
sporophyll: modified leaves that bear one or more sporangia.
stamen: a reproductive structure of a flower, composed of a filament and an anther, where pollen grains are produced.
starch: carbohydrate molecule synthesized from photosynthetic products; common food storage substance in many plants.
stele: the vascular tissue and pith or ground tissue at the central core of a root or stem.
stigma: the upper portion of the pistil of a flower.
stoma: an opening in a plant leaf through which gas exchange takes place.
stomach: a pouch-like digestive organ between the esophagus and the duodenum.
style: the long slender portion of the pistil of a flower.
submucosa: a layer of supportive connective tissue that underlies a mucous membrane.
succession: the sequence of ecological stages by which a particular biotic community gradually changes until there is a community of climax vegetation.
sucrose: a disaccharide (double sugar) consisting of a linked glucose and fructose molecule; the principal transport of sugar in plants.
surfactant: a substance produced by the lungs that decreases the surface tension within the pulmonary alveoli.
suture: a type of immovable joint articulating between bones of the skull.
symbiosis: a close association between two organisms in which one or both species derive benefit.
sympathetic: pertaining to that part of the autonomic nervous system concerned with activites antagonistic to the parasympathetic.
synapse: a minute space between the axon terminal of a presynaptic

neuron and a dendrite of a postsynaptic neuron.
syngamy: union of gametes in sexual reproduction; fertilization.
synovial cavity: a space between the two bones of a synovial joint, filled with synovial fluid.
system: a group of body organs that function together.
systole: the muscular contraction of the ventricles of the heart during the cardiac cycle.
systolic pressure: arterial blood pressure during the ventricular systolic phase of the cardiac cycle.

taproot: a plant root system in which a single root is thick and straight.
target organ: the specific body organ that a particular hormone affects.
tarsus: pertaining to the ankle; the proximal portion of the foot that contains the tarsal bones.
taxonomy: the science of describing, classifying, and naming organisms.
tendo calcaneous: the tendon that attaches the calf muscles to the calcaneous bone.
tendon: a band of dense regular connective tissue that attaches muscle to bone.
testis: the primary reproductive organ of a male, which produces spermatozoa and male sex hormones.
tetrapod: a four-appendaged vertebrate, such as amphibian, reptile, bird, or mammal.
thoracic: pertaining to the chest region.
thorax: the chest.
thymus gland: a bilobed lymphoid organ positioned in the upper mediastinum, posterior to the sternum and between the lungs.
tissue: an aggregation of similar cells and their binding substances, joined to perform a specific function.
tongue: a protrusible muscular organ on the floor of the oral cavity.
toxin: a poisonous compound.
trachea: a tubule in the respiratory system of some invertebrates; the airway leading from the larynx to the bronchi in the respiratory system of vertebrates; also called the *windpipe*.
tract: a bundle of nerve fibers within the central nervous system.
trait: a distinguishing feature studied in heredity.
transpiration: the evaporation of water from a leaf, which pulls water from the roots through the stem of a leaf.
tricuspid valve: the heart valve between the right atrium and the right ventricle; also called *right atrioventricular valve*.
turgor pressure: osmotic pressure that provides rigidity to a cell.
tympanic membrane: the membranous eardrum positioned between the outer ear and middle ear; also called the *tympanum*, or the *eardrum*.

umbilical cord: a cord-like structure containing the umbilical arteries and vein that connects the fetus with the placenta.
umbilicus: the site where the umbilical cord was attached to the fetus; also called the *navel*.
ureter: a tube that transports urine from the kidney to the urinary bladder.
urethra: a tube that transports urine from the urinary bladder to the outside of the body.
urinary bladder: a distensible sac in the pelvic cavity that stores urine.
uterine tube: the tube through which the ovum is transported to the uterus and where fertilization takes place; also called the *oviduct* or *fallopian tube*.
uterus: a hollow, muscular organ in which a fetus develops; located within the female pelvis between the urinary bladder and the rectum.
uvula: a fleshy, pendulous portion of the soft palate that blocks the nasopharynx during swallowing.

vacuole: a fluid-filled organelle.
vagina: a tubular organ that leads from the uterus to the vestibule of the female reproductive tract and receives the male penis during coitus.
vascular tissue: plant tissue composed of xylem and phloem; functions in transport of water, nutrients, and photosynthetic products throughout the plant.
vegetative: plant parts not specialized for reproduction; asexual reproduction.
veins: blood vessels that convey blood toward the heart.
ventral (anterior): toward the front surface of the body.
vertebrate: an animal that possesses a vertebral column.
vestibular folds: the supporting folds of tissue for the vocal folds within the larynx; also called *vocal cords*.
vestibular window: a membrane-covered opening in the bony wall between the middle and inner ear into which the footplate of the stapes fits; also called *oval window*.
viscera: the organs within the abdominal or thoracic cavities.
vitreous humor: the transparent gel that occupies the space between the lens and retina of the eye.
vocal folds: folds of the mucous membrane in the larynx that produce sound as they are pulled taut and vibrated; also called *vocal cords*.
vulva: the external genitalia of the female that surround the opening of the vagina; also called the *pudendum*.

wood: interior tissue of a tree composed of secondary xylem.

zoospore: a flagellated or ciliated spore produced asexually by some protists.

zygote: a fertilized egg cell formed by the union of a sperm and an ovum.

Index

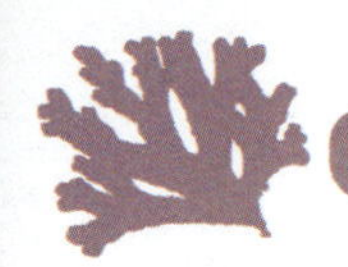

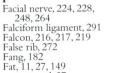